AF88677Ø

Range Science Dept.

USU

Vegetation science applications for rangeland analysis and management

Handbook of vegetation science

FOUNDED BY R. TÜXEN

H. LIETH, EDITOR IN CHIEF

Volume 14

Vegetation science applications for rangeland analysis and management

Edited by

P. T. TUELLER

Kluwer Academic Publishers
DORDRECHT – BOSTON – LONDON

Library of Congress Cataloging in Publication Data

Vegetation science applications for rangeland analysis and management
/ [edited by] Paul T. Tueller.
p. cm. -- (Handbook of vegetation science ; 14)
Includes index.
ISBN 9061931959
1. Range management. 2. Range ecology. 3. Range plants.
I. Tueller, Paul T., 1934- . II. Series.
QK911.H3 pt.14
[SF85]
581 s--dc19
[633.2'02] 88-14846

ISBN 90-6193-195-9

Kluwer Academic Publishers incorporates the publishing programmes of Dr. W. Junk Publishers, MTP Press, Martinus Nijhoff Publishers, and D. Reidel Publishing Company.

Distributors

for the United States and Canada: Kluwer Academic Publishers, 101 Philip Drive, Norwell, MA 02061, USA
for all other countries: Kluwer Academic Publishers Group, P.O. Box 322, 3300 AH Dordrecht, The Netherlands

Copyright

© 1988 by Kluwer Academic Publishers, Dordrecht.

All rights reserved. No part of this publication may be reproduced, stored in a retrieval system, or transmitted in any form or by any means, mechanical, photocopying, recording, or otherwise, without the prior written permission from the copyright owners.

PRINTED IN THE NETHERLANDS

Contents

VEGETATION DISTRIBUTION AND ORGANIZATION

VEGETATION SCIENCE RANGELAND APPLICATIONS

Series editor's preface

Natural grasslands, pastures and meadows are among the vegetation types most frequently investigated with phytosociological methods. This was one of the reasons why volume 13, Application of vegetation science to grassland husbandry and agriculture, edited by W. Krause, appeared as one of the first volumes of this handbook. It appeared under the chief editorship of Prof. R. Tüxen and in his time main emphasis of the handbook was placed on Zürich-Montpellier methods and the European vegetation.

When we redesigned the handbook we felt the need to include other methods and aims of grassland analyses as well as a more global coverage of grasslands. Especially the natural dry and semidry areas of the world needed to be covered.

I was very fortunate in getting Prof. Tueller of the University of Reno Nevada as an editor for this volume. He and the colleagues he motivated to compile volume 14 on Application of vegetation science to rangeland analysis and management have created a truly global coverage of the topics interesting for vegetation analyses in natural grasslands. Since volume 13 covered the problems of anthropogenically created grasslands, this topic was not expressly treated in order to avoid duplication. For the same reason no specific attempt was made to get more papers from Europe and the temperate forest region in general.

The cooperation with Dr. Tueller has been very rewarding for me. Being interested in vegetation problems in developing countries for all my life I find it satisfying to have a volume in the handbook of special importance for the tropical and subtropical regions of the world.

Especially useful for future users of the book will be the combination of basic vegetation science elements with papers devoted to problems of vegetation distribution and the application to range management. It is easy to forecast that this volume will become a major reference source for future work in vegetation science; not only for work in grassland areas but also for general work in vegetation science.

I believe I speak for the whole profession when I express Dr. Tueller and all his authors my gratitude for the solid work they have done. I hope it will be used by professionals and students alike.

Osnabrück, July 1988 H. LIETH

Foreword

In early 1986 I began a correspondence with Professor Helmut Lieth relative to the possible publication of a short book on aridland remote sensing. This led to discussions concerning a contribution to the Handbook of Vegetation Science Series on the subject of rangelands. Previously books in this series had been prepared on Forestry, Grassland Husbandry and Agriculture and Application to Landscape Management. A chapter on Rangeland Analysis and Management had been suggested as an important contribution but the actual preparation had not begun. It has been my pleasure to begin the task and bring it to fruition.

The authors of this chapter are to be commended for the excellent and timely material that they have brought to this volume. They represent an outstanding cross section of expertise on rangeland ecology and management. Only three are from outside of the United States. This is a somewhat unfortunate circumstance related to the fact that I am a U.S. rangeland scientist and have familiarity with the work here. My experience in Australia and South Africa led me to have at least one contributor from each of these areas. Also one author is a Canadian. In spite of this potential inbreeding the subject matter content represents well the contributions of vegetation science to rangeland analysis and management.

Rangelands have often been an almost forgotten resource in the sense that their management, on a worldwide basis, has been rather piecemeal and has followed behind the management of other vegetation resources such as timber and crops. Of course, there have been many reasons for this. The primary reason has been what might be termed the extensive rather than intensive management of the world's rangelands and a feeling that range vegetation resources will somehow take care of themselves. The vegetation approaches to range management discussed in this volume will quickly dispell this notion. It will become clear that the wise management of rangeland is not by happenstance but is based on sound scientifically based management of the vegetation and soil resources. I only hope that this work will provide information and incentive toward better rangeland resources management throughout the world.

Reno, Nevada P. T. TUELLER

List of contributors

B. H. Allen, Assistant Professor of Range Ecology, Department of Forestry and Resource Management, University of California, Berkeley, California 94720.

A. W. Bailey, Professor of Range Ecology and Management, Department of Plant Science, University of Alberta, Edmonton, Alberta, Canada.

O. J. H. Bosch, Head of Department of Plant Sciences, Potchefstroom University for Christian Higher Education, Potchefstroom, Republic South Africa.

C. A. Call, Assistant Professor, Department of Range Science, Utah State University, Logan, Utah 84322-5230.

E. J. DePuit, Associate Professor, Range Management Department, University of Wyoming, Laramie, Wyoming 82071.

K. T. Harper, Professor, Department of Botany and Range Science, Brigham Young University, Provo, Utah 84602.

R. H. Hart, Range Scientist, U.S. Department of Agriculture, Agriculture Research Service, High Plains Grasslands Research Station, Cheyenne, Wyoming 82009.

S. L. Hatch, Associate Professor, Department of Range Science, Texas A and M University, College Station, Texas 77843.

K. C. Hodgkinson, CSIRO, Division of Wildlife and Rangelands Research, Deniliquin, N.S.W. 2710.

D. A. Jameson, Professor, Department of Range Science, Colorado State University, Fort Collins, Colorado 80523.

J. G. Kie, Principal Research Wildlife Biologist, Pacific Southwest Forest and Range Experiment Station, USDA Forest Service, Fresno, California 97310.

D. R. Klein, Unit Leader, Alaska Cooperative Wildlife Research Unit, University of Alaska, Fairbanks, Alaska 99701.

P. C. Lent, U.S. Bureau of Land Management, Reno, Nevada 89512.

S. G. Leonard, Range Scientist, USDA, National Soil Range Team, Department of Range, Wildlife and Forestry, University of Nevada Reno, Reno, Nevada 89512.

J. R. Marble, Department of Botany and Range Science, Brigham Young University, Provo, Utah 84602.

E. D. McArthur, Supervisory Research Geneticist, USDA Forest Service, Intermountain Research Station, Shrub Sciences Laboratory, Provo, Utah 84601.

R. L. Miles, Soil Scientist, USDA, SCS National Soil Range Team, Department of Range, Wildlife and Forestry, University of Nevada Reno, Reno, Nevada 89512.

R. A. Nicholson, Professor, Department of Biological Science, Fort Hays State University, Hays, Kansas 67601.

J. C. Noble, CSIRO, Division of Wildlife and Rangelands Research, Deniliquin, N.S.W. 2710.

B. E. Norton, Associate Professor of Range Ecology, Department of Range Science, Utah State University, Logan, Utah 84322.

R. D. Pieper, Professor of Range Science, Department of Animal and Range Sciences, New Mexico State University, Las Cruces, New Mexico 88003.

K. A. Platou, Ecologist, Inyo National Forest, Bishop, CA 93514.

E. F. Redente, Associate Professor, Range Science Department, Colorado State University, Fort Collins, Colorado 80523.

B. A. Roundy, Assistant Professor, School of Renewable Natural Resources, University of Arizona, Tucson, Arizona 85721.

E. L. Smith, Chairman of Range Resources, Division of Range Management, University of Arizona, Tucson, Arizona 85721.

R. E. Sosebee, Professor of Range Management, Department of Range and Wildlife Management, Texas Tech University, Lubbock, Texas 79409.

S. R. Swanson, Assistant Professor, Department of Range, Wildlife, and Forestry, University of Nevada Reno, Reno, Nevada 89512.

N. M. Tainton, Head of Department of Grassland Science, University of Natal, Peitermaritzburg, Republic of South Africa.

J. W. Thomas, Chief Research Wildlife Biologist, Pacific Northwest Forest and Range Experiment Station, USDA Forest Service, La Grande, Oregon 97850.

P. T. Tueller, Professor of Range Ecology, Department of Range, Wildlife and Forestry, University of Nevada Reno, Reno, Nevada 89512.

C. Wan, Graduate Research Assistant, Gansu Grassland Ecological Research Institute, Lanzhou, Gansu, China.

N. E. West, Professor, Range Science Department, Utah State University, Logan, Utah 84322-5230.

A. D. Wilson, Senior Principal Research Scientist, CSIRO, Division of Wildlife and Rangelands Research, Deniliquin, N.S.W. 2710.

M. K. Wood, Associate Professor, Department of Animal and Range Sciences, New Mexico State University, Las Cruces, New Mexico 88003-0003.

J. A. Young, Range Scientist, U.S. Department of Agriculture, Agriculture Research Service, Reno, Nevada 89512.

I. Introduction

PAUL T. TUELLER

1.1. Abstract

Rangelands are a valuable worldwide resource. The wise use and management of these lands is strongly dependent upon the vegetation sciences including botany, physiology, ecology, taxonomy, morphology, genetics and demography. Closely intertwined with the plant sciences and contributing to further clarification and understanding of rangelands for improved management are various related sciences such as agronomy, forestry, geology/soil science, bioclimatology, hydrology, animal science, wildlife ecology, mathematics, statistics, ecosystem modelling, economics and sociology. In addition, biotechnology will be an important future contributor to range management. The objective of this volume is to describe the role of the various plant sciences in furthering our understanding of rangeland ecosystems and their management.

1.2. Introduction

Vegetation is the carpet of plant life covering the ground. Whether of a square meter or an entire continent, its properties are only those of its component species. These components are modified within the vegetation by interspecific competition and by the biotic and abiotic environment. Grazing and browsing by domestic livestock and wild herbivores has been an important controlling influence on a large part of the vegetation of the world's uncultivated land (Miles 1979), mostly rangelands. Vegetation science is the *sine qua non* of range management. Without an understanding of the plant species and communities that characterize the world's rangelands we would not be able to manage these vegetation resources.

Rangelands are those areas of the world, which by reason of physical limitations — low and erratic precipitation, rough topography, poor drain-

Tueller, P. T. (ed.), Vegetation science applications for rangeland analysis and management.
ISBN 90-6193-195-9.
© 1988, *Kluwer Academic Publishers, Dordrecht, Boston, London.*

age, or cold temperatures — are unsuited to cultivation and which are a source of forage for free-ranging native and domestic animals, as well as a source of wood products, water and wildlife (Stoddart *et al.* 1975). Range is land which supports vegetation useful for grazing on which routine management of that vegetation is through manipulation of grazing rather than cultural practices (RISC 1983). Range is a kind of land that has many uses and should not be thought of as a single use, e.g., livestock grazing. In addition society may elect to maintain some lands technically suited for agriculture as rangelands, e.g., as national parks, ecological preserves and wildlife refuges. Red meat has been a traditional product from rangelands and historically and prehistorically wildlife has been harvested. Presently there is increasing interest in non-consumptive uses of rangeland such as aesthetics and recreation.

Range management refers to the wise use of these lands for the sustained yield of various resources. It is the "science and art of optimizing the returns from rangelands in those combinations most desired by and suitable to society through the manipulation of range ecosystems" (Stoddard *et al.* 1975). Even when rangelands are improved by manipulative management such as plowing and seeding, prescribed burning and water spreading, the management of the vegetation is still based on naturally applied ecological and physiological principals.

If forested ranges and natural vegetation such as tropical savannah and tundra areas are included as rangeland, the total land area may be as high as 47% of the global land surface (Williams *et al.* 1968). On a world wide basis there are about 3.7×10^6 ha of rangeland (Stiles 1984) occurring on all continents of the world, except antarctica, including Africa, Asia, Europe, South America, North America and Australia.

1.3. Vegetation science

The vegetation sciences have contributed immensely to our understanding of rangelands. Range management is based on these sciences which include botany, physiology, ecology, taxonomy, genetics, morphology and demography. Vegetation science principles learned by agronomists and foresters have also proven useful to range management.

Of all the vegetation sciences, plant ecology has thus far been the most influential for range management. It will likely remain important for many years because range management is in many ways applied plant ecology. What is applied plant ecology? It is the use of vegetation or plant ecological principles in such a way as to cause changes to take place on the landscape that are beneficial to man or the animals he wishes to manage. It also involves the maintenance of vegetation cover and the integrity of surface soil consistent with the use of the rangelands for any important value.

Examples of management activities on rangelands based on applied ecology are numerous. Selection of species, seedbed preparation and seeding rate and dates for optimum germination, emergence, establishment and survival are now based on learned ecological principles. Grazing management and the reaction of individual species to grazing intensity, duration, season and type of grazing animal are based on an understanding of applied ecology. Rangeland monitoring methodologies used to evaluate the relative success or failure of management practices are based on ecological principles. Prescribed burning of rangeland vegetation is based on an understanding of the influence that fire intensity, duration, season and other factors will have on the resulting vegetation.

Of profound importance to the management of rangeland vegetation is knowledge of the principles of plant succession (Smith 1988) developed by ecologists and demographers. Range managers have felt for many years that they have a good understanding of these successional principles. However, now there is reason to believe that we must reevaluate our thinking relative to successional ecology as it applies to rangeland management. Management is closely related to an understanding of how vegetation changes in relation to both natural forces such as wildfire, climate changes and flooding as well as manmade forces such as prescribed grazing, prescribed burning, fencing, seeding, water development, wood harvesting and the harvesting of specific numbers from a population of wild animals.

Synecologists have, in many ways, paved the way for range management by providing excellent classifications of rangeland vegetation. These classifications have strongly influenced rangeland management. Fortunately many of the early range scientists had excellent powers of observation and were able to define or describe classifications that proved very useful for mapping and managing units of rangeland that were reasonably homogeneous with respect to grazing and browsing responses. Now modern approaches using multivariate statistical approaches are providing new insight into range vegetation classification and relationships to management. What is needed today is a classification of ecosystems, not just vegetation (West 1988).

Botanists, ecophysiologists and plant morphologists help us to understand how various important species function within a range ecosystem. Energy transformations, water relations, photosynthesis, respiration and gas exchange, nutrient requirements and other related factors of range plant species have been and are being studied by ecophysiologists. Their research and knowledge has improved range management by providing information on how various species react to grazing or browsing. This knowledge is essential for the successful implementation of intensive and improved grazing management systems.

Taxonomists and systematic botanists have provided range managers an improved understanding of the populations of species that make up the plant communities. Taxonomic knowledge allows stronger more refined interpretation of plant synecological units on the range landscapes and provides information useful for the interpretation of successional changes based on floristics. Further knowledge of these populations will be accomplished by plant demographers studying plant population ecology in order to determine how populations of range plants react to various random events (Noble 1984). Knowledge of ecotypes and their genetic makeup is related to the environment in which they are found, which potentially provides interpretive information for range condition and trend (vegetation succession on rangelands) assessment. New biochemical procedures have shed light on many taxonomic groups and provided insights relative to the ecological placement and adaptation of species for many range sites. Practical use of taxonomic information includes the identification of range weeds, of plants suitable for revegetation uses, of poisonous plants and of species from which other useful products can be derived.

Perhaps the most common tool in rangeland management for maintenance of suitable plant communities is the manipulation of wild and domestic animal populations. In some cases the manipulation of range landscapes using the principles of applied ecology is also practiced but often the costs are prohibitive.

1.4. Related sciences

Within the purview of range management there are numerous contributing sciences. Knowledge from the vegetation sciences must be coupled with an understanding of geology/soil science, climatology including micrometeorology, hydrology, statistics, mathematical modelling, animal science including animal behavior, economics, sociology and various related sciences. These sciences interact with the vegetation sciences and contribute to successful range vegetation and soil management.

Geology and soil science play an important role by providing an understanding of the substrate and physical and chemical characteristics of the physical support medium for plant growth. Likewise bioclimatology is a science that integrates additional aspects of the abiotic (temperature, water relations, light) environment and relates them to species and community growth and development. Related to this are hydrological principles where runoff, sedimentation, erosion and other parameters on rangelands are directly related to the kind and amount of vegetation.

Ecosystem modelling, statistics and related mathematics are used to analyze, synthesize and provide strong objective interpretation of range management/vegetation science questions. Range ecosystem modelers assist in the management of rangelands by modelling the processes that take place and by simulating the potential influences of certain kinds and levels of management activities. Modelling efforts potentially allow range ecosystem evaluations to be made over the short term without costly long-term empirical studies.

Understanding the range animals through studies of animal husbandry, animal nutrition, wildlife management, population dynamics and the behavior of animals in the range ecosystem is important because of the complex relationships between these factors and the range vegetation. Questions of forage palatability, preference and overgrazing must be looked at in the context of the animals using the range.

Finally, economics, sociology and politics play an important role in range management because the management and use of range vegetation involves strong human and social interactions. This is especially true in countries with complex land ownership and strong tradition as well as in countries with considerable public lands.

1.5. Biotechnology and the future

Today vegetation science is strongly tempered by what may be generally called biotechnology. Biotechnology is the use of vegetation science principles of genetics, cellular and molecular biology and physiology coupled with newly developed and developing plant propagation and tissue culture procedures to produce new or highly adapted species to fill certain needs. Tissue culture and an understanding of cellular totipotency (all living cells in a plant body, irrespective of their ploidy level and the form of specialization, can potentially give rise to whole plants) has become an important aid in the field of experimental botany (Bhojwani and Razdan 1983). The spin-off from this technology will soon be important to rangelands but for now the efforts are going into high-value cash crops, horticultural varieties, forest regeneration, hydroponics, gardening and crops grown for pharmaceuticals. Biotechnology applications for the improvement of rangeland or wildlife habitat species have, up until now, been minimal.

Plant geneticists have not worked a great deal with range plants. Work on plants adapted to semi-arid rangelands based on ecological rather than agronomic principles is a relatively new effort (McArthur 1988). For the foreseeable future we can expect genetic engineering to place emphasis on vegetation other than rangeland species. Resources for research on range plant species will be modest and progress will be slow. Most of the resources will go to crop plants that primarily provide food directly to man rather that through the production of red meat.

Cellular and molecular biology will lead to a better understanding of how plant systems work which will in turn lead to an improved understanding of how various species react to the environment and to various levels and kinds of range management. Expert systems based on new computer technology and artificial intelligence will provide range scientists and managers with valuable new tools for managing data and interacting with other fields of expertise (Noble 1987).

On rangelands as well as on other resource lands there must be a consideration of population control, resource exhaustion, environmental pollution and endangered species. It is of interest that modern ecologists are called upon by conservationists to assist in the prevention of extinction of plant and animal species while at the same time they must guide scientists who synthesize essentially new species for introduction to the environment.

The most trying problems found in rangeland ecosystems deal with unraveling and managing the abberations of competition created within the ecosystem by humans. Such problems are often caused by the intervention of man with technology into otherwise slowly evolving and dynamically balanced biotic and abiotic systems (De Santo 1978). We can expect that as scientific understanding of range ecosystems increases because of these many considerations, management tools to compensate for man's often inept intervention will be forthcoming and successful, at least we can hope so.

1.6. Scope of this book

The effort of rangeland scientists, pastoral scientists, range ecologists, range conservationists, veld management specialists or any others who create new knowledge related to rangelands is enormous taken in its entirety. Ecologists, ecophysiologists, systematic botanists, geneticists, population biologists, quantitative ecologists, cellular biologists, biochemists and those in the peripheral sciences suggested above all contribute to this knowledge. Based on this knowledge we create management objectives and management options that are beneficial to the vegetation and soil on rangelands.

Within this volume the authors have described vegetation science knowledge and understanding for rangelands. First has been the description of basic vegetation science contributions. Synecology, ecophysiology, taxonomy and plant breeding have been described along with special considerations of successional concepts, the role of cryptogams and seed ecology. This has been followed by one chapter dealing with modelling of rangeland ecosystems.

Secondly, this volume delves into vegetation distribution and organization on rangelands. Various authors have considered vegetation-soil relationships, land-type vegetation relationships in Australia, rangeland (veld) vegetation in South Africa, forest-range relationships, tundra vegetation, shrubland/woodlands, grasslands and riparian vegetation.

Finally, the volume completes the picture of vegetation science and rangelands by taking a closer look at how vegetation science is actually used on rangelands. Applications of vegetation science to the art and science of range management are both varied and complex. The chapters here include rangeland vegetation productivity and biomass, rangeland vegetation-hydrology interactions, grazing management and vegetation response, understanding fire ecology on rangelands, mined land reclamation on rangelands, range vegetation as wildlife habitat and contributions to rangeland revegetation.

Plant species and their rangeland environments must be well understood from all vegetation science viewpoints before there can be higher expectations for improved management. The management of rangelands throughout the world must be based on sound vegetation science principles. This will lead to wise use, sustained yield and the prevention of desertification and other forms of land degradation. These problems are still all to prevalent, leading to decreased productivity, loss of topsoil, reduced carrying capacity, decimation of wildlife species and a reduced ability to produce wood fuels, water and other requirements for man. These in turn have led to declining well-being, malnutrition and even directly to starvation of human populations. Thus it is our hope that this volume can make a contribution toward improved management of planet earth's rangelands.

1.7. References

Bhojwani, S. S., and M. K. Razdan. 1983. Plant tissue culture: theory and practice. Elsevier, Amsterdam.

De Santo, R. S. 1978. Concepts of applied ecology. Springer-Verlag, New York.

Miles, J. 1979. Vegetation dynamics. Chapman and Hall, London.

McArthur, E. D. 1988. New plant development in range management. This volume.
Noble, J. C. 1986. Plant population ecology and clonal growth in arid rangeland ecosystems. p. 16—19. *In*: Rangelands: a resource under siege. Proceeding of the second international rangeland congress. Australian Academy of Science, Canberra.
Noble, I. R. 1987. The role of expert systems in vegetation science. Vegetatio 69: 115—121.
Range Inventory Standardization Committee. 1983. Report presented to the Board of Directors, Society for Range Management, Albuquerque, New Mexico, published by the Society for Range Management.
Smith, E. L. 1988. Successional concepts in relation to range management. This volume.
Stiles, D. 1984. Desertification: a question of linkage. Desertification Control Bull. 11: 1—6.
Stoddard L. A., A. D. Smith, and T. W. Box. 1975. Range Management. McGraw-Hill Book Company. New York.
West, N. E. 1988. Plant synecology in the service of rangeland management. This volume.
Williams, R. E., B. E. Allred, R. M. Denio, and H. A. Poulsen, Jr. 1968. Conservation, development, and use of the world's rangelands. Jour. Range Manage. 21: 355—360.

PART ONE

Basic Vegetation Science Contributors

2. Plant synecology in the service of rangeland management

NEIL E.WEST

2.1 Abstract

Plant synecology deals with the classification and/or ordination of vegetation and the relationships of plant communities to environments. Understanding these aspects of rangelands is important for stratifying inventories, extrapolating research, transferring experience, and effectively evaluating and applying management alternatives. Many different approaches to classifying and mapping vegetation have been used across the rangelands of the world. Each of the approaches has underlying assumptions about the dynamics and organization of vegetation. Vegetation is but one of the most easily seen portions of ecosystems — the level of nature that inevitably responds to management. We are, however, far from agreement on how to classify ecosystems. Thus, vegetation has and will likely remain a major way we stratify land. We should not, however, be satisfied with only a classification scheme for the vegetation of a given area because recognition of how vegetation relates to environmental patterns is needed to explain why certain management approaches succeed or fail.

2.2 Introduction

Vegetation, being composed of primary producers, is the basis of the food web in rangeland ecosystems. In addition to forage, vegetation also provides shade and shelter from adverse weather for livestock and habitat for native animals. Living vegetation and its litter also play important roles in modifying soil chemical and physical structure, protecting it against accelerated erosion, and intermediating in the hydrologic cycle. Range managers must therefore be cognizant of spatial and temporal variation in vegetation and its relationships to environment if they are going to sustain production from their land holdings.

Tueller, P. T. (ed.), Vegetation science applications for rangeland analysis and management.
ISBN 90-6193-195-9.
© 1988, *Kluwer Academic Publishers, Dordrecht, Boston, London.*

Vegetation is the part of ecosystems which can be most easily manipulated to reach certain management objectives. Repetition of successes and avoidance of failures in these manipulations at additional locations cannot be achieved, however, unless the comparability of ecological conditions are known. One commonly used means of assessing similarity of biologically effective environmental conditions is to compare the similarity of vegetation. Many different ways of using this 'phytometer' (Clements and Goldsmith 1924) or indicator idea have been practiced around the world. Methods have grown increasingly sophisticated over time. It is therefore easiest to discuss this topic on a historical basis.

2.3 Definitions

Plant sociology, and particularly the sub-discipline now called syntaxonomy (van der Maarel 1984), deals with the classification and nomenclature of vegetation. Plant synecology, in the European sense, deals only with the study of relationships between plant communities and their environments. Americans, however, have traditionally included both activities under the label, plant synecology. Mueller-Dombois and Ellenberg (1974) propose the new term 'vegetation ecology' to encompass both of the topics plus changes in vegetation over time. I will use plant synecology in the American sense. Vegetation dynamics is considered separately by Smith (this volume). After discussion of some general principles, a review of syntaxonomy will be given, followed by a shorter consideration of environmental relationships.

2.4 General principles

2.4.1 Classification and mapping by dominant plants

Plant community types have been typified on many different bases. The initial stratifications have usually been based on the most conspicuous growth forms and/or species. Height, cover, or weight dominance are the usual variables noted. Cover type maps have been produced as the usual initial stratifications for forage production and grazing capacity determinations. Erection of these categories is usually first seen during the adjudication process that takes place when governments first become involved in allocating grazing commons to individual graziers (West 1982).

Height and cover dominants are the most easily identified components of vegetation. Their use in stratifying landscapes persists to the present day because even sophisticated remote sensors pick up mainly the major overstory forms and species. This simplistic approach has two shortcomings. First, widespread cover dominants are likely to possess considerable ecotypic variation and secondly, widespread cover dominants may fill different successional roles on different sites.

2.4.2 Vegetation as a part of ecosystems

It is the ecosystem that responds to management action or inaction, not just vegetation, soils, micro-climate, etc. The land manager would thus like to recognize where one ecosystem stops and another begins and how similar or different an ecosystem is from other areas where he has observed success or failure of a particular management approach. Ecosystems, however, include all biota (including man), soils and atmospheric influences, as well as the interactions between these components. The possibilities for combining these features are enormous (West and Shute 1978).

Quantitatively describing animal communities, soils, and environmental variables is expensive and time consuming. Vegetation, however, is abundant and sessile and indicates, through the presence and vigor of growth of key species, the integrated influences of climate, soils, herbivores, fires and other catastrophic environmental events. When we know how plant species respond to these environmental complexes, then we can read the effects of past environments. If we rely on the indicator value of only one or a few species, however, we are ignoring other information available in the total community.

2.4.3 Ecotypic and subspecific variation

Widespread, dominant species typically display considerable intraspecific variation. That is, ecotypes have evolved to fit local environments. This means that the same species indicates different environments in different parts of its range. For instance, little bluestem (*Schizachyrium scoparium*) responds differently to temperature and photo-period in northern Texas than in the grasslands of the northern Great Plains (McMillan 1959). Thus, this mid-grass will play different roles in the diverse communities in which it occurs across its total range of distribution. A single description of its indicator value thus cannot be written. Because ecotypic phenomena were not widely recognized before about 1950, one has to be careful of statements about what a given species indicates written in older literature. For instance, the Range Plant Handbook (Forest Service 1937) states that winterfat [*Ceratoides* (*Eurotia*) *lanata*] is indicative of saline soils. There are, however, winterfat ecotypes that are found on non-halomorphic soils (Workman and West 1969). Ecotypic variation can normally be identified only by experimental means; usually by growing these races in common environments and/or reciprocal transplant gardens. Although Quinn (1978) has concluded that plant ecotypes are neither good basic evolutionary or ecological units, the utility of the concept remains for those in applied fields.

Harper (1982) emphasized the problems of using the species as an ecological unit and the "failure of taxonomic categories to fit ecological categories." Nevertheless, range scientists have found subspecific designations of major species are useful ways to recognize differences in range sites. When based on morphology or simple chemical tests, subspecific designations can be done with relative ease. For instance, prior to about 1960, big sagebrush (*Artemisia tridentata*) was commonly considered to be the same kind of plant across its entire 100 million ha range in the western U.S. Scientists (e.g., Beetle 1960), however, began to explore the variations within this species. Over the next 25 years, subspecies were recognized and related to environmental variables. Keys were written so that managers could more easily identify the principal subspecies and use them to compare environments. Although much more remains to be discovered about this and other sagebrush taxa, we now do more sophisticated planning and management because of this knowledge. For instance, (Winward 1983) was recently able to show that different revegetation procedures should be applied where wildfire had covered a large portion of a mountain range occupied by different sub-species of big sagebrush. Johnson (1986) demonstrated how attention to the taxonomy and autecology of other North American *Artemisia* can point out opportunities, as well as allow avoidance of problems that coarser-grained management would create.

Something similar may need to be done for mulga (*Acacia aneura*), bladder saltbrush (*Atriplex vesicaria*), kangaroo grass (*Themeda triandra*), and spear grass (*Heteropogon contortus*) in Australia. As it now stands, these widespread, important species respond very differently to distant environments in that country. Considerable morphological variation in these taxa and the initial work done by Tothill (1977) hint that some of the confusion may begin to be sorted out if sub-species and ecotypes were identified.

Because of genetic variation, environmental compensations, different mixes of companion species and the facilitation and/or competition they provide, a given species may fit into successional pathways in quite different ways. We thus cannot uniformly label a widespread species as being pioneer (early seral), mid-seral or late seral (climax). A plant's successional role depends on ecological contexts and circumstances (Noble and Slatyer 1980). For instance, galleta (*Hilaria jamesii*), a grass species occurring in many parts of the American Southwest and southern Great Plains, is a decreaser with cattle grazing on 12 range sites identified by the Soil Conservation Service, an increaser on 75 other sites and an invader on 3 sites (West 1972). Obviously, a simple description of how this species responds to livestock grazing cannot be written. Thus, its presence and abundance means radically different things in different places.

2.5 Short historical review of syntaxonomy

2.5.1 Zurich-Montpellier School

Species of more restricted distribution have fewer of the problems outlined above, but others of their own. The phenomenon of species restrictions to particular kinds of environments is called "fidelity". This concept was the basis for the earliest development of phytosociology in southern Europe (the Zurich-Montpellier or SIGMA School), beginning in the nineteenth century (Whittaker 1962). Some of the first demonstrations of this approach were on the grazed alpine meadows of the Swiss Alps (Braun-Blanquet 1913). Thus, rangelands have figured prominently in the development of vegetation science since its earliest days.

Phytosociologists have been concerned with describing and classifying vegetation in a manner analogous to the way that classical taxonomists operated. Species that were restricted to particular circumstances (diagnostic species) were searched for in the same way that taxonomists searched for key characters to separate similar taxa. Even though some of the analytic and synthetic methods used in this approach have been widely criticized as being too subjective and unrepeatable (Mueller-Dombois and Ellenberg 1974), the matter remains that more of the world's vegetation has been classified by these means than any other approach. In certain parts of the world the only published literature on vegetation that exists has been done in this way. For instance, virtually the only work on North African rangeland vegetation has been done in the manner of the Zurich-Montpellier School. Werger (1977, 1981) has also demonstrated the utility of this approach in southern Africa. Regardless of one's opinion of the quality of science behind it, one may be forced to rely on these papers, if no other vegetation data exist.

The southern European approach to phytosociology has rarely been taken for rangelands beyond Europe and Africa. Some exceptions are

Bridgewater (1981) and Beadle (1980) in Australia, Looman (1969, 1982) in the prairie provinces of Canada, and Willard (1979) in the alpine of Colorado. Ibrahim *et al.* (1972) tried the Zurich-Montpellier approach for salt desert shrub vegetation in eastern Utah. Because of the low diversity and high fidelity of the cover dominants for different segments of topographic and edaphic gradients, the results were similar to the cover dominance approach favored by British, Australian, New Zealand and American workers. Danin (1983) has had a similar experience trying to adapt the Southern European approaches to the deserts of Israel and the Sinai. Since diagnostic, characteristic or differential species need not be the abundant or dominant ones, a rare species may be the prime basis of this classification. Unless one is well trained in plant taxonomy and willing to search beyond a few small plots, these less abundant, faithful species would be missed.

2.5.2 Monoclimax-based approaches

The lack of concern for successional contexts is also a feature of the earlier work under the Southern European approach that puts off many present day users of phytosociological information. Most American and Soviet vegetation scientists see understanding of successional dynamics of vegetation as crucial to both basic and management issues. American scientists contributed early to the development of the succession concept (Cowles 1901). Clements (1916) was immensely influential in stressing the role of climate in influencing vegetation development. This paralleled Dokuchaev's (1898) stress on climate in soil development. Zonal soils and vegetation were viewed as those in equilibrium with regional climate. The vegetation and soils that could develop without disturbances on undulating uplands became points of reference for classification schemes.

Clements and his students lived and worked mainly on the Great Plains and were much involved with the development of the range science profession during the 1920's and 30's. It is thus inevitable that the monoclimax theory would become imbedded in the earliest approaches to classifying U.S. rangeland. Indeed Dysterhuis (1949) was successful in transferring Clement's theory to every day usage by range conservationists in the Soil Conservation Service (Shiflet 1975).

2.5.3 Polyclimax-based approaches

Clement's point of reference was an ideal perhaps obtainable only on a large spatial and long time scale. Many modifiers to climax had to be used to explain present day deviations from the ideal. A terminological jungle was created. Tansley (1935) saw environmental factors beside climate as preventing further vegetational development. This is the key idea behind the poly-climax concept. Daubenmire (1942, 1952, 1955, 1976) was influential in translating this poly-climax concept, along with other Northern European and Soviet ideas, into a classification scheme for which the 'habitat type' is the key unit. The methods of Daubenmire and Braun-Blanquet are similar in some respects (Ratliff and Pieper 1982), but Daubenmire emphasizes dominance, stratal unions, and a poly-climax base, whereas Braun-Blanquet emphasizes fidelity, even if rare species are involved, and ignores successional phenomena.

The habitat type is a unit of land over which a certain kind of climax vegetation already occurs or could supposedly return. Daubenmire (1970) has had many followers in the Pacific Northwest of the U.S.A., particularly in the Forest Service (Mueggler and Stewart 1980, Pfister 1981, Tew 1983).

During the past few years there have been efforts to relate the land classifications based on habitat types, range sites and community types to each other (Fischer 1986, Hironaka 1986) and thus make it possible to provide the land manager with a more comprehensive information storage

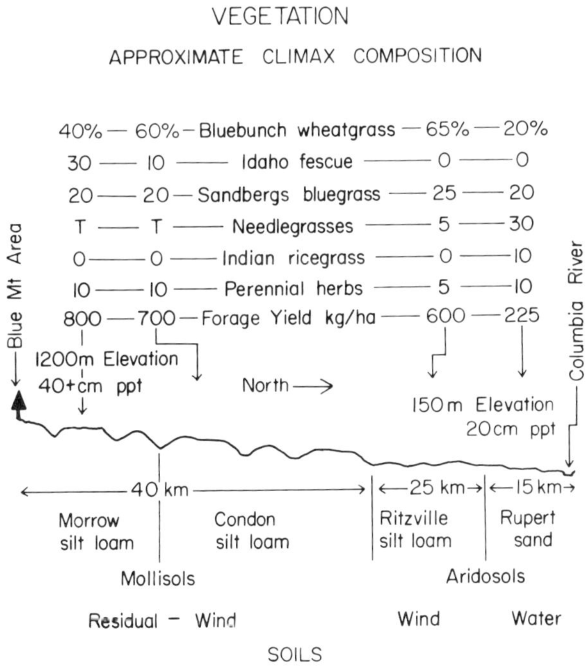

Figure 1. General north-south profile section of the Columbia Basin area of eastern Oregon showing general soil-plant relationships as influenced by changing soils, elevation, precipitation, geographic position and other climatic and physiographic factors (Anderson, 1956).

and retrieval system. This is desirable because the three classification schemes are not duplicative and all three are needed to classify our rangelands in a more adequate way. The habitat type classification provides the hierarchical, ecologically and geographically based framework for classification. The range site designation specifies productivity and relationships to soils and climate, e.g., (Fig. 1). The community type data deals with the present vegetation. We must begin with the extant vegetation to understand successional trajectories under various regimens of disturbance by nature and man.

2.5.4 Newer approaches

The three prevailing phytosociological approaches (monoclimax, polyclimax, Zurich-Montpellier) used on American and other rangelands are underpinned by older assumptions or theories of how vegetation is organized in space and time. Briefly, these are that vegetation forms natural, internally homogeneous entities and that species interact primarily through modifying their neighbors' environment, facilitating certain sequences of species replacement over time. Climates are considered either stable or unimportant under the classical approaches to classifying vegetation. Furthermore, advocates of the habitat type and range site approaches assume that a stable, self-perpetuating climax is possible and describable enough to be the basis of classification of land. All of these assumptions have been questioned and rejected by many (e.g., Slatyer 1977, Randall 1978, Miles 1979, Austin 1986, Roberts 1987, Smith, this volume).

Range scientists should be aware that many basic ecologists no longer believe in the reality of communities. Some regard this level of organization as being composed of only temporary assemblages of species from an available pool. They contend that unless functional interactions between sympatric species can be demonstrated, the community is only a unit of man's convenience. The debate is thus over the strength of interactions and how they contribute to community structure. Another focus for this debate is whether higher order, "emergent" properties (not readily derivable from an analysis of constituent species) occur (MacIntosh 1985, Committee on the Applications of Ecological Theory to Environmental Problems 1986).

Applied ecologists know that some species do indeed interact, often in indirect ways that can be understood only in the context of the functional structure of the whole community. They are also aware that ecological communities have numerous properties that transcend those of their constituent species and require study themselves — e.g., synusia, diversity, food webs, ecotones, successional trends etc. We will thus, probably always operationally retain some of these community level concepts. After all, physicists operate comfortably with both wave and particulate theories of light. Ecologists thus will have to get used to the opposing views of communities. Until more adequate and universally accepted theories of vegetation are developed (Austin 1986), we are left with *ad hoc* ways of solving our problems.

Some of the newer, alternative ways of describing vegetation patterns (numerical classification, ordination; Gauch 1982) involve assumptions of at least some non-connected species response, i.e., individualistic distributions of species in space and time, unique successional trajectories driven by a variety of life history features, many possible biotic interactions only intermittently operative, and a large element of chance, especially in the initial floristic composition.

While many range managers regard grazing disturbance as a simple reversal of succession, many vegetation scientists now reject the notion of a single, equilibrium state and thus the automatic return to the pristine condition once proximate disturbing influences are removed. As Westoby (1979—80) points out, when grazing pressures are removed or relaxed in semi-arid environments, the systems often remain stuck where they are. If soil erosion has ensued, original species have been lost from the seedbank, invading species have become naturalized, and genotypes, climates and fire regimens have changed, one can hardly expect vegetation to return to what it was previously. While it is possible to accept these new notions and still construct an ecological framework for land management via ordination and gradient modelling (Kessell 1979), it is usually more complex and difficult to explain, as well as being more costly.

New kinds of interpretations are possible when computerized, multivariate approaches are used. For instance, Werger *et al.* (1979) and Tueller and Eckert (1987) have recently compared their interpretation of rangeland vegetation patterns via classical and modern "objective" approaches. The major conclusions were that the larger features of the patterns were similar; however, the multivariate approaches allowed more subtle relationships at smaller spatial scales to emerge. With very large data sets, computer analysis may allow syntheses that few human minds can visually interpret from only tabular data. Such levels of sophistication may, however, be unnecessary for current needs. For instance, Shute and West (1982) tested the relative merit of using general versus specific strata, in both discrete and continuous modes, to predict forage production in shrub steppes of Utah. They failed to find the expected increases in accuracy of prediction by using full floristic-based vegetation and environmental variables in continuous modes. These findings should not be unquestioningly extrapolated to other contexts. They do, however, suggest methodology that could help explore the gains of predictive accuracy provided by alternative land stratification schemes.

Use of almost any ecological stratification scheme is going to automatically increase accuracy and precision in prediction of product availability over randomly or regularly spaced inventories (Poulton 1959, West 1964). What is now needed are examples of how to relate productivity levels and management constraints and responses to the preferred strata, whether they be cover types, habitat types, range sites, plant community types or various ordination schema. Much of this could be done rather speedily using expert systems (Walker *et al.* 1985). Foresters and wildlife biologists are far ahead of the range professionals in doing this type of work.

We now also need to recognize that what is really needed for wildland management is a classification of ecosystems or landscapes, not just vegetation. Vegetation varies greatly in its time dimension, whereas land forms and soils are more stable. Land forms and soils however, reflect more about prior rather than current conditions, especially on older landscapes (Anderson 1986). We need to work together with geologists, soil scientists and climatologists to develop integratable systems of land classification (Bailey 1984). There has been much effort expended on this task during the past few decades in the developed nations. Two fundamentally different approaches have been used. The first approach is based on intrinsic properties of the land, with usually a hierarchically upward aggregation of units. The other approach is a top-down one, e.g., the broadest, largest scale differences are first recognized and then successively sub-divided, if finer-grained differences are needed. This latter approach inevitably means that fine-grained units are primarily related to features involving spatial proximity. The most basic units can be the same, but they are combined by different criteria at higher categories.

The rules for integrating characteristics of land cannot always be consistently explained. This is largely due to intrinsic differences in ecosystem components; they naturally sort out on different scales of time and space. For instance, in Figure 1, "hand in glove" interrelationships of vegetation, soils, and climate are implied. Such is not always the case, particularly in more rugged topography where numerous compensations occur (Jensen 1984). That is, one cannot expect that one plant community type will be linked with only one soil family since slope, exposure and disturbance history (e.g., fire) also influence vegetation development. Tightness of correlations of vegetation with present or potential land uses diminish with increased modifications made by man (Havel 1981). Different users also prefer to use favored components in particular combinations of scale in time and space. For instance, a watershed scientist may favor use of more detailed information on slope and soil properties and only rather coarse-grained levels of the vegetation classification to predict erosional losses with certain kinds of management manipulations. The developers of such a land classification, thus, cannot anticipate all the applications of the classification system.

The author was part of a team which developed a wildland classification system called ECOSYM (Davis and Henderson 1977). This employed multiple, hierarchical levels for components of bedrock, regolith, topography, climate, soils, current and potential vegetation (Table 1). A geographic information system was available via computer. Any user could choose the components and levels within components that he felt were relevant to answering his inventory or project planning needs. We wrote a wide variety of examples of how this system could be used to solve problems land managers typically face. This system is not in use, however, mainly because of human attitudes and institutional constraints: the Forest Supervisor could not find a diplomatic way to put a young, computer-aware generalist on his staff without threatening the older personnel responsible for various aspects of national forest operation. Thus, the reductionist trends in training of managers and structuring of institutions mitigates against using such an holistic view (Davis 1980).

Despite 15 years of effort, it has been impossible for the U.S. Forest Service to agree on using one hierarchical classification of wildland ecosystems across all its lands nationwide (Bailey *et al.* 1985). The lack of a uniform classification scheme has impeded aggregation of information on the state of resources in the national forests and grasslands across the nation. Agreement between federal and state resource agencies on a wildlands classification scheme is necessary if we are ever to obtain comparable data on resources (Driscoll 1981). Obtaining such agreement is unlikely, however, because legislatively limited institutional mission and maintenance of its pride mitigates against a willingness to compromise and fill data gaps. These fundamentally human traits are in evidence in other countries also. We will thus likely continue to see competing systems of land classification be discussed, but choices will always involve political influences as well as scientific merit and practicality. This is inevitable because recognition of what constitutes undesirable or preferred uses depends on policies which usually place social and economic criteria ahead of biophysical ones (Havel 1981). Vegetation scientists will always be at the forefront of such debates because vegetation is such an important part of wildland ecosystems and how they change under various use policies.

Table 1. The ECOSYM Classification Framework (Davis and Henderson 1977).

COMPONENTS	BEDROCK GEOLOGY	REGOLITH	TOPOGRAPHIC			CLIMATE		SOIL	CURRENT VEGETATION	POTENTIAL VEGETATION
			ELEVATION	SLOPE	ASPECT	TEMPERATURE	PRECIP.			
	Order (Igneous)	Order (Deposi-tional)	Megarelief (nearest 1000 ft)	Megaincline (nearest 10%)	Megadirection (90° segments)	Megatherm (nearest 10°C)	Megaprecip. (nearest decimeter)	Order (Mollisol)	Physiognomic type (Forest)	Physiognomic type (Forest)
								Suborder (Boroll)		
	Suborder (Course-crystalline)	Class (Colluvium)	Macrorelief (nearest 100 ft.)	Macroincline (nearest 1%)	Macrodirection (45° segments)	Macrotherm (nearest 1°C)			Formation (Subalpine forest)	Series Group (Subalpine forest)
							Macroprecip. (nearest centimeter)	Great Group (Cryoboroll)	Cover type (Abies lasiocarpa)	Series (Abies lasiocarpa)
Hierarchical Level with Illustrative Class Names	Class (Granitoid)	Type (Fall-Talus)	Minirelief (nearest 10 ft.)	Miniincline (nearest .1%)	Minidirection (22.5° segments)	Minitherm (nearest .1°C)		Subgroup (Typic Cryoboroll)	Community type (Abies lasiocarpa/ Berberis repens)	Habitat type (Abies lasiocarpa/ Berberis repens)
	Type (Quartzolite)						Miniprecip. (nearest millimeter)		Phase (Symphoricarpos oreophilus)	Phase (Symphoricarpos oreophilus)
									Condition Class	
		Locale (Rainier Ash)	Microrelief (nearest 1 ft.)					Family (Fine mixed)	Community	Community
		Conditional Series — applies to any level			Microdirection (nearest 1°)			Series (Kings Peak)	Sample	Sample
								Pedor	Site	Site

2.6 Relationships of plant communities to environments

In addition to its use in erecting land or ecosystem classificatory schemes, study of environmental relationships with plant communities is important in devising successful management strategies. For instance, the subtropical savannas of northeastern Australia appear physiognomically similar over a huge area. All of this region has an overstory of *Eucalyptus* trees with a grass dominated understory. The primary way European colonists have used this area is via cattle grazing. Management has been greatly intensified there during the past few decades. Some portions have easily tolerated increased usage whereas other portions have been seriously degraded. These differing responses indicated that even though vegetation structure was grossly similar, the ecosystems involved were not. Pastoralists, encouraged by researchers and extension specialists, had extrapolated results from a few, rather arbitrarily selected stations to locations where the expected responses did not materialize and instead damage was done. Researchers in the C.S.I.R.O. Division of Tropical Pastures realized that other bases of subdivision and comparison within the subtropical savanna region were needed besides phytosociological differences. Perry (1960) showed that understory and overstory components behaved quite independently of each other complicating conventional usage of phytosociology. Others consequently undertook studies of the geographical patterns of climate and soils. They also worked out how climate and soil interacted to influence growing seasons (McCown 1981). Distinctly different subregional patterns emerged (Walker and Gillison 1982, Mott and Tothill 1984, Mott *et al.* 1985). Differences in forage palatability, nutrient content and susceptibilities to fire and fertilization responses were then more readily explained.

The above is but one example of why the range scientist or manager cannot restrict his use of vegetation science only to syntaxonomy; the relationships of vegetation to its environment are frequently more essential to devising and then explaining why certain management strategies succeed or fail. Leonard *et al.* (this volume) give other examples of how vegetation-soil relationships are useful in range science and management. Bailey (this volume) shows how fire influences rangeland vegetation. Smith (this volume) emphasizes how fluctuations in vegetation structure due to climatic variation are often difficult to separate from the truly retrogressional or progressional aspects of vegetation dynamics.

Observation of undesirable impacts of large domestic herbivores on vegetation, along with other problems (droughts, economic depression, etc.), led to the creation of the range management profession. While means of reducing environmental damage was our first and continuing concern, we now need to shift some of our focus to how to reach a sustainable interaction between livestock and other uses of the land. Using plant synecological information more effectively is an important part of the desiderata for attaining this goal.

2.7 References

Anderson, D. J. 1986. Ecological succession. p. 269—285. *In*: J. Kikkawa, and D. J. Anderson (eds.) Community ecology: Pattern and process. Blackwell Sci. Publ., London.

Anderson E. W. 1956. Some soil-plant relationships in eastern Oregon. J. Range Manage. 9: 171—175.

Austin, M. P. 1986. The theoretical basis of vegetation science. Trends in Research on Ecology and Evolution. 1: 161—164.

Bailey, R. G. 1984. Integrating ecosystem components. p. 181—188. *In*: J. Bockheim (ed.) Proc. Forest Land Classification Symp., Univ. Wisconsin, Madison.

Bailey, R. G., S. C. Zoltai, and E. B. Wilken. 1985. Ecological regionalization in Canada and the United States. Geoforum 16: 265—275.

Beadle, N. C. W. 1980. The vegetation of Australia. Cambridge Univ. Press, Cambridge.

Beetle, A. A. 1960. A study of sagebrush — The Section Tridentatae of *Artemisia*. Wyo. Agric. Expt. Sta. Bull. 368.

Braun — Blanquet, J. 1913. Die Vegetations verhaltnisse der Schneestufe in den Ratisch-Lepontischen Alpen. Ein Bild des Pflanzenlebens an seinen ausserten Grenzen. Schweiz. Naturf. Gesell. Neue Denkschr. 48: 1—347.

Bridgewater, P. B. 1981. Potential application of the Zurich-Montpellier system of vegetation classification in Australia. p. 1—9. *In*: A. N. Gillison, and D. J. Anderson (eds.) Vegetation Classification in Australia. Austr. National Univ., Canberra.

Clements, F. E. 1916. Plant succession: an analysis of the development of vegetation. Carnegie Instit. Wash. Publ. 242. Wash., D.C.

Clements, F. E., and G. W. Goldsmith. 1924. The phytometer method in ecology. Carnegie Instit. Wash. Publ. 356 Wash., D.C.

Committee on the Applications of Ecological Theory to Environmental Problems 1986. Ecological knowledge and environmental problem solving: concepts and case studies. National Academy of Sciences, Washington, D.C.

Cowles, H. C. 1901. The physiographic ecology of Chicago and vicinity: a study of the origin, development, and classification of plant societies. Bot. Gaz. 31: 73—108.

Danin, A. 1983. Desert vegetation of Israel and Sinai. Cana Publ. House, Jerusalem.

Daubenmire, R. 1942. An ecological study of the vegetation of southeastern Washington and adjacent Idaho. Ecol. Monogr. 12: 53—79.

Daubenmire, R. F. 1952. Forest vegetation of northern Idaho and adjacent Washington and its bearings on concepts of vegetation classification. Ecol. Monogr. 22: 301—330.

Daubenmire, R. F. 1955. The use of vegetation to indicate grazing potentials of forest land. Proc. Soc. Amer. Foresters.

Daubenmire, R. 1970. Steppe vegetation of Washington. Wash. Agric. Expt. Sta. Tech. Bull. 62.

Daubenmire, R. 1976. The use of vegetation in assessing the productivity of forest lands. Bot. Rev. 42: 115—143.

Davis, L. S. 1980. Strategy for building a location specific, multipurpose information system for wildland management. J. Forestry 78: 402—408.

Davis, L. S., and J. A. Henderson. 1977. Many uses and many users: Some desirable characteristics of a common land and water classification system. p. 13—40. *In*: A. Marmelstein (ed.) Classification, inventory, and analysis of fish and wildlife habitat: The Proceedings of a National Symposium. Fish and Wildlife Service. U.S. Dept. Interior, Biol. Services Prog. FWS/OBS — 78/76.

Dokuchaev, V. V. 1898. Writings (in Russian), 6 vols. Akad. Nauk., Moscow.

Driscoll, R. S. 1981. The quest for a uniform classification system for land use allocation. p. 179—185. Proc. 1980. Convention of the Society for Amer. Foresters.

Dysterhuis, E. J. 1949. Condition and management of rangeland based on quantitative ecology. J. Range Manage. 2: 104—115.

Fischer, N. T. 1986. Vegetation ecology, sample adequacy, and determination of reclamation success. p. 189—215. *In*: C. C. Reith, and L. D. Potter (eds.) Principles and methods of reclamation science with case studies from the arid Southwest. Univ. New Mexico Press, Albuquerque.

Forest Service. 1937. Range plant handbook. Govt. Print. Office, Wash., D.C.

Gauch, H. G. Jr. 1982. Multivariate analysis in community ecology. Cambridge Univ. Press, New York.

Harper J. L. 1982. After description. p. 11—26. *In*: E. I. Newman (ed.) The plant community as a working mechanism, Special Publications Series of the British Ecological Society, No. 1, Blackwell Sci. Publ., Oxford.

Havel, J. J. 1981. Vegetation classification as a basis for land use planning. p. 219—226. *In*: A. N. Gillison, and D. G. Anderson (eds.) Vegetation classification in Australia. Austr. National Univ. Press, Canberra.

Hironaka, M. 1986. Habitat type, range site and community type. p. 15—18. *In*: E. D. McArthur, and B. L. Welch (compilers) Proc-Symp. on the biology of *Artemisia* and *Chrysothamnus*. Provo, Utah. USDA Forest Service. Gen. Tech. Report INT-200. Intermountain Research Station, Ogden, Utah.

Ibrahim, K. M., N. E. West, and D. L. Goodwin. 1972. Phytosociological characteristics of perennial *Atriplex* dominated vegetation of southeastern Utah. Vegetatio. 24: 13—22.

Jensen, M. E. 1984. Some edaphic relations of southern Idaho wildlands. Great Basin Nat. 44: 265—271.

Johnson, K. L. 1986. Sagebrush types as ecological indicators to integrated pest management (IPM) in the sagebrush ecosystem of western North America. p. 1—10. *In*: J. A. Onsager (ed.) Integrated pest management in rangeland: State of the art in the sagebrush ecosystem. USDA Agric. Research Service. ARS-50.

Kessell, S. R. 1979. Gradient modeling: resource and fire management. Springer-Verlag, New York.

Looman, J. 1969. The fescue grasslands of western Canada. Vegetatio 19: 128—145.

Looman, J. 1982. Grasslands of western North America: Fescue grasslands. p. 209—221. *In*: A. C. Nicholson *et al.* (eds.) Grassland ecology and classification. Ministry of Forests, Province of British Columbia.

Maarel, van der E. 1984. Dynamics of plant populations from a synecological viewpoint. p. 66—82. *In*: R. Dirzo, and J. Sarukhan (eds.) Perspectives in plant population ecology. Sinauer Assoc., Sunderland, MA.

MacIntosh, R. P. 1985. The background of ecology: concept and theory. Cambridge Univ. Press, Cambridge.

McCown, R. L. 1981. The climate potential for beef cattle production in tropical Australia. II. Variation in the cessation and duration of the green season. Agric. Syst. 7: 163—178.

McMillan, C. 1959. The role of ecotypic variation in the distribution of the central grasslands of North America. Ecol. Monogr. 29: 285—308.

Miles, J. 1979. Vegetation dynamics. Chapman and Hall, London.

Mott, J. J., and J. C. Tothill. 1984. Tropical and subtropical woodlands. p. 255—269. *In*: G. N. Harrington, A. D. Wilson, and M. D. Young (eds.) Management of Australia's rangelands. C.S.I.R.O., Canberra.

Mott, J. J., J. Williams, M. H. Andrew, and A. N. Gillison. 1985. Australian savanna ecosystems. p. 56—82. *In*: J. C. Tothill, and J. J. Mott (eds.) Management and ecology of the world's savannas. Aust. Acad. Sci., Canberra.

Mueggler, W. F., and W. L. Stewart. 1980. Grassland and shrubland habitat types of western Montana. USDA, Forest Service Genl. Tech Rep. INT-66. Intermountain Res. Sta., Ogden, Utah.

Mueller-Dombois, D., and H. Ellenberg. 1974. Aims and methods of vegetation ecology. John Wiley & Sons, New York.

Noble, I. R., and R. O. Slatyer. 1980. The use of vital attributes to predict successional changes in plant communities subject to recurrent disturbances. Vegetatio 43: 5—21.

Perry, R. A. 1960. Pasture lands of the Northern Territory, Australia. Comm. Sci. and Ind. Res. Organ. Land Research Ser. No. 5, Canberra.

Pfister, R. D. 1981. Habitat type classification for managing western watersheds. p. 59—67. *In*: D. M. Baumgartner (ed.) Interior West Watershed Management. Proc. Symp., Spokane, WA, Wash. St. Univ. Coop. Extn.

Poulton, C. E. 1959. Soil-vegetation research and surveys in multiple use management of western ranges. p. 359—370. *In*: H. B. Spraque (ed.) Grasslands. Amer. Assoc. Adv. Sci., Wash., D.C.

Quinn, J. A. 1978. Plant ecotypes: ecological or evolutionary unit. Bull. Torrey Bot. Club 105: 58—64.

Randall, R. E. 1978. Theories and techniques in vegetation analysis. Oxford Univ. Press, Oxford.

Ratliff, R. D., and R. D. Pieper. 1982. Approaches to plant community classification for the range manager. Soc. for Range Manage., Sci. Mongr. 1., Soc. Range Manage., Denver, Colorado.

Roberts, D. W. 1987. A dynamical systems perspective on vegetation theory. Vegetatio 69: 27—33.

Shiflet, T. N. 1975. Range sites and soils in the United States. p. 23—33. *In*: D. N. Hyder (ed.) Arid Shrublands Proc. 3rd Workshop of the US/Australia Rangelands Panel. Soc. Range Mgt., Denver.

Shute, D. A., and N. E. West. 1982. Two basic methodological choices in wildland vegetation inventories: Their consequences and implications. J. Appl. Ecol. 19: 249—262.

Slatyer, R. O. (ed.) 1977. Dynamic changes in terrestrial ecosystems: patterns of change, techniques for study and applications to management. MAB Tech. Notes No. 4. UNESCO, Paris.

Tansley, A. G. 1935. The use and abuse of vegetation concepts and terms. Ecology 16: 284—307.

Tew, R. K. 1983. Management implications associated with land stratification and habitat typing. p. 2—5. *In*: S. B. Monsen, and N. Shaw (compilers) Managing Intermountain Ragelands — Improvement of Range and Wildlife Habitats. U.S.D.A. Forest Service Genl. Tech. Rep. INT-157. Intermountain Forest and Range Experiment Sta. Ogden, Utah.

Tothill, J. C. 1977. Seed germination studies with *Heteropogon contortus*. Aust. J. Ecol. 2: 477—484.

Tueller, P. T., and R. E. Eckert Jr. 1987. (*Artemisia tridentata vaseyana*) and longleaf snowberry (*Symphoricarpos oreophilus*) plant associations in northeastern Nevada. Great Basin Naturalist 47: 117—131.

Walker, J., J. R. Davis, and A. M. Gill. 1985. Towards an expert system for fire management at Kakadu National Park. C.S.I.R.O., Inst. Biol. Res. Dept. Water and Land Resources, Canberra, Tech. Memo. 85/2.

Walker, J., and A. N. Gillison. 1982. Australian savannas. p. 1—24. *In*: B. J. Huntley, and B. H. Walker (eds.) Ecology of tropical savannas. Ecol. Stud. 42. Springer-Verlag, Berlin.

Werger, M. J. A. 1977. Applicability of Zurich-Montpellier methods in African tropical and sub-tropical rangelands. p. 125—145. *In*: W. Krause (ed.) Application of Vegetation Science to Grassland Husbandry. Handbook Veg. Sci. Vol. 13, Dr. W. Junk, The Hague, Netherlands.

Werger, M. J. A. 1981. Phytosociological inventory as a basis for resource evaluation in aridlands. p. 338—341. *In*: H. G. Lund *et al.* (compilers) Arid land resource inventories: Developing cost-efficient methods. U.S. Dept. Agric., Forest Service, Gen. Tech. Rep. WO-28. Wash., D.C.

Werger, M. J. A., J. W. Morris, and J. M. W. Louppen. 1979. Vegetation-soil relationships in the southern Kalahari. Documents Phytosociologiques (N. S.) 4: 967—981.

West, N. E. 1964. Contributions of plant synecology to pure and applied biology. Biologist 46: (3—4): 73—80.

West, N. E. (ed.) 1972. Galleta: Taxonomy, ecology and management of *Hilaria jamesii* on western rangelands. Utah, Agric. Expt. Sta. Bull. 487.

West, N. E. 1982. Approaches to synecological characterization of wildlands in the intermountain West. p. 633—643. *In*: T. C. Brann (ed.) In-place resource inventories: Principles and practices. A national workshop, Univ. Maine, Orono. Society of American Foresters, McClean, Virginia.

West, N. E., and D. A. Shute 1978. Alternatives for ecosystem classification and their implications for rangeland inventory. p. 174—176. *In*: D. N. Hyder (ed.) Proc. First Int. Rangeland Congress, Society for Range Manage., Denver, Colo.

Westoby, M. 1979—80. Elements of a theory of vegetation dynamics in arid rangelands. Israel J. Bot. 28: 169—194.

Whittaker, R. H. 1962. Classification of natural communities *Bot. Rev.* 28: 1—239.

Willard, B. E. 1979. Plant sociology of alpine tundra, Trail Ridge, Rocky Mountain National Park, Colorado. Colo. School of Mines Quarterly No. 74, Golden, Colorado.

Winward, A. H. 1983. Using sagebrush ecology in wildland management, p. 15—19. *In*: K. L. Johnson (ed.) First Utah Shrub Ecology Workshop. College of Natural Resources, Utah State University, Logan.

Workman, J. P., and N. E. West. 1969. Ecotypic variation of *Eurotia lanata* populations in Utah. Bot. Gaz. 130: 26—35.

3. Ecophysiology of range plants

RONALD E. SOSEBEE AND CHANGGUI WAN

3.1. Introduction

Plant "ecophysiology" is often used interchangeably with "environmental physiology" and "physiological ecology" depending upon one's background (Tracy and Turner 1982). Plant ecophysiology is primarily concerned with "the physiology of plants as it is modified by fluctuating external influences" (Larcher 1980). Emphasis of physiological relationships in plant ecology is not new. Schimper (1898) was one of the first to suggest that ecological relationships be approached from a physiological viewpoint. Tansley (1914) suggested that the most fundamental problems of plant ecology are physiological, and called for the formation of a School of Physiological Ecology to address the basic issues of ecology.

The physiological approach to ecological problems seems to have been largely abandoned until the 1960's. In 1962 an international symposium was held at Montpellier, France, to address ecophysiological problems of arid zone research (Eckardt 1965). This symposium served as the springboard for renewed interest in plant ecophysiology or physiological ecology. Since the mid 1960's, there has been a major emphasis on plant ecophysiology.

It is unfortunate that plant ecologists and plant physiologists generally have worked in restricted niches with limited exchange of information between disciplines. Such a sharing of information would be helpful to workers in both research arenas. For example, the diversity of species exploiting a particular environment reflects specific adaptations of the species to "normal" and "extreme" ranges of evironmental factors which vary in time and space. The nature of these adaptations must be examined in view of the plant's physiological ability to respond to the environment (Fitter and Hay 1981). Favorable environmental conditions are not necessarily determined by regional climates or even microclimates. Rather, adaptability is determined by the environment in which the vital processes

Tueller, P. T. (ed.), Vegetation science applications for rangeland analysis and management.
ISBN 90-6193-195-9.
© 1988, *Kluwer Academic Publishers, Dordrecht, Boston, London.*

take place within the protoplasm (Walter 1973). A plant's adaptability to specific environments largely depends upon its enzyme and membrane responses to the internal conditions created by the external environment. Thus, the field plant ecologist studying species distributions in relation to environmental gradients would be well-advised to develop an appreciation for the "physiological perception of the plant" to its environment.

There are many excellent books and reviews that address various facets of ecophysiology. Therefore, this chapter will not attempt a complete coverage of ecophysiology, but it will attempt to discuss specific physiological responses of rangeland ecosystems.

3.2 Energy relationships

Radiant energy supports all life and drives every ecosystem. The intensity of radiation at the outer limits of the earth's atmosphere (about 30 km above the earth's surface) is defined as the *solar constant* and is equivalent to 1360 W m^{-2}, or 1.95 cal cm^{-2} min^{-1} (Gates 1980). Only 47% of the incoming radiation reaches the earth's surface. The amount of radiation

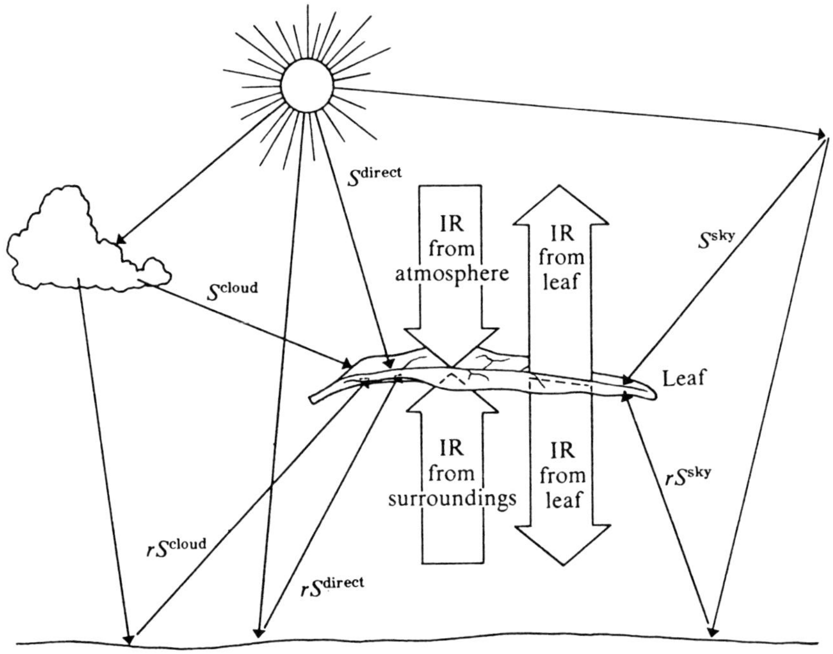

Figure 1. Schematic illustrations of eight forms of radiant energy incident on an exposed leaf, and the infrared radiation emitted from its two surfaces. (From BIOPHYSICAL PLANT PHYSIOLOGY AND ECOLOGY by Park S. Nobel. Copyright 1970, 1974, 1983. W. H. Freeman and Company. Reprinted with permission.)

that reaches a plant community depends upon latitude, altitude, atmospheric conditions, slope exposure, and season of year. Radiation reaching the ground or a plant community consists of direct sunlight, diffuse radiation from clouds or cloud light, and diffuse radiation from the sky or skylight (Fig. 1). Noon radiation at sealevel at intermediate latitudes is approximately 900 W m^{-2} or 1.3 cal cm^{-2} min^{-1} (Larcher 1980).

Incident radiation reaching the earth's atmosphere is composed of various wavelengths. About 5% of the photons are in the ultraviolet (UV, < 390 nm) wavelength, 28% in the visible (390—760 nm), and 67% in the infrared (IR, > 740 nm). Most of the UV entering the earth's upper atmosphere is absorbed by ozone and is prevented from striking the earth's surface. Nearly all wavelengths < 300 nm are eliminated by the ozone. Much of the infrared radiation from the sun is absorbed by water vapor and CO_2 (Fig. 2). Thus, about 2% of the photons impinging upon the earth's surface are in the UV, 45% in the visible, and 53% in the infrared wavelengths. The short wavelengths reaching the earth's surface peak at 680 nm (Fig. 3) which coincides with the red absorption band of chlorophyll (Nobel 1974).

Radiant energy is transferred by photons or discrete packages of energy that travel at the speed of light. The wavelengths of solar radiation are related to the energy of the photons (Table 1) according to Planck's quantum law:

$$e = hc/\lambda$$

where:

e = energy of the wavelength, joules
h = Plack's constant, 6.62×10^{-34} J s
λ = wavelength, m
c = speed of light, 3×10^{8} m sec^{-1}

All objects, including plants, with a temperature above absolute zero (0°K, −273°C) emit energy. Likewise, all objects are capable of absorbing energy. The amount of energy radiated by a body is a function of the fourth power of its absolute temperature according to Stefan-Boltzmann law of radiation:

$$R = \sigma (T + 273)^4$$

where:

R = amount of energy radiated, W m^{-2}
σ = Stefan-Boltzmann constant, 5.67×10^{-8} W m^{-2} T^{-4}
T = surface temperature, °C

Any object that is capable of perfectly emitting its energy is called a

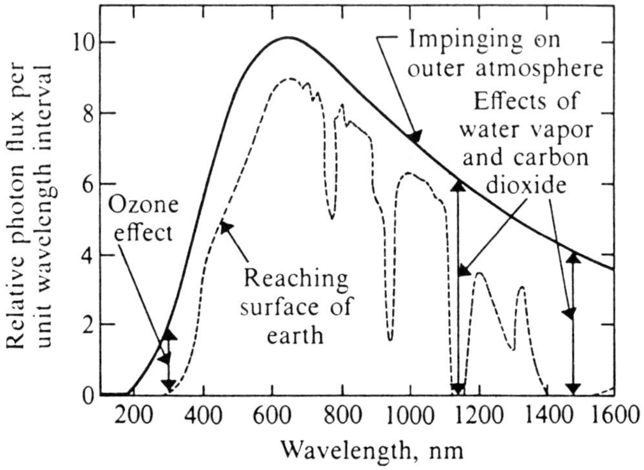

Figure 2. Wavelength distributions of the sun's photons incident on the earth's atmosphere and its surface. The curve for the solar radiation on the atmosphere is an idealized one based on Planck's radiation distribution formula. The spectral distribution and amount of solar radiation reaching the earth's surface depends on clouds, other atmospheric condi-, tions, altitude, and the sun's angle in the sky. The pattern indicated by the lower curve is appropriate at sea level on a clear day with the sun nearly overhead. (From BIOPHYSICAL PLANT PHYSIOLOGY AND ECOLOGY by Park S. Nobel. Copyright 1970, 1974, 1983. W. H. Freeman and Company. Reprinted with permission.)

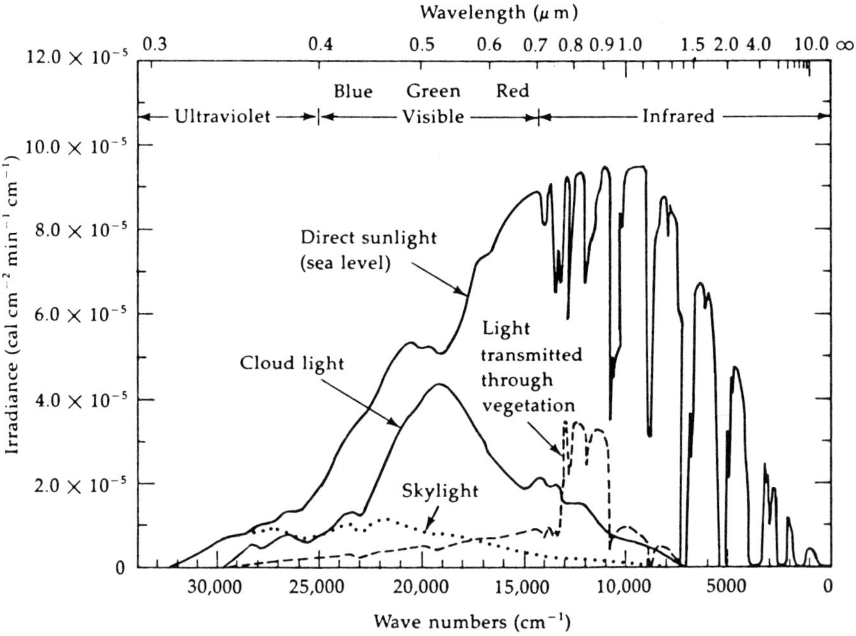

Figure 3. Spectral distribution of direct sunlight, skylight, cloudlight, and sunlight penetrating a stand of vegetation. Each curve represents the energy incident on a horizontal surface. (From Gates 1965. Meteorological Monographs 6: 1—26. The American Meteorological Society. Reprinted with permission.)

Table 1. Definitions and characteristics of the various wavelength regions of light. The ranges of wavelengths leading to the sensation of a particular color are somewhat arbitrary and vary with individuals. Both frequencies and energies in the table refer to the particular wavelength indicated in column 3 for each wavelength interval. Wavelength magnitudes are those in a vacuum. (From BIOPHYSICAL PLANT PHYSIOLOGY AND ECOLOGY by Park S. Noble. Copyright 1970, 1974, 1983. W. H. Freeman and Company. Reprinted with permission.)

Color	Approximate wavelength range (nm)	Representative wavelength (nm)	Frequency (cycles/sec) (Hertz)	Energy (ev per photon)	(kcal/"mole" of photons)
Ultraviolet	below 400	254	11.80×10^{14}	4.88	112.5
Violet	400—425	410	7.31×10^{14}	3.02	69.7
Blue	425—490	460	6.52×10^{14}	2.70	62.2
Green	490—560	520	5.77×10^{14}	2.39	55.0
Yellow	560—585	580	5.17×10^{14}	2.14	49.3
Orange	585—640	620	4.84×10^{14}	2.00	46.2
Red	640—740	680	4.41×10^{14}	1.82	42.1
Infrared	above 740	1400	2.14×10^{14}	0.88	20.4

"blackbody". A blackbody is a convenient idealization describing any object that uniformly absorbs all wavelengths and emits them according to Planck's radiation distribution law (Nobel 1974). This distribution law is a good approximation for describing radiation from the sun whose surface temperature is approximately 5800°K. A perfect emitter has an emissivity $(\varepsilon) = 1.0$; it is also a perfect absorber $(a = 1.0)$.

In nature there are no perfect blackbodies, but rather gray bodies. There are, however, objects that are "black" for certain wavelengths. The amount of energy absorbed or emitted by a plant is related to the emissivity of the surface of the plant according to the following modification of the Stefan-Boltzmann law of radiation:

$$R = \sigma \varepsilon T^4$$

where:

ε = emissivity of the object
T = surface temperature, °K

Emissivity is always equivalent to absorptivity (Campbell 1977) and represents the amount of energy absorbed or emitted by the surface of the plant or plant part.

Radiation emitted by a plant (or any other object) will have a wavelength distribution which is related to its surface temperature since wave propagation is a function of molecular oscillation that increases in frequency with an increase in temperature (Etherington 1974). Wavelengths of radiation are related to velocity of light and frequency of molecular oscillation according to the following equation:

$$\lambda = c/v$$

where:

λ = wavelength, cm
c = velocity of light, cm sec^{-1}
v = frequency, cycles sec^{-1}

An increase in surface temperature increases the frequency of wave propagation. Therefore, a plant with a higher surface temperature will emit a shorter wavelength than a plant with a lower surface temperature. The peak of wavelength distribution is related to temperature by Wien's displacement law:

$$\lambda \max = 2864/T$$

where:

λ = maximum wavelength emitted, μm
T = surface temperature, °K

As the temperature of a plant increases, energy emitted is shifted to shorter wavelengths, thereby dissipating high quantities of energy.

Plants do not absorb all of the radiation incident upon them. Some of the incident radiation is reflected (r; albedo) at the surface, some is absorbed, and some is transmitted (t) (Fig. 4). All of the energy incident upon a plant must be partitioned into these three categories; therefore, $a + r + t = 1.0$. For a blackbody $a = 1$, so $r = t = 0$. The amount reflected, absorbed or transmitted is dependent upon the wavelength and the surface characteristics.

Leaves reflect 70% of the perpendicular incident infrared radiation and 6 to 12% in the visible wavelength. Within the visible wavelengths, green is most strongly reflected (10—20%), less orange and red (3—10%) are reflected. Generally, very little UV is reflected (3%); most of it is absorbed by waxy, cuticular, and suberized epidermis.

The amount of radiation transmitted by leaves depends upon their structure and thickness. Soft, flexible leaves transmit 10—20% of the incident radiation whereas, very thin leaves transmit 40% (Larcher 1980). Thick, solid leaves might not transmit any radiation. Radiation transmitted through foliage is rich in wavelengths 500 and > 800 nm (Larcher 1980). Red and green wavelengths prevail beneath a canopy; and in the depths of a forest with a closed canopy, only far-red and infrared are present.

Absorption of the visible wavelengths (largely blue and red) is determined by chloroplast pigments. About 70% of the photosynthetically

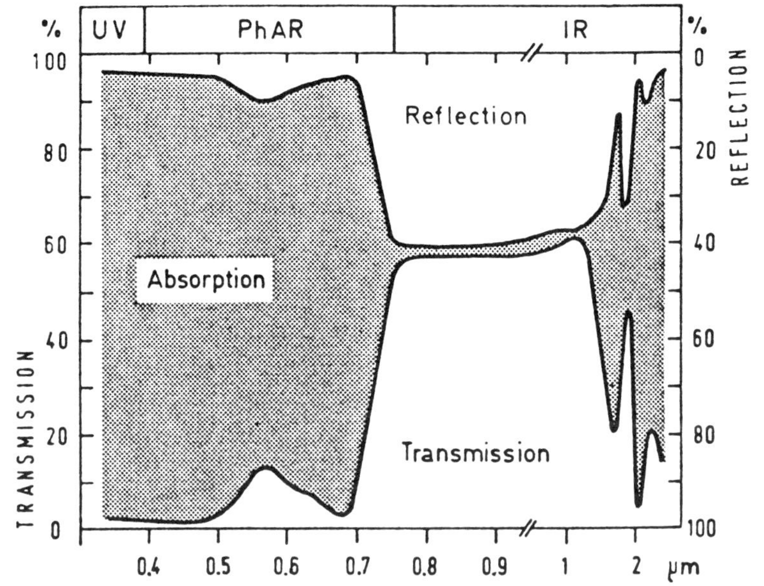

Figure 4. Relative reflection, transmission and absorption of a popular (*Populus deltoides*) leaf as a function of the wavelength of the incident radiation (UV, ultraviolet; PhAR, photosynthetic active radiation; IR, infrared). (From PHYSIOLOGICAL PLANT ECOLOGY by W. Larcher. Copyright 1975, 1980. Springer-Verlag New York. Reprinted with permission.)

active radiation (PhAR) is absorbed by the chloroplast in mesophyll cells. Only small amounts of infrared below 2000 nm are absorbed. However, approximately 97% of the wavelengths > 7000 nm are absorbed (Larcher 1980). Therefore, plants behave like blackbodies with respect to heat radiation.

The importance of these energy concepts becomes clear when it is realized that the heat load of a plant adjusts to that of the ambient environment until the energy input is equivalent to the output according to the following equation:

$$M + Q_a = R + C + LE + G + X$$

where:

M = rate at which metabolic energy is generated
Q_a = amount of energy absorbed by surface of the plant
R = radiation emitted by surface of the plant ($W\ m^{-2}$)
C = energy transferred by convection ($W\ m^{-2}$)
LE = energy transferred by evaporation or condensation ($J\ m^{-2}\ sec^{-1}$)
G = energy transferred through conduction, and
X = energy put into or taken out of storage.

The amount of energy generated by metabolism contributes to the energy budget of a plant under certain conditions, but it is usually a negligible amount; therefore, the M-term is eliminated. Likewise, the amount of energy stored is negligible and also eliminated from the equation. The net amount of energy transferred to and from the plant via the soil approaches zero diurnally; therefore, G is eliminated. The energy budget of a plant is reduced to the following equation:

$$Q_a = R + LE + C.$$

3.2.1 Energy relations of an individual plant

The energy budget of a plant (or specifically an individual leaf) is extremely important to that plant. Energy (heat) is important because it influences all metabolic processes, including photosynthesis, respiration, transpiration and reproduction. It also is important because of its impact on the cytological state of individual cells. For example, protoplasmic streaming (membrane integrity) and proteins (enzymes) are very temperature dependent. Plants must be able to withstand the warm temperatures of the summer and, yet, tolerate the cold of winter. Every plant or leaf absorbs a certain fraction of the incident radiation and partitions this energy into three outgoing streams: reradiation, convection, and transpiration (Gates 1980).

Energy is radiated by a leaf according to the fourth power of its absolute temperature. The emissivity usually ranges between 0.92 and 0.96. Energy is removed from the surface of a leaf by convection, even with rates as minute as eddy currents. Transpirational water loss is an important mechanism of energy dissipation for plants when they transpire. However, CAM plants and xerophytes can endure extended periods of high heat-loads without transpiring. Therefore, transpiration or latent energy transfer may account for very little energy dissipation. Gates (1980) calculated temperatures of leaves that absorbed 419 (low heat load), 698 (typical of normal midday conditions), and 977 $W\,m^{-2}$ (approximately the maximum amount possible in the natural world) as they would be influenced by radiation, radiation plus convection, and radiation plus convection and transpiration (Table 2). Leaves exposed to low amounts of radiation (e.g. 419 $W\,m^{-2}$) would actually be warmed by convection with an air temperature of 30°C. Leaves exposed to high amounts of radiation are significantly cooled by convection, but an increase in wind speed from 1.0 to 5.0 $m\,s^{-1}$, or more, does not significantly lower the temperature of a leaf (Gates 1980). Transpirational cooling of leaves exposed to moderate radiation loads and some wind is relatively low. The temperature of leaves exposed to large radiation loads

Table 2. Temperatures (in Degrees Centigrade) of an idealized leaf under various conditions. (From BIOPHYSICAL ECOLOGY by David M. Gates. Copyright 1980. Springer-Verlag New York. Reprinted with permission.)

		Reradiation and convection			Reradiation, convection and transpiration		
	Radiation only	V = 0.1 m s^{-1}	V = 1.0 m s^{-1}	V = 5.0 m s^{-1}	V = 0.1 m s^{-1}	V = 1.0 m s^{-1}	V = 5.0 m s^{-1}
Radiation absorbed							
419 W m^{-2}	20.0	27.0	28.7	29.4	23.7	26.0	27.8
698 W m^{-2}	60.0	41.0	34.6	32.3	34.1	31.3	30.7
977 W m^{-2}	89.2	55.0	40.5	35.1	44.5	36.8	34.4

The leaf has a characteristic dimension of 0.05 m and an internal resistance to water loss of 200 m s^{-1}. Air temperature is 30°C and relative humidity is 50%. V is wind velocity.

and in still air might be significantly lowered by transpiration (10—12°C). Generally, evaporative cooling is more important in large leaves with low rates of convectional heat transfer. Transpirational cooling is particularly effective when the air temperature is high, the humidity is low, and the soil water is not limiting (Larcher 1980). Evaporative cooling can lower the leaf temperature several degrees below the air temperature. Under steppe conditions, one fourth of the energy dissipated can be accounted for by transpiration (Rashke 1956). Evaporative cooling is less important for small leaves because convective heat transfer is more important.

The temperatures within which plants live range from −5°C to 90°C. However, most metabolic activities and vital processes take place within about 5°C of the optimum temperature for specific plants or ecotypes and each particular process. The temperature range for most terrestrial vascular plants extends from −5 to 55°C, but most dry matter is produced and growth occurs between 5° and 40°C. There are some algae (e.g. *Chlamydomonas nivalis*) that are specifically adapted to freezing temperatures and some bacteria that colonize hot water (90°C) near geysers.

The temperature range for photosynthesis in most plants is only about 5°C narrower than the maximum and minimum temperatures that injure or kill the leaves. Optimum temperatures for photosynthesis are species dependent, but usually have a range of 10—15°C. Shoot growth and development are closely related to photoperiod as well as temperature. Temperature optima for shoot growth of most tropical and sub-tropical species are between 30 and 40°C and for temperate zone plants the optima lie between 15 and 30°C (Larcher 1980). Root growth can occur at soil temperatures as low as 2—5°C. Growth of endogenous species will reflect their evolutionary history and the prevailing environment of their native habitat.

Plants endogenous to high temperature regimes are remarkably well-adapted and dissipate energy very efficiently. However, heat injury occurs and the temperature at which it occurs depends largely upon the species, its internal water content, the time to which the plant is exposed to high temperatures, and the phenological stage of the plant at the time of exposure to high temperatures. Quiescent or dormant plants are less susceptible to heat injury than actively growing plants. Seedlings are much more susceptible than older plants because of their lack of protective tissue at the soil surface where temperatures commonly reach 50—75°C in some environments. Heat tolerance among plants is inversely related to plant water content; the higher the water content, the lower temperature at which heat injury occurs (Levitt 1972). Heat injury results primarily from damage to cellular membranes and denaturization of proteins.

Likewise, the plasma membrane is the tissue damaged by freezing and chilling; it is also the site where acclimation to cold occurs (Steponkus 1984). Cold acclimation and winter hardiness have been attributed to a number of factors including high concentration of sugars and proline, absence of sulfhydryl groups in enzymes, extracellular rather than intra-cellular freezing, increase in soluble proteins, increase in ribonucleic acid (RNA), increase in phospholipids, and increase in the degree of un-saturation of fatty acids. The literature is inundated with details of the effects of freezing and thawing on plants but most textbooks do not address the issue.

Plants exposed to temperatures colder than their optimum range of adaptations often have an increase in soluble proteins. Two of the more common changes are enzymes concerned with synthesis of phenolic compounds, phenylalanine ammonialyase and hydroxycinnamoyl CoA quinate hydroxycinnamoyl transferase (Graham and Patterson 1982). Most enzymes are more stable at lower temperatures, but some are inactivated by cold. RNA increases along with soluble and membrane-bound proteins. However, deoxyribonucleic acid (DNA) is usually not affected. Levitt (1974) proposed that an adaptive feature of plants to cold temperatures is the lack of enzymes with sulfhydryl groups (SH) that tend to form disulfide bonds upon dehydration caused by chilling and freezing; however, this remains a controversial issue.

Very often there is an increase in lipids, especially phospholipids, and the degree of unsaturation of fatty acids in plants adapted to cold climates (Graham and Patterson 1982), which increases fluidity of the cell membranes. Plants native to colder climates also generally have higher respiratory rates at a given temperature than plants native to warmer climates.

Non-hardened (tender) plants exposed to a "killing frost" (−1 to −3°C) in the fall are injured or killed from rapid intracellular freezing that causes irreparable damage, notably destruction of membrane continuity (Burke *et al.* 1976). Burke *et al.* (1976) also noted injury that is caused in tropical species exposed to temperatures 0 to 14°C that induce lipid structure transitions which disrupt membrane proteins.

Most plants that survive freezing temperatures usually can tolerate some ice formation in their tissues. Herbaceous plants such as winter wheat (*Triticum aestivum*) and turf grasses may survive winter temperatures to about −25°C (Burke *et al.* 1976). Often air temperatures fall below the level at which tissues are damaged or killed, but the buds and meristems are at the soil surface or belowground where the temperature is ameliorated. Soil water content can influence hardiness and survival because it affects tissue hydration and soil temperature. Hydrated tissues seem to be more tolerant of low temperatures and supercooling than dry tissues. Ice crystals do not spread uniformly in desiccated tissues; therefore, some cells are supercooled several degrees and then freeze suddenly with injurious effects (Burke *et al.* 1976). Ice formation seems to propagate throughout extracellular spaces and not intracellular spaces of tolerant herbaceous plants.

Deciduous woody plants and flower buds of azalea (*Rhododendron* sp.), blueberries (*Vaccinium* sp.), apricot (*Prunus armenica*), peaches (*Prunus persica*), cherries (*Prunus* sp.), and plums (*Prunus domestica*) avoid freezing injury by "deep supercooling" to temperatures as low as −40°C (Burke *et al.* 1976). Supercooling apparently occurs because there are no nucleating substances necessary for ice initiation. However, the lowest point of supercooling is considered to represent the homogeneous ice nucleation temperatures of the cellular solutions of the tissues involved (Burke *et al.* 1976).

Neither coniferous nor deciduous trees native to the Boreal Forest of North America supercool. However, they can tolerate a greater degree of dehydration than other woody plants. Paper birch (*Betula papyrifera*), red-ozier dogwood (*Cornus stolonifera*), willow (*Salix* sp.), and quaking aspen (*Populus tremuloides*) can survive experimental freezing to −196°C when they are fully acclimated (Burke *et al.* 1976).

3.3 Water relations

Water is one of the most important environmental factors influencing rangeland vegetation. Ecologically, availability of water determines plant distribution more than any other single factor (Brown 1977). Water is the principal constituent of protoplasm, comprising more than 80% of the fresh weight of herbaceous plants and more than 50% of woody plants.

Water is a unique chemical compound. It is a universal solvent and a medium in which all physiological processes occur. It is the solvent in which gases and minerals enter plant cells and the medium in which organic and inorganic compounds are translocated throughout the plant. It is the reagent in which photosynthesis, respiration, and other metabolic processes occur. Water is essential for cell turgidity and enlargement as well as plant growth and development.

One of the most unique properties of water is its specific heat. It is higher than that of any other similar chemical compound except liquid ammonia; 1 cal of heat is required to raise the temperature of 1 g of water 1°C at 14.5°C. Closely associated with specific heat are latent heat of fusion and latent heat of vaporization. The amount of energy required to melt 1 g of ice at 0°C (latent heat of fusion) is 80 cal. Conversion of water to steam (latent heat of vaporization) at 100°C requires 580 cal. Bodies of water and plant communities near water change temperature slowly because of the amount of energy required to either raise or lower the temperature of water 1°C. Plants that have high water contents (e.g., cacti) are less affected by dramatic changes in temperature than plants with lower water contents.

Viscosity or density of water varies with temperature; its greatest density occurs at 4°C. Availability and uptake of water are affected by soil temperature and the viscosity of water. Cohesion of water molecules creates an unusually high tensile strength resulting in high surface tensions. Surface tension is important in transpiration (boundary layer effect) and in preventing passage of air bubbles through pores in cell walls.

Water relationship terminology has evolved through several modifications. Terms such as diffusion pressure deficit, suction force, pF, etc., arc common in older literature, but since 1960 an effort has been made to unify terminology of water relationships. Water potential has been adopted from the thermodynamics of chemical potential of water and is defined as the difference in chemical potential of water in a specified system and that of pure, free water at the same temperature and pressure.

Rate and direction of movement of water through the soil-plant-atmosphere continuum (SPAC) are determined by the energy relationships of the various components within the SPAC gradient. Water potential has been defined by the universal gas law as:

$$\Psi = RT \ln (e/e_o)/\overline{V}_w$$

where:

Ψ = water potential (Pa; 1 Pa = 1 N m^{-2} = 10^{-5} bar)
R = universal gas constant (8.3 1 atm °K^{-1} mol^{-1})
T = temperature (°k)
e = vapor pressure of water in the system (mm Hg)
e_o = vapor pressure of pure, free water (mm Hg)
$\overline{V}_w$ = molar volume of water (0.018 l/mol)

Water potential of pure, free water is defined as zero. Addition of a solute, the presence of a porous capillary matrix, or gravitational pull will reduce the potential (make more negative). Pressure within a cell caused by the force of water exerted against the cell wall increases (makes less negative) the potential.

Plant water potential is determined by matric potential (or intercellular space, fiberous cell walls, etc.), solute (or osmotic) potential (result of dissolved solutes within plants), and turgor potential (or pressure):

$$\Psi \text{ plant} = \Psi\tau + \Psi\pi + \Psi p$$

where:

Ψ plant = plant water potential
$\Psi\tau$ = matric potential
$\Psi\pi$ = osmotic or solute potential
Ψp = pressure potential

Matric potential is derived from forces of capillarity, absorption, and hydration acting on water molecules. Spanner (1972) describes these as Newtonian forces that “seize water molecules” and pull them into matric surfaces. Capillary forces exist mainly in intercellular spaces and at the outer surface of cell walls. Within the cell wall, forces of adsorption or imbibition predominate and may also predominate in the colloidal matrix of the cytoplasm. Forces of hydration or adsorption are responsible for orienting layers of water around hydrated proteins and other macromolecules (Warren Wilson 1967a, Brown 1977). Matric potential is the primary force influencing water potential of cell walls and in dry seeds. Forces acting on water molecules by matric surfaces reduce free energy of the water below that of pure water; therefore, matric potential lowers water potential.

Osmotic potential also lowers free energy of water because of the presence of dissolved solutes in solution. Their presence increases entropy of the total system thereby, lowering water potential. Salts and sugars lower the osmotic potential of a solution by reducing the kinetic energy of the water molecules (Brown 1977). A 0.5% NaCl solution has a water potential (Ψ) = −0.42 MPa, $\Psi_{1\%\ \text{NaCl solution}}$ = −0.83 MPa, and $\Psi_{3\%\ \text{NaCl solution}}$ = 2.0 MPa (approximately equal to sea water). Sugars and other macromolecules do not lower osmotic potential per mole as much as salts because the salts are more highly dissociated. Osmotic potential contributes minimally to water potential of the xylem, but is very important in cell vacuoles, cytoplasm, and phloem.

Pressure potential (or turgor pressure) results from forces of hydrostatic pressure in excess of atmospheric pressure acting outward on the cell walls and internal membranes (Brown 1977). It includes only forces resulting from turgor and does not include any forces originating from capillary forces (Warren Wilson 1967b). An increase in pressure potential increases free energy of water; thereby, increasing water potential.

3.3.1 Transpiration

About 99% of water taken up by the plant is lost through transpiration. Evaporation of water vapor from the leaf surface is a diffusion process. It consists of cuticular and stomatal transpiration. The former refers to the evaporation from the outer wall of the epidermis; in the latter case, the liquid water evaporates from the inner wall of the epidemis into the leaf air spaces, and then water vapor diffuses out through stomatal pores. The water vapor eventually escapes from the leaf surface to the open air via the boundary layer.

According to Fick's law, transpiration (E) can be expressed as $E = \Delta W\, g$, where:

W = water vapor pressure difference between leaf and air
g = conductance

The water vapor pressure is a function of both humidity and temperature. As leaf temperature rises, ΔW becomes steeper and transpiration tends to increase. The conductance has three components. Cuticular conductance is usually very small, because the outer surface of epidermis cell wall is cutinized and forms a hydrophobic barrier to water vapor. Cuticular conductance in most range plants is two orders of magnitude lower than the stomatal conductance; and cuticular transpiration is 3—10% in desert shrubs and in succulents only 1—2% (Larcher 1980). Boundary layer conductance is a function of geometry of the leaf surface and wind velocity. Small leaves have a high boundary layer conductance. In open country or on rangeland, high wind velocities usually make boundary layer conductance higher than the stomatal conductance.

The stomata serve as a valve in regulating water loss, therefore conductance depends primarily on the aperture of stomata. Stomata can respond directly to light to open and the response is indirectly mediated by intercellular CO_2 concentration. Efficient operation of photosynthesis depletes intercellular CO_2 and demands stomata to open so that ambient CO_2 can diffuse inward. It also demands an increase in stomatal conductance as light intensity and temperature approach the optimum for CO_2 assimilation. Therefore, the higher rate of transpiration is an indication that the plant is actively photosynthesizing. When photosynthesis is reduced, stomata will close so extra water loss can be prevented. Stomata are sensitive to vapor pressure difference between the leaf and air. This sensitivity is a mechanism of desiccation avoidance. It has the effect of reducing water loss when evaporative demand is high. Stomata function to regulate leaf temperature to keep it near the optimum for photosynthesis and prevent heat stress in the leaf tissue through transpirational cooling.

Transpiration also depends on the hydraulic resistance to water flow in the soil-root-plant system which in turn is determined by soil water availability. In arid and semiarid rangeland, plants transpire much less water than irrigated crop species. Transpiration rate is proportional to the water potential gradient from the soil to the plant. Under water stress conditions, plants which can maintain a steeper water potential gradient through osmotic adjustment will transpire more water providing the soil has extractable water within reach of the root system.

In arid and semiarid rangelands, and latent heat transfer by transpiration may have certain influence on the leaf energy balance in the rainy season, nevertheless it is a minor component affecting energy balance in the summer. A continuous water flow from apoplast to symplast to maintain a relatively constant cell volume is a major determinant of metabolic activity. The volume change (dV/dt) is given by difference between transpiration (E) and root uptake (R) (Sinclair and Ludlow 1985). A plant which can keep dV/dt close to zero will have better water balance and higher relative water content (RWC). Otherwise, RWC will decline if root uptake cannot equal transpiration. When RWC drops to a critical level, plants will die from desiccation.

In some species, stomatal conductance is related to leaf water potential (or turgor potential). Stomata close when turgor potential approaches zero. Diurnal transpiration rate may show a linear increase and then level off in the dry air. This is referred to as negative feedback control (Cowan 1977). Minimum water potential occurs when the plant reaches a constant maximum transpiration rate. In other species, stomatal conductance is determined by epidermal water potential. With increased VPD, transpiration would reach a peak and then decrease. This is referred to as feed-forward control (Farquhar 1978). It can prevent leaf water potential from declining. When transpiration rate begins to decline at high VPD, leaf water potential actually increases.

3.3.2 Soil water

Although the amount of precipitation that an area receives annually is important, the amount that infiltrates and is stored in the soil is of paramount importance. The level at which soil water is available has been and continues to be a nebulous point.

Field capacity is defined as that point where water has been removed from the macropores of the soil but remains in the micropores. Generally, tension with which water is held at field capacity is −0.03 MPa. The wilting coefficient is the point where plants have difficulty removing water from the soil in sufficient quantities to maintain turgidity. This point is usually represented by −1.5 MPa. The point at which temporary and permanent wilting occurs in native plants is often difficult to define. Wilting occurs within native, wildland plants at different soil water potentials. Nearly all plants wilt temporarily because of excessive atmospheric demands (e.g., during a summer day), but they will regain turgor when the excessive demands are removed (e.g., at night). Permanent wilting is defined as the point beyond which a plant has been damaged physiologically and cannot regain turgor.

Very few native range plants wilt permanently at −1.5 MPa. Plants adapted to arid and semiarid environments tend to wilt at much lower soil water potentials, whereas those plants adapted to more mesic environments will tend to wilt at higher soil water potentials (e.g., −1.5 MPa). Brown (1977) defined wilting on the basis of a plant's response to its environment. Physiological wilting represents the point at which the pressure potential reaches 0 MPa and incipient wilting occurs. As the plant water potential decreases to a minimum, the protoplast collapses and the pressure potential again tends toward 0 MPa which represents the lethal threshold for a plant (Fig. 5). The lethal threshold or "permanent wilting point" for range plants often occurs at less than −2.0 MPa soil water potential. Minimum water potential for native plants are indicative of their habitat, but they are much lower than −2.0 MPa for xerophytes (Larcher 1980).

Stomatal conductance (and transpiration) as well as plant water potentials are determined by soil water potentials (Schulze 1986). Availability of water also depends upon soil texture. Water is held more tenaciously by soils with a fine texture than those with a coarse texture.

Salinity significantly alters soil water availability, often dictating the kind of vegetation that will grow in an area. Many regions of the western

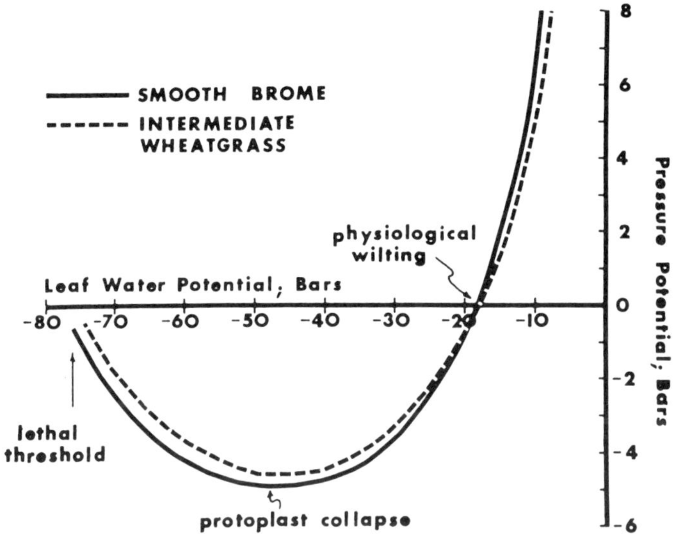

Figure 5. The relationship between leaf pressure potential and total leaf water potential for smooth brome (*Bromus inermis*) and intermediate wheatgrass (*Agropyron intermedium*). The points of physiological wilting, protoplast collapse, and the lethal threshold for water stress under this environmental condition are similar for both species. (From Water Relations of Range Plants by R. W. Brown. *In*, RANGELAND PLANT PHYSIOLOGY. Copyright 1977. Society for Range Management. Reprinted with permission.)

and southwestern United States have high concentrations of salts in and on the surface of soils because of limited precipitation and high rates of evaporation. Salts increase the osmotic potential of these soils, rendering them less productive. Only halophytes occupy saline areas where salt concentrations are high and soil water potentials are very low (−5.0 to −7.0 MPa).

3.3.3 Germination and seedling establishment

Water is necessary to initiate the germination process. Seeds of native wildland plants have a very low water content (<15%) and will remain dormant until the environment provides adequate water to assure the germination process (McDonough 1977).

The amount of water required for germination varies among species depending upon their adaptations (see Young, this volume). However, there is a minimum threshold in water content that must be surpassed before germination will begin in native wildland plants. After germination is initiated, a minimum amount of water must be maintained in the environment to insure completion of the germination process and establishment of the seedling.

The seedling stage of plant growth is the most vulnerable stage in the life cycle of a plant. Survival of the seedling initially depends upon the ability of the radicle to grow in wet soil which is very critical since the surface soil often dries very rapidly. It is difficult for the radicle or young root to grow fast enough to keep up with the drying front of surface soil. Seedlings are very susceptible to water stress because of lack of development of anatomical and morphological features that conserve water.

3.3.4 Flooding

Although flooding is not usually considered a problem with plants, it causes physiological stress to those plants intolerant of flooding. Non-adapted plants subjected to extended periods of flooding generally succumb to anaerobiosis (oxygen-deficit stress). Secondary affects associated with oxygen deficits can include accumulation of toxic levels of manganous and ferrous ions or other toxic intermediate compounds resulting from anaerobic respiration. Often death occurring from anaerobiosis results from increased levels of CO_2 rather than decreased levels of O_2. Also NH_3, which is very toxic to plants and plant tissues, is produced in flooded or waterlogged soils.

Plants that are adapted to extended periods of flooding, such as rice, have the ability to compensate for low oxygen levels by employing a specialized fermentation system for anaerobic respiration. Reduction in translocation of plant growth regulators or hormones also results from flooding.

Plants adapted to extended periods of inundation also have anatomical characteristics that enable them to tolerate oxygen deficits in the environment. Many of these plants have large cavities in which oxygen can accumulate.

3.4 Photosynthesis

Photosynthesis is a process by which radiant energy is converted into chemical energy contained in organic compounds. Photosynthesis can be summarized by the chemical equation:

$$6CO_2 + 6H_2O \xrightleftharpoons[\text{chlorophyll}]{\text{light}} C_6H_{12}O_6 + 6O_2,$$

where $C_6H_{12}O_6$ represents glucose, a simple sugar. Photosynthesis includes the temperature independent light reaction (photochemical process) and temperature dependent dark reaction (enzymatic process). In green plants, photons (400—700 nm) absorbed by chlorophyll molecules are used to remove electrons from water and raise them to a high energy status. Through a subsequent electron transport process, the electrons are donated to nicotinamide adenine dinucleotide phosphate ($NADP^+$) to form NADPH which then serves to reduce carbon dioxide. In the process, a proton concentration gradient as well as electrical potential gradient is created across the thylakoid membrane in the chloroplast. This potential energy is used by a coupling factor enzyme to convert adenine diphosphate (ADP) and phosphate to adenine triphosphate (ATP), a high energy compound used in carbon reduction and many other biochemical reactions.

The photochemical process involves a cooperation of photosystem I and photosystem II. The former has a high ratio of chlorophyll a to chlorophyll b and is maximally excited by light at longer wavelengths (700 nm). The electron transfer in photosystem I leads to reduction of NADP. The latter has larger proportions of chlorophyll b and maximal absorption at 680 nm. The main function of photosystem II is to split water and to provide the electrons for photophosphorylation of ADP and reduction of NADP. As a result of photolysis of water, O_2 is released and enters the photosynthetic gas exchange process.

In the dark reaction, a CO_2 molecule is fixed by a pentose sugar, ribulose bisphosphate (RuBP) via the enzyme ribulose bisphosphate carboxylase oxygenase (RuBisCO) and forms an unstable six-carbon intermediate compound which hydrolyzes to give two molecules of 3-phosphoglyceric acid (PGA). This C_3 acid is phosphorylated and reduced to triose phosphate at the expense of NADPH and ATP. The triose phosphate then undergoes a series of reactions in which five molecules of triose phosphate are rearranged and phosphorylated to regenerate three molecules of RuBP, the CO_2 acceptor. This is termed the Calvin-Benson cycle, photosynthetic carbon reduction (PCR) cycle or the reductive pentose phosphate cycle (Fig. 6) (Bassham and Calvin 1962). This is the only known carbon reduction mechanism. All green plants, whether they are C_3, C_4 (Fig. 7) or CAM (Fig. 8) plants, possess the PCR cycle.

Most cool season grasses and legumes have the C_3 photosynthetic pathway in which the first product of CO_2 fixation is a three carbon acid (PGA). Most warm season grasses have the C_4 photosynthetic pathway in which the first product of CO_2 fixation is oxaloacetic acid (OAA), a four carbon dicarboxylic acid. Some plants have Crassulacean Acid Metabolism (CAM), so termed because it was first discovered in the family Crassulaceae. Most succulent plants such as cacti are characterized by this photosynthetic pathway. In C_4 photosynthesis, the reactions are separated in space: CO_2 is fixed by phosphoenopyruvate (PEP) via PEP carboxylase in the mesophyll cell; the resulting C_4 dicarboxylic acid is transported to the bundle sheath cells where the C_4 acid is split into CO_2 and pyruvate, a three carbon acid. CO_2 enters the PCR cycle and pyruvate is transported back to the mesophyll cell and phosphorylated to regenerate PEP. In CAM, the reactions are separated in time: the stomata open at night, thus CO_2 is fixed into malate and stored in the vacuole; during the day, the stomata are closed and malate is decarboxylated, and released CO_2 is assimilated through the PCR cycle.

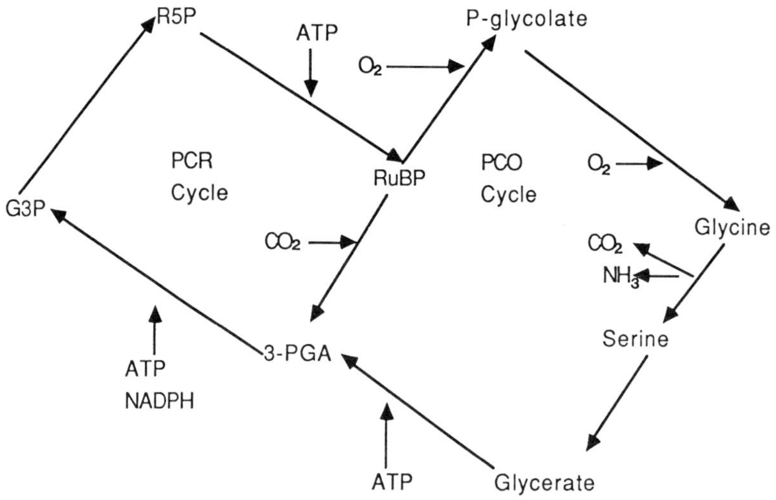

Figure 6. Photosynthetic carbon reduction (PCR) and photorespiratory carbon oxidation (PCO) cycles.

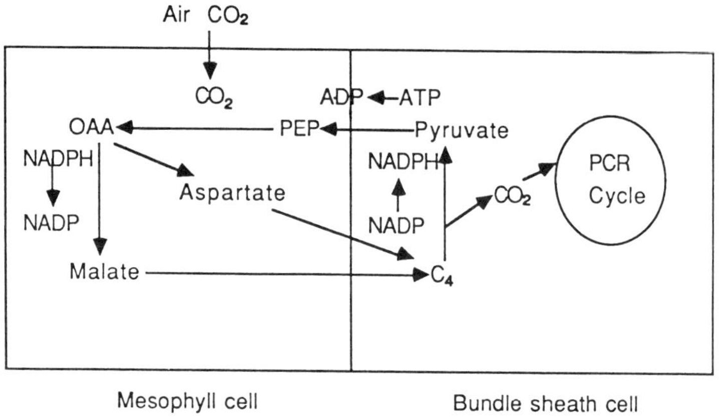

Figure 7. C_4 photosynthetic pathway involving carboxylation via PEP carboxylase in the mesophyll cells and decarboxylation in the bundle sheath cells. Photosynthesis proceeds by the PCR cycle in the bundle sheath cell.

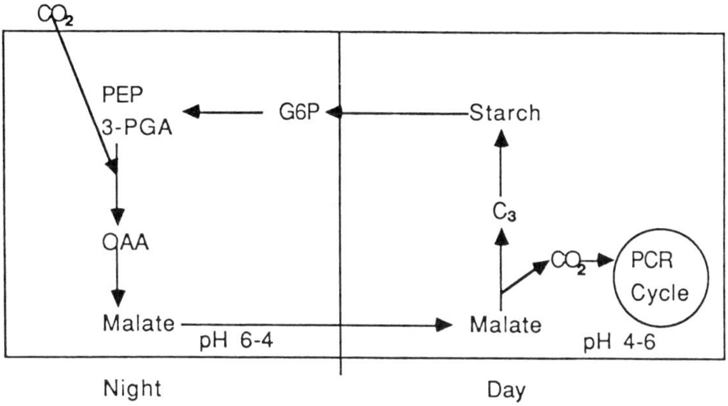

Figure 8. Crassulacean acid metabolism (CAM) photosynthetic pathway involving carboxylation or acidification via PEP carboxylase at night and decarboxylation or deacidification during the day. Photosynthesis proceeds by the PCR cycle.

The most important regulatory enzyme of the PCR cycle is RuBisCO. It can function as a carboxylase as well as an oxygenase, depending on the CO_2/O_2 ratio. When intercellular CO_2 partial pressure is high, it acts mainly as a carboxylase, and PCR cycle is favored. When O_2 partial pressure is high, and CO_2 pressure is low, RuBisCO functions as a oxygenase, and the PCR cycle is suppressed. In C_3 plants growing at ambient air (340 μbar CO_2 and 210 mbar O_2), a considerable amount of carbon flows through the photorespiratory carbon oxidation (PCO) cycle in which RuBP is oxidized into PGA and phosphoglycolate and CO_2 is released in the subsequent reaction (Fig. 6).

One of the advantages of C_4 over C_3 photosynthesis lies in its CO_2-concentrating mechanism. Carboxylation occurs in mesophyll cells and affinity to CO_2 (in the form of HCO_3) is 10 times higher in PEP carboxylase than in RuBP carboxylase. CO_2 molecules are efficiently trapped by PEP carboxylase in the mesophyll and then they are effectively "pumped" into the bundle sheath cells through C_4 acid decarboxylation and, thus, a high CO_2 concentration is maintained in the bundle sheath cells which favors the PCR cycle and reduces photorespiration (all respiratory activity in the light in which CO_2 is released or O_2 consumed). Even if some CO_2 molecules escape from the bundle sheath cells, they will be refixed in the mesophyll cells because the bundle sheath cells are essentially surrounded by PEP carboxylase. Therefore, photorespiration is usually not detected in C_4 plants, while it may be near 40% in C_3 plants under normal conditions (Bjorkman 1966, Schnarrenberger and Fock 1976).

High CO_2 concentration in the bundle sheath cells can compensate for a much lower RuBP carboxylase in C_4 plants. RuBP carboxylase comprises up to 50% of soluble leaf protein of C_3 plants. Therefore, C_4 plants can use leaf nitrogen more efficiently. Efficient CO_2 entrapment by PEP carboxylase allcws C_4 plants to use less RuBP carboxylase. In addition, C_4 photosynthesis requires coordination between the C_4 acid and the PCR cycle. It also requires that carboxylation and decarboxylation take place within different compartments through extensive metabolite transport. This makes it efficient in nitrogen utilization. C_3 plants invest 40—60% of their leaf nitrogen supply in RuBP carboxylase in contrast to C_4 plants which only invest 10—25%. On the average, C_4 plants can produce about two-fold greater dry matter per unit of leaf nitrogen than C_3 plants can. This gives C_4 plants an advantage in poor- nitrogen habitats.

Photosynthetic capacity (A) is positively correlated with leaf conductance to water vapor (g), but the slope of the regression line (A/g) is different for C_3 and C_4 plants (Korner *et al.* 1979, Wong *et al.* 1979, Schulze and Hall 1982). C_4 plants usually have a higher A/g than C_3 plants; i.e., to achieve the same photosynthetic capacity, C_4 plants would have their stomata more tightly closed than C_3 plants. At air temperatures above 25°C, C_4 plants are superior to C_3 plants in photosynthetic performance with the same stomatal conductance. Under favorable environmental conditions, stomata function to maintain an intercellular CO_2 partial pressure (C_i) at 100—150 μbar in C_4 plants and 220—260 μbar in C_3 plants (Wong, Cowan and Farquhar 1978, 1979, Korner *et al.* 1979, Werk *et al.* 1983). Due to the CO_2 concentrating mechanism, C_4 plants become CO_2 saturated at 100—150 μbar pressure, but it would require C_i much greater than 220—260 μbar for C_3 plants to reach CO_2 saturation. Therefore, lower C_i associated with lower g in C_4 plants does not inhibit photosynthesis because of efficient carboxylation of CO_2 by PEP carboxylase, but it restricts transpirational water loss (E). Water use efficiency (WUE) can be expressed as $A/E = (C_a - C_i)/1.6$ VPD; where C_a is the ambient CO_2 partial pressure, VPD is vapor pressure deficit, 1.6 is the diffusion ratio of water vapor to CO_2. Under the same VPD environment, C_4 plants have greater WUE. Since a lower C_i in C_4 plants means greater CO_2 diffusion gradient from the ambient to intercellular environment, C_4 plants have a greater A/E.

CAM plants can fix CO_2 at night rather than during the day, so they are even more conservative in water usage than C_4 plants. Leaf conductance is extremely low in CAM plants; a minimum g of 0.01 cm s^{-1} is not uncommon under water stress conditions. Low stomatal frequency and small stomatal aperture are both responsible for low conductance. The WUE are about 10—20 μg CO_2 gained per mg water transpired in CAM plants, 3.4—4 in C_4 plants, and 2—2.5 in C_3 plants (Edwards and Walker 1983). CAM and C_4 plants are well adapted to hot, dry environments where the evaporation potential is extremely great because of a high VPD.

Small conductances enable CAM and C_4 plants to more efficiently utilize the limited water resources. Since transpirational cooling is less, these plants must be capable of tolerating high temperature and heat stress.

The optimum temperature range for photosynthesis in C_4 plants (30—40°C) is considerably higher than that of C_3 plants (20—30°C). As temperature increases from 20 to 40°C, the CO_2/O_2 ratio at the carboxylation site is reduced, because the solubility of CO_2 is more affected by high temperature. For C_3 plants, this reduced CO_2/O_2 ratio will increase O_2 inhibition on photosynthesis by increasing RuBP oxygenase activity and enhanced photorespiratory CO_2 evolution. In C_4 plants, the CO_2 concentrating mechanism will help to keep the CO_2/O_2 ratio higher and reduce O_2 inhibition of photosynthesis to a minimum. If ambient CO_2 pressure is increased to 1000 μbar such that even C_3 plants could become CO_2 saturated, then photosynthesis would show strong temperature dependent characteristics which resemble that of C_4 plants (Osmond *et al.* 1980). Quantum yield of C_3 plants at normal air (340 μbar CO_2) is, therefore, temperature-dependent and becomes temperature independent at CO_2 saturation. Low quantum yield of C_3 plants at high temperature is probably related to stimulation of RuBP oxygenase activity, whereas, the quantum yield of C_4 plants is independent of temperature even at normal air.

High temperature also results in substantial decline in the rate of electron transport and the dark reaction is limited by the reduced rates of photophosphorylation and/or NADP reduction and by insufficient RuBP supply. Some plants are able to maintain high rates of photosynthesis at high temperatures, because their photosynthetic apparatus has great stability and rate of electron transport is faster at high temperature. In some C_4 plants such as *Tidestromia oblongifolia*, the optimum temperature for photosynthesis is as high as 47°C (Bjorkman *et al.* 1978). This is primarily due to high thermostability of the photosynthetic machinery evolved from high temperature acclimation. But this thermostability is not absolutely linked to the C_4 photosynthetic pathway. Some heat-adapted C_3 plants such as *Prosopis glandulosa* can maintain positive carbon balance at leaf temperatures of 42.5°C, indicating high thermostability. The reduced photosynthetic rate of *Prosopis* at high temperatures is mainly caused by O_2 inhibition of RuBP carboxylase. In high temperature and low humidity environments, C_4 plants are superior to C_3 plants in photosynthetic performance. In cold and mesic environments, however, C_3 plants are superior to C_4. In C_4 photosynthesis, regeneration of PEP from pyruvate is catalyzed by pyruvate phosphate dikinase. This enzyme is very sensitive to low temperature. The enzyme activity will decline drastically when low temperature prevails. At low temperature, however, photosynthesis tends to become CO_2 saturated in C_3 plants. Thus, C_3 plants are well adapted to cool season. Although the average photosynthetic rates of C_4 plants are 50% greater than those of C_3 plants (Osmond *et al.* 1982) and WUE is approximately 100% higher, C_3 plants may achieve greater photosynthetic rates than C_4 plants and the same WUE value as C_4 plants during the cool season (Caldwell *et al.* 1977). In the Sonoran desert, the C_3 photosynthetic pathway is associated with all winter ephemerals, whereas the C_4 pathway is largely found in summer ephemerals (Shreve and Wiggins 1964, Mulroy and Rundel 1977). Most plants have the ability to acclimate to their growth-temperature regimes. Plants usually show a seasonal shift in temperature optima:maxima photosynthetic rates and the optimum temperature increases from spring to summer and declines in the autumn. The optimum temperature also decreases with increasing site elevation.

Since photosynthesis is a light-driven process, net photosynthesis parallels light intensity until the saturation point (Is) is reached, providing other environmental factors are not limiting CO_2 exchange. Photosynthesis in C_4 plants has a very strong light dependence curve, since the PCR cycle is nearly CO_2 saturated in normal air and light becomes an important rate-limiting factor. In addition, 5 ATP and 2 NADPH (instead of 3 ATP and 2 NADPH in C_3 plants) are used to reduce a CO_2 molecule. C_4 plants must require more energy from their photochemical reaction than C_3 plants, hence a strong light dependence curve for photosynthesis. Many C_4 plants are not light saturated until full sunlight is reached. In C_3 plants, the carboxylation site is far from CO_2 saturating, so it usually requires only 1/3 to 1/2 of full sunlight to reach Is; i.e., on bright sunny days, photosynthesis in C_3 plants is more restricted by the CO_2 fixation process than by the photochemical process. If sun leaves of C_3 plants are CO_2 saturated, then photosynthesis will have a light dependence curve more like that of C_4 plants. At temperatures of 30°C and C_i of 325 μbar, O_2 inhibition reduces quantum yield by 30% in C_3 plants; whereas in C_4 plants, the quantum yield is not inhibited by 21% O_2, even at low CO_2 pressures (Bjorkman *et al.* 1970, Ehleringer and Bjorkman 1977).

Plants are able to acclimate to different light intensities. Shade leaves utilize low light better than do the sun leaves and they reach their light saturation point at lower intensities, but when shade leaves are exposed to high light, photoinhibition occurs. This involves a series of damaging effects to the photosynthetic apparatus such as chlorophyll bleaching and reduction of photosynthesis. When the capacity of CO_2 fixation is relatively low compared to the light absorption by chlorophyll, only a fraction of absorbed light can be utilized in photosynthesis; a large excess of excitation energy will cause damage to the photosynthetic machinery.

Under water stress conditions, leaf bleaching can be frequently observed. In a dense canopy, only the top layer of leaves can achieve high photosynthetic rates at light saturation levels and the lower leaves usually have reduced photosynthetic rates at low light intensity. In an alfalfa field, the top layer of the canopy displays photosynthetic rates 33-fold greater than the lower layer even in high sunlight (Larcher 1978).

The light intensity at the mid-point of the canopy is only about 10% of the full illumination. Low-light-grown plants usually have low light saturation photosynthesis and lower stomatal conductance. When photosynthesis is mostly restricted by low light intensity, stomata should be adjusted to maintain a low C_i to match the low photosynthesis. Higher stomatal conductance at low light intensities would not enhance photosynthesis, since the photochemical process is limiting CO_2 fixation; however, it would increase transpirational water loss and thus lower WUE. Photosynthesis of high-light-grown plants is saturated at high light intensities and high stomatal conductance. The increased conductance, therefore, is a result of increased demand for CO_2 supply in the mesophyll cells.

Photosynthesis is more affected by water deficit than any other environmental factor. The effect of water stress on photosynthesis involves two major aspects: CO_2 transport and the biochemical processes of photosynthesis. Stomatal conductance of CO_2 will decline in some species under water stress conditions, causing decreased intercellular CO_2 concentration which may reduce CO_2 fixation. On the other hand, water stress can also reduce photosynthesis by influencing photosynthetic machinery directly. Water stress may result in alteration of chloroplast membranes and losses in structural integrity of chloroplasts (Kurkova and Motorina 1974). The reactions in the chloroplasts, especially the light reactions, are inhibited by severe stress (Hsiao 1973). Maize (*Zea mays*) suffers substantial loss of chlorophyll when subjected to stress (Giles *et al.* 1974). Water stress causes conformational changes in membrane-bound electron carriers and enzymes and reduces the capacity for electron transport. Inhibition of photosystem I and II activities by water stress leads to uncoupling of phosphorylation from electron flow (Keck and Boyer 1974). The coupling factor enzyme loses its activity in rapidly stressed spinach (*Spinacea oleracea*) leaves (Younis *et al.* 1979). Quantum yields are also reduced by stress (Mohanty and Boyer 1978). Activity of RuBP carboxylase is decreased as water stress develops (Jones 1973, Johnson *et al.* 1974).

Photosynthesis has been thought to be primarily affected by stomatal conductance (Jones 1973, Setter *et al.* 1980, Lange *et al.* 1982). Midday depression of assimilation rate was attributed to midday stomatal closure for several desert species (Lange *et al.* 1982, Schulze and Hall 1982). The curvilinear relationship between A and g does not necessarily mean that a change in g causes A to change. Midday stomatal closure may cause C_i to decline which would affect CO_2 uptake. But this stomatal limitation of photosynthesis has been shown to be of minor importance. It usually constitutes 15—25% of total limitation of photosynthesis (Mooney *et al.* 1977, Farquhar and Sharkey 1982, Hutmacher and Krieg 1983). If stomata had exerted a major limitation on photosynthesis then stomatal limitation on photosynthesis would have been greater under water stress conditions because of a much reduced stomatal conductance. However, Mooney (1977) found that inhibition on photosynthesis in *Larrea divaricata* caused by stomatal closure was actually decreased as water potential declined. At low C_i, assimilation responds almost linearly to C_i, since effect of increasing C_i is the activation of RuBP carboxylase. In this linear region, the enzymes are saturated with respect to the substrate RuBP. As C_i increases, a point is reached where RuBP regeneration becomes rate limiting and assimilation will no longer be a linear function of C_i. The capacity of RuBP regeneration depends on capacity for electron transport, which in turn depends on absorbed light (Farquhar and Sharkey 1982). Water stress can reduce photosynthesis by affecting the capacity of electron transport and/or the activity of RuBP carboxylase independently. It is also unlikely that a single factor such as C_i which is regulated by stomatal movement would be rate-limiting across all conditions. In some experiments, the initial effect of water stress is reduction in the capacity of RuBP regeneration (von Caemmerer and Farquhar 1981). Water stress reduces photosynthesis substantially even at CO_2 saturation (Mooney 1977), indicating that potential photosynthetic capacity is inhibited by stress. In some species the intercellular CO_2 concentration increases rather than decreases under water stress conditions, indicating that the mesophyll has reduced its ability to fix CO_2 and, thus, causes C_i to rise. Stomata may close in response to increased C_i, thus preventing water loss without concurrent reduction in CO_2 fixation. When the mesophyll suffers water stress, a drop in pH within the mesophyll cells could result in an increase in abscissic acid (ABA) in the apoplast. ABA then diffuses to guard cells and causes the stomata to close. Stomata also respond to signals generated by the process of assimilation in the mesophyll and transmitted to the guard cells via the epidermis (Wong *et al.* 1978). Farquhar and Sharkey (1982) concluded that stomatal conductance may respond to changes in photosynthetic rate to maintain C_i near saturation.

Photosynthesis is very sensitive to changes in osmotic potential. Rate of photosynthesis depends on the maintenance of proton and electrical potential gradient created across the thylakoid membrane, which requires a pH of 8.3 at the stroma and a pH of 5 at intrathylakoid space. With increasing desiccation, there is a reduced hydration of protoplast which could cause reduction in osmotic potential. In response to water stress, a net accumulation of solutes will occur. This process involves breakdown of macromolecules and will lead to stromal acidification which tends to diminish the potential gradient and reduce electron transport.

Photosynthesis is also very sensitive to changes in protoplast volume. Reduction of the maximal protoplast volume below 55% decreased photosynthesis in leaf tissue of hydro-, meso- and xerophytes (Kaiser 1982). It is suggested that volume change and subsequent increase in concentration due to osmotic dehydration not only inactivate the enzymes, but also causes disturbance of compartmentalization.

Many investigations have shown that photosynthesis at CO_2 saturation decreases at low air humidity. Resemann and Raschke (1984) found that midday depression in stomatal and photosynthesis activity of *Arbutus unedo* is caused by a large vapor pressure differences between leaf and air. The CO_2 saturated assimilation rate of some species is reduced by high transpiration rates and the sensitivity of mesophyll capacity to transpiration rate or VPD change is greater at high C_i than at low C_i (Sharkey 1982). It appears that at low C_i, activity of RuBP carboxylase is rate-limiting, whereas at high C_i, limitation induced by high transpiration rate is due to RuBP regeneration (Farquhar and von Caemmerer 1982).

Stomata not only respond to internal physiological processes, but also to external environmental conditions such as light, temperature, relative humidity and VPD. The optimization hypothesis proposed by Cowan and Farquhar (1977) suggests that stomata function to minimize water cost of carbon gain. The optimum behavior of stomata is that they tend to open when environmental conditions favor CO_2 fixation and to close when environmental conditions promote transpirational water loss; therefore, their operation has the effect of reducing the average rate of water loss relative to the average rate of carbon fixation (Cowan, 1982). If assimilation (A) is a benefit to the plant and transpiration (E) is a cost, the curve relating A and E has a slope $\partial E/\partial A$, termed unit marginal cost or gain ratio which should remain constant during a day. Because a plant maximizes its water use efficiency by maintaining a constant gain ratio ($\partial E/\partial A$), it has an adaptive advantage in the sense that when water is available, it tends to take up and lose water rapidly, but it can also fix as much carbon as possible relative to water lost. When water is less available and VPD is high, stomata tend to close which conserves water, but the assimilation rate is also reduced; thus, the overall water cost is kept minimal with respect to carbon gained. The sensitive response of stomata to VPD provides a means by which stomata guard against excessive water loss and prevents leaf water status from dropping to levels which may cause damage to the tissue.

Intrinsic water use efficiency (A/g) has ecological significance in plant adaptation to its habitat. A CO_2 concentrating mechanism allows C_4 plants to achieve an equal or greater rate of CO_2 uptake than C_3 plants with more tightly closed stomata and at a lower water cost of CO_2 fixation. The slope relating maximum photosynthesis (Am) and stomatal conductance of Am has been suggested as being different among species and similar within a species (Schulze and Hall 1982). Greater slopes of regression lines, indicating higher intrinsic water use efficiency, exist in C_4 species and some C_3 species that are adapted to more arid habitats (Schulze and Hall, 1982). However, the slopes do not have to be the same within the same species. Development of a cultivar or ecotype within a native range species on different habitats may result in significant differences in the slopes. Usually the regression lines of A versus g of plants native to a more xeric habitat have greater slopes than those of plants native to a more mesic habitat, thus yielding greater water use efficiency. This variation is derived from adaptation to the particular habitat. Under selective pressure, genetic diversity may emerge. Large genotype and environment interactions may cause the slope to change. The changes in slopes not only occur within a given species, but also occur during different seasons within a plant. According to Mooney *et al.* (1977), the slope may double from March to May. This increase in slope has an adaptive advantage in semiarid areas because it can counteract the effect of increased VPD that develops in summer induced by high leaf temperatures. Consequently, actual WUE (A/E) could be maintained at a relatively constant level as it is in spring, or in some cases, it might only decline slightly.

This increase in slope is due to more rapid decline in conductance than photosynthetic rate. However, this does not lead to the conclusion that stomata are solely responsible for the reduced photosynthetic rate, because under high VPD environment, stomatal conductance, to some extent, is affected by high VPD. Response of stomatal conductance to reduced photosynthetic rates is somewhat masked by its response to increased VPD. Nonstomatal limitation of photosynthesis still plays a major role even in the case of increased slope of some range species. *Prosopis glandulosa* has shown about two-fold increase in its slope from spring to summer at different levels of soil water potential. The plant under mild soil water stress has an assimilation rate two-fold higher than the plant under severe soil water stress, even though they both have the same average intercellular CO_2 partial pressure.

High photosynthesis and conductance are usually associated with active root growth. Activity of the root system may affect synthesis and transport of cytokinins which are biosynthesized in an active root meristem. There is a growing body of evidence indicating that cytokinins increase photosynthesis by stimulating chloroplast formation and protein synthesis and at the same time promotes stomatal opening (Meidner 1967, Jewer and Incoll 1980). According to Meidner, the increased stomatal opening in Kinetin-treated leaves is a reduction of intercellular CO_2 pressure due to increased assimilation rate. Roles of hormones in regulating photosynthesis, transpiration and conductance have been recently reviewed by Schulze (1986). It has been shown that stomatal closure is mediated by ABA. Increased ABA synthesis was found to occur at a threshold close to zero turgor (Pierce and Raschke 1980). Raschke and Hedrich (1985) indicated that ABA has simultaneous and independent effects on stomata and photosynthetic apparatus in whole leaves. However, active plant growth may occur at low turgor while stomata remain open. This indicates that the effect of ABA is reduced or eliminated in some way, probably via the effect of other hormones. Blackman and Davies (1983) found that response of stomata to ABA can be modified by cytokinins. It seems that photosynthetic capacity and stomatal conductance both respond to a signal transmitted from roots. The signal is related to metabolic activity of roots and probably related to metabolism of cytokinins, which can counterbalance the effect of ABA (Schulz 1986).

3.5 Physiological applications in range ecology and range management

A fundamental aspect of range ecology is the study of species coexistence. A physiological approach can be very useful. Wieland and Bazzaz (1975) presented an excellent example of plant physiological adaptations to their habitats. Three codominant annuals characterize old field successions: *Setaria faberii* (C_4) is shallow rooted, *Polygonum pensylvanicum* (C_3) is deep rooted, and *Abutilon theophrasti* (C_3) is intermediate in rooting depth. Wieland and Bazzaz (1975) investigated the physiological consequences accompanying the spatial segregation of root systems which apparently allows for coexistence among these three species. Leaf water potential was most negative in *S. faberii* and least negative in *P. pensylvanicum.* Stomata began to close at −0.2 Mpa in *P. pensylvanicum,* −0.8 Mpa in *A. theophrasti*, and −1.0 Mpa in *S. faberii.* However, photosynthesis was less affected by stomatal closure in *S. faberii* because of low CO_2 compensation point (5 ppm at 27°C) as compared to the other two species (> 44 ppm). Maximum photosynthetic rates were 38, 24, and 18 mg CO_2 dm^{-2} h^{-1}, respectively, for *S. faberii, A. theophrasti*, and *P. pensylvanicum. S. faberii* photosynthesizes at 30% of maximum rate at leaf water potential of −2.4 Mpa (and recovered from −5.0 Mpa) while photosynthesis in *A. theophrasti* was trivial at −2.6 Mpa. *P. pensylvanicum* died at −2.0 Mpa. Although *P. pensylvanicum* had roots accessing a deep water source unavailable to the other species, this advantage was offset by susceptibility of its photosynthetic machinery to water stress and by lower photosynthetic capacity.

In this study, it is reasonably clear that physical root system separation allows coexisting species to exploit different soil depths. More importantly, however, this physical niche separation is accompanied by physiological consequences; and it it through the study of these physiological adaptations to habitat that a complete understanding of factors enabling species coexistence can be attained. The shallow-rooted *S. faberii* may be able to pre-empt incoming precipitation by virtue of concentrating roots in the upper 15 cm of soil. The physiological consequence of this morphological adaptation to habitat is the development of very negative leaf water potentials. However, *S. faberii* can both maintain high photosynthetic rates at low leaf water potentials, and recover from leaf water potentials exceeding −5.0 Mpa. *S. pensylvanicum*, on the other hand, exploits deeper soil water. Photosynthesis of this species is very sensitive to leaf water stress. However, deep roots insuring a relatively continuous supply of water, compete with stomatal sensitivity, maintaining leaf water potentials at levels near optimum for photosynthesis. These species seem to have evolved different physiological adaptations to function effectively under the conditions of their habitat; and whereas root separation may provide a ready means for coexistence, it does not explain *how* these species are mutually adapted to their environment.

Since *S. faberii* cannot successfully compete with *P. pensylvanicum* for deep water, it has to develop some adaptive strategies to survive. It tends to conserve water by narrowing its stomatal aperture and its photosynthetic machinery becomes more tolerant to lower leaf water potentials. Similar physiological adaptation of C_4 plants to nutrient stress is also observed in unproductive habitats. Plants indigenous to nutrient deficient habitats are generally not necessarily more efficient in the uptake of these nutrients when they are present in low concentrations. It is not competition for limited water and nutrients that is the most important factor; rather, it is important to be able to conserve the resources which have been captured, or to use the resources more efficiently. *Tidestromia oblongifolia* (C_4) grown in hot desert environments had a high photosynthetic rate at 45°C, but its RuBP carboxylase comprised only 4—8% of soluble leaf protein (Osmond *et al.* 1980). This value is 10—25% in most C_4 plants and 40—60% in C_3 plants. It seems that high water use efficiency and high nitrogen use efficiency are the main factors that make C_4 plants successful in the harsh desert environments.

Somewhat different patterns exist in the North American shortgrass plains. In a west Texas *Bouteloua-Buchloe-Prosopis* community, deep-rooted *Prosopis glandulosa* (C_3) coexists with relatively shallow-rooted grasses (often C_4). In this semiarid rangeland, *Bouteloua* cannot out-compete *Prosopis*, despite high photosynthetic rates [maximum photosynthesis of *B. gracilis* was as high as 75 mg CO_2 dm^{-2} h^{-1} (Detling*et al.* 1978)]. Although *P. glandulosa* is a C_3 plant, its physiological activities closely resemble those of C_4 plants. Its highly thermostable photosynthetic machinery allows it to carry on photosynthesis at rates similar to C_4 plants (maximum net photosynthesis 32—42 mg CO_2 dm^{-2} h^{-1}) under optimum leaf temperatures of 27—31°C. It is very drought tolerant and recovers rapidly from severe water stress. It can photosynthesize at a rate of 25 mg CO_2 dm^{-2} h^{-1} even when xylem water potential was −4.6 Mpa.

Water use efficiency of *Prosopis* at the leaf level (2.5—3 μmol CO_2/mmol water) also resembles that of C_4 plants. In contrast, net photosynthetic rate of *B. gracilis* was very sensitive to leaf water potential and decreased exponentially with decreasing leaf water potential under high and low growth temperature regimes (Kemp and Williams 1980). Superior performance of *Prosopis* in the short grass plains over C_4 grasses may be correlated with its phreatophytic characteristics and high water use efficiency. Coexistence of these species in this ecosystem can be explained by their physiological adaptations to their respective niches and is not necessarily related to their photosynthetic pathways. Although C_4 plants are more adapted to hot and dry environments, this need not be true in all circumstances.

Kemp and Williams (1980) showed a niche separation pattern between *Agropyron smithii* (C_3) and *B. gracilis* (C_4) in the same habitat of the Colorado short grass prairie. Cool-season *A. smithii* produced 2-fold more dry matter when grown at cool (20°/15°C) compared to warm (35°/15°C) temperatures. Dry matter production of *B. gracilis* was 2.7-fold greater when grown at warm compared to cool temperatures. This difference is apparently caused by their photosynthetic pathways which allows occupation of different niches through temporal partitioning of resources in a given habitat. *A. smithii* occupies the habitat during the early part of the growing season by having the capacity for early spring growth while *B. gracilis* becomes dominant during summer from its ability to grow rapidly at high temperature regimes.

Generally, C_3 plants have lower light saturation points than C_4 plants. Therefore, mesophytes and cool-season plants can carry on photosynthesis at sufficient rates for optimum growth at lower light intensities than xerophytes. Bazzaz (1979) showed a similar relationship between plants characteristic of particular successional stages to adaptation to light intensity: late successional plants become light saturated at lower intensities than early successional plants. One could ask, "In mesic environments and closed habitats where light is a limiting factor, are early successional plants typically C_4 species and climax plants typically C_3 species?" In open habitats of the western U.S. the above relationship is not necessarily true because climax plants are not usually shade tolerant species. C_3 plants *P. glandulosa* and *Larrea tridentata*, dominant in the southwestern desert, are good examples.

Dry matter production of a range plant community is not totally determined by maximum photosynthetic rates because environmental conditions favorable for maximum photosynthesis rarely occur in the field. According to Schulze and Hall (1982), daily average photosynthesis of several species was 32—68% below maximum photosynthesis. Nowak and Caldwell (1986) compared photosynthetic characteristics of two bunch grass species. Photosynthetic rates and stomatal conductances of senescent late-season blades on bluebunch wheatgrass (*Agropyron spicatum*) were higher than those of crested wheatgrass (*A. desertorum*) at photon flux densities greater than 0.8 mmol $m^{-2} s^{-1}$ and at all foliage temperatures between 18 and 41°C. However, the difference in net photosynthetic rates between late-season blades on two bunch grasses diminished at photon flux density below 0.8 mmol $m^{-2} s^{-1}$. During 30% of the summer, light intensity within the bunchgrass canopy was below 0.8 mmol photon $m^{-2} s^{-1}$, which reduced the significance of higher photosynthetic rate of bluebunch wheatgrass. Furthermore, leaf area of bluebunch wheatgrass was 43% less than that of crested wheatgrass. Therefore, higher photosynthetic rates and stomatal conductances of bluebunch wheatgrass at the single leaf level do not result in greater photosynthetic carbon gain and water loss for individual tillers.

Since photosynthetic capacity depends on light interception surface area and phenological stage of leaves, and since microenvironmental gradients may exist within a canopy, photosynthetic rates at the leaf level may or may not reflect photosynthetic capacity of the whole plant. Furthermore, canopy photosynthesis appears to be more meaningful than single leaf photosynthesis in an ecophysiological context. When a dense canopy is opened up by defoliation, photosynthetic surface, carbon dioxide and water vapor profiles within the canopy change as well as light and temperature gradients. All of these changes influence carbon allocation pattern, energy flow, water use efficiency, and primary production of the canopy. Therefore, physiological response to defoliation bears an ecological significance in understanding trophic level dynamics of a range ecosystem.

There is a curvilinear response between leaf area indices (LAI) and growth rate. Once LAI exceeds 4—5, about 95% of incident radiation is intercepted, which corresponds to the breaking point from the linear portion to the plateau of the response curve. Canopy density should be managed to optimize economic production; thus a LAI slightly below optimum (4—5) should be maintained. If LAI exceeds the optimum, light interception stays the same, but canopy maintenance respiration increases and biomass production decreases. If LAI falls far below optimum after heavy defoliation, light interception decreases. In addition, considerable advection energy will arise from bare soil and cause increased vapor pressure deficits between leaf and air, which tends to inhibit photosynthesis.

Detling *et al.* (1979) showed in blue grama, the net photosynthetic rate of remaining tissue immediately after defoliation was only 40% of pre-clipping rate, and it increased to a maximum three days after clipping. From 3 to 10 days following defoliation, net photosynthetic rates in the remaining leaves were maintained at a constant level of about 21% greater than those unclipped controls. A more recent study by Painter and Detling (1981) found similar trends in *A. smithii.* Although the net photosynthetic rate of remaining tissue following defoliation may increase, the magnitude of this increase may not compensate for reduction in biomass production due to decreased light interception capacity. Detling *et al.* (1979) indicated that aboveground biomass of control blue grama plants doubled during a 10-day study period, while biomass of clipped plants increased only 67%. Therefore, plant species characterized by rapid replacement of green foliage following defoliation will be desirable in a grazing system (Dahl and Hyder 1977). The plants with a greater capability to allocate a large proportion of current photosynthate to production of new leaf blades are those more tolerant to grazing. Partially defoliated blue grama plants allocated over 50% new growth to production of new leaf blades and only 18% to new roots, while only 33% of new growth of unclipped plants was allocated to new leaf blades but 29% went to roots (Detling *et al.* 1979). These changes in carbon allocation patterns and enhanced net photosynthetic rates of the remaining leaves following moderate defoliation results in improved efficiency of photosynthetic carbon fixation. One could expect an increased total photosynthetic carbon gain for a range plant community if proper grazing management is used. On the other hand, studies in defoliation effect on photosynthetic carbon gain in a range plant community will provide valuable information in developing better grazing systems.

In moderately grazed rangeland, annual primary production was greater than standing crop production at the end of growing season (Wiegert and Evans 1964, McNaughton 1979). This is due to facilitation of energy flow by partial defoliation as suggested by McNaughton (1976). Moderate defoliation by grazing has been shown to be especially beneficial to tallgrass prairie communities (Sims and Singh 1978). Mean annual aboveground net primary production for grazed tallgrass prairie sites was 422 g m^{-2}, 28% higher than that for ungrazed sites. This increase is apparently related to increased canopy photosynthesis resulting from improved light penetration within the canopy and postponement of leaf senescense. In ungrazed areas, plants tend to increase height as result of continued elongation of dense reproductive shoots. In grazed areas, however, moderate grazing stimulates vigorous tillering which produces dense green foliage (increased photosynthetic surface area) and improved forage quality. In grazing systems, effect of defoliation on canopy photosynthesis and production should be carefully studied at various degrees of intensities and frequencies in developing appropriate grazing strategies.

Physiological adaptation of plant species to grazing is a valuable trait for maintaining stability and high productivity of a range ecosystem. Caldwell *et al.* (1981) showed great variation in photosynthetic capacity and carbon allocation pattern after defoliation in two wheatgrass (*Agropyron*) species. *A. spacanum* was able to photosynthesize at greater rates than *A. desertorum*, but replacement of new tillers and leaf blades following defoliation was more rapid for *A. desertorum. A. desertorum* produced 65% more foliage elements than *A. spacanum* and invested less nitrogen and biomass in each element. *A. desertorum* was able to produce photosynthetic surface area 3 to 5 times greater than *A. spacanum* following defoliation, partly due to lower cost of green foliage element. Greater flexibility of resource allocation in *A. desertorum* allows greater interception of solar radiation, resulting in a higher total photosynthetic carbon gain of regrowth foliage as compared with *A. spacanum.* Because of rapid reestablishment of photosynthetic surface, *A. desertorum* is a more desirable species in a grazing system.

Since relatively less carbon is translocated to the root system after defoliation, plant vigor might be affected, especially during drought periods when root extension rate is not enough to keep roots ahead of the drying front. Therefore, root carbohydrate replenishment pattern of species after spring "drawdown" is another physiological consideration in grazing management. Plants less affected by defoliation are those with narrow V-shaped root TNC curves, while those with U-shaped and flat or extended V-shaped root TNC curves are more sensitive to defoliation (Menke and Trlica 1981). Blue grama is able to grow rapidly during the spring rainy season and to replenish TNC reserves quickly. It has a narrow V-shaped root TNC curve. Because of rapid root TNC recharge, blue grama tends to minimize the period of low TNC reserve levels and is able to tiller rapidly when favorable growing conditions exist. Western wheatgrass (*A. smithii*) has a flat V-shaped root TNC curve. It takes a longer time for reserve replenishment, thus, it is more affected by defoliation (Buwai and Trlica 1977). The narrow V-shaped reserve curve of blue grama is probably related to high photosynthetic rates and rapid foliage development so sufficient photosynthates are available for rapid downward translocation. The high growth rate of blue grama is accompanied by rapid root and crown TNC depletion, indicating great root activity. Rapid leaf blade growth will not be possible unless sufficient root growth provides adequate water and nutrient uptake. Therefore, carbon allocation to roots after defoliation could be as important as carbon allocation to leaf tissue. Menke and Trlica (1981) suggested that the shape of the seasonal TNC curve could serve as a screening tool for assessing the relative effects of defoliation on different plant species.

There is some evidence that shape of root TNC curve may have certain physiological importance in chemical control of woody plants which sprout from the crown and roots. It is well known that mesquite trees growing on sands or sandy loams are easier to kill by aerially applied herbicides than those growing on heavier soils (Fisher 1955, Dahl *et al.* 1971). We recently found that mesquite trees on a sandy loam site are characterized by narrow V-shaped TNC curve, while those on a clay loam site had an extended V-shaped reserve curve during late May through early July. Since narrow V-shaped TNC curves indicate rapid carbohydrate translocation to the roots, sufficient herbicides are expected to be transported to the basal crown and bud zone. Four out of five herbicides applied from May to early July showed a significant difference in mortality between the two mesquite populations on the two sites, with greater mortality in the sandy loam population than in the clay loam population (M. Dumesnil unpublished data). Rabbitbrush (*Chrysothamus* sp.) is very difficult to control chemically, which may be related to its flat root TNC curve. Adequate soil moisture and adequate shoot growth are primary conditions for greater than 90% control in *C. nauseousus* (Mohan 1973). Under these conditions, rate of carbohydrate translocation to the roots is expected to be high.

Plant ecophysiology has many applications to range ecology and range management. The more we can learn about ecophysiological relationships, the better managers we can become. This information will enhance our abilities in grazing management, revegetation of drastically disturbed sites (including seeding of rangeland), and control of noxious (and sometimes poisonous) plants. It will also aid in our understanding of ecological changes on rangelands.

3.6 References

Bassham, J. A. 1982. The reduction pentose phosphate cycle. Encycl. Plant Physiol. Vol. 12B: 10—30. Springer-Verlag. New York.

Bazzaz, F. A. 1979. The physiological ecology of plant succession. Ann. Rev. Ecol. Syst. 10: 351—371.

Bjorkman, O. 1966. The effect of oxygen on photosynthesis in higher plants. Physiol. Plant. 19: 618—633.

Bjorkman, O., M. Badger, and P. A. Armond. 1978. Thermal acclimation of photosynthesis: Effect of growth temperature on photosynthetic characteristics and components of the photosynthetic apparatus in *Nerium oleander*. Carnegie Inst. Wash. Yearbook. 77: 262—282.

Bjorkman, O., E. Gauhl, and M. A. Nobs. 1970. Comparative studies of Atriplex species with and without B. carboxylation photosynthesis. Carnegie Inst. Wash. Yearbook. 68: 620—633.

Blackman, P. G., and W. J. Davies. 1983. The effects of cytokinins and ABA on stomatal behavior of maize and *Commelina*. J. Exp. Bot. 34:1619—26.

Bradford, K. J., and T. C. Hsiao. 1982. Physiological responses to moderate water stress. Encycl. Plant Physiol. Vol. 12B: 264—324. Springer-Verlag. New York.

Brown, R. W. 1977. Water relations of range plants. p. 97—140. *In*: R. E. Sosebee (ed.) Rangeland plant physiology. Range Science Series No. 4. Soc. for Range Manage., Denver, Colo.

Burke, M. J., L. V. Gusta, H. A. Quamme, C. J. Weiser, and P. H. Li. 1976. Freezing and injury in plants. Ann. Rev. Plant Physiol. 27: 507—528.

Buwai, M., and M. J. Trlica. 1977. Multiple defoliation effects on herbage yield, vigor, and total non-structural carbohydrates of five range species. J. Rage Manage. 30: 164—171.

Caldwell, M. M., J. H. Richards, D. A. Johnson, R. S. Nowak, and R. S. Dzurec. 1981. Coping with herbivory: photosynthetic capacity and resource allocation in two semiarid *Agropyron* bunchgrasses. Oecologia 50: 14—24.

Caldwell, M. M., R. S. White, R. T. Moore, and L. B. Camp. 1977. Carbon balance, productivity, and water use of cold-winter desert shrub communities dominated by C_3 and C_4 species. Oecologia 29: 275—300.

Campbell, G. S. 1977. An introduction to environmental biophysics. Springer-Verlag. New York.

Cowan, I. R. 1977. Stomatal behavior and environment. Adv. Bot. Res. 4: 117—228.

Cowan, I. R. 1982. Regulation of water use in relation to carbon gain in higher plants. Encycl. Plant Physiol. Vol. 12B: 589—613. Springer-Verlag. New York.

Cowan, I. R., and G. D. Farquhar. 1977. Stomatal function in relation to leaf metabolism and environment. p. 471—505. *In*: Jennings (ed.) Integration of activity in the higher plant. SEB Symp. 31. Cambridge, Cambridge University Press.

Dahl, B. E., and D. N. Hyder. 1977. Developmental morphology and management implication. p. 257—290. *In*: R. E. Sosebee (ed.) Rangeland plant physiology. Range Science Series No. 4. Society for Range Management. Denver, Colo.

Dahl, B. E., R. B. Wadley, M. R. George, and J. L. Talbot. 1971. Influence of site on mesquite mortality from 2, 4, 5-T. J. Range Manage. 24: 210—215.

Detling, J. K., M. I. Dyer, and D. T. Winn. 1979. Net photosynthesis, root respiration, and regrowth of *Bouteloua gracilis* following stimulated grazing. Oecologia 41: 127—134.

Detling, J. K., W. J. Parton, and H. W. Hunt. 1978. An empirical model for estimating CO_2 exchange of *Bouteloua gracilis* (H.B.K.) Lag. in the shortgrass prairie. Oecologia 33: 137—147.

Eckardt, F. E. (ed.) 1965. Methodology of plant eco-physiology. Montpellier Symp., Proc. UNESCO.

Edwards, G., and D. Walker. 1983. C_3, C_4: mechanisms and cellular and environmental regulation, of photosynthesis. Univ. of California Press. Berkeley.

Ehleringer, J., and O. Bjorkman. 1977. Quatum yields for CO_2 uptake in C_3 and C_4 plants. Plant Physiol. 59: 86—90.

Etherington, J. R. 1975. Environment and plant ecology. John Wiley & Sons. New York.

Farquhar, G. D. 1978. Feedforward responses of stomata to humidity. Aust. J. Plant Physiol. 5: 787—800.

Farquhar, G. D., and T. D. Sharkey. 1982. Stomatal conductance and photosynthesis. Ann. Rev. Plant Physiol. 33: 317—345.

Farquhar, G. D., and S. von Caemmerer. 1982. Modelling of photosynthetic response to environmental conditions. Encycl. Plant Physiol. Vol. 12B: 549—587. Springer-Verlag. New York.

Fliter, A. H., and R. K. M. Hay. 1981. Environmental physiology of plants. Academic Press. New York.

Fisher, C. E., C. H. Meadors, and R. Behrens. 1956. Some factors that influence the effectiveness of 2, 4, 5-Trichlorophenoxyacetic acid in killing mesquite. Weeds 4: 139—147.

Gates, D. M. 1965. Radiant energy, its receipt and disposal. Meteorol. Monogr. 6: 1—26.

Gates, D. M. 1980. Biophysical ecology. New York. Springer-Verlag. New York.

Giles, K. L., M. F. Beardsell, and D. Cohen. 1974. Cellular and ultrastructural changes in mesophyll and bundle-sheath cells of maize in response to water stress. Plant Physiol. 54: 208—212.

Graham, D., and B. D. Patterson. 1982. Responses of plants to low, nonfreezing temperatures: Proteins, metabolism, and acclimation. Ann. Rev. Plant Physiol. 33: 347—372.

Hutmacher, R. B., and D. R. Krieg. 1983. Photosynthetic rate control in cotton. Plant Physiol. 73: 658—661.

Hsiao, T. C. 1973. Plant responses to water stress. Ann. Rev. Plant Physiol. 24: 519—570.

Jewer, P. C., and L. D. Incoll. 1980. Promotion of stomatal opening in the grass *Anthephora pubescens* Nees by a range of natural and synthetic cytokinin. Planta 150: 218—221.

Johnson, R. R., N. M. Frey, and D. N. Moss. 1974. Effect of water stress on photosynthesis and transpiration of flag leaves and spikes of barley and wheat. Crop Sci. 14: 728—731.

Jones, H. G. 1973. Limiting factors in photosynthesis. New Phytol. 72: 1089—1904.

Kaiser, W. M. 1982. Correlation between changes in photosynthetic activity and changes in total protoplast volume in leaf tissue from hydro-, meso- and xerophytes under osmotic stress. Planta 154: 538—545.

Keck, R. W., and J. S. Boyer. 1974. Chloroplast response to low leaf water potentials. III. Differing inhibition of electron transport and photophosphorylation. Plant Physiol. 53: 474—479.

Kemp, P. R., and G. J. Williams (III). 1980. A physiological basis for niche separation between *Agropyron smithii* (C_3) and *Bouteloua gracilis* (C_4). Ecology 61: 846—858.

Korner, C. H., J. A. Sheel, and H. Bauer. 1979. Maximum leaf diffusive conductance in vascular plants. Photosynthetica 13: 45—82.

Kurkova, E. B., and M. V. Motorina. 1974. Chloroplast ultrastructure and photosynthesis at different rates of dehydration. Sov. Plant Physiol. 21: 28—31.

Lange, O. L., J. D. Tenhunen, and M. Braun. 1982. Midday stomatal closure in Mediterranean type sclerophylls under simulated habitat conditions in an environment chamber. I. Comparison of the behavior of various European Mediterranean species. Flora 172: 563—557.

Larcher, W. 1978. Physiological plant ecology. Springer-Verlag. New York.

Larcher, W. 1980. Physiological plant ecology. (Second ed.). Springer-Verlag. New York.

Levitt, J. 1972. Response of plants to environmental stresses. Academic Press. New York.

McDonough, W. T. 1977. Seed physiology. p. 155—184. *In*: R. E. Sosebee (ed.) Rangeland plant physiology. Range Science Series No. 4, Soc. for Range Manage. Denver, Colo.

McNaughton, S. J. 1975. Serengeti Migratory Wildebeast: Facilitation of energy flow by grazing. Science 92—94.

McNaughton, S. J. 1979. Grazing as an optimization process. Grass ungulate relationship in the Serengeti. Amer. Nat. 113: 691—703.

Meidner, H. 1967. The effect of Kinetin on stomatal opening and the rate of intake of carbon dioxide in mature primary leaves of barley. J. Exp. Bot. 18: 556—561.

Menke, J. W., and M. J. Trlica. 1981. Carbohydrate reserve, phenology, and growth cycles of nine Colorado range species. J. Range Manage. 34: 269—277.

Mohan, J. M. 1973. Fourteen years of rabbitbrush control in central Oregon. J. Range Manage. 26: 448—451.

Mohanty, P., and J. S. Boyer. 1976. Chloroplast response to low leaf water potentials. IV. Quantum yield is reduced. Plant Physiol. 57: 704—709.

Mooney, H. A., O. Bjorkman, and G. J. Collatz. 1977. Photosynthetic acclimation to temperature and water stress in the desert shrub, *Larrea divaricata*. Carnegie Inst. Wash. Yearbook 76: 328—335.

Mulroy, T. W., and P. W. Rundel. 1977. Annual plants: adaptations to desert environments. BioScience 27: 109—114.

Nobel, P. S. 1974. Biophysical plant physiology. W. H. Freeman and Co. San Francisco. California.

Nowak, R. S., and M. M. Caldwell. 1986. Photosynthetic characteristics of crested wheatgrass and bluebunch wheatgrass. J. Range Manage. 39: 443—450.

Osmond, C. B., O. Bjorkman, and D. J. Anderson. 1980. Physiological processes in plant ecology. Springer-Verlag. New York.

Painter, E. L., and J. L. Detling. 1981. Effects of simulated grazing on net photosynthesis and regrowth of western wheatgrass. J. Range Manage. 34: 68—71.

Pearcy, R. W., and J. Ehleringer. 1984. Comparative ecophysiology of C_3 and C_4 plants. Plant Cell and Environ. 7: 1—13.

Pierce, M., and K. Raschke. 1980. Correlation between loss of turgor and accumulation of absusic acid in detached leaves. Planta 148: 174—182.

Raschke, K. 1956. The physical relationships between heat transfer coefficients, radiation exchange, temperature, and transpiration of a leaf. Planta 48: 200—239.

Raschke, K., and R. Hedrich. 1985. Simultaneous and independent effects of abscisic acid on stoma and the photosynthetic apparatus in whole leaves. Planta 163: 105—118.

Resemann, H., and K. Raschke. 1984. Midday depression in stomatal and photosynthesis activity of leaves of *Arbutus unedo* are caused by large water vapor pressure differences between leaf and air. Plant Physiol. (Suppl.) 75: 66.

Schimper, A. F. W. 1898. Pflanzengeographie auf physiologischer grundlage. Jena, Fischer. Transl. by Wm. A. Fischer. 1903. Plant geography upon a physiological basis. Clarendon Press, Oxford.

Schulze, E. D. 1986. Carbon dioxide and water vapor exchange in response to drought in the atmosphere and in the soil. Ann. Rev. Plant Physiol. 37: 247—274.

Schulze, E. D., and A. E. Hall. 1982. Stomatal responses, water loss and CO_2 assimilation rates of plants in contrasting environments. Encycl. Plant Physiol. Vol. 12B: 181—230. Springer-Verlag. New York.

Setter, T. L., W. A. Grun, and M. L. Brenner. 1980. Effect of obstructed translocation on leaf abscisic acid, and associated stomatal closure and photosynthesis decline. Plant Physiol. 65: 1111—1115.

Sharkey, T. D. 1984. Transpiration-induced changes in the photosynthetic capacity of leaves. Planta 160: 143—150.
Shreve, F., and I. L. Wiggins. 1964. Vegetation and flora of the Sonoran Desert. Stanford Univ. Press. Stanford, California.
Sims, P. L., and J. S. Singh. 1978. The structure and function of ten western North American grasslands. III. New primary production, turnover and efficiencies of energy capture and water use. J. Ecol. 66: 573—597.
Sinclair, T. R., and M. M. Ludlow. 1985. Who taught plants thermodynamics? The unfulfilled potential of plant water potential. Aust. J. Plant Physiol. 12: 213—217.
Spanner, D. C. 1972. Plants, water, and some other topics. p. 29—39. *In*: R. W. Brown and B. P. van Haveren (eds.) Psychrometry in water relations research. Utah Agr. Exp. Sta. Utah State Univ. Logan.
Steponkus, P. L. 1984. Role of the plasma membrane in freezing injury and cold acclimation. Ann. Rev. Plant Physiol. 35: 543—584.
Tansley, A. G. 1914. Presidential address. J. Ecol. 2: 194—202.
Tracy, C. R., and J. J. Turner. 1982. Commentary: What is physiological ecology? Ecological Society of America, Bull. 63: 340—341.
von Caemmerer, S., and G. D. Farquher. 1981. Some relationships between the biochemistry of photosynthesis and the gas exchange of leaves. Planta 153: 376—387.
Walter, H. 1973. Vegetation of the earth in relation to climate and the eco-physiological conditions. The English Universities Press Ltd. London.
Warren-Wilson, J. 1967a. The components of leaf water potential. I. Osmotic and matric potentials. Aust. J. Biol. Sci. 20: 329—347.
Warren-Wilson, J. 1967b. The components of leaf water potential. II. Pressure potential and water potential. Aust. J. Biol. Sci. 20: 349—357.
Wiegert, R. G., and F. C. Evans. 1964. Primary production and the disappearance of dead vegetation on an old field in southeastern Michigan. Ecology 45: 49—63.
Wieland, N. K., and F. A. Bazzaz. 1975. Physiological ecology of three codominant successional annuals. Ecology 56: 681—688.
Werk, K. S., J. Ehleringer, I. N. Forseth, and C. S. Cook. 1983. Photosynthetic characteristics of Sonoran desert winter annuals. Oecologia (Berlin) 59: 101—105.
Wong, S. C., I. R. Cowan, and G. D. Farquhar. 1979. Stomatal conductance correlates with photosynthetic capacity. Nature 282: 424—426.
Younis, H. M., J. S. Boyer, and Bovindjee. 1979. Conformation and activity of chloroplast coupling factor exposed to low chemical potential of water in cells. Biochem. Biophys. Acta 548: 328—340.

4. Rangeland plant taxonomy

STEPHAN L. HATCH

4.1. Abstract

Rangeland plant taxonomy is primarily concerned with the accurate identification of plants as influenced by the basic processes of plant taxonomy. The processes of characterization, classification, and nomenclature of plants influence the identification of species. Changes in Latin names are necessary when new information dictates a reclassification of the organism. Just as important as identification is the use of the correct name. Voucher specimens are an important part of the documentation for all range vegetation studies. The proper preparation techniques are discussed for the collection, pressing, and drying of plant specimens.

4.2. Introduction

Daubenmire (1968) stated the following, "The synecology of a particular area can never rise above the quality of the taxonomic treatment available for that area, nor the taxonomic competence of the synecologist." It is evident from this statement that a high degree of taxonomic accuracy is required in all areas of rangeland work.

Precise identifications are required at times when the diagnostic characters of plants are not available. Therefore, vouchers of immature as well as mature specimens become important. The problem of inaccurate identification is most frequently associated with the sight identification rather than critical examination and verification of specimens.

Rangeland plant taxonomy is based on the fundamentals of plant taxonomy as these basic processes influence the identification and naming of plants. Common and scientific names are important for range professionals, each with their respective strengths and weaknesses. Just as important as the correct identification is the correct name.

Tueller, P. T. (ed.), Vegetation science applications for rangeland analysis and management.
ISBN 90-6193-195-9.
© 1988, *Kluwer Academic Publishers, Dordrecht, Boston, London.*

Documentation of plant materials is accomplished by proper preparation of plant specimens as vouchers. Range plants are best preserved for identification by collection, pressing, and drying.

4.3. Fundamentals of plant taxonomy

The fundamentals of plant taxonomy are characterization, classification, nomenclature, and identification. These processes are utilized in the exploration of any vegetation regions. Catalogues of species are developed that are eventually replaced by manuals describing the species. This provides a basis for the ecologist to gather datas to quantify densities, and composition of individual species within a site.

Characterization is the data collection and descriptive phase of taxonomy. For the non-professional taxonomist characterization pertains to gross morphology. The number of petals, sepals, number of stamens and their position relative to one another constitute a few of the important characters of a plant species. Data from this phase are used in identification as well as in the writing of descriptions for new species. These characters are also used to place plants within the taxonomic hierarchy i.e., division, class, order, family, tribe, genus, and species.

Classification is the placement of plants in taxonomic categories following a particular system. Similarities and differences of organisms are used to group plants and indicate or determine how closely or distantly organisms are related. Using this process, one may place a new plant taxon in the proper rank, i.e., genus or family. Modern classification is a result of the synthesis of data from the study of gross morphology, anatomy, physiology, biochemistry, cytology, plant breeding, and ecology. It is believed that a large numbers of characters from different fields will provide a more accurate indication of relationships than the traditional gross morphology.

Plants that have been characterized and partially or fully classified are ready for the nomenclature process. Nomenclature deals with the application of Latin names to plant species in accordance with rules located in the *International Code of Botanical Nomenclature* (1983). The purpose of these rules is to maintain order in the usage of plant names. The Code governs the scientific names of plants but does not deal with common or vernacular names.

Identification is the matching of an unknown plant with a known identity. Nomenclature is naming or attaching a name to a specimen. When a plant is collected it is compared with one or more of the following: mental pictures; line drawings in taxonomic works; diagnostic keys; botanical descriptions; and herbarium or reference collections of known specimens. The visual comparison of the unknown to the known specimen, if needed, may be accomplished with the aid of magnification for fine details. The unknown is either very similar (or the same) to or different from the known in one or more characters. When the unknown matches with the known an identification has been made. At that time the name of the known specimen is assumed to be the name of the previously unknown plant.

Generally people working in range management are not concerned with the first three fundamentals of plant taxonomy. All four processes are the domain of the professional plant biosystematist or taxonomist. However, the latter process of identification is the important aspect for all users of taxonomy.

4.4. Use of common versus scientific names

Range conservationists, and ecologists work with a clientele, the public, landowners, or lessees of federal lands, that communicate in common names. It is a necessary element for a professional range trained person to communicate with most ranchers in common names. The best policy is to learn the plants in an area by the scientific or Latin name. This can be accomplished by collecting and identifying the dominant plant species of an area. Next learn and use the common names of that location. This process may be repeated several times if you move to other locations for employment.

Common names are not desirable for communicating scientific information for four reasons: (1) common names are not universal, they are applicable in a single language and over a small area; (2) relatively few species of the world have common names; (3) common names are applied indiscriminantly to several categories i.e., genera, species, or varieties; (4) two or more unrelated species are frequently known by the same common name (Benson 1962).

Junegrass is an example of a common name which can lead to confusion without the accompanying Latin name. In Kentucky, junegrass is *Poa pratensis* L. (Kentucky bluegrass), whereas in Utah it is either *Bromus tectorum* L. (cheatgrass brome or downy chess) or *Koeleria pyramidata* (Lam.) Beauv., in Maine it is *Danthonia spicata* (L.) Beauv. ex Roem. & Schult. (povertygrass) and in Colorado and Texas it is *Koeleria pyramidata*. Examples like this could be given for many plants such as: bitterweed, sneezeweed, sourgrass, sweetgrass, and johnsongrass.

Scientific names allow people to obtain information such as forage values, seed or seeding characteristics, poisonous compounds, habitat preference, and distribution for a particular species. This information can be gathered from throughout the distribution of the species including other countries.

There is only one correct scientific name for each plant. The scientific name consists of three parts. The first part is the genus, which is capitalized and the word underlined or italicized demarcating it as a foreign word. Second is the specific epithet which follows the first part. It is italicized but not capitalized. Together these names are called the Latin name, binomial, or scientific name, or species name for a plant. Following the Latin name is an abbreviation or whole name of the person or persons who first described the species. An example is *Bromus mollis* L. This brome was described by Carl Linnaeus and that is indicated by the abbreviation for Linnaeus, which is L. When changes in scientific names occur the name or name of people making the new combinations follow the name(s) of the original describer. *Sporobolus cryptandrus* (Torr.) Gray is an example of a species names that has changed because of a new interpretation of plant morphology. This species was originally classified and described as *Vilfa cryptandra* Torr. by John Torrey. Later Asa Gray determined the placement of this taxon should be in the genus *Sporobolus.* He then published the new name combination with Torr. for Torrey (in parenthesis for the original description) followed by Gray who was responsible for making the new name combination. This means that whenever a species name is followed by one or more authority names in parenthesis the species has been reclassified and a new name combination given.

4.5. Modern systematics and ecology

Traditionally, plant taxonomy has been based on gross morphology. This was exemplified by Hitchcock's (1951) *Manual of the Grasses of the United States.* In recent times, plant biosystematics is becoming an important tool in understanding phylogenetic relationships within plant groups. Gould's (1968) application of a biosystematic approach to grass taxonomy involved the following five areas: (1) gross morphology; (2) anatomy and histology; (3) cytology and genetics; (4) biochemistry; and (5) ecology and plant distribution. Radford *et al.* (1974), recognized 15 areas of study in biosystematics. Those listed which were different from Gould's (1968) were sub-divisions of criteria. Regardless of the number of areas, it is important to note that numerous non-morphological criteria are presently used in addition to morphology. A high correlation among these characters is expected to elucidate relationships better and thus evolutionary history.

The experimental approach of Clausen *et al.* (1940) helped to emphasize the importance of systematics to taxonomy. This study also noted the importance of the ecological aspects of variation in plant speciation. Their 200 mile transect in California extended through many ecological zones. As a part of that study Clausen and Hiesey (1958a) recognized five ecotypes of *Potentilla glandulosa* Lindl. that were studied for morphological, cytological, and ecological variations. A common garden (Clausen and Hiesey 1958b) was used to evaluate the effect of the environment on plant species. These studies on plant variations indicated that genotype and habitat produce a recognizable unit, the ecotype.

Biochemical techniques which correlate highly with ecology have been used in classifying a taxon below species rank, subspecies and ecotypes of *Artemisia*. Kelsey *et al.* (1976) studied the shrubby species of *Artemisia* (section *Tridentatae*) in Montana using sesquiterpene lactones as taxonomic markers. Lactones are compounds that can be easily extracted from aerial plant parts. Analysis of the lactones by thin layer chromatography supports the classification based upon the morphological and ecological data.

Historically the species of *Artemisia* (sagebrush) were identified vegetatively using leaf morphology. Shultz (1986) reported on the leaf morphology and anatomy of species of *Artemisia* grown in an experimental garden in Wyoming. Employing multiple regression analysis she developed a predictive model relating ecology to leaf structure. There was a trend in leaf morphology for a reduction in size progressing from cold, moist habitats, to hot, dry habitats (Shultz 1986). However, a large amount of variation in leaf morphology existed between species.

There are many other biochemical techniques that are currently being used in plant systematics. One such technique is refraction fragment pattern analysis of chloroplast DNA which is used to estimate the probable percentage of differences in the base sequences among species. These types of comparisons will provide new insights to ecology and taxonomy of important range plant species.

4.6. Why Latin names change

Only one correct scientific name exists for a plant. However, the Latin name for a particular plant will change if information is obtained that causes the plant to be reclassified or if it is discovered that an earlier name was validly published. The assignment of rank i.e., genus, and position in the classification categories is taxonomic opinion. Users of plant taxonomy are frequently annoyed by these name changes. However, one must remember that taxonomy is a dynamic science in which new data and its interpretation brings about changes in the classifications and thus in nomenclature.

An example of a Latin name change is that of little bluestem, *Schizachyrium scoparium* (Michx.) Nash. Formerly this species was classified as *Andropogon scoparius* Michx. Data from micromorphology, anatomy, and gross plant morphology support the recognition of several genera within the genus *Andropogon* as previously recognized by Hitchcock (1935, 1951). If that interpretation of several recognizable genera in a genus like *Andropogon* is accepted, then the name of the species must change when the taxon is reclassified. In this case names like *Andropogon scoparius* Michx. become a synonym of *Schizachyrium scoparium* (Michx.) Nash.

4.7. Voucher specimens

The importance of voucher specimens to the study of vegetation on rangelands cannot be over emphasized. A voucher specimen is a plant specimen that has been collected, accurately identified, and then placed in a herbarium. Placement of specimens in a public herbarium allows others to re-examine specimens and check identifications. Voucher specimens are important for ecological impact statements as prepared for long term baseline data. These data are used to document changes in species, and species composition over long periods of time. These specimens become part of the documentation. Similar studies on nutritional values of forages, herbicide use on plants, and plant poisoning of livestock should also have a voucher specimen of the plants involved.

An example of a rangeland taxonomy problem in central Texas is the herbicide control of annual broomweed *Xanthocephalum dracunculoides* (DC.) Shinners. When researchers do not accurately check the species identification the resulting data may be difficult to interpret. Three species of *Xanthocephalum* occur in central Texas that are very similar at a casual glance. *Xanthocephalum dracunculoides, X. texana* (DC.) Shinners, and *X. sphaerocephalum* (Gray) Shinners can occur in the same area. Frequently *X. dracunculoides* and *X. texana* occur together. Treatment of these species with herbicide may give ambiguous results possibly due to the differing effect of the herbicide on the particular species. These mistakes are usually the result of sight identification rather than a careful examination followed by verification.

4.8. Preparing plant collections

4.8.1. Why collect and press specimens

Plant specimens are collected primarily for the purpose of identification. Mounted and labeled specimens serve the following purposes: (1) identification of plant materials, (2) voucher specimens, (3) classroom study, (4) records of distribution, (5) records of phenology, and (6) anatomical study. Further the identification of plants is not always certain and need verification in the laboratory. One of the best ways to learn plants is to collect and determine the identification (Harrington and Durrell 1957). If help with an identification is needed or verification of an identification is desired, most herbaria are willing to help. In exchange for their identification or verification it is customary to provide the herbarium with a specimen for their collection. It should be a good specimen with complete and dependable data attached. Hence, it is preferable to collect several specimens of a plant species.

Voucher specimens should be prepared not only for floristic study, but also for plant breeding experiments, plant disease analysis, and determination of poisonous plants. Voucher specimens are the plants collected that will become a permanent record of the plant species in a study. If any questions arise on the identification or any aspect of the study in which identification of the specimen is needed or involved the voucher specimen provides the data. These specimens are usually donated to a herbarium to be stored as permanent record. In this way, everyone has access to the material. Floras and checklists use herbarium specimens to document or determine the distributions of plants (Gould and Shaw 1983). Recently, voucher specimens are being used to document the host plant of many insects, smuts, rusts, fungi, lichens, and algae (Gould and Shaw 1983).

Pressed specimens are an essential tool in teaching plant identification and morphological characters representative of species, genera, and other taxa (Gould and Shaw 1983).

Pressing and drying are the best ways of processing most range plants. However, most of the nonvascular plants such as lichens, mosses, liverworts, algae, and fungi require special techniques and storage facilities. Bulky plant structures (pine cones, bark, large stem sections, and large cacti) are not pressed when dried. Pressing not only promotes retention of forms close to their natural shape but prevents plant structures from shriveling and distortion. Drying prevents the specimen from molding. The pressed and dried specimen looks different from the live specimen because it has been removed from its natural three dimensional state and pressed into a two dimensional state. If stored and handled properly; pressed, dried and mounted plants can last several hundred years. Plant specimens that are dried without prior pressing will not last very long, because the specimen is very brittle and breaks apart with handling.

4.8.2. Collecting plants

Various tools can be used to collect specimens. The following is a list of the equipment needed to collect and press specimens: (1) plant press, (2) tools for collecting: (a) digging tool, (b) pruning shears, and (c) knives, (3) plastic bag and moistened paper towels to place in bag, (4) rubber bands, or ties to close bag, (5) tags or small pieces of paper, (6) field notebook and pencil, (7) maps (State and County) showing distances and elevations, (8) camera (optional) — used to photograph plant when only a small portion is being collected, and (9) plant identification manuals and floras, and a 10× hand lens to aid in field identification (Forest Handbook 1962).

A plant press consists of a pair of frames, corrugated cardboard (driers), felt blotter pads, and single sheets of newspaper. The dimension of the frames should be 12 × 18 inches (30.2 × 45.3 cm) and be made of 1/4—3/8 inch (0.63 cm — 0.95 cm) plywood, or hardwood slats 3/4 inch (1.9 cm) wide and 1/4 inch (0.63 cm) or 1/2 inch (1.27 cm) thick. These slats can be held together with copper or aluminum rivets, or contact cement and woodscrews (use 1/2 inch thick slats when using woodscrews). A pair of nylon webbed straps 4 to 8 feet long and 1—1 1/4 inches wide are ideal for applying pressure to the press. Straps with a positive lock parachute buckle work the best, but conventional tooth buckles do a satisfactory job. The corrugated cardboard and blotter pads should also be 12 × 18 inches in dimension. The corrugated cardboard allows air to pass through the press. The blotter absorbs the plant moisture. Single sheets of newspaper folded crosswise contain the plant that is being pressed. If grasses and other plants with a relatively low moisture content are being collected then the blotter pads are optional. For plants with a high moisture content (i.e., wildflowers, succulents etc.) blotters are essential.

Collect plants that are representatives of the population, while avoiding diseased plants. If you do collect a specimen that varies from the population note the variation in your field notebook. Collect mature plants that are in flower and beginning to set fruit. This will assure that you have all the structures necessary for keying. Collect enough material to make three good mounts, plus extra flowers and fruits for the identification process. Collect a complete specimen (roots, stems, leaves, flowers and/or fruits) when collecting herbaceous plants (i.e., grass and wildflowers). Collect a branch or limb from woody species. Grasses and grass-like plants may be collected and placed in a large plastic bag with a moist paper towel and then tie the bag. Plants can then be pressed at a later time.

A conventional herbarium mounting sheet is 11 1/2 by 16 1/2 inches and specimens need to be folded to a size than 12 to 14 inches in length as they are pressed (Gould and Shaw 1983). Trees and other woody plants should be 12—14 inches in length to insure fitting the mounting sheet. Cut leafy branches that show typical leaf variation. Be aware that the flowers can be male or female and occur on separate plants. Collect fruits or cones (acorns, pods, nuts and berries) whenever possible. Always collect your specimens from the same individual plant to avoid mixtures (Forest Service 1962).

4.8.3. Pressing and drying plants

Pressing plant specimens may be accomplished in the following sequence: place a sheet of corrugated cardboard on one of the frames; place a single folded sheet of newspaper on the cardboard and place the specimen within; and place a blotter (if being used) on the newspaper. Continue to build up the press with this series of cardboard, newspaper and blotter. When all the plants have been placed in the press place a cardboard on top of the blotter and add the other frame. Thus, one press frame is on the top and one on the bottom. Place straps around the press and then tighten as much as possible. Best results are obtained if you can get one or two people to stand on the press while tightening.

Once the plants are in the tightened press they are ready to be dried. The secret to producing specimens with natural-colored foliage and flowers is in drying them as rapidly as possible without artificial heat (Forest Service 1962). This prevents any chance of mold growth, much of the discoloration and may prevent the abscission of leaves. The amount of time required for drying specimens depends on three major factors: (1) moisture content in plant, (2) the amount of humidity in the atmosphere and, (3) whether you are drying the specimens with or without artificial heat.

Specimens with a low moisture content (such as grasses) can be dried with a fan in two-three days. Other plants will take longer depending on their moisture content. The blotters and newspapers should be changed once a day, sometimes twice, when using a fan on succulents and other plants with a high moisture content. The specimens are well-dried when they feel crisp and dry to the touch, not dry if they feel cool and soft.

Collecting, pressing, and drying succulent plants (cacti) requires a different technique. Collect the entire plant when it is under 14 inches tall. If the plant is taller, collect several stem sections of joint-stemmed cacti (cholla or prickley pear). If the cacti is a ribstemmed (barrel cactus) type, collect stem sections including several ribs that show spine type. Once the specimen is collected the spongy tissue in the plant should be removed. Split the cactus in half and remove the spongy tissue. Press the half without spongy tissue. With care the specimen will dry very quickly and will make a good herbarium mount.

Drying plants with artificial heat is recommended in humid climates. This not only decreases drying time but also the need for changing cardboard and blotters. Instructions on building a stationary or a collapsible drying box are given in Forest Service Handbook 4084.5 (1962). Place the press so that the corrugation in the cardboard is parallel to the flow of air. Do not use too much heat; about 110°F is the optimum temperature. Excessive heat causes the plants to become very brittle (Forest Service 1962). Periodic checking and tightening of the press results in higher quality specimens. The press straps should be tightened at least four hours after the plants are pressed and then again in eight hours. After that the press should be checked at least twice a day and tightened whenever the straps are loose.

4.8.4. Field or collection book and collection number

Plants of the same species collected at the same time in a given area are entered in the field book under one number (Gould and Shaw 1983). Letters can be added to the number to give special recognition to individual plants (i.e., 501a, 501b, etc.). The field or collection notebook is essential for the preparation of good specimens. All data should be kept in this notebook with only the collection number being entered on the newspaper in which the plant is placed. The field book should be a pocket size, permanently bound, hardcover notebook. If possible the pages should be made of at least 50% rag to prevent water damage. The data should be entered in pencil or permanent (waterproof) ink to prevent loss of data due to moisture. Essential information to be recorded is: (1) an exact geographical location for the collection; This should include distances from the closest city, the highway number, county/parish, state and country (if not in the U.S.). The longitude and latitude may also be given; (2) the habitat or local conditions under which the plants are growing: roadside, prairie, wooded area, cropland, field etc.; soil (sand, clay) and/or rock type, (limestone etc.); shaded or sunlight area; xeric, mesic, or hydric site; (3) date of collection; and (4) the collector's collection number. Other information that should be included: longevity, plant height, flower color, growth habit, abundance, odor, presence of other characters (i.e., milky juices or stinging hairs). All of the information should be recorded at the time of collection.

The information in the field notebook will be put onto a plant label that will be glued on the bottom right hand corner of the herbarium sheet (Fig. 1 for an example of a herbarium label). The plant label is an important part

S. M. TRACY HERBARIUM
PLANTS OF TEXAS

ASTERACEAE

Croptilon divaricatus (Nutt.) Raf.

7 mi. E. of Hwy. 6 on Greens Prairie Rd.; S. side of road. Deep sandy mesic arroya cutting through on oak forest. Flowers : rays (yellow), and disc (yellow). Ca. 1 m. in height. Associated species: Indian grass, purpletop, goldenrod, longspike tridens and knotroot bristlegrass.

BRAZOS COUNTY

S. Jones 455 27 SEP 1987

Figure 1. Example of a herbarium label.

of a plant specimen. The minimum amount of information that should appear on the label includes scientific name, exact location (including county and state), the collector's name, collection number, and date. Additional information such as habitat data, height, flower color, abundance etc. is desirable. The more information included in a label; the more valuable that specimen.

4.9. References

Benson, L. 1962. *Plant taxonomy*. New York. Ronald Press Co.

Clausen, J., D. D. Keck, and W. M. Hiesey. 1940. Experimental environments on western North American plants. Washington D.C. Carnegie Inst. Wash. Publ. 520.

Clausen, J., and W. M. Hiesey. 1958a. Experimental studies on nature of species, IV Genetic structure of ecological races. Washington D.C. Carnegie Inst. Wash. Publ. 615.

Clausen, J., and W. M. Hiesey. 1958b. Phenotypic variation of genotypes in contrasting environments. Report, Scot. Pl. Breeding Station, 1958: 41—51.

Daubenmire, R. 1968. Plant communities. New York. Harper & Row Publ.

Forest Service. Handbook 4084.5. 1962. Handbook on Forest Service plant collections. U.S.D.A., Forest Service.

Gould, F. W., and R. B. Shaw. 1983. Grass systematics. 2nd ed. College Station: Texas A&M University Press.

Harrington, H. D., and L. W. Durrell. 1957. How to identify plants. Chicago. Swallow Press.

Hitchcock, A. S. 1935. Manual of the grasses of the United States. U.S.D.A. Misc. Pub. 200.

Hitchcock, A. S. 1951. Manual of the grasses of the United States. Revised by Agnes Chase. U.S.D.A. Misc. Pub. 200.

Kelsey, R. G., M. S. Morris, and F. Shafizadeh. 1976. The use of sesquiterpene lactones as taxonomic markers in the shrubby species of *Artemisia* (section *Tridentatae*) in Montana. J. Range Management 29(6): 502—505.

Radford, A. E., W. C. Dickison, J. R. Massey, and C. R. Bell. 1974. Vascular plant systematics. New York. Harper & Row Publ.

Shultz, L. M. 1986. Comparative leaf anatomy of sagebrush: Ecological considerations. p. 253—264. *In*: E. D. McArthur, and B. L. Welch (compilers). Proceedings Symposium on the biology of *Artemisia* and *Chrysothamnus*.

Voss, E. G. (ed.) 1983. International code of botanical nomenclature. The Hague. Dr. W. Junk, Publishers.

5. New plant development in range management

E. DURANT McARTHUR

5.1 Abstract

Developing plants for semiarid rangelands ($\leqslant$ 500 mm precipitation and management based on ecological rather than agronomic principles) is a relatively new effort supported in the main by public resources. Development of plant materials is a two step process with discovery and characterization followed, when warranted, by selection and genetic manipulation. Biotechnological techniques hold promise for long-term future semiarid land plant improvement, but for the present and foreseeable future traditional methods are much more likely to form the basis for progress. Grasses, grasslike plants, forbs, and shrubs are all found on semiarid rangelands. Grasses are preeminent in grazing management systems, but forbs and shrubs are also important rangeland plants. Grasses, shrubs, and forbs, often have complementary rangeland values. Many naturally occurring rangeland plant populations are used in rangeland rehabilitation projects. However, each of these life forms includes species that are being developed to improve the productivity of rangelands. For western North American rangelands some 60 species of grasses encompassing 130 cultivars are listed in this chapter. Forbs and shrubs have about 50 species apiece listed with a few more than 50 forb cultivars and about a dozen shrub cultivars. Examples of the development through various means are given for rangeland grasses, forbs, and shrubs. A chronic problem in range rehabilitation projects is to obtain plant materials of known (and high) quality. Concurrent with the ongoing effort to discover superior natural populations and develop improved cultivars, efforts are under way to improve and standardize the quality of seed and planting stock.

5.2 Introduction

This chapter is restricted to treatment and review of plant materials for use

Tueller, P. T. (ed.), Vegetation science applications for rangeland analysis and management.
ISBN 90-6193-195-9.
© 1988, *Kluwer Academic Publishers, Dordrecht, Boston, London.*

on rangelands that are not intensively managed for cultivated or irrigated lands of any type or for pastures of favorable moisture regimes ($\geqslant 500$ mm precipitation). The plant materials considered are those for rangelands that Schwarz *et al.* (1976) defined as lands suitable for grazing or browsing and as "more suitable for management by ecological rather than agronomic principles." For discussion of rangeland sites with more favorable water relationships, e.g., riparian and meadow habitats, I recommend Hafenrichter *et al.* (1968), Plummer *et al.* (1968), Thomas (1979), Monsen and Shaw (1983), and Johnson *et al.* (1985).

Rangelands are usually valued at a lesser rate than agronomic or forested lands. As a corollary, development of plant materials for rangelands has lagged behind development of plant materials for agronomy and forestry. Only in recent decades and in a few locations have efforts been made to develop plant materials for rangelands (Campbell and Herbel 1975, McIvor and Bray 1983, Carlson and McArthur 1985). Harlan (1983) said the problems of forage plant improvement, including range forage plants, were difficult because the available development resources (monies and personnel) are extremely limited in the face of biological resources requiring substantial study for adequate use.

In the United States, Canada, Australia, and in other countries much of the rangeland resource is under public ownership (Smith and Walker 1977, Walker 1979). As a consequence, much of the development effort for plants is institutional and government funded. To illustrate this, I cite the affiliations of contributing authors to three representative publications: Carlson and McArthur (1985), Heath *et al.* (1985), and Johnson (1986). Of the 175 separate authors (a few authored more than one contribution) in these publications, 97 (55.4%) had primary affiliations with public (= state) supported universities, 69 (39.4%) with national government agencies, 4 (2.3%) with private industry, 3 (1.7%) with state agencies, and 2 (1.1%) with private universities. Public sector involvement in rangeland plant development is underscored by the fact that virtually all new plant materials for rangelands in the United States are made available directly or cooperatively through U.S. Department of Agriculture agencies (Soil Conservation Service Plant Materials Centers and Agricultural Research Service Germplasm Enhancement Projects) or various State Agricultural Experiment Station Systems (Stein *et al.* 1986, USDA Soil Conservation Service 1987). In Canada that role is played by Agriculture Canada, a federal agency (Smith and Walker 1977, Stein *et al.* 1986).

Plant material development for rangeland has two distinct phases. The first is to discover and characterize (evaluate) plant germplasms that are useful for particular management objectives. This phase is dependant upon the existence and discovery of natural plant populations and the use of those populations in rangeland management practices. Two important factors must be considered at the outset: (1) the plants must be relatively common in at least some vegetative communities so that there is an expectation that they can be successfully established and contribute substantially to the productivity of reconstructed plant communities, and (2) successful propagation and establishment techniques must be on place for the plants under consideration. Many of these plants are indigenous to the particular country of use, but others are introductions from other parts of the globe (e.g., see Dewey and Plummer 1980, McIvor and Bray 1983, Monsen and Shaw 1983, Carlson and McArthur 1985).

The second phase of development is manipulation of plant materials obtained in the first phase processes by selection and other genetic procedures. Products of the second phase may be "cultivars," which are plant accessions with demonstrated superior characteristics and which are ordinarily made available for commercial seed production for germplasm increase and wider distribution. Effectiveness in the second phase is dependent not only upon the internal rigor used in that phase but also upon the quality of plant materials obtained from the first phase (Cameron 1983). For example, our group has outlined a development plan for wildland shrubs based on management needs as follows (Welch *et al.* 1986b, McArthur *et al.* 1987):

> A description of management needs (DMN) based on the range use categories, soil stabilization, animal habitat, and esthetics . . . (is developed). Evaluation criteria for soil stabilization center around a precise DMN and include root characteristics, potential disease and insect problems, ease and method of establishment, and longevity. Animal habitat needs include forage and cover with an emphasis on nutritional and cover requirements of those animal species designated for special consideration. Depending on the DMNs, candidate accessions are then subjected to a battery of greenhouse, laboratory, and field evaluations. When characteristics need to be combined between or among accessions then hybridization and cycles of selection . . . (are) . . . implemented.

Some effort, usually limited, for development of plants for rangelands is under way for all the major continents. In this chapter, I emphasize developments in western North America, but I also use some examples from a worldwide perspective. Plant material development activities and achievements are divided according to plant life form (grasses, forbs, and shrubs). I also discuss problems and prospects of plant material availability and quality control.

5.3 Plant materials

Plant materials development for rangelands encompasses the plant life forms of grasses, forbs (broad-leafed herbaceous plants), and shrubs. Grasslike plants may also occur on rangelands, but they have received little attention in plant development. The relative importance of each class of plants may be inferred by the number of each class that Stubbendieck *et al.* (1986) considered important. Of the 200 plants they considered to be the most important North American range plants, 94 (47%) were grasses, 2 (1%) were grasslike plants, 41 (20.5%) were forbs, 59 (29.5%) were shrubs (in the broad sense including suffrutescent shrubs and plants that may be large shrubs or small trees), and 4 (2%) were trees. Likewise, Nechaeva (1985) showed that shrubs, forbs, and grasses were all important for vegetative improvement of desert ranges in Soviet Central Asia.

All of these plant classes except the grasslike plants have representatives that have been subjected to intense conventional plant development (breeding and selection) programs for traditional agricultural or silvicultural purposes (Wright 1962, 1976, Simmonds 1976). To a lesser extent and much more recently each major plant class has been receiving some biotechnological or genetic engineering attention as well (Miksche 1976, Hansen *et al.* 1986, Namkoong 1986, Carmen *in press*, Wurtele *in press*). However, relatively little of either (conventional, biotechnological) kind of plant development has been accorded range plants. Most of what has been and is being accomplished is of the conventional type and is directed to grasses and legumes (Allen and Allen 1981, Estes *et al.* 1982, McIvor and Bray 1983).

Two quotes from literature help put in perspective the priority accorded different plant classes. Jonathan Swift in *Gulliver's Travels* said, "Whoever could make . . . two blades of grass, to grow upon a spot of ground where only one grew before, would deserve better of mankind, and do more essential service to his country, than the whole race of politicians put together" (Bartlett 1953). But Mark Twain in *Roughing It* said, "Sagebrush is a very fair fuel, but as a vegetable it is a distinguished failure" (Twain 1984).

5.3.1 Grasses

Grasses are, of course, the single most important class of plants in grazing management systems (Knight 1983, Heath and Kaiser 1985). Many of the 10,000 grass species that occur in the world contribute to natural grazing, but only 40 species receive significant use as sown pasture plants (Knight

1983). The indigenous distribution of these 40 species is centered on Eurasia with 24 species, east Africa with 8 species, and South America with 4 species. Most of these 40 species do not have application in the semiarid rangelands that we are considering in this chapter. Those that do, e.g., orchardgrass (*Dactylis glomerata*), smooth brome (*Bromus inermis*), and fescue (*Festuca* spp.), are genotypes that are adapted to drier climatic regimes than are ordinary for their respective taxa (Knight 1983, Ensign 1985, Monsen and Stevens 1985). Grasses are important not only for grazing but also for soil stability. In both of these functions, the grass contribution is more significant, in temperate climates at least, than contributions of other plant life forms.

Table 1 lists about 60 grass species (representing seven tribes) of current interest to plant developers in western North America. Although the 131 released range grass cultivars and the additional potential cultivars under evaluation demonstrate that progress is being made, these numbers pale in comparison to hundreds of individual cultivars available for some single agronomic species (Heath *et al.* 1985). In addition to the cultivars, many "common" sources are commercially available, and unimproved native or naturalized stands are collected and placed on the market (Carlson 1982, Young 1984). The listings in Table 1 are confined to species that (1) include populations or cultivars adapted to semiarid sites as defined here ($\leqslant$500 mm of annual precipitation) and (2) have demand as use in revegetation practice (Carlson and McArthur 1985, Heath *et al.* 1985). Grass groups excluded from consideration under (1) include the bluestems (*Andropogon* spp.) of the new and old worlds (Voigt and MacLauchlan 1985, Jacobson *et al.* 1985), many bluegrasses (*Poa* spp.) (Duell 1985), timothy (*Phleum pratense*) (Childers and Hanson 1985), tufted hairgrass (*Deschampsia caespitosa*) (Ward 1969), and others (Carlson and McArthur 1985, Heath *et al.* 1985). Groups excluded under (2) are such taxa as saltgrass (*Distichlis stricta*) and several weedy species, e.g., quackgrass (*Elytrigia repens*), cheatgrass (*Bromus tectorum*), and medusahead (*Taeneatherum caput-medusae*) (Holmgren and Holmgren 1977, Asay and Knowles 1985b). Work on grass improvement for semiarid environments is also going on in other places besides western North America. Two areas of such activity are southern Africa and Australia (McIvor and Bray 1983). In Australia, for example, orchardgrass and hardinggrass (*Phalaris aquatica*) have been under development.

To illustrate the development of a range grass I review briefly the history of crested wheatgrass in North America. This Eurasian introduction, called the "golden grass of the West" (Johnson 1986), is recognized as being composed of three separate species and consists of a polyploid

Table 1. Grasses under development for rangelands in western North America [1].

Tribe [2] Common Name [2]	Scientific Name [2]	Area of Natural Distribution	Bunch (B) or Sod (S)	*Cultivars* [3] No.	Examples
Aveneae					
Hardinggrass (Koleagrass)	*Phalaris aquatica* (*P. tuberosa*)	Africa	S/B	3	Perla Wintergreen
Chlorideae					
Sideoats grama	*Bouteloua curtipendula*	N. Amer.	S/B	12	El Reno Niner Pierre
Black grama	*B. eriopoda*	N. Amer.	S	2	Sonora Nogal
Blue grama	*B. gracilis*	N. Amer. S. Amer.	S/B	2	Hachita Lovington
Buffalograss	*Buchloe dactyloides*	N. Amer.	S	3	Mesa Sharp's improved Texoka
Curlymesquite	*Hilaria belangeri*	N. Amer.	S	0	
Galleta	*H. jamesii*	N. Amer.	S	1	Viva
Tobosa	*H. mutica*	N. Amer.	S	0	
Big gallenta	*H. rigida*	N. Amer.	S	0	
Green spangletop	*Leptochloa dubia*	N. Amer.	B	0	
Eragrosteae					
Prairie sandreed	*Calamovilfa longifolia*	N. Amer.	S	1	Goshen
Weeping lovegrass	*Eragrostis curvula*	Africa	B	4	Morpa Renner
Boer lovegrass	*E. c.* var. *conferta*	Africa	B	2	A-84 Catalina
Plains lovegrass	*E. intermedia*	N. Amer.	B	0	
Lehmann lovegrass	*E. lehmanniana*	Africa	S	2	Kuivato Puhuima
Wilman lovegrass	*E. superba*	Africa	B	1	Palar
Sand lovegrass	*E. trichoides*	N. Amer.	B	3	Nebraska-27 Bend Mason
Spike muhly	*Muhlenbergia wrightii*	N. Amer.	B	2	El Vado
Alkali sacaton	*Sporobolus airoides*	N. Amer.	B	1	Salado Saltalk
Sand dropseed	*S. cryptandrus*	N. Amer.	B	0	
Big sacaton	*S. wrightii*	N. Amer.	B	0	

Table 1. (Continued)

Tribe[2] Common Name[2]	Scientific Name[2]	Area of Natural Distribution	Bunch (B) or Sod (S)	*Cultivars*[3] No.	Examples
Paniceae					
Buffelgrass	*Cenchrus ciliaris*	Africa Asia	B/S	3	Llano Nueces
Arizona cottontop	*Digitaria californica*	N. Amer.	B(?)	0	
Blue panic	*Panicum antidotale*	Asia	S	1	A-130
Kleingrass	*P. coloratum*	Africa	S	3	Verde
Vine mesquite	*P. obtusum*	N. Amer.	S	0	
Plains bristlegrass	*Setaria leucopila*	N. Amer.	B	0	
Poeae (Festuceae)					
California brome (Mountain brome)	*B. carinatus* (*B. marginatus*)	N. Amer.	B	1	Bromar
Smooth brome	*B. inermis*	Eurasia	S	15	Homesteader Lancaster Lincoln Manchar Southland
Orchardgrass	*Dactylis glomerata*	Eurasia	B	5	Berber Palistine Paiute
Sheep fescue	*Festuca ovina*	N. Amer. Eurasia	B	2	Covar Durar
Arizona fescue	*F. arizonica*	N. Amer.	B	1	Redondo
Idaho fescue	*F. idahoensis*	N. Amer.	B	2	Joseph Nezpurs
Greenleaf fescue	*F. viridula*	N. Amer.	B	0	
Big bluegrass	*Poa ampula*	N. Amer.	B/S	1	Sherman
Bulbous bluegrass	*P. bulbosa*	Europe	B	1	P-4874
Canby bluegrass	*P. canbyi*	N. Amer.	B	1	Canbar
Sandberg bluegrass	*P. secunda*[2] (*P. sanbergii*)	N. Amer. S. Amer.	B	0	
Stipeae					
Indian ricegrass	*Oryzopsis hymenoides* (*Stipa hymenoides*)	N. Amer.	B	2	Nezpar Paloma
Smilograss	*O. mileacea*	Eurasia/ Africa	B	1	Smilo
Green needlegrass	*Stipa viridula*	N. Amer.	B	2	Green stipagrass Lodorm

Table 1. (Continued)

Tribe[2] Common Name[2]	Scientific Name[2]	Area of Natural Distribution	Bunch (B) or Sod (S)	*Cultivars*[3] No.	Examples
Needle-and-thread	*S. comata*	N. Amer.	B	0	
Columbia needlegrass	*S. columbiana*	N. Amer.	B	0	
Texas needlegrass	*S. leucotricha*	N. Amer.	B	0	
Purple needlegrass	*S. pulchra*	N. Amer.	B	0	
Triticeae (Hordeae)					
Crested wheatgrass[4]	*Agropyron cristatum*	Eurasia	B/S[4]	4	Ephraim Fairway Parkway Ruff
	A. desertorum	Eurasia	B	3	Nordan Standard[5] Summit
	A. cristatum x *A. desertorum*	—	B	1	Hycrest
	A. fragile	Asia	B	1	Siberian
Canada wildrye	*Elymus canadensis*	N. Amer.	B	1	Mandan
Bottlebrush squirreltail	*E. elymoides*	N. Amer.	B	0	
Blue wildrye	*E. glaucus*	N. Amer.	B	0	
Thickspike wheatgrass (Streambank wheatgrass)	*E. lancelatus* (*A. dasystachyum, A. riparium*)	N. Amer.	S	3	Critana Elbee Sodar
Bearded wheatgrass	*E. trachycaulus* ssp. *subsecundus* (*A. subsecundum*)	N. Amer.	B	0	
Slender wheatgrass	*E. t.* ssp. *trachycaulus*	N. Amer.	B	3	San Luis revenue
Altai wildrye	*Leymus angustus* (*E. angustus*)	Asia	B	1	Prairieland
Basin wildrye	*L. cinerius* (*E. cinerius*)	N. Amer.	B	1	Magnar
Mammoth wildrye	*L. racemosus* (*E. giganteus*)	Eurasia	S	1	Volga
Beardless wildrye	*L. triticoides* (*E. triticoides*)	N. Amer.	S	1	Shoshone
Western wheatgrass	*Pascopyrum smithii* (*A. smithii*)	N. Amer.	S	6	Arriba Baeton Rodan Rosana

Table 1. (Continued)

Tribe[2] Common Name[2]	Scientific Name[2]	Area of Natural Distribution	Bunch (B) or Sod (S)	*Cultivars*[3] No.	Examples
Russian wildrye	*Psathyrostachys junceus* (*E. junceus*)	Eurasia	B	6	Bozoisky select Cabree Swift Vinall
Bluebunch wheatgrass	*Pseudoroegneria spicata* ssp. *spicata* (*A. spicata*)	N. Amer.	B	1	Secar
Beardless wheatgrass	*P. s.* ssp. *inerme* (*A. inerme*)	N. Amer.	B/S	1	Whitmar
Mountain rye	*Secale montanum*	Eurasia	B	0	
Pubescent wheatgrass	*Thinopyrum intermedium* ssp. *barbulatum* (*A. trichophorum*)	Eurasia	S	4	Greenleaf Luna Mandan 759 Topar
Intermediate wheatgrass	*T. i.* ssp. *intermedium* (*A. intermedium*)	Eurasia	S	7	Amur Chief Clarke Greenar Oahe Tegmar
Tall wheatgrass	*T. ponticum* (*A. elongatum*)	Eurasia	B	6	Alkar Jose Largo Orbit Platt

[1] Compiled from Wasser (1972), Sours (1983), Adams and Kay (1985), Asay and Knowles (1985a), (1985b), Hafenrichter *et al.* (1968), Ensign (1985), Hassel and Barker (1985), Heizer and Hassell (1985), Voight and MacLauchlan (1985), Voight and Oaks (1985), Vogel *et al.* (1985), Monsen-personal communication (1987), Asay (1987), and USDA Soil Conservation Service (1987).

[2] Names in parentheses are synonyms. Scientific names from the Triticeae follow the Barkworth and Dewey (1985) recent revision. Otherwise taxonomy follows Holmgren and Holmgren (1977), and the individual contributors of Carlson and McArthur (1985) and Heath *et al.* (1985).

[3] Cultivar data are from references in footnote 1. There are cultivars of some of these species from other locations, e.g., South Africa, Australia, southwestern United States (Carlson and McArthur 1985, Heath *et al.* 1985).

[4] The three species listed and their hybrids are considered crested wheatgrass (Asay and Knowles 1985b). They are ordinarily bunch grasses, but the cultivar Ephraim has sodding tendencies.

[5] Not a cultivar but a broad recognized "generic" class.

complex (x = 7) (Table 1, Dewey 1986). Crested wheatgrass was first introduced into North America in 1898 and 1906. The first introduction was unsuccessful, but the second led to its ultimate establishment of 12.5 million acres in North America (Sharp 1986). All three crested wheatgrass species of Table 1 (*Agropyron cristatum*, *A. desertorum*, and *A. fragile*) were successfully introduced in 1906 (Dillman 1946, Dewey 1986). A selection of *A. cristatum* was termed 'Fairway', but that name came to be applied generically so that all diploid *A. cristatum* with broad pectinate spikes are considered Fairway crested wheatgrass. Strictly speaking, however, not all *A. cristatum* is Fairway-like. An example is the cultivar 'Ephraim', which is a tetraploid *A. cristatum* with sod forming tendencies (Stevens and Monsen 1985). Another early variety with a generic connotation is 'Standard'. The name Standard crested wheatgrass was applied to a plant introduction (P.I.) line of *A. desertorum* that other accessions of that species were measured against (Dewey 1986). It became conventional to call any non-Fairway type crested wheatgrass Standard crested wheatgrass. Standard crested wheatgrass is not a formally recognized cultivar per se but is a generic class of crested wheatgrass. *A. desertorum* is tetraploid with cylindrical spikes and imbricated spikelets. 'Siberian' crested wheatgrass is the common name for *A. fragile*; a cultivar for this taxon is 'P-27'. This species is characterized by narrow spikes and awnless lemmas. A recent development is the release of 'Hycrest' crested wheatgrass. This cultivar is the product of hybridization between an artificially induced tetraploid Fairway crested wheatgrass and Standard crested wheatgrass (naturally tetraploid), followed by selection for fertility, seedling vigor, and growth performance on semiarid ranges (Asay and Knowles 1985a). Other hybrids in the Triticeae also show promise for semiarid rangeland improvement (Asay and Knowles 1985a, 1985b, Asay 1987).

The use of crested wheatgrass as an example of development of rangeland grass cultivars is appropriate because crested wheatgrass is tolerant of grazing and is of Eurasian origin; both conditions are common denominators of most grasses that are artificially seeded on semiarid temperate rangelands (Platou and Tueller 1985). Native American grasses of the semiarid region did not evolve under and can not generally withstand the intensive grazing pressures as can many Eurasian grasses adapted to similar climatic regimes (Leopold 1950, Young *et al.* 1979, Platou and Tueller 1985).

Each cultivar listed in Table 1 has selection criteria that are documented in that table's references. In addition to these cultivars many other collected accessions and wild populations are useful for planting and will provide the basis for additional range grass development programs (Wasser 1982, Carlson and McArthur 1985, Heath *et al.* 1985).

5.3.2 Forbs

Forbs, important components of rangeland plant communities, provide much of the ecological and botanical diversity on rangeland sites (Shaw and Monsen 1983, Rumbaugh and Townsend 1985, McArthur *et al.* 1987). Rangeland forb values include: (1) preferential palatability (of some taxa) for a variety of animals; (2) enhancement of the forage value of rangeland plant communities, especially in the spring when forbs, in general, have higher protein and phosphorus levels than grasses and shrubs; some forbs also have elevated levels of protein and phosphorus in the fall; (3) nitrogen fixation by leguminous forbs enriches range soils; (4) erosion control and ground cover on unstable sites; (5) low-maintenance, aesthetically pleasing landscape plants; and (6) resistance to the spread of wildfire (Wasser 1982, Shaw and Monsen 1983, Rumbaugh and Townsend 1985).

Development of forb plant materials for arid rangelands has received little attention in comparison to grasses. For example, Wasser (1982) in his handbook treating selected species of use in revegetating disturbed western lands listed 48 grass taxa (species or subspecies) and 29 forb taxa. He listed 98 grass cultivars (2.05 per taxon) excluding those with many cultivars, e.g., Kentucky bluegrass with 33 cultivars and timothy with 25 cultivars, but only 17 forb cultivars (0.58 per taxon) excluding the cultivars of alfalfa and white clover. Wasser's (1982) treatment, while not restricted to low precipitation sites, gave primary emphasis to those areas.

Table 2 lists about 50 species of forbs adapted to semiarid conditions (< 500 mm ppt.) from 10 plant families of current interest to plant developers in western North America. While there are some 54 cultivars, two-thirds (38) represent only two introduced species (sweetclover and alfalfa). Most of the species listed in Table 2 are used as direct collections from wildland stands or from seed increased as noteworthy wildland ecotypes. Uses and demands for many of these species are increasing. I believe that many additional cultivars will be developed and made available for use over the course of the next two decades.

Forb plant materials have received little attention from rangeland plant developers. Most efforts that have been expended relate to leguminous forbs (88% of the cultivars of Table 2 are legumes). Legumes are valuable forage plants that have been consciously used as forage plants by pastoralists for over 11,000 years (Mathison 1983). The family, universally recognized as valuable for its forage, fiber, industrial chemical, and medicine production (Allen and Allen 1981), also includes some poisonous plants, most notably some species of the large genus *Astragalus* (Williams and Barneby 1977a, 1977b). The legume family or Fabaceae

Table 2. Forbs under development for rangelands in western North America [1].

Family Common Name [2]	Scientific Name [2]	Area of Natural Distribution	Rhizo-matous	*Cultivars* [3] No.	Examples
Apiaceae (= Umbelliferae)					
Nineleaf lomatium	*Lomatium triternatum* ssp. *platycarpum*	N. Amer.	No	0	
Nuttall lomatium	*L. nuttallii*	N. Amer.	No	0	
Sweetanise	*Osmorhiza occidentalis*	N. Amer.	No	0	
Asteraceae (= Compositae)					
Yarrow	*Achillea millefolium*	Eurasia	Yes	0	
Fringed sagewort	*A. frigida*	N. Amer., Eurasia	No	0	
Louisiana sagewort	*A. ludoviciana*	N. Amer.	Yes	1	Summit
Mugwort wormwood	*A. vulgaris*	Eurasia	Yes	0	
Pacific aster	*Aster chilensis* ssp. *adcendens*	N. Amer.	Yes	0	
Engelman aster	*A. engelmannii*	N. Amer.	Yes	0	
Blueleaf aster	*A. glaucodes*	N. Amer.	Yes	0	
Hairy balsamroot	*Balsamorhiza hirsuta*	N. Amer.	No	0	
Cutleaf balsamroot	*B. macrophylla*	N. Amer.	No	0	
Arrowleaf balsamroot	*B. sagittata*	N. Amer.	No	0	
Oneleaf helianthella	*Helianthella uniflora*	N. Amer.	No	0	
Showy goldeneye	*Heliomeris multiflora* var. *multiflora* (*Viguiera multiflora* var. *multiflora*)	N. Amer.	No	0	
Nevada goldeneye	*H. multiflora* var. *nevadensis* (*Viguiera multiflora* var. *nevadensis*)	N. Amer.	No	0	
Boraginaceae					
Tall bluebell	*Mertensia arizonica* ssp. *leonardi*	N. Amer.	No [4]	0	
Chenopodiaceae					
Rillscale	*Atriplex suckleyi*	N. Amer.	No	0	

Table 2. (*Continued*)

Family Common Name[2]	Scientific Name[2]	Area of Natural Distribution	Rhizo-matous	*Cultivars*[3] No.	Examples
Fabaceae (Leguminosae)					
Cicer milkvetch	*Astragalus cicer*	Eurasia	Yes	3	Lutana Monarch Oxley
Sicklepod milkvetch	*A. falcatus*	Eurasia	Yes	0	
Utah sweetvetch	*Hedysarum boreale*	N. Amer.	No[4]	0	
Sulla sweetvetch	*H. coronarium*	Eurasia, Africa	Yes	0	
Mongolian sweetvetch	*H. mongolicum*	Asia	Yes	0	
Sickle-keeled lupine	*Lupinus albicaulis*	N. Amer.		1	Hederma
Alfalfa (Lucerne)	*Medicago falcata*	Eurasia, Africa	Yes	—[5]	
	Medicago sativa	Eurasia, Africa	Yes	15[5]	Drylander Ladak Nomad Ranger Rambler Roamer Spredor Teton Travois
Medics	*Medicago* spp.	Eurasia	No	—[6]	
Sweetclover	*Meliotus alba*	Eurasia	No	23[7]	Norgold
	M. officinalis	Eurasia	No	—[7]	Polara
Sainfoin	*Onobrychis viciaefolia*	Eurasia	No	6	Eski Melrose Nova Onar Remont Renumex
Kura clover	*Trifolium ambiguum*	Asia	Yes	—[6]	
Rose clover	*Trifolium hirtum*	Eurasia	No	1[6]	Wilton
Subterranean clover (subclover)	*Trifolium subterraneum*	Eurasia, Africa	No	—[6]	
American vetch	*Vicia americana*	N. Amer.	Yes	0	
Geraniaceae					
Richardson geranium	*Geranium richardsonii*	N. Amer.	Yes	0	

Table 2. (*Continued*)

Family Common Name[2]	Scientific Name[2]	Area of Natural Distribution	Rhizo-matous	*Cultivars*[3] No.	Examples
Sticky geranium	*G. viscosissimum*	N. Amer.	Yes	0	
Linaceae					
Lewis flax	*Linum perenne* ssp. *lewisii*	N. Amer.	No	1	Appar
Malvaceae					
Scarlet globemallow	*Sphaeralcea coccinea*	N. Amer.	Yes	0	
Gooseberryleaf globemallow	*S. grossulariaefolia*	N. Amer.	No[4]	0	
Munro globemallow	*S. munroana*	N. Amer.	No[4]	0	
Rosaceae					
Small burnet	*Sanguisorba minor*	Eurasia, Africa	No	1	Delar
Scrophulariaceae					
Wasatch penstemon (Wasatch beardtongue)	*Penstemon cyananthus*	N. Amer.	No	0	
Scabland penstemon (Scabland beardtongue)	*P. deustus*	N. Amer.	No	0	
Eaton penstemon (Eaton beardtongue)	*P. eatonii*	N. Amer.	No	0	
Low penstemon (Low beardtongue)	*P. humilis*	N. Amer.	Yes	0	
Palmer penstemon (Palmer beardtongue)	*P. palmeri*	N. Amer.	No	1	Cedar
Rydberg penstemon (Rydberg beardtongue)	*P. rydbergii*	N. Amer.	Yes	0	
Rocky Mountain penstemon (Rocky mountain beardtongue)	*P. strictus*	N. Amer.	Yes	1	Bandera

Table 2. (Continued)

[1] Compiled from Hafenrichter *et al.* (1968), Schuman and Howard (1978), McKell *et al.* (1979), Wasser (1982), Shaw and Monsen (1983), Voorhees *et al.* (1984), Heath *et al.* (1985), Rumbaugh and Townsend (1985), Stevens *et al.* (1985b), and Cooperrider and Bailey (1986), McArthur *et al.* (1987), and USDA Soil Conservation Service (1987).
[2] Names in parentheses are synonyms. Scientific names are those of the contributors of footnote 1.
[3] Cultivar data are from references in footnote 1.
[4] Spreads vegetatively but not by rhizomes.
[5] Only dryland cultivars are counted, see Rumbaugh and Townsend (1985). These cultivars are most often hybrids between *M. falcata* x *M. sativa.*
[6] Each of these species has had several Australian cultivars developed. Several of these are adapted to North American situations (Barnes and Sheaffer 1985, Carlson *et al.* 1985, Knight and Hoveland 1985, and Rumbaugh and Townsend 1985).
[7] In addition to Norgold (*M. alba*) and Polara (*M. officinalis*) 21 cultivars of both species in combination have been developed (Hoveland and Townsend 1985).

(= Leguminosae) is an enormous family with a worldwide distribution and some 16,000 species in 750 genera (Allen and Allen 1981). The family includes trees, shrubs, vines, and annual and perennial forbs. All of these life forms have representation in semiarid habitats, but several of them, most introduced, are important forage plants in western North America (Table 2). Currently, the most widely used rangeland legumes are genotypes and cultivars of alfalfa, milkvetch, and sweetclover adapted to drier conditions than the great mass of their respective species arrays.

The development of rangeland alfalfa in western North America will serve as an illustration of the development of a rangeland legume. Alfalfa is the common name of two perennial species of *Medicago* (*M. sativa* and *M. falcata*). These species are related, can hybridize, and are sometimes divided into subspecific taxa (Lesins and Lesins 1979, Barnes and Sheaffer 1985, Rumbaugh and Townsend 1985). Their center of origin and diversification is in central and western Asia, although the genus as a whole evolved in the Mediterranean region (Lesins and Lesins 1979, Rumbaugh and Townsend 1985). The species evolved in their Asian homeland under grazing in rangeland ecosystems, but most plant development and selection have been under favorable moisture regimes (> 500 mm ppt. or irrigation) and fertile soils (Rumbaugh and Townsend 1985). Consequently, of the 400 cultivars recognized in North America only 15 are adapted to dryland situations (Barnes and Sheaffer 1985, Rumbaugh and Townsend 1985). These are a legacy of the South Dakota horticulturalist N. E. Hansen (Rumbaugh 1979). In 1911 Hansen stated, "If we could cloth our naked hillsides with these wild Siberian alfalfas we could increase . . . carrying capacity . . . seven to eight times." And in 1913, Hansen promoted introduction of wild alfalfa "into the native ranges" of the North American West "where they will probably be able to hold their own with any plant now found there" (Hansen 1913, Rumbaugh and Townsend 1985).

Much of the germplasm incorporated in dryland alfalfa cultivars is derived from some of Hansen's early collections. These dryland grazing alfalfas are all interspecific hybrids between *M. falcata* and *M. sativa*. The former species primarily contributes characteristics of drought resistance, winterhardiness, and tolerance to grazing, whereas the latter is more responsible for disease and insect resistance along with increased forage and seed yield (Rumbaugh and Townsend 1985). Prominent among these dryland cultivars are 'Teton', 'Travois', 'Spredor', 'Rambler', 'Drylander', and 'Nomad' (Table 2). Other legumes are, of course, important in many wildland plantings. Medics or burclovers (*Medicago* spp.) are annual relatives of alfalfa. They have a Mediterranean basin origin but have been most useful in Australia where they have undergone rigorous selection (Crawford 1983, Rumbaugh and Townsend 1985). These plants are currently being tested for adaptation in other semiarid environments. Milkvetches (*Astragalus* spp.), sweetvetches (*Hedysarum* spp.), lupine (*Lupinus* spp.), sweetclover (*Meliotus* spp.), sainfoin (*Onobrychis visiifolia*), clover (*Trifolium* spp.), and American vetch (*Vicia americana*) all have realized value or potential promise in enhancing the productivity of semi-arid rangelands (Table 2).

Nonleguminous forbs are receiving increasing attention as a revegetation resource (Wasser 1982, Shaw and Monsen 1983, Stevens *et al.* 1985b). Most of this plant material now comes from wildland stands or first generation cultivated field stands that have undergone little directed selection. Only five nonleguminous forbs are available as named cultivars for semiarid western North America (Table 2). All of these have been released for commercial seed production recently by public agencies (Shaw and Monsen 1983, McArthur *et al.* 1987). All are first releases for their respective taxa and have specific superior attributes for specific management purposes. However, these traits are naturally occurring and have not been enhanced by advanced generation selection. The five cultivars are 'Appar' Lewis flax (released for its ornamental value and ability to grow with grasses), 'Bandera' Rocky Mountain penstemon (released for ornamental landscaping and soil stabilization), 'Cedar' Palmer penstemon (released for its ability to provide forage diversity in seeded mixtures and for its ornamental value), 'Delar' small burnet (released for its forage value in mixed seedings), and 'Summit' Louisiana sagewort (released for its ability to stabilize disturbed soils).

5.3.3 Shrubs

Shrub dominance often defines rangelands, i.e., lands not intensively managed by agronomic or forestry principles. Shrubs are widely distributed in vegetation communities dominated by other life forms (Küchler 1964) but are dominants, for the most part, only in habitats that place plants under considerable stress (McArthur 1984). Stebbins (1972, 1975) listed the conditions, often interacting, that promote shrubby habitats as drought or aridity, nutrient poor soils, fire, shade, poor soil aeration, winter cold, short growing seasons, and wind. Most of these conditions are best met in arid or semiarid continental climates that also characterize most of the world's rangelands (Walter 1973). Shrub values on rangelands include: (1) extending the grazing season because shrubs, especially evergreens, provide more available and higher quality forage in the fall and winter seasons than do herbaceous plants (Cook 1972, Welch 1983) and because of deep roots, many species of shrubs are a reliable forage source during droughts (Graetz and Howes 1979); (2) substantially increasing plant community diversity and, as a corollary, wild and domestic animal community diversity (Shelford 1963, Zimmerman 1980); (3) providing materials for low maintenance landscaping (Schmutz *et al.* 1973, Schmidt 1980); (4) supplying materials for fiber and fuel and for medicinal and industrial chemicals (Foster *et al.* 1983, McArthur and Sanderson 1984, Kelsey 1986, Hegerhorst *et al.* 1987); (5) assisting ecosystem stability by providing ground cover, erosion control, nutrient cycling, water harvesting, and nitrogen fixation (Klemmedson 1979, Carlson and McArthur 1985, Laycock and Shoop 1986).

In terms of formal plant improvement, programs for rangeland shrubs trail even the poorly developed ones of rangeland grasses and forbs. Woody plant improvement programs require a longer time investment than herbaceous plants. There are, however, successful forest tree and shrub crop improvement programs (Wright 1962, 1976, Simmonds 1976). Although naturally occurring populations provide most of the use made of shrubs in the world's arid and semiarid regions (Table 3, McKell *et al.* 1972, McKell *in press*), some rangeland shrub improvement programs have begun (Carlson and McArthur 1985). Table 3 lists nearly 60 species of shrubs (including suffrutescents and lianas) adapted to western North American rangelands that have been subjected to comparative intraspecific performance evaluation, at least to a limited extent. Many other shrub species of the genera listed (Table 3 footnotes) and other genera (e.g., *Acacia, Ambrosia, Clematis, Coleogyne, Gutierrezia, Holodiscus, Physocarpus, Prunus, Sarcobatus, Spiraea, Tetradymia*) grow in western North America. As pressures on rangeland increase because of expanding resource needs, I expect that more intensive development of the rangeland shrubs listed above and in Table 3 will occur. My treatment emphasizes western North America, but parallel work is being done elsewhere; e.g., Olivares and Gasto (1981) and Contreras *et al.* (1984) in Chile on

Table 3. Shrubs under development for rangelands in western North America[1].

Family Common Name[2]	Scientific Name[2]	Area of Natural Distribution	Resprouts after fire[3]	*Cultivars*[4] No.	Examples
Anacardiaceae					
Smooth sumac	*Rhus glabra*	N. Amer.	Yes	0	
Skunk bush sumac	*R. trilobata*	N. Amer.	Yes	2	Autumn amber Bighorn
Asteraceae (= Compositae)					
Oldman wormwood	*Artemisia*[5] *abrotanum*	Europe	Yes	0	
Bigelow sagebrush	*A. bigelovii*	N. Amer.	No	0	
Silver sagebrush	*A. cana*	N. Amer.	Yes	0	
Sandsage	*A. filifolia*	N. Amer.	Yes	0	
Black sagebrush	*A. nova*	N. Amer.	No	0	
Budsage	*A. spinescens*	N. Amer.	No	0	
Basin big sagebrush	*A. tridentata* ssp. *tridentata*	N. Amer.	No	0	
Mountain big sagebrush	*A. t.* ssp. *vaseyana*	N. Amer.	No	1	Hobble Creek
Wyoming big sagebrush	*A. t.* ssp. *wyomingensis*	N. Amer.	No	0	
Spreading rabbitbrush	*Chrysothamnus*[5] *linifolius*	N. Amer.	No	0	
Rubber rabbitbrush	*C. nauseosus*	N. Amer.	No	0	
Parry rabbitbrush	*C. parryi*	N. Amer.	No	0	
Low rabbitbrush	*C. viscidiflorus*	N. Amer.	Some forms	0	
Guayule	*Parthenium argentatum*	N. Amer.	No	0	
Caprifoliaceae					
Tatarian honeysuckle	*Lonicera tartarica*	Eurasia	No	0	
Longflower snowberry	*Symphoricarpos longiflorus*	N. Amer.	Yes	0	
Western snowberry	*S. occidentalis*	N. Amer.	Yes	0	
Mountain snowberry	*S. oreophilus*	N. Amer.	Yes	0	
Chenopodiaceae					
Fourwing saltbush	*Atriplex*[6] *canescens*	N. Amer.	Some forms	4	Marana Rincon Santa Rita Wytana

Table 3. (Continued)

Family Common Name[2]	Scientific Name[2]	Area of Natural Distribution	Resprouts after fire[3]	*Cultivars*[4] No.	Examples
Shadscale	*A. confertifolia*	N. Amer.	No	0	
Castle Valley saltbush	*A. cuneata*	N. Amer.	No	0	
Gardner saltbush	*A. gardneri*	N. Amer.	No	0	
Quailbush	*A. lentiformis*	N. Amer.	Yes	1	Casa
Cattle saltbush	*A. polycarpa*	N. Amer.	Yes	0	
Australian saltbush	*A. semibaccata*	Australia	No?	1	Corto
Trident saltbush	*A. tridentata*	N. Amer.	Yes	0	
Common winterfat	*Ceratoides lanata*	N. Amer.	Some forms	1	Hatch
Pamirian winterfat	*C. latens*	Asia	No	0	
Spineless hopsage	*Grayia brandegei*	N. Amer.	No	0	
Spiney hopsage	*G. spinosa*	N. Amer.	Some forms	0	
Gray molley	*Kochia americana* ssp. *vestita*	N. Amer.	No	0	
Forage kochia	*K. prostrata*	Eurasia	Some forms	1	Immigrant
Elaeagnaceae					
Silver buffaloberry	*Shepherdia argentea*	N. Amer.	Yes	1	Sakakawea
Roundleaf buffaloberry	*S. rotundifolia*	N. Amer.	Yes	0	
Ephedraceae					
Nevada mormon tea	*Ephedra nevadensis*	N. Amer.	Some forms	0	
Green mormon tea	*E. viridis*	N. Amer.	Some forms	0	
Fabaceae (Leguminosae)					
Leadplant amorpha	*Amorpha canescens*	N. Amer.	Yes	0	
Siberian peashrub	*Caragana arborescens*	Asia	Yes	0	
Oleaceae					
New Mexico forestiera	*Forestiera neomexicana*	N. Amer.	No	1	Jemez
Opuntiacea (Cactaceae)					
Prickly pear	*Opuntia* spp.	N. Amer.	Yes	0	
Polygonaceae					
California buckwheat	*Eriogonum fasciculatum*	N. Amer.	No	1	Duro
Snow buckwheat	*E. niveum*	N. Amer.	No	0	

Table 3. (*Continued*)

Family Common Name[2]	Scientific Name[2]	Area of Natural Distribution	Resprouts after fire[3]	*Cultivars*[4] No.	Examples
Sulfur buckwheat	*E. umbellatum*	N. Amer.	No	1	Sierra
Rhamnaceae					
Fendler ceanothus	*Ceanothus*[7] *fendleri*	N. Amer.	Yes	0	
Martin ceanothus	*C. martinii*	N. Amer.	Yes	0	
Rosaceae					
Saskatoon serviceberry	*Amelanchier alnifolia*	N. Amer.	Yes	0	
Utah serviceberry	*A. utahensis*	N. Amer.	Yes	0	
Curlleaf mountain mahogany	*Cercocarpus ledifolius*	N. Amer.	No	0	
Birchleaf mountain mahogany	*C. montanus*	N. Amer.	Some forms	1	Montane
Stansbury cliffrose	*Cowania*[8] *stansburiana*	N. Amer.	No	0	
Apache plume	*Fallugia paradoxa*	N. Amer.	Yes	0	
Squawapple	*Peraphyllum ramosissimum*	N. Amer.	Yes	0	
Desert bitterbrush	*Purshia glandulosa*	N. Amer.	Yes	0	
Antelope bitterbrush	*P. tridentata*	N. Amer.	Some forms	1	Lassen
Cherries	*Prunus* spp.[9]	N. Hemisphere	Some forms	—[9]	
Saxifragaceae					
Golden current	*Ribes aureum*	N. Amer.	Yes	0	
Simmondsiaceae					
Jojoba	*Simmondsia chinensis*	N. Amer.	Yes (?)	0	

[1] Compiled from Blauer *et al.* (1975, 1976), McArthur *et al.* (1979), McKell *et al.* (1979), Stubbendieck *et al.* (1986), Robertson (1982), Wasser (1982), Ferguson (1983), Foster *et al.* (1983), Stutz (1983), Tiedemann and Driver (1983), Carlson (1984, personal communication), McArthur *et al.* (1985), Monsen and Davis (1985), Stutz and Carlson (1985), Patton *et al.* (1986), Monsen (1987, personal communication), McArthur *et al.* (1987), and USDA Soil Conservation Service (1987).

[2] Names in parentheses are synonyms. Scientific names are those of contributors of footnote 1.

[3] From shrub bases. Many but not all of the fire adapted plants have extensive lateral root or rhizome systems. Conditions (yes, no, some forms) from references in footnote 1 and from personal observation.

[4] Cultivar data are from references in footnote 1 and include only those from rangelands. Some selections have been made for guayule and jojoba in agronomic situations.

[5] Several other *Artemisia* and *Chrysothamnus* species are useful rangeland plants but are generally managed in place and are not being developed or evaluated — see McArthur *et al.* (1979, 1985).
[6] Several other *Atriplex* are useful rangeland plants being managed in place although some development has been suggested — see Blauer *et al.* (1976), Stutz (1983), Carlson (1984), and Stutz and Carlson (1985).
[7] Important and widespread genus with numerous other useful species — see Nobs (1963) and Conard *et al.* (1985).
[8] Some authors suggest reducing *Cowania* to synonymous with *Purshia* (e.g., Welsh 1986, Schaack 1987). There are several other species of *Cowania* — see McArthur *et al.* (1983).
[9] Several species were recommended for use by Plummer *et al.* (1968). None of the released cultivars are rangeland types.

Atriplex, Alimov and Amirkhanov (1980) in the Soviet Union on *Kochia*, Nechaeva (1985) also in the Soviet Union on several shrub species, Kaul (1986) in India on *Haloxylon*, and Sankary and Barbour (1972) in Syria on *Salsola*.

Examples of rangeland shrub development are the 16 cultivars listed in Table 3. These, without exception, are early generation selections from naturally occurring populations. All are native except ‘Immigrant’ forage kochia (*Kochia prostrata*) (Stevens *et al.* 1985a). Each was released for production of commercial seed because it fulfilled particular management needs. For example, ‘Rincon’ fourwing saltbush (*Atriplex canescens*) is widely adapted with excellent growth characteristics (McArthur *et al.* 1984); ‘Hobble Creek’ mountain big sagebrush (*Artemisia tridentata* ssp. *vaseyana*) is a vigorous, nutritious plant that is palatable to mule deer (*Odocoileus hemionus*) and domestic sheep (*Ovis aries*) (Welch *et al.* 1986a); and ‘Lassen’ antelope bitterbrush (*Purshia tridentata*) has a vigorous upright growth habit and is widely adapted (Shaw and Monsen 1986). Seed production of many of these shrub species poses special problems because of the difficulty of harvesting seed with conventional equipment and providing seed orchard designs optimizing plant cultural care and seed quality and production (Plummer 1980, Briggs 1984, Carlson *et al.* 1984, Noller *et al.* 1984). Some researchers have proposed, and in some cases initiated, more intensive selection on rangeland shrub taxa including artificial hybridization or discovery of natural hybrids (Stutz 1982, Carlson 1984, McArthur *et al.* 1985, 1987, *in press*).

5.4 Availability and quality control

A serious continuing problem in improving the quality of rangelands is that of obtaining quality plant materials for many rehabilitation projects

(Plummer 1984). Although bare root and containerized stock is sometimes useful (McKell *et al.* 1979, Murphy 1984), seed is ordinarily the most important means of establishing plants in rehabilitation plantings of arid and semiarid lands (Plummer 1980, McArthur *et al.* 1987, Young and Evans 1987). Availability of quality seed can sometimes be a problem. Grass seed in particular and many types of forb seed can be grown and harvested by standard agronomic techniques, but shrub seed production often requires innovative techniques (Plummer 1980, Monsen 1985). To a degree some seed of all three classes (grass, forb, and shrub) are collected from native or naturalized stands. In the United States the Association of Official Seed Certifying Agencies (AOSCA) has source identification standards for tree seed and is considering expanding these standards to include native collected grasses, forbs, and shrubs (AOSCA 1986). The states of Colorado and New Mexico have functioning source identification programs for wildland-collected seeds other than tree seed (Colorado Seed Certification Standards 1979, Official Seed Certification Handbook for New Mexico 1984). Released cultivar varieties of rangeland grasses, forbs, and shrubs are eligible for seed certification as specified by the requirements and standards of individual states such as Utah (Young 1984) under the umbrella of AOSCA (AOSCA 1987).

A continuing challenge for the manager is to find seed sources. I offer six references as guides to help find rangeland seed suppliers in western North America: Crofts and McKell (1977), Brown *et al.* (1980), Slayback (1980), Everett (1981), Long (1981), and USDA Soil Conservation Service (1982). Other sources of information include various public agencies such as the U.S. Department of Agriculture, Soil Conservation Service, State Agricultural Experiment Stations, National and State Seed Certifying Agencies, and Agriculture Canada (Carlson 1982, Stein *et al.* 1986, USDA 1987), and the annual reports of the Vegetative Rehabilitation and Equipment Workshop (VREW) (available from the USDA Forest Service, Equipment Development Center, Missoula, Montana). Many of the cultivars listed in Tables 1, 2, and 3 are not universally available, and consequently, effort is required on the part of the seed-use manager to locate supplies. I believe such an effort is warranted.

Another problem, exacerbated by the often uncontrolled conditions inherent in wildland seed collection, is that of seed quality. Efforts are currently under way to standardize evaluation techniques for various plant classes so that seed quality tests are standard and broadly interpretable (Allen and Meyer 1986, Stein *et al.* 1986).

Work on plant materials for rangelands is slowly accelerating from a relatively recent and hard earned foundation. Obtaining the best plant materials available is generally a difficult job. However, I believe it is a job worth doing well — the seeker after plant materials should make a thorough, energetic search.

5.5 Acknowledgements

I thank many of my colleagues with whom I have shared the adventure of documenting genetic variation in plant populations both for the intrinsic interest and for possible use in increasing land and plant productivity. Foremost among these is my (and many of my colleagues') mentor, A. Perry Plummer. Those colleagues, among others, are A. Clyde Blauer, Jack R. Carlson, James N. Davis, D. Carl Freeman, Wendell G. Hassell, Gary L. and Kent R. Jorgensen, Sherel Goodrich, Susan E. Meyer, Stephen B. Monsen, Gary L. Noller, David L. Nelson, Rosemary L. Pendleton, Stewart C. Sanderson, Nancy Shaw, Richard Stevens, Sam E. Stranathan, Howard C. Stutz, Bruce L. Welch, and Stanford A. Young. Work that I accomplished that contributed, in part, to this review was facilitated by cooperative efforts in wildlife habitat restoration funded partially through Pittman-Robertson Project W82-R (cooperators: U.S. Department of Agriculture, Forest Service, Intermountain Research Station and Utah Department of Natural Resources, Division of Wildlife Resources).

5.6 References

Adams, T. E., Jr., and B. L. Kay. 1985. Phalaris, orchardgrass, fescue and selected minor grasses. Part I: Perennial grasses for California rangelands. p. 128—132. *In*: J. R. Carlson, and E. D. McArthur (eds.) Proceedings, selected papers presented at the 38th Annual Meeting of The Society for Range Management. III. Range plant improvement, Society for Range Management, Denver, Colorado.

Alimov, E., and Zh. Amirkhanov. 1980. Breeding of *Kochia prostrata* Schard. in deserts of southeastern Kazakhstan. Problems of Desert Development 3: 57—61.

Allen, O. N., and E. K. Allen. 1981. The Leguminosae, a source book of characteristics, uses and nodulation. The University of Wisconsin Press, Madison.

Allen, P., and S. Meyer. 1986. Developing procedures for testing wildland seed. AOSA Newsletter 60(3): 26—28.

AOSCA, Association of Official Seed Certifying Agencies. 1986. Sixty-eighth annual report of the Association of Official Seed Certifying Agencies. Raleigh, North Carolina.

AOSCA, Association of Official Seed Certifying Agencies (*In Press*). 1987. Sixty-ninth annual report of the Association of Official Seed Certifying Agencies. Raleigh, North Carolina.

Arnow, L. A. 1987. Gramineae A. L. Juss. p. 684—788. *In*: S. L. Welsh, N. D. Atwood, S. Goodrich, and L. C. Higgins (eds.) A Utah Flora, Great Basin Naturalist Memoirs No. 9. Brigham Young University, Provo, Utah.

Asay, K. H. 1987. Revegetation in the sagebrush ecosystem. p. 19—27. *In*: J. A. Onsager (ed.) Integrated pest management on rangeland, state of the art in the sagebrush ecosystem, USDA Agricultural Research Service ARS-50. U.S. Government Printing Office, Washington, D.C.

Asay, K. H., and B. P. Knowles. 1985a. Current status and future of introduced wheatgrasses and wildrye for rangeland improvement. p. 109—116. *In*: J. R. Carlson, and E. D. McArthur (eds.) Proceedings, selected papers presented at the 38th Annual Meeting of The Society for Range Management. III. Range plant improvement. Society for Range Management, Denver, Colorado.

Asay, K. H., and R. P. Knowles. 1985b. The wheatgrasses. p. 166—176. *In*: M. E. Heath , R. F. Barnes, and D. S. Metcalf (eds.) Forages, the science of grassland agriculture (4th ed.). Iowa State University Press, Ames.

Barnes, D. K., and C. C. Sheaffer. 1985. Alfalfa. p. 89—97. *In*: M. E. Heath, R. F. Barnes, and D. S. Metcalf (eds.) Forages, the science of grassland agriculture (4th ed.). Iowa State University Press, Ames.

Barkworth, M. E., and D. R. Dewey, 1985. Genomically based genera in the perennial Triticeae of North America. Amer. J. Bot. 72: 767—776.

Bartlett, J. 1953. The shorter Bartlett's familiar quotations. Pocket Books, Inc., New York.

Cameron, D. F. 1983. To breed or not to breed. p. 237—250. *In*: J. G. McIvor, and R. A. Bray (eds.) Genetic resources of forage plants. Commonwealth Scientific and Industrial Research Organization, East Melbourne, Australia.

Campbell, R. S., and C. H. Herbel. 1975. Improved range plants. Range symposium series No. 1. Society for Range Management, Denver, Colorado.

Carlson, G. E., P. B. Gibson, and D. D. Baltensperger. 1985. White clover and other perennial clovers. p. 118—127. *In*: M. E. Heath, R. F. Barnes, and D. S. Metcalfe (eds.) Forages, the science of grassland agriculture (4th ed.). Iowa State University Press, Ames.

Carlson, J. R. 1982. The role of SCS Plant Materials Centers in western mined-land rehabilitation. p. 8—15. *In*: K. E. Evans (tech. coordinator) Proceedings of western mined-land rehabilitation research workshop. USDA Forest Service, Intermountain Forest and Range Experiment Station, Ogden, Utah.

Carlson, J. R. 1984. *Atriplex* cultivar development. p. 176—182. *In*: A. R. Tiedemann, E. D. McArthur, H. C. Stutz, R. Stevens, and K. L. Johnson (compilers) Proceedings — symposium on the biology of *Atriplex* and related chenopods. USDA Forest Service Gen. Tech. Rep. INT-172. Intermountain Forest and Range Experiment Station, Ogden, Utah.

Carlson, J. R., and E. D. McArthur (eds.) 1985. Range plant improvement in North America. p. 107—220. *In*: Proceedings, selected papers presented at the 38th Annual Meeting of The Society for Range Management. III. Society for Range Management, Denver, Colorado.

Carlson, J. R., J. G. Scheetz, and W. R. Oaks. 1984. Seed production techniques of two chenopods: Gardner saltbush and winterfat. p. 191—195. *In*: A. R. Tiedemann, E. D. McArthur, H. C. Stutz, R. Stevens, and K. L. Johnson (compilers) Proceedings — symposium on the biology of *Atriplex* and related chenopods. USDA Forest Service Gen. Tech. Rep. INT-172. Intermountain Forest and Range Experiment Station, Ogden, Utah.

Carman, J. G. (*In Press*) Cellular and molecular approaches to the ecophysiology and genetic improvement of shrubs. *In*: A. Wallace, E. D. McArthur, and M. R. Haferkamp (compilers) Proceedings — symposium on shrub ecophysiology and biotechnology. USDA Forest Service Gen. Tech. Rep. INT-. Intermountain Research Station, Ogden, Utah.

Childers, W. R., and A. A. Hanson. 1985. Timothy. p. 217—223. *In*: M. E. Heath, R. F. Barnes, and D. S. Metcalfe (eds.) Forages, the science of grassland agriculture (4th ed.). Iowa State University Press, Ames.

Colorado Seed Certification Standards. 1979. Colorado Seed Growers Association. Department of Agronomy, Colorado State University, Ft. Collins.

Conard, S. G., A. E. Jaramillo, K. Cromack, Jr., and S. Rose (compilers) 1985. The role of the genus *Ceanothus* in western forest ecosystems. USDA Forest Service Gen. Tech. Rep. PNW-182. Pacific Northwest Forest and Range Experiment Station, Portland, Oregon.

Contreras, D., P. Azocar, G. Covarrabias, and G. Soto. 1984. Use of forage shrubs in the arid land of Chile. p. 237—242. *In*: A. R. Tiedemann, E. D. McArthur, H. C. Stutz, R. Stevens, and K. L. Johnson (compilers) Proceedings — symposium on the biology of *Atriplex* and related chenopods. USDA Forest Service Gen. Tech. Rep. INT-172. Intermountain Forest and Range Experiment Station, Ogden, Utah.

Cook, C. W. 1972. Comparative nutritive values of forbs, grasses and shrubs. p. 303—310. *In*: C. M. McKell, J. P. Blaisdell, and J. R. Goodin (eds.) Wildland shrubs — their biology and utilization. USDA Forest Service Gen. Tech. Rep. INT-1. Intermountain Forest and Range Experiment Station, Ogden, Utah.

Cooperrider, A. Y., and J. A. Bailey. 1986. Fringed sagebrush (*Artemisia frigida*) — a neglected forage species of western ranges. p. 46—54. *In*: E. D. McArthur, and B. L. Welch (compilers) Proceedings — symposium on the biology of *Artemisia* and *Chrysothamnus*. USDA Forest Service Gen. Tech. Rep. INT-200. Intermountain Research Station, Ogden, Utah.

Crawford, E. J. 1983. Selecting cultivars from naturally occurring genotypes: evaluating annual *Medicago* species. p. 203—215. *In*: J. G. McIvor, and R. A. Bray (eds.) Genetic resources of forage plants. Commonwealth Scientific and Industrial Research Organization, East Melbourne, Australia.

Crofts, K. A., and C. M. McKell. 1977. Sources of seeds and planting materials in western states for land rehabilitation projects. Agricultural Experiment Station Land Rehabilitation Series No. 4. Utah State University, Logan.

Dewey, D. R. 1986. Taxonomy of the crested wheatgrasses (*Agropyron*). p. 31—41. *In*: K. R. Johnson (ed.) Crested wheatgrass: it values, problems and myths; symposium proceedings. Utah State University, Logan.

Dewey, D. R., and A. P. Plummer. 1980. New collections of range plants from the Soviet Union. J. Range Manage. 33: 89—94.

Dillman, A. C. 1946. The beginnings of crested wheatgrass in North America. J. Amer. Soc. Agron. 38: 237—250.

Duell, R. W. 1985. The bluegrasses. p. 188—197. *In*: M. E. Heath, R. F. Barnes, and D. S. Metcalfe (eds.) Forages, the science of grassland agriculture (4th ed.). Iowa State University Press, Ames.

Ensign, R. D. 1985. Phalaris, orchardgrass, fescue, and selected minor grasses. Part II: The fescues — perennial western rangeland grasses. p. 133—136. *In*: J. R. Carlson, and E. D. McArthur (eds.) Proceedings, selected papers presented at the 38th Annual Meeting of The Society for Range Management. III. Range plant improvement. Society for Range Management, Denver, Colorado.

Estes, J. R., R. J. Tyrl, and J. N. Brunken (eds.) 1982. Grasses and grasslands, systematics and ecology. University of Oklahoma Press, Norman.

Everett, W. 1981. Sources of seed and planting stock. USDA Forest Service Equipment Development Center, Missoula, Montana.

Ferguson, R. B. 1983. Use of rosaceous shrubs for wildland plantings in the Intermountain West. p. 136—149. *In*: S. B. Monsen, and N. Shaw (compilers) Managing Intermountain rangelands — improvement of range and wildlife habitats, proceedings of symposia. USDA Forest Service Gen. Tech. Rep. INT-157. Intermountain Forest and Range Experiment Station, Ogden, Utah.

Foster, K. E., M. M. Karpiscak, J. G. Taylor, and N. G. Wright. 1983. Guayule, jojoba,

buffalo gourd, and Russian thistle: plant characteristics, products and commercialization potential. Desert Plants 5: 112—117, 126.
Graetz, R. D., and K. M. W. Howes (eds.) 1979. Studies of the Australian arid zone. IV. Chenopod shrublands. Commonwealth Scientific and Industrial Research Organization, Melbourne, Australia.
Hafenrichter, A. L., J. L. Schwendiman, H. L. Harris, R. S. MacLauchlan, and H. W. Miller. 1968. Grasses and legumes for soil conservation in the Pacific Northwest and Great Basin States. USDA Soil Conservation Service Agriculture Handbook 339. U.S. Government Printing Office, Washington, D.C.
Hansen, M., L. Busch, J. Burkhardt, W. B. Lacy, and L. R. Lacy. 1986. Plant breeding and biotechnology. BioScience 36: 29—39.
Hansen, N. E. 1913. Cooperative tests of alfalfa from Siberia and European Russia. Bulletin 141. p. 33—157. South Dakota Agricultural Experiment Station, Brookings.
Harlan, J. R. 1983. The scope for collection and improvement of forage plants. p. 3—14. *In*: J. G. McIvor, and R. A. Bray (eds.) Genetic resources of forage plants. Commonwealth Scientific and Industrial Research Organization, Melbourne, Australia.
Hassell, W. G., and R. E. Barker. 1985. Relationships and potential development of selected needlegrasses and ricegrasses for western North American rangelands. p. 122—127. *In*: J. R. Carlson, and E. D. McArthur (eds.) Proceedings, selected papers presented at the 38th Annual Meeting of The Society for Range Management. III. Range plant improvement. Society for Range Management, Denver, Colorado.
Heath, M. E., R. F. Barnes, and D. S. Metcalfe (eds.) 1985. Forages, the science of grassland agriculture (4th ed.). Iowa State University Press, Ames.
Heath, M. E., and C. J. Kaiser. 1985. Forages in a changing world. p. 3—11. *In*: M. E. Heath, R. F. Barnes, and D. S. Metcalfe (eds.) Forages, the science of grassland agriculture (4th ed.). Iowa State University Press, Ames.
Hegerhorst, D. F., D. J. Weber, and E. D. McArthur. 1987. Resin and rubber content in *Chrysothamnus*. Southwestern Nat. 32: 475—482.
Heizer, R. B., and W. G. Hassell. 1985. Improvement of the gramas and other shortgrass prairie species. p. 171—177. *In*: J. R. Carlson, and E. D. McArthur (eds.) Proceedings, selected papers presented at the 38th Annual Meeting of The Society for Range Management. III. Range plant improvement. Society for Range Management, Denver, Colorado.
Holmgren, A. H., and N. H. Holmgren. 1977. Poaceae. Volume 6, p. 175—464. *In*: A. R. Cronquist, A. H. Holmgren, N. H. Holmgren, J. L. Reveal, and P. K. Holmgren (eds.) Intermountain flora. Columbia University Press, New York.
Hoveland, C. S., and C. E. Townsend. 1985. Other legumes. p. 146—153. *In*: M. E. Heath, R. F. Barnes, and D. S. Mecalfe (eds.) Forages, the science of grassland agriculture. Iowa State University Press, Ames.
Jacobsen, E. T., C. M. Taliaferro, C. L. Dewald, D. A. Tober, and R. J. Haas. 1985. New and old world bluestems. p. 148—158. *In*: J. R. Carlson, and E. D. McArthur (eds.) Proceedings, selected papers presented at the 38th Annual Meeting of The Society for Range Management. III. Range plant improvement. Society for Range Management, Denver, Colorado.
Johnson, K. L. (ed.) 1986. Crested wheatgrass: Its values, problems and myths; symposium proceedings. Utah State University, Logan.
Johnson, R. R., C. D. Ziebell, D. R. Patton, P. F. Ffolliott, and R. H. Hamre (technical coordinators) 1985. Riparian ecosystems and their management: reconciling conflicting uses. USDA Forest Service Gen. Tech. Rep. RM-120. Rocky Mountain Forest and Range Experiment Station, Ft. Collins, Colorado.
Kaul, A. 1986. *Haloxylon salicornicum* — an arid land shrub: its ecology and potential. Annals of Arid Zone 25: 31—43.

Kelsey, R. R. 1986. Foliage biomass and crude terpenoid productivity of big sagebrush (*Artemisia tridentata*). p. 375—388. *In*: E. D. McArthur, and B. L. Welch (compilers) Proceedings — symposium on the biology of *Artemisia* and *Chrysothamnus*. USDA Forest Service Gen. Tech. Rep. INT-200. Intermountain Research Station, Ogden, Utah.

Klemmedson, J. O. 1979. Ecological importance of actinomycete-nodulated plants in the western United States. Bot. Gaz. 140: S91—S96.

Knight, R. 1983. Mediterranean and temperate forage grasses. p. 47—61. *In*: J. G. McIvor, and R. S. Bray (eds.) Genetic resources of forage plants. Commonwealth Scientific and Industrial Research Organization, East Melbourne, Australia.

Knight, W. E., and C. S. Hoveland. 1985. Arrowleaf, crimson, and other annual clovers. p. 136—145. *In*: M. E. Heath, R. F. Barnes, and D. S. Metcalfe (eds.) Forages, the science of grassland agriculture. Iowa State University Press, Ames.

Küchler, A. W. 1964. Potential natural vegetation of the conterminous United States map and manual. (map scale = 1 : 3,168,000). Special Publication 36. American Geographical Society, New York.

Laycock, W. A., and M. C. Shoop. 1986. Shrubs for living snow fences on the central Great Plains. p. 431—457. *In*: H. Stepphun, and W. Nicholaichuk (eds.) Snow management for agriculture symposium. Great Plains Agricultural Council Publication No. 120. Water Studies Institute, Saskatoon, Saskatchewan.

Leopold, A. S. 1950. Deer in relation to plant succession. J. Forestry 48: 675—678.

Lesins, K. A., and I. Lesins. 1979. Genus *Medicago* (Leguminosae): a taxogenetic study. W. Junk Publishers, The Hague.

Long, S. G. 1981. Characteristics of plants used in western reclamation (2nd ed.). Environmental Research and Technology, Inc., Ft. Collins, Colorado.

Mathison, M. J. 1983. Mediterranean and temperate forage legumes. p. 63—81. *In*: J. G. McIvor, and R. A. Bray (eds.) Genetic resources of forage plants. Commonwealth Scientific and Industrial Research Organization, East Melbourne, Australia.

McArthur, E. D. 1984. Natural diversity of western range shrubs. p. 193—209. *In*: J. L. Cooley, and J. H. Cooley (eds.) Natural diversity in forest ecosystems, proceedings of the workshop. Institute of Ecology, University of Georgia, Athens.

McArthur, E. D., A. C. Blauer, A. P. Plummer, and R. Stevens. 1979. Characteristics and hybridization of important Intermountain shrubs. III. Sunflower family. USDA Forest Service Res. Pap. INT-220. Intermountain Forest and Range Experiment Station, Ogden, Utah.

McArthur, E. D., S. B. Monsen, and B. L. Welch. 1987. Shrubs and forbs for revegetation plantings in the sagebrush ecosystem. p. 28—39. *In*: J. A. Onsanger (ed.) Integrated pest management on rangeland, state of the art in the sagebrush ecosystem. USDA-Agricultural Research Service ARS-50. U.S. Government Printing Office, Washington, DC.

McArthur, E. D., and S. C. Sanderson. 1984. Distribution, systematics, and evolution of Chenopodiaceae: an overview. p. 14—24. *In*: A. R. Tiedemann, E. D. McArthur, H. C. Stutz, R. Stevens, and K. L. Johnson (compilers) Proceedings — symposium of the biology of *Atriplex* and related chenopods. USDA Forest Service Gen. Tech. Rep. INT-172. Intermountain Forest and Range Experiment Station, Ogden, Utah.

McArthur, E. D., S. E. Stranathan, and G. L. Noller. 1984. 'Rincon' fourwing saltbush — proven for better forage and reclamation. Rangelands 6: 62—64.

McArthur, E. D., H. C. Stutz, and S. C. Sanderson. 1983. Taxonomy, distribution, and cytogenetics of *Purshia, Cowania*, and *Fallugia* (Rosoideae, Rosaceae). p. 4—24. *In*: A. R. Tiedemann and K. L. Johnson (compilers) Proceedings — research and management of bitterbrush and cliffrose in western North America. USDA Forest Service Gen. Tech. Rep. INT-152. Intermountain Forest and Range Experiment Station, Ogden, Utah.

McArthur, E. D., B. L. Welch, and D. L. Nelson. 1985. Developing improved cultivars of sagebrushes and other composite shrubs. p. 188—196. *In*: J. R. Carlson, and E. D. McArthur (compilers) Proceedings, selected papers presented at the 38th Annual Meeting of The Society for Range Management. III. Range plant improvement. Society for Range Management, Denver, Colorado.

McArthur, E. D., B. L. Welch, and S. C. Sanderson (*In Press*). Natural and artificial hybridization between big sagebrush (*Artemisia tridentata*) subspecies. J. Heredity.

McIvor, J. G. and R. A. Bray (eds.) 1983. Genetic resources of forage plants. Commonwealth Scientific and Industrial Research Organization, East Melbourne, Australia.

McKell, C. M. (ed.) (*In Press*). Biology and utilization of shrubs. Academic Press, Inc., New York.

McKell, C. M., J. P. Blaisdell, and J. R. Goodin (eds.) 1972. Wildland shrubs — their biology and utilization. USDA Forest Service Gen. Tech. Rep. INT-1. Intermountain Forest and Range Experiment Station, Ogden, Utah.

McKell, C. M., G. A. Van Epps, S. G. Richardson, J. R. Barker, C. Call, E. Alvarez, and K. A. Crofts. 1979. Selection, propagation, and field establishment of native plant species on disturbed arid land. Utah Agricultural Experiment Station Bulletin 500. Institute for Land Rehabilitation, Utah State University, Logan.

Miksche, J. P. (ed.) 1976. Modern methods in forest genetics. Springer-Verlag, Berlin.

Monsen, S. B. (chairman) 1985. Seed harvesting. p. 37—41. *In*: R. R. Hall (ed.) 39th annual report, Vegetative Rehabilitation and Equipment Workshop. Washington Office, USDA Forest Service, Washington, DC.

Monsen, S. B. 1987. Shrub selections for pinyon-juniper plantings. p. 316—329. *In*: R. L. Everett (compiler) Proceedings — Pinyon-juniper conference. USDA Forest Service Gen. Tech. Rep. INT-215. Intermountain Research Station, Ogden, Utah.

Monsen, S. B., and J. N. Davis. 1985. Progress in the improvement of selected western North American rosaceous shrubs. p. 201—209. *In*: J. R. Carlson, and E. D. McArthur (eds.) Proceedings, selected papers presented at the 38th Annual Meeting of The Society for Range Management. III. Range plant improvement. Society for Range Management, Denver, Colorado.

Monsen, S. B., and N. Shaw (compilers) 1983. Managing Intermountain rangelands — improvement of range and wildlife habitats, proceedings of symposia. USDA Forest Service Gen. Tech. Rep. INT-157. Intermountain Forest and Range Experiment Station, Ogden, Utah.

Monsen, S. B., and R. Stevens. 1985. 'Paiute' orchardgrass — forage species for semi-arid range and wildland sites. Rangelands 7: 114—115.

Murphy, P. M. (compiler) 1984. The challenge of producing native plants for the Intermountain area, proceedings: Intermountain Nurseryman's Conference. USDA Forest Service Gen. Tech. Rep. INT-168. Intermountain Forest and Range Experiment Station, Ogden, Utah.

Namkoong, G. 1986. Genetics and the forests of the future. Unasylva 38: 3—18.

Nechaeva, N. T. (ed.) 1985. Improvement of desert ranges in Soviet Central Asia. Harwood Academic Publishers, Chur, Switzerland.

Nobs, M. A. 1963. Experimental studies on species relationships in *Ceanothus*. Publication 623. Carnegie Institution of Washington, Washington, DC.

Noller, G. L., S. E. Stranathan, and E. D. McArthur. 1984. Establishment and initial results from a 'Rincon' fourwing saltbush orchard. p. 196—204. *In*: A. R. Tiedemann, E. D. McArthur, H. C. Stutz, R. Stevens, and K. L. Johnson (compilers) Proceedings — Symposium of the biology of *Atriplex* and related chenopods. USDA Forest Service Gen. Tech. Rep. INT-172. Intermountain Forest and Range Experiment Station, Ogden, Utah.

Official Seed Certification Handbook for New Mexico. 1984. New Mexico Crop Improvement Association, New Mexico State University, Las Cruces.

Olivares E., A., and J. Gasto C. 1981. *Artiplex repanda*, organizacion y manejo de ecosistemas con arbustos forrajeros. Ciencias Agricolas No. 7. Universidad de Chile, Santiago.

Patton, D. R., C. E. Gonzales V., A. L. Medina, L. A. Segura T., and R. H. Hamre (technical coordinators) 1986. Management and utilization of arid land plants: symposium proceedings. USDA Forest Service Gen. Tech. Rep. RM-135. Rocky Mountain Forest and Range Experiment Station, Ft. Collins, Colorado.

Platou, K. A., and P. T. Tueller. 1985. Evolutionary implications for grazing management systems. Rangelands 7: 57—61.

Plummer, A. P. (chairman) 1980. Seed harvesting. p. 25—27. *In*: T. V. Russell (ed.) 34rd annual report, Vegetative Rehabilitation and Equipment Workshop. Washington Office, USDA Forest Service, Washington, DC.

Plummer, A. P., D. R. Christensen, and S. B. Monsen. 1968. Restoring big-game range in Utah. Publication No. 68—3. Utah Division of Fish and Game, Salt Lake City.

Plummer, M. 1984. Considerations in selecting chenopod species for range seedings. p. 183—186. *In*: A. R. Tiedemann, E. D. McArthur, H. C. Stutz, R. Stevens, and K. L. Johnson (compilers) Proceedings — symposium on the biology of *Atriplex* and related chenopods. USDA Forest Service Gen. Tech. Rep. INT-172. Ogden, Utah.

Robertson, J. H. 1982. Pamirian winterfat (*Eurotia ceratiodes* [L.] C. A. M.). Mentzelia 5: 33—36.

Rumbaugh, M. D. 1979. N. E. Hansen's contributions to alfalfa breeding in North America. Agriculture Experiment Station B 665. South Dakota State University, Brookings.

Rumbaugh, M. D., and C. E. Townsend. 1985. Range legume selection and breeding in North America. p. 137—147. *In*: J. R. Carlson, and E. D. McArthur (eds.) Proceedings, selected papers presented at the 38th Annual Meeting of The Society for Range Management. III. Range plant improvement. Society for Range Management, Denver, Colorado.

Sankary, M. N. and M. G. Barbour. 1972. Autecology of *Salsola vermiculata* var. *villosa* from Syria. Flora 161: 421—439.

Schaack, C. 1987. The status of *Purshia subintegra* (Kearney) Hendrickson and a new *Purshia* for Arizona. J. Arizona-Nevada Academy of Science 22: 13.

Schmidt, M. G. 1980. Growing California native plants. University of California Press, Berkeley.

Schmutz, E. M., Y. H. Havens, J. S. Horton, M. S. James, J. E. Jay, C. W. Renney, and A. P. Thatcher. 1973. Landscaping with native Arizona plants. University of Arizona Press, Tucson.

Schuman, G. E., and G. S. Howard. 1978. *Artemisia vulgaris* L.: an ornamental plant for disturbed land reclamation. J. Range Manage. 31: 392—393.

Schwarz, C. F., E. C. Thor, and G. H. Elsner. 1976. Wildland planning glossary. USDA Forest Service Gen. Tech. Rep. PSW-13. Pacific Southwest Forest and Range Experiment Station, Berkeley, California.

Sharp, L. A. 1986. Crested wheatgrass: its values, problems, and myths. p. 3—6. *In*: K. L. Johnson (ed.) Crested wheatgrass, its values, problems, and myths, symposium proceedings. Utah State University, Logan.

Shaw, N., and S. B. Monsen. 1983. Nonleguminous forbs for rangeland sites. p. 123—131. *In*: S. B. Monsen, and N. Shaw (compilers) Managing Intermountain rangelands — improvement of range and wildlife habitats, proceedings of symposia. USDA Forest Service Gen. Tech. Rep. INT-157. Intermountain Forest and Range Experiment Station, Ogden, Utah.

Shaw, N., and S. B. Monsen. 1986. 'Lassen' antelope bitterbrush: a browse plant for game and livestock ranges. Rangelands 8: 122—124.

Shelford, V. E. 1963. The ecology of North America. University of Illinois Press, Urbana.

Simmonds, N. W. (ed.) 1976. Evolution of crop plants. Longman Group Limited, London.
Slayback, R. D. 1980. Plant materials 1980, commerical sources of plant materials. Technical Note Plant Materials 10. USDA Soil Conservation Service, Davis, California.
Smith, L. B., and M. Walker (eds.) 1977. Public property? The Fraiser Institute, Vancouver.
Sours, J. M. 1983. Grasses for arid western rangelands. p. 90—94. *In*: S. B. Monsen, and N. Shaw (compilers) Managing Intermountain rangelands — improvement of range and wildlife habitats, proceedings of symposia. USDA Forest Service Gen. Tech. Rep. INT-157. Intermountain Forest and Range Experiment Station, Ogden, Utah.
Stebbins, G. L. 1972. Evolution and diversity of arid-land shrubs. p. 111—120. *In*: C. M. McKell, J. P. Blaisdell, and J. R. Goodin (eds.) Wildland shrubs — their biology and utilization. USDA Forest Service Gen. Tech. Rep. INT-1. Intermountain Forest and Range Experiment Station, Ogden, Utah.
Stebbins, G. L. 1975. Shrubs as centers of adaptation and evolution. p. 120—140. *In*: H. C. Stutz (ed.) Proceedings of symposium and workshop on the occasion of dedication of USDA Forest Service Shrub Sciences Laboratory. Brigham Young University Printing Service, Provo, Utah.
Stein, W. I., R. Danielson, N. Shaw, S. Woolff, and D. Gerdes. 1986. Users guide for seeds of western trees and shrubs. USDA Forest Service Gen. Tech. Rep. PNW-193. Pacific Northwest Forest and Range Experiment Station, Portland, Oregon.
Stevens, R., K. R. Jorgensen, E. D. McArthur, and J. N. Davis. 1985a. 'Immigrant' forage kochia. Rangelands 7: 22—23.
Stevens, R., and S. B. Monsen. 1985. 'Ephraim' crested wheatgrass — a rhizomatous grass. Rangelands 7: 163—164.
Stevens, R., N. Shaw, and C. G. Howard. 1985b. Important nonleguminous forbs for Intermountain ranges. p. 210—220. *In*: J. R. Carlson, and E. D. McArthur (eds.) Proceedings, selected papers presented at the 38th Annual Meeting of The Society for Range Management. III. Range plant improvement. Society for Range Management, Denver, Colorado.
Stubbendieck, J., S. L. Hatch, and K. J. Hirsch. 1986. North American range plants (3rd ed.). University of Nebraska Press, Lincoln.
Stutz, H. C. 1982. Broad gene pools required for disturbed lands. p. 113—118. *In*: E. F. Aldon, and W. R. Oaks (eds.) Reclamation of mined lands in the Southwest. New Mexico Chapter, Soil Conservation Society of America, Albuquerque.
Stutz, H. C. 1983. Some promising chenopods for use on disturbed lands. p. 132—135. *In*: S. B. Monsen, and N. Shaw (compilers) Managing Intermountain rangelands — improvement of range and wildlife habitats, proceedings of symposia. USDA Forest Service Gen. Tech. Rep. INT-157. Intermountain Forest and Range Experiment Station, Ogden, Utah.
Stutz, H. C., and J. R. Carlson. 1985. Genetic improvement of saltbush (*Atriplex*) and other chenopods. p. 197—200. *In*: J. R. Carlson, and E. D. McArthur (eds.) Proceedings, selected papers presented at the 38th Annual Meeting of The Society for Range Management. III. Range plant improvement. Society for Range Management, Denver, Colorado.
Thomas, J. W. (ed.) 1979. Wildlife habitats in managed forests, the Blue Mountains of Oregon and Washington. USDA Forest Service Agriculture Handbook No. 553. U.S. Government Printing Office, Washington, D.C.
Tiedemann, A. R., and C. H. Driver. 1983. Snow eriogonum; a native halfshrub to revegetate winter game ranges. Reclam. Reveg. Res. 2: 31—39.
Twain, M. 1984. The innocents abroad (and) roughing it. Literary Classics of the United States, Inc., New York (annotated by G. Cardwell).
USDA Soil Conservation Service. 1982. Sources of native seeds and plants. Ankey, Iowa.
USDA Soil Conservation Service. 1987. Improved plant materials cooperatively released by SCS through December 1986. National Plant Materials Center, Beltsville, Maryland.

Vogel, K. P., C. L. Dewald, H. J. Gorz, and F. A. Haskins. 1985. Improvement of switchgrass, indiangrass, and eastern gamagrass: current status and future. p. 159—170. *In*: J. R. Carlson, and E. D. McArthur (eds.) Proceedings, selected papers presented at the 38th Annual Society for Range Management. III. Range plant improvement. Society for Range Management, Denver, Colorado.

Voorhees, M. E., S. W. Uresk, and R. M. Hansen. 1984. *Atriplex suckleyi*: a native annual plant for revegetating bentonite mine spoils. p. 306—309. *In*: A. R. Tiedemann, E. D. McArthur, H. C. Stutz, R. Stevens, and K. L. Johnson (compilers) Proceedings — symposium on the biology of *Atriplex* and related chenopods. USDA Forest Service Gen. Tech. Rep. INT-172. Intermountain Forest and Range Experiment Station, Ogden, Utah.

Voight, P. W., and R. S. MacLauchlan. 1985. Native and other western grasses. p. 177—186. *In*: M. E. Heath, R. F. Barnes, and D. S. Metcalfe (eds.) Forages, the science of grassland agriculture (4th ed.). Iowa State University Press, Ames.

Voight, P. W., and W. R. Oaks. 1985. Lovegrasses, dropseeds, and other desert and subtropical grasses. p. 178—187. *In*: J. R. Carlson, and E. D. McArthur (eds.) Proceedings, selected papers presented at the 38th Annual Meeting of The Society for Range Management. III. Range plant improvement. Society for Range Management, Denver, Colorado.

Walker, B. H. (ed.) 1979. Management of semi-arid ecosystems. Elsevier Scientific Publishing Co., Amsterdam.

Walter, H. 1973. Vegetation of the earth in relation to climate and the eco-physiological conditions. Springer-Verlag, Berlin.

Ward, R. T. 1969. Ecotypic variation in *Deschampsia caespitosa* (L.) Beauv. from Colorado. Ecology 50: 519—522.

Wasser, C. H. 1982. Ecology and culture of selected species useful in revegetating disturbed lands in the west. USDI FWS/OBS-82/56. Fish and Wildlife Service, Washington, D.C.

Welch, B. L. 1983. Improving the nutritive value of winter range forage. p. 158—164. *In*: S. B. Monsen, and N. Shaw (compilers) Managing Intermountain rangelands — improvement of range and wildlife habitats. USDA Forest Service Gen. Tech. Rep. INT-157. Intermountain Forest and Range Experiment Station, Ogden, Utah.

Welch, B. L., E. D. McArthur, D. L. Nelson, J. C. Pederson, and J. N. Davis. 1986a. 'Hobble Creek' — a superior selection of low-elevation mountain big sagebrush. USDA Forest Service Res. Rap. INT-370. Intermountain Research Station, Ogden, Utah.

Welch, B. L., D. L. Nelson, and E. D. McArthur. 1986b. Improvement of wildland shrubs through selections and breeding. p. 467—468. *In*: P. J. Joss, P. W. Lynch, and O. B. Williams (eds.) Rangelands: a resource under siege. Australian Academy of Science, Canberra.

Welsh, S. L. 1986. New taxa and combinations in the Utah flora. Great Basin Nat. 46: 254—260.

Williams, M. C., and R. C. Barneby. 1977a. The occurrence of nitro-toxins in North American *Astragalus* (Fabaceae). Brittonia 29: 310—326.

Williams, M. C., and R. C. Barneby. 1977b. The occurrence of nitro-toxins in Old World and South American *Astragalus* (Fabaceae). Brittonia 29: 327—331.

Wright, J. W. 1962. Genetics of forest tree improvement. FAO Forestry Products Studies No. 16. Food and Agriculture Organization of the United Nations, Rome.

Wright, J. W. 1976. Introduction to forest genetics. Academic Press, New York.

Wurtele, E. S. (*In Press*) Biotechnology improvement: possibilities and western shrubs, focus on *Atriplex*. *In*: A. Wallace, E. D. McArthur, and M. R. Haferkamp (compilers) Proceedings — symposium on shrub ecophysiology and biotechnology. USDA Forest Service Gen. Tech. Rep. INT-. Intermountain Research Station, Ogden, Utah.

Young, J. A., R. E. Eckert, Jr., and R. A. Evans. 1979. Historical perspectives regarding the sagebrush ecosystem. p. 1—13. *In*: The sagebrush ecosystem: a symposium. Utah State University, Logan.

Young, J. A., and R. A. Evans. 1987. Technology for seeding on sagebrush rangelands. p. 15—18. *In*: J. A. Onsanger (ed.) Integrated pest management, state of the art in the sagebrush ecosystem. USDA Agricultural Research Service ARS-50. U.S. Government Printing Office, Washington, D.C.

Young, S. A. 1984. Seed certification in Utah. Utah Science 45: 80—83.

Zimmerman, E. A. 1980. Desert ranching in central Nevada. Rangelands 2: 184—186.

6. Successional concepts in relation to range condition assessment

E. LAMAR SMITH

6.1 Abstract

Range managers have generally accepted that assessment of range condition should be related to the successional status of vegetation on a given range site. The traditional concept of succession employed has been based on Clementsian ecology. Succession has been seen as an orderly process of ecosystem development leading, in the absence of abnormal disturbance, to a stable and predictable endpoint (climax) which is used as a standard for measuring condition. Regression from climax was viewed as a reversible, linear process related to the nature, degree and duration of disturbance (including livestock grazing). This approach has worked well in some grasslands but has not been adequate in situations where invasion of shrubs and exotic species is not spontaneously reversible, and it is not adequate for multiple use management or to report the real state of range management on public ranges. The climax approach is untenable in view of modern ecological theory which stresses the importance of disturbance and chance occurrence of rare events as they interact with life histories of plants and animals. Vegetation change may take multiple pathways and enter multiple steady states. New approaches to range condition assessment, more in line with modern ecology, are being developed to meet the needs of managers.

6.2 Introduction

"Speculating about what vegetation would be like if it were to develop under specified conditions is the basis of scientific land management, and therefore obviously worthwhile . . ." (Zedler 1981). In the United States interest in applying scientific management to rangelands began near the turn of the century. From about 1865—1885 many thousands of cattle and

Tueller, P. T. (ed.), Vegetation science applications for rangeland analysis and management.
ISBN 90-6193-195-9.
© 1988, *Kluwer Academic Publishers, Dordrecht, Boston, London.*

sheep were stocked on the vast open rangelands of the West. Livestock operators were attracted by the abundant, high-quality forage (which was also free). Blizzards and drought, however, periodically caused widespread starvation of livestock, and it became evident that the "condition" of the ranges had deteriorated due to heavy and uncontrolled grazing. This situation, along with the destruction of forests by selective logging and man-caused forest fires, led to establishment of the National Forests and, later, to government control of grazing on the public domain.

Most of the early range managers were foresters, botanists or agronomists by training, rather than animal scientists or veterinarians as in other parts of the world. As such, their prime interest was in conservation of the land rather than livestock production *per se*. They observed decreased vigor and abundance of desirable forage plants, increase of ephemerals and woody plants, and accelerated soil erosion by wind and water. These changes were attributed to improper grazing, drought, lack of fire, and possible climatic changes. Their objective was to determine sustainable stocking rates which would halt range deterioration and reverse the changes which had already occurred if possible (For example, see Smith 1895; Wooten 1908; Thornber 1910; Griffiths 1902).

It was probably not mere coincidence that the foundations of "dynamic ecology" were begun at this same time by Frederic E. Clements at the University of Nebraska. Clements travelled widely in the West and recognized the value of his writings on succession, disturbance and plant indicators for use in range and forest management. Range managers, faced with management of millions of acres of rangeland, and lacking the resources or inclination to manage "agronomically," adopted an "ecological" approach to management based heavily on Clements' ideas of succession. Westoby (1980) stated that "Insofar as range management has a theoretical basis, it is the concept of range succession." That may be an overstatement, but it is true that range management, especially in the U.S., has been heavily influenced by successional theories, and the whole concept of range condition is inextricably intertwined with succession.

The purpose of this chapter is to summarize how current methods of assessing range condition have incorporated successional concepts; to describe some of the problems created by the successional concepts used; and to explore some of the implications of modern concepts of succession relative to the use and interpretation of range condition assessments.

6.3 Development of range condition concepts in the U.S. and Canada

The history of the development of range condition concepts and methodology in the U.S. has been reviewed several times (e.g., Tueller 1973; Smith 1978, 1984, and in press; Pitt 1984; Wilson and Tupper 1982; West 1985) thus only a general summary will be included here. Pitt (1984) described development of approaches used in western Canada. Canadian and American workers have been closely associated and, for all practical purposes, the conceptual basis has been identical in the two countries.

Sampson (1917, 1919) is usually credited as the first to suggest that changes in vegetation and soil caused by heavy grazing were analogous to stages in "retrogressive succession." He felt that recognition of the successional status of vegetation could be used to evaluate deterioration of range productivity and, implicitly at least, to predict recovery of the range vegetation if grazing pressure were reduced or eliminated. The emphasis of the various government agencies from about 1910 until the late 1940's was to bring livestock stocking in line with carrying capacity and to apportion grazing rights among lessees on public range. Range surveys during this time were based on present vegetation types, i.e., the kind, amount, and forage value of the vegetation present irrespective of its successional status. Several authors (e.g., Talbot 1937; Costello 1939; Clements 1920), and instructions for "grazing examiners" in the Forest Service and other agencies described indicators of range condition and trend. The landmark publication, Senate Document 199 The Western Range (U.S. Forest Service 1936), reported the estimated extent of reduction in carrying capacity due to degradation of the rangelands. However, during this period the concept of range condition was not clearly defined. For example, heavy utilization by livestock was sometimes listed as an indicator of poor condition and the various interactions of management, weather and site potential were often not separated as to their effects on vegetation and soil cover.

During the 1940's and early 1950's a number of people, mainly in the Forest Service and Soil Conservation Service, began to develop methods for assessing or classifying range condition and trend. In the process it was necessary to attempt better definition of condition and the criteria used to classify it. At least two basic approaches were developed, the productivity approach and the climax approach (Hacker 1973).

Proponents of the productivity approach maintained that range condition should be evaluated as the amount of forage currently produced in relation to what was attainable under good management. Humphrey (1945, 1947, 1949) was the main proponent of this concept, although numerous others expressed similar ideas (e.g., Ellison 1949; Costello 1964). A number of authors recognized that some species of plants decrease and some increase, at least in relative amounts, in response to grazing by specific kinds of animals. While range condition was defined in terms of forage production, and thus closely related to carrying capacity, it was often stated that good range condition was often, but not necessarily, typified by higher stages of succession. Some writers also recognized that range condition could be better for one kind of animal than another due to their differing selectivity of forage species (Humphrey 1949).

The climax approach related range condition directly to successional status of the vegetation. The procedure outlined by Dyksterhuis (1949), often called the Quantitative Climax Method (QCM), has been adopted with some modifications by the Soil Conservation Service, Bureau of Land Management and some other agencies in both the U.S. and Canada (Ross 1985). As presently used, the composition of the climax vegetation for a given site (land type) is estimated through study of relic or protected areas. The vegetation composition occurring whereever the same site conditions exist can then be compared to the climax community for the site. The percentage similarity of present vegetation to the climax community determines the range condition score, which is usually divided into four arbitrary classes: 76—100% Excellent; 51—75% Good; 26—50% Fair; 0—25% Poor (Soil Conservation Service 1976).

In the QCM, range condition is based solely on composition of the vegetation. Plant vigor, soil erosion, plant reproduction, population age structure or other factors are not used to rate condition but are noted as indicators of trend. Forage values or other utilitarian attributes of plants are, supposedly, not considered. Therefore, the condition rating is purely "ecological" and independent of land use objectives. The basic assumption of this approach is that the climax or near climax composition represents the ultimate in biological productivity of the system and provides maximum possible conservation of the soil (Dyksterhuis 1958a, 1958b).

Both the "productivity" and the "climax" approaches depend on recognition of differences in site potential (Dyksterhuis 1949; Humphrey 1945). The concept of site goes back at least to Korstian (1919), but was not specifically related to range condition assessment until considerably later (Renner and Johnson 1942; Renner 1948; Humphrey 1945). A range site (also called an ecological site; see RISC 1983) is a classification of land according to its potential to produce a particular kind and amount of vegetation. It is similar in concept to the habitat type of Daubenmire (1968). Although the importance of site potential has long been recognized, site differences (i.e., soil, topography, climate) have often not been adequately considered with the result that range condition ratings may confound management effects with site differences. For example, an area rated as in poor condition may reflect improper management, or low site potential, or both (Smith 1978).

The two general approaches described above both relate to interpretations made of data on the amount and composition of the vegetation. Some people have held that range condition classification should not be based solely on vegetation composition but should include a separate and independent evaluation of soil condition (Ellison 1949; Packer 1951; Parker 1954; Smith 1978). Such ratings of the soil have been included in several scorecard methods for evaluating range condition and trend, e.g., the Parker 3-Step Method (Parker 1954) and the Deming 2-Phase Method (Deming 1957). Proponents of the QCM maintain that because soil changes are the result of changes in vegetation (and thus lag behind them) vegetation is a more reliable and sensitive indicator of change than are soil characteristics (Soil Conservation Service 1976).

6.4 Approaches in Australia and South Africa

Efforts to develop and apply methods to evaluate range condition outside North America have, apparently, been limited mainly to South Africa and Australia. Until recently, these efforts have consisted mostly of testing or modifying North American techniques. In South Africa the QCM was adapted for use in some grass veldt types (Foran, Tainton and Booysen 1978; Tainton, Foran and Booysen 1978; Tainton, Edwards and Mentis 1980). The best condition was considered to be a fire subclimax. With undergrazing and fire exclusion the vegetation composition became less palatable and nutritious for livestock, and when overgrazed the composition shifted to lower value plants and more bare soil. Range condition was judged by the relative composition of Decreaser (most abundant in the fire subclimax), Increaser I (increase without fire or grazing) and Increaser II (increase with overgrazing) species. Soil cover also was included in the score. Later modifications included adding an Increaser III (species which increase on selectively grazed range) and separation of soil ratings from vegetation composition scores. The QCM was considered to be useful without modification in the drier, shortgrass areas where grazing values reached a maximum in the climatic climax community.

Although the basic QCM was followed in South Africa, the purpose of rating condition was clearly related to productivity for livestock. In fact, the method was adapted to be used in estimating grazing capacity (Tainton, Edwards, and Mentis 1980). The approach used was, then, a "productivity" approach rather than a purely "ecological" approach as is claimed for the QCM in the U.S.

Australian workers have also tested the QCM as well as the Parker 3-Step and Deming methods (Wilson, in press). The Parker method was criticized because of lack of repeatability among observers and the Deming Method for its subjectivity and lack of independence among factors rated.

These same criticisms have been made in the U.S. In general, the QCM has, with some modifications, been found reasonably useful (Cunningham 1976; Lendon and Lamacraft 1976). However, it appears that relatively little large-scale effort to classify range condition has been done in Australia. More emphasis has been placed on monitoring changes in vegetation and soil attributes rather than trying to arrive at a condition "score" (Wilson, in press).

6.5 Practical problems of range condition assessment

A number of problems have been recognized with regard to use and interpretation of range condition assessments. These problems have intensified in the U.S. in recent years because legislation requires land management agencies to report periodically to the public on the condition of rangelands both publicly and privately owned. Increased emphasis on multiple use of public lands has increased involvement of interest groups which may question or misinterpret range condition data. Some of these problems are as follows:

1 Lack of adequate consideration of site potential. Some agencies have set up guides or scorecards which specify criteria (such as species composition, productivity, ground cover, erosion rates) with little regard for site potential. Thus, a given scorecard may be used for broad vegetation types or life zones (e.g., pinyon-juniper, mountain bunchgrass) with little or no adjustment for soil type, slope, or topographic position. Sites with low potential may be rated as being in poor condition (Smith 1978). Even when range sites are recognized and correlated between range and soil scientists there is considerable internal variation in potential (Hyder *et al.* 1966). Using a better-than-average situation as potential will tend to reduce condition scores, and vice-versa (Wilson, in press).

2 Exotic species are often not considered. Strict proponents of the climax approach argue that introduced species of plants were not part of the climax or pristine situation and thus their presence should reduce the condition rating (Ross 1985). This means that a different approach must be used on ranges where introduced species have been seeded even though these lands are similar in nature and management to adjacent native rangelands. In some cases exotic species have become naturalized and show no sign that succession will exclude them from a stand, yet they too detract from the condition rating even though they may be valuable plants for soil protection or forage (Smith 1979).

3 The climax approach does not work well on ranges where the climax

is forest or dense woodland. In such situations understory forage plants generally decline as succession to forest or dense woodland proceeds (Dyksterhuis 1958b). If the intent of range managers were truly to measure ecological status (similarity to climax) without regard to management objectives, this situation would pose no theoretical problem. However, since "range condition" in this case would be inversely related to forage production for grazing animals, it does not seem to useful to range managers. Consequently, the Soil Conservation Service (1976) (for example) uses a productivity approach on forest and woodlands. Condition may then decrease either due to overgrazing of understory or to increase of trees, or both (Hall 1978). Foran, Tainton and Booysen (1978) selected the fire subclimax rather than the "climatic climax" to represent excellent condition because it was more productive for livestock. In effect, a similar approach has been taken in western U.S. where mesquite, juniper or other shrub and woodland species have invaded what was formerly grassland, considered by some to have been kept free of shrubs and trees by fire.

4 A "climax type" of vegetation is not necessary for conservation of the site nor is it usually the management objective. Range managers have often stated that vegetation characteristic of seral communities often meets management objectives as well or better than the climax vegetation type, i.e., they may deliberately manage ranges for "fair" or "good" condition rather than "excellent" (climax) (Costello 1964). Since a prime objective in managing renewable resources is sustained yield, managing for good or fair condition implies a conviction that communities other than the climax type are capable of protecting the site against accelerated soil erosion or other processes which lead to diminished productivity and that these communities may meet management goals better than a climax type community. If there can be several types of communities which are equal to or exceed the climax in terms of biological productivity and site protection, then a simple measure of similarity to "climax" does not furnish any information about productivity or site protection. Especially where land is managed for multiple uses, vegetation changes may increase values for one use while decreasing another. The desired composition of the vegetation depends, therefore, on management objectives. Similarity to climax does not indicate whether management is succeeding or failing to achieve its goals (Comptroller General 1982).

5. Public reports of range condition on a national or local basis are confusing and misleading. Because of the different approaches used by different agencies, or by one agency in different vegetation types, it is difficult to combine data from different agencies. Although the same terminology, e.g., good, fair, poor, is used the meanings may be quite different for different data sets. With present methods, there is no way for

the general public to separate ranges which are in, say, poor condition because:

a. Site potential is poor but condition has been judged with standards from a more productive site; it is impossible to improve the condition.

b. The present vegetation is being deliberately maintained because it meets management objectives even though it is of "lower" successional status.

c. Vegetation composition was altered by excessive grazing, fire control, etc. at some time in the past. Although it is presently well-managed, it will not return to something like the "original" condition without technological intervention such as shrub control or land treatment. Such treatments may be economically unfeasible or unacceptable to certain political interests. Thus, although present vegetation is not optimum for multiple use objectives, the manager has no intention of doing anything to change the situation.

d. Present vegetation is undesirable but improving in response to management, i.e., the trend is up.

e. Present vegetation is undesirable for management objectives and present management is causing further deterioration or at least preventing improvement.

Only in the last case is there any reason to criticize the management of the land, and even here there is no necessary implication of land deterioration, only that vegetation is changing away from climax or to a less useful type for some particular use (depending on the approach taken in evaluating it).

In order to reduce some of these problems the Society for Range Management formed the Range Inventory Standardization Committee (RISC) in 1978. RISC (1983) recommended the following with respect to range condition:

1. Site potential always should be considered in rating range condition. The concept of range sites used by the Soil Conservation Service was endorsed, although RISC recommended the term "ecological sites," to signify their applicability in any type vegetation.

2. The term "range condition" should only be used in a general sense to refer to the status of range ecosystems.

3. The similarity of present vegetation on a site to the potential natural vegetation (Tuxen 1956) for the site would represent a rating of "ecological status" of the vegetation. Potential natural vegetation was suggested rather than climax because it should accommodate for past modifications

of the system by human or natural causes, e.g., naturalized exotic species, fundamental changes in community structure, e.g., through extinction or altered fire history.

4. Soil condition should be evaluated independently. Present erosion rates, or indicators of erosion hazard, should be compared to "natural" erosion for the site.

5. Resource value rating(s) should be developed to express the value of present vegetation for a specific use or uses, e.g., cattle forage, deer cover. The present mix of resource value ratings could be compared to the desired mix to meet management objectives.

The recommendations of RISC have been implemented to some extent. They have the potential to help solve some of the difficulties enumerated above. However, the successional concepts underlying the RISC recommendations are not different from those traditionally employed in range condition assessment.

6.6 Successional concepts underlying traditional approach to range condition

Regardless of whether an ecological (climax) approach or a use-oriented (productivity) approach is taken to evaluating range condition the basic successional concepts underlying the evaluation of range condition have been rooted in Clementsian theory. Some of Clements' views have been modified. For example, the monoclimax theory was replaced by a polyclimax view (Dyksterhuis 1949). The climax vegetation for a range site represents the combined influences of climate, topographic position, and soils (topo-edaphic climax). Also, the Clementsian view of plant communities as discrete units has been replaced by acceptance that species composition of plant communities changes in continuous fashion in both space and time (Dyksterhuis 1958a). However, the basic concepts of Clements and "neo-Clementsian" ecologists (e.g., Odum 1969) have been, and largely still are, accepted by most range managers and other natural resource professionals in North America. Some of these concepts employed in approaches to range condition are presented below.

Secondary succession after disturbance is viewed as an autogenic process involving an improvement of nutrient status and amelioration of extremes of temperature and moisture by action of plants and animals. Such reaction of early succession species on the site makes it possible for later, more demanding species to establish. The later species are not only more demanding but may be more specialized and/or efficient in extracting resources and, thus, more competitive than earlier species under the improved conditions. These processes involve certain aspects of both the "facilitation model" and the "tolerance model" of Connell and Slatyer (1977). The successional process continues in the absence of abnormal disturbance until the climax is reached (no further directional change occurs). The climax represents the highest development, i.e., the most efficient use of resources possible. Because the climax species have evolved into a complex system to effectively exploit resources of the site, it is unlikely that exotic species can successfully enter the climax community. It is thought that climax is not static but fluctuates around a mean in response to variations in weather and other characteristics of the natural environment. The climax ecosystem is more productive than seral communities and provides the best possible protection of site potential. When disturbed, climax vegetation will retrogress to a different species composition characteristic of lower successional stages. The disturbance may be abrupt and catastrophic, as by plowing or intense fire, or it may be a continued response to a stress like heavy grazing or prolonged drought. Chronic or frequently repeated disturbance may result in a relatively stable condition in equilibrium with the frequency and intensity of disturbance occurring, e.g., a grazing or fire "disclimax." The nature of the change in species composition from the climax (i.e., which species increase, decrease or invade) depends on the type of disturbance occurring, therefore a variety of seral communities may occur on a given site depending on the type of disturbance applied (Huschle and Hironaka 1980). However, when the disturbance is removed, or lessened, secondary succession will inevitably and "irresistibly" (Dyksterhuis 1958a) proceed to re-establish the original climax vegetation. The exception would be where soil degradation has been sufficient to create a situation where the original climax can no longer exist until new soil formation (primary succession) has occurred. In this case a new climax or site potential has been created for practical purposes (Ellison 1949; Soil Conservation Service 1976).

In the traditional viewpoint, succession is seen as a predictable process. No matter what the nature or extent of disturbance, in the absence of disturbance, vegetation will progressively change back to the climax. The time required may depend on many factors such as degree of disturbance, weather, longevity of seral species, size of area disturbed (seed source) etc., but the end result is predictable. The pathways of retrogression can also be predicted insofar as knowledge is available about species reactions to specific natural or man-caused influences. Substitution of the concept of potential natural vegetation (PNV) for climax (or original) vegetation as RISC (1983) suggested does not lessen the predictability substantially, but only recognizes that species introduction or extinctions may produce a somewhat different "climax" than existed in pristine conditions.

The view that climax (or PNV) furnishes the only objective standard by which range condition (ecological status) can be measured (Dyksterhuis 1958b) is based on two arguments. One is that the climax is the most stable and productive vegetation and provides the best soil protection of any vegetation capable of growing on the site. The second is that many seral communities can occur on a site varying greatly in composition and other characteristics. The climax is the only community which, presumably, can be accurately specified and which is independent of land use. It, therefore, is the only objective standard on which to rate range condition. Using climax as a standard does not mean that it is the management goal.

6.7 Alternative theories of succession and their implications for range condition assessment

For many years some ecologists (e.g., Gleason 1926; Egler 1954) have taken a population-centered approach to succession rather than the community/ecosystem approach of Clements and his successors (McIntosh 1980). More recently, a number of ecologists have criticized the Clementsian model of succession (e.g., McCormick 1968; Drury and Nisbet 1973) and have proposed alternative mechanisms of succession mainly based on plant life-histories and interactions among plant populations or individual plants (Connell and Slatyer 1977; Cattelino *et al.* 1970; Peet and Christensen 1980; Holling 1973; Horn 1974; Miles 1979; Noble and Slatyer 1980; Walker *et al.* 1981; Westoby 1980; White 1979; Huston and Smith 1987; Pickett, Colling and Armesto 1987). A few range scientists fairly early raised questions about the relation of range condition to succession (Love 1961; Jameson 1970), but Clementsian concepts were so firmly entrenched in the range and other natural resource professions that their correctness and their utility in evaluating range conditions were not widely questioned until recently (e.g., Smith 1978, 1984; Wilson and Tupper 1982; Tueller and Platou 1986; Mentis 1986; West *et al.* 1984; Norton 1978; Anderson 1986). As a result, in the U.S., some rangeland management agencies have begun to modify their approaches to range condition assessment, but considerable resistance to change has been expressed (Ross 1985; Dyksterhuis 1985). While the Clementsian "paradigm" has been largely rejected by ecologists there is not yet any one generally accepted model to replace it (McIntosh 1980). Some question whether there will be. The following discussion points out some of the alternative views of succession which relate to the evaluation of range condition.

Rather than viewing succession as a process occurring in the absence of "abnormal" disturbance, some emphasize that disturbance is characteristic of natural systems (White 1979). The nature of disturbance, its intensity, its frequency and regularity influence changes in vegetation and soil which are observed. Which disturbances are considered abnormal is a matter of opinion and may be influenced by considerations of time and spatial patterns. Allogenic (off-site) factors, such as arroyo cutting, sediment deposition, proximity to seed sources, or availability of water or cover for herbivores, may outweigh the autogenic processes emphasized in classical succession. In arid regions abiotic influences may exceed biotic interactions in shaping plant communities (Noble 1984). Viewed in this context the importance of distinguishing between "natural disturbance" and management (man-caused disturbance) is diminished.

Early colonizers of a disturbed site are those for which a residual seed bank exists, or which can quickly spread seed into the site, and which encounter conditions favorable for establishment. Rather than facilitating the establishment of late succession species, the early plants may inhibit this process (Connell and Slatyer 1977). Which species establish first may depend more on the existence of a propagule source (e.g., the nature and duration of the disturbance or the size of area affected) than ability of the propagules to establish new plants on the open site. Succession may then take multiple pathways which can only be understood and predicted by knowing the life history characteristics of the species involved and their relation to weather conditions, herbivory and other influences (Cattelino *et al.* 1979). Average conditions (e.g., of precipitation, temperature, incidence of disease and insects, grazing) may be less useful in explaining changes than extreme (abnormal) conditions. Changes may be effected largely by "the fortuitous co-occurrence of infrequent events" (Noble 1984), and once changes have occurred, they may persist for considerable time. The present state of the vegetation reflects its own history of such events and its future direction depends on those yet to come. Prediction of future changes depends on our knowledge of how populations will react to management prescriptions as these interact with natural events. Predictability for specific situations is limited by lack of knowledge about species' life histories, the stochastic nature of "infrequent events," and their interaction with management prescriptions. Succession does not appear to be the orderly, deterministic process envisioned by Clements.

According to the concept of multiple steady states advanced by Holling (1973), Walker *et al.* (1981), Noy-Meir (1975), Noy-Meir and Walker (1986), vegetation changes induced by imposition of a disturbance or stress may not be reversed when the stress is reduced or removed. Shrubs and trees have increased in many former grasslands, woodlands and savannahs in many parts of the world. This change may be due to cessation of burning, introduction of heavy livestock grazing, or both. However, once occupied by woody plants, removal of livestock may not reverse the process. It may be impossible to re-introduce fire because of inadequate fuel. Westoby (1980) points out that competitive ability among species may differ depending on which one is mature and which one is a seedling. Well-established perennial grasses may effectively exclude seedlings of woody plants from a community. Reduction of grass competition by grazing or drought may allow shrub seedlings to establish. Once established they are superior competitors to the grass and perpetuate the open understory required for their own seedlings to establish.

The use of climax composition as the standard for range condition or ecological status has been based on the view that climax is the only stable point in a "natural" system, and that when disturbance is removed or reduced disturbed vegetation will inevitably change back to the climax. Thus climax represents the only objective benchmark and the only basis for predicting change by "natural" succession. The multiple steady state hypothesis indicates that the classical view may work in special cases but is not adequate as a general model.

6.8 Current directions in evaluating range condition

"The philosophical concepts which form the basis of range condition assessment are now in a period of considerable ferment and change." This statement by Australian researchers (Foran, Bastin and Shaw 1986) provides a succinct conclusion to the topic addressed in this chapter. Yet, these same authors point out the vital importance of rangelands for production of meat, milk, fiber and other values in both developed and developing countries. Because of their usual relatively low production per unit area, cultural and economic factors influencing their management and the erratic climatic influences which drive their biological productivity, these lands often suffer degradation. It is very important to have objective methods to assess their productivity and the effectiveness of management applied to these rangelands (Foran, Bastin and Shaw 1986). Given the theoretical and practical problems discussed above of present range evaluation approaches, what are the likely alternatives?

Australian and South African range scientists have rejected the traditional theories of plant succession and climax used as a basis for range condition assessment as scientifically inadequate and subjective in application (Westoby 1980; Wilson, in press; Mentis 1986; Foran, Bastin and Shaw 1986). Wilson and Tupper (1982) pointed out there are many attributes of vegetation and soil which can be measured as a basis for inventory and monitoring of range condition. Selection of appropriate attributes to measure will depend on both the type of vegetation involved and the intended land use (West 1983). It is unrealistic to suppose that one attribute (e.g., composition by weight) will be appropriate in all situations.

Several Australian workers are pursuing the use of multivariate techniques for monitoring range condition. Hacker (1983) used a reciprocal averaging technique to ordinate vegetation stands on a gradient of range condition and range sites. Wilson (1986) and Hacker (1986) described multivariate approaches that could be used to objectively quantify range site differences and vegetation changes over time. They pointed out that multivariate techniques may help clarify which attributes are important indicators of site or conditions and to sort out the various influences of grazing, fire, weather and other factors. Foran, Bastin and Shaw (1986) used the Information Statistic (Grunow and Lance 1969) to classify range land "types" using presence and absence of perennial species and soil data. Within each range type, vegetation stands were classified and ordinated based on vegetation composition using the Bray-Curtis metric. Ordination along principal coordinates allowed identification of major variables associated with condition classes. Plotting of change over years in position of each stand on these vectors aided in detecting those stands where changes were management caused, rather than weather-related. These authors claim that such techniques remove some of the subjectivity of previous approaches which placed a premium on experience and judgment of the range examiner.

In the U.S.A. range managers have begun to realize that traditional approaches to range condition assessment do not provide all the information needed for multiple-use management in many vegetation types. The Bureau of Land Management (BLM) has followed the recommendations of RISC (1983) by developing "Resource Value Ratings" to measure utility of vegetation for specific uses in addition to an index of "Ecological Status" based on similarity to potential natural vegetation (Eshelman 1985). Similar recommendations were made for the Ministry of Forests in British Columbia (Pitt 1984). The Forest Service is considering independent ratings of soil condition and resource value ratings for the major planned land uses. If either is "unsatisfactory" the range condition is unsatisfactory (Schlatterer 1986). This approach is basically the same as suggested by Ellison, Croft and Bailey (1951).

Some difficulties have been encountered in the approach suggested by RISC (1983) (Eshelman 1985). There may be several, or many, resource value ratings applied to one stand of vegetation, some high, some low, depending on the number and type of uses they refer to. These may be helpful in management but do not provide a basis to report to the public whether the range is in "satisfactory" condition or not. The BLM has reported range condition as ecological status. Although the concept of potential natural vegetation, instead of climax (or original) vegetation, is used there is little to distinguish this rating from that of the QCM. Environmental critics of BLM still equate a rating of "mid-seral" (25—49% similarity to PNV) as "fair" condition and evidence of unsatisfactory management (Wald and Alberswerth 1985). It is not apparent what management decision can be made on the basis of "ecological status."

The BLM is presently considering reporting range condition as similarity of present vegetation to the *desired* vegetation for that range site (Eshelman 1987). The desired plant community would be the type of plant community, capable of being produced on a given range site, which most nearly meets the management objectives of the agency. These objectives are decided by considering conservation needs, public desires, and economic/legal constraints or requirements. Range condition would be reported as satisfactory if present vegetation meets or is changing toward that prescribed as desirable. This approach seems to have merit since it does not depend on any particular concept of succession or climax, does not distinguish among native and exotic species, and should be usable in any vegetation type, including seeded stands.

American range managers have been much more reluctant than their Australian and South African colleagues to recognize the inadequacy of traditional concepts of succession as a basis for evaluating range conditions. That is understandable considering how thoroughly the traditional views have been established in the training, textbooks, research and land management agency procedures in the U.S.

6.9 Summary and conclusions

It is apparent that classical successional theory has had enormous impact on our approaches to evaluating range condition. The most widely used methods to evaluate range condition have equated condition with similarity to composition of the climax vegetation. This approach worked reasonably well when livestock grazing was the principal land use and where the "original" vegetation was productive for livestock use. Succession was seen as a natural process to improve the stability and productivity of the range (even if the process had to be encouraged by use of herbicides, bulldozers, or reseeding in some cases). Where succession clearly did not increase forage production, a productivity approach to range condition was substituted. In multiple use management of rangelands this approach has proven inadequate. This is so because any given stand of vegetation has different values for different uses and, further, the values may relate to different attributes of the vegetation, e.g., cover, structure (West 1985). Therefore, it is unlikely that a single rating of condition based only on composition will be useful for many management objectives (Wilson 1984).

Modern ecologists do not accept the classical (Clementsian) view of succession and climax as a general model for explaining vegetation changes. The predictable, orderly and reversible process of community succession to a undisturbed climax has been replaced by emphasis on multiple stable states, life histories of plant and animal population, and on co-occurrence of infrequent events (i.e., "disturbance"). The acceptance of various disturbances as normal processes in nature, varying in intensity, scale, frequency or periodicity, and the probabilistic nature of complex interactions among these disturbances makes the selection of any particular state as climax arbitrary. Therefore, the "climax" on any given site seems no more objective as a standard than any other community, and since undisturbed (i.e., unmanaged) vegetation is rarely the management goal, it seems inherently less useful. The unpredictability of environmental extremes limits our ability to predict successional pathways. Predictability may be improved by intensive management intervention which may override or dampen natural disturbances to some extent.

Range managers should follow the admonitions of Love (1961), Slatyer (1973), Costello (1974), Tueller and Platou (1986), Smith (1987) and others to take the responsibility to decide the type of vegetation which most closely meets management goals (including the fundamental goal of site protection) and use that as their standard for condition. Condition should be based on those attributes pertinent to projected uses of the land. In this approach the distinction among successional stages, native and exotic species, or natural versus man-caused "disturbance" is irrelevant.

It has been pointed out (Neil West, personal communication) that the change to "desired" vegetation as the focus for range monitoring will lead to moving targets as the desires of society and managers change over time, raising concern that agencies will use this as an excuse for lack of good trend data. However, range condition and trend are interpretations of field data characterizing attributes of the plant/soil system. The basic data collected should provide a record of trends in those attributes even though the values placed on them may change over time.

The contribution of successional theory to range management is not so much as a basis for evaluating condition as to explain past changes or to predict future responses to management. In order to do either, knowledge is needed of the life histories and competitive relations among plant and animal populations and how they interact with the abiotic environment and management actions. Even if such knowledge were considerably greater than at present, predictability may be limited. Range (or other natural resource) management is not likely to become an exact science.

The quote of Zedler (1981) which introduced this paper continues as follows: "but there seems to be no reason justifiable from a theoretical point of view for the inordinate concern for 'disturbance' free vegetation development. There are strong arguments for discarding the climax concept."

That seems like good advice.

6.10 References

Anderson, J. E. 1986. Development and structure of sagebrush steppe plant communities. p. 10—12. *In*: P. J. Joss, P. W. Lynch, and O. B. Williams (eds.) Rangelands: A Resource Under Siege — Proceedings of the Second International Rangeland Congress. Australian Academy of Science Canberra.

Cattelino, P. J., I. R. Noble, R. O. Slatyer, and S. R. Kessell. 1979. Predicting the multiple pathways of plant succession. J. Environ. Mgmt. 3(1): 41—50.

Clements, F. E. 1920. Plant indicators: the relation of plant communities to process and practice. Carnegie Inst. Wash. Publ. 290: 1—388.

Comptroller General. 1982. Public rangeland improvement — a slow, costly process in need of alternate funding. Report to the Congress. GAO/RCED-83-23 (October 14, 1982). General Accounting Office, Washington, D.C.

Connell, J. H., and R. O. Slayter. 1977. Mechanisms of succession in natural communities and their role in community stability and organization. Am. Nat. 111: 1119—1144.

Costello, D. F. Range ecology. U.S. Forest Service Rocky Mt. For. and Range Expt. Sta., Ft. Collins, Colorado.

Costello, D. F. 1964. Range dynamics control — an ecological urgency. p. 91—107. *In*: D. J. Crisp (ed.) Grazing in terrestrial and marine environments. Blackwell Sci. Publ., Oxford.

Cunningham, G. M. 1976. Some concepts for range assessment in Australia. Aust. Rangeland J. 1(1): 60—63.

Daubenmire, R. 1968. Plant Communities. Harper & Row. New York.

Deming, M. H. 1957. Two-phase range condition surveys. U.S.D.I. Bureau of Land Mgmt. Mimeo.

Drury, W. H., and J. C. Nisbet. 1973. Succession. J. of Arnold Arboretum 54: 331—368.

Dyksterhuis, E. J. 1949. Condition and management of rangeland based on quantitative ecology. J. Range Mgmt. 2: 104—115.

Dyksterhuis, E. J. 1958a. Ecological principles in range evaluation. Bot. Rev. 24: 253—272.

Dyksterhuis, E. J. 1958b. Range conservation as based on sites and condition classes. J. Soil and Water Cons. 13: 151—155.

Dyksterhuis, E. J. 1985. Follow-up on range sites and condition classes as based on quantitative ecology. Rangelands 7: 172—173.

Egler, F. E. 1954. Vegetation science concepts. I: Initial floristic composition, a factor in old field vegetation development. Vegetatio 4: 412—417.

Ellison, L. 1949. The ecological basis for judging condition and trend on mountain rangeland. J. Forestry 47: 785—795.

Eshelman, K. R. 1985. Vegetation inventory and monitoring on public lands. p. 79—83. *In*: Use of Frequency and Density for Rangeland Monitoring Symposium Proceedings, February 13, 1985, Salt Lake City, Utah. Society for Range Management, Denver, Colorado.

Eshelman, K. R. 1987. Range condition reporting — is it time for a change? Abstracts of 40th Annual Meeting, Society for Range Management. February 8—13, 1987. Boise, Idaho.

Foran, B. D., N. M. Tainton, and P. de V. Booysen. 1978. The development of a method for assessing veld condition in three grassveld types in Natal Proc. Grassld. Soc. South Africa 13: 27—33.

Foran, B. D., G. Bastin, and K. A. Shaw. 1986. Range assessment and monitoring in arid lands: the use of classification and ordination in range survey. J. of Environ. Manage. 22: 67—84.

Gleason, H. A. 1926. The individualistic concept of the plant association. Bull. Torr. Bot. Club 53: 1—20.

Griffiths, D. 1902. Forage conditions in the northern border of the Great Basin. U.S.D.A. Bur. Plant Ind. Bull. No. 15.

Grunow, J. O., and G. B. Lance. 1969. Classification of savannah by information analysis. S. Afr. J. Sci. 35: 341—348.

Hacker, R. B. 1973. The objective of land resource management. Arid Zone Newsletter. p. 112—119. CSIRO, Perth, Australia.

Hacker, R. B. 1986. Multivariate methods in range assessment and monitoring. *In*: P. J. Joss, P. W. Lynch, and O. B. Williams (eds.) Rangelands: A resource under siege — Proceedings of the Second International Rangeland Congress. Australian Academy of Science. Canberra.

Hall, F. C. 1978. Applicability of Rangeland Management concepts to forest-range in the Pacific Northwest. p. 496—499. *In*: D. N. Hyder (ed.) Proceedings of the First International Rangeland Congress. Society for Range Management, Denver, Colorado.

Holling, C. S. 1973. Resiliency and stability of ecological systems. Ann. Rev. Ecol. Syst. 4: 1—23.

Horn, H. S. 1974. The ecology of secondary succession. Ann. Rev. Ecol. Syst. 5: 25—37.

Humphrey, R. R. 1945. Some fundamentals of the classification of range condition. J. For. 43: 646—647.

Humphrey, R. R. 1947. Range forage evaluation by the range condition method. J. For. 45: 10—16.

Humphrey, R. R. 1949. Field comments on the range condition method of forage survey. J. Range Mgmt. 2(1): 1—10.

Huschle, G., and M. Hironaka. 1980. Classification and ordination of some plant communities. J. Range Mgmt. 33: 179—182.

Huston, M., and T. Smith. 1987. Plant succession: life history and competition. Amer. Nat. 130: 168—198.

Hyder, D. N., R. E. Bement, E. E. Remmenga, and C. Terwilliger, Jr. 1966. Vegetation-soils and vegetation-grazing relations from frequency data. J. Range Mgmt. 19: 11—16.

Jameson, D. A. 1970. Land management policy and development of ecological concepts. J. Range Mgmt. 23: 316—321.

Korstian, C. F. 1919. Native vegetation as a criterion of site. Plant World 22: 253—261.

Love, R. M. 1961. The range — natural plant communities or modified ecosystems? J. British Grassl. Soc. 16(2): 89—99.

McCormick, J. 1968. Succession. Via 1: 1—16.

McIntosh, R. P. 1980. The relationship between succession and the recovery process in ecosystems. p. 11—62. *In*: J. Cairns (ed.) The recovery process in damaged ecosystems. Ann Arbor Science Publ., Ann Arbor, Michigan.

Mentis, M. T. 1986. Range dynamics by classical succession and strategic modelling. p. 19—21. *In*: P. J. Joss, P. W. Lynch, and O. B. Williams (eds.) Rangelands: A resource under siege — Proceedings of the Second International Rangeland Congress. Australian Academy of Science. Canberra.

Miles, J. 1979. Vegetation dynamics. New York, John Wiley.

Noble, I. R. 1986. The dynamics of range ecosystems. p. 3—5. *In*: P. J. Joss, P. W. Lynch, and O. B. Williams (eds.) Rangelands: A resource under siege — Proceedings of the Second International Rangeland Congress. Australian Academy of Science. Canberra.

Noble, I. R., and R. O. Slatyer. 1980. The use of vital attributes to predict successional changes in plant communities subject to recurrent disturbances. Vegetatio 43: 5—21.

Norton, B. E. 1978. The impact of sheep grazing on long-term successional trends in salt desert shrub vegetation of southwestern Utah. p. 610—613. *In*: D. N. Hyder (ed.) Proceedings of the First International Rangeland Congress. Society for Range Management, Denver.

Noy-Meir, I. 1975. Stability of grazing systems: an application of predator-prey graphs. J. Ecol. 63: 459—481.

Noy-Meir, I., and B. H. Walker. 1986. Stability and resilience in rangelands. p. 21—25. *In*: P. J. Joss, P. W. Lynch, and O. B. Williams (eds.) Rangelands: A resource under siege — Proceedings of the Second International Rangeland Congress. Australian Academy of Science. Canberra.

Odum, E. P. 1969. The strategy of ecosystem development. Science 164: 262—270.

Packer, P. E. 1951. An approach to watershed protection criteria. J. For. 49: 639—644.

Parker, K. W. 1954. Application of ecology in the determination of range condition and trend. J. Range Mgmt. 7: 14—23.

Peet, R. K., and N. L. Christensen. 1980. Succession: a population process. Vegetatio 43: 131—140.

Pickett, S. T. A., S. L. Collins, and J. J. Armesto. 1987. Models, mechanisms and pathways of succession. Bot. Rev. 53: 335—371.

Pitt, M. D. 1984. Range condition and trend assessment in British Columbia. British Columbia Ministry of Forests Research Report RR84004—HQ.

Renner, F. G. 1948. Range condition: A new approach to the management of natural grazing lands. Proc. Inter-Amer. Conf. on Conser. of Renewable Natural Resources. U.S. State Dept. Publ. 3382: 527—535.

Renner, F. G., and E. A. Johnson. 1942. Improving range condition for wartime livestock production. U.S.D.A. Farmer's Bulletin No. 1921.

RISC (Range Inventory Standardization Committee). 1983. Guidelines and terminology for range inventories and monitoring. Society for Range Management. Denver, Colorado.

Ross, R. L. 1985. Viewpoint: Response to the range inventory standardization committee (RISC). Rangelands 7: 169—170.

Sampson, A. W. 1917. Succession as a factor in range management. J. For. 15: 593—596.

Sampson, A. W. 1919. Plant succession in relation to range management. U.S.D.A. Bull. No. 791.

Schlatterer, E. R. 1986. Background, present status, and future of the evaluation of soil condition in rangeland monitoring in the Forest Service. p. 41—46. *In*: Use of Cover Soils and Weather Data in Rangeland Monitoring Symposium Proceedings. February 12, 1986, Kissimmee, Florida. Society for Range Management, Denver, Colorado.

Slayter, R. O. 1973. Structure and function of Australian arid shrublands. p. 66—73. *In*: Arid Shrublands: Proceedings of the Third United States/Australia Rangeland Panel. Society for Range Management, Denver, Colorado.

Smith, E. L. 1978. A critical evaluation of the range condition concept. p. 266—267. *In*: D. N. Hyder (ed.) Proceedings of the First International Rangeland Congress. Society for Range Management, Denver, Colorado.

Smith, E. L. 1984. Use of inventory and monitoring data for range management purposes. p. 809—842. *In*: Nat. Res. Council/Nat. Acad. Sciences. Developing strategies for rangeland management. Westview Press, Boulder, Colorado.

Smith, L. 1987. Does RISC offer security? p. 19—23. *In*: G. Secrist, and K. Eshelman (coordinators) Rangeland Monitoring Workshop Proceedings. Jan. 12—16, 1987, Golden, Colorado. U.S.D.I. Bureau of Land Management.

Smith, E. L. (*In Press*). Range condition and secondary succession: a critique. *In*: W. K. Lauenroth, and W. A. Laycock (eds.) Secondary succession and the evaluation of rangeland condition. Springer-Verlag, New York.

Smith, J. G. 1895. Forage condition of the prairie regions. p. 309—324. *In*: U.S. Department of Agriculture yearbook. U.S. Government Printing Office, Washington, D.C.

Soil Conservation Service. 1976. National Range Handbook. U.S. Dept. of Agric., Washington, D.C.

Tainton, N. M., B. D. Foran, and P. De V. Booysen. 1978. The veld condition score: an evaluation in situations of known past management. Proc. Grassl. Soc. South Africa 13: 35—40.

Tainton, N. M., P. J. Edwards, and M. T. Mentis. 1980. A revised method for assessing veld condition. Proc. Grassl. Soc. South Africa 15: 37—42.

Talbot, M. W. 1937. Indicators of Southwestern range conditions, U.S. Dept. Agric. Farmer's Bulletin No. 1782.

Thornber, J. J. 1910. The grazing ranges of Arizona. Ariz. Agric. Expt. Sta. Bull. 65.

Tuxen, R. 1956. Die heutige potentielle naturliche Vegetation als Gegenstand der Vegetationskartierung. Stolzenau/Weser. Angewandte Planzensoziologie 13: 5—42.

Tueller, P. T. 1973. Secondary succession, disclimax and range condition standards in desert shrub vegetation. p. 57—65. *In*: D. N. Hyder (ed.) Arid Shrublands: Proceedings of the Third United States/Australia Rangeland Panel. Society for Range Management, Denver, Colorado.

Tueller, P. T., and K. A. Platou. 1986. Range vegetation condition classification problems. p. 540—541. *In*: P. J. Joss, P. W. Lynch, and O. B. Williams (eds.) Rangelands: A resource under siege — Proceedings of the Second International Rangeland Congress. Australian Academy of Science. Canberra.

U.S. Forest Service. 1936. The western range. Senate Document 199. 74th Congress, 2nd Session.

Wald, Johanna and David Alberswerth. 1985. Our ailing public rangelands: condition report — 1985. National Wildlife Federation and Natural Resources Defense Council.

Walker, B. H., D. Ludwig, C. S. Holling, and R. M. Peterman. 1981. Stability of semi-arid savannah grazing systems. J. Ecol. 69: 473—498.

West, N. 1983. Choice of vegetation variables to monitor range condition and trend. p. 636—639. *In*: J. F. Bell, and T. Atterbury (eds.) Proceedings of an International Conference on Renewable Resource Inventories for Monitoring Change and Trends. Oregon State Univ., Corvallis.

West, N. E. 1985. Origin and early development of the range condition and trend concepts. p. 75—78. *In*: Proceedings Selected Papers Presented at the 36th Annual Meeting of the Society for Range Management. Salt Lake City, Utah.

West, N. E., F. D. Provenzia, P. S. Johnson, and M. Keith Owens. 1984. Vegetation change after 13 years of livestock grazing exclusion on sagebrush semidesert in West Central Utah. J. Range Mgmt. 37: 262—264.

West, N. Personal communication.

Westoby, M. 1980. Elements of a theory of vegetation dynamics in rangelands. Israel J. Bot. 28: 169—194.

White, P. S. 1979. Pattern, process and natural disturbances in vegetation. Bot. Rev. 45: 229—299.

Wilson, A. D., and G. J. Tupper. 1982. Concepts and factors applicable to the measurement of range condition. J. Range Mgmt. 35: 684—689.

Wilson, A. D. 1986. The monitoring of changes in range condition a multivariate site potential approach. p. 517—521. *In*: P. J. Joss, P. W. Lynch, and O. B. Williams (eds.) Rangelands: A resource under siege — Proceedings of the Second International Rangeland Congress. Australian Academy of Science. Canberra.

Wilson, A. D. (*In Press*). The development of systems of assessing the condition of rangeland in Australia. *In*: W. K. Lauenroth, and W. A. Laycock (eds.) Secondary succession and the valuation of rangeland condition. Spring-Verlag.

Wooten, E. O. 1908. The range problem in New Mexico. New Mexico Agric. Exp. Sta. Bull. 66.

Zedler, P. H. 1981. Vegetation change in chaparral and desert communities in San Diego County, California. p. 406—430. *In*: D. C. West, H. H. Shugart, and D. B. Botkins (eds.) Forest Succession. Springer-Verlag, New York.

7. A role for nonvascular plants in management of arid and semiarid rangelands

KIMBALL T. HARPER AND JAMES R. MARBLE

7.1 Abstract

Blue-green algae, lichens and mosses are common features on soil surfaces of arid and semiarid rangelands worldwide. Such plants often cover as much or more of the soil surface as vascular plants and make useful contributions to arid and semiarid ecosystems via enhanced soil stability against wind and water erosion and greater availability of nitrogen. In some situations, cryptogamic plants roughen soil surfaces and increase water retention and eventual infiltration. Situations are documented in which cryptogamic covers (particularly those with blue-green algae) have improved establishment and growth of vascular plant seedlings. Responses of various cryptogamic species to a broad array of environmental challenges are discussed. Evidence is presented that cryptogamic plants complement the effects of vascular plants relative to soil stability and water infiltration. It is suggested that cryptogamic cover usually does not develop at the expense of higher plant cover. Data indicate that some range management practices are damaging to cryptogamic plants, while others permit those plants (particularly blue-green algae) to persist in the ecosystem. It is recommended that future management practices on arid rangelands be designed to retain at least blue-green algae and lichens having blue-green phycobionts in the system.

7.2 Introduction

Range managers have traditionally focused attention on vascular plants or plants with true stems and leaves. Vascular plants include the ferns, fern allies, and seed plants, but managers have emphasized seed plants primarily, since they are responsible for most of the forage produced on rangelands worldwide. Nevertheless, rangelands on all continents are

Tueller, P. T. (ed.), Vegetation science applications for rangeland analysis and management.
ISBN 90-6193-195-9.
© 1988, *Kluwer Academic Publishers, Dordrecht, Boston, London.*

known to support numerous species of nonvascular plants (often referred to as cryptogams) that grow on soils and modify those surfaces in significant ways. These are tiny, often microscopic, plants including many species representative of such groups as the algae, bacteria, fungi, lichens, and mosses. The cryptogamic communities of soil surfaces have been studied most often by North American workers (e.g., Cooke 1955, Shields and Drouet 1962, Looman 1964, Cameron and Blank 1966, Worley 1973, Anderson and Rushforth 1976, Nash *et al.* 1977, Rundell 1978a, Nash and Moser 1982, Meyer 1986), but the literature documents the existence of similar nonvascular plant assemblages on soil surfaces of Eurasia (Reichert 1936, Galun 1963, Bourlière 1964, Anantani and Marathe 1974a, Gayel and Shtina 1974), South America (Forest and Weston 1966, Follmann and Redon 1972), Africa (Reichert 1936, Faurel *et al.* 1953), Australia (Rogers and Lange 1971, 1972), Antarctica (Cameron 1969), and islands of both the Atlantic and the Pacific oceans (Flint 1958, Brock 1973, Carson and Brown 1978). Novichkova-Ivanova (1972) has described the algal crusts of the takyr deserts of central Asia. Takyrs are clay deserts of great aridity where the primary source of soil organic matter is from blue-green algae.

7.3 Important cryptogamic groups

The cryptogamic cover on rangeland soils usually includes algae representative of the divisions Chlorophyta (green algae) and Chrysophyta (diatoms and golden-brown algae), lichens, mosses, and a few xerophytic liverworts on more mesic sites (Worley 1973, Friedmann and Galun 1974). The bulk of the algal tissue, however, is usually contributed by the cyanobacteria or blue-green algae on more arid, calcareous soils (Shields and Drouet 1962, Shubert and Starks 1980, Johansen *et al.* 1981, Johansen and Rushforth 1985). A rich diversity of fungi and bacteria occur in association with the photosynthetic cryptogams in North American desert soils (Durrell and Shields 1960, Hunt and Durrell 1966, Vollmer *et al.* 1977, States 1978). Microfungi of the Sahara and Negev Deserts of North Africa and Israel have been studied by Nicot (1960) and Borut (1960) respectively. The cryptogamic covers thus form self sufficient microcommunities in which both primary production and decomposition proceed. Many of the blue-green algae and lichens in which the algal associate is a blue-green alga also fix atmospheric nitrogen and thus enhance the nutrient status of the soil (Snyder and Wullstein 1973, Klubek and Skujins 1980). Other kinds of organisms such as protozoans, nematodes and mites have also been reported from the soil crusts generated by cryptogamic species (Worley 1973). Such crust often form nearly complete covers on soils of arid lands worldwide (Fig. 1).

Figure 1. A heavy cryptogamic cover on a silty clay loam soil in northern Utah. The cover includes a moss, the lichen *Collema tenax* and blue-green algae.

7.4 Cryptogams and rangelands

7.4.1 Effects on soil-water relations

Runoff and Erosion. Workers concerned with the nonvascular plant cover of desert and steppe rangelands have long postulated that such growths retarded erosion by wind and water (Fritsch 1922, Booth 1941, Fletcher and Martin 1948, Rogers and Lange 1971). In more recent years, such qualitative evaluations have received support from better quantified observations (Loope and Gifford 1972, Blackburn 1975, Mackenzie and Pearson 1979, Harper and St. Clair 1985). The positive values of soil cryptogamic communities on soil stability seem to be related to both the effects of layers of living tissue on the surface and to highly stable soil aggregates that form in the presence of both algae (Bailey *et al.* 1973, Anantani and Marathe 1974b, Mackenzie and Pearson 1979) and fungi

(Went and Stark 1968, Sutton and Sheppard 1976). Numerous algal, bacterial, and fungal species of arid environments secrete thick layers of extracellular polysaccharides. Such secretions encapsulate the cells and filaments producing them. Because the polysaccharides are strongly hygroscopic, enough water is apparently retained around cells to significantly enhance survival during dry periods (Shields and Durrell 1964). These mucilaginous secretions have been shown to draw fine soil particles together into aggregates that are stable in water and over periods of at least two years (Martin and Waksman 1940, Bond and Harris 1964, Bailey *et al.* 1973). Schulten (1985) notes that mosses also enhanced soil structure even though mucilaginous secretions were not produced. She further noted that lichen covered soils displayed significantly more large aggregates ($>2{,}000\ \mu m$) than adjacent, bare soils. Others have observed that fungal hyphae often ramify the soil as much as 5 mm from the lichen thallus from which they originate (Ahmadjian *et al.* 1980). Blackburn (1975) found that vesicular surface horizons strongly reduced infiltration rates in Nevada desert. Such horizons rarely occurred under soils described as having Type I or II surfaces. As Eckert *et al.* (1986a) make clear those soil surfaces had significant amounts of cryptogamic cover.

Using an artificial rain device, Loope and Gifford (1972) demonstrated that sediment losses from plots with sandy soils did not differ significantly between those stabilized by grasses and those with a heavy cryptogamic cover. Blackburn (1975) demonstrated that soil surface conditions referred to as Type I and II had infiltration rates of up to 7 cm/hr and lost only about 160 kg/ha of soil, while type III and IV surfaces had infiltration rates of less than 4 cm/hr and lost up to 400 kg/ha of soil under comparable conditions of applied water. A major difference between the two basic groups was abundant vascular (Type I) or cryptogamic cover (Type II) (Eckert *et al.* 1986a). MacKenzie and Pearson (1979) reported that sand covered with a mat of the green alga *Microspora* withstood over twice the wind velocity and 40 times more overland water flow than bare sand before significant soil erosion occurred. Harper and St. Clair (1985) showed that scalped plots lost almost 15 times as much soil as comparable plots with the cryptogamic cover intact when plots were treated with equal amounts of artificial rain (Fig. 2). Booth (1941) demonstrated a similarly large reduction in erosion due to a heavy mat of blue-green algae.

Soil-Moisture. Although some have reported that blue-green algae and lichen crusts impede infiltration (Evanari 1981, Brotherson and Rushforth 1983), other studies demonstrate a significant enhancement of infiltration rate for algal-lichen crusts relative to similar, uncrusted sites (Loope and Gifford 1972, Harper and St. Clair 1985 and Fig. 2). In a recent Utah study, algal-lichen crusts showed infiltration rates that did not differ

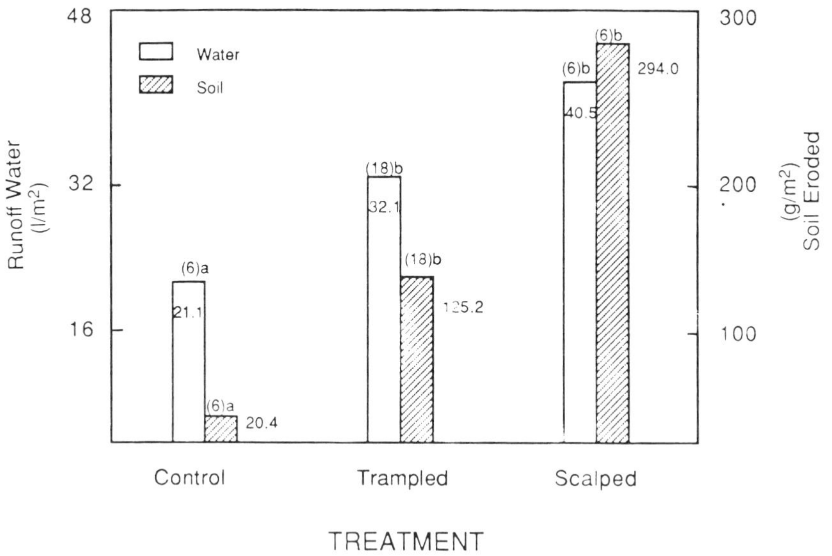

Figure 2. Water and soil losses from 0.25 m^2 plots. All plots initially supported heavy covers of blue-green algae, crustose lichens and moss with minor amounts (less than 4%) of the grass *Poa secunda.* Plots were trampled and treated when dry. The number in parentheses at the top of each bar is the number of replications in that treatment. Scalped plots had the surface 1.0 cm of soil removed. All plots received the equivalent of 5.0 cm of artificial rain in 8 min from a pipette infiltrometer. Bars for water or soil losses that bear a common letter at their top do not differ significantly. Differences were interpreted at the $p \leqslant 0.05$ level.

significantly from rates on scalped plots (Harper and Holmstead, in review).

Well developed cryptogamic crusts often produce roughened surfaces as in Fig. 1, but the effect is particularly conspicuous on sands as shown by photographs in Loope and Gifford (1972) and Brotherson and Rushforth (1983). Such roughening results in large amounts of water being detained in small pools until infiltration can occur during intense storms (Loope and Gifford 1972, Brotherson and Rushforth 1983). Thus even when the crusts significantly slow rates of infiltration, there may be no significant increase in surface flow and sediment transport. Fritsch (1922), Booth (1941), and Fletcher and Martin (1948) all seemed to accept the assumption that the cryptogamic crusts improved soil water relations for vascular plants.

Evaporation. The question of whether cryptogamic species that secrete copious amounts of polysaccharides into the soil can improve soil-water relations by "sealing" the surface has been commented on favorably by Fritsch (1922), Booth (1941), and Brotherson and Rushforth (1983).

Booth (1941) actually reported data supporting that view, but his sample size was so small that the results cannot be considered conclusive. Harper and Holmstead (in review) have tested the effects of a dark colored blue-green algal and *Collema tenax* lichen crust (Fig. 1) on surface evaporation rates. They found that heavily crusted sites in central Utah lost significantly more water from the surface 7.5 cm of soil than intermixed scalped plots (Fig. 3). They concluded that the dark crusts absorbed more solar radiation than did the light colored soil of scalped plots. Temperatures were consistently higher on the crusted plots by as much as 5°C in mid-October.

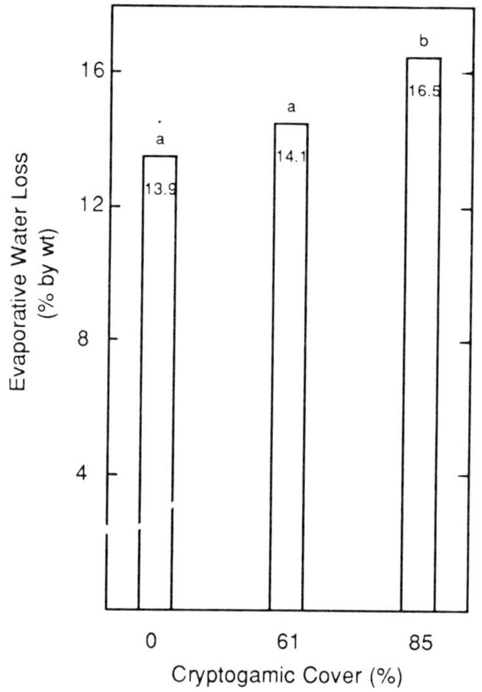

Figure 3. Evaporative water losses over a seven day period from plots covered by variable amounts of cryptogamic cover. The values reported for each category represent averages for cover and evaporative water losses from ten replications. Bars sharing the same letter at their top do not differ significantly ($p \leqslant 0.05$). One of the heavy cover plots is shown in Figure. 1.

7.4.2 Effects on soil nutrients

Nitrogen Fixation. A large body of literature documents the role of several kinds of nonvascular plants (bacteria, blue-green algae and lichens taking a blue-green as the phycobiont) in nitrogen fixation on rangelands. The growth of knowledge in this area can be traced in some detail using a few summary articles (Stokes 1941, Shields 1957, Shields and Durrell 1964, Mayland *et al.* 1966, MacGregor and Johnson 1971, Snyder and Wullstein 1973, Rychert and Skujins 1974, West and Skujins 1977, Rychert *et al.* 1978). Rychert and Skujins (1974) concluded that "nitrogen fixation by algal crusts probably constitutes a major input of nitrogen into the desert ecosystem". West (1981) considered that cool desert ecosystems showed "high rates of nitrogen fixation, primarily due to the surface-incrusting blue-green algae and lichens". As Shields and Durrell (1964) have noted, such crusts do not develop in some warm deserts and the primary nitrogen sources become higher plants that form symbiotic associations with bacteria capable of N-fixation. Algal-lichen crusts are also lacking from some of the more severe deserts of Northeastern Africa (personal observation). Rychert *et al.* (1978) reported that blue-green algal-lichen crusts could fix as much as 100 kg of nitrogen per hectare per year, but their literature survey suggests that 10 kg/ha/yr is a more realistic average fixation rate for desert cryptogamic crusts. West and Skujins (1977) considered crusts to fix about 25 kg/ha/yr in Utah deserts. The blue-green alga *Microcoleus vaginatus* was probably the dominant alga in the crusts tested (Shields *et al.* 1957, Durrell and Shields 1961, Anderson and Rushforth 1976). Such an annual input of available nitrogen is crucial, since vascular plants take up 10—12 kg/ha/yr in above-ground parts in Utah deserts (West and Skujins 1977), and neither symbiotic nor heterotrophic bacterial fixers perform well in cold deserts (West 1981).

Nitrogen fixation has been demonstrated for some species of at least the following blue-green algal genera: *Anabaena, Anabaenopsis, Aulosira, Calothrix, Chlorococcus, Cylindrospermum, Dicothrix, Gloeotrichia, Hapalosiphon, Lyngbya, Nodularia, Nostoc, Plectronema, Scytonema*, and *Tolypothrix* (Fuller *et al.* 1960, Shields and Durrell 1964, Stewart and Lex 1970, Pankratova and Vakhrushev 1971, Ali and Sandhu 1972). Although 18 lichen genera have been reported to fix nitrogen (Millbank 1976), only three genera (*Collema, Heppia* and *Peltigera*) appear to have species that occur on soils in the arid or semiarid zones of North America (Shields *et al.* 1957, Hale 1979). *Collema*'s phycobiont is a *Nostoc* species (Ahmadjian 1967). Four different *Collema* species, each from dry soils of a different continent have been shown to fix nitrogen (Scott 1956, Rogers and Lange 1966, Fogg and Stewart 1968, Sheridan 1979). *Peltigera* is unusual in that its thalloid phycobiont is a green alga, but some species of the genus incorporate colonies of *Nostoc* into surficial structures known as cephalodia. At least *Peltigera aphthosa, P. canina, P. evansiana, P. polydactyla*, and *P. praetextata* have proven nitrogen fixation capabilities attributable to such cephalodia (Millbank 1972, Kershaw 1974). A few species of *Lecanora* and *Lecidea* (lichen genera with many soil dwelling species in deserts) also produce surficial cephalodia that contain blue-green algae (Hale 1974). Rogers and Lange (1966) detected some nitrogen fixing activity for *Lecidea crystalifera* at an Australian site. Sheridan (1979) also reported N-fixation for a species of *Lecidea*. Whether such activity was attributable to cephalodia, adherent and free-living blue-green algae, or to adherent heterotrophic bacterial fixers such as *Azotobacter* is unknown. Snyder and Wullstein (1973) observed some evidence of N-fixation for *Dermatocarpon lachneum* which does not harbor a blue-green phycobiont, but they concluded that the activity had originated with free-living, closely entwined *Nostoc* filaments. That or a similar situation may also account for the weak tendency of *Parmelia adhaerens* to fix nitrogen as reported by Rogers and Lange (1966). Snyder and Wullstein (1973) considered that a positive test for N-fixation for a species of the moss genus *Grimmia* was attributable to heterotrophic bacterial fixers intermixed with rhizoids of the moss.

Although heterotrophic nitrogen fixation is known to occur in deserts (Steyn and Delwiche 1974, Klubek and Skujins 1980), only 2 kg/ha/yr can be expected under natural vegetation in arid environments (Steyn and Delwiche 1974). Klubek and Skujins (1980) considered that the heterotrophic fixers were most active in association with blue-green algae, perhaps because the latter provided an abundance of respirable organic material. They termed such a response "a phycosphere-like effect". Jones (1930) had postulated a similar relationship between blue-green algae and free-living bacteria fixers, but Stokes (1941) presented evidence that such an effect was "unlikely" in nature. Additional studies on heterotrophic N-fixers in algal-lichen crusts would be helpful. West (1981) discounted the importance of heterotrophic N-fixers in deserts because of the scarcity of organic matter on which such bacteria are dependent.

Fogg and Stewart (1968) observed some N-fixation at 0°C in species of *Collema* and *Stereocaulon* on one of the South Orkney Islands at high latitude in the Southern Hemisphere. Maximum fixation rates occurred at temperatures above 10°C. In Utah, Rychert *et al.* (1978) demonstrated maximum N-fixation in blue-green algal-lichen soil crusts at circa 20°C with sharp declines below about 18 and above 23°C.

Kershaw (1974) demonstrated that lichens of xeric environments achieved maximum N-fixation rates at as little as 60% moisture in the thallus, while species in the same genus (*Peltigera*) but indigenous to more mesic environments naturally achieved maximum rates at over 80% moisture. Rychert and Skujins (1975) showed a 50% decline in N-fixation of algal-lichen crusts by the time soil moisture potential had fallen to −1.0

bar. Huss-Danell (1978) reported that night temperatures below 0°C did not reduce nitrogenase activity (the essential enzyme for N-fixation) in the lichen *Stereocaulon paschale* provided temperatures rose to favorable levels on the following day. He did find that lichen thalli collected in winter showed significantly lower nitrogenase activity that thalli collected in summer. That effect was induced artificially by simulating the photoperiod of winter in the laboratory (Huss-Danell 1978).

In respect to light requirements for N-fixation by blue-green algal or blue-green algal-lichen crusts, the evidence demonstrates that full sunlight is strongly inhibitory (Reddy and Giddens 1975, Rychert and Skujins 1975). Using an algal crust including species of *Anabaena, Nostoc, Oscillatoria*, and *Scytonema*, Reddy and Giddens (1975) showed that N-fixation was most rapid at 55% shade; at 80% shade that rate decline by more than 90% and at 30% shade the rate was about 40% of that at 55% shade. Rychert and Skujins (1975) showed that N-fixation by a blue-green algal-lichen (*Collema*) crust was maximal at about 300 micro-einsteins/m^2/sec under incandescent light. That intensity of light would be "somewhat comparable to a day with a heavy grey cloud cover". The last named authors also found that fluorescent light produced greater nitrogen fixation at lower light intensities than incandescent light.

Denitrification. Nitrogen in soils of desert rangelands is maintained at a relatively constant level despite significant inputs from blue-green algae and a few lichens having blue-green phycobionts. Losses of gaseous nitrogen from desert soils due to the process of denitrification are rapid (often exceeding rates of fixation) and occur simultaneously with N-fixation when the soils are periodically wetted by rain or melting snow (MacGregor 1972, Skujins and Klubek 1978, Burns 1983). The denitrifiers apparently maintain themselves by respiring organic matter from the algal-lichen crusts.

Factors Impeding Fixation. Rychert and Skujins (1974) provided good evidence that the major shrub species [*Artemisia tridentata, Atriplex confertifolia*, and *Eurotia* (*Ceratoides*) *lanata*] of the cold deserts of western United States all exerted an allelopathic effect on the N-fixing capacity of associated blue-green algal-*Collema* lichen crusts. The effect was large, apparently reducing N-fixation rates by as much as 75—80%. They were able to demonstrate the effect both in the field under natural conditions and in the laboratory using leachates from foliage of the three shrubs. Rice (1964) has observed a similar suppression of N-fixing bacteria in the presence of photosynthetic seed plants.

Delwiche and Wijler (1956) and Reddy and Giddens (1975) have shown that various forms of mineral nitrogen strongly suppress N-fixation by nonsymbiotic fixers. Thus the fixation process works best where available nitrogen is most needed, in environments severely deficient in fixed nitrogen. Reddy and Giddens (1975) also demonstrated that addition of lime significantly enhanced blue-green algal fixation of nitrogen on acidic soils. Terry and Burns (1987) showed that heavy grazing, wildfire, or cultivation severely reduced N-fixation by cryptogamic crusts in Utah deserts. The latter authors noted that the N-fixing capacity of crusts on grazed and burned soils recovered to control levels within three years.

The nitrogen −15 isotope has been used to demonstrate that nitrogen fixed by blue-green algal crusts can be metabolized by associated higher plants (Mayland and MacIntosh 1966, Stewart 1967). Fuller *et al.* (1960) reported that nitrogen fixed by blue-green algae was "as available to plants as an equivalent amount of nitrate nitrogen".

Phosphorus. Kleiner and Harper (1972, 1977a) noted consistently more phosphorus in surficial (surface 2 cm) soils overlain by cryptogamic crusts relative to adjacent soils where local erosion had been rapid enough to preclude crust development. They concluded that the enhanced phosphorus effect was related to an enrichment in fine soil particles (silt and clay) where cryptogamic crusts existed. Fletcher and Martin (1948) reported a similar enrichment of soil fine particles where nonvascular plant crusts were well developed. They suggested that raindrop splash moved more fine than sand-sized particles, and the gelatinous, roughened surfaces of the "living" crusts trapped and held such dislodge particles with greater efficiency than adjacent, bare areas. One might expect wind erosion, an ever present phenomenon in deserts, to have a similar effect. Thus although the algal-lichen crusts cannot add new phosphorus to the system as they do nitrogen, they may nonetheless create enriched microsites for available phosphorus by selectively trapping soil fines which are much richer in that element than sands (Black 1968).

7.4.3 Effects on seedling establishment

General Observations. Several researchers have reported that seedling success was apparently enhanced by the presence of cryptogamic crusts (Fritsch 1922, Booth 1941, Fletcher and Martin 1948, Shields and Durrell 1964). A few workers have reported positive effects on seedling growth from inoculation of soils with blue-green algae (Breazeale 1929, Watanabe *et al.* 1951, Dadhich *et al.* 1969).

Allelopathy and Antibiosis. Other reports demonstrate that numerous lichen species produce chemicals that strongly inhibit growth of gram positive bacteria (Burkholder *et al.* 1944, Burkholder and Evans 1945). Production of such chemicals probably gives the mycobiont of lichens that synthesize them some advantage by eliminating or reducing the number of bacterial competitors for organic molecules produced by the phycobiont (Hale 1974). Miller *et al.* (1963) showed that extracts from the lichen *Umbilicaria papulosa* inhibited growth of fungal species belonging to nine different genera. Brown and Mikola (1974) found that extracts from four species of the lichen genus *Cladonia* as well as from the lichens *Cetraria islandica* and *Stereocaulon paschale* suppressed growth of 46—83% of 18 fungal species (15 genera) known to form mycorrhizal associations with forest trees.

Pyatt (1967) reported that macerated tissue and sometimes even sand on which the lichen *Peltigera canina* had grown inhibited germination and root growth of seedlings of five of the seven grass species tested. Miller *et al.* (1963) noted that extracts from *Umbilicaria* inhibited cucumber root growth. Brown and Mikola (1974) showed that dense mats of *Cladonia alpestris* significantly inhibited growth of seedlings of the forest trees, *Pinus sylvestris* and *Picea abies.* Brown (1967) presented evidence that *Cladonia cristatella* inhibited germination of *Pinus banksiana.* Cowles (1982) has shown that *Cladina stellaris* strongly suppressed seedlings of *Picea mariana.* Evidence also suggests that the epiphytic lichen *Evernia prunastri* adversely affects photosynthesis of the host tree (in this case, *Quercus rotundifolia*) through disruption of the thylakoid system of chloroplasts (Ascaso *et al.* 1983, this paper cited other European workers who have also documented adverse effects of lichens on seed plants).

Considering the wide array of secondary plant products produced by lichens (Culbertson 1969), it would indeed be surprising if lichen extracts did not suppress at least some of the nonlichen taxa associated with them. Rundel (1978b) has postulated that the secondary products of lichens may benefit the organisms in one or more of four possible ways. He suggests that the secondary plant chemicals may: (1) screen out excessive light, (2) hasten weathering of the substrate and permit more secure attachment, (3) serve as allelopathic and/or antimicrobial compounds, or (4) function as defenses against herbivory.

Schlatterer and Tisdale (1969) found that tissue of *Tortula ruralis*, a moss common in cryptogamic covers in the *Artemisia* steppes of western North America, significantly reduced seedling emergence of *Agropyron spicatum*, a dominant grass in that zone. No lichens were included in their study. Harris *et al.* (1987) considered that “cryptogamic crust forms a barrier to seedling establishment”.

Seedlings in Deserts. The foregoing literature references demonstrate an apparent divergence between the casual observations of workers who have had broad experience in the field and those who have quantitatively investigated the influences of tissue and tissue extracts of lichens and at least one moss on germination and growth of angiosperm seedlings under controlled conditions. We have cited several papers where carefully designed experiments or field surveys have shown that lichens, at least, often exert a significantly negative influence on seedlings of some angiosperms and gymnosperms.

Unfortunately, few experimental studies have considered the influence of cryptogamic plants of arid lands on seed germination and seedling establishment and growth of important range plants. Singh (1950) demonstrated that blue-green algal growth did make an Indian desert soil less alkaline and more favorable for plant growth. St. Clair *et al.* (1984) reported that blue-green algal-*collema* lichen crusts enhanced seedling establishment of range grasses, but the effect was positive for only one species under greenhouse conditions. Under field conditions, all species (*Agropyron elongatum, Elymus cinereus* and *E. junceus*) performed far better where the cryptogamic cover had been destroyed by a wildfire.

Two more recent studies show strong positive effects of cryptogamic crusts on seed plant establishment. In Utah, Harper and St. Clair (1985) planted seed of several species through holes in a template onto 1.0 m^2 plots that were either scalped or left undisturbed on a random basis. Scalped plots had the surface 1.0 cm of soil removed before seeds were planted. One seed was planted into a small hole opened with a metal probe at each of 32 openings in the 1.0 m^2 template. Seeds were covered and compacted before the template was removed. Since two steel pegs were driven into the ground through holes in opposite corners of the template, planting sites could be relocated exactly at the time of future evaluations of the effects of cryptogamic crusts on seedling establishment. At the end of the first growing season, over 2.5 times as many seedlings survived on plots with the cryptogamic cover intact as on scalped plots. At the end of the third growing season, the difference between control and scalped plots had narrowed, but there was still over 1.6 times as many seedlings surviving on cryptogamic covered as opposed to scalped plots (Harper and St. Clair 1985). At the sites considered, the cryptogamic community was dominated by blue-green algae, *Collema tenax* and *Tortula ruralis*. Average cover percentages for those three groups at three planting sites was 14.0, 5.5, and 7.4% respectively. Other cryptogams of regular occurrence on the plots included the moss *Grimmia*, and the lichens *Agrestia hispida, Dermatocarpon lachneum*, and *Fulgensia bracteata*. These and other lesser cryptogamic species combined contributed an average of only 10.0% additional cover on each of the three sites.

In Table 1, we show the number of seedlings of each of five seed plants that survived the first growing season on scalped and unscalped planting sites. It will be noted that four of the five species established more seedlings on plots with intact cryptogamic covers than on scalped plots.

Table 1. Effects of a blue-green algal moss-lichen soil crust on establishment of seedlings of five plant species in Utah. At each plot, 32 seeds were planted through a 1.0 m^2 template. The template location was marked with steel pegs to facilitate relocation. The decision to scalp or leave a plot undisturbed was determined by the toss of a coin prior to planting. Plots were seeded in December 1981 and inventoried for established seedlings in October 1982. Data are from Harper and St. Clair (1985).

Species	Treatment Scalped	Treatment Control	Significance of Difference
Elymus junceus[1,2]			
No. plots	11	11	
No. seedlings	8	2	NS
Kochia prostrata[1,2]			
No. plots	12	9	
No. seedlings	11	14	NS
Lepidium montanum			
No. plots	5	5	
No. seedlings	8	50	$p < 0.005$[3]
Linum perenne			
No. plots	5	5	
No. seedlings	11	16	NS
Sphaeralcea coccinia[1,4]			
No. plots	11	9	
No. seedlings	20	64	$p < 0.005$

[1] Species planted at two different sites. Seedlings showed same preference for scalped or control conditions at each site.
[2] The trend was statistically significant at one site but not the other by Chi-square analysis.
[3] Probability that the observed difference is due to chance.
[4] Seedlings showed a statistically significant preference for control plots at both planting sites.

The difference was statistically significant for two of those species. *Elymus junceus* established more seedlings on scalped than on unscalped plots, but that difference was not statistically significant.

Eckert *et al.* (1986b) planted *Artemisia tridentata wyomingensis, Agropyron desertorum, Bromus tectorum, Poa sandbergii, Sitanion hystrix* and *Stipa thurberiana* onto three kinds of soil surfaces in northcentral Nevada. Those surface types, Types I, II and III, are characterized in the following ways (Eckert *et al.* 1986a). Type I: a roughed surface lightly covered by plant litter and overtopped by shrub canopy. Type II: a soil well covered by cryptogamic growth with a pinnacled surface on which polygonal cracks separated individual pinnacles. This surface occurred at the border between shrub dominated sites and barren interspaces. Type III surfaces were drained and broken into flat-topped polygons separated by shallow cracks. Cryptogamic and vascular plant cover was poorly developed on Type III surfaces. Results showed that emergence and survival of perennial grasses was best on untrampled Type I surfaces and in the polygonal crack microsites of Type II surfaces. *Artemisia* seedlings were most abundant on Type I and III surfaces. Native annual and perennial forb seedlings were most commonly found in crack microsites of Type II and III surfaces. Thus seedlings of all herbaceous species tested established as well or better on Type II surfaces (areas which supported good cryptogamic cover) as on any other surface type. The shrubby *Artemisia* species, however, performed poorly on Type II surfaces.

In Table 2, we show results obtained by Harper and St. Clair (1985) on the effects of blue-green algal, lichen (primarily *Collema*), and moss (*Tortula* and *Grimmia*) cover on establishment of seedlings of three vascular species. The results suggest that blue-green algal cover had a significantly positive effect on establishment of seedlings of two species (*Lepidium* and *Sphaeralcea*) and a weak negative (not statistically significant) effect on a third (*Elymus junceus*). The lichen cover also had a significant positive effect on establishment of seedlings of the former two seed plants and a nonsignificant negative impact on *Elymus* seedlings. Moss cover did not have a significant impact on seedling establishment of any of the three seed plant species, but the correlation coefficient was negative for both *Elymus* and *Lepidium*. The data in Table 2 must be

Table 2. Relative effects of blue-green algae, lichen (*Collema tenax* primarily), and moss (*Tortula ruralis* and *Grimmia* spp.) cover on establishment of seedlings of three seed plant species. Ten 1.0 m^2 plots were seeded with a constant number of seeds (32) for each species. Since seeds were sown through a template, fates of each could be accurately determined. Cryptogamic cover was estimated ocularly using a decimeter square grid. In the analyses, number of seedlings surviving at the end of the first growing season was correlated with percentage cover of each cryptogamic category.

Cover Type	Vascular Species *Elymus cinereus*	*Lepidium montanum*	*Sphaeralcea coccinea*
	r-value (Correlation coefficient)		
Algal	−0.23	0.84**	0.77**
Lichen	−0.40	0.75*	0.87**
Moss	−0.47	−0.14	0.17
All cryptogams combined	−0.31	0.74*	0.85**

** Correlation coefficient significant at the 0.01 probability level.
* Correlation coefficient significant at the 0.05 probability level.

considered as tentative, but such results do seem reasonably in light of reports by Breazeale (1929), Watanabe *et al.* (1951), and Dadhich *et al.* (1969) which show significant positive effects of blue-green algae on certain flowering plant seedlings. It will be recalled that the lichen *Collema tenax* takes a blue-green algal phycobiont and has minimal thallus development, thus minimizing the phycobiont influence. The results in Table 1 are also compatible with Schlatterer and Tisdale's (1969) results in which tissue of the moss *Tortula ruralis* suppressed seedling establishment in two of three perennial grasses.

7.4.4 Environment and cryptogams

Temperature. Any organisms that lives on or near the soil surface in desert environments must be tolerant of high temperatures. Soil surface temperatures of 70°C are not uncommon (Sinclair 1922), and a value of 94°C has been reported for Death Valley, California (Mooney *et al.* 1975). It seems clear that perennial cryptogams (such as lichens and mosses) of soil surfaces must be able to endure the temperature extremes of such surfaces and still resume growth when more favorable conditions reappear. Thus even though controlled experiments have not documented their survival under extreme temperatures (e.g., 60°C), it can be accepted that the organisms are able to tolerate those conditions, since they persist on soils whose surfaces occasionally experience that temperature. No doubt part of their tolerance is related to the fact that both lichens (MacFarlane and Kershaw 1978) and soil algae (Trainor 1982) are known to be more resistant to heat shock when dry.

Direct experimental evidence demonstrates that some soil algae can survive temperature as great as 160°C for one hour provided the soil is dry (about 1% moisture, Trainor 1982). That author also showed that no algal species survive 100°C conditions for even one hour when soil moisture was 17%. Trainor (1982) cultured one or more species of the following genera from a soil containing 1% moisture after three months of incubation at 40°C: *Anabaena, Chlorococcus*-like, *Klebsormidium*, and *Oscillatoria.* Blue-green algae have been shown to survive ten weeks of incubation in dry soil at 75°C (Buzer *et al.* 1985). Green algae endured such conditions at 70°C but not at 75°C. Diatoms survived at 55°C but disappeared at all higher temperatures.

MacFarlane and Kershaw (1978) demonstrated that two subspecies of *Peltigera canina* showed a net gain in carbon from photosynthesis at 45°C. Nash *et al.* (1982a) showed that *Parmelia kurokawae* and *Acarospora schleicheri* showed some net photosynthesis at 31°C provided light intensity was 125 microeinsteins/m^2/sec or greater. Lange (1980) found that the lichen *Ramalina maciformis* also maintained a net gain in carbon at 30°C. Maximum carbon gains were achieved by *Parmelia* and *Acarospora* at about 10°C in the studies of Nash *et al.* (1982a). Lange (1980) reported that *Ramalina* was maximally active photosynthetically at 15°C. Net carbon gains at 25°C were much lower than those at 15°C, and rates at 5°C were intermediate to those at 15 and 25°C.

Physiological activity of soil surface cryptogamic species at low temperatures becomes an important concern in cold and polar deserts. Unfortunately, experimental data concerning response of cryptogams to cold temperatures is not abundant. Since good moisture conditions often coincide with cold weather in the world's cold deserts, there is need for more information on response of major cryptogamic taxa to low temperatures. Such data as do exist, show that lichens (Fogg and Stewart 1968, Lange 1980, Nash *et al.* 1982b) and blue-green algal-*Collema* lichen soil crusts (Rychert *et al.* 1978) show only minimal activity for photosynthesis and N-fixation at temperature of 0°C. Those studies reported maximum physiological activities between 10 and 20°C. At temperatures above 20°C net gains in nitrogen and carbon declined rapidly.

Moisture. As might be expected, desert soil cryptogams are highly resistant to desiccation. Bristol (1919) reported that British agricultural soils stored at 3—10% moisture for over 70 years yielded two blue-green algae (*Nostoc muscorum* and *Nodularia harveyana*) when cultured. Five blue-greens and one green alga were cultured from soils similarly stored for over 59 but less than 70 years. Cultures from soils stored over 47 but less than 59 years yielded two green algal species, one diatom and three blue-green algal taxa. Lipman (1941) successfully cultured a *Nostoc commune* colony from an herbarium specimen eighty-seven years old.

Lichens show maximum rates of carbon assimilation at surprisingly low water contents in thallus tissue. Kershaw (1972) worked with 12 lichen taxa of southeastern Canada. He demonstrated that species of the xeric environments of that region achieved maximum rates of net photosynthesis at tissue water contents as low as 35% of saturation. Even species of more mesic habitats achieved maximum photosynthetic rates at tissue moisture contents as low as 80% of saturation. Lange (1980) reported that *Ramalina maciformis*, a fruticose lichen of the Negev Desert of Israel, achieved its maximum rate of photosynthesis at 15°C and 55% water by weight in the thallus. Lange (1980) reported that dewfall in the desert could bring *Ramalina* tissue to a water content as high 90% of dry weight. At saturation, *Ramalina* tissue held an amount of water equal to 160% of dry weight.

Nash *et al.* (1982a) show that two Sonoran Desert lichens achieved near maximum rates of photosynthesis at about 60% tissue water (dry weight basis). At saturation both lichens contained 120% or more tissue water content. In the Phoenix, Arizona area, lichens were able to obtain enough moisture for net carbon gains via photosynthesis on only about 40 days per year. Those days were almost exclusively associated with precipitation events (Nash *et al.* 1982b). Dew fell in that area on only about 12 days per year, and those events were usually associated with periods of precipitation. Thus in Arizona, lichen growth was confined to only a few days per year (mostly in the winter) when some precipitation fell and temperatures were cool enough to permit lichen thalli to retain moisture adequate for photosynthesis for a few hours (Nash *et al.* 1982b). Since north-facing slopes are more shaded than south-facing slopes in that area, moist conditions persist for a few hours more per precipitation event and cryptogamic growth was correspondingly more lush on north exposures (Nash *et al.* 1977).

In contrast, the lichens of the Negev Desert appear to accumulate as much carbon on the abundant days with dew fall each year (198 days in the year considered by Kappen *et al.* 1979) as on days when rain fell (29 days per year). Cryptogamic growth is accordingly more lush in the Negev than in interior portions of the Sonoran Desert.

Such data support the conclusion that lichens of deserts probably achieve net gains in carbon under conditions of dewfall and/or mere sprinkles of rain. Rundel's (1978a) conclusion that lichens of fog zones can be sufficiently hydrated by water vapor to permit net gains in carbon while most vascular plants cannot seems justified by an abundance of experimental data. It is of interest that photosynthesis in lichens is strongly or even totally inhibited when freewater is adequate to completely saturate the thallus (Kershaw 1972, Lange 1980).

Some evidence suggest that high rates of nitrogen fixation require greater moisture contents in the thallus of lichens having blue-green phycobionts than is needed for maximal rates of photosynthesis (MacFarlane and Kershaw 1978). The data of Rychert *et al.* (1978) also show that N-fixation of algal-lichen crusts declines strongly with minor reductions in soil moisture.

Soil Factors. On the basis of field experience in southern Nevada, Shields and Drouet (1962) concluded that blue-green algae were best developed on fine textured soils. They attributed that preference to the cohesiveness and water retaining properties of soils containing larger percentages of silt and clay-sized particles. Anderson *et al.* (1982a) found a similar preference of cryptogamic species in general (algae, lichens and mosses) for heavier textured soils in Utah. Shields and Durrell (1964) noted that several workers had earlier concluded that soil texture exerted a strong influence on the distribution and development of soil algae. The greater likelihood of finding well developed cryptogamic covers on fine as opposed to coarse textured soils is not easily explained in terms of soil chemistry (Shields and Drouet 1962, Shields and Durrell 1964, Anderson *et al.* 1982a). The advantage of such soils for cryptogamic growths may be related primarily to surface stability as hinted at by Shields and Drouet (1962). Although gypsiferous soils often support nearly complete covers of soil cryptogams (Meyer 1986), careful chemical analyses have failed to show consistent differences between those soils and adjacent soils that supported essentially no cryptogams (McKnight 1980). The most obvious difference is instead a very firm surface cemented by gypsum crystals concentrated there as soil moisture evaporates. Gypsiferous soils worldwide are noted for strongly crusted surfaces in desert environments (Watson 1979).

Soil Chemistry. Stokes (1940) demonstrated that growth of both soil algae and bacteria was inhibited in soils as acidic as pH 4.3. Liming treatments alone that raised pH to about 6.3 were adequate to double algal cell numbers and more than quadruple bacterial cell numbers. Walp and Schopbach's (1942) work on *Nostoc muscorum* showed that that species grew well at pH 5 provided the growth medium did not include nitrogen in the nitrate form. The organism failed to survive at pH 3.5. In contrast, many lichens and mosses thrive at low pH values (Brown and Mikola 1974). Blue-green algae of soils appear to prefer basic conditions, while green algae are better represented under acidic conditions (King and Ward 1977, Durrell and Shields 1961, Shields and Durrell 1964).

Cobalt seems to be essential for blue-green algal growth (Allen 1956). At least two blue-green algae of drier soils (*Anabaena cylindrica* and *Nostoc muscorum*) are known to require sodium for growth (Shields and Durrell 1964). The need for sodium may be general among blue-greens, since they show strong preferences for basic soil conditions in numerous studies. Starks and Shubert (1982) reported significant positive correlations between available iron and both richness in algal species and total algal biomass on newly colonized mine spoils in North Dakota. Those authors also presented evidence that the elements cadmium, lithium, manganese, and strontium may have hampered colonization and growth of algae on mine spoils.

Cryptogamic species in general show greater tolerance to soil salinity than vascular plants of the same general site (Hunt and Durrell 1966). Anderson *et al.* (1982a) reported a significant positive correlation between cryptogamic soil crust cover and soil salinity in the Great Basin. That response for soil surface organisms may be a consequence of the fact that vascular plant cover tends to decline with increasing soil salinity thus reducing shading on the surface photosynthesizers. It also seems likely that salinity of surfaces where cryptogamic species exist would be greatly reduced for at least brief periods by even small precipitation events. It is perhaps relevant that as Rundel (1978a) has reported, some lichens are able to maintain normal rates of photosynthesis at up to −100 bars water potential. Starks and Shubert (1982) observed a negative correlation between soil algal biomass and soil salinity in the early years of invasion of mine spoils.

The consistently higher extractable phosphorus content of soils that support heavy growths of cryptogamic plants, suggests that biologically available P may be locally limiting for some nonvascular species. Higher concentrations of P in soils with well developed versus scanty cryptogamic covers has been reported frequently in western North America (Kleiner and Harper 1972, 1977a, McKnight 1980, Anderson *et al.* 1982a).

Topography. As Nash *et al.* (1977) and Nash and Moser (1982) have abundantly demonstrated, desert cryptogams are strongly influenced by both slope aspect and elevation. Both number of species and amount of tissue produced are far more abundant on north as opposed to south-facing slopes in North America. Nash (1974) and Nash and Moser (1982) showed a tendency for total number of lichen species and cover of soil surface cryptogams to increase with elevation. Unpublished data in our files suggests that soil cryptogamic cover may not always increase with elevation as observed by Nash and Moser (1982). In Utah, soil cryptogamic cover reaches maximum values (both absolute and relative) in low elevation, cold deserts of the Colorado Plateau and Great Basin. Cover of cryptogams runs as high as 42% absolute cover in situations where vascular cover averages only 26% in low elevation deserts that receive significant amounts of summer rain (Kleiner and Harper 1977b). At mid- and higher elevations, cover of soil cryptogams falls to 5% or less, while vascular plant cover rises to over 65%. Thus in North America, cryptogamic plants have their greatest impact on soil surface dynamics in interior cold deserts.

Nash and Moser (1982) summarize the literature that documents the proliferation of cryptogamic organisms (especially lichens) in regions where deserts lie adjacent to the sea. In such zones of high atmospheric humidity, epiphytic cryptogamic species become more important than terricolous taxa.

Brotherson and Masslich (1985) document the tendency of cover of soil cryptogams to decline in an absolute sense from ridge top through midslope positions to floodplain. The decline is associated with a steady increase in total vascular plant cover and a probable increase in mechanical disturbance due to heavier grazing pressure from hooved animals and scouring from floods along the drainage channel. As suggested earlier, heavier vascular plant cover may reduce growth of soil cryptogams by shading.

Grazing Effects. Poulton (1955) appears to be the first to document the adverse effect of grazing animals on soil cryptogamic covers. Pegau (1970) noted the destructive effect of reindeer trampling on lichens, and Daubenmire (1970) documented the devastating effects of domestic grazers on cryptogamic covers of steppe lands in Washington. Rogers and Lange's (1971) work in Australia showed that sheep destroyed cryptogamic crusts around watering stations in desert environments. Work in Utah showed that cryptogamic cover was reduced to less than 15% its probable potential by cattle grazing (Kleiner and Harper 1972). Since that time, many studies of the impact of grazing on soil cryptogams have appeared (Anderson *et al.* 1982a, 1982b, Brotherson *et al.* 1983, Johansen and St. Clair 1986, Andrew and Lange 1986, Terry and Burns 1987). All studies demonstrate unequivocally that excessive grazing severely depletes the soil cryptogamic community. Mosses appear to be more vulnerable than lichens and lichens more vulnerable than blue-green algae (Anderson *et al.* 1982b, Johansen and St. Clair 1986). Grazing during dry seasons seems particularly disruptive (Anderson *et al.* 1982a).

Savory and Parsons (1980) advocate a short duration grazing system for many range types varying from tropical grasslands and jungles to deserts. They suggest that physical animal impact is desirable on arid rangelands, because it breaks up hard surfaces capped by "algae, lichen and moss communities, and allows for greater grass seedling success".

Fire Effects. Because of sparse cover, fire is often not a concern in deserts. However, regions where plant cover is greater (in excess of 40%) because of greater precipitation or invasion of nonindigenous annual species, plant cover may become fire prone during dry periods when humidity is low and winds are strong. Several studies demonstrate that cryptogamic soil communities are temporarily severely damaged by wildfire (Johansen *et al.* 1982, Johansen *et al.* 1984, Callison *et al.* 1985, Johansen and St. Clair 1986). Although at least the blue-green algal components of the crusts may recover rapidly (significant recovery in three years), soils may be exposed to accelerated erosion by water and wind for many months after a fire (Johansen *et al.* 1984 and St. Clair *et al.* 1986).

Although algal biomass decreases significantly after a fire, species composition remains relatively unchanged (Johansen *et al.* 1984). Considering the great tolerance of numerous algal species to high temperatures (Trainor 1982), this observation is not surprising. Lichens and mosses, however, were more seriously impacted by fire and showed only limited recover after five years (Johansen *et al.* 1984). Both cover and species richness of lichens and mosses would appear to require many more than five years for recovery to prefire values.

Other Mechanical Disturbances. Harris *et al.* (1987) noted that 10 mm of fine volcanic ash was sufficient to destroy lichen crusts on the Columbia Plateau. Mosses were somewhat more tolerant of an ash fall, but even moss growth was eliminated by a fall of 20 mm of ash.

Terry and Burns (1987) showed that plowing essentially eliminated the ability of desert soils to fix nitrogen. Blue-green algae were the probable N-fixers involved at their site. Within one year, N-fixing capacity of the plowed soil had recovered to about one-third the fixation rate of the control.

Our observations in Utah suggest that off road vehicular traffic also has a destructive impact on cryptogamic crusts. Motorcycle traffic in the dry season is especially damaging, since recreation cyclists often ascend slopes at a steep angle. In the process, a track is pulverized in the crust. When rain falls or snow melts, runoff water collects in the unstable track and rapidly erodes a gully.

Effects of Atmospheric Contaminants. In recent decades, coal fired electrical power generating plants have appeared in deserts worldwide. Those plants often release large quantities of gaseous oxides of sulfur and nitrogen and other pollutants. In other areas, smelters and other mining operations release noxious gases and heavy metal enriched dusts into the atmosphere. Long experience in more mesic environments has demonstrated that numerous gaseous and dust effluents from such industrial activities can have disastrous effects on regional plant cover (Treshow 1970).

Little is currently known about the impact of industrial effluents on cryptogamic species that grow on soil surfaces. It is widely recognized that lichens are among the more sensitive of photosynthetic plants (Nash 1976a), but most data have been collected in urban areas of mesic climatic zones. Responses of mosses to air contaminants are less well known, but Nash and Nash (1974) suggest that mosses are at least as sensitive as lichens. Responses of soil algae to such challenges are essentially unknown. It seems to be a general conclusion among workers that desert cryptogams will rarely be susceptible to atmospheric pollutants, because they metabolize the contaminants slowly, if at all (Marsh and Nash 1979, Nash and Nash 1974). The former authors suggest that atmospheric pollutant concentrations are commonly relatively low at the soil surface even though emission concentrations are high.

Mosses and SO_2. Nash and Nash (1974) fumigated moist gametophyte tissue of seven moss species for 12 hr with concentrations of SO_2 ranging from 0.5 to 4.0 ppm. Two species (*Hypnum fertile* and *Leucobryum glaucum*) displayed a significant decline in chlorophyll concentration at 1.0 ppm. Two other species (*Dicranum fulvum* and *D. scoparium*) suffered a significant decline in chlorophylls at 2.0 ppm. Three species (*Dicranella heteromalla, Polytrichum ohioense*, and *Pottia truncata*) were not injured by even 4 ppm SO_2. It is important to note, however, that the earliest stages of the protonema (the initial, filamentous stage of the gametophyte) of *Polytrichum ohioense* was killed by as little as 0.2 ppm SO_2.

Lichens and SO_2. Nash (1975) showed that chlorophyll content of three lichens (*Actinogyra muehlenbergii, Cladonia cristatella* and *Parmelia rudecta*) was significantly depressed by 0.5 ppm SO_2 applied in 12 hr fumigations on each of five consecutive days. Marsh and Nash (1979) fumigated eight desert lichen species with SO_2 at concentrations of 0.1 and 0.5 ppm for 8 hr. The lichen thalli were fumigated at the 0.5 ppm level in both the hydrated and air dry condition: the 0.1 ppm treatment was applied only to hydrated thalli. No taxa considered showed a significant reduction in respiration rate at the 0.1 ppm SO_2 level of fumigation. At the 0.5 ppm level, hydrated tissue of five species (*Caloplaca trachyphylla, Dermatocarpon miniatum, Lecanora melanophthalma, Parmelia arseneana*, and *P. mexicana*) experienced major declines of over 75% in rate of respiration. The other three species (*Lecanora alphoplaca, L. christo*, and *L. novomexicana*) also showed a significant decline in respiration when hydrated thalli were fumigated with 0.5 ppm SO_2, but the decline was less than 50%. Respiration rates of air dry thalli of all species were also determined after 8 hr exposures to 0.5 ppm SO_2. In all cases, rates were indistinguishable from those of control tissue.

Sheridan (1979) exposed the arid soil lichen *Collema tenax* to a wetting solution of $NaHSO_3$ at levels ranging from 0.1 to 100 ppm. The nitrogen fixing capacity of *Collema* was reduced by 50% by the 0.1 ppm treatment. Sheridan considered that the 0.1 ppm level of HSO_3 was equivalent to 0.09 ppm SO_2 gas.

Fields and St. Clair (1984) showed that three lichen species (*Parmelia chlorochroa, Collema polycarpon*, and *Lecanora muralis*) showed enhanced electrolyte leachage through membranes after only 4 hours of fumigation with 2.0 ppm SO_2. The effect was significantly enhanced by fumigation periods of 8 and 12 hr.

Henriksson and Pearson (1981) have investigated the effects of SO_2 on the soil lichen *Peltigera canina*. They showed that chlorophyll a content of the thallus was significantly reduced after 14 days of incubation at 5 ppm SO_2. Using values averaged over 14 days, nitrogen fixation rates were not significantly reduced at 5 ppm SO_2, but were significantly depressed at 50 ppm. They used no treatments of less than 5 ppm. In another experiment Henriksson and Pearson (1981) showed that *P. canina* experienced significant reduction in the rate of N-fixation following continuous exposure to 0.1 ppm SO_2 for nine days. That decline continued with increasing loss in ability to fix nitrogen to 10 ppm SO_2. An anomalous increase in N-fixation rate occurred at 30 ppm SO_2 followed by a large decline again at 100 ppm SO_2.

Pearson and Rodgers (1982) demonstrated that short exposures of lichens to elevated levels of SO_2 were sufficient to enhance leachage of electrolytes from thallus tissue. The membrane leachage technique should prove useful as an early warning system by permitting managers to detect adverse effects of atmospheric pollutants to lichen thalli before organisms suffer so much damage that death occurs (Fields and St. Clair 1984).

Lichens and O_3. Ross and Nash (1983) have shown that photosynthesis in *Ramalina menziesii* was significantly depressed after 12 hr fumigations at as little as 200 $\mu g\ O_3/m^3$, a value reached in the air over the Santa Monica Mountains of southern California on between 50 and 100 days per year. In contrast, the lichen *Pseudoparmelia caperata* showed no decline in photosynthesis after being exposed to as much as 1567 $\mu g\ O_3/m^3$ for 12 hr. Nash and Sigal (1979) demonstrated that photosynthetic rate of both *Parmelia sulcata* and *Hypogymnia enteromorpha* was significantly depressed after exposure to 980 $\mu g/m^3$ of O_3 for 3 hr.

Lichens and NO_2. Nash (1976b) has shown that four lichen species (*Anaptychia neoleucomelaena, Lecanora chrysoleuca, Parmelia praesignis,* and *Usnea cavernosa*) experienced significant reduction in thallus chlorophyll content after 6 hr with 4 or 8 ppm NO_2. Concentrations as high as 3.0 ppm total nitrogen oxide have been observed in the atmosphere above the city of Los Angeles, California.

Lichens and F. Sheridan (1974) wetted tissue of *Collema tenax* and *Lecidea* sp. with a solution containing NaF and $PbCl_2$ in a ratio of 5.3:1 over a range of 0—10 ppm NaF. At 0.01 ppm fluorine, the nitrogen fixation rate of *Lecidia* was reduced by about 30%. N-fixation of *Collema* at 0.01 ppm F was reduced by about 75%. As noted elsewhere in this manuscript, the genus *Lecidia* does not take a blue-green algal phycobiont, so the N-fixing capabilities reported are problematic.

Lichens and CO_2. Since CO_2 levels have been slowly rising for several decades on this planet, it seems useful to consider response of cryptogams to that factor. Nash *et al.* (1983) have evaluated the response of six lichen species from widely different habitats and of a variety of morphological forms to elevated levels of CO_2 (ranging up to four times current atmospheric concentrations). The thallus samples were exposed to short (30 s) bursts of CO_2 (enriched with ^{14}C) at various concentrations. All species showed rates of photosynthesis that were maximal at between 0.07 and

0.09% CO_2. At the highest levels tested (0.109%), carbon accumulation rates of all species were essentially the same as those observed for the 0.07—0.09% CO_2 range (i.e., there was no apparent suppression of photosynthesis at the highest CO_2 levels). All species accumulated carbon fastest at CO_2 concentrations above current amounts in the atmosphere (about 0.03%). Work with vascular plants shows that photosynthetic rates accelerate up to about 0.5% CO_2 in bright light, but leaves begin to show injury after two to three weeks of exposure to concentrations that greatly exceed natural ambient levels (Weier *et al.* 1974).

Cryptogams and Heavy Metals. Nash (1975) showed that photosynthesis was depressed at lower concentrations of aqueous solutions of zinc and cadmium than was respiration in the lichens *Lasallia papulosa* and *Cladonia uncialis.* Net photosynthesis of *L. papulosa* fell below zero when tissue content of Zn rose above 482 ppm and Cd rose above 863 ppm. For *C. uncialis,* net photosynthesis dipped below zero when tissue content rose above 450 ppm Zn and 320 ppm Cd. The foregoing concentrations of Zn and Cd in the two lichen species were accumulated from an aqueous solution in which both elements were at a concentration of 2.0 mM. Living lichens were often found in the field with greater Zn and Cd concentrations that the foregoing values. That apparent inconsistency was believed to be due the demonstrated tendency of Zn and Cd when applied together to ameliorate the adverse effects of either alone. Also, when those metals accumulate slowly in the thallus, the effects appear to be less adverse than if they are accumulated rapidly (Nash 1975).

The phycobiont of *Cladonia cristatella, Trebouxia erici,* and the algal associate of *Parmelia caperata, T. anticipata,* were isolated from their lichen thalli and cultured by Nash (1975). In culture, he exposed both algae to variable concentrations of aqueous Zn and Cd. Growth of both algae was significantly depressed by 5×10^{-5} M concentration of either Zn or Cd alone. Thus net photosynthesis and growth of algal cells were suppressed by more dilute solutions of heavy metals than was needed to push net photosynthesis below zero in intact lichens (Nash 1975).

Other Elements. Nash and Sommerfeld (1981) reported elemental content for several populations of six dominant lichen species in the Four Corners area of New Mexico. In total, average and standard error values are reported for 13 elements. For eight of those elements, tissue concentrations are considerably higher than previously reported values for lichens of other regions. Nash and Sommerfeld (1981) attributed the higher values to greater frequency of dust storms in their area and entrapment of mineral fragments in the lichen thalli. The element fluorine occurred in concentrations that are potentially toxic to lichens within 2 km of the power plant.

7.4.5 *Community succession*

Because of their ability to withstand desiccation and to fix nitrogen, blue-green algae are pioneer colonists of denuded lands in arid and semiarid environments worldwide (Booth 1941, Novichkova-Ivanova 1972, Cameron 1972, Anantani and Marathe 1974a). In less arid regions, green algae may become important in such situations (Fritsch 1922). As early as 1972, Cameron described a successional sequence observed in soil cryptogamic communities of Antarctica. He considered that the first colonists of exposed soils were heterotrophic bacteria apparently deriving their energy for life from windblown organic matter. Those organisms were followed by blue-green and coccoid green algae. Yeasts, molds and protozoans quickly colonized the algal mats and were eventually followed in colonization by lichens having blue-green phycobionts and mosses in that order. Schubert and Starks (1979) demonstrated a succession among soil algae occupying North Dakota mine spoils ranging in age from one to 45 years. Blue-green algae accounted for over half the species on newly deposited spoils. Soils long modified by organisms supported more green than blue-green algae. Starks and Shubert focused on a shorter successional sequence on mine spoils in a 1982 paper. They demonstrated a progressive increase in number of algal species encountered during the first three years of colonization of mine spoils.

In 1982b, Anderson *et al.* described a successional process in which soils once denuded of cryptogams by heavy traffic of hooved animals were reinvaded by blue-green algae (and other algae in lesser amounts), lichens and mosses in that order. The entire successional sequence for cryptogams required 15—20 years in the absence of disturbances such as grazing or wheeled traffic. That successional sequence has been confirmed by Johansen and St. Clair (1986) and Johansen *et al.* (1984) for Utah deserts.

7.5 Grazing management and cryptogams

7.5.1 *Value of cryptogams*

The literature suggests that cryptogamic covers on semiarid and arid soils exist on resources unused by vascular plants. Several workers in western United States have shown that vascular plant cover is not significantly greater within ungrazed control areas than it is on adjacent grazed lands where cryptogamic cover has been eliminated by trampling, but cryptogamic cover often equals or exceeds vascular plant cover on ungrazed areas (Kleiner and Harper 1972, Anderson *et al.* 1982a, b, Brotherson *et al.* 1983). Other lines of evidence suggest that not only are the cryptogamic covers not in competition with the vascular plant community, but they may actually improve conditions for those plants. Such benefits may include greater available nitrogen, more stable soils, improved soil moisture relations due to greater infiltration on surfaces "roughened" by cryptogamic growth and, in some situations, improved reproduction of vascular plants. If grazing systems can be devised that retain the cryptogamic components of desert and steppe rangelands reasonably intact, the evidence suggests that such systems would be desirable.

7.5.2 Management considerations

Mack and Thompson (1982) appear to conclude that large grazing animals are incompatible with cryptogamic communities of soil surfaces. Savory and Parsons (1980) advocated grazing management schemes that destroyed those communities in order to improve conditions for vascular plant seedling establishment. Other workers have shown that soil cryptogamic covers have persisted in the face of historic use by grazers (Reichert 1936, Looman 1964, Bourlière 1964, Gayel and Shtina 1974). Anderson *et al.* (1982b) and Johansen and St. Clair (1986) show that blue-green algae maintain significant amounts of cover on ranges that have been grazed for scores of years at intensities that are regularly recommended as "safe" for the major associated vascular plants. Both groups of investigators also found *Collema tenax*, a crustose lichen having a blue-green algal phycobiont, to be common on grazed ranges. In general, however, lichens and mosses were severely depleted by domestic grazing herds. Anderson *et al.* (1982a) were persuaded that significant cover of blue-green algae could be maintained on arid rangelands, if grazing was curtailed during dry periods when such growths would be crushed by the hooves of grazing animals. Marble and Harper (1988) presented evidence that cryptogamic cover can be significantly enhanced by avoidance of dry season grazing. Andrew and Lange (1986) observed severe depletion of desert lichen cover around permanent water tanks within two years of initiation of light sheep grazing in South Australia. Eckert *et al.* (1986a) showed that cryptogamic cover does persist on moderately grazed *Artemisia* steppes in Nevada.

In all probability, cryptogamic crusts that are well developed do offer fewer safe sites for seeds to lodge, germinate and establish vigorous seedlings. As Savory and Parsons (1980) suggested, some "hoof action" on such surfaces may enhance seedling establishment, but Eckert *et al.* (1986b) demonstrated that too much trampling significantly reduced rates of seedling establishment. Even without trampling, some seedlings regularly established in the narrow cracks that separated polygons formed naturally as cryptogamic crusts wet and dry (Eckert *et al.* 1986b).

The recommendation of Anderson *et al.* (1986a) that desert ranges in the Great Basin of North America not be grazed during the dry season in order to preserve some cryptogamic cover is not incompatible with management considerations recommended for higher plants by Platou and Tueller (1985). They suggested that year-long and high intensity grazing schemes that permitted grazing throughout the growing season were incompatible with an arid zone vegetation that received the bulk of its moisture during the dormant season. They advocated a complete rest from grazing during one full growing season "every few years" in order to maintain bunchgrasses in shrub-steppe environments. Such a grazing scheme would also favor cryptogamic species.

7.5.3. Needed research

For those attempting to incorporate cryptogamic species into range management programs, more information would be helpful in several areas. The response of desirable cryptogams to any grazing system is poorly known for any area on earth. Managers of rangelands rarely consider cryptogamic taxa as information on rangelands is collected. In view of the literature considered in this report, more managerial attention to cryptogamic components of rangeland ecosystems seems justified. Those organisms may exert a variety of useful or undesirable influences on rangelands depending upon the organism, the ecosystem, and the process of concern.

It seems especially important that potential linkages between the cryptogamic and vascular plant communities be explored. Connections relative to soil stability, soil moisture dynamics, available nitrogen and other essential elements, and vascular plant growth and reproduction are important and likely to be impacted.

As noted earlier, wildfires and severe mechanical disturbances such as disking are known to temporarily reduce activity of blue-green algae and other cryptogams. It may be possible to enhance soil stability and success of vascular plant seedlings on such perturbed sites by inoculating with blue-green algae and/or lichens that harbor a blue-green alga and are tolerant of seral situations. That such organisms can be successfully inoculated onto open sites has been known since the work of Dean *et al.* (1973). Both Ashley and Rushforth (1984) and St. Clair *et al.* (1986) have reported successful inoculations and rapid establishment of cryptogamic species. All of the workers cited used pulverized cryptogamic crust material. In the later two studies, the commonest organisms extracted from the inoculated site were blue-green algae, except on highly saline-alkali shales where moss protonema were the only colonists observed.

Before inoculation of cryptogamic species onto denuded rangelands is likely to become a common management procedure, three kinds of research seem needed. Firstly, it should be determined whether inoculation will significantly improve soil stability under field conditions. Secondly, it must be determined whether or not inocula rich in blue-green algae will enhance success of vascular plant seedlings on rangelands. Work by Singh (1950) has demonstrated that growth of some blue-green algae does ameliorate both chemical and physical characteristics of barren, alkaline soils and permit subsequent growth of crop plants on those sites. Finally, assuming that cryptogamic inocula are shown to have a significant impact on soil stability and success of vascular plant seedlings, mass culturing processes should be developed for the more promising species. Ideally, the inocula should be available as dry powders that can be dusted over soil surfaces. Fortunately, pure cultures of numerous blue-green algae adapted to arid environments have already been assembled (Lewin 1977). Selection to improve on desirable traits and research on culturing techniques could proceed rapidly.

7.6 References

Ahmadjian, V. 1967. The lichen symbiosis. Blaisdell Publ. Co., Waltham, Mass.

Ahmadjian, V., L. A. Russell, and K. C. Hildreth. 1980. Artificial establishment of lichens. I. Morphological interactions between the phycobionts of different lichens and the mycobionts of *Cladonia cristatella* and *Lecanora chrysoleuca*. Mycologia 72: 73—89.

Ali, S., and G. R. Sandhu. 1972. Blue-green algae of the saline soils of the Punjab. Oikos 23: 268—272.

Allen, M. B. 1956. Photosynthetic nitrogen fixation of blue-green algae. Sci. Mon. 83: 100—106.

Anantani, Y. S., and K. V. Marathe. 1974a. Observations on algae of some arid and semi-arid soils of Rajasthan. J. of the Univ. of Bombay 41(68): 88—93.

Anantani, Y. S., and K. V. Marathe. 1974b. Soil aggregation effects of some algae occurring in the soil of Kutch and Rajasthan. J. of the Univ. of Bombay 41(68): 94—100.

Anderson, D. C., K. T. Harper, and R. C. Holmgren. 1982a. Factors influencing development of cryptogamic soil crusts in Utah deserts. J. Range Manager. 35: 180—185.

Anderson, D. C., K. T. Harper, and S. R. Rushforth. 1982b. Recovery of cryptogamic soil crusts from grazing on Utah winter ranges. J. Range Manage. 35: 355—359.

Anderson, D. C., and S. R. Rushforth. 1976. The cryptogamic flora of desert soil crusts in Utah deserts. Nova Hedwigia 28: 691—729.

Andrew, M. H., and R. T. Lange. 1986. Development of a new piosphere in arid chenopod shrubland grazed by sheep. I. Changes to the soil surface. Aust. J. Ecol. 11: 395—409.

Ascaso, C., I. Orus, and P. Estevez. 1983. Chloroplast crystalloids and other alterations in response to lichen substances. Photosynthetica 17: 198—203.

Ashley, J., and S. R. Rushforth. 1984. Growth of soil algae on topsoil and processed oil shale from the Uintah Basin, Utah, U.S.A. Reclam. and Reveg. Res. 3: 49—63.

Bailey, D., A. P. Mazurak, and J. R. Rosowski. 1973. Aggregation of soil particles by algae. J. Phycol. 9: 99—101.

Black, C. A. 1968. Soil-plant relationships, 2nd ed. John Wiley and Sons, Inc., New York.

Blackburn, W. H. 1975. Factors influencing infiltration rate and sediment production of semiarid rangelands in Nevada. Water Resources Res. 11: 929—937.

Bond, R. D., and J. R. Harris. 1964. The influence of the microflora on physical properties of soils. I. Effects associated with filamentous algae and fungi. Aust. J. Soil Res. 2: 111—122.

Booth, W. E. 1941. Algae as pioneers in plant succession and their importance in erosion control. Ecol. 22: 38—46.

Borut, S. 1960. An ecological and physiological study of soil fungi of the northern Negev (Israel). Bull. Res. Counc. Isr., Sect. D 8: 65—80.

Bourlière, F. 1964. The land and wildlife of Eurasia. Time Inc., New York.

Breazeale, J. M. 1929. Algae and their effect upon the growth of citrus seedlings. Master's Thesis, Univ. of Arizona, Tucson.

Bristol, B. M. 1919. On the retention of vitality by algae from old stored soils. New Phytol. 18: 92—107.

Brock, T. D. 1973. Primary colonization of Surtsey, with special reference to the blue-green algae. Oikos 24: 239—243.

Brotherson, J. D., and W. J. Masslich. 1985. Vegetation patterns in relation to slope position in the Castle Cliffs area of southern Utah. Great Basin Nat. 45: 535—541.

Brotherson, J. D., and S. R. Rushforth. 1983. Influence of cryptogamic crusts on moisture relationships of soils in Navajo National Monument, Arizona. Great Basin Nat. 43: 73—78.

Brotherson, J. D., S. R. Rushforth, and J. R. Johansen. 1983. Effects of long-term grazing on cryptogam crust cover in Navajo National Monument, Arizona. J. Range Manage. 36: 579—581.

Brown, R. T. 1967. Influence of naturally occurring compounds on germination and growth of jack pine. Ecol. 48: 542—546.

Brown, R. T., and P. Mikola. 1974. The influence of fruticose soil lichens upon the mycorrhizae and seedling growth of forest trees. Acta For. Fennica 141: 5—22.

Burkholder, P. R., and A. W. Evans. 1945. Further studies on the antibiotic activity of lichens. Bull. Torrey Bot. Club 72: 157—164.

Burkholder, P. R., A. W. Evans, I. McVeigh, and H. K. Thornton. 1944. Antibiotic activity of lichens. Proc. Nat. Acad. Sci. 30: 250—255.

Burns, S. J. 1983. Nitrogen fixation and transformations in cryptogamic soil crusts as affected by disturbance. Master's thesis, Brigham Young University, Provo, Utah.

Buzer, J. S., R. A. Dohmeier, and D. R. DuToit. 1985. The survival of algae in dry soils exposed to high temperatures for extended time periods. Phycol. 24: 249—251.

Callison, J., J. D. Brotherson, and J. E. Bowns. 1985. The effects of fire on the blackbrush (*Coleogyne ramosissima*) community of southwestern Utah. J. Range Manage. 38: 535—538.

Cameron, R. E. 1969. Abundance of microflora in soils of desert regions. Nat. Aeronautics and Space Admin., Jet Propulsion Lab., Tech. Report No. 32-1378. California Inst. of Tech., Pasadena.

Cameron, R. E. 1972. A comparison of soil microbial ecosystems in hot, cold, and polar desert regions. p. 185—192. *In*: L. E. Rodin (ed.) Eco-physiological foundation of ecosystems productivity in arid zone. Publishing House "Nauk", Leningrad, USSR.

Cameron, R. E., and G. B. Blank. 1966. Desert algae: soil crusts and diaphanous substrata as algal habitats. Nat. Aeronautics and Space Admin., Jet Propulsion Lab., Tech. Report No. 32-971. California Inst. of Tech., Pasadena.

Carson, J. L., and R. M. Brown, Jr. 1978. Studies of Hawaiian freshwater and soil algae. II. Algal colonization and succession on a dated volcanic substrate. J. Phycol. 14: 171—178.

Cooke, W. B. 1955. Fungi, lichens and mosses in relation to vascular plant communities in eastern Washington and adjacent Idaho. Ecol. Monogr. 25: 119—180.

Cowles, S. 1982. Preliminary results investigating the effect of lichen ground cover on the growth of black spruce. Naturaliste Can. (Rev. Ecol. Syst.) 109: 573—581.

Culbertson, C. F. 1969. Chemical and botanical guide to lichen products. Univ. of N. Carolina Press, Chapel Hill.

Dadhich, K. S., A. K. Varma, and G. S. Venkataraman. 1969. The effect of *Calothrix* inoculation on vegetable crops. Plant and Soil. 31: 377—379.

Daubenmire, R. 1970. Steppe vegetation of Washington. Agric. Exp. Stn. Tech. Bull. 62.

Dean, K. C., R. Havens, K. T. Harper, and J. B. Rosenbaum. 1973. Vegetative stabilization of mill mineral wastes. p. 119—136. *In*: R. J. Hutnik, and G. Davis (eds.) Ecology and reclamation of devastated land, vol. 2. Gordon and Breach, Science Publ. Ltd., London.

Delwiche, C. C., and J. Wijler. 1956. Nonsymbiotic nitrogen fixation in soil. Plant and Soil 7: 113—129.

Durrell, L. W., and L. M. Shields. 1960. Fungi isolated in culture from soils of the Nevada test site. Mycologia 52: 636—641.

Durrell, L. W., and L. M. Shields. 1961. Characteristics of soil algae relating to crust formation. Trans. Amer. Microsc. Soc. 80: 73—79.

Eckert, R. E., Jr., F. F. Peterson, and J. T. Belton. 1986a. Relation between ecological-range condition and proportion of soil-surface types. J. Range Manage. 39: 409—414.

Eckert, R. E., Jr., F. F. Peterson, M. S. Meurisse, and J. L. Stephens. 1986b. Effects of soil-surface morphology on emergence and survival of seedlings in big sagebrush communities. J. Range Manage. 39: 414—420.

Evenari, M. 1981. Synthesis. p. 555—591. *In*: D. W. Goodall, R. A. Perry, and K. M. W. Howes (eds.) Arid-land ecosystems: structure, functioning and management. Cambridge Univ. Press, Cambridge, U.K.

Faurel, L., P. Ozenda, and G. Schotter. 1953. Les lichens du Sahara Algerien. Res. Counc. Israel Spec. Publ. 2: 310—317.

Fields, R. D., and L. L. St. Clair. 1984. A comparison of methods for evaluating SO_2 impact on selected lichen species: *Parmelia chlorochroa, Collema polycarpon* and *Lecanora muralis*. Bryol. 87: 297—301.

Fletcher, J. E., and W. P. Martin. 1948. Some effects of algae and moulds in the rain crust of desert soils. Ecol. 29: 95—100.

Flint, E. A. 1958. Biological studies of some tussock-grassland soils. IX. Algae: preliminary observations. New Zealand J. of Agric. Res. 1: 991—997.

Fogg, G. E., and W. D. P. Stewart. 1968. In situ determinations of biological nitrogen fixation in Antarctica. Brit. Antarctic Survey Bull. 15: 39—46.

Follmann, G., and J. Redon. 1972. Ergaenzungen zur flechtenflora der nordchilenischen nebeloasen fray jorge und talinay. Willdenowia 6: 431—459.

Forest, H. S., and C. R. Weston. 1966. Blue-green algae from the Atacama Desert of Northern Chile. J. Phycol. 2: 163—164.

Friedmann, E. I., and M. Galun. 1974. Desert algae, lichens, and fungi. p. 165—212. *In*: G. W. Brown, Jr. (ed.) Desert biology. Academic Press, New York.

Fritsch, F. E. 1922. The terrestrial algae. J. of Ecol. 10:220—236.

Fuller, W. H., R. E. Cameron, and N. Raica. 1960. Fixation of nitrogen in desert soils by algae. Working Papers of Seventh Congr. Intern. Soc. of Soil Sci. for 19 August. Madison, Wisconsin.

Galun, M. 1963. Autecological and synecological observations on lichens of the Negev, Israel. Israel J. of Bot. 12: 179—187.

Gayel, A. G., and E. A. Shtina. 1974. Algae on the sands of arid regions and their role in soil formation. Soviet Soil Sci. 6(3): 311—319.

Hale, M. E., Jr. 1974. The biology of lichens, 2nd ed. American Elsevier Publ. Co., Inc., New York.

Hale, M. E., Jr. 1979. How to know the lichens, 2nd ed. W. C. Brown Co. Publ., Dubuque, Iowa.

Harper, K. T., and G. Holmstead. *In review.* Effects of cryptogamic plants on some aspects of the water relations of Utah desert soils.

Harper, K. T., and L. L. St. Clair. 1985. Cryptogamic soil crusts on arid and semiarid rangelands in Utah: effects on seedling establishment and soil stability. Final report Bureau of Land Management, Utah State Office, Salt Lake City.

Harris, E., R. N. Mack, and M. S. B. Ku. 1987. Death of steppe cryptogams under the ash from Mount St. Helens. Amer. J. Bot. 74: 1249—1253.

Henriksson, E., and L. C. Pearson. 1981. Nitrogen fixation rate and chlorophyll content of the lichen *Peltigera canina* exposed to sulfur dioxide. Amer. J. Bot. 68: 680—684.

Hunt, C. B., and L. W. Durrell. 1966. Distribution of fungi and algae. U. S. Geol. Surv., Prof. Pap. 509: 55—66.

Huss-Danell, K. 1978. Seasonal variation in the capacity for nitrogenase activity in the lichen *Stereocaulon paschale.* New Phytol. 81: 89—98.

Johansen, J. R., A. Javakul, and S. R. Rushforth. 1982. Effects of burning on the algal communities of a high desert soil near Wallsburg, Utah. J. Range Manage. 35: 598—600.

Johansen, J. R., and S. R. Rushforth. 1985. Cryptogamic soil crusts: seasonal variation in algal populations in the Tintic Mountains, Juab County, Utah. Great Basin Nat. 45: 14—21.

Johansen, J. R., S. R. Rushforth, and J. D. Brotherson. 1981. Subaerial algae of Navajo National Monument, Arizona. Great Basin Nat. 41: 433—439.

Johansen, J. R., and L. L. St. Clair. 1986. Cryptogamic soil crusts: recovery from grazing near Camp Floyd State Park, Utah. Great Basin Nat. 46: 632—640.

Johansen, J. R., L. L. St. Clair, B. L. Webb, and G. T. Nebeker. 1984. Recovery patterns of cryptogamic soil crusts in desert rangelands following fire disturbance. Bryol. 87: 238—243.

Jones, J. 1930. An investigation into the bacterial associations of some Cyanobacteria, with special reference to their nitrogen supply. Ann. Bot. 44: 721—740.

Kappen, L., O. L. Lange, E. D. Schulze, M. Evenari, and U. Buschbom. 1979. Ecophysiological investigations on lichens of the Negev Desert. VI. Annual course of the photosynthetic production of *Ramalina maciformis* (Del.) Bory. Flora 168: 85—108.

Kershaw, K. A. 1972. The relationship between moisture content and net assimilation rate of lichen thalli and its ecological significance. Can. J. Bot. 50: 543—555.

Kershaw, K. A. 1974. Dependence of the level of nitrogenase activity on the water content of the thallus in *Peltigera canina, P. evansiana, P. polydactyla*, and *P. praetextata.* Can. J. Bot. 52: 1423—1427.

King, J. M., and C. H. Ward. 1977. Distribution of edaphic algae as related to land usage. Phycol. 16: 23—30.

Kleiner, E. F., and K. T. Harper. 1972. Environment and community organization in the grasslands of Canyonlands National Park. Ecol. 53: 229—309.

Kleiner, E. F., and K. T. Harper. 1977a. Soil properties in relation to cryptogamic groundcover in Canyonlands National Park. J. Range Manage. 30: 202—205.

Kleiner, E. F., and K. T. Harper. 1977b. Occurrence of four major perennial grasses in relation to edaphic factors in a pristine community. J. Range Manage. 30: 286—289.

Klubek, B., and J. Skujins. 1980. Heterotrophic nitrogen fixation in arid soil crusts. Soil Biol. Biochem. 12: 229—236.

Lange, O. L. 1980. Moisture content and CO_2 exchange of lichens. I. Influence of temperature on moisture-dependent net photosynthesis and dark respiration in *Ramalina maciformis.* Oecologia 45: 82—87.

Lewin, R. A. 1977. The use of algae as soil conditioners. CIBCASIO Trans. 3: 33—35. (Publ. of Centros de Investigacion de Baja California, Scripps Inst. f Oceanography, Lajolla, Calif.)

Lipman, C. B. 1941. The successful revival of *Nostoc commune* from an herbarium specimen eighty-seven years old. Bull. Torrey Bot. Club 68: 664—666.

Looman, J. 1964. The distribution of some lichen communities of the prairie provinces and adjacent parts of the Great Plains. Bryol. 67: 209—224.

Loope, W. L., and G. F. Gifford. 1972. Influence of a soil microfloral crust on selected properties of soils under pinyon-juniper in southeastern Utah. J. Soil and Water Conserv. 27: 164—167.

MacFarlane, J. D., and K. A. Kershaw. 1978. Thermal sensitivity in lichens. Sci. 201: 739—740.

MacGregor, A. N. 1972. Gaseous losses of nitrogen from freshly wetted desert soils. Soil Sci. Soc. Amer. Proc. 36: 594—596.

MacGregor, A. N., and D. E. Johnson. 1971. Capacity of desert algal crusts to fix atmospheric nitrogen. Soil Sci. Soc. Amer. Proc. 35: 843—844.

Mack, R. N., and J. N. Thompson. 1982. Evolution in steppe with few large, hooved mammals. Amer. Nat. 119: 757—773.

MacKenzie, H. J., and H. W. Pearson. 1979. Preliminary studies on the potential use of algae in the stabilization of sand wastes and wind blow situations. British Phycol. J. 14: 126.

Marble, J. R., and K. T. Harper (*In Press*). 1988. The effect of timing of grazing on soil-surface cryptogamic communities in a Great Basin low-shrub desert: a preliminary report. Great Basin Nat.

Marsh, J. E., and T. H. Nash, III. 1979. Lichens in relation to the Four Corners Power Plant in New Mexico. Bryol. 82: 20—28.

Martin, J. P., and S. A. Waksman. 1940. Influence of microorganisms on soil aggregation and erosion. Soil Sci. 50: 29—47.

Mayland, H. F., and T. H. McIntosh. 1966. Availability of biologically fixed atmosphere nitrogen-15 to higher plants. Nature 209: 421—422.

Mayland, H. F., T. H. McIntosh, and W. H. Fuller. 1966. Fixation of isotopic nitrogen in a semi-arid soil by algal crust organisms. Soil Sci. Amer. Proc. 30: 56—60.

McKnight, K. B. 1980. Factors influencing size and hyphal pigmentation of soil microfungal populations: a study from gypsiferous soils of a Utah desert. Master's thesis, Brigham Young University, Provo, Utah.

Meyer, S. E. 1986. The ecology of gypsophile endemism in the eastern Mojave Desert. Ecol. 67: 1303—1313.

Millbank, J. W. 1972. Nitrogen metabolism in lichens. IV. The nitrogenase activity of the *Nostoc* phycobiont in *Peltigera canina*. New Phytol. 71: 1—10.

Millbank, J. W. 1976. Aspects of nitrogen metabolism in lichens. p. 441—555. *In*: D. H. Brown, D. L. Hawksworth, and R. H. Bailey (eds.) Lichenology: process and problems. Academic Press, London.

Miller, E. V., R. Greene, A. S. Cancilla, and C. Curry. 1963. Antimetabolites in lichens. A preliminary report. Penn. Acad of Sci. 37: 104—108.

Mooney, H. A., O. Bjorkman, and J. Berry. 1975. Photosynthetic adaptation to high temperatures. p. 138—151. *In*: N. F. Hadley (ed.) Environmental physiology of desert organisms. Dowden, Hutchinson and Ross, Stroudsberg, Penn.

Nash, T. H., III. 1974. Lichens of the Page environs as potential indicators of air pollution. J. Ariz. Acad. Sci. 9: 97—101.

Nash, T. H., III. 1975. Influence of effluent from a zinc factory on lichens. Ecol. Mongr. 45: 183—198.

Nash, T. H., III. 1976a. Lichens as indicators of air pollution. Naturwissenschaften 63: 364—367.

Nash, T. H., III. 1976b. Sensitivity of lichens to nitrogen dioxide fumigations. Bryol. 79: 103—106.

Nash, T. H., III, O. L. Lange, and L. Kappen. 1982a. Photosynthetic patterns of Sonoran desert lichens. II. A multivariate laboratory analysis. Flora 172: 419—426.

Nash, T. H., III, and T. J. Moser. 1982. Vegetational and physiological patterns of lichens in North American deserts. J. Hattori Bot. Lab. No. 53: 331—336.

Nash, T. H., III, T. J. Moser, C. C. Bertke, S. O. Link, L. L. Sigal, S. L. White, and C. A. Fox. 1982b. Photosynthetic patterns of Sonoran desert lichens. I. Environmental considerations and preliminary field measurements. Flora 172: 335—345.

Nash, T. H., III, T. J. Moser, S. D. Link, L. J. Ross, A. Olafsen, and U. Matthes. 1983. Lichen photosynthesis in relation to CO_2 concentration. Oecologia 58: 52—56.

Nash, T. H., III, and E. A. Nash. 1974. Sensitivity of mosses to sulfur dioxide. Oecologia 17: 257—263.

Nash, T. H., III, and L. L. Sigal. 1979. Gross photosynthetic response of lichens to short-term ozone fumigations. Bryol. 82: 280—285.

Nash, T. H., III, and M. R. Sommerfeld. 1981. Elemental concentrations in lichens in the area of the Four Corners Power Plant, New Mexico. Environ. and Experi. Bot. 21: 153—162.

Nash, T. H., III, S. L. White, and J. E. Marsh. 1977. Lichen and moss distribution and biomass in hot desert ecosystems. Bryol. 80: 470—479.

Nicot, J. 1960. Some characteristics of the microflora of desert soils. p. 94—97. *In*: D. Parkinson and J. S. Ward (eds.) International Symposium on the ecology of soil fungi. Liverpool Univ. Press, Liverpool, U.K.

Novichkova-Ivanova, L. N. 1972. Soil algae of middle Asia deserts. p. 180—182. *In*: L. E. Rodin (ed.) Eco-physiological foundation of ecosystems productivity in arid zone, Publishing House "Nauka", Leningrad, USSR.

Pankratova, Ye. M., and A. S. Vakhrushev. 1971. Field determination of the fixation of atmospheric nitrogen by blue-green algae using N^{15}. Soviet Soil Sic. 3: 726—733.

Pearson, L. C., and G. A. Rodgers. 1982. Air pollution damage to cell membranes in lichens. III. Field experiments. Phyton 22: 329—337.

Pegau, R. E. 1970. Effect of reindeer trampling and grazing on lichens. J. Range Manage. 23: 95—97.

Platou, K. A., and P. T. Tueller. 1985. Evolutionary implications for grazing management systems. Rangelands 7: 57—61.

Poulton, C. E. 1955. Ecology of non-forested vegetation in Umatilla and Morrow Counties, Oregon. Ph.D. diss., Washington State Univ., Pullman.

Pyatt, F. B. 1967. The inhibitory influence of *Peltigera canina* on the germination of graminaceous seeds and the subsequent growth of the seedlings. Bryol. 70: 326—329.

Reddy, G. B., and J. Giddens. 1975. Nitrogen fixation by algae in fescuegrass soil crusts. Soil Sci. Soc. Amer. Proc. 39: 654—656.

Reichert, I. 1936. Steppe and desert in the light of lichen vegetation. Proc. of the Linnean Soc. of London 149: 19—23.

Rice, E. L. 1964. Inhibition of nitrogen-fixing and nitrifying bacteria by seed plants. Ecol. 45: 824—837.

Rogers, R. W. 1977. Lichens of hot arid and semi-arid lands. p. 211—252. *In*: M. R. D. Seaward (ed.) Lichen ecology. Academic Press, New York.

Rogers, R. W., and R. T. Lange. 1971. Lichen populations on arid soil crusts around sheep watering places in South Australia. Oikos 22: 93—100.

Rogers, R. W., and R. T. Lange. 1966. Nitrogen fixation by lichens of arid soil crusts. Nature 209: 96—97.
Rogers, R. W., and R. T. Lange. 1971. Lichen populations on arid soil crusts around sheep watering places in South Australia. Oikos 22: 93—100.
Rogers, R. W., and R. T. Lange. 1972. Soil surface lichens in arid and subarid south-eastern Australia. I. Introduction and floristics. Aust. J. Bot. 20: 197—213.
Ross, L. J., and T. H. Nash III. 1983. Effect of ozone on gross photosynthesis of lichens. Environ. and Experi. Bot. 23: 71—77.
Rundel, P. W. 1978a. Ecological relationships of desert fog zone lichens. Bryol. 81: 277—293.
Rundel, P. W. 1978b. The ecological role of secondary lichen substances. Biochem. Syst. and Ecol. 6: 157—170.
Rychert, R. C., and J. Skujins. 1974. Nitrogen fixation by blue-green algae-lichen crusts in the Great Basin desert. Soil Sci. Soc. Amer. Proc. 38: 768—771.
Rychert, R., J. Skujins, D. Sorensen, and D. Porcella. 1978. Nitrogen fixation by lichens and free-living microorganisms in deserts. p. 20—30. *In*: N. E. West and J. Skujins (eds.) Nitrogen in desert ecosystems. Dowden, Hutchinson & Ross, Inc., Stroudsburg, Penn.
Savory, A., and S. D. Parsons. 1980. The Savory grazing method. Rangelands 2: 234—237.
Schlatterer, E. F., and E. W. Tisdale. 1969. Effects of litter of *Artemisia, Chrysothamnus* and *Tortula* on germination and growth of three perennial grasses. Ecol. 50: 869—873.
Schulten, J. A. 1985. Soil aggregation by cryptogams of a sand prairie. Amer. J. Bot. 72: 1657—1661.
Scott, G. D. 1956. Further investigation of some lichens for fixation of nitrogen. New Phytol. 55: 111—116.
Scheridan, R. P. 1979. Impact of emissions from coal-fired electrical generating facilities on N_2-fixing lichens. Bryol. 82: 54—58.
Shields, L. M. 1957. Algal and lichen floras in relation to nitrogen content of certain volcanic and arid range soils. Ecol. 38: 661—663.
Shields, L. M., and F. Drouet. 1962. Distribution of terrestrial algae within the Nevada Test Site. Amer. J. Bot. 49: 547—554.
Shields, L. M., and L. W. Durrell. 1964. Algae in relation to soil fertility. Bot. Rev. 30: 92—128.
Shields, L. M., C. Mitchell, and F. Drouet. 1957. Alga- and lichen stabilized surface crusts as nitrogen sources. Amer. J. Bot. 44: 489—498.
Shubert, L. E., and T. L. Starks. 1979. Algal succession on orphaned coal mine spoils. p. 661—669. *In*: M. K. Wali (ed.) Ecology and coal development, Pergamon Press, New York.
Shubert, L. E., and T. L. Starks. 1980. Soil-algal relationships from surface mined soils. British Phycol. J. 15: 417—428.
Sinclair, J. G. 1922. Temperatures of the soil and air in a desert. Monthly Weath. Rev. 49: 142—144.
Singh, R. N. 1950. Reclamation of "Usar" lands in India through blue-green algae. Nature 165: 325—326.
Skujins, J., and B. Klubek. 1978. Nitrogen fixation and denitrification in arid soil cryptogamic crust microenvironments. p. 543—552. *In*: Environmental biogeochemistry and geomicrobiology, Vol. 2. Ann Arbor Publ., Ann Arbor, Mich.
Snyder, J. M., and L. H. Wullstein. 1973. The role of desert cryptogams in nitrogen fixation. Amer. Midl. Nat. 90: 257—265.
Starks, T. L., and L. E. Shubert. 1982. Colonization and succession of algae and soil-algal interactions associated with disturbed areas. J. Phycol. 18: 99—107.

States, J. S. 1978. Soil fungi of cool-desert plant communities in northern Arizona and southern Utah. Arizona-Nevada Acad. Sci. 13: 13—17.

St. Clair, L. L., J. E. Johnsen, and B. L. Webb. 1986. Rapid stabilization of fire-disturbed sites using a soil crust slurry: Inoculation studies. Recl. and Reveg. Res. 4: 261—269.

St. Clair, L. L., B. L. Webb, J. R. Johansen, and G. T. Nebeker. 1984. Cryptogamic soil crusts: Enhancement of seedling establishment in disturbed and undisturbed areas. Reclam. and Reveg. Res. 3: 129—136.

Stewart, W. D. P. 1967. Transfer of biologically fixed nitrogen in a sand dune slack region. Nature 214: 603—604.

Stewart, W. D. P., and M. Lex. 1970. Nitrogenase activity in the blue-green alga *Plectronema boryanum* strain 594. Archiv. Mikrobiol. 73: 250—260.

Steyn, P. L., and C. C. Delwiche. 1970. Nitrogen fixation by nonsymbiotic microorganisms in some California soils. Environ. Sci. and Tech. 4: 1122—1128.

Stokes, J. L. 1940. The influence of environmental factors upon the development of algae and other microorganisms in soil. Soil Sci. 49: 171—184.

Stokes, J. L. 1941. The relation of algae to the nitrogen economy of the soil. Chronica Bot. 6: 202—203.

Sutton, J. C., and B. R. Sheppard. 1976. Aggregation of sand-dune soil by endomycorrhizal fungi. Can. J. of Bot. 54: 326—333.

Terry, R. E., and S. J. Burns. 1987. Nitrogen fixation in cryptogamic soil crusts as affected by disturbance. p. 369—372. *In*: Proceedings — pinyon-juniper conference. USDA For. Serv. Gen. Tech. Report INT-215.

Trainor, F. R. 1982. Survival of algae in soil after high temperature treatment. Phycol. 22: 201—212.

Treshow, M. 1970. Environment and plant response. McGraw-Hill Book Co., New York.

Vollmer, A. T., F. Au, and S. A. Bamberg. 1977. Observations on distribution of microorganisms in desert soil. Great Basin Nat. 37: 81—86.

Walp, L., and R. Schopbach. 1942. Influence of pH on the proliferation of *Nostoc muscorum*. Growth 6: 33—37.

Watanabe, A., A. S. Nishigake, and C. Konishi. 1951. Effect on nitrogen-fixing blue-green algae on growth of rice plants. Nature 168: 748—749.

Watson, A. 1979. Gypsum crusts in deserts. J. Arid Environ. 2: 3—20.

Weier, T. E., C. R. Stocking, and M. G. Barbour. 1974. Botany — an introduction to plant biology. 5th ed. John Wiley & Sons, New York.

Went, F. W., and N. Stark. 1968. The biological and mechanical role of soil fungi. Proc. Nat. Acad. Sci. 60: 497—504.

West, N. E. 1981. Nutrient cycling in desert ecosystems. p. 301—324. *In*: D. W. Goodall, R. A. Perry, and K. M. W. Howes (eds.) Arid-land ecosystems: Structure, functioning and management. Cambridge Univ. Press, Cambridge, U.K.

West, N. E., and J. Skujins. 1977. The nitrogen cycle in North American cold-winter semi-desert ecosystems. Oecol. Plant. 12: 45—53.

Worley, I. A. 1973. The "black crust" phenomenon in upper Glacier Bay, Alaska. Northwest Sci. 47: 20—29.

8. Seedbeds as selective factors in the species composition of rangeland communities

JAMES A. YOUNG

8.1. Abstract

On the uncultivated rangelands, seeds, as the reproductive reserves of the plant community, often occur as a thin, discontinuous layer on the soil suface. The distribution of this layer is mandated by local microtopography and the locations of plant litter deposition such as subcanopy litter accumulations. Despite the fact that range managers rarely cultivate, irrigate, or fertilize this surface layer containing seeds, it is subject to manipulation through burning, trampling, and by controlling the rate of litter deposition through grazing management. The principles discussed in this chapter apply to rangelands in general, but most of the examples are taken from *Artemisia*/bunchgrass rangelands of western North America.

8.2. Importance of seeds in temperate desert eco-systems

For many of the plant communities in the *Artemisia*/bunchgrass rangelands of western North America, recruitment of new plants to the community is dependent on seed production, germination, and seedling establishment. Most of the woody species of *Artemisia* that are the overstory dominants of millions of hectares of rangelands do not regenerate from crown or root buds following burning of the aerial portion of the plant in wildfires. Burning in wildfires is the primary form of stand renewal for most *Artemisia*/bunchgrass plant communities (Blaisdell 1953, Young and Evans 1978).

There are plant communities in the complex of *Artemisia*/grass vegetation where rhizomatous species play an important role in the herbaceous portion of community structure, but they do not predominate (Billings 1951, Daubenmire 1970, Franklin and Dyrness 1973). The adventive flora of the *Artemisia*/grasslands is nearly totally dominated by annuals,

Tueller, P. T. (ed.), Vegetation science applications for rangeland analysis and management.
ISBN 90-6193-195-9.
© 1988, *Kluwer Academic Publishers, Dordrecht, Boston, London.*

placing the ultimate test of adaptation on the seed production, germination, and establishment sequence (Young *et al.* 1972).

8.3. Concept of safe sites for germination

The plant communities of the *Artemisia*/grasslands of western North America may well reproduce with exquisite precision in the minds of ecologists, but most of the plants that constitute the structure of the communities regenerate through seed production, germination, and seedling establishment. The inherent potential of the seeds of the component species of *Artemisia* dominated plant communities to germinate is only a portion of the equation to determine the species composition of the communities. The remaining factors in the equations are furnished by the seedbeds and are inherent to the physical environment of the site, and biological interactions. The concept of seedbed potential was first proposed by J. L. Harper and his students (e.g., Harper *et al.* 1965).

8.4. Analogy between range & agronomic seedbeds

When a farmer undertakes to establish a stand of small seeded legumes, he goes through several steps in seedbed preparations. Mechanical tillage is employed to mechanically uproot competing vegetation and to bury the seeds of competing weeds so deep in the surface soil profile that they can not emerge. If the seedbank of weeds in the soil is sufficient to ensure excessive competition even after mechanical tillage, a selective pre and/or post emergence herbicide may be used. The rough seedbed left by mechanical tillage is worked with implements to break down the soil aggregates to ensure a firm, fine textured seedbed to give optimum contact between the seeds and soil particles in the seedbed. The goal of the farmer is an even stand across the field so the mechanical operations are aimed toward uniformity. In a final step in stand renewal the farmer uniformly meters out the seeds with a drill that opens a drill row in the seedbed, places the uniformly spaced seed and closes the row over the seed, repacking the surface soil. The seed may be pretreated with a protective coating of pesticides, fertilizer, and/or inoculum of beneficial symbiotic bacteria. The post planting care of the seeds and resulting seedlings may include irrigation and application of selective pesticides.

8.5. Biology & physical structure of *Artemisia*/bunchgrass seedbeds

In the case of *Artemisia*/bunchgrass plant communities, natural stand regeneration is goverened by the same physical and biological laws that are manipulated in the farmer's field, but the mode of action of the interacting players in this biophysical symphony are quite different.

There is no mechanical or herbicidal reduction of competition in the natural seedbed. There is no dispersal of seeds in the soil profile other than that accomplished through the actions of soil macro-organisms and surface disturbance by animals or physical processes of the soil such as cracking. Seeds are largely distributed across the surface of the seedbed to gather in windrows about some micro obstruction or depressions in the microtopography.

8.5.1. Physical site

The biological and physical structure of the seedbeds of *Artemisia*/ bunchgrass communities are far from uniform. The more arid the site the greater the tendency for deflation of interspaces and mound building beneath shrub canopies. This process may be accentuated by grazing which reduces the herbaceous species established between the shrubs (Wood *et al.* 1978 and 1982). Virtually the entire litter fall and subsequent cycling of nutrients takes place beneath the shrub canopies (Charley and West 1975). This cycling, of especially nitrogen, is fueled from the litter dropped by the shrubs, the accumulation of litter from between shrubs which is deposited under the canopies by wind action and by the favorable micro-environment provided by the shrub canopy. The subcanopy micro-environment favors decomposition by micro organisms as well as the growth of mosses which provide a moisture receptive and evaporation protective surface, and lichens that may fix atmospheric nitrogen, respectively. The bottoms of the interspaces between shrubs may become mini-playas with vesicular soil crusts and polygonal cracking. The end product of such mound-interspace evolution is a site with two soils distinct in structure, organic matter and nutrient content in the surface horizons. This polarizaiton of the site produces a gradient in seedbed from the shrub stem to the interspace. Close to the stem the accumulation of *Artemisia* leaves and associated litter is sufficiently deep (2.5 to 5.0 cm) to keep seeds from contact with the mineral soil. Around the margins of the shrub canopy, the tops of soil prisms tend to be exposed in minature badlands. The mini canyons of the microstructure have crumbling walls or leafy crusts of lichens that provide a great opportunity for natural seed burial. In the center of the interspaces, the only opportunity for soil burial may be the polygonal cracks. The surface of the polygons is a hard, bare, biological wasteland.

8.5.2. Biological site

Accumulations of herbaceous litter on the surface of seedbeds can drastically alter the microenvironment (Evans and Young 1970). Extremes in high and low temperatures are moderated and relative humidity is increased beneath the litter. Litter accumulates seeds that otherwise could not find purchase on the smooth surface of interspace soils and stops their dispersal flight. The accumulation of *Artemisia* leaves beneath shrub canopy forms an overlapping, indurate layer that virtually deflects seeds as opposed to herbaceous litter that is a porous media to receive seeds.

8.6. Germination characteristics

The surface microtopography of seedbeds also contributes to the potential of the seedbed to support germination of seeds of various species (Evans and Young 1972). The prismatic-badlands around shrub mounds, as mentioned, provide opportunities for the self burial of seeds. Artificial structures such as hoof prints are dependent on the clay content of the surface soil and moisture content when the imprint was made for their duration. A great deal of trampling of dry soils may not be equal to a few animals walking across the site when deep tracks are left in a moist soil. Surface microtopography of seedbeds is obviously greatly influenced by the amount of natural surface stone cover. Rocks not only cast shadows and break up surface wind patterns, they also absorb and retain heat.

Soils such as those derived from decomposing granite, where surface particle size distribution tends to lead to wind sorting, can lead to the burial of small seeds despite their inhospitable appearance (unpublished research ARS-Reno, NV). Such soils can provide burial for a very specific size and density of seeds.

8.6.1. Artemisia tridentata

Despite the aspect characterization of *Artemisia tridentata*, little is known of the seed and seedbed ecology of this species. Flowering occurs in late September when the soil profiles are often virtually completely free of moisture normally considered available for plant growth. Anthesis, fertilization, and the subsequent development of the fertilized egg occurs with diurnal fluctuations in temperature from −15°C to 25°C (unpublished research, ARS-Reno). Seeds usualy begin to dehisce from the inflorescences by December 1st. By the time the seeds of the *A. tridentata* subsp. *vaseyana* begin to dehisce the canopies are often partially covered with snow.

The seed production per plant by *A. tridentata* is highly variable. Dense, maturing stands may not flower on drier than normal seasons. Open grown plants invading an area of disturbance may produce several thousand seeds (achenes) the second year after establishment (unpublished research, ARS-Reno). Plants of *A. tridentata* subsp. *tridentata* growing in the burrow pits along roads to a height of 1.5 m may produce several million viable achenes.

At the time of seedfall from *A. tridentata* plants, seed reserves in the litter and soil are virtually zero (unpublished research, ARS-Reno, NV). A tiny fraction of the previous year's crop of seeds are viable, but it is nearly impossible to detect these seeds through bioassay of samples from the seedbed. Seeds of most collections of *A. tridentata* will germinate as soon as they are collected, although afterripening requirements have been reported for some sources (McDonough and Harniss 1974). Usually there is a 6 to 8 week delay before seedbed temperatures are sufficiently warm to permit germination (Evans *et al.* 1970). The seeds of big sagebrush are relatively small (0.5 by 1.0 mm) and are dispersed to find sites in the surface of the seedbed that will support germination. These seeds are not going to germinate at the surface of a flat, bare seedbed unless relative humidity of the surrounding atmosphere is near saturation.

The *A. tridentata* seeds that have a chance to become established seedlings germinate in late winter. Years may pass without a successful seedling becoming established in a stand. A period of warm rain in February may condition the sudden germination of hundreds of *A. tridentata* seeds in a limited area. Barring such unusual moisture events, the chances of *A. tridentata* seedling establishment is directly related to the potential of the seedbed as defined by the number of sites with temperature and moisture relations capable of supporting germintion (Table 1).

Considering the mound interspace polarization of most big sagebrush sites one would assume the area with the greatest potential for germination would be on the mounds. This is not necessarily so because of the moisture competition posed by the established fruiting shrubs. Where studies have been made of the progeny of isolated *A. tridentata* plants the seedlings are distributed in an elliptical pattern reaching out from the shrub canopy in the lee of the prevailing winds. The dispersal pattern is

Table 1. Estimated percent germination and confidence interval for seeds of *Artemisia tridentata* subsp. *tridentata* grown in a garden at Medell Flat near Reno, Nevada.[1]

Cold Period °C—16 hrs	Warm period °C—8 hrs: 0	2	5	10	15	20	25	30	35	40
	%									
0	51(9)	53(7)	56(6)	59(5)	60(6)	59(6)	56(6)	51(6)	44(7)	35(10)
2		54(7)	57(6)	60(4)	62(5)	61(5)	59(5)	54(5)	48(5)	39(8)
5			56(6)	60(4)	63(4)	63(4)	61(4)	57(4)	52(4)	44(7)
10				37(6)	61(4)	62(4)	62(4)	59(4)	55(4)	48(6)
15					55(7)	58(5)	58(4)	57(4)	54(5)	48(7)
20						49(7)	51(5)	51(4)	49(5)	44(7)
25							39(7)	40(5)	39(5)	36(7)
30								26(7)	26(5)	24(7)
35									9(8)	8(8)
40										0(12)

[1] Confidence interval at the 0.01 level of probability. Means underlined are defined as optimum values. They are not lower than the maximum and its confidence interval. The estimates of mean germination are derived from a quadratic response surface developed by using multiple regression techniques (Evans *et al.* 1982). Data for this profile is unpublished material, ARS-Reno, Nevada (unpublished data, ARS-USDA, Reno).

usually oriented from the northeastern side of the shrub. The length of the colony is rarely more than a meter in length (Frischknecht and Bleak 1957).

The accumulation of hard leaf litter near the base of *A. tridentata* plant is not a desirable site for big sagebrush seed germination and seedling establishment. The terraced, prismatic margins of the mound provides ideal sites for seed germination, but competition from the fruiting big sagebrush plant may inhibit seedling establishment through moisture competition. When an established shrub dies it would seem logical that *A. tridentata* seedlings would become established on the mound because it would be the most productive site in the community. As an *A. tridentata* plant nears senescence, decay of the litter accumulation may accelerate as moisture competition is reduced. The decay of litter and subsequent enrichment in nitrogen through nitrification provide an ideal habitat for seedlings of the natural successional species of grass *Sitanion hystrix* or the alien species *Bromus tectorum*. Either of these species in sufficient density can inhibit the establishment of big sagebrush seedlings. Competition forces sagebrush seedlings to the interspaces. This sequence of events is rare in *A. tridentata* communities because stand renewal is usually catastrophic through burning in wildfires rather than individual shrubs reaching senescence. This sequence is the rule in *Artemisia nova* communities where there is insufficient herbaceous vegetation to support wildfires (unpublished research, ARS-Reno).

This discussion of the establishment of *A. tridentata* seedlings has direct application to the management of grazing and range improvement on sagebrush-grass rangelands. The economic feasibility of range improvement practices on sagebrush rangelands is often based on the effective life of the treatment. The effective life of the treatment is directly influenced by the reestablishment of seedlings of *A. tridentata*.

8.6.2. Chrysothamnus viscidiflorus

The subdominant shrubs in *A. tridentata* communities mainly sprout after having the aerial portion of the plants burned in wildfires. This sprouting can be from crown buds as in the case of *Chrysothamnus* sp. or from roots as in the case of *Tetradymia* sp. *Chrysothamnus viscidiflorus* illustrates this kind of sprouting stand renewal (Young and Evans 1974). About 20 percent of the *Chrysothamnus viscidiflorus* plants that exist in an *A. tridentata* dominated community will be killed when the community is burned in wildfires. The percent kill is highly variable depending on the fire conditions and 20 percent mortality would be near the maximum loss. The plants that sprout have the advantage of reduced competition and increased nutrients made available by the fire. The sprouts flower profusely the first year and produce large numbers of viable seeds. The seedling establishment of *C. viscidiflorus* the second year following burning of the community in a wildfire may fully occupy the site. Land managers often look at the results and attribute the shrub stand to sprouting after the fire. The sexual reproduction and seedling establishment are the critical points in this stand renewal by *C. viscidiflorus*. The small achenes of *C. viscidiflorus* have abundant pappus that aids in wind dispersal across burned plant communities. The halos of ash around burned big sagebrush stems are largely sterile the first season following burning and provide a receptive seedbed for the *C. viscidiflorus* seeds (Young and Evans 1978). There is considerable variability among the species, subspecies, and forms of *Chrysothamnus* in the colonizing ability of seedlings (Young *et al.* 1984). Despite the tremendous numbers of seeds produced by *C. viscidiflorus* plants, seed viability is often very low (Table 2). Many embryos apparently abort or otherwise fail to develop. The seeds are also subject to attack by insect larva that hatch from eggs laid in flowers.

Table 2. Estimated percent germination and confidence interval for seeds of *Chrysothamnus viscidiflorus*.[1]

Cold period °C—16 hrs	Warm period °C—8 hrs 0	2	5	10	15	20	25	30	35	40
	%									
0	23(11)	24(16)	37(21)	56(12)	74(8)	52(8)	37(5)	18(6)	7(16)	0(11)
2		37(11)	39(3)	69(10)	78(9)	64(11)	28(9)	18(11)	16(10)	6(11)
5			68(11)	74(6)	68(11)	73(9)	65(19)	51(13)	28(17)	11(10)
10				70(8)	74(9)	75(11)	57(14)	58(11)	33(16)	10(9)
15					57(10)	54(12)	29(11)	26(10)	17(11)	5(10)
20						34(6)	39(7)	21(14)	10(9)	2(7)
25							26(9)	21(5)	9(6)	5(8)
30								17(6)	10(5)	5(9)
35									0(10)	0(11)
40										0(16)

[1] Confidence interval at the 0.01 level of probability. Maeans underlined are defined as optimum values. They are not lower than the maximuma and its confidence interval (unpublished data, ARS-Reno, Nevada).

8.6.3. *Purshia tridentata*

A. tridentata is not a highly preferred browse species, although certain forms are more preferred than others. Most of the valuable browse species that occur in higher elevation *A. tridentata* communities are members of the rose family. *Purshia tridentata* is one of the most important browse species. The achenes of *P. tridentata* are borne in persistent flower parts. These flower parts contain a water soluble germination inhibitor (Hormay 1943). The occurrence of the germination inhibitor is not significant in the germination ecology of the species. The seeds are almost entirely collected by various species of heteromyid rodents. When the rodents collect the seeds they remove the papery covering, thus negating the inhibitor. The seed of *P. tridentata* still will not germinate. The coat of the achene apparently interferes with oxygen transfer to the embryo (Nord 1965). This inhibitor in the coat is overcome by cool-moist stratification (Young and Evans 1976). Rodents cache the seeds in groups of 15 to 30 seeds 2.5 cm deep in the soil. The optimum temperature for stratification ranges from −2 through 5°C (stratification at 2°C is recommended by proposed rules of the Association of Official Seed Analysts) (Young and Evans 1976). This range of temperatures is characteristic of conditions found under continuous snow cover. *Purshia* seeds are sensitive to moisture stress during the stratification period. If the seeds are placed under moisture stress during stratification, the duration of stratification can be greatly extended. Under optimum conditions the cool-moist stratification can be satisfied in as little as 2 weeks (Table 3).

The seeds of *Purshia* are subject to predators while they are in the cache and after the seedlings emerge from the cache. *Purshia tridentata* is an example of a seed whose germination ecology is nearly totally interrelated with animal activities (Young and Evans 1981).

8.6.4. Perennial Folbs

At higher elevations there are a host of perennial forbs native to big sagebrush communities. The germination requirements of plants such as species of *Castilleja* are virtually unknown. The coarse forbs, *Balsamorhiza sagittata* and *Wyethia* sp. have been determined to have extremely long cool-moist stratification requirements (Table 4) (Young and Evans 1979). These coarse forbs tend to be more abundant at higher elevations where snow cover is more consistent. The herbaceous and shrub diversity of *A. tridentata* communities drops at lower elevations as available soil moisture is reduced, but the mechanism of this reduction may be through the failure to consistently satisfy cool-moist stratification requirements.

Table 3. Estimated percent germination and confidence interval for seeds of *Purshia tridentata*. Seeds stratified for 2 weeks at 5°C before incubation.[1]

Cold period °C—16 hrs	Warm period °C—8 hrs									
	0	2	5	10	15	20	25	30	35	40
	%									
0	0(20)	38(10)	43(8)	48(7)	51(7)	50(7)	46(7)	34(7)	30(8)	17(11)
2		40(10)	44(7)	50(5)	52(6)	52(6)	48(6)	51(5)	32(6)	19(10)
5			46(8)	51(5)	54(5)	53(5)	56(5)	43(5)	33(5)	20(8)
10				50(8)	53(6)	52(5)	48(5)	42(5)	36(6)	19(8)
15					47(4)	46(6)	43(6)	36(6)	27(6)	14(8)
20						36(8)	33(6)	26(5)	17(6)	4(9)
25							18(9)	12(7)	2(7)	11(10)
30								8(12)	17(17)	0(22)
35									0(20)	0(24)
40										0(29)

[1] Confidence interval at the 0.01 level of probability. Means underlined are defined as optimum values. They are not lower than the maximum and its confidence interval (Young and Evans 1981).

Table 4. Estimated percent germination and confidence interval for seeds of *Balsamorhiza sagittata*. Seeds cool-moist stratified at 5°C for 12 weeks.[1]

Cold period °C—16 hrs	Warm period °C—8 hrs 0	2	5	10	15	20	25	30	35	40
	%									
0	6(21)	2(20)	69(14)	68(13)	76(16)	72(21)	47(16)	28(17)	0(14)	0(13)
2		36(19)	80(14)	67(17)	69(17)	43(19)	27(21)	30(17)	15(14)	0(27)
5			87(22)	86(20)	65(16)	57(18)	43(31)	50(29)	3(27)	0(18)
10				69(21)	34(13)	43(17)	35(16)	21(17)	3(21)	0(18)
15					36(17)	20(16)	8(14)	19(18)	0(4)	0(7)
20						27(19)	9(11)	6(17)	0(4)	0(3)
25							4(17)	3(18)	1(7)	0(10)
30								6(13)	0(11)	0(19)
35									0(18)	0(17)
40										0(14)

[1] Confidence interval at the 0.01 level of probability. Means underlined are defined as optimum values. They are not lower than the maximum and its confidence interval (Young and Evans 1979).

8.6.5. Perennial grasses

The dominant perennial grass species of the *Artemisia*/bunchgrass communities generally belong to either the Poaceae tribes Triticeae or Stipeae. This is not to discount the importance of the tribe Poaceae with *Festuca* and *Poa*, but for much of the moderate and lower elevation *A. tridentata*/bunchgrass rangeland the first grass mentioned in order of dominance will be *Elytrigia* or *Stipa*. *Elytrigia spicata*, which has the S genome of the Triticeae (Dewey 1983), characterizes vast areas of landscape from the interior of the Pacific Northwest down through eastern Oregon, northeastern California, Idaho, and northern Utah. This species is also found on the higher potential portions of Nevada in the Great Basin. Obviously such a widely distributed species contains variable ecotypes, but in general the seeds will germinate over a fairly wide range of temperatures (Table 5). Optimum germination of seeds of the cultivar Secar of *Elytrigia spicata* occurs at relatively warm temperatures compared to other species in the communities.

The members of the tribe Stipeae are represented in *Artemisia* communities from the margin of the salt desert to the mountain brushlands. The diverse species of *Stipa* are generally characterized by highly dormant seeds (e.g. see Rogler, 1960 for *S. viridula*). Seed of *S. speciosa* will

Table 5. Estimated percent germination and confidence interval for seeds of 'Secar' *Elytrigia spicata*.[1]

Cold period °C—16 hrs	Warm period °C—8 hrs									
	0	2	5	10	15	20	25	30	35	40
	%									
0	3(13)	9(11)	17(9)	27(8)	33(8)	36(9)	35(9)	30(8)	22(10)	16(14)
2		21(11)	29(8)	39(7)	46(7)	49(8)	48(7)	44(7)	36(8)	24(12)
5			44(9)	55(6)	62(6)	66(6)	65(6)	61(6)	54(6)	43(10)
10				73(10)	81(6)	85(6)	86(6)	83(6)	76(6)	66(9)
15					90(10)	95(7)	97(6)	94(6)	88(7)	78(10)
20						95(10)	97(7)	96(6)	90(7)	81(10)
25							88(10)	87(7)	82(7)	74(10)
30								68(10)	65(7)	57(10)
35									37(12)	30(12)
40										0(17)

[1] Confidence interval at the 0.01 level of probability. Means underlined are defined as optimum values. They are not lower than the maximum and its confidence interval.

germinate (Young and Evans 1980), but are not highly viable. *Stipa thurberiana* is the dominant perennial grass on vast areas of alluvial fans spreading out from the bases of desert mountain ranges. The seeds of this species are highly dormant (Table 6). Apparently the nature of the dormancy is related to light quality requirements, but can not be consistently overcome with near red light (unpublished research, ARS-Reno, NV).

The light requirement for germination of seeds of *S. thurberiana* is surprising because most of the caryopses of the *Stipa* species have geniculate awns that coil and uncoil in response to changing relative humidity. These awns apparently provide a system of self burial for the caryopses similar to those reported for species of *Erodium* (Young *et al.* 1975). Self burial enhances the moisture relations of seeds for germination, but limits the effectiveness of light quality as a germination stimulator.

Among the native herbaceous species there is a group of short lived perennial species such as *Poa sandbergii* and *Elymus elymoides*. Both of these species are highly variable, but generally produce seeds that will germinate at a relatively large number of temperatures (Young and Evans 1977, Evans *et al.* 1977).

The abundant anual species in *Artemisia*/bunchgrass communities are largely alien species which have been introduced during this century (Young *et al.* 1970). Much has been written about how these species have preempted environmental potential in these disturbed environments (Mack 1981, Young and Evans 1985). The first species to become established in highly disturbed areas in *Salsola australus* (equals *S. iberica*) (Evans and Young 1982). Seeds of *S. australus* mature late in the fall and have germination greatly restricted by temperature related afterripening requirements. At maturity these seeds will only germinate at

Table 6. Estimated percent germination and confidence interval for seeds of *Stipa thurberiana*.[1]

Cold period °C—16 hrs	Warm period °C—8 hrs 0	2	5	10	15	20	25	30	35	40
	%									
0	2(5)	3(4)	4(3)	5(3)	5(3)	5(3)	4(3)	2(3)	0(4)	0(5)
2		7(5)	8(3)	8(2)	8(3)	8(3)	6(3)	4(3)	1(3)	0(5)
5			13(3)	13(2)	12(2)	11(2)	9(2)	6(2)	3(2)	0(4)
10				21(3)	19(2)	17(2)	13(2)	10(2)	5(2)	0(3)
15					25(4)	21(3)	17(2)	12(2)	7(3)	0(4)
20						25(4)	20(3)	14(2)	7(3)	0(4)
25							22(4)	15(2)	7(3)	0(4)
30								15(4)	6(3)	0(4)
35									5(4)	0(4)
40										0(6)

[1] Confidence interval at the 0.01 level of probability. Means underlined are defined as optimum values. They are not lower than the maximum and its confidence interval.

Table 7. Estimated percent germination and confidence interval for seeds of *Sitanion hystrix*.[1]

Cold period °C—16 hrs	Warm period °C—8 hrs 0	2	5	10	15	20	25	30	35	40
	%									
0	0(18)	0(14)	0(12)	16(18)	35(14)	20(9)	21(11)	10(8)	2(10)	0(18)
2		42(14)	59(18)	67(17)	68(16)	71(11)	67(9)	55(11)	2(14)	0(15)
5			64	78(12)	81(9)	83(10)	79(11)	72(6)	21(14)	1(16)
10				87(5)	86(11)	90(12)	89(12)	83(13)	39(9)	10(11)
15					87(10)	88(14)	91(13)	82(8)	81(14)	46(14)
20						82(11)	81(11)	77(14)	52(11)	57(16)
25							72(13)	74(12)	40(9)	50(21)
30								47(18)	28(21)	29(14)
35									2(10)	2(11)
40										0(10)

[1] Confidence interval at the 0.01 level of probability. Means underlined are defined as optimum values. They are not lower than the maximum and its confidence interval (Young and Evans 1977).

Table 8. Estimated percent germination and confidence interval for seeds of *Poa sandbergii*.

Cold period °C—16 hrs	Warm period °C—8 hrs: 0	2	5	10	15	20	25	30	35	40
	%									
0	0(19)	0(18)	12(14)	36(18)	53(14)	58(8)	32(11)	0(8)	0(11)	0(8)
2		0(9)	44(11)	48(14)	57(8)	65(16)	30(8)	22(19)	0(8)	0(11)
5			74(14)	75(11)	88(18)	90(14)	84(8)	22(9)	0(8)	0(16)
10				95	96(12)	91(8)	96(12)	48(8)	10(11)	0(9)
15					82(9)	88(11)	79(10)	56(8)	12(10)	0(10)
20						65(9)	64(11)	17(12)	61(10)	0(12)
25							48(12)	8(8)	17(14)	0(14)
30								0(14)	0(18)	0(20)
35									0(11)	0(18)
40										0(19)

[1] Confidence interval at the 0.01 level of probability. Means underlined are defined as optimum values. They are not lower than the maximum and its confidence interval (Young and Evans 1977).

very restricted temperatures. Over the winter the afterripening requirements gradually break down and by spring the seeds will germinate at extreme diurnal fluctuations in temperature (Table 9). Seeds of Russian thistle can germinate and become established on bare seedbeds under proper environmental conditions due to their extremely rapid rate of germination. Establishment can occur during a moisture event lasting a few hours.

The next step in secondary succession on disturbed *Artemisia* sites is dominance by annual members of the mustard family. Several species in this group have mucilaginous seedcoats (Young *et al.* 1970). The mucilage improves the moisture relations of the seeds so they can germinate on the surface of bare seedbeds (Young and Evans 1973).

8.6.6. Bromus tectorum

The final stage in secondary succession on disturbed areas is dominance by *Bromus tectorum*. This annual grass provides abundant caryopses that require some form of coverage for germination to occur. Burial in depressed microtopography or litter or the caryopses themselves gathering in small windrows provides enough coverage for some seeds of *B. tectorum* to germinate. When initially produced, the seeds of *B. tectorum* lack dormancy (Table 10). Left in the seedbed in the field over the winter the seeds acquire a dormancy that breaks down slowly over time

Table 9. Estimated percent germination and confidence interval for seeds of *Salsola australia.*[1]

Cold period °C—16 hrs	Warm period °C—8 hrs 0	2	5	10	15	20	25	30	35	40
	%									
0	60(11)	66(9)	79(7)	83(7)	89(7)	91(7)	89(7)	82(7)	72(8)	58(12)
2		63(9)	71(7)	81(5)	88(6)	90(6)	88(6)	82(6)	73(7)	59(10)
5			65(8)	77(5)	84(5)	87(5)	86(5)	81(5)	72(5)	59(8)
10				66(8)	74(5)	79(5)	80(5)	76(5)	69(5)	57(9)
15					61(8)	67(6)	69(5)	67(5)	61(6)	51(8)
20						52(9)	55(6)	54(5)	50(6)	41(9)
25							37(8)	38(6)	34(6)	27(9)
30								17(8)	15(6)	9(8)
35									0(10)	0(10)
40										0(14)

[1] Confidence interval at the 0.01 level of probability. Means underlined are defined as optimum values. They are not lower than the maximum and its confidence interval.

Table 10. Estimated percent germination and confidence interval for seeds of *Bromus tectorum.*[1]

Cold period °C—16 hrs	Warm period °C—8 hrs 0	2	5	10	15	20	25	30	35	40
	%									
0	26(9)	35(8)	48(6)	65(6)	78(6)	87(6)	92(6)	92(6)	87(7)	79(10)
2		40(8)	52(6)	69(5)	82(5)	91(6)	95(5)	95(5)	91(6)	82(9)
5			57(7)	74(9)	87(4)	95(5)	99(5)	99(4)	94(5)	85(7)
10				79(7)	92(5)	100(4)	100(4)	100(4)	97(5)	87(7)
15					92(7)	99(5)	100(5)	100(5)	95(5)	85(7)
20						95(8)	98(5)	96(5)	90(5)	79(8)
25							89(7)	86(5)	80(5)	69(7)
30								73(7)	66(5)	54(7)
35									47(9)	36(9)
40										13(12)

[1] Confidence interval at the 0.01 level of probability. Means underlined are defined as optimum values. They are not lower than the maximum and its confidence interval.

(Young *et al.* 1969). The dormant seeds of *B. tectorum* respond to enrichment of the seedbed with nitrate (Evans and Young 1975).

8.7. Synthesis of community characteristics of seedbed ecology

Comparisons of the germination characteristics of the various groups of species offers some clues about the structure of *Artemisia*/bunchgrass communities (Table 11). The optimum temperature for the germination of seeds of the shrub species presented ranges from 10 to 0°C cold period temperature alternating with 5 to 30°C days. Roughly two thirds of the optimum temperatures for germination are in the self-stratifying range or at 5°C or below during the cold period, although only *Purshia tridentata* has a precise cool-moist stratification requirement for seed germination. None of the three shrubs have large banks of seed reserves in the soil. In the total absence of rodents and ants, it would be interesting to determine if naturally dispersed seeds of *P. tridentata* would establish.

The herbaceous perennial species for illustration the partitioning of habitat through germination requirements (Table 11). Seed of *Balsamorhiza sagittata* require such a long cool-moist stratification period that plants are restricted to the higher, more prolonged snow covered portion of the *Artemisia*/bunchgrass environment. Seeds of *Stipa thurberiana* with their high level of unresolved dormancy mask an important part of the seedbed ecology of the drier portion of the *Artemisia*/bunchgrass environment. The cultivar Secar of *Elytriga spicata* illustrates that high germination can be selected for in seeds of native perennial grasses, but seedling vigor in

Table 11. Summary of the characteristics of optimum regimes for germination for selected shrubs, long-lived perennial herbaceous species, short-lived perennial herbaceous species, and woody annuals that are prominent in the structure of *Artemisia*/bunchgrass plant communities.[1]

	Germination Parameters				
	Mean	Maximum	Optimal Regimes	Maximum Optimal Regime	Minimum Optimal Regime
	-----------------	%	-----------------	------------- °C—16/8 hr	-------------
Shrubs	63	64	22	10/30	0/10
Long lived herbaceous species	64	69	13	20/35	0/5
Short lived herbaceous species	91	94	15	10/30	5/10
Alien annuals	90	100	22	15/35	0/15

[1] Optimal regimes refers to these incubation temperatures that produced germination not significantly ($P = 0.01$) lower than the maximum germination observed. This is expressed as a percentage of the total number of regimes tested (55), ranging from 0 through 40°C.

the face of competition for moisture does not necessarily follow. Note that the optimum temperature regimes for the two long lived perennial grasses do not overlap with those for the native shrubs. Optima for the native perennial grasses are at higher temperature regimes.

It is little wonder the *Sitanion hystrix* and *Poa secunda* are the native secondary successional species in many *Artemisia* plant communities. The optima for these two short-lived grasses have high average and maximum germination at a wide range of temperatures (Table 11). Only two profiles were present for these species and one must remember that both taxons are extremely variable.

For a host of alien annuals only two profiles were present to illustrate the bottom and top of current secondary succession. Seeds of *Salsola australus* illustrate phenological partitioning through temperature related afterripening requirements. The seeds of *B. tectorum* with their highly evolved dormancy system illustrate a physiological system adapted to variable environmental conditions. Seed production, seed germination, and seedling establishment provide land managers with options to manipulate plant populations if they understand the critical points in the seed and seedbed processes that lead to or prevent seedling establishment.

8.8. References

Billings, W. D. 1951. Vegetation zonation in the Great Basin of Western North America. p. 101—122. *In*: Les Bases Ecologigues de la Regeneration de la Vegetation des zones Arides Union Int. Sci. Bio. Ser. Bull. No. 9. Par. 15.

Blaisdell, J. P. 1953. Ecological effects of planned burning of sagebrush-grass range on the upper Snake River Plains. Tech. Bull. 1075. U.S. Dept. Agric., Washington, D.C.

Charley, J. L., and N. E. West. 1975. Plant-induced soil chemical patterns in some shrub-dominated semi-desert ecosystems of Utah. Journal of Ecology 63: 945—964.

Daubenmire, R. 1970. Steppe vegetation of Washington. Tech. Bull. 62. Washington Agric. Expt. Stn., Pullman, WA.

Dewey, D. R. 1983. Historical and current taxonomic perspectives of *Agropyron, Elymus*, and related genera. Crop Science 23: 637—642.

Evans, R. A., D. A. Easi, D. N. Book, and J. A. Young. 1982. Quadratic response surface analysis of seed-germination trials. Weed Science 30: 411—416.

Evans, R. A., H. R. Holbo, R. E. Eckert, Jr., and J. A. Young. 1970. Functional environment of downy brome communities in relation to weed control and revegetation. Weed Science 18: 89—97.

Evans, R. A., and J. A. Young. 1970. Plant litter and establishment of alien annual species in rangeland communities. Weed Science 18: 697—703.

Evans, R. A., and J. A. Young. 1972. Microsite requirements for establishment of annual rangeland weeds. Weed Science 20: 350—356.

Evans, R. A., and J. A. Young. 1975. Enhancing germination of dormant seeds of downy brome. Weed Science 23: 354—357.

Evans, R. A., and J. A. Young. 1982. Russian thistle and barbwire Russian thistle seed and seedbed ecology. ARR-W-25. Agric. Res. Serv., U.S. Dept. of Agric., Berkeley, CA.

Evans, R. A., J. A. Young, and B. A. Roundy. 1977. Seedbed requirements for germination of Sandberg bluegrass. Agronomy Journal 69: 817—820.

Franklin, J. F., and C. T. Dyrness. 1973. Natural vegetation of Oregon and Washington. Gen. Tech. Report NW-8. Pacific Northwest Forest and Range Experiment Station. Forest Service, U.S. Dept. of Agric., Portland, OR.

Frischknecht, N.C., and A. T. Bleak. 1957. Encroachment of big sagebrush on seeded range in northeastern Nevada. J. Range Manage. 10: 165—170.

Harper, J. L., W. T. Williams, and G. R. Sagar. 1965. The behavior of seeds in soil. Part I. the heterogeniety of soil surface and its role in determining the establishment of plants from seed. J. Ecology 53: 273—286.

Hormay, A. L. 1943. Bitterbrush in California. Research Note 9. California Forest and Range Expt. Sta. Forest Service, U.S. Dept. of Agric., Berkeley, CA. 13 p.

Mack, R. M. 1981. Invasion of *Bromus tectorum* L. into western North America: An ecological chronicle. 7: 145—165.

McDonough, W. T., and R. O. Harniss. 1974. Seed dormancy in *Artemisia tridentata* Nutt. subsp. *vaseyana* Rydb. Northwest Science 48: 17—20.

Nord, E. C. 1965. Autecology of bitterbrush in California. Ecological monographs 35: 193—194.

Rogler, G. A. 1960. Relation of seed dormancy of green needlegrass (*Stipa viridula* trin.) to age and treatment. Agron. J. 53: 467—469.

Wood, M. K., R. E. Eckert, Jr., W. H. Blackburn, and F. F. Peterson. 1982. Influence of crusting soil surfaces on emergence and establishment of crested wheatgrass, squirreltail, Thurber needlegrass, and fourwing saltbush. J. of Range Manage. 35: 282—287.

Wood, M. K., W. H. Blackburn, R. E. Eckert, Jr., and F. F. Peterson. 1978. Interrelations of the physical properties of coppice dune and vesicular dune interspace soils with grass seedling emergence. J. of Range Manage. 31: 250—253.

Young, J. A., and R. A. Evans. 1973. Mucilaginous seedcoats. Weed Science 21: 52—54.

Young, J. A., and R. A. Evans. 1974. Population dynamics of green rabbitbrush in disturbed big sagebrush communities. J. of Range Manage. 27: 127—132.

Young, J. A., and R. A. Evans. 1976. Stratification of bitterbrush seeds. J. of Range Manage 29: 421—425.

Young, J. A., and R. A. Evans. 1977. Squirreltail seed germination. J. Range Manage. 30: 33—36.

Young, J. A., and R. A. Evans. 1978. Population dynamics after wildfires in sagebrush grasslands. J. Range Manage. 31: 283—289.

Young, J. A., and R. A. Evans 1979. Arrowleaf balsamroot and mules ear seed germination. J. of Range Manage. 32: 71—74.

Young, J. A., and R. A. Evans. 1980. Germination of desert needlegrass. 5: 40—46.

Young, J. A., and R. A. Evans. 1981. Germination of seeds of antelope bitterbrush, desert bitterbrush, and cliffrose. ARR-W-17. Sci. Education Admin. U.S. Dept. of Agric., Berkeley, CA.

Young, J. A., and R. A. Evans. 1985. Demography of *Bromus tectorum* in *Artemisia* communities. pp. 489—502. J. White, ed. The Population Structure of Vegetation: Dr. W. Junk, Publisher, Dordrecht.

Young, J. A., R. A. Evans, and R. E. Eckert, Jr. 1981. Temperature profiles for germination of bluebunch and beardless wheatgrass. J. of Range Manage. 34: 84—89.

Young, J. A., R. A. Evans, and B. L. Kay. 1975. Dispersal and germination dynamics of broadleaf filaree. *Erodium botrys* (Cav.) Betrol. Agron. J. 67: 54—57.

Young, J. A., R. A. Evans, and B. L. Kay. 1984. Persistence and colonizing ability of rabbitbrush collections in a common garden. J. of Range Manage. 37: 373—377.

Young, J. A., R. A. Evans, and J. Major. 1972. Alien plants in the Great Basin. J. of Range Mange. 25: 194—201.

Young, J. A., R. A. Evans, and R. E. Ecket, Jr. 1969. Population dynamics of downy brome. Weed Science 17: 20—26.

Young, J. A., R. A. Evans, R. O. Gifford, and R. E. Eckert, Jr. 1970. Germination characteristics of three species of Cruciferae. Weed Science 18: 48—48.

9. Modelling rangeland ecosystems for monitoring and adaptive management

DONALD A. JAMESON

9.1. Abstract

Major topics covered in this paper are (1) the statistical analyses of model-based monitoring systems, (2) model-based analyses for determining optimal management strategies, and (3) system attributes that dictate model structure. Each topic is discussed in relationship to models for management of rangeland ecosystems.

The statistics of model-based monitoring in the engineering and aerospace literature need modifications for rangeland and vegetation systems. These modifications make calculations of appropriate variances for model-based estimates straightforward for linear and linear-in-a-neighborhood systems.

Simulation models may be coupled with optimization models within the A matrix of a linear programming formulation, in the right hand side vector (b) of a linear programming formulation, or in the recursion equation of a dynamic programming formulation.

Thus both statistical and optimal strategy procedures are well in hand for linear and linear-in-a-neighborhood problems. If these formulations are acceptable, incremental management may be appropriately guided by a suitable predictive model. However, linearity-in-a-neighborhood methods are not appropriate when discontinuities are critical to behavior of the system. Fold and cusp catastrophe manifolds express discontinuities of equilibria, but are seldom used in ecological models.

9.2. Introduction

9.2.1. The status of modelling in vegetation management

Modelling in ecological studies has become commonplace and continuing

Tueller, P. T. (ed.), Vegetation science applications for rangeland analysis and management.
ISBN 90-6193-195-9.
© 1988, *Kluwer Academic Publishers, Dordrecht, Boston, London.*

changes in computer technology have had an impact on types and implementation of models. The number of individuals building and using computer models of range and other ecological systems is continuing to increase. However, there are still serious questions about "why model"? When ecosystem models first became popular in the late 1960's, the main objective in modelling often was to demonstrate that we could build such models, and that they mimicked the ecological systems that they purported to represent. Such an objective is naturally shortlived. Since these early days of model induced euphoria, we have come to ask more serious questions about model objectives.

For the purposes of this paper, modelling efforts will be categorized as follows:

1. Research guidance models. These incorporate fundamental questions of system behavior as a guide to relevant research questions. Since its purpose is to help direct research in biological or physical phenomena, it is important that such models faithfully represent what is known about fundamental biological or physical processes involved. It is not necessary that the "entire system" be included, only those aspects relevant to the critical research questions.

2. Systems behavior models. The fundamental behavior of many systems can only be examined by modelling. Thus, important aspects of model behavior may rest on mathematical or computational principles as well as on biological or physical principles.

3. Management oriented models. If a management model is to be useful, it must be predictive because managerial decisions are always about the future. It matters little about biological or mathematical elegance; what does matter is the ability to predict future states of the system.

In this paper management oriented models will be emphasized. Ecologically based management disciplines, such as range science, are more concerned with such models. In addition, such models can be evaluated for their contribution to management decisions. Research guidance models, in particular, have more specialized audiences, and it is more difficult to evaluate such models. An appropriately elegant mathematical model of an ecological phenomena might become a useless abstraction when applied to a managerial system. For reviews of model building methods appropriate to many aspects of range science, see Williams (1985), Wilson & Bennett (1985), Hazell & Norton (1986), and Starfield & Bleloch (1986).

The categories listed above explicitly exclude the large "include everything ecosystem model". Such models may help coordinate a large modelling and research effort, but have little other utility. There is also no evidence that such models are better at mimicking ecosystem behavior than smaller and simpler models that focus on a portion of the system under study.

The test of a predictive model is the comparison of previous predictions to sample data. Such comparisons of any model soon lead to the conclusion that no model is perfect. At best, ecological models are a close approximation to sample data. Because of sampling error and other uncertainties, data is often not much better. Given a relatively poor model and relatively poor data, where should we put our trust? The answer: put a little bit of trust in each, so that the combined result is more reliable than either the data or the model alone.

Thus, evaluation of a predictive model is based on previous experiences with model predictions and sampled data values. These comparisons allow computation of model prediction errors (even for a deterministic model). Reliability of a sample mean is evaluated by its variance; similarly, the appropriate confidence to be placed in a model prediction can be estimated from the variance of its prediction errors.

For the purposes of this paper, monitoring is the repetitive measurement of some system state or condition to estimate the state or condition within a specified limit of error. An optimal monitoring procedure achieves this goal at the lowest possible cost; an optimal monitoring system almost always requires some model. A corollary of optimal monitoring is the optimal management strategy; such a strategy requires a predictive model, and a series of measurements to keep the system close to the optimal strategy.

Consideration of optimal monitoring facilitates development of suitable predictive models. If model development proceeds without consideration of monitoring requirements, it is difficult to focus on the modelling objectives.

This paper covers three general topics:

* The statistics of model-based monitoring systems.
* Model based analysis for determining optimal management strategies.
* System attributes that dictate model structure.

9.2.2. The role of monitoring and adaptive management in range management

Monitoring should provide not only baseline data, but also information feedback so that managers can take corrective action. Monitoring design should consider sample adequacy, cost effectiveness, repeatability of measurements, and timeliness. Monitoring information should relate directly to management actions. Managers not only may use data collected at a given time, but also may make predictions about how the ecological system will change in relationship to this baseline data. Comparisons of the relative value of predictions and baseline data should be part of the monitoring scheme.

Adaptive management requires that some measurements or other observations are taken at intervals, and a predictive model is used to supply information between measurements. Anything that we can say about the system between these measurement periods comes from our concept or "model" of how the system will perform during this time. The simplest possible model predicts that everything remains the same between measurements. Without a model we would have to measure continuously to know what's going on, and would have little opportunity for advance planning. To keep our understanding about the system at a satisfactory level, we can use either better models or better measurements. This combination of models and measurements is monitoring:

Monitoring: a combination of repeated predictions and sequential measurements to provide information for management.

As a result of monitoring, needed changes in management can be made in an adaptive management scheme:

Adaptive management: management in which periodic changes are made to adjust to newly acquired information about the system being managed.

Sampling to acceptable levels of statistical error may be a difficult and costly task. However, a combination of model results and field data frequently yields an estimate with satisfactory statistical errors. The combination of two sources of information requires calculation of weighting factors that are proportional to the reliability of the information. The variances required for the calculations are computed from sampling error and from the differences between model results and field data over a series of time steps. The appropriate variance when both model predictions and measurements are used will always be less than the measurement variance, indicating an improvement due to combination of measurements and model results. A "good" model will result in a considerable reduction in the variance, but even a "poor" model will result in at least some reduction.

Model based predictions are an acceptable basis for management if:

1. The statistical errors in measurements and models are not strongly skewed from a normal distribution.

2. The cost or penalty for a given degree of management error in one direction is about the same as for the same degree of management error in the opposite direction, and

3. The changes in system performance in one direction follow the same time path as changes in the reverse direction.

The conditions described above seldom apply to ecological systems. For example, responses that depend heavily on current rainfall most likely will not be normally distributed, but will have a skewed distribution because periods below mean rainfall are more common than periods above mean rainfall. The penalty for underestimating ecological carrying capacity is much less than the penalty for overestimating carrying capacity; a minor inconvenience caused by misjudging an ecological response in one direction might be a disaster when the response is misjudged by a similar amount but in the opposite direction. The time required to shift from a climax biotic community to an early successional stage may be much less than the time required to shift from an early successional stage back to the climax community. Although the short-term monitoring and adaptation approach may help, the successful manager should not neglect these cases where it will not work.

Much of the current activity in systems research and risk management research concentrates on deriving strategies when conditions for suitable short-term management do not apply. A very useful approach in these cases is to use a stochastic dynamic programming model that considers all of the uncertainty, possible management corrections, and measurements that will be encountered throughout the length of the planning period (2 years, 200 years?), and compare results to a decision model in which future measurements and uncertainty of future events is ignored.

In general, the appropriateness of ignoring uncertainty of future events requires the same three conditions as for the simpler approaches to adaptive management. If a computational approach that considers uncertainty of future events yields the same decision as a simpler approach that ignores uncertainty, we then know that the simpler approach is appropriate, and "certainty equivalence" applies. When future uncertainty can be suitably ignored, the computational load ranges from trvial to reasonable, and could be done under many management situations. If the three required conditions do not apply and more complicated approaches must be used, the computational load ranges from uncomfortable to unbearable. A research program can be designed to determine if deterministic and stochastic approaches yield the same or different results, and research direction can focus on the differences.

9.3. Statistics of model-based management systems

9.3.1. Combining observations with different variances

Much of the basics of model evaluation can be demonstrated by considering two separate measurements of an ecological or natural resource system. Generally, the arithmetic mean of two measurements would be calculated. However, computation of a simple mean assumes that two measurements come from the same population; i.e., the variance relative to each measurement is the same. On the other hand, if one measurement comes from a population with much lower variance than the other, a different approach is required (Carande & Jameson, 1986).

Suppose that ecological measurements are taken by two different field crews. Crew 1 consistently collects data carefully; consequently, their measurements have a low variance. Crew 2, on the other hand, usually reports data with a higher variance. Rather than discard data collected by Crew 2, it seems reasonable to weight data collected by each crew with some appropriate weighting factor to calculate an appropriate mean. The measurements are:

$$y_1 = x + v_1 \quad \text{and} \tag{2.1.1}$$

$$y_2 = x + v_2 \tag{2.1.2}$$

where y_1 and y_2 are measurements of Crew 1 and Crew 2, respectively, x is the "true" state of the system, and v_1 and v_2 are measurement errors. The best estimate of the true state of the system $\hat{x}$ is:

$$\hat{x} = g_1 y_1 + g_2 y_2 \tag{2.1.3}$$

where g_1 and g_2 are appropriate weighting factors. The estimation error resulting from those weights is:

$$\tilde{x} = \hat{x} - x \tag{2.1.4}$$

The problem is to find values of g_1 and g_2 to provide the "best" estimate of x. The usual "least squares" criteria is to minimize $\tilde{x}^2$; a series of n values of $\tilde{x}$ would allow the expression

$$\text{minimize} \sum_{i=1}^{n} (\tilde{x}_i)^2 \tag{2.1.5}$$

i.e., to find the minimum variance.

The expected value of a random variable w is denoted by E[w]. For normally distributed populations the expected value is the true, but unknown, population mean; i.e., if the population mean is zero the sample size tends to zero as the sample size become increasingly large:

$$\text{if } E[w] = 0, \text{ then } \sum_{i=1}^{n} w_i/n \rightarrow 0 \text{ as } n \rightarrow \infty \tag{2.1.6}$$

The expected variance of w is the sums of squares of individual values w_i of a sample of the population as the sample size becomes large:

$$\text{if } E[w^2] = \sigma w, \text{ then } \sum_{i=1}^{n} (w_i)^2/n \rightarrow \sigma_w \text{ as } n \rightarrow \infty \tag{2.1.7}$$

where σ indicates population variance (*ordinarily, σ^2 indicates a variance and σ a standard deviation, but to be consistent with other notation in this* *paper σ will be used for a variance and $\sqrt{\sigma}$ for a standard deviation*). For small samples, division is by (n − 1) instead of by n.

Substitute equations (2.1.1) and (2.1.2) into equation (2.1.3) to yield:

$$\hat{x} = g_1 (x + v_1) + g_2 (x + v_2) \tag{2.1.8}$$

Then substitute equation (2.1.8) into equation (2.1.4):

$$\tilde{x} = g_1 (x + v_1) + g_2 (x + v_2) - x \tag{2.1.9}$$

The expected values of both sides of equation (2.1.9) are:

$$E[\tilde{x}] = E[g_1 (x + v_1) + g_2 (x + v_2) - x] \tag{2.1.10}$$

If measurements represent unbiased estimates of a population, $E[\tilde{x}] = 0$, $E[v_1] = 0$, and $E[v_2] = 0$. However, since x is nonrandom, $E[x] = x$, and for constants $E[g_1] = g_1$ and $E[g_2] = g_2$. Equation (2.1.10) can be rewritten:

$$0 = g_1 x + g_1 E[v_1] + g_2 x + g_2 E[v_2] - x \tag{2.1.11}$$

$$0 = g_1 x + g_2 x - x \tag{2.1.12}$$

$$g_1 = 1 - g_2 \tag{2.1.13}$$

Since g_1 is determined by g_2, and vice versa, only one of these values need be computed. Henceforth, g_1 will be designated by g, and g_2 will designated by $1 - g$.

It has already been stated that $E[v_1] = 0$ and $E[v_2] = 0$. A major assumption in development of optimal estimation theory is zero correlation between v_1 and v_2; i.e., $E[v_1 v_2] = 0$. In the example of two crews, the errors of one crew are not related to the errors of the other. To find the expected value of $\tilde{x}^2$, substitute $g = g_1$ and $(1 - g) = g_2$ into equation (2.1.9), square both sides and take the expected value:

$$E[\tilde{x}^2] = E[\{g(x + v_1) + (1 - g)(x + v_2) - x\}^2]. \tag{2.2.14}$$

Expanding and reducing equation (2.1.14) gives:

$$E[\tilde{x}^2] = E[(gv_1 + v_2 - gv_2)^2] \tag{2.1.15}$$

Squaring terms within the parentheses in equation (2.1.15) and omitting all terms with an expected value of 0 (i.e., v_1, v_2 and $v_1 v_2$) yields:

$$E[\tilde{x}^2] = E[g^2(v_1)^2 + (v_2)^2 + g^2(v_2)^2 - 2g(v_2)^2] \tag{2.1.16}$$

By the definition of equation (2.1.7), the expected value of the square of a zero mean variable is its variance and equation (2.1.16) can be rewritten:

$$E[\tilde{x}^2] = g^2 \sigma_1 + \sigma_2 + g^2 \sigma_2 - 2g \sigma_2 \tag{2.1.17}$$

where σ indicates a variance, as previously defined. To find the minimum value of $E[\tilde{x}^2]$ in equation (2.1.17), find the first derivative with respect to g and set the result to zero:

$$2g\sigma_1 + 2g\sigma_2 - 2\sigma_2 = 0 \qquad (2.1.18)$$

Divide this result by 2, collect terms containing g, and factor g:

$$g(\sigma_1 + \sigma_2) = \sigma_2 \qquad (2.1.19)$$

Therefore:

$$g = \sigma_2/(\sigma_1 + \sigma_2) \qquad (2.1.20)$$

Thus, g as expressed by equation (2.1.20) is the weighting factor for combining the measurement of Crew 1 with the measurement of Crew 2. The best estimate of x is:

$$\hat{x} = gy_1 + (1 - g)y_2 \qquad (2.1.21)$$

The variance of $(\hat{x} - x)$ is $E[\tilde{x}^2]$. Factor equation (2.1.17) as follows:

$$E[\tilde{x}^2] = g^2\sigma_1 + (1 - 2g + g^2)\sigma_2 \quad \text{thus} \qquad (2.1.23)$$

$$E[\tilde{x}^2] = g^2\sigma_1 + (1 - g)^2\sigma_2 \qquad (2.1.24)$$

Equations (2.1.20) and (2.1.24) are fundamental in combining two sources of information, and will be the basis of calculating model variances in subsequent sections.

9.3.2 Prediction error in linear dynamic models

9.3.2.1 Development of variances for models. Assume that a prediction has been made about the state or condition of a natural resource or ecological system. After some interval of time, a measurement is made. Just as in the case of two measurements in the previous section, both the prediction and the measurement contain random errors and require computation of appropriate weighting factors.

Perhaps the simplest predictive model is:

$$x_{k+1} = x_k \qquad (2.2.1)$$

which simply states that x will be the same at some future time k + 1 as at time k. Of course, the relationship of equation (2.2.1) will not exactly occur, and more correctly:

$$x_{k+1} = x_k + w_k \qquad (2.2.2)$$

where w_k is a random prediction (process) error with $E[w_k] = 0$. The random variable w is assumed to be normally distributed with mean zero, and the variance of w is q. These properties are indicated by:

w is N(0, q)

This expression is read "w is normally distributed with mean zero and variance q." There is no correlation assumed between w_k and w_{k+1}, i.e., $E[w_k w_{k+1}] = 0$. The measurement y at time k is:

$$y_k = x_k + v_k \tag{2.2.3}$$

where v is a random measurement error with $E[v_k] = 0$ and v is N(0, r).

To distinguish between a "true" value, a predicted value, and the "best estimate", the following notation will be used:

x_k = true but unknown value at time k,

$x_{k|k}$ = the best estimate of x given information at time k (with variance $p_{k|k}$),

$x_{k+1|k}$ = the predicted value of x for time k + 1 given information at time k (with variance $p_{k+1|k}$), and

$x_{k+1|k+1}$ = the updated best estimate of x for time k + 1 given information at time k + 1 (with variance $p_{k+1|k+1}$).

Because the prediction error w is not known and $E[w] = 0$:

$$x_{k+1|k} = x_{k|k} \tag{2.2.4}$$

but the actual or true values are:

$$x_{k+1} = x_k + w_k \tag{2.2.5}$$

Subtracting equation (2.2.5) from equation (2.2.4) yields:

$$x_{k+1|k} - x_{k+1} = x_{k|k} - x_k - w_k \tag{2.2.6}$$

and departure of the estimated value from the true value of x is:

$$\tilde{x}_{k+1} = \tilde{x}_k - w_k \tag{2.2.7}$$

Taking the expected value of the square of equation (2.2.7) yields:

$$E[(\tilde{x}_{k+1})^2] = E[(\tilde{x}_k - w_k)^2] \tag{2.2.8}$$

It is assumed that $E[\tilde{x}_k w_k] = 0$; therefore, in expansion of equation (2.2.8), the cross product term is omitted:

$$E[(\tilde{x}_{k+1})^2] = E[(\tilde{x}_k)^2 + (w_k)^2] \quad \text{and} \tag{2.2.9}$$

$$E[(\tilde{x}_{k+1})^2] = E[(\tilde{x}_k)^2] + E[(w_k)^2] \tag{2.2.10}$$

As defined in equation (2.1.7), the expected value of the square of a zero mean random variable is the variance of the variable. The following definitions will be used:

$$E[(\tilde{x}_k)^2] = p_{k|k} \tag{2.2.11}$$

$$E[(\tilde{x}_{k+1})^2] = p_{k+1|k} \tag{2.2.12}$$

$$E[w^2] = q, E[w] = 0 \quad \text{and} \tag{2.2.13}$$

$$E[v^2] = r, \text{and } E[v] = 0 \tag{2.2.14}$$

With definitions (2.2.11) to (2.2.14), substituting into equation (2.2.10) gives:

$$p_{k+1|k} = p_{k|k} + q \tag{2.2.15}$$

Thus, the variance of the prediction for time k + 1, given information at time k, is $p_{k+1|k}$. Equating this variance to σ_2 of the previous section and the variance r to σ_1, equation (2.1.20) can be expressed:

$$g = p_{k+1|k}/(r + p_{k+1|k}) \tag{2.2.16}$$

The weighting factor g is the "Kalman gain", and the procedure of updating variance estimates using the Kalman gain is the Kalman filter. The updated best estimate of x, from equation (2.1.21), is:

$$x_{k+1|k+1} = gy_{k+1} + (1-g)x_{k+1|k} \quad \text{and} \tag{2.2.17}$$

$$x_{k+1|k+1} = x_{k+1|k} + g(y_{k+1} - x_{k+1|k}) \tag{2.2.18}$$

Thus, the difference between measurement y and the prediction is $(y_{k+1} - x_{k+1|k})$; this difference is called an innovation or residual. A series of values of $(y_{k+1} - x_{k+1|k})$ is termed an innovation sequence, and the variance of this sequence is designated by s. Define i_{k+1} as the innovation $(y_{k+1} - x_{k+1|k})$, and add and subtract x_{k+1} to the right side of the equation:

$$i_{k+1} = y_{k+1} - x_{k+1|k} + (x_{k+1} - x_{k+1}) \tag{2.2.19}$$

$$i_{k+1} = (y_{k+1} - x_{k+1}) - (x_{k+1|k} - x_{k+1}) \tag{2.2.20}$$

$$E[(i_{k+1})^2] = E[\{(y_{k+1} - x_{k+1}) - (x_{k+1|k} - x_{k+1})\}^2] \tag{2.2.21}$$

If measurement errors are not correlated with prediction errors:

$$E[(y_{k+1} - x_{k+1})(x_{k+1|k} - x_{k+1})] = 0 \tag{2.2.22}$$

Therefore, expansion of equation (2.2.21) becomes:

$$E[(i_{k+1})^2] = E[(y_{k+1} - x_{k+1})^2] + E[(x_{k+1|k} - x_{k+1})^2] \tag{2.2.23}$$

Thus the variance s of the innovation sequence i_k is:

$$E[(i)^2] = s = r + p_{k+1|k} \tag{2.2.24}$$

Once a new measurement y is made at time k + 1, an updated estimate of the variance $p_{k+1|k+1}$ of the combined information can be calculated. The variance update equation can be developed from equation (2.1.24). Replacing σ_1 with r, σ_2 with $p_{k+1|k}$, and $E[\tilde{x}^2]$ with $p_{k+1|k+1}$, this equation becomes:

$$p_{k+1|k+1} = g^2 r + (1 - g)^2 p_{k+1|k} \tag{2.2.25}$$

This is the "Joseph" form of the variance update equation.

Thus far in this section the prediction equation has stated that $x_{k+1} = x_k + w_k$. More generally, a prediction can be written:

$$x_{k+1} = \phi x_k + w_k \tag{2.2.26}$$

where the multiplier ϕ is a state transition multiplier that describes how the state of the system changes between time k and k + 1. With this form the state x is not constant but is multiplied by the factor ϕ. However, the state transition multiplier is not necessarily constant over time and can be indicated by

$$x_{k+1} = \phi_k x_k + w_k \tag{2.2.27}$$

which indicates that ϕ need only be constant over the interval k to k + 1. Equation (2.2.9) becomes

$$E[(\tilde{x}_{k+1})^2] = E[\phi^2(\tilde{x}_k)^2] + E[(w_k)^2] \tag{2.2.28}$$

and equation (2.2.15) becomes:

$$p_{k+1|k} = \phi^2 p_{k|k} + q \tag{2.2.29}$$

9.3.2.2. Summary of computations. Given the system:

$$x_{k+1} = \phi_k x_k + w_k, \text{ w is N(0, q)} \tag{2.2.30}$$

$$y_{k+1} = x_{k+1} + v_k, \text{ v is N(0, r)} \tag{2.2.31}$$

Because $E[w_k] = 0$, the system can be modelled as:

$$x_{k+1|k} = \phi_k x_{k|k} \tag{2.2.32}$$

Variances are:

$$p_{k+1|k} = \phi^2 p_{k|k} + q \tag{2.2.33}$$

$$g = p_{k+1|k}/(p_{k+1|k} + r) \tag{2.2.34}$$

The state variable x is updated by:

$$x_{k+1|k+1} = gy_{k+1} + (1-g)x_{k+|k} \tag{2.2.35}$$

or in innovation form:

$$x_{k+1|k+1} = g(y_{k+1} - x_{k+1|k}) + x_{k+1|k} \tag{2.2.36}$$

The updated variance is given in Joseph form by:

$$p_{k+1|k+1} = (1-g)^2 p_{k+1|k} + g^2 r \tag{2.2.37}$$

or in the simpler but numerically equivalent standard form by:

$$p_{k+1|k+1} = (1-g)p_{k+1|k} \tag{2.2.38}$$

To advance to the next time step, $p_{k+1|k+1}$ is redesignated as $p_{k|k}$, $x_{k+1|k+1}$ as $x_{k|k}$, and processing returns to equation (2.2.32). These equations can be found in Jazwinski (1970), Gelb (1974), Anderson and Moore (1979), and Maybeck (1979).

9.3.3 Calculation of appropriate variances

9.3.3.1 The measurement multiplier. A measurement multiplier h is either a 0 or 1 in the gain equation:

$$g = hp_{k+1|k}/(p_{k+1|k} + r) \tag{2.3.1}$$

Thus, if h = 0, g = 0 (i.e., if no measurement is taken, no weight is given to the measurement) and

$$x_{k+1|k+1} = x_{k+1|k} \tag{2.3.2}$$

Equation (2.3.2) merely states that the best estimate of x at time k + 1 with no measurements is simply the predicted value $x_{k+1|k}$. Also, if h = 0, g = 0, and

$$p_{k+1|k+1} = (1-0)p_{k+1|k} \tag{2.3.3}$$

$$p_{k+1|k+1} = p_{k+1|k} \tag{2.3.4}$$

Thus, the updated variance without measurement is simply the variance the estimated state at time k plus the variance of the prediction from k to k + 1. In recursive calculation of the variance p, $p_{k+1|k+1}$ is redesignated as $p_{k|k}$, thus, if h = 0 (i.e., no measurement):

$$p_{k+1|k+1} = \phi^2 p_{k|k} + q \tag{2.3.5}$$

$$p_{k+2|k+2} = \phi^2 p_{k+1|k+1} + q \quad \text{and} \tag{2.3.6}$$

$$p_{k+3|k+3} = \phi^2 p_{k+2|k+2} + q \tag{2.3.7}$$

9.3.3.2 The sampling variance. The sampling variance r can be found by repeated measurements at a single point in time. Since the "true value" of x at a given time is the same for all measurements, the square of the differences between the true value of x and measurement values y will yield the variance σ:

$$\sigma = \sum_{i=1}^{n} (y_i - x)^2/n \tag{2.3.8}$$

However, since the true value of x is not known, an estimate of the variance r′ is computed with the mean value of y as an estimate of x:

$$r' = \sum_{i=1}^{n} (y_i - \bar{y})^2/(n-1) \tag{2.3.9}$$

For computation Equation (2.3.9) is usually written as:

$$r' = \left[\sum_{i=1}^{n} (y_i)^2 - \left(\sum_{i=1}^{n} y_i \right)^2 \Big/ (n) \right] \Big/ (n-1) \tag{2.3.10}$$

Equations (2.3.9) and (2.3.10) give the variance or error appropriate to a single measurement. Multiple measurements at a single time yield a mean value at a point in time; the variance of these mean values is the square of the standard error of the mean as defined in elementary statistics, and:

$$r = r'/n \tag{2.3.11}$$

The result of Equation (2.3.11) is the measurement error of the sample.

Most of the literature on optimal estimation assumes that the dimension represented by the subscripts k and k + 1 refers to time. However, there is no particular reason that this should be always so. In studies of ecological pattern, k may refer to a location along a transect. In hydrology, k may refer to depth (stage) of water. In any event, sampling variance refers to multiple measurements taken at a given k, and if k does not refer to time, measurements may actually be at different times.

9.3.3.3 The variance of the innovation sequence. A series of differences between measurements and predictions or innovation sequence is:

$$i_{k+1} = y_{k+1} - x_{k+1|k} \tag{2.3.12}$$

If multiple measurements are made at a single time, then y_{k+1} of equation

(2.3.12) is actually the mean value of measurements at that time. The variance of the sequence of i's is designated by s:

$$s = \left[\sum_{k=1}^{m} (i_k)^2 - \left(\sum_{k=1}^{m} i_k \right)^2 \Big/ m \right] \Big/ (m-1) \qquad (2.3.13)$$

where k indicates the time step and m indicates the length of the sequence.

9.3.3.4 The initial value of the system variance. The calculation of the variance update in equations (2.2.33), (2.2.37), and (2.2.38) converges quickly; in applications with a long sequence of measurements any initial value of system variance (indicated by $p_{o|o}$) is satisfactory. However, in many range management applications, the measurement sequence will be short and a more precise value of $p_{o|o}$ would be helpful. Without benefits of the update equations described here, knowledge of the system would be described only by the measurement variance r; thus an estimate of r from a prior sample would be an acceptable starting value $p_{o|o}$. With advantages of the update equations, $p_{k|k}$ will eventually be less than the measurement variance r, indicating an improvement due to the recursive procedure.

9.3.3.5 The model or prediction variance. The variance q of Equation (2.2.33) is the process or prediction variance appropriate to the model. This variance is not common in natural resource management and ecological literature, but is extremely important in evaluating models because it is the appropriate "model variance". The variance can be derived from other available information as described below.

In the variance update calculations $p_{k+1|k+1}$ replaces $p_{k|k}$ at each iteration. When this procedure has converged, $p_{k+1|k+1} = p_{k|k}$. Jameson (1985) has presented development of a method to find the variance q at convergence:

$$q = \phi^2(s-r)^2/s + (s-r) - \phi^2(s-r) \qquad (2.3.14)$$

If $\phi = 1$, Equation (2.3.14) reduces to

$$q = (s-r)^2/s \qquad (2.3.15)$$

Thus, q can be computed from s and r, provided that $s > r$. If $s \leqslant r$, divide r by a sufficiently large value of n (sample size) so that $s > r$. This value of n will become the basic sample size for measurements, and subsequent calculation will indicate multiples of this basic sample size.

The variances have the following relationship:

$$s = \phi^2 p + q + r \qquad (2.3.16)$$

This equation holds at convergence when p attains its minimum value.

This, the minimum value of p, for a given s, will result in the maximum value of q. With q having a large value, the prediction is relatively poor and more measurements will be required.

An alternative for finding the variance q is to compare a detailed (possibly nonlinear) mechanistic model of the system to a linear approximation. The difference between the two models is identified as model error w, and the variance of w is q. This approach allows a comparison of two models, but universally yields a numerical value of q which is too small.

9.3.4 Sample size and monitoring cost

The state of knowledge about the environmental system following a measurement can be computed recursively using equations (2.2.30) through (2.2.38), which describe the well known Kalman filter. In this computation, relative values or "gain" of the predicted value and the sampled value are computed at each time interval. However, if the prediction is extended over a sufficient number of intervals without measurements, usefulness of the prediction approaches zero, and knowledge about the system, when it is measured, can be assumed to be expressed solely by r. This assumption will introduce a slightly conservative bias in calculation of sample sizes, but greatly simplifies the following calculations.

If the variable in question is not predicted to change, $\phi = 1$ and uncertainty about the system at the next interval will increase by q. Thus,

$$p_{1|0} = r + q \tag{2.4.1}$$

and at the next time step

$$p_{2|1} = p_{1|0} + q \tag{2.4.2}$$

Without measurements the system variance increases by q at each time step. Efficient monitoring delays measurements until the model prediction is nearly useless; a measurement returns the system variance $p_{k|k}$ to the value of the variance r. The effect of periodic measurements on system variance is shown in Figure 1. Sisler & Jameson (1983) and Jameson (1986b) present methods of using information in this figure to calculate a least cost monitoring procedure.

9.3.5 Adaptive monitoring and prediction testing

9.3.5.1 Linear — Quadratic — Gaussian (LQG) models. If ϕ in previous sections is a simple multiplier and not a nonlinear function, the result is a *linear system.* If $\phi > 1$, values of x will increase over time, and if $\phi < 1$, values of x will decrease. Since the random effect w in such systems is

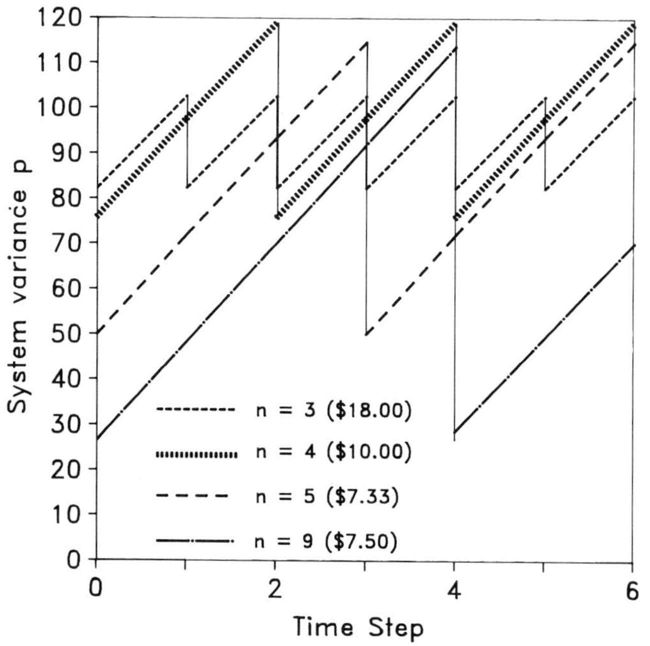

Figure 1. Example monitoring schemes with sampling at one, two, three and four intervals result in sample sizes of n = 3, n = 4, n = 5 and n = 9 respectively.

normally distributed, the systems are called Gaussian (a synonym for normally distributed). The word Quadratic refers to the penalty for departing from a desired pathway; if a departure is indicated by d, and the penalty for departure is proportional to d^2, the system is said to have a quadratic cost function.

The Linear-Quadratic-Gaussian (LQG) conditions include standard assumptions for monitoring conditions used in this paper. Although it is unlikely that any natural resource management or ecological system is strictly LQG, conditions that are approximately LQG often can be treated by standard monitoring procedures with good results. Systems that are "linear-in-a-neighborhood" can be treated as linear over a few time steps. For example, a linear approximation might be made for x_k, x_{k+1}, x_{k+2}, x_{k+3}, and a new approximation made for x_{k+1}, x_{k+2}, x_{k+3}, x_{k+4}, etc. In this example, a linear projection is made over the next three time steps.

9.3.5.2 Adaptive calculation of variances. Suppose that a continuing series of predictions and measurements are being made as time progresses. Thus, new information to calculate the variance s becomes available at every time step, and perhaps older information is becoming less appropriate. In this situation it would be useful to calculate a new value of s at each time step, using the last m values (m may indicate the last 3 to 20 values). Since s is recalculated at each time step, a new value of q is also calculated. Thus, an adaptive monitoring procedure or "adaptive filter" can be constructed by this method. Since m steps are always used, this method is a "fixed memory filter".

Development of a suitable adaptive monitoring procedure has been a troublesome area in the monitoring literature (Chin 1979). The procedure outlined in equations (2.3.14) and (2.3.15) is useful, but may give erratic values of q. To smooth the values of q, a "fading memory" device can be used:

1. Compute q by equation (2.3.14) or (2.3.15)
2. Retrieve the last value q_{k-1} from storage
3. Compute a new value $q_k = (q + 5q_{k-1})/6$.

9.3.5.3 Tests of an adaptive monitoring procedure. An efficient monitoring procedure should possess the property that i_k is uncorrelated with i_{k+1}. To test this condition, autoregression analysis of the innovation sequence is appropriate. High "t" values of autoregression indicate that assumptions made in the monitoring procedure are suspect: small "t" values indicate that the monitoring procedure is performing satisfactorily. If the innovation sequence is autocorrelated, a procedure based on differences between consecutive measurements can remove the autocorrelation (Jazwinski, 1970). Alternatively, methods of Soeda & Yoshimura (1973) may be useful.

9.3.5.4 Monitoring procedures for nonlinear models. Thus far, monitoring procedures discussed in this paper have used simple linear models. However, many models expressing natural resource or ecological systems are not of this type. To develop a monitoring system in these cases, it is useful to assume "linearity-in-a-neighborhood". Since $x_{k+1} = \phi x_k$, divide both sides of this equation by x_k to yield:

$$\phi = x_{k+1}/x_k \qquad (2.5.1)$$

as an appropriate linear state transition multiplier for the interval k to k + 1. Then find process error q by equation (2.3.14) and project the system error through several time steps as described in equations (2.3.5), (2.3.6), and (2.3.7). The model can be assumed to be adequate as long as the system variance p is less than an allowable value; when p exceeds this value the model has been used beyond its satisfactory prediction limits and a measurement must be taken.

Although the procedure described in this section is a linear approximation to a more complicated model, results will generally be more satisfactory than an attempt to develop a procedure based on nonlinear theory *unless the theoretical basis is in itself an important part of the analysis.* The cases for which the linear projection approach is satisfactory include LQG conditions described at the beginning of this section. Conditions for which the approach is not satisfactory are described later.

9.3.6 Multivariable monitoring

In the previous section, design of optimal monitoring procedures was presented with the assumption that for each time step the single variable would either be measured or not measured, and that sample size would be adjusted so that the variable was sampled to an acceptable level of uncertainty. In this section, multivariable states are considered, and the possibility of measuring only some of the variables at a time step is introduced. Methods of this section are also used when the system contains states that can never be measured: these "artificial" states may occur as a result of linearization of a nonlinear system. Although examples of this situation are common it the literature, the approach of "linearity-in-a-neighborhood" may be more useful (Emara-Shabaik and Leondes, 1981).

Operations for multivariable monitoring are essentially the same as for a single variable, with vectors and matrices used instead of scalar values. Multiplication by the matrix inverse is used instead of scalar division, and the order of matrix multiplication must be correct. The total number of state variables is designated as n, and the number of variables measured at each time step is m.

The linear state-space system is:

$$x_{k+1} = \phi_k x_k + w_k,\ w \text{ is } N(0, Q) \tag{2.6.1}$$

$$v_{k+1} = x_{k+1} + v_{k+1},\ v_{k+1} \text{ is } N(0, R) \tag{2.6.2}$$

Since $E[w_k] = 0$, the system is modelled as:

$$x_{k+1|k} = \Phi_k x_{k|k} \tag{2.6.3}$$

The covariances are:

$$P_{k+1|k} = \Phi P_{k|k} \Phi^T + Q \tag{2.6.4}$$

and, if every variable is measured, the Kalman gain matrix is:

$$G = P_{k+1|k}(P_{k+1|k} + R)^{-1} \tag{2.6.5}$$

The term $(P_{k+1|k} + R)$ is the covariance matrix S of the innovation sequence (see equation (2.2.24)).

The state vector x is updated by:

$$x_{k+1|k+1} = x_{k+1|k} + G(y_{k+1} - x_{k+1|k}) \tag{2.6.6}$$

and the system covariance is updated by:

$$P_{k+1|k+1} = (I - G)P_{k+1|k} \tag{2.6.7}$$

in standard form, or

$$P_{k+1|k+1} = (I - G)P_{k+1|k}(I - G)^T + GRG^T \tag{2.6.8}$$

in Joseph form, where I is the identity matrix. If there are no measurements, the covariance is projected by

$$P_{k+1|k+1} = \Phi P_{k|k} \Phi^T + Q \tag{2.6.9}$$

and the predicted state vector by

$$x_{k+1|k+1} = \Phi x_{k|k} \tag{2.6.10}$$

9.3.6.1 The measurement matrix. The measurement matrix, indicated by H, is a device to include measured variables in the procedure that are updated according to equation (2.6.3) through (2.6.7), and to relegate unmeasured variable to equations (2.6.9) and (2.6.10). The measurement matrix, in the simple case discussed here, is a matrix with 1's on the main diagonal and 0's otherwise. If all variables are measured, the measurement matrix $H = I$ (the identity matrix) and multiplication by the measurement matrix in this case has no effect. However, if the measurement matrix is not square, use of the measurement matrix has the effect of reducing the size of the covariance matrix. Equation (2.6.5) is modified as:

$$G = P_{k+1|k} H^T(HP_{k+1|k} H^T + R)^{-1} \tag{2.6.11}$$

Matrix P is of size n, n, where n is the total number of state variables in the system. Note that the matrix $(HP_{k+1|k+1} H^T + R)$ is square, and of the same dimension as R. Since R is the covariance matrix of measurement error, the dimension m of R represents those variables measured at this time step. The operation $HP_{k+1|k} H^T$ reduces the $P_{k+1|k}$ matrix to the same size as R. The gain matrix G is rectangular, with n rows (i.e., as many rows as state variables), and m columns (i.e., as many columns as measured variables). The update equation is:

$$P_{k+1|k+1} = (I - GH)P_{k+1|k} \tag{2.6.12}$$

or

$$P_{k+1|k+1} = (I - GH)P_{k+1|k}(I - GH)^T + GRG^T \tag{2.6.13}$$

where the operation GH^T results in a square matrix of dimension n, and the full number of state variables is restored.

The state variables are updated by:

$$x_{k+1|k+1} = x_{k+1|k} + G(y_{k+1} + Hx_{k+1|k}) \tag{2.6.14}$$

where the operation $Hx_{k+1|k}$ shortens the vector of predicted states to correspond to the number of measured variables.

The operations with the H matrix require that the first m variables are the measured variables. Thus all of the measured variables must be in the first subvector of length m of the vector x, and the first submatrix of size m, m of the matrices. If the items to be measured change from one time step to the next, it will be necessary to resort the vectors and matrices so that the measured items are in the first m positions.

Some authors also adjust the covariance matrix Q in a similar fashion with a diagonal rectangular matrix Γ and the operation rQr^{T}. If the dimension of Q are not n, n this operation is necessary. However, this adjustment is seldom needed and will not be discussed further.

9.3.6.2 Summary of definitions and computations for multivariable monitoring. The vectors and matrices, and their dimensions and definitions, in a multivariable monitoring problem are:

n: number of state variables
m: number of measured variables
x: vector of state variables, dimension n, 1
P: system covariance matrix, dimension n, n
Q: process covariance matrix, dimension n, n
H: measurement matrix, dimension m, n
y: measurement vector, dimension m, 1
R: measurement covariance matrix, dimension m,m
G: gain matrix, dimension n, m.

To clarify the matrix dimensions, the update equations will be shown below with dimensions. To do this, time subscripts will be omitted and $P_{k+1|k}$ will be designated as $\hat{p}$. Equations (2.6.3) through (2.6.8) become:

$$x_{n,1} = \Phi_{n,n}\, x_{n,1} \tag{2.6.15}$$

$$\hat{p}_{n,n} = \Phi P_{n,n}\, \Phi^{\mathrm{T}} + Q_{n,n} \tag{2.6.16}$$

$$G_{n,m} = \hat{p}_{n,n}\, H^{\mathrm{T}}_{n,m}(H_{m,n}\,\hat{p}_{n,n}\, H^{\mathrm{T}}_{n,m} + R_{m,m})^{-1} \tag{2.6.17}$$

$$x_{n,1} = x_{n,1} + G_{n,m}(y_{m,1} - H_{m,n}\, x_{n,1}) \quad \text{and} \tag{2.6.18}$$

$$P_{n,n} = (I_{n,n} - G_{n,m}\, H_{m,n})\, \hat{p}_{n,n} \tag{2.6.19}$$

or

$$P_{n,n} = (I_{n,n} - G_{n,m}\, H_{m,n})\, \hat{p}_{n,n}(I_{n,n} - G_{n,m}\, H_{m,n})^{\mathrm{T}} + G_{n,m}\, R_{m,m}\, G^{\mathrm{T}}_{m,n} \tag{2.6.20}$$

9.3.6.3 Linear approximations with multivariable filters. If linear approximations to nonlinear systems are used to design monitoring systems, a suitable state transition multiplier Φ must be derived. In the scalar case:

$$x_{k+1|k}/x_{k|k} = \phi_k \tag{2.6.21}$$

However, matrix division is not defined. Therefore, Φ must have only diagonal elements according to equation (2.6.21). Naturally, a nonlinear model will not be approximated well with such a simple expression of Φ over more than a very few time steps.

9.4 Model based analysis for determining optimal management strategies

9.4.1 Activity scheduling and passive adaptive management

In earlier sections the basic linear system has been given as:

$$x_{k+1} = \Phi_k x_k + w_k \tag{3.1.1}$$

Equation (3.1.1) represents a "free" system, i.e., one without any controls. To represent a controlled system:

$$x_{k+1} = \Phi_k x_k + Bu_k + w_k \tag{3.1.2}$$

where u_k is a vector of controls (i.e. additions to and subtractions from state vector x_k). The matrix B is a diagonal rectangular (possibly square) matrix used to adjust vector u to the same length as x. If u is already the same length as x, $B = I$ and can be omitted.

Because $E[w_k] = 0$, and if u is the same length as x_k:

$$x_{k+1|k} = \Phi_k x_{k|k} + u_k \tag{3.1.3}$$

To express a natural resource management situation as a control problem, consider a plant growth model in which plants can be harvested. Plant growth in this example has two components; one depends on current amount of plant material present and the other is a random event. The control that is applied is the amount of plant material harvested. The system can be represented by:

$$x_{k+1} = \Phi_k x_k + (-u_k) + w_k \tag{3.1.4}$$

where x is amount of plant material at time k. u_k is amount harvested, Φ_k is the "growth multiplier", and w_k is the random growth component. Note that u_k is amount harvested and has a negative sign; if u_k were a generic control, it might be more convenient to use a positive sign as in equation (3.1.3). The conditions at time $k = 0$ (initial conditions) must be stated.

Equation (3.1.4) can be rewritten

$$x_{k+1} - \Phi_k x_k + u_k = w_k \tag{3.1.5}$$

but since E[w] = 0, the system can be modelled as:

$$x_{k+1} - \Phi_k x_k + u_k = 0 \tag{3.1.6}$$

A series through several time steps can be expanded (for single variables) as:

$$\begin{array}{lllllllllll}
x_0 & & & & & & & & & = x(0) \\
-\phi_0 x_0 & + x_1 & + u_0 & & & & & & & = 0 \\
 & -\phi_1 x_1 & & + x_2 & + u_1 & & & & & = 0 \\
 & & & -\phi_2 x_2 & & + x_3 & + u_2 & & & = 0 \\
 & & & & & -\phi_3 x_3 & & + x_4 & + u_3 & = 0
\end{array} \tag{3.1.7}$$

If variables x_k and u_k in equation (3.1.7) are redesignated as a vector of x's, and only coefficients are shown, a matrix of coefficients results. In addition, a vector c showing benefits of harvests can be added with the result as shown below:

state-space variables:

$$x_0 \quad x_1 \quad u_0 \quad x_2 \quad u_1 \quad x_3 \quad u_2 \quad x_3 \quad u_3$$

linear programming variables:

$$\begin{array}{llllllllll}
x_1 & x_2 & x_3 & x_4 & x_5 & x_6 & x_7 & x_8 & x_9 & \\
1 & & & & & & & & & = b_1 \\
-\phi_0 & 1 & 1 & & & & & & & = b_2 \\
 & -\phi_1 & & 1 & 1 & & & & & = b_3 \\
 & & & -\phi_2 & & 1 & 1 & & & = b_4 \\
 & & & & & -\phi_3 & & 1 & 1 & = b_5 \\
 & & c_3 & & c_5 & & c_7 & & c_9 & = z\ ,
\end{array} \tag{3.1.8}$$

where z is total benefits. The initial condition is b_1, and E[w_k] = 0 constitutes the other elements of the b vector (commonly called the right hand side). The matrix of coefficients is commonly called the A matrix, and coefficients of x and u of the original dynamic equation are included in the A matrix. The solution of the redesignated x that maximizes harvest is found by maximizing the scalar $z = c^T x$, called the objective function. In matrix notation:

$$\begin{array}{l} \text{maximize } c^T x \\ \text{subject to } Ax = b,\ x \geqslant 0 \end{array} \tag{3.1.9}$$

which is a standard linear programming problem.

The solution of equation (3.1.8) is appropriate for forage harvest problems in which growth is a linear function of remaining plant material. Of course, a linear plant growth model cannot be valid for more than a few time steps. At the end of a series of time steps, a new initial condition can be set for b_1 and a new optimal path computed. The multiplier ϕ_k may be updated if the new series comes from a different time window of the plant growth model. If forage growth of one period is not expressed as a function of forage of the previous period (i.e., $\phi = 1$), but instead is an explicit statement of the amount of forage growth independent of previous forage, the right hand side or *b* vector must contain the amount of new forage at each period (Jameson & Bartlett 1987).

Many dynamic management systems can be adequately treated by serialized or dynamic linear programming formulations (Bartlett *et al.*, 1974; Propoi, 1979; Propoy, 1985). As outlined, the formulation does not explicitly allow for updating information and is called "open-loop". Because of uncertainties involved in natural resource management, an open-loop approach should be used with caution. If used without updating based on new measurements, open-loop control would eventually lead to solutions in which values of state variables in solution depart appreciably from the true state.

More realistically, information is updated at intervals by new measurements, and new predictions made where appropriate (such as new forage growth predictions). This is a "feedback" approach. Since the formulation does not contain explicit plans to acquire new information, the feedback is "passive adaptive".

Although control problems shown thus far in this section can be stated as serialized or dynamic linear programming problems, the formulation becomes rather tedious if more than a few time steps and state variables are involved. These cases are in the realm of dynamic programming, the subject matter of the next section. Note that serial or *dynamic linear programming*, treated in this section, is a different subject than *dynamic programming.*

9.4.2 Stochastic dynamic programming

The control problems of the previous section can be stated as serialized or dynamic linear programming problems, but it may be easier to generalize to a recursive formulation as:

$$\begin{aligned}&\text{maximize } z = f(x_k, u_k),\ k = 1, 2 \ldots N\\&\text{subject to } x_{k+1} = \Phi_k x_k + u_k + w_k\end{aligned} \quad (3.2.1)$$

where $f(x_k)$ is some (possibly nonlinear) function of x. The vector u_k has now been generalized to include all inputs and outputs, and N is the number of time steps in the planning period. The random effects w_k are unknown, but if $E[w] = 0$, and updating measurements are made, the system can be modelled as:

$$\begin{aligned} &\text{maximize } z = f(x_{k|k}, u_k), k = 1, 2 \ldots N \\ &\text{subject to } x_{k+1|k} = \Phi_k x_{k|k} + u_k \end{aligned} \qquad (3.2.2)$$

Equation (3.2.2) implies that some measurements are taken to give estimates $x_{k|k}$ as specified in preceding sections. At each time step $x_{k|k}$ is updated, and the system is re-optimized over future time steps. Re-optimization based on updated information is the basis of adaptive management.

The recursive formulation of equation (3.2.2) cannot be incorporated into linear programming solution procedures, but is exactly the kind of formulation needed for dynamic programming. There is no restriction (in dynamic programming) that Φ_k be a simple matrix of constants, but can be a matrix of functions. Thus nonlinear relationships can readily be handled with dynamic programming. However, dynamic programming is often computationally expensive, and detailed state transition functions may therefore not be satisfactory.

There are several assumptions so far in this section that influence the appropriate approach to solution of measurement and control problems. The simplest solutions are possible when:

(1) Φ is a matrix of numbers, rather than of functions, i.e. the system is linear.

(2) The cost of control is assessed according to the square of control required, i.e. the system has a quadratic cost function.

(3) $E[w] = 0$ and w is $N(0, Q)$, i.e., the noise (error) is normally distributed (Gaussian).

Systems that possess these linear — quadratic — Gaussian, (LQG) properties have the following important characteristics:

(1) Although the system may be stochastic, the optimal path (solution) is the same as though the system were deterministic. This is the "certainty equivalence" (CE) property. If new measurements are made at intervals, but the system is "certainty equivalent" between intervals, the approach is "updated certainty equivalence" (Taylor & Chavas 1980).

(2) The accompanying measurement procedure can be designed separately from any decision about control vector u. This is known as the separation theorem.

Although strict LQG conditions described above are rare in nature, close approximations may mean that certainty equivalence and the separation theorem can still be applied. Approximations to LQG are marked by:

(1) The system is at least "smooth enough" that it can be perceived as "linear-in-a-neighborhood", i.e. the system can be perceived as linear between measurements.

(2) The cost (or benefit) of control is approximately the same for an upward control as it is for a downward control.

(3) The statistical errors are additive, even if not strictly Gaussian (Bar-Shalom & Tse 1976).

The control portion of systems that possess these properties can be treated as a linear programming problem (or perhaps more conveniently as a deterministic dynamic programming problem), and the optimal monitoring system designed according to procedures of earlier sections. Combinations of linear and dynamic programming may also be useful (Casti, 1983; Fisher, 1985; Williams, 1985).

In contrast to conditions of certainty equivalence are conditions that lead to the "dual-effect". This name arises because the control not only influences state variables, but also uncertainty, i.e.

$$w_k = f'(u_k) \tag{3.2.3}$$

which merely states that the process noise is not additive, but instead depends on the control (Fel'dbaum 1965). This condition holds if the control action influences uncertainty, i.e., deliberate learning is part of the control process (Bar-Shalom & Tse 1976). Also, the certainty equivalence or LQG approach cannot be used if there are significant discontinuities in the system and "linearity-in-a-neighborhood" is not appropriate. In these situations some method must be found to "anticipate" across the discontinuity. Walters (1986) has an excellent presentation of difficulties encountered in natural resource management with this class of problems.

Another case of non-applicability of the CE approach occurs when the penalty function is so strongly skewed that results of overachievement are not comparable to the results of underachievement. In natural resource management, underutilization may not be serious and overutilization may be disastrous. In any of these cases a more conservative strategy than CE is required (Bar-Shalom 1981). An objective function such as max-min (i.e., the minimum value of the objective function over time is maximized) is appropriate when a more conservative strategy is desired (Hof *et al.* 1985). Commonly no other constraints are needed for a solution to be suitably cautious when a max-min objective function is used.

A review of applications of dynamic programming in resource management has been completed by Kennedy (1981), and an excellent elementary discussion of how dynamic programming works is presented by Dreyfus & Law (1977).

9.5 System attributes that dictate model structure and performance

In this paper we have studied various *dynamic systems*, particularly *linear dynamic systems*, or at least systems that could be suitably studied as *linear-in-a-neighborhood.* The theoretical properties of systems that can be managed incrementally are well established (Bar-Shalom & Tse 1976) and have been discussed in terms of biological and ecological systems (Jameson 1986a). It has been well documented that fixed schedule plans for management of ecological systems are not satisfactory, and traditional approaches to analyses of uncertain systems emphasize the importance of using a stochastic model (Karp & Pope, 1984). However, the few studies that have considered LQG properties have concluded that a stochastic model is not always helpful for optimal management (Taylor & Chavas, 1980; Went & Bar-Shalom, 1980).

9.5.1 Elementary catastrophe theory

It was not until recently that we understood why fixed schedule systems so often failed. If model equilibria demonstrate properties of certain of the catastrophe classes, the deterministic or fixed approach will not even approximate the optimal strategy. An important consideration of study of such systems is their equilibria, i.e., to what states does the system migrate because of its own properties (Gilmore, 1981). Some extremely helpful aids in addressing the concept of equilibria are found in the area of catastrophe theory:

> Elementary catastrophe theory is the study of how equilibria of a dynamic system change as control parameters change.

In this section we will look at various *catastrophe manifolds.* These manifolds are system responses surfaces that show response of state variables x resulting from application of various controls u. Although the response has a time dimension, in the catastrophe manifold time trace of the system is not explicitly shown. From the plot of x against u, however, the impact of assumptions of linearity are readily apparent. The basic concepts of both adaptive management and mathematical catastrophies were developed in the 1960's. However, it was not until a decade later that these two concepts were incorporated into studies of ecological systems (Zeeman 1976, Bar-Shalom & Tse 1976, Jones 1977), and were combined even later (Casti 1980).

Several natural resource systems have been modelled as fold catastrophes (McMurtrie & Wolf 1983, Noy-Meir 1982, Walker & Noy-Meir 1982, Walker *et al.* 1981). Jones (1977) presented a model of spruce budworm outbreaks as a cusp catastrophe. Jameson (1987) presented the application of a cusp catastrophe to a grassland-savanna ecotone. Loehle (1985) published a theoretical paper on application of catastrophe theory to grazing, but no concrete examples were included. Johnson & Parsons (1985) studied an example pasture system to collect data on a fold catastrophe response. To date, there has not been a published result that has experimentally examined biological response of a natural resource system to see if cusp catastrophe conditions were statistically discernable.

9.5.2 A coolseason-warmseason grassland example

This example applies to a mixture of coolseason (C_3) and warmseason (C_4) plants. The basic treatments considered will be early season and late season partial harvesting. Typically, we would expect a shift toward warmseason species and away from coolseason species with early season harvesting and vice versa with late season harvesting. Although these reversals are not documented in the scientific literature, they seem entirely reasonable and have been frequently observed in range management. In mountain grasslands, a shift to cool season grasses with late season grazing is commonplace. Shifts from mixed communities to warmseason grasses are frequently observed in mixed prairie as well; the reverse shift to coolseason species is less frequent.

What is not known is whether a shift toward warmseason plants, due to early season harvesting, can be reversed to a shift toward coolseason plants by comparable degree of late season harvesting. Naturally, if harvesting is slight, the impact will be difficult to determine; the impact of severe harvesting will be easier to determine. The expected response surface is shown in Figure 2.

The back (folded) edge of Figure 2 represents three of four properties of a catastrophe model listed by Jones (1977):

1. Bimodality: the system tends to be either in a coolseason state or in a warmseason state; intermediate values cannot be reached directly along the grazing season axis and tend not to occur.
2. Discontinuity (catastrophe): as the controlling factor (grazing season) moves toward either extreme, a point is reached where response can no longer move smoothly, the jump to a different level at this discontinuity is what gives catastrophe theory its name.
3. Hysteresis: The time path that the coolseason/warmseason response

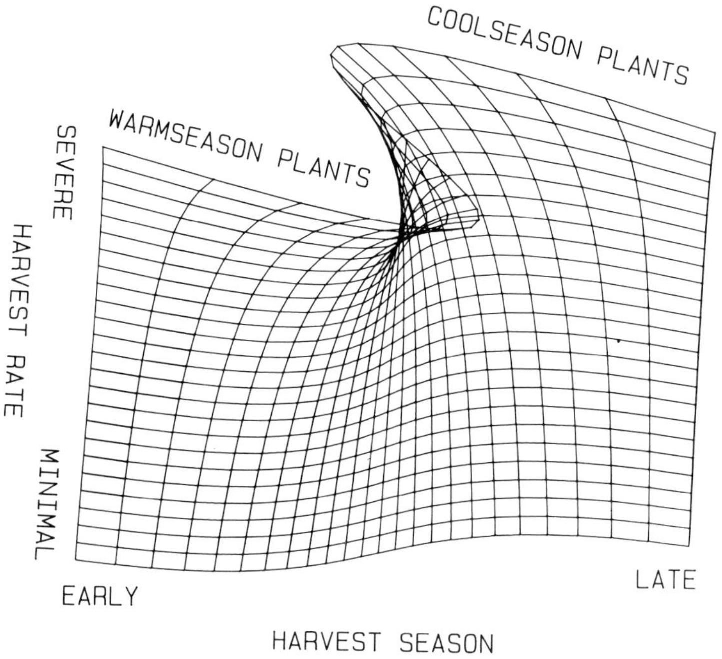

Figure 2. Harvesting effects on a coolseason/warmseason plant community perceived as a cusp catastrophe.

must make as grazing season control moves to the right is different than the response as control moves to the left.

The fourth property of a cusp catastrophe, divergence, is represented in a movement from the front edge toward the back edge of Figure 3.

In pastures managed as a single species, LQG conditions described earlier are reasonable, and the incremental adaptive approach is generally acceptable. Under these conditions, we can use simple observations of vegetation and animal performance to make adjustments to grazing schedules and stocking rates; the major problem is in knowing what to observe. In some vegetation types, such as the shortgrass plains, this method will be very useful as long as the system is constrained to the "warmseason" lower plateau of Figure 2. It is only when control extends beyond the discontinuity that cusp behavior becomes critical.

Almost any natural resource system could be perceived and modelled as LQG, or conversely could be modelled as non-LQG. The real question is, does it make any difference? A useful approach is to model the system as non-LQG, then use experiments on the model to determine if non-LQG properties of the model lead to management decisions that are different from those reached using LQG assumptions. This has been done in several natural resource management studies (Rodriguez & Taylor,

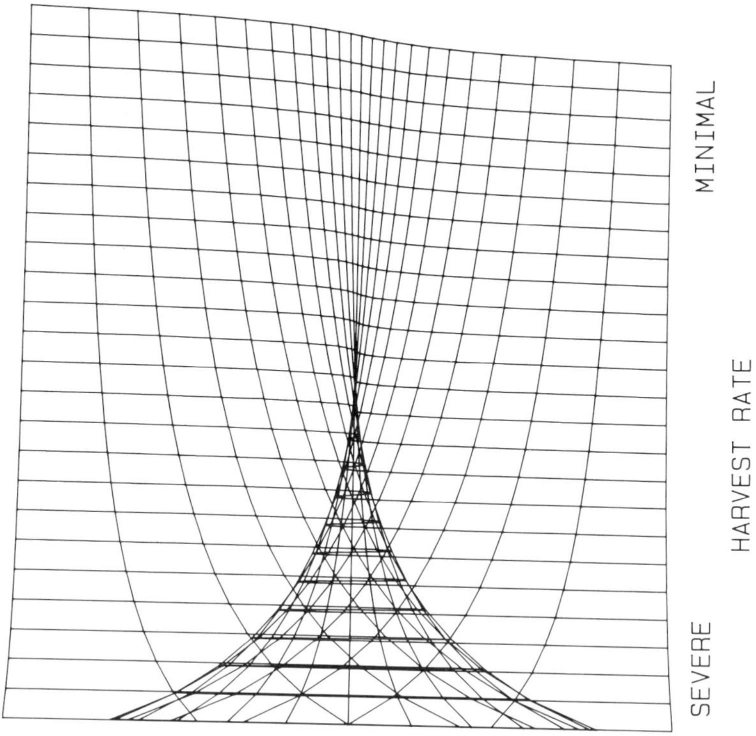

Figure 3. An overhead view of Figure 2. Lines represent the path of plant communities during a change from little harvest impact to severe harvest impact. Overlapping lines indicate the "bifurcation" where the pathways move into the folded area of Figure 2.

1988), and will continue to be used in many applied problems, but has the obvious disadvantage that it depends on structure of the model rather on measured behavior of the natural system.

An alternative approach is to directly examine the behavior of the natural system to see if responses are such that a non-LQG approach must be used. For some systems, the LQG approach may lead only to inefficiencies rather than to "catastrophes"; for other systems the consequences are more severe. The latter include fold and cusp catastrophes described earlier in this section. If such properties can be found experimentally, the system will respond catastrophically if managed incrementally (i.e., with a passive adaptive or LQG approach). For grazing management, lack of "cusp catastrophe" behavior implies that satisfactory management corrections could be based on observations of plant and animal responses without leading to plant community destruction. If "cusp catastrophe" conditions are found, then errors in grazing scheduling cannot be corrected merely by reversing the harvest season and changing harvest rates.

9.5.3 Algebraic models v. mechanistic models

The models described in this section are only conceptual. The surface for the figures was generated from:

$$-(x^3 + u_1 x + u_2) = 0 \qquad (4.3.1)$$

where:

x is coolseason/warmseason response,
u_1 is control along the grazing intensity axis, and
u_2 is control along the grazing season axis.

The algebraic model also has some other interesting properties in addition to those mentioned previously. Note that in the zone between the edges of the fold in Figure 2, the equation gives three solutions. However, for a given response and position along the response axis, there is a single input or control. Thus it may be possible to reconstruct from ecological evidence the causes leading to a particular response (i.e., solve equation (4.3.1) for u_1 or u_2 rather than for x). However, prediction of the response may not be possible without knowing the ecological history of the community.

Obviously the algebraic equation does not represent any ecological reality, but is the simplest mathematical form that will generate the desired surface (Zeeman 1976, Jones 1977, Fararo 1978, Gilmore, 1981). On the other hand, most models in the literature that were cited earlier in this section (see especially Johnson & Parsons, 1985; Noy-Meir, 1982) appear to be more biologically realistic and may have adequate detail to demonstrate catastrophe properties. However, these appealing models may contain parameters not readily identifiable with known biological states or processes, and thus may not be experimentally determinable. Highly mechanistic models, especially if they contain many components, may be even more satisfying biologically. However, the larger and more complicated the model, the more difficult it is to observe important attributes of system behavior that make the catastrophe model necessary for management in the first place. If catastrophic system attributes are not identifiable, then the same management conclusions will be reached with a simpler linear-in-a-neighborhood approach.

9.6 Summary

The major topics covered in this paper are (1) statistics of model-based monitoring systems, (2) model based analyses for determining optimal management strategies, and (3) system attributes that dictate model structure. Techniques for construction of simulation models are not included, but can be found in the references.

The statistics of model-based monitoring are covered extensively in engineering and aerospace literature, and are referenced here in several texts. However, modifications to these approaches are needed for limited data bases typical of natural resource systems, and these modifications have been presented in this paper. Calculations of appropriate variances for model-based estimates are straightforward for linear and linear-in-a-neighborhood systems.

Selection of optimal management strategies using linear programming or dynamic programming is referenced in several works. This paper includes methods for coupling simulation models with these optimization models. The coupling may be within the A matrix of a linear programming formulation, as the b or right hand side vector of a linear programming formulation, or in the recursion equation of a dynamic programming formulation. Linear programming is restricted to linear and linear-in-a-neighborhood problems, but dynamic programming can treat nonlinear functions equally as well.

Thus both statistical and optimal strategy selection questions are well in hand for linear and linear-in-a-neighborhood problems. If these formulations are acceptable, incremental management guided by a suitable predictive model is also acceptable. Although biological and physical scientists may be more comfortable expressing phenomena of their studies with nonlinear functions, usually even complicated nonlinear functions can be treated as linear-in-a-neighborhood for monitoring and management purposes. The most critical cases where these methods are not appropriate is when discontinuities are critical to behavior of the system. Fold and cusp catastrophe manifolds express these conditions, but are seldom recognized in ecological models.

This paper concludes that straightforward procedures are available for modelling both monitoring and adaptive management aspects of range management as long as the system can be adequately treated as linear or linear-in-a-neighborhood. If the system cannot be so modelled because it behaves as a fold or cusp catastrophe, then the model should explicitly address catastrophe behavior.

10.7 References

Anderson, B. D. O., and J. B. Moore. 1979. Optimal filtering. Prentice-Hall, New York.

Bar-Shalom, Y. 1981. Stochastic dynamic programming: caution and probing. IEEE Trans. Automat. Contr. AC-26: 1184—1195.

Bar-Shalom, Y., and E. Tse. 1976. Concepts and methods in stochastic control. Control and Dynamic Systems 12: 99—172.
Bartlett, E. T., G. Evans, and R. Bement. 1974. A serial optimization model for ranch management. J. Range Manage. 27: 233—239.
Carande, V., and D. A. Jameson. 1986. Combination of weight estimates with clipped sample data. J. Range Manage. 39: 88—89.
Casti, J. L. 1980. Bifurcations, catastrophes and optimal control. IEEE Trans. Automat. Contr. AC-25: 1008—1011.
Casti, J. 1983. Forest harvesting and harvesting policies. Appl. Math. and Computation 12: 19—48.
Chin, L. 1979. Advances in adaptive filtering. Control and Dynamic Systems 15: 277—350.
Dreyfus, S. E., and A. M. Law. 1977. The art and theory of dynamic programming. Ch 1. Academic Press, New York.
Emara—Shabaik, H. E., and C. T. Leondes. 1981. Non-linear filtering — the link between Kalman an extended filters. Internat. J. Control 34: 1207—1214.
Fararo, T. J. 1978. An introduction to catastrophes. Behavioral Sci. 23: 291—317.
Fel'dbaum, A. A. 1965. Optimal control systems. Academic Press. New York.
Fisher, I. H. 1985. Derivation of optimal stocking policies for grazing in arid regions. I. Methodology. Appl. Math. and Computation 17: 1—35.
Gelb, A. 1974. Applied optimal estimation. MIT Press, Cambridge, Massachusettes.
Gilmore, R. 1981. Catastrophe theory for scientists and engineers. Wiley, Chichester.
Hazell, P. P. R., and R. D. Norton. 1986. Mathematical programming for economic analysis in agriculture. MacMillan, New York.
Hof, J., J. Pickens, and E. T. Bartlett. 1985. A maxmin approach to nondeclining yield timber harvest scheduling problems. Forest Science 32: 653—666.
Jameson, D. A. 1985. A priori analysis of allowable interval between measurements as a test of model validity. Appl. Math. and Computation 17: 93—105.
Jameson, D. A. 1986a. Models vs measurements in grazing systems analysis. Proceedings of Society for Range Management Special Symposium, Feb. 1986, Kissimee, FL.
Jameson, D. A. 1986b. Sampling intensity for monitoring of environmental systems. Appl. Math. and Computation 18: 71—75.
Jameson, D. A. 1987. Climax v alternative steady states in woodland ecology. Proceedings of Woodland Ecology Conference, Jan. 1986, Reno, Nevada.
Jameson, D. A., and E. T. Bartlett. 1987. Selection of optimal management strategies based on stochastic dynamic ecological models. Ecological Modelling 36: 5—13.
Jazwinski, A. H. 1970. Stochastic processes and filtering theory. p. 269—272. Academic Press. New York.
Johnson, I. R., and A. J. Parsons. 1985. A theoretical analysis of grass growth under grazing. J. Theor. Biol. 112: 345—367.
Jones, D. 1977. Catastrophe theory applied to ecological systems. Simulation 29: 1—15.
Karp, L., and A. Pope. 1984. Range management under uncertainty. Am. J. Agr. Econ. 66: 437—446.
Kennedy, J. O. S. 1981. Applications of dynamic programming to agriculture, forestry and fisheries. Rev. Marketing and Agr. Econ. 49: 141—173.
Loehle, C. 1985. Optimal stocking for semi-desert range: a catastrophe theory model. Ecological Modelling 27: 285—297.
McMurtrie, R., and L. Wolf. 1983. A model of competition between trees and grass for radiation, water and nutrients. Ann. Bot. 52: 449—458.
Maybeck, P. S. 1979. Stochastic models, estimation, and control. Vol. 1. Academic Press, New York.

Noy-Meir, I. 1982. Stability of plant-herbivore models and possible application to savanna. Ecol. Stud. 42: 591—609.
Propoi, A. 1979. Dynamic linear programming models for livestock farms. Behavioral Sci. 24: 200—207.
Propoy, A. I. 1985. Problems and methods of dynamic linear programming. Engineering Cybernetics 21: 109—123.
Rodriguez, A., and G. Taylor. 1988. Livestock and forage management as a stochastic control problem. Am. J. Agr. Econ. 70: 121—132.
Sisler, J. F., and D. A. Jameson. 1983. Optimal multi-observation monitoring systems, a practical approach. p. 235—240. *In*: W. K. Lauenroth, G. V. Skogerboe and M. Flug (eds.) Analysis of ecological systems: state of the art in ecological modeling, Elsevier.
Soeda, T., and T. Yoshimura. 1973. A practical filter for systems with unknown parameters. J. Dynamic Systems, Measurements and Control 95: 396—401.
Starfield, A. M., and A. L. Bleloch. 1986. Building models for conservation and wildlife management. MacMillan, New York.
Taylor, C., and J. Chavas. 1980. Estimation and control of an uncertain production process. Amer. J. Agr. Econ. 62: 675—680.
Walker, B. H., D. Ludwig, C. S. Holling, and R. M. Peterman. 1981. Stability of semi-arid savanna grazing systems. J. Ecol. 69: 473—498.
Walker, B. H., and I. Noy-Meir. 1982. Aspects of the stability and resilience of savanna ecosystems. Ecol. Stud. 42: 556—590.
Walters, C. 1986. Adaptive management of renewable resources. MacMillan, London.
Wenk, C., and Y. Bar-Shalom. 1980. A multiple model adaptive dual control algorithm for stochastic systems with unknown parameters. IEEE Trans. Automat. Contr. AC-25: 703—710.
Williams, B. K. 1985. Optimal management strategies in variable environments: stochastic optimal control methods. J. Environ. Manage. 21: 95—115.
Wilson, A. G., and R. J. Bennett. 1985. Mathematical models in human geography and planning. John Wiley & Sons, Chichester.
Zeeman, E. 1976. Catastrophe theory. Sci. Am. 234(4): 65—83.

PART TWO

Vegetation Distribution and Organization

10. Vegetation-soil relationships on arid and semiarid rangelands

S. G. LEONARD, R. L. MILES AND P. T. TUELLER

10.1 Abstract

Vegetation-soil relationships on rangelands, or any other lands must include all ecosystem components. That soil is a function of climate, organisms, topography, parent materials, and time establishes the basic model. Rangeland plant community distribution and species composition are known to be related to specific soil properties such as soil climate (moisture and temperature), texture, depth, structure, fertility, pH, salinity and toxic influences. These properties relate to spatial variability of vegetation responses ranging from broad geographic distributions to landscape influences to specific site characteristics.

Vegetation-soil relationships are discussed relative to classification systems, soil morphology and chemistry, and data extrapolation for management interpretations.

10.2 Background

The subject of vegetation-soil relationships is not new. Re'aumer recognized over 250 years ago that the development of most plants, animals and insects was affected by various environmental factors such as soil characteristics, nutrient levels, and atmospheric variables (Richardson and Leonard 1981). The Russian soil scientist, Dokuchaev, in 1898, described soil as a function of climate, vegetation and animals, geologic substratum and relative age. This initial formulation has developed into the five soil forming factors of climate, organisms, topography, parent material and age (Jenny 1983). Jenny further elaborates on total ecosystem relationships through "state factor analysis"; the state factors being synonymous with the soil forming factors.

A. W. F. Schimper in 1898 introduced the word edaphon (soil) into

Tueller, P. T. (ed.), Vegetation science applications for rangeland analysis and management.
ISBN 90-6193-195-9.
© 1988, *Kluwer Academic Publishers, Dordrecht, Boston, London.*

the botanical literature (Jenny 1983). Since then, botanists and plant ecologists, including such names as Clements, Daubenmire, Dyksterhuis, Kuchler, Tuxen and many others have included discussions of vegetation-soil relationships even though often emphasizing climatic relationships or geographic distribution. The fact that vegetation, soil and climate are intimately related in ecosystem functions is well established.

Present day range ecologists direct their energy toward improving the established concepts and models as they relate to rangeland ecosystems. The study of vegetation-soil relationships necessarily includes a combination of synecology and autecology. The kind and amount of research done on rangeland soils and vegetation has not approximated the work done for domestic crops and commercial forests, in part because of the relatively low economic value per unit area. The degree of variability and the vast expanses involved has led most range ecologists to generalize at some level in order to handle the large amounts of data required to understand range ecosystems. This generalization has been accomplished through various kinds of vegetation and soil classification schemes.

10.3 The relationship of vegetation classification to soil classification

Classification systems in general are developed as a tool for communication, for aggregation of information into logical units, for interpretation, and for the extrapolation of interpretations and information among units with similar properties. The utility of one system when compared to another and the intensity or scale of classification often hinges on an individuals perspective in relation to intended use and interpretation.

It is often assumed that plant communities can be consistently correlated with soils (classification units) because of their shared relationship in the five soil forming factors (Tiedeman and Terwilliger 1979). This assumption is erroneous because soil classification does not necessarily parallel vegetation classification even though relationships exist between soil properties and vegetation communities. Soil classifications are based on observable or measurable soil characteristics. Since soils may vary continuously over the landscape, classifications often set arbitrary boundaries based on interpretational needs that may or may not include natural vegetation communities. Likewise, vegetation classifications must often set arbitrary boundaries to separate continua for use and management interpretations. It is no wonder that soil and vegetation classifications seldom exhibit a 1 : 1 relationship. However, since many soil boundaries are based on landform boundaries, a better correlation often exists between soil and vegetation map units than classification units.

Shantz stated in 1938 "In the classification of soils, only soil characteristics should be considered, and in the classification of vegetation, only characteristics strictly limited to vegetation. These independent groupings can then be compared." This approach is still valid if applied correctly. Vegetation-soil relationships documented in the United States classification systems have generally used this approach. In contrast, Tiedeman and Terwilliger have successfully used a "phyto-edaphic" classification that integrates both soil and vegetation characteristics. The phyto-edaphic unit (PEU) is management oriented rather than strictly ecological.

Most nations develop systems of soil classification that reflect their own preferences and needs. The United States has developed a comprehensive Soil Taxonomy (USDA 1975) integrating both soil genesis and soil morphology into a hierarchical system of six categorical levels. Soil Taxonomy has become the standard in the U.S. but is far from static. The use of Soil Taxonomy over a 10 year period has prompted over 150 changes, agreed upon through the efforts of the National Cooperative Soil Survey (NCSS), an organization with representatives from various agencies and other users of Soil Taxonomy. Consistency in terminology, classification and correlation is maintained through this cooperative effort.

Brief descriptions of the six categories of the soil taxonomy follows: (From Driscoll *et al.* 1983)

Orders. There are 10 orders defined by the presence or absence of identifiable horizons that reveal the major soil-forming processes, such as accumulation of organic matter.

Suborders. There are 44 suborders defined by important properties that influence soil development and plant growth, such as wetness and major kinds of parent material.

Great Groups. There are 186 great groups defined by similarities in kind, arrangement, and distinctiveness of horizons as well as close similarities in soil, moisture, temperature regimes, and base status.

Subgroups. There are 987 subgroups defined by characteristics already described for the higher classes. These characteristics are subordinate because their presence modifies the dominant soil-forming processes.

Families. There are 5,417 families defined by physical and chemical characteristics that affect soil use and response to management. Particle size, mineralogy, temperature regime, and depth of root penetration are examples. This level in the hierarchy contains features that are most closely correlated with the distribution of natural vegetation.

Series. There are about 17,000 series recognized by the kind, thickness, number, and arrangement of horizons. The range of properties, such as particle size, used to define series may be more restrictive than those used at the family level.

Phases are not part of the soil taxonomic hierarchy but are utilitarian created to serve specific purposes in individual soil surveys. They are not part of the soil taxonomic hierarchy. Any property or combination of properties that does not duplicate class limits for a taxon can be used to differentiate phases. The choices of properties and limits are determined by the purpose of the particular soil survey and by how consistently the phase criteria can be applied. In practice, phases can be defined for any class of any category (USDA 1980). Some of the features commonly used in defining phases in soil surveys are texture of the surface layer; presence of rock fragments, such as gravel or stone; slope; climatic characteristics, such as moisture and temperature; salinity; kind of substratum; erosion or flooding potential. Thus phases at any level in the hierarchy may also delimit certain vegetation parameters.

To date no universally recognized classification or taxonomy has evolved for organizing and evaluating knowledge about vegetation communities. However, a few major systems have evolved within U.S. government agencies and the academic community. The data used involve different vegetation parameters and/or attributes depending upon the kind of community (grass, grass-shrub; forest, etc.) and the desired interpretation (forage, watershed, wildlife habitat, etc.) This lack of consistency has seriously impaired the ability to extrapolate data for use and management interpretations and applied research activities.

The susceptibility of vegetation to change in short periods of time as compared to soils has presented a dilemma for vegetation classification as well as the establishment of vegetation-soil relationships. Succession and retrogression characteristics due to both management and natural perturbations and species interactions within communities result in very different plant communities occurring on the same or similar soils. On the other hand, many seral species occur over a wide range of soil properties and other environmental factors resulting in apparently similar plant communities on contrasting soils capable of producing a different potential plant community. This dilemma has led to two primary strategies in vegetation classification; one based on present vegetation attributes and one based on potential or climax attributes.

The classifications most extensively used systems used on rangelands in the United States are range sites (Shiflet 1973, USDA 1976) and habitat types (Daubenmire 1970, Hironaka *et al.* 1983, Mueggler and Stewart 1980). Both are based on the potential or climax plant community, at least as perceived by the classifiers. The most obvious differences between the two appear to be the vegetation attributes (species production vs structure and cover) used for classification and level of detail. Both depend heavily on a prior knowledge of vegetation-soil relationships (as well as climate and topographic relationships) to classify landscapes in the absence of the potential or climax vegetation. As stated by Passey *et al.* (1982), "Remnants of climax vegetation are scarce in most western rangeland areas. Thus, approximation of the probable plant-community potential for many sites has necessarily been empirical (i.e. speculative)." The range site and habitat type are closely related to the ecotope classification used in South Africa (Mentis) and the pasture, as used in Australia, except that the potential implied is South Africa and Australia is the management potential, which may not be ecological climax.

Other vegetation classifications, e.g., the International Classification and Mapping of Vegetation (UNESCO 1973) and the closely related Vegetation Classification System applied to Southern California (Paysen *et al.* 1980) emphasize present vegetation communities either independently or in conjunction with soils, climate and topographic attributes. Paysen and co-workers state that an ecological relationship defining site potential will eventually be determined.

In Australia Perry (1960) and Mabbutt *et al.* (1963) have described general land classifications. Mabbutt describes a land system classifications in which each unit is "an area or group of areas throughout which there is a recurring pattern of topography, soils and vegetation". Present vegetation is emphasized in relation to pastoral interpretations.

A 1 : 1 relationship between soil classification and vegetation classification is not apparent in any of the strategies discussed. However, all strategies provide an initial framework for determining vegetation-soil relationships as indicated by Shantz (1938). We must realize that the classification provide only an initial basis for establishing relationships that must be quantified and refined through additional information and research. Relating specific vegetation communities and species attributes to soil properties can provide a reliable basis for extrapolation of both ecological and management induced community changes. The authors suggest that the relationship between vegetation and soil classifications can be conceptually portrayed (Figure 1) using the United States Soil Taxonomy as the soil base. The soil family defines general physical and chemical characteristics and is further separated into series which define specific limits of individual soil attributes. In Figure 1, a unique assemblage of plants representing a potential or climax plant community may have a narrower range of associated soil properties than a seral community dominated by one or two species with broad ecological amplitude.

10.4 Soil-climate relationships

Soil moisture regimes and soil temperature regimes are collectively and informally referred to as soil-climate as opposed to ambient climate. When

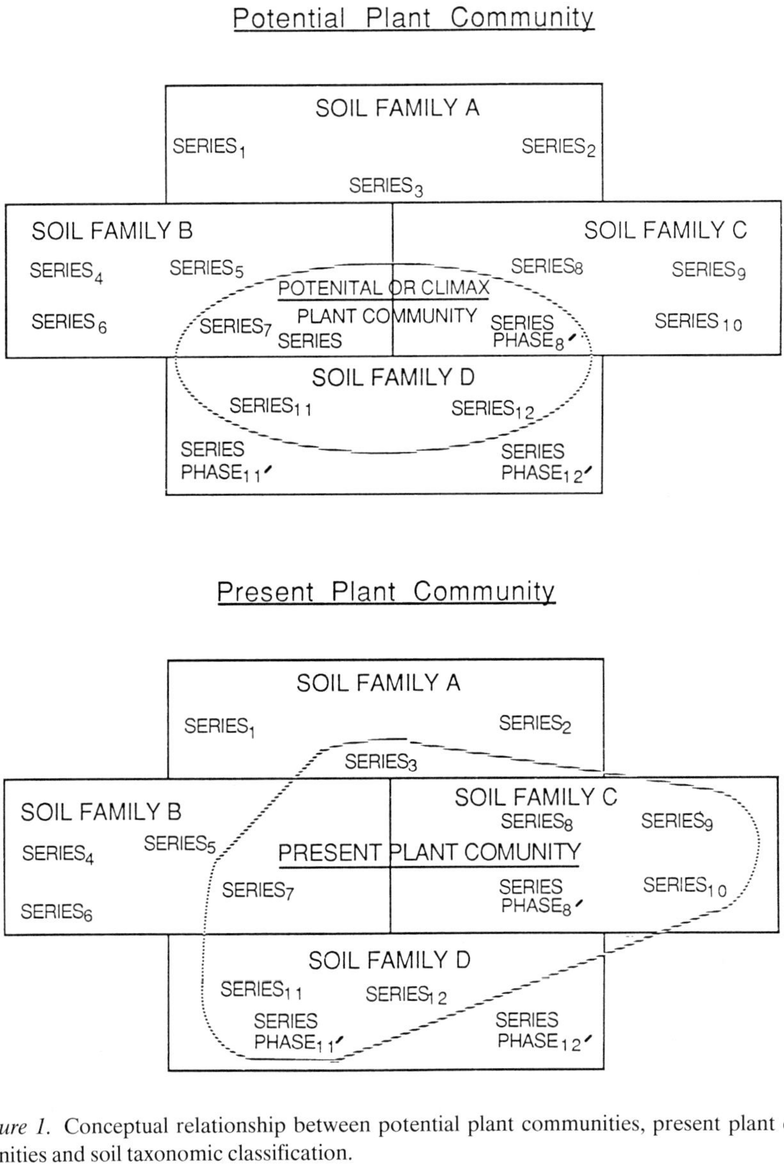

Figure 1. Conceptual relationship between potential plant communities, present plant communities and soil taxonomic classification.

one considers the complex interaction between soil, vegetation and ambient climate, it is not surprising that geographic differences occur in soil-climate and vegetation relationships. Several authors have discussed plant community evolution and migrations through the various geological periods and epochs (Daubenmire 1978). Reveal (1979) and Stutz (1983) have discussed this process as it relates to portions of the Intermountain

Western U.S. The various glacial epochs have evidently led to the movement of northern species to the south with each glacial advancement and the movement of southern species to the north with each recession. These processes may have led to an evolution of ecotypes within species as well as new species, each adapting to slightly different soils and climates. According to Stutz (1983), members of the genus *Atriplex* have especially undergone an evolutionary explosion of hybridization and polyploidy since the recession of pleistocene lakes in the western U.S. to take advantage of "new" habitats. This migration and genetic adaptation is a continuing process.

Major vegetation biomes (Odum 1971) or provinces (Daubenmire 1978) have been associated with climatic influences, especially temperature and precipitation in association with geographic and topographic positions. Climographs plotting ambient mean monthly temperature with mean monthly precipitation have been used to characterize major biomes (Kormandy 1969). Temperature and moisture ultimately control all biological processes in the soil and are important soil characteristics (Smith *et al.* 1964). Soil moisture and temperature are directly influenced by the ambient climate characteristics and further modified by vegetation cover and topography and soil properties.

Soil moisture and soil temperature are collectively referred to as soil-climate. The effects of soil-climate are essentially the major determining criteria separating rangeland ecosystems from other natural ecosystems (e.g., forests and true deserts). In addition, the soil-climate influences, at least indirectly, all soil-plant relationships within rangeland ecosystems. Rooting depth, water potentials, nutrient uptake and even nutrient distribution are all influenced by the amount and timing of soil moisture availability in relationship to the critical temperatures at which root activity occurs.

Soil-climate is used for defining at least some classes at all levels of the U.S. system of Soil Taxonomy from the major soil-forming processes associated with the soil orders to "moist", "dry", "cool" or "warm" phases of individual soil series for specific interpretations. Of the major soil orders in the U.S., Aridisols and Mollisols are associated with the majority of native rangeland in the western U.S. Aridisols are mostly assoicated with the desert shrub and grass-shrub ecosystems of the Great Basin and Intermountain ranges while Mollisols are associated with the transition to forest in the mid elevations of the west, prairie grasslands in the midwest and transition to the forests of the eastern U.S. Other soil orders are included in less extensive but important range ecosystems such as riparian areas (the wetter components of almost any soil order) and many flood plains (commonly Entisols in the arid portion of the western U.S.).

10.4.1 Soil moisture and temperature regimes

Specific soil moisture and temperature regimes are defined in soil taxonomy by annual and seasonal distribution of temperature class limits and moisture availability. At the family level of soil classification combinations of moisture and temperature regimes are incorporated that have proven useful in predicting plant community distribution and in some cases individual species distribution on rangelands.

Moisture regimes reflect the time and distribution of precipitation (or water table for aquic) because they are based on potential wetting depth and period of time the soil is moist during the growing season. Soil taxonomy recognizes five broad categories for moisture regimes (aquic, udic, ustic, xeric and aridic). These are listed from the wettest to driest situation. Some subgroups describe gradations between the categories.

Soil temperature classes are: hyperthermic, thermic, mesic, frigid, cryic and pergelic. These are listed from the hottest to coldest condition. The regimes have a difference of more than 5°C between mean summer and mean winter soil temperatures except for cryic and pergelic and unless an organic layer or water table is present. If the prefix so becomes precedes frigid, mesic, thermic or hyperthermic, the difference between mean summer and mean winter soil temperature is less than 5°C. Soil Taxonomy (USDA 1975) should be consulted for exact definition.

Hironaka *et al.* (1983) describe soil-climate regimes in association with other soil features for sagebrush (*Artemisia* spp.) — grass habitat types in southern Idaho (Figure 2). Soil climate regimes are also discussed by Mueggler and Stewart (1980) in describing grassland and shrubland habitat types in western Montana. Sasich and Nielsen (1984) described specific plant species as indicators of soil-climate in Montana. Among the grassland species they found to indicate soil-climate were: (1) Idaho fescue (*Festuca idahoensis*) = cryic/ustic; (2) Columbia needlegrass (*Stipa columbiana*) and richardson's needlegrass (*Stipa richardsonii*) = cryic/udic; and (3) needleandthread (*Stipa comata*) = aridic and/or frigid. Associated species that are used as temperature regime indicators include big sagebrush (*Artemisia tridentata* var. wyomingensis) = frigid temperature regime and big sagebrush (*Artemisia tridentata* var. vaseyana) and low sagebrush (*Artemisia arbuscula*) = cryic temperature regimes.

An analysis of soil survey data in the Great Basin (Platou *et al.* 1986) indicates both similarities and differences in soil climate relationships with several grass species common to southern Idaho habitat types, Montana habitat types and the Great Basin region of Nevada and Utah. The differences found indicate that there may be variation between geographic areas and other associated soil properties and climate variables.

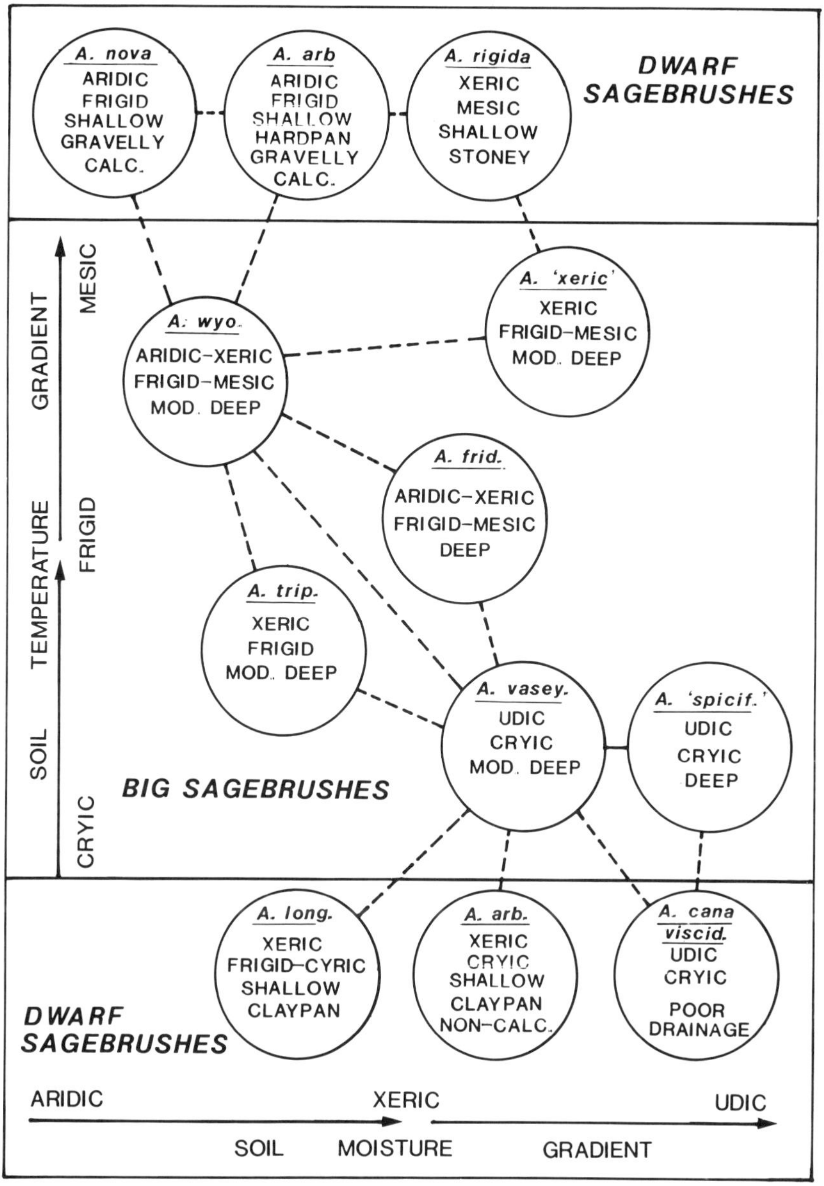

Figure 2. Conceptual relationship of sagebrush species and subspecies based on soil moisture, temperature and other soil characteristics.*

* From Sagebrush-Grass Habitat Types of Southern Idaho, Hironaka *et al.* (1983) with permission.

Ecotypic variation within species may be one explanation for these geographic variations. Another possible explanation is that studies conducted in local geographic areas may represent one extreme or the other of the adaptive range for individual plant species. We must also remember that moisture and temperature regimes are soil taxonomic criteria with specified limits that may not always correspond with specific physiologic responses of range plants.

10.4.2 Soil-climate gradients

Gradients in soil temperature and moisture from the surface through the effective root zone over time influence a plants ability to be metabolically active and extract moisture and nutrients at various depths. Adaptation of range plants to respond to variability in temperature and moisture gradients between soils and within soils over time affects species distribution, their dynamics within plant communities and even their relationship to seral status or range condition.

The soil temperature regime is based on mean annual temperature and mean summer temperature (measured at 50 cm in the U.S.). Plants are responding to daily or even hourly temperature fluctuations within the root zone and to other environmental factors. The mean temperatures used for classification may not reflect extreme events that affect plant species.

Daily soil temperature fluctuations are affected by clouds, vegetation, length of day, soil color, soil slope, soil moisture, air circulation near the ground and temperature of any rain that falls. Daily changes in air temperature have a significant effect on the soil temperature, particularly in soils of dry climates where the daily range in temperature at 2.5 cm may approach 55°C. At the other extreme, under melting snow, the temperature at the soil surface may be constant throughout the day (USDA 1975).

Soil moisture regimes are based impart on the identification of a "moisture control section." The upper and lower boundaries of this control section are determined by the depth to which a dry (tension 1.5×10^6 Pa) soil will be moistened by 2.5 cm of water within 24 hours and 7.5 cm of water within 48 hours respectively (USDA 1975). Other restrictions apply and it is recognized that uneven moistening occurs for a variety of reasons. The amounts of water governing the moisture control section may not occur in the time periods required by plants and many native plants do not "recognize" the 1.5×10^6 Pa tension as too dry to maintain turgor. Therefore, we must look beyond the taxonomic limits of soil moisture regimes to the actual gradients of soil moisture both with depth over time and laterally as influenced by vegetation and topography.

Soil moisture gradients can vary markedly between soils of similar climate as well as within individual soils over time. Soil properties that can directly affect the moisture gradient include texture (and kind of clays), horizon boundaries, structure, soluble salts, rock fragments, depth and temperature. Topographic variables such as slope, aspect and elevation affect moisture gradients, other factors being similar. Microrelief characteristics such as surface roughness and depressions are also factors. Vegetation further modifies the gradients through interception, cover of both live material and residual litter, root distribution and transpiration characteristics.

Many authors, too numerous to cite here, have documented germination characteristics of rangeland species in relation to temperature and moisture at or near the soil surface. Many soils exhibit conditions when either temperature or moisture are favorable for germination of one species or another. However, Jordan (1983) states that both temperature and moisture may be favorable for germination for no more than 3 to 7 days each year. In addition to germination, establishment and persistence over time are also dependent on soil temperature and moisture.

Blaisdell (1958) indicates that in the upper Snake River plains, *Poa secunda* begins sprouting at soil temperatures about 0°C or as soon as snow disappears, followed by *Poa nevadensis* and *Koeleria cristata*, then for *Agropyron spicatum*, *A. dasystachyum* and *Stipa comata* progressively later. Pearson (1979) found that *Oryzopsis hymenoides* commenced growth when soil temperature stayed at 4°C for at least 3 or 4 days (about the same as for *Stipa comata*) and he correlated maximum plant height with soil temperature at 20 cm depth.

The stratification of species response to temperature in relation to soil moisture can account for annual differences in total production and in the production of individual species as found by Passey *et al.* (1982). Differences as high as 1200% between years have been found for range ecosystems in the Great Basin (Tueller and Monroe 1975).

A plants ability to withstand moisture stress is undoubtedly a tremendous factor in community dynamics as is its ability to occupy different sites. Barker and McKell (1986) found that soil-water potentials decreased faster in soils occupied by Wyoming big sagebrush (*Artemisia tridentata* var. *wyomingensis*) than in contiguous soils occupied by basin big sagebrush (*Artemisia tridentata* var. *tridentata*). Leaf-water potentials were also significantly lower for Wyoming big sagebrush than for basin big sagebrush from mid-June to September. Their results supported earlier suggestions that Wyoming big sagebrush plants tolerate drier soil conditions than basin big sagebrush.

In a study between big sagebrush/galleta grass (*Hilaria jamesii*) and shadscale (*Atriplex confertifolia*)/galleta grass communities, Everett *et al.*

(1977) found that soil-water potentials varied among depths and communities but differences were not consistent. However, plant water potentials of big sagebrush and shadscale were consistently lower than those of galleta or rabbitbrush (*Chrysothamnus viscidiflorus*). They concluded that big sagebrush and shadscale were physiologically capable of maintaining low water potentials for extended periods while rabbitbrush and galleta grass reflected current conditions. Also, early morning soil water gradients indicated water movement into the zone of greatest root concentration in both communities. This information suggests that as soil water gradients change between sites, different root systems are favored, and a species may become more or less adapted if other factors remain the same.

Range condition as reflected by the kind, amount and proportion of species within a community is directly related to water use differences exhibited between species. The effects of soil-water depletion by undesirable species on forage production have been addressed by Sturges (1977), Ueckert (1979) and Sosebee *et al.* (1982) to name a few.

Plants directly influence the effectiveness of precipitation in relation to soil moisture retention and spatial distribution. Rauzi *et al.* (1968) documented decreasing water intake rates and accompanying declines in herbage production as range condition declined on mixed and true prairie rangelands in the midcontinental U.S. These declines were attributed to a combined effect of the kind and amount of range cover as well as changes in the surface structure of the soil.

Plant height and community structure are important factors influencing transport or retention of blowing snow and its contribution to soil moisture (Sturges and Tabler 1981). Some plants such as mulga (*Acacia aneura*) in Australia intercept virtually all rain falling on them resulting in extreme spatial variation in wetting depth and distribution. Deep moisture penetration occurs near the stem while the soil remains dry under the canopy and intermittently moist and dry in the plant interspaces (USDA 1975).

10.4.3 Soil-climate models

Regardless of the direct implications of soil temperature and moisture dynamics within and between plant communities, the fact remains that meaningful information is often difficult and always time consuming to acquire in the detail necessary to establish credible soil-plant relationships on rangelands. Mathematical modeling has provided a tool for predicting vegetation responses from more easily measured ambient climate data and soil physical and chemical properties generally available from soil surveys.

Many attempts to model moisture relationships on rangelands correlated annual production with annual or seasonal precipitation, bypassing soil characteristics altogether. More complex models such as the Ekalaka Rangeland Hydrology and Yield Model (ERHYM) (Wight and Neff 1983) and Soil-Plant-Air-Water (SPAW) models (Saxton et al. 1974) and the Simulation of Production and Utilization (SPUR) (USDA 1983) model involve calculation of soil moisture input characteristics over time in conjunction with evapotranspiration characteristics. These more complex models generally provide more satisfactory results for production estimates as well as additional information; however, they are not well suited to evaluate species dynamics either within or between plant communities.

The heat unit approach to plant development has been in use for over two centuries (Wang 1960), especially on croplands. Phenological development, height and productivity relationships of *Agropyron cristatum* and *Hilaria jamesii* to heat unit accumulation, soil moisture and day length response (Richardson and Leonard 1981) indicate this approach may have application to modeling plant community dynamics on rangelands.

Many vegetation relationships with other soil properties are either directly or indirectly related also to soil moisture and temperature. In areas with similar climate, it's often easier to make direct associations between chemical and physical soil characteristics and vegetation attributes. This may also be true along climatic gradients, but the relationships often become more empirical.

10.5 Relationships to soil morphology

By soil morphology, we are primarily referring to texture, structure, depth, rock fragment content (stoniness), bulk density, consistence, color and horizonation in relation to these factors. Soil surface features can also be included. With the exception of bulk density, the morphological characteristics are easily observed and measured or estimated when a pedon is described. In soil taxonomy, some characteristics become diagnostic at the subgroup or higher level, but most specific criteria useful for vegetation-soil relationships are included at the family and series levels of classifications. General descriptions of soil texture, depth, stoniness and color are often included with vegetation studies and descriptions for various purposes. In many studies, a distinct vegetation relationship with morphological characteristics is obvious while in others there appears to be no relationship. The relationships are most often distinct when the landscapes involved contribute to a pattern of soils with distinctly different moisture infiltration and retention characteristics. As the landscapes become more homogeneous, the physical differences contributing to plant community distribution become more subtle and are often masked by differences in soil chemistry or vegetation interactions.

As climate gradients shift geographically or topographically, plant responses in relation to available moisture, as influenced by the soil characteristics, are also likely to shift. Shantz (1938) remarked that the drainage of a sandy soil is so rapid in more moist regions that plants may respond similarly to those on sandy soils of arid regions because the scant moisture enters the soil rapidly, penetrates deeply and is protected from evaporation. The nearly identical soil moisture availability in these two situations results in rather humid species occurring far into the semidesert types on sandy soils and semidesert species likewise pushing into the more humid regions on sands.

When considering moisture relations affected by soil morphology, combinations of factors such as climate-soil interactions stated above or different combinations of characteristics often compensate for one another and thus may result in similar plant responses. Single factor analysis (i.e. texture, depth, etc. taken independently) often provide misleading if not erroneous interpretation for vegetation-soil relationships. Actually, this is the case for all soil properties and not restricted just to morphology characteristics. In addition, the properties or combinations of properties must be analyzed in relation to the physiological constraints of individual species as well as in relation to the categorical constraints of man-made definitions. In other words, we must learn to "think like a plant" in the conceptual sense.

To illustrate this point, soil depth is defined in Soil Taxonomy as depth to bedrock or cemented pan. Plants, however, respond to other soil properties which restrict soil moisture (e.g. rock fragments, salts, contrasting textures, structures, etc.). Depending on concentrations and depth to such soil features, plants may "think" the soil is more shallow than taxonomic criteria would indicate. Soil depth (shallow, mod. deep, deep) criteria associated with sagebrush (*Artemisia*) species (Hironaka *et al.* 1983) have been used successfully as rule-of-thumb by soil and range scientists in some areas of the western U.S., although exceptions were always known to occur. Miles and Leonard (1984) tested the criteria for four of the subspecies using a large national soil database over a wide geographic distribution. As shown in Figure 2, black sagebrush (*A. nova*) and low sagebrush (*A. arb.*) are generally thought to occur on shallow soils while Wyoming big sagebrush (*A. wyo.*) and mountain big sagebrush (*A. vasey.*) are generally thought to occur on moderately deep soils. However, little or no consistency with soil depth was found when the standard criteria of "depth" referred to bedrock or cemented pan as the restricting element. A second analysis including physiologically restrictive layers from the moisture availability standpoint as well as the physically restrictive pan or bedrock supports the concept that black sagebrush and low sagebrush do indeed grow predominantly on shallow soils as determined by root restricting criteria. The additional restricting criteria evaluated in this case were depth to a layer with greater than 35% clay with an abrupt textural change or greater than 60% rock fragments.

As the analysis of vegetation-soil relationships becomes more detailed and involves the combination of species within specific communities rather than individual dominants of vegetation types, the associated soil characteristics become more difficult to quantify. Compensating soil and climate factors combined with species interactions confound our ability to delimit contributing soil factors.

Eckert (1957) described soil characteristics associated with several sagebrush dominated habitat types in southern Oregon. He found several soil series associated with an individual habitat type, but also found several habitat types associated with one soil series. Eckert suggested that although similar in gross morphological characteristics, the soils differed in minor but important compensating characteristics which allowed distinctly different vegetational units to exist and suggested moisture-holding capacity as one of those characteristics.

Because of the nearly unlimited combinations of different horizon characteristics that might have the "compensating" effects Eckert mentioned, the correlation between soil series and their phases and different habitat types or range sites has been mostly empirical as judged from vegetation responses. Lentz (1984), working in the same area as Eckert, used an aggregating strategy of super-horizon-categories to avoid this problem and statistically analyzed twenty-eight morphological parameters associated with sagebrush habitat types. Lentz determined that combinations of soil texture, subhorizon sequence, rock fragment content, color, horizon thickness, ped size, and structure type are, in that order of relative importance and criteria, useful in differentiating the sagebrush habitat types. Depth of mollic epipedon, surface rock fragment cover, and thickness and average dry consistence of the argillic horizon also become important as soil-landscape units become more complex. Lentz suggests additional soil series could be differentiated based on the soil properties described or at least phases could be consistently associated with sagebrush habitat types.

The distribution of chemical characteristics in the soil are influenced by physical morphology and vegetation, so the interrelationships must be considered or at least concurrently evaluated. In South Africa, Brendenkamp *et al.* (1983) quantitatively analyzed 27 soil variables, including both morphological and chemical characteristics in relation to distribution of plant communities. The soil characteristics were aggregated into relatively homogeneous "soil habitat classes" which are characterized by sets of variables rather than single attributes. The soil habitat classes were significantly correlated with recognized plant communities. The super-horizon-categories used by Lentz appear to be conceptually similar to the soil habitat classes described by Brendenkamp.

West and Ibrahim (1967) found that morphological characteristics such as profile depth, surface horizon depth, and practicle size (sand, silt, clay) as well as chemical characteristics of base exchange capacity, exchangeable potassium, and exchangeable sodium vary significantly between four Atriplex habitat types in southeastern Utah. Also in southeastern Utah, Kleiner and Harper (1977) concluded that sites dominated by the grass *Hilaria jamesii* are characterized by finer texture, slightly warmer average temperature, and higher surface potassium and organic matter compared to sites dominated by the grass *Stipa comata*. Fairchild and Brotherson (1980) found that microhabitat characteristics beneath six major shrubs in northern Arizona were more indicative of site classification than characteristics of the shrub interspaces. Soil depth was determined to be the most important factor, but other factors of importance included slope, clay content, topographic position, aspect and soil pH.

The presence and thickness of a mollic epipedon has been positively correlated with average range forage production (Cannon and Nielsen 1984, Jensen 1984). The mollic epipedon is an organically-enriched dark-colored surface horizon determined in the field primarily by color and thickness although other specific criteria apply (USDA 1975). The mollic epipedon is generally present only in the more moist rangeland soils and contains greater than 1% organic matter content.

Soil surface features are also important characteristics influencing vegetation-soil relationships. Walker (1979) states that the development of a surface cap or crust on exposed soils is an important consideration in management of semi-arid ecosystems. The caps can form an effective seal, reducing infiltration and increasing run-off and erosion. Rauzi *et al.* (1968) indicate that even highly permeable sand range sites have reduced infiltration where poor surface structure has resulted from livestock trampling, splash erosion, or sediment deposition. The morphological properties associated with vesicular soil crusts have been shown to impair seedling emergence and plant establishment (Wood *et al.* 1982).

The physical surface crusting just described is apparently always undesirable; however, a different morphology and crusting characteristic associated with cryptogamic floral components in the soil surface may be either desirable or undesirable. Walker also states that caps compounded by the development of algae and/or lichens can enhance soil water properties due to high water retention and in others can be unfavorable because of hydrophobic characteristics.

In arid and semi-arid rangelands of Utah and Nevada, the presence of cryptogamic soil crusts appear to have favorable implications, especially the blue-green algae and lichen components. Kleiner and Harper (1972) found that cryptogamic soil crusts are a major cover component in the vascular plant interspaces of a pristine grassland and discussed favorable aspects of soil stabilization and nutrient cycling compared to a similar grazed site with less cryptogamic composition. Anderson *et al.* (1982a, 1982b) suggest that cryptogamic crusts have been reduced by grazing and that soil variables such as texture, salinity and phosphorus influenced the establishment or recovery of the cryptogamic crust cover. These authors also suggest that the encouragement of cryptogamic crusts in the areas studied will not necessarily result in less vascular plant production. The soil surface features described for well developed cryptogamic crusts is quite similar to the soil surface morphological type II described by Eckert *et al.* (1986a, 1986b) for sagebrush-grass vegetation types in Nevada.

Eckert and co-workers describe four surface types associated with microtopographic position. Type I and II surfaces are associated with coppice dune accumulations around shrubs and grasses and have a canopy and litter cover (Type I) or a surface stabilized by cryptogams (Type II). Types III and IV are characterized by durable vesicular crusts and low organic matter. Type III is on the intercoppice microplain microtopographic position, a transition to the playette microtopographic position where water often ponds after rain or snowmelt and the Type IV surface forms. Type III and IV surfaces are mostly barren, except for partial cryptogam cover on large polygons formed by surface cracks or around polygon margins. Grass seedling emergence and establishment is much better on Type I and II surfaces than types III and IV. These authors also suggest that relative proportions of these surface morphological types may be indicative of range condition.

10.6 Soil chemistry relationships

A high proportion of the worlds rangelands are arid or semi-arid. Semi-arid soils are characteristically low in organic matter but generally quite fertile (except in much of Australia and with other local exceptions). Lack of leaching and high evaporation rates lead to a fairly common occurrence of saline soils (Walker 1979).

Often the strong chemical patterning that occurs in native rangelands that is dependent on the interaction between the inherent soil properties, both chemical and physical, climate and vegetation. Vegetation composition and spatial distribution within communities can have a profound influence in the chemical patterns (and vice versa over time). Relationships must include the microfloral components of the vegetation community within the soil profile and at the soil surface as well as the higher plants. In addition to spatial patterning, the vegetation community characteristics affect the availability of nutrients and the cycling rate.

10.6.1 Soil nutrients

Soil macronutrients essential for plant growth include nitrogen, phosphorus, potassium, calcium, magnesium and sulfur. Essential micronutrients include iron, manganese, copper, zinc, boron, molybdenum, chlorine and cobalt (Buckman and Brady 1969). The role of these soil nutrients is generally understood as it relates to plant growth in the common sense. Understanding nutrient dynamics and interrelationships within and between natural rangeland communities is only rudimentary. Nutrient relationships as well as others have been intensively studied for some species in specific locations, such as desert plants at the Nevada Test Site (Wallace and Romney 1972). Ranges of nutrient characteristics associated with geographic distribution of communities and community dynamics are yet to be determined in most cases. For many species and communities, information is often sketchy or absent.

The relationship and importance of microbial symbiotic nitrogen fixing plants to community productivity has been the subject of past and continuing research. Of equal importance is the availability of organic N over time as influenced by the relative amounts incorporated in living biomass and plant residues and the nitrification rate of organic matter and the relative amounts in living biomass as compared with plant residues. A comprehensive study of nitrogen in desert ecosystems is presented by West and Skujins (1978). Without reiterating the basic nitrogen cycle and related research on rangelands, we hope the reader will already realize the ultimate importance while we move on to some possibly less known, but also important phenomenon.

Only recently (the past decade or a little more) has the symbiotic relationships of vesicular-arbuscular mycorrhizal (VAM) fungi within rangeland communities have been explored. VAM fungi have been shown to improve water absorption as well as assist plants in mineral nutrition, especially phosphorus. At some levels of phosphorus availability, however, they may apparently act as parasites, depressing plant growth. Call and McKell (1984) reported that VAM inoculated fourwing saltbush (*Atriplex canescens*) are more effective in the uptake of water and phosphorus than noninoculated plants on processed shales deficient in both nitrogen and phosphorus.

The influence of VAM fungi may vary both geographically among vascular plants of the same species as well as between species in the same plant community. Both similarities and differences were observed between two geographically isolated species of the VAM Glomus and the water relationships of western wheatgrass (*Agropyron smithii*) (Stahl and Smith 1984). Zajicek *et al.* (1986) suggest that VAM symbiosis may play a role in the survival of deep rooted forbs in the tallgrass prairie. Although not conclusive, the deep rooted habit of the forbs may be a survival mechanism reducing competition for water and nutrients with shallower rooted, fast-growing grasses. Soil phosphorus levels were consistently low and the VAM fungi were suspected to enhance P translocations as well as water relations.

Levels of mycorrhizal fungi are affected by soil disturbance (Doerr *et al.* 1984, Call and McKell 1984) and by grazing (Bethlenfalvay and Dakessian 1984). These factors associated with the obligate requirement of VAM fungi with vascular plants led the authors to speculate that the associations and coinciding nutrient relationships are both site specific (in the context of range site, habitat-type, etc.) and dependent on seral status. Much more needs to be researched in this area.

The ratio between nutrients in the soil is another factor influencing rangeland vegetation responses, the serpentine soils presenting one of the most striking examples. The serpentine soils have an unfavorably high magnesium to calcium ratio coupled with other unfavorable nutrient characteristics including low levels of nitrogen, phosphorus, potassium and possibly molybdenum and toxic levels of chromium and nickel (Jenny 1980, Rosiere and Vaughn 1986). The serpentine sites are characterized by low productivity and a number of endemic species evidently able to secure calcium at low degrees of Ca-saturation (Jenny 1980).

Shock *et al.* (1984a, 1984b) report that nitrogen phosphorus and sulfur commonly restrict productivity of California annual rangelands. The addition of sulfur alone enhanced N : S ratios in subclover (*Trifolium subterraneum*), strongly enhanced nitrogen fixation by subclover and improved forage yields.

The potassium-magnesium ratio within the soil has been suggested by Jensen (1984) as a possible indicator between shrub dominated sites and grass dominated sites. The root cation exchange capacity (CEC) for the shrubs was over twice that of the grasses and would apparently favor Mg absorption by shrubs rather than K and vice versa for grasses. The relations suggest sites with high K-Mg ratios would favor shrub growth while low K-Mg ratios favor grass production.

10.6.2 Soil reaction (pH)

Soils of arid and semiarid rangelands are generally in the slightly acid to alkaline range while the more humid forests tend to have from slightly alkaline to acid reactions. The soil pH may influence nutrient absorption and plant growth either through direct effect of the hydrogen ion or indirectly through on nutrient availability and presence of toxic ions, the latter being of greater significance (Buckman and Brady 1969). Several essential elements such as iron, manganese and zinc tend to become less available as pH increases from the moderately acid to slightly alkaline. At very high pH values, the bicarbonate ion may interfere with uptake of other ions. Also, the availability of HPO_4 or H_2PO_4 ions of phosphate is dependent upon soil pH (Buckman & Brady 1969). Some research indicates non-conclusive relationships between soil pH and rangeland community characteristics. Many researchers recognize general relationships of pH in conjunction with other chemical and physical soil characteristics, particularly salinity.

10.6.3 Salinity

Salinity is the total salt content of all salts more soluble than gypsum as measured by the electrical conductivity (EC) of a saturation extract. Sodicity is the proportion of exchangeable sodium, commonly estimated from the sodium to calcium plus magnesium ratio of a saturation extract (USDA 1980). In older literature, the term sodicity was commonly referred to as alkali.

Salinity affects plant growth by decreasing osmotic potentials, producing ion toxicity and affecting nutrient uptake. General relationships between plant communities and saline soils such as by Shantz and Piemeisal (1940), Billings (1949) and others have been documented worldwide. Most research, however, has concentrated on individual species or even varietal relationships, especially as they might apply to range improvement projects or reclamation.

Roundy (1983, 1985a, 1985b), for instance, has made an extensive study on the establishment and growth characteristics of basin wildrye (*Elymus cinereus*) and tall wheatgrass (*Agropyron elongatum*) for range improvement on saline soils. Roundy suggests high salinity values are limiting for plant reestablishment and growth and that arid sites require supplemental irrigation for successful resegregation. Vegetation conversion trials from black greasewood (*Sarcobatus vermiculatus*) to basin wildrye by Wilson (1986) confirmed suggestions in a report by Leonard *et al.* (unpublished 1984) that soils with salinities above 0.775 S/m and pH above 8.3 would be marginal for vegetation conversion.

Hodgkinson (1987) made a direct correlation of six saltbrush species in Arizona and New Mexico with SAR, ECe and pH. Species ranked from highest to lowest adaptability to SAR, ECe and pH were: mat saltbush (*Atriplex corrugata*), mound saltbush (*A. obovata*), Castle Valley clover (*A. cuneata*), sickle saltbush (*A. falcata*), shadscale (*A. confertifolia*) and fourwing saltbush (*A. canescens*).

The extrapolation of these relationships to other geographic areas, however, may have limitations. Billings (1949), for instance, indicates that shadscale is not necessarily a good indicator of subsoil salinity. Stutz (1983) has more recently determined several polyploid forms of shadscale. The diploid chromosome type is apparently more often associated with sites above the old pleistocene lake Bonneville level while various polyploid varieties occupy the more saline sites exposed by lake recession. In another study by Potter *et al.* (1986), different ecotype collections of fourwing saltbush seed exhibited different responses in germination to osmotic stress as well as pH.

10.6.4 Chemical patterning

Vegetation relationships with chemical patterning in the soil can be considered in two ways — the distribution of vegetation as a result of chemical patterning or chemical patterning occurring as a result of vegetation distribution. Chemical patterning involves plant nutrients, salts, mineral elements and organic compounds which ultimately may result in chemical differences of individual plant species between sites as well as the distribution of species within and between sites.

In an analysis of a saltgrass meadow (Bowman *et al.* 1985), vegetation cover and species composition were related to salinity, sodicity, fertility and physical characteristics (horizonation). Although not explicitly stated, the authors seem to attribute patterning to species occurrence and cover attributes with soil properties. The occurrence of blue grama (*Bouteloua gracilis*) was positively correlated with thickness of A horizon, depth to salinity and depth to C horizon but decreases with higher salt levels near the surface. Saltgrass (*Distichlis stricta*) is dominant where shallow saline horizons occur but is negatively correlated to nitrogen and positively to sodicity. Western wheatgrass (*Agropyron smithii*) and alkali sacaton (*Sporobolus airoides*) are not strongly influenced by salinity but western wheatgrass is positively correlated with N in the surface soil and does not occur on sodic sites.

Stein and Ludwig (1979) state that present vegetation and soil patterns are clearly related between creosotebush (*Larrea tridentata*) dominated communities and three different associated grassland communities. Soils of the four communities studied differed significantly in pH, magnesium, organic carbon and total nitrogen. Creosotebush apparently accumulates nitrogen under the canopy at the expense of interspace areas so that N may become limiting for grass establishment in the interspaces without fertilization (Cox *et al.* 1983).

Brotherson and Osayande (1980) compared mineral concentrations of both soils and plants between true mountain mahogany (*Cercocarpus montanus*) and Utah juniper (*Juniperus osteosperma*) communities. Zinc, manganese, and phosphorus are higher in the soil substratum for juniper while calcium and magnesium are in greater concentration in mountain mahogany soils. Nitrogen concentrations are significantly higher beneath plant canopies in both communities than in the interspaces.

Both vertical and horizontal plant-induced chemical gradients have been documented in semi-desert shrub ecosystems of Utah by Charley and West (1975). These authors cite similar research in Australia by both Charley and Cowling indicating that gradients are so severe that they could result in serious soil survey errors and plant community identification if not considered. Nitrogen, carbon, available organic and total phosphorus are concentrated toward the soil surface at all locations. Nitrogen, carbon and organic phosphorus regularly showed horizontal patterning and a species influence was apparent for pH and salinity.

Allelopathy is technically a biochemical interaction between all types of plants including microorganisms and covers both inhibitory and stimulatory effects (Rice 1984). Since allelopathic affects are commonly transmitted through the soil as root exudates or water soluble leachates of organic compounds, we felt that it should at least be mentioned in the discussion of vegetation-soil relationships. Rice states that there is increasing evidence that allelopathy probably plays a role in most, if not all, spatial distributions of plants and devotes an entire chapter on its role in the patterning of ecosystems with native vegetation, including some rangelands. Unfortunately, there are few references that suggest a possibility of allelopathy such as West *et al.* (1984) or address chemical effects of weedy species such as halogeton (*Halogeton glomeratus*) (Kinsinger and Eckert, Jr. 1961), the work by Rice represents the most comprehensive information available. There appears to be a great need for additional research on the role of allelopathy in regard to rangelands in general.

10.7 Data extrapolation for management

Vegetation and soil (and climate at the earths surface, at least) are mutually dependent. In spite of this, the quantification of precise relationships is difficult due to spatial and temporal variability associated with natural ecosystems. Moreover, the classification systems designed to describe and communicate attribute information about soils and vegetation seldom exhibit a 1 : 1 relationship. So, why do we persist in identifying relationships at all?

With the term range management, we imply action; either maintaining a current state of productivity (sustained yield) or altering the current productivity by manipulation. Range management is dependent on the inherent natural productivity of the land since expensive land treatments such as leveling, irrigation or even fertilization are not feasible in most cases. Mechanical or chemical treatments used on rangelands are often feasible only because of an expected long term benefit. In many areas, livestock control is the only economical management alternative. Whatever the management alternative is, there is a predicted or at least desired result. Vegetation-soil relationships are an essential piece of knowledge required for accurately and consistently predicting management responses on rangelands. The desired result must be within the capability of the site.

Similar and repeatable soil-vegetation units provide a basis for making more consistent interpretations. During the process of data collection and evaluation, ranges of characteristics can be identified that help range managers determine reasonable expectations. Where probabilities have been determined such as with seeding trials, risks can be evaluated. Past management successes and failures as well as those of new management trials and new research results can be extrapolated based on associated vegetation-soil relationships.

The extrapolation of vegetation-soil relationships and associated management implications is still best accomplished through soil and vegetation classification even though the 1 : 1 relationship may not exist. Neither soil nor vegetation classification by itself provides the necessary information for extrapolation until a correlation between them is achieved. The correlation may identify combinations of soil properties that split soil taxonomic units or relate to additional properties not normally considered in soil classification. Soil identification then provide a stable classification unit with which to extrapolate management interpretations.

This kind of correlation can be done with either present or potential vegetation or different categorical levels of soils classification. The authors prefer to correlate of soil taxonomic units and their phases at the family or series level with ecological units perceived as having the same climax or potential vegetation. Present vegetation communities can then be effectively interpreted as to their seral status and management opportunities for change can be identified. The classifications identifying climax or potential vegetation have often been criticized in that most rangelands have been so extensively disturbed that climax is rarely known. Even the methodology for determining a climax plant community is not universally agreed upon. However, the climax concept can be used to relate the various ecosystem components into ecological units with the same use and management potential for vegetation, even if we're not absolutely sure of the exact community. We must base our perception of climax on actual observations. Our perception of the climax community may change with additional information; but the susceptibility to change should not detract from the utility of the concept.

Once the ecosystem components are correlated within an ecological unit, the units become a reliable source of information. These units provide a basis for prediction and extrapolation of vegetation responses due to management or other influences. Since these units are subject to change, either because of additional information or interpretation, the process of correlation must be continued.

10.8 References

Anderson, D. C., K. T. Harper, and R. C. Holmgren. 1982a. Factors influencing development of cryptogamic soil crust in Utah deserts. J. Range. Manage. 35: 180—185.

Anderson, D. C., K. T. Harper, and S. R. Rushforth. 1982b. Recovery of cryptogamic soil crusts from grazing on Utah winter ranges. J. Range. Manage. 35: 355—359.

Barker, J. R., and C. M. McKell. 1986. Differences in Big Sagebrush (Artemisia tridentata) plant stature along soil-water gradients: Genetic Components. J. Range Manage. 39: 450—454.

Bethlenfalvay, G. J., and S. Dakessian. 1984. Grazing effects on mycorrhizal colonization and floristic composition of the vegetation on a semiarid range in northern Nevada. J. Range Manage. 37: 312—316.

Blaisdell, J. P. 1958. Seasonal development and wild of native plants on the upper Snake River plains and their relation to certain climatic factors. U.S. Dep. Agr. Tech. Bull. 1190: 68.

Billings, W. D. 1949. The shadescale vegetation zone of Nevada and eastern California in relation to climate and soils. The Amer. Midland Naturalist. 42: 870—109.

Bowman, R. A., D. M. Mueller, and W. J. McGinnies, 1985. Soil and vegetation relationships in a central plains saltgrass meadow. J. Range Manage. 38: 325—328.

Brotherson, J. D., and S. T. Osayande, 1980. Mineral concentrations in true mountain mahogany and Utah juniper and in associated soils. J. Range Manage. 33: 182—185.

Brendenkamp, G. J., G. K. Theron, and D. R. J. Van Vuuren, 1983. Ecological interpretation of plant communities by classification and ordination of quantitative soil characteristics. p. 692—699. Bothalia 14, 3 & 4.

Buckman, H. O., and N. C. Brady. 1969. The nature and properties of soils. Seventh Edition. The Macmillan Co./Collier-Macmillian Limited, London.

Call, C. A., and C. M. McKell. 1984. Field establishment of fourwing saltbrush in processed oil shale and distributed native soil as influenced by vesicular-arbuscular mycorrhizae. Great Basin Naturalist. 44: 363—371.

Cannon, M. E., and G. A. Nielsen. 1984. Estimating production of range vegetation from easily measured soil characteristics. Soil Sci. Soc. of Amer. J. 48: 1393—1397.

Charley, J. L., and N. E. West. 1975. Plant-induced soil chemical patterns in some shrub dominated semi-desert ecosystems of Utah. J. of Ecology. 63: 945—962.

Cox, J. R., J. M. Parker, and J. L. Strochlein, 1983. The initial growth of range grasses on nonfertilized and fertilized soils collected from creasotebush communities in the southwestern United States. J. Range Manage. 36: 726—729.

Daubenmire, R. 1970. Steppe vegetation of Washington. Wash., Agr. Exp. St., Tech. Bull. 62.

Daubenmire, R. 1978. Plant geography with spatial reference to North America. Academic Press, New York.

Doerr, T. B., E. F. Redente, and F. B. Reeves. 1984. Effects of soil disturbance on plant succession and levels of mycorrhizal fungi in a sagebrush-grassland community. J. Range Manage. 37: 135—139.

Driscoll, R. S., D. L. Merkel, J. S. Hagihara, and D. L. Radloff. 1983. A component land classification for the United States: Status report. U.S. Dep. Int. BLM. Tech. Note 360.

Eckert, R. E. Jr. 1957. Vegetation-soil relationships in some Artemisia types in northern Hainey and Lake counties, Oregon. (PHD thesis) Oregon State Univ, Oregon.

Eckert R. E. Jr., F. F. Peterson, M. S. Meurisse, and J. L. Stephens. 1986a. Effects of soil-surface morphology on emergence and survival of seedlings in big sagebrush communities. J. Range. Manage. 39: 414—420.

Eckert R. E. Jr., F. F. Peterson, and J. T. Belton. 1986b. Relation between ecological range condition and properties of soil-surface type. J. Range Manage. 39: 409—414.

Everett, R. L., R. O. Meewig, P. T. Tueller, and R. A. Evans. 1977. Water potential in sagebrush and shadscale communities. Northwest Sci. 51: 271—281.

Fairchild, J. A., and J. D. Brotherson. 1980. Microhabitat relationships of six major shrubs in Navajo National Monument, Arizona. J. Range Manage. 33: 150—156.

Hironaka, M., M. A. Fosberg, and A. H. Winward. 1983. Sagebrush-grass habitat types of southern Idaho. Forest, Wildlife and Range Experiment Station Bull. 35.

Hodgkinson, H. S. 1987. Relationship of saltbush species to soil chemical properties. J. Range Manage. 40: 23—26.

Jenny, H. 1983. The Soil Resource Origin and Behavior. Ecol. Studies 37. Springer-Verlag, New York.

Jensen, M. 1984. Soil moisture regimes on some rangelands of southern Idaho. Soil Science Soc. Amer. 48: 1328—1330.

Jordan, G. L. 1983. Planting limitations for arid, semiarid and salt-desert shrublands. p. 11—16. *In*: Managing intermountain rangelands — Improvement of Range and Wildlife Habitats. USDA Gen. Tech. Rpt. IMT-157.

Kinsinger, F. E., and R. E. Eckert Jr. 1961. Emergence and growth of annual and perennial grasses and forbs in soils altered by halogeton leachate. J. Range Manage. 14: 194—197.

Kleiner, E. F., and K. T. Harper. 1972. Environment and community organization in Grassland of Canyonlands National Parks. Ecology. 53: 299—309.

Kleiner, E. F., and K. T. Harper. 1977. Occurrence of four major perennial grasses in relation to edaphic factors in a pristine community. J. Range Manage. 30: 286—289.

Kormandy, E. J. 1969. Concepts of ecology. Concepts of Modern Biology Series. Prentice-Hall, Inc., New York.

Lentz, R. D. 1984. Correspondence of soil properties and classification units with sagebrush communities in southeastern Oregon. (MS Thesis) Oregon State Univ.

Leonard, S. G., R. L. Miles, and J. W. Burkhardt. 1984. Unpublished report: Comparison of soil properties associated with basin wildrye and black greasewood in the Great Basin region.

Mabbutt, J. A., W. H. Litchfield, N. H. Speck, J. Sofanlis, D. G. Wilcox, J. M. Arnold, M. Brookfield, and R. L. Wright. 1963. General report on lands of the Wiluna-Meekatharra Area, Western Australia. *In*: Land Research Series No. 7. Commonwealth Sci. and Ind. Research Org., Australia.

Mentis, M. T. 1984. Monitoring in South African grasslands. South Africa Nat. Sci. Programmers Report No. 9.

Miles, R. L., and S. G. Leonard. 1984. Documenting soil-plant relationships of selected sagebrush species using the soil resource information system. Soil Survey Horizons. 25: 22—26.

Miles, R. L. *et al.* (*Unpublished*) Greasewood.

Muegller, W. F., and W. H. Stewart. 1980. Grassland and Shrubland Habitat Types of Western Montana. U.S. Dept. Agr. Forest Service. Gen. Tech. Rep. INT-66.

Odum, E. P. 1971. Fundamentals of ecology. 3rd Ed. W. B. Saunders Company. Philadelphia. London. Toronto. Lib. Cong. No. 76—81826.

Passey, H. B., V. K. Hugie, E. W. Williams, and D. E. Ball. 1982. Relationships between soil, plant community and climate on rangelands of the intermountain west. U.S.D.A., S.C.S. Tech. Bull. 1669.

Paysen, T. E., J. A. Derby, H. Black, V. Gleich, and J. W. Mincks. 1980. A vegetation classification system applied to southern California. Pacific Southwest Forest and Range Experiment Station. Gen. Tech. Rept. PSW-45.

Perry, R. A. 1960. Pasture lands of the northern territory, Australia. Land Research Series No. 5. Commonwealth Sci. and Ind. Research Org., Australia.

Pearson, L. C. 1979. Effects of temperature and moisture on phenology and productivity of Indian ricegrass. J. Range Manage. 33: 127—134.

Platou, K. A., P. T. Tueller, S. G. Leonard, and R. L. Miles. 1986. Soil properties associated with six common grasses of the Great Basin. J. of Soil & Water Conser. 41: 417—421.

Potter, R. L., D. W. Ueckert, J. L. Peterson, and M. L. McFarland, 1986. Germination of fourwing saltbush seeds: Interaction of temperature, osmotic potential and pH. J. Range Manage. 39: 43—46.

Rauzi, F., C. L. Fly, and E. J. Dyksterhuis. 1968. Water intake on midcontinental rangelands as influenced by soil and plant cover. U.S. Dept of Agr. ARS and SCS in coop. w/Wyo Agr. Exp. St.

Reveal, J. L. 1979. Mentzelia (Biogeography of the Intermountain Region). J. of No. NV. Native Plant Soc. No. 4.

Rice, E. L. 1984. Allelapathy. Physiological Ecology. Academic Press, Inc., Orlando.

Richardson, E. A., and S. G. Leonard. 1981. Climatic modeling of winter rangelands in Utah. Extended abstracts. 15th Conf. on Agriculture and Forest Meteorology and 5th Conf. on Biometeorology. p. 182—185. American Meteorological Soc., Anaheim, Cal.

Rosiere, R. E., and C. E. Vaughn. 1986. Nutrient content of sheep diets on a serpentine barrens range site. J. Range Mange. 39: 8—13.

Roundy, B. A. 1983. Response of basin wildrye and tall wheatgrass seedlings to salination. Agro. J. 75: 67—71.

Roundy, B. A. 1985a. Root penetration and short elongation of tall wheatgrass and basin wildrye in relation to salinity. Canadian J. of Pl. Sci. 65: 335—343.

Roundy, B. A. 1985b. Emergence and establishment of basin wildrye and tall wheatgrass in relation to moisture and salinity. J. Range Manage. 38: 126—131.

Sasich, J., and G. A. Nielsen. 1984. Indicators of soil climate used by Montana soil classifiers: A survey. Montana Agr. Exp. St. Special Report 4.
Saxton, K. E., H. P. Johnson, and R. H. Shaw. 1974. Modeling evapotranspiration and soil moisture. Trans. Amer. Soc. Agr. Eng. 17: 673—677.
Shantz, H. L. 1938. Plants as soil indicators. p. 835—859. *In*: Soils and men. U.S. Dep. Agr. Yearbook of Agr.
Shantz, H. L., and R. L. Piemeisal. 1940. Types of vegetation in Escalante Valley, Utah, as indicators of soil conditions. U.S. Dept. of Agr. Tech. Bulletin 713.
Shiflet, T. N. 1973. Range sites and soils in the United States. p. 26—33. *In*: Arid Shrublands-Proceedings of the Third Workshop of the USA/Australia Rangelands panel.
Shock, C. C., W. A. Williams, M. B. Jones, D. M. Center, and D. A. Phillips. 1984a. Nitrogen fixation by sub clover associations fertilized with sulfur. Plant and Soil. 83: 323—333.
Shock, C. C., W. A. Williams, M. B. Jones, D. M. Center, and D. A. Phillips. 1984b. Completion for S and N by associations of three annual range species in lysimeters. Plant and Soil. 81: 322—321.
Smith, G. D., F. Newhall, L. H. Robinson, and D. Swanson. 1964. Soil-temperature regimes — Their characteristics and predictability. U.S. Dept. Agr. SCS-TP-144.
Sosebee, R. E., F. M. Churchill, and C. W. Green. 1982. Soil water depletion by Yucca. J. Range Mange. 36: 774—776.
Stahl, P. D., and W. K. Smith. 1984. Effects of different geographic isolates of Glomuson the water relations of *Agropyron smithii*. Mycologia. 76: 261—267.
Stein, R. A., and J. A. Ludwig. 1979. Vegetation and soil patterns on a Chihuahuan desert bajada. Amer. Midland Naturalist. 101—102: 28—37.
Sturges, D. L., and R. D. Tabler. 1968. Management of blowing snow on sagebrush rangelands. J. of Soil and Water Conser. 36: 287—292.
Sturges, D. L. 1977. Soil water withdrawal and root characteristics of big sagebrush. The Amer. Midl. Natur. 98: 257—274.
Stutz, H. C. 1983. Some promising chenopods for use on disturbed lands. p. 132—135. *In*: Managing intermountain rangelands — Improvement of range and wildlife habitates. USDA. Gen. Tech. Rpt. IMT-157.
Tiedeman, J. A. and C. Terwilliger Jr. 1979. Phyto-edaphic classification for rangeland inventory and management. *In*: Abstract of papers presented at the 32nd annual meeting. Society for Range Management.
Tueller, P. T., and L. A. Monroe. 1975. Management guidelines for selected deer habitats in Nevada. Nevada Agri. Exp. Sta. Publ. No. R104.
Ueckert, D. W. 1979. Effect on shortgrass forage production and soil water depletion. J. Range Manage. 32: 216—220.
United Nations Educational, Scientific, and Cultural Organization. 1973. International classification and mapping of vegetation, Series 6, Ecology and Conservation. Paris, France.
U.S. Dept. of Agr. 1975. Soil Conservation Service. Soil Taxonomy. Agr. Handbk. 436. Washington, D.C.
U.S. Dept. of Agr. S.C.S. 1976. National Range Handbook. Section 300. Washington, D.C.
U.S. Dept. of Agr. 1980. Soil Conservation Service. Soil Survey Manual Rev. Draft. Chapt. 4—92 and Chapt. 5—10. Washington, D.C.
U.S. Dept. of Agr. 1983. SPUR — Simulation of production and utilization of rangelands. (A rangeland model for management and research). Agr. Research Service. Misc. Pub. No. 1431.
Walker, B. H. 1979. Management of semi-arid ecosystems. (Management principles for semi-arid ecosystems) Elsevier Scientific Publishing Co. Amsterdam-Oxford-New York.
Wallace, A., and E. M. Romney. 1972. Radioecology and ecophysiology of desert plants at

the Nevada Test site. United States Atomic Energy Commission of Information Services TID-25954.

Wang, J. Y. 1960. A critique of the heat unit approach to plant response studies. Ecology 41: 785—789.

West, N. E., and K. I. Ibrahim. 1967. Soil-vegetation relationships in the shadscale zone of southern Utah. Ecology. 49: 445—456.

West, N. E., and J. Skujins. 1978. Nitrogen in desert ecosystems. US/IBP Synthesis Series 9. Dowden, Hutchinson & Ross, Inc. Stroudsburg. 244 pp.

West, N. E., F. D. Provenza, P. J. Johnson, and M. K. Owens. 1984. Vegetation change after 13 years of livestock grazing exclusion on sagebrush semidesert in west central Utah. J. Range Mange. 37: 262—264.

Wight, J. R. and E. L. Neff. 1983. Soil-vegetation-hydrology studies. Volume II. A user manual for ERHYM: The Ekalka rangeland hydrology and yield model. USDA, Agr. Research Ser. ARR-W-29.

Wilson, J. S. 1986. Controlling black greasewood with fire and tebuthiuron on depleted great basin wildrye sites in northwestern Nevada. (M. S. Thesis) Univ. of Nev. Reno.

Wood, M. K., W. H. Blackburn, R. E. Eckert, Jr., and F. F. Peterson. 1982. Influence of crusting soil surfaces on emergence and establishment of crested wheatgrass, squirreltail, thurber needlegrass and fourwing saltbush. J. Range Manage. 35: 282—287.

Zajicek, J. M. Daniels, B. A. Hetrick, and C. E. Owensby, 1986. The influence of soil depth on mycorrhizal colonization of forbs in the tallgrass prairie. Mycologia. 78: 316—320.

11. Vegetation attributes and their application to the management of Australian rangelands

A. D. WILSON, K. C. HODGKINSON AND J. C. NOBLE

11.1 Abstract

This chapter presents the attributes of species and communities in Australian rangelands and describes how these may be used to manipulate vegetation composition towards desired objectives. The characteristics of ephemerals, herbaceous perennials and woody perennials are outlined, including the flux of their populations in relation to variation in climate and the occurrence of rare climatic events. The vital attributes of species are used to construct the impact of various fire regimes and the response of communities to grazing is used to develop a theory of grazing management. Where the vegetation exhibits a periodic response to a variable climate, these vectors of change can be used to determine the origin of the current vegetation or to direct the course of its management.

11.2 Introduction

The rangelands of Australia cover a vast area of some 6,300,000 km^2 and include a wide range of vegetation types. Nevertheless they form a single entity of natural vegetation that is used primarily for grazing by domestic animals. The botanical composition has been changed to various degrees by this use, but it is still almost wholly comprised of native species, whose populations are determined primarily by the soil and climate and only secondarily by land use. Management must therefore be ecological; based on understanding the natural relationship between the plant community and its environment and how this might be modified by management. The options for management are the various compositions of flora that are possible on each land type. The tools of management are the grazing and fire regimes that may be applied, including the frequency, timing and intensity of their application. The management goal is the plant community

Tueller, P. T. (ed.), Vegetation science applications for rangeland analysis and management.
ISBN 90-6193-195-9.
© 1988, *Kluwer Academic Publishers, Dordrecht, Boston, London.*

which gives the greatest animal production in the long-term, with the proviso that the productive potential of the soil is maintained. This is not necessarily synonymous with the 'original' vegetation composition before settlement, partly because the original vegetation was not lightly grazed and partly because 'original' has little meaning in a variable environment. For each range type one combination of the available species will be more desirable for grazing than another. It may be that it has a higher production of edible forage, a higher quality of forage in terms of protein content or digestibility of its energy, or that its production varies less from year to year.

The rangelands of Australia are all warm to hot environments. Climatically the vegetation is controlled by aridity, the season of rainfall and rainfall variability. The season of rainfall changes gradually from a summer dominance in the north to a winter dominance in the south, but many range types may receive rainfall in either season (Austin and Nix 1973).

Variability in both amount and season of rainfall is a dominant characteristic arising from El Nino-Southern Oscillation (ENSO) phenomena (Vines and Tomlinson 1985). The Southern Oscillation is a see-saw of change in atmospheric pressure between the Pacific and Indian Oceans with associated changes in rainfall, temperature and winds, particularly in equatorial regions (Philander 1983). It is possible that these oscillations are related to such meteorological phenomena as the southward monsoonal excursions which result in significant summer rainfall events across much of the arid zone. If such summer weather phenomena also precede or follow major cool-season rainfall events across arid Australia, then substantial herbage stocks will be generated to fuel widespread wildfires (Noble *et al.* 1980, Griffin *et al.* 1983).

The variation in rainfall patterns mentioned above suggests that certain years, or series of years, will be dominated by cool-season rainfall and other periods by warm-season rainfall. In non-seasonal rainfall areas, there are quite distinct plant species guilds associated with both cool- and warm-season patterns (e.g., Mott 1972). Analysis of rainfall data from Pooncarie in western New South Wales, where on average rainfall is evenly distributed throughout the year showed that, on a 10—12 year cycle, cool-season (April—September) rainfall was generally out-of-phase with warm-season (October—March) rainfall (Noble and Vines, unpublished data). Major pluvial periods arise on the rare occasions when the two occur in sequence, as in 1973/74.

The landscapes of the Australian rangelands are predominantly flat. Nevertheless water runoff and redistribution on a local scale is a common feature. At a larger scale, water movement from the slopes to the flats forms the mosaics of land units that differ in composition and production, and may need to be managed separately. At a finer scale, infiltration is often greater in gilgais (depressions formed in cracking clays) or on contours, producing a banded or clumped vegetation distribution. Such patterned landscapes have different plant production — rainfall and animal production — forage availability relationships to non-patterned landscapes. The soils have been subject to extensive weathering and many have been transported and redeposited as sandy dunes, aeolian A-horizons or alluvial clays. These contrasts in soil types and their interaction with climate are the primary determinants of the species composition of the vegetation.

In this chapter we discuss the attributes of key species and communities of three major range types, to illustrate how knowledge of species biologies and their interactions with climate, fire and grazing, can be used in management. We will attempt to give an overview of the science and its application, rather than provide precise management prescriptions. Those who seek more detail are referred to appropriate chapters in the book by Harrington *et al.* (1984) on the management of Australia's rangelands.

11.3 Vegetation types

In this chapter examples will be drawn primarily from three vegetation types — the chenopod low shrublands, the *Eucalyptus* and *Acacia* semi-arid woodlands and the Mitchell grasslands whose continental distribution is shown in Figure 1.

The chenopod low shrublands is one of the major vegetation types in the winter rainfall zone, with rainfall varying from 125 to 350 mm. The saltbush (*Atriplex* spp.) and bluebush (*Maireana* spp.) shrubs are mostly edible, albeit of relatively low palatability. They have a cover of 5—20% depending on aridity and are usually clumped; one or more individuals growing in a low soil hummock originating from either the entrapment of wind-blown soil or the natural gilgai-forming processes in certain clays (Noble *et al.* 1984).

Ephemeral and perennial herbaceous species grow between the clumps. The cover and species composition of this layer varies widely from almost completely bare to almost a complete cover according to the amount of autumn-winter rainfall (Noble 1977).

Soil texture has a strong influence on species composition. *Atriplex vesicaria* is the dominant species on clay soils, *Maireana sedifolia* on medium textured soils and *M. pyramidata* on calcareous sands and loams. However, a wide array of other shrub species including *M. astrotricha* and *M. brevifolia*, and sub-shrubs from the genera *Maireana* and *Sclerolaena*,

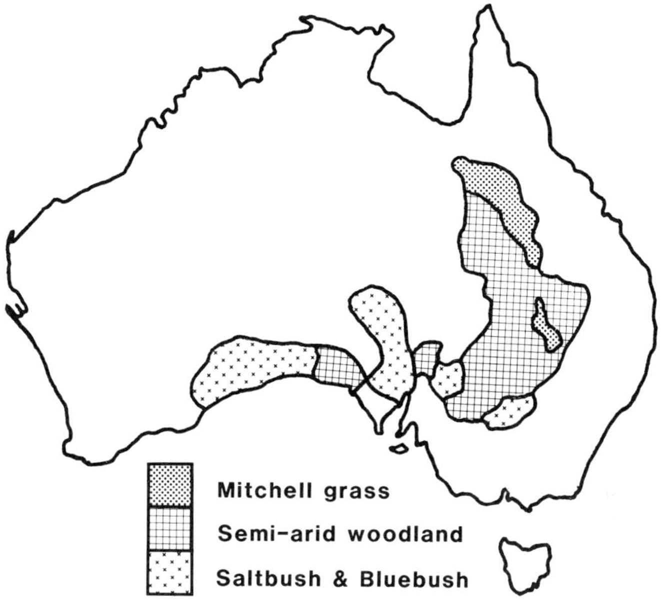

Figure 1. The location of the range types discussed in this chapter.

are also important. A knowledge of the biology of these plants, and their response to grazing, provides the key to management. They are the elements that dominate the landscape and protect the soil from erosion. They also provide a reserve of forage for the grazing animal (Graetz and Wilson 1984).

The semi-arid woodlands of eastern Australia have a dominant tree or tree-shrub layer containing many species that are unpalatable to herbivores, and an understorey of perennial and annual herbs. Rain falls at any time of the year, trending from winter effectiveness in the south (250—300 mm) to summer effectiveness in the north (300—600 mm). The soils vary in texture from sandy loams to clay loams.

Tree species vary widely including the summer-growing *Eucalyptus populnea, E. dumosa* and *Acacia aneura.* The shrub layer is similarly diverse including widespread species such as *Eremophila mitchellii, E. sturtii* and *Dodonaea attenuata.* A common characteristic is a patterned vegetation in which the shrubs are clumped beneath the canopy of the larger trees (Harrington *et al.* 1981). The soils within these clumps have both higher fertility and higher permeability because of litter accumulation. Other more drought resistant trees, such as *A. aneura*, may grow in the inter-thicket areas. Grass species also vary with position in the thicket/inter-thicket catena. The herbaceous layer includes perennial (e.g., *Thyri-*

dolepis mitchelliana and *Eragrostis eriopoda*) and biennial (e.g., *Stipa* spp. and *Aristida jerichoensis*) and annual species, whose abundance depends on both rainfall and tree-shrub cover. In its natural state, trees dominate the landscape because they are taller and live longer. When the overstorey cover is high (e.g., 25—30% canopy), grasses are suppressed in all but the wettest years. When the overstorey canopy is reduced, by either fire or mechanical clearing, grass production is greatly enhanced (Beale 1973, Noble 1988), but the increase is not sustained unless the treatment is repeated. Competition between the two layers, and its manipulation by fire and grazing, constitute the core of range management in these communities.

The Mitchell grasslands provide another contrast. They are dominated by various species of the tussock grass *Astrebla* and are found on clay soils in summer or predominantly summer rainfall (250—550 mm) areas. Trees and shrubs are sparse to rare, so that the grass is dominant and usually remains so even when grazed. Heavy grazing reduces the cover and abundance of *Astrebla* but with the exception of communities at the extremity of its distribution it rebounds again in favourable seasons. Since communities are generally resilient to both grazing and fire, management may be directed towards manipulating the composition of the herbage layer to improve animal production (Orr and Holmes 1984).

11.4 Population dynamics

The three vegetation types described differ markedly in floristics, with few species in common. This is mainly attributable to major differences in soil texture, aridity and seasonality of rainfall. However, we are not concerned here with why species grow where they do, but rather with the manner in which community composition and productivity respond to climatic variation. In this chapter we will discuss such climatic influences at the level of the plant growth-form, because this classification can be applied to all three vegetation types. The conclusions also have greater generality and the implications of a change in composition at this level are fundamental to both production and management.

Each rangeland plant community may contain up to 100 species, most of which are herbaceous. These present a continuum of perenniality from the short-lived ephemeral to those shrubs and trees that live for 300+ years. For convenience we may classify these into ephemerals, perennial herbs and trees and shrubs, even though this may at times disguise some important differences between species of similar life form. These different forms co-exist in each rangeland community because of different temporal niches arising from intermittent and variable moisture regimes (Westoby 1979–80). When rain falls, perennial species commence growth almost immediately and have an advantage in both start-up time after major rainfall and in response to minor storms. They also grow longer into the dry period because they are able to draw on moisture at depth (Walker and Noy-Meir 1982). Because of this persistence they are successful competitors with ephemeral forms but their absolute density is limited by the amount of moisture available for survival through dry periods. Hence they are unable to utilize all the surface moisture available after rain. This provides a niche for ephemerals, which grow more slowly at first because of the time taken in germination, but subsequently grow faster than perennials because they do not invest energy in perennating tissue. Since the length of wet and dry periods is variable, the relative prominence of each group changes with time. The density of perennial herbs builds up during pluvial phases and declines during droughts. Ephemerals, on the other hand, are most prominent after drought breaking rains when both moisture and space are plentiful. The most productive community is one that contains a range of growth forms, including species with different temperature optima, able to exploit contrasting moisture regimes.

Trees and shrubs usually live longer than most perennial herbs, and current population densities reflect a balance between the more extreme wet and drought phases of the past. They are generally more persistent through moderate droughts than perennial herbs and in the absence of fire become the dominant species in semi-arid communities. They are less persistent in arid areas where the frequency of long dry periods is high. Usable production is invariably low because of a high investment in woody tissue. Periods of growth and their relative growth rates are illustrated in Figure 2.

The growth-form compositions of the three vegetation types referred to earlier are shown in Table 1. The response of different forms to climate variation is presented in the ensuing sections.

11.4.1 Cool-season ephemerals

These occur in all three vegetation types and their germination is genetically programmed to respond to certain evironmental cues (Mott and Groves 1981) so that their subsequent growth and reproduction is completed within the cool season. In a chenopod low shrubland of the Koonamore Vegetation Reserve, Noble and Crisp (1979/80) found that for a 30 year period, there was frequent germination of ephemerals in autumn and winter, but rarely in spring. All ephemeral species in this region of non-seasonal rainfall, can germinate and establish over the full range of

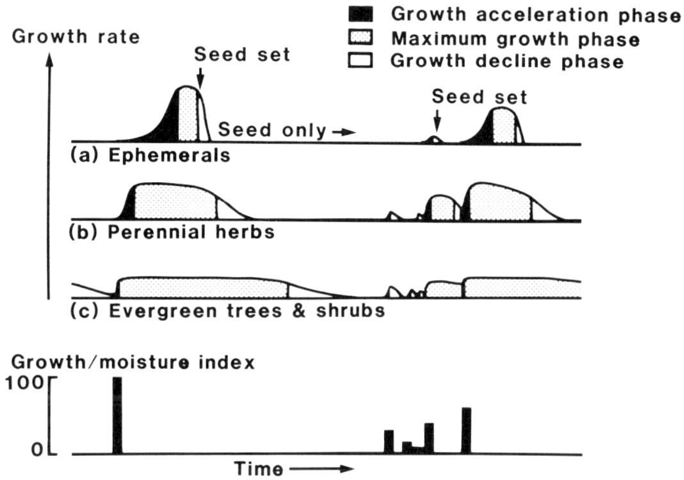

Figure 2. The growth response of three plant growth-forms to a sequence of rainfall events (after Westoby 1979—80).

Table 1. Plant growth-form composition of the three plant communities referred to in this chapter.

Plant growth form	Chenopod low shrublands	Semi-arid woodlands	Mitchell grasslands
Ephemerals			
Cool season	X	X	X
Warm season	O	O	X
Herbaceous perennials			
Cool season	X	X	O or X
Warm season	X	X	X
Woody perennials			
Shrubs	X	X	O
Trees	O or X	X	O or X

X = present
O = absent

temperature regimes that occur from summer to winter (Nicholson 1986); there are no exclusively winter or summer ephemeral floras as there are in the northern Mitchell grasslands (Orr and Holmes 1984). The life cycle of ephemerals is completed within four to five months of germination irrespective of soil moisture conditions. Their pulse of growth is therefore short-lived and once reproduction has occurred they rapidly disintegrate.

11.4.2 Warm-season ephemerals

With increased dominance of summer rainfall, a second group of ephemerals appears that germinates and grows only during the warm season. In addition to higher temperature requirements for germination and growth, they may also require cool winter temperatures to break seed dormancy (Mott 1972).

Both cool- and warm-season ephemerals are slower than perennials to achieve peak growth rates because they start from seed rather than perennating organs. They have a longer "start-up" time covering the transition from heterotroph to autotroph. Regrowing perennials, especially herbs, therefore pose considerable competition for ephemerals, and it is only in the chenopod low shrublands and overgrazed semi-arid woodlands that ephemerals come to periodically dominate the herbaceous layer.

Cool and warm season ephemerals contribute significantly to forage production in Australian rangelands. Nevertheless their populations are largely regulated by the season and amount of rainfall, with grazing pressure and fire having little influence except insofar as they reduce the density of overstorey species. In some species recruitment is an annual event, but the amount of rainfall determines the size of the cohort. This applies especially to the widespread species of *Stipa* (Williams and Roe 1975). The response is sufficiently predictable to be modelled quite accurately using temperature and soil moisture budgets (Noble and Crisp 1979/80). The total biomass of short-lived species is correlated with rainfall received during the previous 12 months (Noble 1977) due no doubt to the fact that many species so classified, such as *Stipa nitida, Salsola kali*, and species of *Sclerolaena* and *Zygophyllum*, are facultative perennials and may survive for more than one year given favourable rainfall. Given two successive seasons of above-average rainfall, which occurs about once in 15 years, *Stipa* becomes the dominant herbaceous component and provides abundant fuel for extensive wildfires that usually follow such sequences (Noble *et al.* 1986).

Some species, for example *Haloragis odontocarpa* and *Swainsona colutoides* (Noble 1988), depend on long-lived seed banks in the soil to only appear briefly after a disturbance such as fire. Others may resemble some coastal dune annuals in having no reserve of dormant seeds and depend wholly on annual seed production to maintain their populations (Harper 1977).

11.4.3 Cool-season herbaceous perennials

This group of plants constitutes a major source of palatable forage in semi-arid rangelands, except in the north where the reduced winter rainfall provides an unsuitable climate for their recruitment and growth. Many species in this group are grasses but forbs may abound in some areas. They generally possess the C_3 photosynthetic pathway, as do cool-season ephemerals, and therefore optimal temperatures for plant growth are 5 to 10° lower than those for C_4 plants. The proportion of herbaceous perennials that are of the cool-season type generally declines with decreasing latitude. Hattersley (1983) estimated that in eastern semi-arid rangelands the proportion of C_3 species in the grass flora declines progressively from about 60% in the south to 0 to 10% in the north.

A widespread and often locally abundant species is the grass *Danthonia caespitosa*. Extensive demographic and ecophysiological studies have been conducted on this species because of its forage importance. Populations have evolved that are able to grow and reproduce in a variety of environments ranging from cool temperate environments in Tasmania to warm semi-arid environments in the semi-arid woodlands (Quinn and Hodgkinson 1983, 1984). Notable genetically-based differences between populations include a latitudinal gradient in relative growth rates and photoperiodic and different temperature requirements for floral initiation (Hodgkinson and Quinn 1976, 1978). In southern temperate environments floral initiation is programmed for day length and temperature so that flowering and seed maturity coincide with a predictable growing season. In the hotter semi-arid environment this contrast is relaxed, permitting opportunistic reproduction whenever soil moisture and temperature permit growth. The success of this species and others with a wide latitudinal distribution in semi-arid environments is therefore achieved by a rapid response in growth and reproduction to irregular rainfall events.

11.4.4 Warm-season herbaceous perennials

The timing of germination in this group is dependent on rainfall and temperature. It takes place principally in the spring or autumn, but not during the extremes of winter or summer (Silcock 1973). This is probably because of sub- or supra-optimal soil temperatures although dormancy may also be involved. Brown (1982) demonstrated seed dormancy in *Aristida armata*, a common grass species of semi-arid woodlands. Seed production of this and other C_4 grass species occurs in summer or autumn and the seed dormancy mechanism prevents seedling establishment until the following spring or autumn. Another mechanism aiding establishment of warm-season grasses is the ability of partially germinated seed to remain viable over extended dry periods until suitable establishment conditions return (Watt 1982).

The growth rates of warm-season species given adequate soil moisture differs amongst *Astrebla* species (Jozwick 1970), and between genera (Christie 1975a), but are no higher than C_3 species at optimum temperatures. Differences between species in growth rates are less related to photosynthetic capacity than they are to the proportion of photosynthate partitioned into leaf surface (Christie 1975b, Doley and Trivett 1974).

As depicted in Figure 2, the herbaceous perennials are able to more effectively utilise available soil moisture, especially from small rainfall events, than the ephemerals because they can immediately respond to raised soil moisture. Their deep roots also enable them to utilise water out of reach of the shallower rooted ephemerals, and so they continue to grow after ephemerals cease. In both the chenopod low shrublands and the semi-arid woodlands, there generally are both cool- and warm-season species present which extend the growing season of pastures. Elimination of one or both groups of herbaceous perennials from pasture by heavy grazing will therefore reduce pasture production by narrowing the season of response to rainfall and growth from small rainfall events.

The seedling stage is obviously the most critical phase in the life history of arid perennial plants. Silcock (1977) studied the fate of more than 2,100 seedling populations of the four most numerous taxa in semi-arid woodlands over a 2.5 year period. He found that more than 50% of the seedlings died within a fortnight and only 7% of the original seedling population ultimately flowered.

In the principal habitat of a species, recruitment may be regular and almost annual, but in habitats that are more marginal for the species (rainfall lower or out-of-season) it may be quite irregular. *Astrebla* for instance, establishes regularly in the geographical centre of its distribution in Queensland (Orr and Holmes 1984), but infrequently at its southern distribution (Williams and Roe 1975, Roe 1987). This is related to the distribution of suitable rainfall events for germination, namely a fall of *c.* 120 mm or more in the late spring-summer period. The response of *Astrebla* populations to grazing is similarly related to habitat, with an increase in density in principal habitats but a decrease in marginal ones. *Danthonia caespitosa* is another example. It establishes regularly and is highly resistant to grazing within its principal habitat (Williams 1968), but establishes only episodically and has a reduced persistence under sheep grazing in more marginal habitats of lower winter moisture availability (Graetz 1980). Nevertheless there are other grasses within the semi-arid *Danthonia* habitat that are more episodic in establishment and are sensitive to grazing. These include *Themeda australis, Stipa aristiglumis* and *Enteropogon acicularis* (Figure 3).

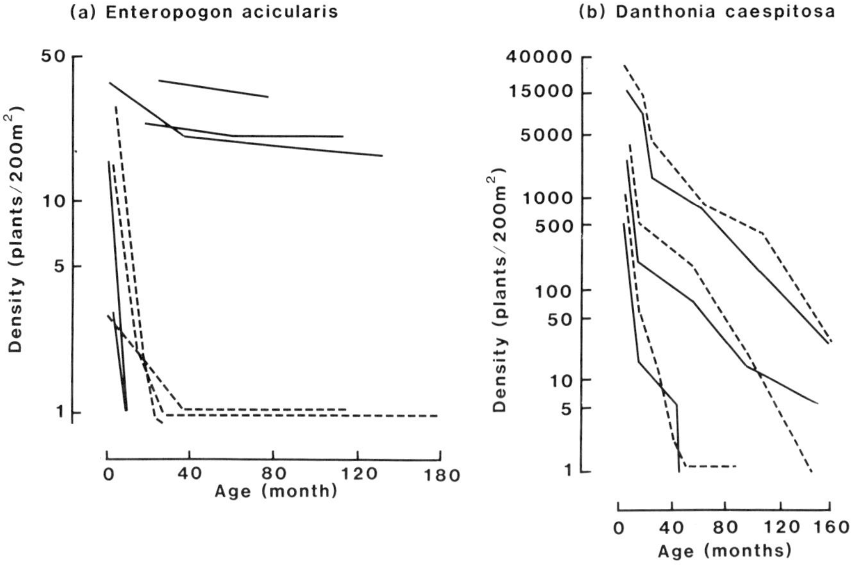

Figure 3. Survivorship curves for grazed (———) and ungrazed (—) populations of (a) *Enteropogon acicularis* and (b) *Danthonia caespitosa* in a disclimax grassland at Deniliquin (after Williams 1970).

11.4.5 Perennial trees and shrubs

The woody perennials that live from 10—300+ years represent a quite different category of rangeland plants. Apart from *Atriplex* species, all trees and shrubs are C_3 plants. Growth can occur at any time of the year although their highest growth rates are in the warmer periods (Hodgkinson and Beeston 1982, Noble 1977). All species are evergreen except for the winter-deciduous *Nitraria billardieri*, so they can very rapidly commence photosynthesis after drought-breaking rains, even more rapidly than perennial herbs which have to reestablish a leaf surface from quiescent meristems. Furthermore trees and shrubs have a more extended growth period than perennial herbs because their extensive roots have a monopoly on soil moisture at depth (Walker and Noy-Meir 1982). Mature trees often have access to water beyond the reach of many shrubs.

The recruitment of trees and shrubs is less frequent than for ephemerals and herbaceous perennials but the frequency and level of recruitment is both species- and rainfall-dependent. Temperature optima for maximum germination of all the species tested so far lie between 20 and 30°C except for *Atriplex vesicaria* which is between 15 and 20°C (Hodgkinson 1979, Burbidge 1945). *Atriplex vesicaria* has been found to establish every 2 to

Plate 1. Chenopod low shrubland (*Atriplex vesicaria*). The shrub is edible, but of lower palatability than the sparse grasses and forbs that grow between the shrubs.

Plate 2. Mulga (*Acacia aneura*) semi-arid woodland. The maintenance of a suitable balance between the tree-shrub and herbage layers is the primary requirement of vegetation management.

5 years, although germination may occur more than once a year in seasons of high rainfall (Crisp 1978, Graetz 1980). Recruitment of species examined so far is episodic and infrequent; *Acacia aneura* on average every 10 years (Preece 1971, Crisp 1978), *Maireana sedifolia* and *M. pyramidata* every 50 years or more (Crisp 1978) *following periods* of above-average rainfall. Mallee seedling recruitment is highly sporadic in the absence of fire (Wellington and Noble 1985) but may be extensive after fire, especially spring fire (Noble 1982). The lack of recruitment in most years has been attributed to seed harvesting by ants. After fire, large amounts of seed are released from the fruit within the canopy which provides a degree of satiation of seed predators (O'Dowd and Gill 1984). Hence certain rainfall sequences are often a prerequisite for successful reproduction. They need to be such as to produce an abundant crop of viable seed, suitable conditions for germination the next year, and then further rainfall to establish seedlings. If any one of these requirements in sequence do not occur then recruitment is thwarted. In semi-arid woodlands, cohorts of shrubs recruited infrequently are readily seen in the stepped size distributions of mature plants of different species. Furthermore different seasonal sequences favour episodic recruitment of different perennial species (Noble 1986).

The studies by Milthorpe (1978) on the chenopod shrub *Maireana pyramidata* show three essential features of woody plant demography (see Figure 4). The first is that some sites present a safer environment for seedling survival than others with the important features of a site varying from species to species. The second is that mortality in the early months after germination is extremely high (Deevy III survivorship). It is logical that only a proportion of seedlings will attain sufficient root extension to withstand the ensuing dry periods. In fact only a small proportion *need* to establish to maintain the population of such long-lived plants. The third is that grazing, in this case by rabbits, affects the degree of mortality in these early months. The mortality of adult shrubs is more or less constant (Deevy II) and not affected by grazing after this establishment period.

The competition provided by herbaceous species during the seedling stage is a further indirect effect of grazing. For instance the survival of seedlings of *Nitraria billardieri* is greatly reduced by the concurrent growth of annuals (Noble and Whalley 1978) and such interspecific competition is probably significant in the establishment of all shrubs and trees. Competition from annuals also reduces the seedling establishment of *Sclerolaena birchii* (Auld 1976), *Dodonaea attenuata* (G. N. Harrington, personal communication) and *Atriplex vesicaria*. Reduction in competition by herbs, especially perennial grasses, is one of the reasons contributing to increases in unpalatable shrubs in the heavily grazed semi-arid woodlands.

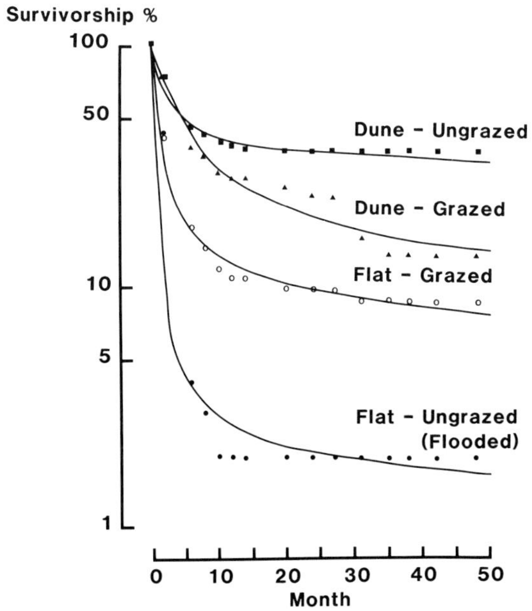

Figure 4. Survivorship of *Maireana pyramidata* seedlings in four contrasting situations (adapted from Milthorpe 1978).

The other reason is reduced fire frequency, for shrub seedlings are susceptible to defoliation by fire. The influence of fire is dealt with more fully at a later stage.

Populations of long-lived shrubs and trees, such as *Maireana sedifolia, Acacia sowdenii* and *Heterodendrum oleifolium*, show a very low mortality of adults. *M. sedifolia* populations showed a negative exponential survivorship with a half-life of more than 150 years (Crisp 1978). Such species establish only rarely in arid areas, with establishment related to the occurrence of rare events or the co-occurrence of two infrequent events. The absence of juveniles during the lifetime of a research worker does not necessarily mean that the population is headed for extinction.

In general, recruitment of perennial grasses, trees and shrubs in Australian rangelands is highly episodic and linked with major pluvial events. These are interspersed with long drought periods which may result in substantial mortality of established perennial plants. Cohorts established during such major recruitment events are therefore usually large because of the availability of drought-induced gaps and survivorship of such large cohorts of perennials is often considerably longer than small cohorts (Williams 1978). The wide fluctuations in amount and seasonal distribution of rainfall in these semi-arid rangelands, coupled with the genetically based differences between species in terms of their recruitment, growth, reproduction and mortality responses to climatic variation, result in constant fluctuations in both biomass and botanical composition of the herbaceous layer. Whilst biomass production is largely governed by rainfall, it is the genetic diversity that enables the full productive potential of the rainfall to be realised. Reduction in genetic diversity of these plant communities through mismanagement, may therefore reduce the productive potential of vegetation.

11.5 Response to fire

Fire is a naturally occurring phenomenon in Australia's rangelands and all three vegetation types are prone to periodic fire at frequencies ranging from 10 to 100 years (Hodgkinson and Griffin 1982). Despite this low frequency if has far-reaching consequences for the composition of the vegetation and its value for grazing animals. In this section we show how the attributes of species effect their response to fire and how these responses contribute to changes in community composition under various fire regimes.

The prerequisites for the occurrence of fire are sufficient dry herbaceous fuel, an ignition source and weather conditions conducive to the spread of fire. The quantity of fuel is directly related to recent grass growth and hence to rainfall in the previous 6—9 months, except where the herbaceous layer contains *Triodia* spp. and perennial fuel such as litter accumulates from year to year. Before European settlement, Aborigines used fire on a regular basis for hunting and stimulation of food plants, producing a mosaic of fire ages across the landscape. Lightning is now the main ignition agent, although in the semi-arid woodlands, fires are now deliberately lit to modify the composition of vegetation thereby raising the level and efficiency of animal production. Some of the effects of fire are similar to that of grazing. In both cases plants are defoliated but fire is unique in that defoliation is non-selective and instantaneous and the transient heat pulse induces a variety of plant responses (plants and seeds are killed while seed of some species is promoted to germinate) and longer term successional changes.

There has undoubtedly been a substantial reduction in wildfire frequency throughout most of Australia's rangelands following European settlement last century (Leigh and Noble 1981). Only when there is a sequence of above-average rainfall events do large-scale wildfires now occur in these areas, as during 1974/75 and to a lesser extent in 1984/85. One of the more serious consequences of this decline in fire frequency has been a significant increase in the density of undesirable shrub species, particularly species of *Acacia, Cassia, Dodonaea* and *Eremophila,* throughout the semi-arid woodlands (Hodgkinson 1979, Walker *et al.* 1981).

The longer-term effects of fire on plant populations and the structure and composition of a community are best considered within the framework of the fire regime (Gill 1975) rather than considering the effects of a single fire alone. The fire regime is characterised by the intensity, season and frequency of fire and species have different responses to particular fire regimes. Fire has been a significant factor in the Australian environment for at least the last 15 million years (Kemp 1981), and species have evolved a variety of adaptations to survive fire (Gill 1975, Hodgkinson and Griffin 1982), each having its own range of fire regimes under which their populations persist. Altering the fire regime can have a significant effect on the composition of vegetation so that prescribed fires used either for shrub or wildfire control, are both legitimate options and powerful vegetation management tools (Hodgkinson *et al.* 1984).

11.5.1 Vital attributes of species

Following the occurrence of a fire in any plant community, a series of successional changes occur in the species composition, as well as community structure depending on the post-fire rainfall regime. Immediately after a fire the species composition is dependent upon propagules being present, or which immigrate to the area from unburnt surrounds, or upon vegetative resprouting from surviving plant organs. There is an initial pulse of recruitment and regrowth under conditions of little competition for space and other resources before recruitment slows. Once an individual is established it is difficult to displace. This means that subsequent recruitment of additional species is unlikely and in the absence of further fires, long-lived species and those that can regenerate in the presence of adults will finally become dominant.

A major contribution to understanding the responses of plant communities to fire and other disturbances has been the development by Noble and Slatyer (1980) of a framework, based on vital attributes of a species, for examining vegetation replacement sequences. They considered the three most important attributes to be the method of arrival or persistence of the species during and after a fire, the ability to establish and grow to maturity in the developing community, and the time taken for the species to reach critical life stages.

All species in the three vegetation types have the potential to persist through fire disturbance as seed, but only a few survive fire as vegetative individuals. Seed longevity in the soil bank varies between species. This is an important factor governing the quantity and presence of viable seed of a species in the soil since species that do not produce seed regularly are at risk of extinction if seeds are viable for only a short time.

Most perennials are able to regrow from protected meristematic tissue but differences in level of survival varies greatly amongst species, especially the trees and shrubs, and within species according to age/size class (Figure 5). At one end of the continuum are species which die if all their leaves

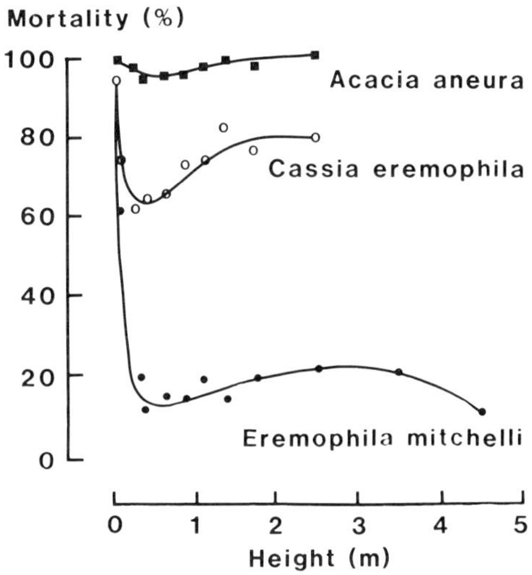

Figure 5. Fire induced mortality of three woody species in a semi-arid woodland, in relation to plant height (age) (from Hodgkinson 1986).

are killed (*Acacia aneura, Atriplex vesicaria* and *Callitris columellaris* are examples) either by direct heat or indirectly through the death of stem cambium. Conversely some species nearly always resprout after fire, such as mallee eucalypts. Many species have intermediate survival capability varying with size of plant and thickness of bark. Plant survival following a single fire is not significantly influenced by fire intensity within the normal range characteristic of each community, nor season of fire or physiological status of individuals (Hodgkinson 1986). However, resprouting species such as mallee eucalypts are vulnerable to a regime of multiple fires involving at least two fires in the autumn over a 10 year period. Noble (1982) recorded 90% mallee mortality after three fires in 10 years, of which the last two were in the autumn, compared with only 10% mortality for the same number of spring fires.

In contrast to shrubs and trees, nearly all perennial grasses are able to vegetatively regenerate from basal meristems (these are well insulated from heat damage by soil and sheaths). Typically, mortality of grass plants ranges between 0—10% but there are exceptions. For example *Eragrostis eriopoda*, a weakly stoloniferous grass with tiller bases exposed above the soil surface, will survive fire if there has been recent rain and the stem bases are green but if the plants are droughted, survival will be low. Presumably under these circumstances the fire consumes more of the base of the plant and the transfer of heat to basal meristems is higher.

The second group of vital attributes describe the conditions under which a species can establish and grow to maturity. All species in the three vegetation types appear to be able to establish and grow immediately after a fire. However, the chance of successful establishment of a species reduces with the length of time after fire, because of increasing competition. Grazing reduces the competitive effect of perennial grasses and allows a larger cohort of woody species to survive to maturity (Harrington 1986a). Ephemerals are often abundant during the first year after fire provided there is cool season rainfall.

Some species of shrub produce seeds which require weathering or heat to crack the testa to allow water entry for germination. This fire-stimulated germination response is common in the genera *Acacia, Cassia* and *Dodonaea* and often leads to mass germination of seedlings. Seed of some grass species such as *Triodia irritans* also appears to have a heat and scarification requirement. Suitable soil moisture conditions are also needed to initiate germination and sustain establishment. This phenomenon increases the chance of juvenile establishment but does not guarantee replacement of fire-killed adults since rainfall is a prerequisite.

The third vital attribute group deals with the timing of the life stages and comprises the time taken for a species to reach reproductive maturity after a disturbance, the lifespan of individuals and the longevity of propagules. Ephemerals and herbaceous perennials generally set viable seed within 1 or 2 years of a fire but shrubs and trees take much longer, in the order of 4 to 10 years, depending on the species. The frequency of fire is therefore an important determinant of the species composition of the shrub layer. If fuel is available, frequent fires can be used to radically reduce shrub populations especially obligate reseeders with low levels of viable seed in the soil bank.

11.5.2 Plant community response to fire

Community response to a fire is governed by both the vital attributes of the dominant species and the post-fire rainfall pattern. Figure 6 is a

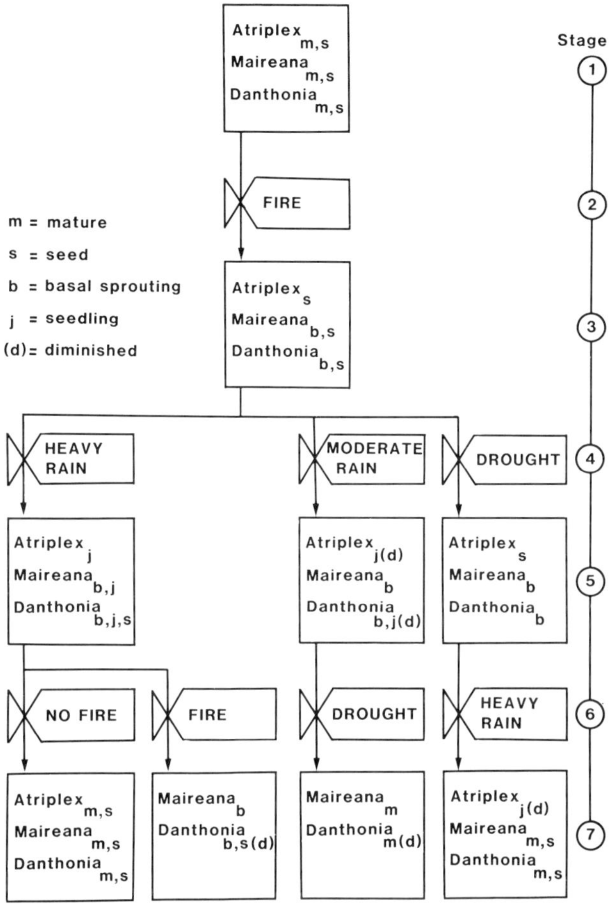

Figure 6. Model of changes in an *Atriplex vesicaria — Maireana aphylla — Danthonia caespitosa* community near Deniliquin, as influenced by fire and rainfall.

summary of how a chenopod shrubland community near Deniliquin, N.S.W., would react to a variety of fire and post-fire rainfall combinations. If a mature community is burnt by wildfire (prescribed burning is not practised or recommended), all the juvenile and mature *Atriplex vesicaria* will be killed. Some of the mature *Maireana aphylla* and most of the mature *Danthonia caespitosa* will regrow. All species would also be present in the soil as seed. Differences in post-fire rainfall (Stage 4 and 6) would influence community composition. Drought followed by heavy rainfall would result in recruitment of all species but the density of *Atriplex vesicaria* would be diminished because of its low seed longevity (Stage 7). Moderate rainfall followed by drought would lead to local extinction of *Atriplex vesicaria* because of poor persistence of juveniles under drought. Heavy rainfall soon after the fire would, in time, restore the original structure of the community but a second wildfire fuelled by abundant herbage growth, would result in local extinction of *Atriplex vesicaria* and diminished populations of *Danthonia caespitosa*. This model is not comprehensive but illustrates the way fire and rainfall may interact to produce structural and floristic differences in plant communities.

In contrast to the rapid demise of *Atriplex vesicaria* from chenopod shrublands given certain fire and rainfall sequences, semi-arid *Triodia*/mallee communities are floristically stable under a range of fire and rainfall regimes. Applying Noble and Slatyer's (1980) vital attribute scheme to this community gives a replacement sequence shown in Figure 7. The mallee is a resprouting species (species type V) which must pass through a juvenile phase before achieving reproductive maturity, whereas the few herbaceous species have a long-lived seed bank in the soil (species type S). Species intolerant of competition (I species) can establish as seedlings only immediately after disturbance while competition tolerant species (T species) establish and reach maturity both immediately after disturbance and also for an indefinite time afterwards. The scheme depicts succession in the absence of fire with the predicted demise of mallee and *Triodia* after about 400 years. Recurrence of fire at various seral stages

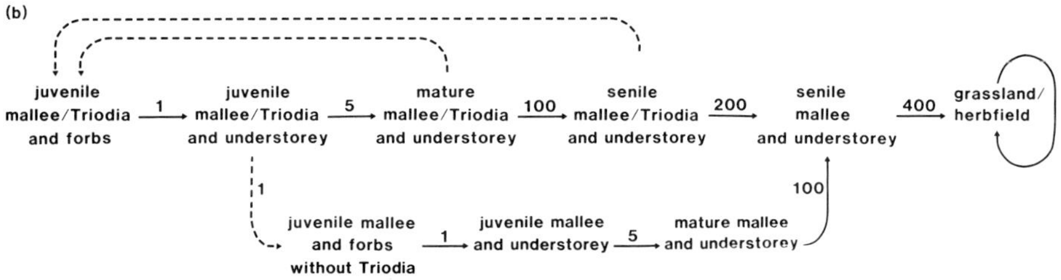

Figure 7. Vegetation replacement sequence for a mallee semi-arid woodland. Fig. 7a summarizes the vital attributes of the species where m = time to reproductive maturity, l = longevity of mature plant, e = longevity of seed bank, V = vegetative resprouter, S = persists through seed, I = intolerant of competition (establishes only after fire) and T = tolerant of competition. Fig. 7b shows the community transitions after periods without fire (solid lines) or fire (broken lines).

will return the composition to juvenile *Triodia*/mallee and juvenile herbs except for two fires, 1 or 2 years apart, when the *Triodia* may be lost from the community or at least in those habitats carrying sufficient ephemeral *Stipa* fuel. In general the community is resilient to fire with the initial pre-fire species being reproduced in post-fire communities. The same general population response is expected from semi-arid *Acacia aneura* and *Eucalyptus populnea* woodlands except that there would be no reduction in shrub and tree density following a long fire-free period because all species have the ability to regenerate from seed at any successional stage.

The short-lived herbage pulse following a spring fire in *Triodia*/mallee is portrayed in Figure 8. Forbs, and particularly *Haloragis odontocarpa*, dominate the herbage flora in the first year providing there is adequate autumn rainfall, while other short-lived forms including sub-shrubs such as *Sclerolaena divaricata* and the facultative perennial *Stipa* spp. dominate thereafter. A winter burn applied in the same mallee community had a quite different effect on the herbage flora, with only a small forb component present in the first year after fire. Replicate prescribed fires applied during the 1982/83 drought demonstrated the overriding importance of rainfall in mediating ephemeral herbage productivity.

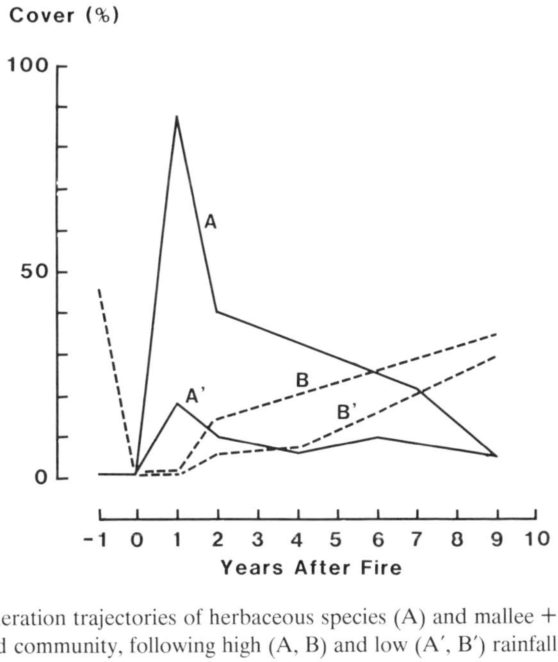

Figure 8. Regeneration trajectories of herbaceous species (A) and mallee + *Triodia* (B) in a mallee woodland community, following high (A, B) and low (A′, B′) rainfall in the first 1—3 years after fire (after Noble 1988).

Plate 3. Mitchell (*Astrebla* spp.) grassland. The community is resilient to both grazing and fire.

Plate 4. Mallee (*Eucalyptus* spp.) semi-arid woodland with a hummock porcupine grass (*Triodia irritans*) understorey. The composition of this community may be manipulated by both the frequency and timing of fire.

Mitchell grasslands are also highly resilient to fire. Botanical composition and long-term herbage production are not altered significantly, nor for long after a fire, irrespective of rainfall (Scanlan 1980). This is because no or at most few shrubs or trees are present in the grasslands and all the herbaceous species have seed stored in the soil. New growth of *Astrebla* after fire is increased above that of unburnt plants provided rainfall during the growing season exceeds 500 mm (Scanlan 1980, 1983). Seed production is up to 10 times greater on recently burnt than unburnt pastures. Since the general effects of fire on Mitchell grassland are benign, prescribed burning is not recommended.

Unlike the chenopod shrublands and Mitchell grasslands, there is a case for periodic prescribed burning of semi-arid woodlands to maintain the woody overstorey at an acceptable level for profitable pastoralism (Hodgkinson and Harrington 1985, Noble 1986). Prior to the advent of pastoralism, widespread shrub establishment events and wildfires were closely coupled so that the shrub populations and biomass were regularly checked by fire. Following European settlement grazing by domestic herbivores has reduced the competitive stress imposed by herbaceous plants on shrub seedlings (Harrington 1986a) and has also reduced herbage fuel thereby removing fire as the main lethal agent. To check shrub increase, government agencies are now recommending prescribed fire following uncommon periods of high rainfall when there is widespread germination and establishment of shrubs and abundant grass is available as fuel. In addition to prescribed fire, prudent grazing is being advocated to foster an increase in the herbaceous perennial component of the vegetation.

11.6 Response to grazing

Grazing is the major disturbing force in rangeland vegetation, and the reaction of species and communities to this force is one we seek to understand and manage. It is made more complex by the fact that it is also the process by which we reap our economic return from the rangelands. The topic will be considered here under three headings: the response of plants, the nature of grazing and the response of communities, and the concept of grazing management.

11.6.1 Plant response to grazing

Rangeland plants differ in their ability to withstand defoliation by grazing. At low levels of defoliation the survival of all species is unaffected, however at some threshold level of defoliation (defined by both severity and/or frequency and which is presumably different for each species) plant survival is reduced; as is the productive vigour of those individuals that survive (Hodgkinson and Mott 1987). Grazing can alter the botanical composition of rangeland vegetation directly by accelerating mortality of some species and indirectly, by influencing the recruitment levels of species into the gaps created by plant death.

Amongst the shrubs of chenopod shrublands and semi-arid woodlands there is a wide range of tolerance to grazing. At the low end are species such as *Atriplex vesicaria* which cannot tolerate a complete defoliation (Leigh and Mulham 1971) and at the other are species such as *Eremophila mitchellii* which readily resprout after complete defoliation by caterpillars or fire (Hodgkinson 1979). Mortality of *A. vesicaria* is increased when leaf number per bush is reduced below 50, with mortality rate rising in proportion to further leaf reduction (Leigh and Mulham 1971). This mortality response is not altered by alleviating water stress or defoliating in different seasons, suggesting that its basis is morphologic rather than physiologic. Sustained heavy defoliation by grazing will also kill high proportions of *Acacia aneura, Dodonaea viscosa* and *Eremophila bowmanii* (Harrington 1979). Although the evidence is equivocal many of the edible shrubs, irrespective of mature height, appear not to cope with a complete defoliation.

In contrast all, or nearly all, established perennial grasses survive a single complete defoliation, whether by clipping or heavy grazing (Hodgkinson 1976, Williams 1970, Brown 1986). Grasses, because of their basally located meristems, are better adapted to close grazing than the shrubs with their predominantly aerial meristems. Some grasses, although resistant to grazing when established, succumb to defoliation at the juvenile phase. Williams (1970) found that seedlings of *Enteropogon acicularis* but not *Danthonia caespitosa*, are easily killed by grazing (see Figure 3). Both species persisted under moderate to heavy grazing by sheep when established. Seedlings of *Thyridolepis mitchellianna*, but not *Aristida armata*, fail to survive severe defoliation (Brown 1987). These studies suggest that grass species differ in the rate at which they acquire tolerance to grazing following germination.

Reasons for species differences in grazing tolerance amongst Australia's rangeland have not been clearly established. Recently Hodgkinson *et al.* (1986) compared the physiological and morphological responses of *Themeda australis*, a ubiquitous but grazing sensitive Australian grass with *Cenchrus ciliaris*, an exotic species sometimes sown to form pastures in eastern parts of semi-arid woodlands. It was found that *T. australis* was unable to cope with increasing defoliation stress because of an inability to form nodal tillers and hence maintain an adequate basal leaf area to meet the carbon requirements for plant growth and respiration. This led to temporary impairment of the photosynthetic capacity of leaves produced immediately after defoliation and a reduced allocation of photosynthate into leaves during the first three weeks of regrowth. As a consequence, leaf initiation and expansion of individual leaves was impaired, thereby lowering the efficiency of light interception of leaves and the rate of shoot regrowth. This study indicates that species differences in tolerance to grazing have complex bases.

The grazing pressure exacted on individual plants in the field fluctuates within and between years, even with animal numbers constant. This follows unpredictable variation in supply of water to plants and hence plant growth. Grazing pressure is expected to be greatest on an individual plant after growth is arrested by drought or soon after drought breaking rains, when the amount of forage on offer to grazing animals is very low. The significance of simultaneous pressures of water stress and defoliation for subsequent plant survival and growth have not been adequately studied. A study of *Danthonia caespitosa* (Hodgkinson 1976) indicated that growth was slightly increased by frequent severe defoliation of summer-droughted plants but greatly reduced for irrigated plants. Recovery growth after good autumn rains indicated that defoliation during active growth supresses vigour of plants more than for the droughted plants (see Table 2). Although this study was limited to one species the results support the theory that the critical time for rangeland grasses, in terms of maintenance of vigour and possibly their survival, is grazing pressure immediately after rather than during drought.

Grazing has little or no direct influence on the germination and early growth of seedlings but may do so indirectly by influencing the composition and amount of seed in the soil seed pool and the suitability of the surface soil for recruitment processes. Inflorescences may also be removed by grazing so that seed output may be reduced. The decline in seed production is not necessarily linearly related to grazing pressure as shown for *Astrebla* species (D. M. Orr personal communication). Further studies are required to show if this curvilinear response is common to all perennial grasses.

11.6.2 The nature of grazing and the response of communities

At the plant level, herbivores prefer the soft and "sweet" components. This preference involves the selection of green plants or green components of plants, before dry components, and leaf before stem (Arnold 1981). In the growing season (particularly after cool-season rainfall) the softest and greenest plants are the ephemerals and these may dominate the diet while they are available (Wilson *et al.* 1969; Lorimer 1978; Harrington 1986b). Ephemerals are of high nutritive value while they last and hence are valuable components of the community. In the dry season the pressure moves to the perennials because these retain green leaf longer than the annuals. The highest grazing pressure falls on those perennials growing out of phase with others in the community. Examples are the sparse winter-growing legumes found in Mitchell grass communities and the summer-growing perennial grass *Enteropogon acicularis* growing in an essentially temperate *Danthonia caespitosa* grassland (Williams 1970). The intensity with which sheep search out green forage is shown in the work of Leigh and Mulham (1966) who found that 50% of the summer diet in an *Atriplex vesicaria* shrubland was composed of the grass *Sporobulus caroli*, comprising only 1% of the plant biomass on offer. Such species whose vegetative phenology differs from the main body of the community are heavily defoliated and are likely to be the first species removed by grazing.

As the length of the dry period increases, animals turn more and more to browse for their forage requirements. With browse, preference for soft and green is often modified by the presence of secondary compounds such as tannins, alkaloids, essential oils and salt. These are common in shrubs and represent their major defence against herbivory. The relative unpalatability of chenopod shrubs such as *Atriplex vesicaria* (Wilson *et al.* 1969), is the major reason for its persistence in grazed communities. Whilst sensitive to defoliation (Leigh and Mulham 1971), it remains the dominant plant over wide areas of the Riverine Plain of New South Wales. In other communities it is readily removed, particularly where its palatability is higher than other species such as the unpalatable shrub *Maireana pyramidata*. Paradoxically browse shrubs of low palatability (e.g., *M. pyramidata* in the chenopod shrublands, *Acacia aneura* in the semi-arid woodlands), are valuable in the short-term for the maintenance of stock through droughts, but detrimental in the long-term by sustaining grazing pressure on palatable grasses. This is a simple extension of predator-prey theory which holds that a herbivore cannot eliminate the major species on which it depends for its food, but can eliminate a secondary more preferred species because the grazing animal is sustained by forage contributed by the less preferred species (Caughley 1982).

Some species become less palatable as they age through increasing fibrousness or spinyness. Such plants are well protected from high grazing pressure during periods of forage shortage. There may also be changes in chemical constituents. For instance the tall shrub *Eremophila mitchellii* has a high content ($\simeq 1\%$) of essential oils in the foliage of established plants protecting it from grazing by all herbivores, including goats (Harrington 1979). Consequently it has become a serious weed in the semi-arid woodlands. However coppice regrowth after disturbance has a relatively low content of essential oils and is eaten by goats. Goats may therefore be used to control the regrowth of shoots after clearing operations even though they avoid adult shrubs. Possibly a combination of fire (to defoliate the shrubs) and goats (to consume the low oil regrowth) would achieve a greater reduction in shrub density than either fire or goats used separately, although this has not been tested in the field.

The distribution of grazing across a landscape is another important component of grazing pressure. Watering points are always foci of high grazing pressure, particularly in saltbush communities where the dominant shrub *Atriplex vesicaria* is high in soluble salts. Water intakes may exceed 10 l/day for sheep and 130 l/day for cattle (Wilson and Graetz 1980). The effects of these differential grazing pressures are obvious along radial transects from water sources. The analysis of these gradients, referred to as piospheres (Barker and Lange 1969), shows the response of the communities to high grazing pressure near watering points. Andrew and Lange (1986) showed increased defoliation of the palatable shrubs *Maireana sedifolia* and *A. vesicaria* near water, increased mortality of *A. vesicaria* and increased density of less palatable and shorter-lived species such as *Sclerolaena obliquicuspis*. Grazing effects are therefore an integration of animal preferences, resistance to defoliation and place in the landscape.

Different levels of selectivity may also be found between herbivore species and these may be exploited in management. Generally, cattle have a lower *ability* than sheep to harvest small grass tussocks, so that these have a lower degree of defoliation and mortality than in sheep grazed areas (Graetz 1980). Such differences in selectivity may be exploited in management in two ways. The first is in the management of vegetation composition. Goats are sometimes used to remove a plant that is preferred by them but avoided by sheep (e.g., *Dodonea attenuata*). However if the weed is eaten only sparingly by goats and then only at high grazing pressure, such as *Dodonaea viscosa*, the desirable perennial grasses will also be removed along with the weed and further degradation has been induced (Harrington 1979, 1986b). In this case fire is the preferred management option because it is non-selective. The second application to management is when animal productivity may be increased by grazing a mixed population of herbivores. It is often assumed from diet separation studies that this is a widespread occurrence. For instance, sheep have a preference for forbs and kangaroos for grasses and they may therefore be seen as complementary grazers. However, such differences diminish as grazing pressure increases and they become almost directly competitive (Short 1985). Differences in preference must be maintained at high grazing pressure before advantage accrues from mixed species grazing. While this situation is rare, animal species do differ in their ability to exert grazing pressure on vegetation. This is a function of both their ability to graze close to the ground and also to browse high shrub foliage. It is least for cattle and greatest for goats. Where domestic herbivores are grazed in competition at high grazing pressure, sheep obtain a nutritional advantage over cattle, and in turn goats over sheep. Animal survivorship is the converse of plant survivorship (Graetz and Wilson 1980, Wilson and Mulham 1980).

Rabbits differ from other herbivores in that not only do they defoliate plants by direct grazing of foliage but they may also induce plant mortality, even of quite tall woody species, by ringbarking stems. Accordingly, several workers (e.g., Lange and Graham 1983, Friedel 1985, Silander 1983) have suggested that adequate regeneration of some of the more conspicious shrubs and trees of arid Australia is seriously threatened by rabbit predation. Foran *et al.* (1985) found that whilst some grass and shrub species (e.g., *Acacia kempeana*) were reduced by rabbit grazing, establishment of other species appeared to be promoted by rabbits. Whether or not direct predation by rabbits of threatened species is minimised significantly by episodic mass seedling recruitment following rare climatic sequences, has yet to be determined. In areas of higher rainfall the effect is less apparent, presumably because of predator satiation. Such frequency-dependant changes in relative consumption are often the key to understanding why grazing has a major regulatory function in some communities and not in others (Westoby 1979/80). Thus shrubs such as *Acacia aneura* and *Cassia nemophila* have declining populations in arid regions while increasing in the semi-arid.

In summary the response of a plant to grazing is complex. The outcome is dependent on its ability to withstand defoliation, its avoidance of grazing through unpalatability, the relative palatability of its neighbours, its place in the landscape, and the type of herbivore.

11.6.3 Animal production

There is a general assumption that herbivores forage in an optimal manner, seeking the greatest amount of nutrients from the forage for the least expenditure of energy. In the rangelands forage is sparsely distributed over much of the landscape and with high nutrient species and plant components (leaves) interspersed amongst those of low nutrient status. At very low biomass levels, total intake may be limited directly by scarcity. At high levels, it is limited indirectly by quality, or more precisely by sparseness of high quality components. With pasture swards intake of the grazing animal is reduced when availability falls below 1000 kg/ha (Arnold 1981). However with rangeland plants that grow as spaced tussocks, intake is not reduced until much lower levels of biomass. Threshold levels have been measured at 300 kg/ha for *Danthonia* grassland (Leigh *et al.* 1968), 200 kg/ha for a grass-chenopod community (Short 1985) and only 50 kg/ha for a semi-arid woodland (Leigh *et al.* 1979). Most species in the communities discussed in this chapter have an erect habit and may be reduced to as low as 20 kg/ha by grazing during droughts (Short 1985).

In the short term, many compositional changes have little influence on animal productivity. This is because species within the one plant growth form (e.g., ephemerals, herbaceous perennials) have similar forage values. Furthermore, domestic stock are able to select the most nutritious material and it is usually amount rather than quality which limits animal production. Differences between growth-forms however are important. In the Mitchell grasslands the growth of cool-season ephemerals, which occurs in some years, may almost double the number of lambs weaned (Orr and Holmes 1984). Ephemerals are not only more digestible than perennial grasses, but C_3 plants and plants growing in cool conditions are more digestible than those growing in warm conditions (Minson and Wilson 1980). After winter rain the annual flora is predominantly dicotylodenous, whereas after summer rainfall it is predominantly monocotylodenous (Mott 1972, Orr and Homes 1984). These contrasting suites have sharply different forage characteristics. The monocots provide a greater bulk of low quality forage that persists for many months after senescence. The dicots on the other hand, are not persistent but provide high quality forage when available and give a significant boost to animal production. However, as stated earlier, the productivity and composition of these ephemerals are almost entirely dependent on rainfall and are beyond management control except for those following prescribed fire.

The two major forage attributes of a community, quality and quantity, are integrated within the production-stocking intensity relationship shown in Figure 9. This is known as the Jones-Sandland model (Jones and Sandland 1974, Jones 1981) which, in all pastures studied, shows a linear decline in production per animal (y) as stocking rate (x) is increased at most practical stocking rates. This is expressed in the form $y = a-bx$. Theoretically we may expect a pasture composed of perennial grasses, which are characterised by high persistence but moderate quality, to have a production-stocking rate function that has a low 'b' value (high carrying capacity) and moderate 'a' value. A pasture composed of a mixture of perennial grass and ephemerals will have reduced carrying capacity because of lower persistence during dry seasons, but higher average quality. It will have a production-stocking rate relationship with a moder-

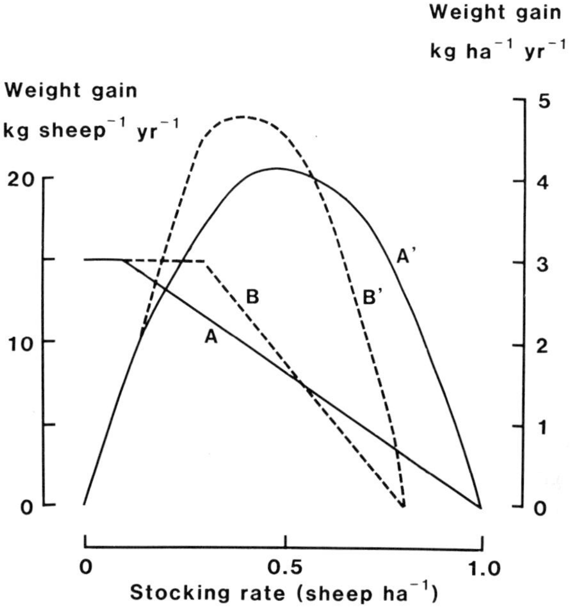

Figure 9. Jones-Sandland model of weight gain per sheep (A) and per hectare (A′) for a perennial grassland. The situation (B) and (B′) represents the same model for a theoretical change in composition to a mixed perennial and annual grassland with higher quality but lower carrying capacity.

ate 'b' value and high 'a' value. The mixture therefore has greater total animal production per hectare. The grass dominant 'A' situation shown in Figure 9 is found in years of good summer rainfall in the Mitchell grasslands and the sandplain mulga woodlands with an *Eragrostis eriopoda* understorey. Animal production is higher in years of more average and seasonally distributed rainfall, when the composition is more diverse and quality higher, as represented by situation 'B' in Figure 9.

11.6.4 Grazing management

Changes in botanical composition induced by grazing are a widespread phenomena. Sometimes the change is due directly to increased vigour of defoliated species. Seedlings of the perennial grass *Astrebla*, both establish and grow more vigorously under moderate grazing compared with ungrazed communities (Orr and Holmes 1984). New tiller formation is inversely related to old tiller density suggesting that an optimal management strategy is periodic heavy grazing to remove old tillers thereby stimulating new tiller formation. This release from apical dominance was observed by Hall and Lee (1980) following heavy cattle grazing and light sheep grazing of *Astrebla* pastures. Orr (1980b) also suggested that moderate grazing of *Astrebla lappacea* was beneficial in promoting both seed production and leaf growth.

More commonly however, changes arise from the death or loss of vigour of grazed species with a consequent increase in availability of resources, particularly soil water, to the remaining species. The local extinction of a species through grazing may be due to two reasons. Either that species is more sensitive to defoliation than others in an evenly grazed community, or it is no more sensitive to defoliation than its neighbours but receives a higher grazing pressure. The latter may arise from either a higher palatability or because it maintains green leaves when other species are senescent.

By far the most important factor in grazing management is the selection of optimum grazing pressure. The response of communities to grazing must therefore be viewed as a function of grazing pressure rather than as a simple matter of grazing or resting. In semi-arid woodlands a safe grazing pressure is considered to be that which will utilise 20—30% of pasture biomass present at the end of the main growing season (March). In this case pasture biomass refers to the total herbage mass, green and dry, and it is considered that one sheep will consume 400 kg of forage in a year (Beale *et al.* 1986). Economic returns are maintained because each sheep has high individual productivity and the variable costs of production are the same as for sheep of low productivity. The stocking rates are thus low on the Jones-Sandland scale mentioned earlier and allow the maintenance of pasture vigour. They also increase the potential for prescribed burning to control any increase in density of undesirable trees and shrubs.

Such a system represents a rule of thumb approach to the control of grazing pressure. It is adapted to breeding enterprises requiring relatively stable animal population by setting the utilisation rate at an average of 20% and allowing the actual utilisation rate to rise and fall between years of low and high rainfall. It has the advantage of requiring few pasture measurements in contrast to more intensive systems of defoliation control. On more *resilient* pastures the utilisation rate may be higher. In the Mitchell grasslands research has shown that 30% is optimal for animal production and at this rate there is no damage to the pasture (Orr 1980b).

The challenge for research is to develop grazing systems allowing the maintenance of less resilient pasture types under higher utilisation rates. The application of grazing systems involving formal rests and rotations, has failed to demonstrate any advantage in long-term animal production in Australia (Wilson 1977). Such systems have been imported on an *ad hoc* basis from other countries and applied without regard to the actual requirements of key pasture species. The only successful system was developed in the Pilbara region of Western Australia and is based on summer burning followed by grazing deferment (Suijdendorp 1980). This system has led to a change in dominance from *Triodia pungens* to *Eragrostis eriopoda* and was developed specifically for this purpose. It is essential that in future grazing systems be developed from the study of species response to defoliation and survivorship under various grazing pressures. Systems will then be developed that are specific to the particular seed production, establishment and defoliation requirements of individual key species. This detailed research is only justified for pasture systems that show a lack of resilience under normal grazing pressures (i.e., pressures that optimize short-term animal production), such as the semi-arid woodlands. It is important to realize that grazing systems do not improve animal production directly (i.e., without first improving pasture composition or production) and will not be of value in resilient pasture types, such as the Mitchell grasslands (Wilson 1986).

It is also essential that they be tested across a range of grazing pressures. Bransby and Tainton (1986) in South Africa have shown that rotational grazing systems change the production-stocking rate relationship in the Jones-Sandland model, such that carrying capacity is increased but production per animal is decreased (see Figure 10). In the case shown, for a Coastgrass-Bermudagrass pasture, the optimum stocking rate was almost doubled by changing from continuous to rotational grazing but there was little increase in total productivity per hectare. It would appear that highest economic return would arise from a lightly stocked-continuously grazed pasture. This further illustrates the requirement to integrate studies

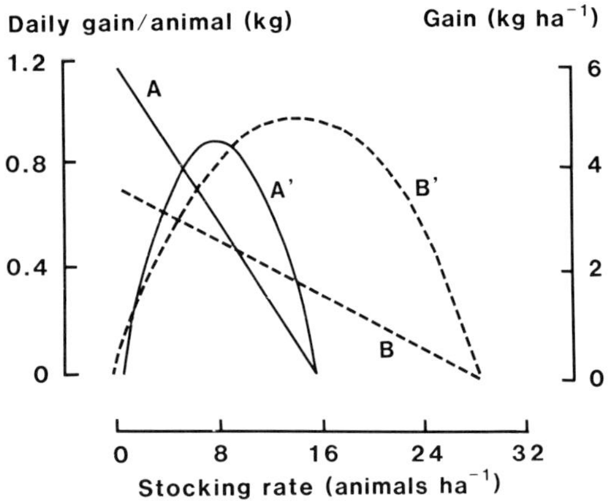

Figure 10. Relationship between weight gain and stocking rate for continuously grazed (A) and rationally grazed (B) systems (after Bransby & Tainton 1986).

on plant demography, animal production and economics before deciding on a grazing management strategy, for it is never safe to assume a direct relationship between grass production and economic return.

In the Australian rangelands, where pasture production is highly variable and seedling establishment an episodic event, grazing management decisions are also likely to be episodic. In this context a study on *Themeda triandra* in South Africa found that pasture growth was not enhanced by deferment unless rain fell at the appropriate time (Walker *et al.* 1986). In this context an appropriate grazing management strategy is to manage according to the needs of the plants at those times that are critical to plant survival or growth, and at other times to manage according to the needs of the animals.

Earlier we mentioned that a species will establish more frequently and at a greater density within its principal habitat than in more marginal ones. Consequently a species will be more resilient to grazing within its principal habitat than in a more marginal one and the retention of a species in a marginal habitat will require more careful grazing management. Grazing systems may therefore vary in their application between range sites even for the one species. *A. vesicaria* for example, remains essentially unaffected by continuous grazing at the centre of its distribution but has been largely removed by grazing in both the more arid and humid parts of its distribution.

11.7 Patterns in space and time

This chapter has so far been concerned with the response of plant species and communities to climate, fire and grazing. Species attributes are important in determining population response to any of these environmental events, but alone do not determine the response. At the population and community level the outcome is mediated by direct competition from other species and the indirect influence of neighbouring plants on fire regime and grazing pressure. Similarly the response of a community is influenced by neighbouring communities. At the landscape level there are patterns of water runoff and runon, fire mosaics, and grazing preferences which also need to be understood if we are to apply vegetation science to management at the scale of the animal's habitat or of a commercial grazing enterprise (the habitat of man!).

Much of the variation within a landscape arises from patterns of water runoff and runon. At a scale of 20—30 km this pattern defines the major land systems that are observed in arid areas within the one land holding. An example is the Burt Plain near Alice Springs. Water is shed from the hill land system, runs across the foothill fans and is deposited on the plains beyond. This is the basis of the different land units on which management is based. Animals concentrate their grazing on those units with the most nutritious species, in this case the alluvial fans (Low *et al.* 1981). Here grazing has its biggest impact on species survival and these units should be the focus for management research. There is also a finer scale pattern. Within these alluvial fans, there is a pattern of erosion, transfer and sink zones, formed in series as water and sediment are moved downslope (Pickup 1985). Erosion zones will have a poor soil cover and relatively sparse vegetation whilst sink zones will have the deepest soil and the densest cover of both herbaceous and tree species. These zones move downslope in time. The sink zones are more favourable for plant growth and easiest to manage. In contrast the erosion zones are easy to damage by grazing because the vegetation already has a precarious hold and erosion is easily accelerated. Those who are assessing landscape erosion arising from grazing must be aware that there is already natural erosion within the landscape which affects vegetation composition and its trajectory in time.

Pattern is also a feature in determining the effects of fire. In the mallee woodlands there is often a complex of dunes and swales. Forage growth is more abundant in the swales and hence fire frequency is greater (Noble 1984). Fire can therefore be used to change the density of the trees on the swales but not on the dunes. When fire occurs naturally it will only carry across the dunes about once every 20 years and this is an important feature in the control of prescribed fires in such woodlands. It is also important in reducing post-fire wind erosion which may occur on dunes. A mosaic of fire ages is regarded as desirable in such intrinsically flammable landscapes. This not only limits the spread of natural fires that may occur at times of high fire danger, but provides a diversity of successional stages in the vegetation. Grazing by both sheep and kangaroos is likely to be more intense on small burnt patches, with the probability of higher grass mortality from overgrazing whereas large prescribed fire can act to produce predator satiation as described previously.

Grazing pattern arises from a preference by sheep and cattle for land units that are flat and open. There is also a preference for units with short grasses, within the limitations imposed by fences and water supply. If these units are sensitive to grazing then complex grazing regimes or paddock subdivision may be required to achieve satisfactory production and utilization levels. For instance Foran *et al.* (1985) recommend grazing of degraded units only in the late dry season for the "red calcareous" land units in the Victoria River District. In contrast to sheep and cattle, grey kangaroos prefer to graze in areas with high tree cover. Therefore, with the exception of periods of forage stress, there is a degree of separation of the grazing by sheep and kangaroos. The separation serves to provide a more even grazing across the landscape than would be provided by either herbivore alone.

Grazing is also more severe near water supply, natural camp site and in the direction of the prevailing wind (Stafford Smith 1984, Orr 1980a). The effects of this uneven distribution are most pronounced in the chenopod shrublands where the major species are sensitive to grazing. In these communities concentric circles of species change may be recognized, with the more sensitive species removed at greater distances from water (Graetz and Ludwig 1978) and increaser species becoming prominent near water (Barker and Lange 1969) as mentioned earlier. It may appear that grazing pattern is an intrinsic characteristics of the landscape that is not amenable to management. However, since the placement of both water and fences are under man's control there is scope for management. Knowledge of water requirements and movement behaviour can be applied in mathematical models that predict the outcome of new fences or watering points (Stafford-Smith 1984). Furthermore, watering points can be placed on those parts of the landscape that are resilient to grazing and of low erosion potential (Pickup 1985).

The final factor creating pattern is the time that a continuous disturbance, such as grazing, has been active, or the time since a discontinuous disturbance such as fire. This may be described as a trajectory of a land unit within species space, as revealed by ordination, which is synonymous with trend or succession (Austin *et al.* 1981, Foran *et al.* 1986). If comparative sites of known fire age and grazing history are available, the relative contribution of these factors and climate can be separated. This allows the prediction of the future direction of a community when one of these factors is varied, as well as explaining the various vegetation compositions that are seen at one time. This method has been used to separate variation due to grazing and climate for grasslands in central Australia (Foran *et al.* 1986).

11.8 Conclusion

The vegetation of the Australian rangelands may be portrayed as dynamic ecosystems in which plant cover and composition respond to cycles of drought and rain, continuous disturbance by grazing and recurrent disturbance by fire. In drought years the density and size of most plant populations are reduced. Some species will suffer more severely than others so that changes in botanical composition will be evident. An inexperienced observer may reach the erroneous conclusion that a permanent change has been wrought on the landscape. Such was the conclusion of Chippendale (1963) who wrote of *Acacia aneura* in central Australia that replacement seedlings were rare in stocked country. Subsequently the long drought sequence was replaced by a pluvial phase and abundant regeneration of this species followed in grazed and ungrazed areas alike. Fire then became the centre of management interest rather than grazing (Griffin and Friedel 1985). New populations of plants, particularly of long-lived species that need only periodic replacement, are established during the pluvial phase and determine the composition of the vegetation for many years ahead. Rains also vary in seasonality. The vegetation is not stable, even in the absence of human intervention but varies in response to a variable environment.

Fire and grazing provide the other vectors for change. Fire is an infrequent event in arid areas (frequency of 1:10 to 1:100 years) but it nevertheless has a powerful influence on populations of long-lived species. It occurs mainly in association with very wet phases and is therefore concurrent and interactive with the establishment of new seed populations. In one sense it is non-selective, defoliating all species in its path. But in another sense it is partially selective, for species vary in their tolerance to fire (sprouting ability) and facility with which they re-establish after the fire. Grazing is definitely a selective process. It will always eliminate some species, particularly those that are sensitive to defoliation, or those that remain palatable when others are dormant. Unlike fire, which is linked in its effect with wet cycles, grazing influences are linked with drought. It is only in dry periods that dominant species receive high grazing pressure and hence it is the interaction of drought and grazing on species survival that is responsible for shifts in vegetative composition. The species composition of any one site therefore occupies an ordination space that is defined by the vectors acting on the vegetation and the length of time they operate (Foran *et al.* 1986).

The long-lived components of the vegetation possess 'inertia'; populations reflecting the effects of significant past events (Noble 1986). M. Westoby (personal communication) has combined these observations into a model of management for non-equilibrium rangelands, which he refers to as the state and transition model. This terminology arises from the various states (compositions) which are possible on the one land unit and the mechanisms required to move from one state to another. Critical decisions relating to vegetation management will often be episodic; applied only at those times where there is an interaction with a climatic extreme and associated with either plant establishment or plant death.

We have endeavoured to show in this chapter how the management of rangeland vegetation is essentially an ecological process. In the final analysis, however, rangeland management also includes animal production. It therefore is concerned with both exploitation and conservation or more specifically with maintaining a balance between the immediate needs of the manager and society's long-term goals.

11.9 References

Andrew, M. H., and R. T. Lange. 1986. Development of a new piosphere in arid chenopod shrubland grazed by sheep. 2. Changes to the vegetation. Aust. J. Ecol. 11: 411—424.

Arnold, G. W. 1981. Grazing behaviour. p. 79—104. *In*: F. H. W. Morley (ed.) Grazing Animals. Elsevier, Amsterdam.

Auld, B. A. 1976. The biology of *Bassia birchii* (F. Muell.) F. Muell. Weed Res. 16: 323—330.

Austin, M. P., O. B. Williams, and L. Belbin. 1981. Grassland dynamics under sheep grazing in an Australian Mediterranean type climate. Vegetatio 47: 201—211.

Austin, M. P., and H. A. Nix. 1978. Regional classification of climate and its relation to the Australian rangeland. p. 9—17. *In*: K. M. W. Howes (ed.) Studies of the Australian arid zone. 111. Water in rangelands. CSIRO, Melbourne.

Barker, S., and R. T. Lange. 1969. Effects of moderate sheep stocking on plant populations of a black oak — bluebush association. Aust. J. Bot. 17: 527—537.

Beale, I. F. 1973. Tree density effects on yields of herbage and tree components in south-west Queensland mulga (*Acacia aneura* F. Muell.) scrub. Trop. Grassl. 7: 135—142.

Beale, I. F., D. M. Orr, and J. R. Mills. 1986. Pastoral impacts on the Mulga Ecosystem. *In*: P. S. Sattler (ed.) The mulga lands. Royal Society of Queensland, Brisbane.

Bransby, D. I., and N. M. Tainton. 1986. Management of grazing systems: research proposals for the future with reference to stocking rate and rotational grazing. p. 211—213. *In*: P. J. Joss, P. W. Lynch, and O. B. Williams (eds.) Rangelands: A resource under siege. Proc. 2nd Inter. Rangel. Congr., Adelaide, 1984. Aust. Acad. Sci/Camb. Univ. Press.

Brown, R. F. 1982. Seed dormancy in *Aristida armata*. Aust. J. Bot. 30: 67—73.

Brown, R. F. 1986. The effects of burning, fertilizing and clipping on populations of *Aristida armata, Thyridolepis mitchelliana* and *Monachather paradoxa* in a mulga woodland pasture. Aust. Rangel. J. 8: 4—10.

Brown, R. F. 1987. Defoliation regrowth, and dry matter partitioning in the two arid zone grasses, *Aristida armata* and *Thyridolepis mitchelliana*. Aust. J. Agric. Res. 38: 881—893.

Burbidge, N. T. 1945. Germination studies of Australian chenopodiaceae with special reference to the conditions necessary for regeneration. I. *Atriplex vesicaria* Heward. Trans. R. Soc. South Aust. 69: 73—84.

Caughley, G. 1982. Vegetation complexity and the dynamics of modelled grazing systems. Oecologia 54: 309—312.

Chippendale, G. M. 1963. Pasture deterioration in central Australia. J. Aust. Inst. Agric. Sci. 29: 84—89.

Christie, E. K. 1975a. Physiological responses of semi-arid grasses III growth in relation to temperature and soil water deficit. Aust. J. Agric. Res. 26: 447—457.

Christie, E. K. 1975b. Physiological responses of semi-arid grasses IV photosynthetic rates of *Thyridolepis mitchelliana* and *Cenchrus ciliaris* leaves. Aust. J. Agric. Res. 26: 459—466.

Crisp, M. D. 1978. Demography and survival under grazing of three Australian semi-desert shrubs. Oikos 30: 520—528.

Doley, D., and N. B. A. Trivett. 1974. Effects of low water potentials on transpiration and photosynthesis in Mitchell grass (*Astrebla lappacea*). Aust. J. Plant Physiol. 1: 539—550.

Foran, B. D., G. Bastin, and B. Hill. 1985. The pasture dynamics and management of two rangeland communities in the Victoria River district of the Northern Territory. Aust. Rangel. J. 7: 107—113.

Foran, B. D., G. Bastin, and K. A. Shaw. 1986. Range assessment and monitoring in arid lands: the use of classification and ordination in range survey. J. Environ. Manage. 22: 67—84.

Friedel, M. H. 1985. The population structure and density of central Australian trees and shrubs, and relationships to range condition, rabbit abundance and soil. Aust. Rangel. J. 7: 130—139.

Gill, A. M. 1975. Fire and the Australian flora. Aust. For. 38: 4—25.

Graetz, R. D. 1980. The influence of grazing by cattle and sheep on the population dynamics of bladder saltbush (*Atriplex vesicaria*) and white top (*Danthonia caespitosa*). Aust. Rangel. J. 2: 208—213.

Graetz, R. D., and J. A. Ludwig. 1978. A method for the analysis of piosphere data applicable to range assessment. Aust. Rangel. J. 1: 126—136.

Graetz, R. D., and A. D. Wilson. 1980. Comparison of the diets of sheep and cattle grazing a semi-arid chenopod shrubland. Aust. Rangel. J. 2: 67—75.

Graetz, R. D., and A. D. Wilson. 1984. Saltbush and bluebush. p. 209—222. *In*: G. N. Harrington, A. D. Wilson, and M. D. Young (eds.) Management of Australia's rangelands. CSIRO, Melbourne.

Griffin, G. F., and M. H. Friedel. 1985. Discontinuous change in central Australia: some implications of major ecological events for land management. J. Arid Environ. 9: 63—80.

Griffin, G. F., N. F. Price, and H. F. Portlock. 1983. Wildfires in the central Australian rangelands, 1970 to 1980. J. Environ. Manage. 17: 311—323.

Hall, T. J., and G. R. Lee. 1980. Response of an *Astrebla* spp. grassland to heavy grazing by cattle and light grazing by sheep in north-west Queensland. Aust. Rangel. J. 2: 183—193.

Harper, J. L. 1977. Population Biology of Plants. Academic Press, London.

Harrington, G. N. 1979. The effects of feral goats and sheep on the shrub populations in a semi-arid woodland. Aust. Rangel. J. 1: 334—345.

Harrington, G. N. 1986a. Critical factors in shrub dynamics in eastern mulga lands. *In*: P. S. Sattler (ed.) The mulga lands. Royal Society of Queensland, Brisbane.

Harrington, G. N. 1986b. Herbivore diet in a semi-arid *Eucalyptus populnea* woodland. 2. Feral goats. Aust. J. Exp. Agric. Anim. Husb. 26: 423—429.

Harrington, G. N., G. T. Dawes, and J. A. Ludwig. 1981. An analysis of the vegetation pattern in a semi-arid *Eucalyptus populnea* woodland in northwest New South Wales. Aust. J. Ecol. 6: 279—287.

Harrington, G. N., A. D. Wilson, and M. D. Young. 1984. Management of Australia's rangelands. CSIRO, Melbourne.

Hattersley, P. W. 1983. The distribution of C_3 and C_4 grasses in Australia in relation to climate. Oecologia 57: 113—128.

Hodgkinson, K. C. 1976. The effects of frequency and extent of defoliation, summer irrigation and fertilizer on the production and survival of the grass *Danthonia caespitosa* Gaud. Aust. J. Agric. Res. 27: 755—767.

Hodgkinson, K. C. 1979. The shrubs of poplar box (*Eucalyptus populnea*) lands and their biology. Aust. Rangel. J. 1: 280—293.

Hodgkinson, K. C. 1986. Responses of rangeland plants to fire in water-limited environments. p. 437—441. *In*: P. J. Joss, P. W. Lynch, and O. B. Williams (eds.) Rangelands: A resource under siege. Proc. 2nd Inter. Rangel. Congr., Adelaide, 1984. Aust. Acad. Sci./ Cambridge Univ. Press.

Hodgkinson, K. C., and J. A. Quinn. 1976. Adaptive variability in the growth of *Danthonia caespitosa* Gaud. populations at different temperatures. Aust. J. Bot. 24: 381—396.

Hodgkinson, K. C., and J. A. Quinn. 1978. Environmental and genetic control of reproduction in *Danthonia caespitosa* populations. Aust. J. Bot. 26: 351—364.

Hodgkinson, K. C., and G. R. Beeston. 1982. The biology of Australian weeds 10. *Eremophila mitchellii* Benth. J. Aust. Inst. Agric. Sci. 48: 200—208.

Hodgkinson, K. C., and G. F. Griffin. 1982. Adaptation of shrub species to fires in the arid zone. *In*: W. R. Barker and P. J. M. Greenslade (eds.) Evolution of the flora and fauna of arid Australia. Peacock Publications, Adelaide.

Hodgkinson, K. C., and G. N. Harrington. 1985. The case for prescribed burning to control shrubs in eastern semi-arid woodlands. Aust. Rangel. J. 7: 64—74.

Hodgkinson, K. C., G. N. Harrington, G. F. Griffin, J. C. Noble, and M. D. Young. 1984. Management of vegetation with fire. *In*: G. N. Harrington, A. D. Wilson, and M. D. Young (eds.) Management of Australia's rangelands. CSIRO, Melbourne.

Hodgkinson, K. C., J. J. Mott and M. M. Ludlow. 1986. Coping with grazing: A comparison of two savanna grasses differing in tolerance to defoliation. p. 1089—1091. *In*: Proc. 15th Inter. Grassld. Congr. Kyoto.

Hodgkinson, K. C., and J. J. Mott. 1987. On coping with grazing. *In*: F. P. Horn, J. Hodgson, J. J. Mott, and R. W. Broughton (eds.) Grazing lands research at the plant-animal interface. Winrock International.

Jones, R. J. 1981. Interpreting fixed stocking rate experiments. p. 419—430. *In*: J. L. Wheeler and R. D. Mochrie (eds.) Forage evaluation: Concepts and techniques. CSIRO, Melbourne.

Jones, R. J., and R. L. Sandland. 1974. The relation between animal gain and stocking rate. Derivation of the relation from the results of grazing trials. J. Agric. Sci. Camb. 83: 335—342.

Jozwick, F. X. 1970. Response of mitchell grasses (*Astrebla* F. Muell.) to photoperiod and temperature. Aust. J. Agric. Res. 21: 395—405.

Kemp, E. M. 1981. Pre-quaternary fire in Australia. p. 3—22. *In*: A. M. Gill, R. H. Groves, and I. R. Noble (eds.) Fire in the Australian Biota. Aust. Acad. Sci., Canberra.

Lange, R. T., and C. R. Graham. 1983. Rabbits and the failure of regeneration in the Australian arid zone *Acacia*. Aust. J. Ecol. 8: 377—381.

Leigh, J. H., and W. E. Mulham. 1966. Selection of diet by sheep grazing semi-arid pastures on the Riverine plain. 1. A bladder saltbush (*Atriplex vesicaria*) — cotton bush (*Kochia aphylla*) community. Aust. J. Exp. Agric. Anim. Husb. 6: 460—467.

Leigh, J. H., and W. E. Mulham. 1971. The effect of defoliation on the persistence of *Atriplex vesicaria*. Aust. J. Agric. Res. 22: 239—244.

Leigh, J. H., and J. C. Noble. 1981. The role of fire in the management of rangelands in Australia. *In*: A. M. Gill, R. H. Groves, and I. R. Noble (eds.) Fire and the Australian Biota. Australian Academy of Science, Canberra.

Leigh, J. H., A. D. Wilson, and W. E. Mulham. 1968. A study of Merino sheep grazing a cotton-bush (*Kochia aphylla*) — grassland (*Stipa variabilis* — *Danthonia caespitosa*) community on the Riverine plain. Aust. J. Agric. Res. 19: 947—961.

Leigh, J. H., A. D. Wilson, and W. E. Mulham. 1979. A study of sheep grazing on belah (*Casuarina cristata*) — rosewood (*Heterodendrum oleifolium*) shrub woodland in western New South Wales. Aust. J. Agric. Res. 30: 1223—1236.

Lorimer, M. S. 1978. Forage selection studies. 1. The botanical composition of forage selected by sheep grazing *Astrebla* spp. pasture in north-west Queensland. Trop. Grassl. 12: 97—108.

Low, W. A., M. L. Dudzinski, and W. J. Muller. 1981. The influence of forage and climatic conditions on range community preference of shorthorn cattle in central Australia. J. Appl. Ecol. 18: 11—26.

Milthorpe, P. L. 1978. Some factors affecting establishment and growth of black bluebush (*Maireana pyramidata* (Benth.) P. G. Wilson) in western New South Wales. p. 362—365. *In*: D. N. Hyder (ed.) Proc. 2nd Inter. Rangel. Congr., Denver, 1978. Soc. Range Manage, Denver.

Minson, D. J., and J. R. Wilson. 1980. Comparative digestibility of tropical and temperate forage — a contrast between grasses and legumes. J. Aust. Inst. Agric. Sci. 46: 247—249.

Mott, J. J. 1972. Germination studies on some annual species from an arid region of Western Australia. J. Ecol. 60: 293—304.

Mott, J. J., and R. H. Groves. 1981. Germination strategies. *In*: J. S. Pate and A. J. McComb (eds.) The Biology of Australian plants. Univ. W. A. Press, Nedlands.

Nicholson, K. P. 1986. Spatial and temporal patterns of herbaceous species at Middleback Station, South Australia. Ph.D. Thesis, Univ. Adelaide.

Noble, I. R. 1977. Long-term biomass dynamics in an arid chenopod shrub community at Koonamore, South Australia. Aust. J. Bot. 25: 639—653.

Noble, I. R. 1986. The dynamics of range ecosystems. p. 3—5. *In*: P. J. Joss, P. W. Lynch, and O. B. Williams (eds.) A resource under siege. Proc. 2nd Inter. Rangel. Congr., Adelaide, Aust. Acad. Sci./Cambridge Univ. Press.

Noble, I. R., and M. D. Crisp. 1979/80. Germination and growth models of short-lived grass and forb populations based on long term photo-point data at Koonamore, South Australia. Isr. J. Bot. 28: 195—210.

Noble, I. R., and R. O. Slatyer. 1980. The use of vital attributes to predict successional changes in plant communities subject to recurrent disturbances. Vegetatio 43: 5—21.

Noble, J. C. 1982. The significance of fire in the biology and evolutionary ecology of mallee *Eucalyptus* populations. p. 153—159. *In*: W. R. Barker and P. J. M. Greenslade (eds.) Evolution of the flora and fauna of arid Australia. Peacock Publications, Adelaide.

Noble, J. C. 1984. Mallee. p. 223—240. *In*: G. N. Harrington, A. D. Wilson, and M. D. Young (eds.) Management of Australia's rangelands. CSIRO, Melbourne.

Noble, J. C. 1986. Prescribed fire in mallee rangelands and the potential role of aerial ignition. Aust. Rangel. J. 8: 118—130.

Noble, J. C. *In Press*. 1988. Fire studies in mallee (*Eucalyptus* spp.) communities of western New South Wales: the effects of fires applied in different seasons on herbage productivity and their implications for management. Aust. J. Ecol.

Noble, J. C., G. M. Cunningham, and W. E. Mulham. 1984. Rehabilitation of degraded land. p. 171—186. *In*: G. N. Harrington, A. D. Wilson, and M. D. Young (eds.) Management of Australia's rangelands. CSIRO, Melbourne.

Noble, J. C., G. N. Harrington, and K. C. Hodgkinson. 1986. The ecological significance of irregular fire in Australian rangelands. p. 577—580. *In*: P. J. Joss, P. W. Lynch, and O. B. Williams (eds.) Rangelands: A resource under siege. Proc. 2nd Inter. Rangel. Congr., Adelaide, 1984. Aust. Acad. Sci./Cambridge Univ. Press.

Noble, J. C., A. W. Smith, and H. W. Leslie. 1980. Fire in the mallee shrublands of western New South Wales. Aust. Rangel. J. 2: 104—114.

Noble, J. C., and R. D. B. Whalley. 1978. The biology and autecology of *Nitraria* L. in Australia. II. Seed germination, seedling establishment and response to salinity. Aust. J. Ecol. 3: 165—177.

O'Dowd, D. J., and A. M. Gill. 1984. Predator satiation and site alteration following fire: mass reproduction of alpine ash (*Eucalyptus delegatensis*) in southeastern Australia. *Ecology* 65: 1052—1066.

Orr, D. M. 1980a. Effects of sheep grazing *Astrebla* grassland in central western Queens-

land. I. Effects of grazing pressure and livestock distribution. Aust. J. Agric. Res. 31: 797—806.
Orr, D. M. 1980b. Effects of sheep grazing *Astrebla* grassland in central western Queensland. II. Effects of seasonal rainfall. Aust. J. Agric. Res. 31: 807—820.
Orr, D. M., and W. E. Holmes. 1984. Mitchell grassland. p. 241—254. *In*: G. N. Harrington, A. D. Wilson, and M. D. Young (eds.) Management in Australia's rangelands. CSIRO, Melbourne.
Philander, S. G. H. 1983. El Nino/Southern Oscillation phenomena. Nature (Lond.) 302: 295—301.
Pickup, G. 1985. The erosion cell — a geomorphic approach to landscape classification in range assessment. Aust. Rangel. J. 7: 114—121.
Preece, P. B. 1971. Contributions to the biology of mulga. II. Germination. Aust. J. Bot. 19: 39—49.
Quinn, J. A., and K. C. Hodgkinson. 1983. Population variability in *Danthonia caespitosa* (Gramineae) in responses to increasing density under three temperature regimes. Amer. J. Bot. 70: 1425—1431.
Quinn, J. A., and K. C. Hodgkinson. 1984. Plasticity and population differences in reproductive characters and resources allocation in *Danthonia caespitosa* (Gramineae). Bull. Torrey Bot. Club 111: 19—27.
Roe, R. 1987. Recruitment of *Astrebla* spp. in the Warrego region of south-west Queensland. Trop. Grassld. 21: 91—92.
Scanlan, J. C. 1980. Effects of spring wildfires on *Astrebla* (Mitchell grass) grasslands in north-west Queensland under varying levels of growing season rainfall. Aust. Rangel. J. 2: 162—168.
Scanlan, J. C. 1983. Changes in tiller and tussock characteristics of *Astrebla lappacea* (curly Mitchell grass) after burning. Aust. Rangel. J. 5: 13—19.
Short, J. 1985. The functional response of kangaroos, sheep and rabbits in an arid grazing system. J. Appl. Ecol. 22: 435—447.
Silander, J. A. 1983. Demographic variation in the Australian desert cassia under grazing pressure. Oecologia (Berlin) 60: 227—233.
Silcock, R. G. 1973. Germination responses of native plant seeds to rainfall in south-west Queensland. Trop. Grassld. 7: 99—104.
Silcock, R. G. 1977. A study of the fate of seedlings growing on sandy red earths in the Charleville district, Queensland. Aust. J. Bot. 25: 337—346.
Stafford-Smith, M. 1984. Behavioural ecology of sheep in the Australian arid zone. Ph.D. thesis. Aust. Nat. Univ.
Suijdendorp, H. 1980. Pastoral development and research in the Pilbara region of Western Australia. Aust. Rangel. J. 2: 115—123.
Vines, R. G., and A. I. Tomlinson. 1985. The Southern Oscillation and rainfall patterns in the Southern Hemisphere. S. Afr. J. Sci. 81: 151—156.
Walker, B. H., and I. Noy-Meir. 1982. Aspect of the stability and resilience of savanna ecosystems. p. 556—590. *In*: B. J. Huntley and B. H. Walker (eds.) Ecology of tropical savannas. Springer-Verlag, Berlin.
Walker, B. H., D. A. Matthews, and P. J. Dye. 1986. Management of grazing systems — existing versus an event-orientated approach. S. Afr. J. Sci. 82: 172.
Walker, J., R. W. Condon, K. C. Hodgkinson, and G. N. Harrington. 1981. Fire in pastoral areas of poplar box (*Eucalyptus populnea*) lands. Aust. Rangel. J. 3: 12—23.
Watt, L. A. 1982. Germination characteristics of several grass species as affected by limiting water potentials imposed through a cracking black clay soil. Aust. J. Agric. Res. 33: 223—231.

Wellington, A. B., and I. R. Noble. 1985. Seed dynamics and factors limiting recruitment of the mallee *Eucalyptus incrassata* in semi-arid, south-eastern Australia. J. Ecol. 73: 657—666.

Westoby, M. 1979/80. Elements of a theory of vegetation dynamics in arid rangelands. Isr. J. Bot. 28: 169—194.

Williams, O. B. 1968. Studies in the ecology of the Riverine Plain. IV. Basal area and density changes of *Danthonia caespitosa* Gaudich. in a natural pasture grazed by sheep. Aust. J. Bot. 16: 565—578.

Williams, O. B. 1970. Population dynamics of two perennial grasses in Australian semi-arid grassland. J. Ecol. 58: 869—875.

Williams, O. B. 1978. Plant demography of Australian arid rangeland and implications for management, research and land policy. p. 185—186. *In*: D. N. Hyder (ed.) Proc. 1st Inter. Rangel. Congr., Denver, 1978. Soc. Range Manage., Denver.

Williams, O. B., and R. Roe. 1975. Management of arid grasslands for sheep: plant demography of six grasses in relation to climate and grazing. Proc. Ecol. Soc. Aust. 9: 142—156.

Wilson, A. D. 1977. Grazing management in the arid areas of Australia. p. 83—92. *In*: The impact of herbivores on arid and semi-arid rangelands. Australian Rangeland Society: Perth.

Wilson, A. D. 1986. Principles of grazing management systems. p. 221—225. *In*: P. J. Joss, P. W. Lynch, and O. B. Williams (eds.) Rangelands: A resource under siege. Proc. 2nd Inter. Rangel. Congr., Adelaide, 1984. Aust. Acad. Sci./Cambridge Univ. Press.

Wilson, A. D., and R. D. Graetz. 1980. Cattle and sheep production on an *Atriplex vesicaria* (bladder saltbush) community. Aust. J. Agric. Res. 31: 369—378.

Wilson, A. D., J. H. Leigh, and W. E. Mulham. 1969. A study of Merino sheep grazing a bladder saltbush (*Atiplex vesicaria*) — cottonbush (*Kochia aphylla*) community on the Riverine Plain. Aust. J. Agric. Res. 20: 1123—1136.

Wilson, A. D., and W. E. Mulham. 1980. Vegetation changes and animal productivity under sheep and goat grazing an arid belah (*Casuarina cristata*) — rosewood (*Heterodendrum oleifolium*) woodland in western New South Wales. Aust. Rangel. J. 2: 183—188.

12. The ecology of shrubland/woodland for range use

KAREN A. PLATOU AND PAUL T. TUELLER

12.1 Abstract

Shrubland/woodland vegetation communities throughout the world are considered to be rangelands and are used for range products. They often represent ecotonal or transitional areas between grassland and forest but are usually managed differently than either. Great variability is inherent in this rangeland vegetation resource because of this transitional nature influenced by such factors as fire, disease, grazing, climate and the activities of man. Management is often constrained by a lack of local ecological knowledge concerning the vegetation.

12.2 Introduction

Shrublands and woodlands represent the largest percentage of the world's rangelands, somewhat over 60 million km^2 (Mckell *et al.* 1971). These shrub and woodland ranges include such categories as mixed pine and oak woodland, woodland, open woodland, heath, steppe, steppe woodland, semi-desert, desert tall thicket matorral, matorral (chaparral, maquis), open shrubland (desert chaparral, margnal maquis) phrygana (coastal sage) (Shmida 1985), mountain brush, pinyon-juniper woodlands, acacia woodlands, and tropical savannahs. Terminology is varied as exemplified by the above and a listing of some of the names given throughout the world to mediterranean climate shrublands — maquis, maccia, fynbos, garrique, chaparral, tomillares and mallee. Deserts, except for annual desert, are dominated by woody species.

Definitions also vary widely. Steppe, for example, is dominated by dwarf-shrubs which cover more than 10% of the ground. Perennial grasses and forbs can grow between the suffrutescent shrubs, but they are not dominant. Shrublands are areas dominated by shrubs (more than 0.5 m and up to 2 m high) which cover more than 30% of the ground. If shrubs

Tueller, P. T. (ed.), Vegetation science applications for rangeland analysis and management.
ISBN 90-6193-195-9.
© 1988, *Kluwer Academic Publishers, Dordrecht, Boston, London.*

are sparser (3 to 30% cover) they may be called open shrublands. Open arboreal formations in which trees (and sometimes shrubs) are higher than 2 m and cover ranges between 20 and 80% are woodlands. If the woodland becomes sparse, less that 20% cover, they should be called open woodland. Savannas are open tropical woodlands in which trees are usually thorny and have facultatively deciduous microphylly. Usually the savanna has a grass layer between the trees (especially in Africa), but not always (e.g., in the Sonoran Desert). The shrubs can be evergreen or deciduous. If most of them are broad-leaved evergreens, the vegetation usually corresponds to mediterranean matorral. If the shrubs are leptophyll evergreen, the vegetation generally corresponds to heath (Shmida 1985). Western United States cool high elevation shrub communities are referred to as mountain brush. And so it goes.

Despite their wide distribution and varied names and definitions, these vegetation types hold certain characteristics in common. Generally they are found in areas with limited growing conditions. Precipitation is not adequate during the growing season for forest or grassland and/or the length of the growing season is too short. Often, soils are shallow, poorly drained, or excessively stony. Adapting to, or evolving with these poor growing conditions, the vegetation has taken on a distinct physiognomy. These are woody life-forms consisting of small trees (chamaephytes) or shrubs (nano-, or microphanerophytes) (Raunkiaer 1934). Trees are often considered to be woody plants with a single stem and reaching some minimum height while shrubs are generally perceived to be shorter and with multiple stems arising at or near the ground.

In addition to the vast acreages which continually support woodland and shrubland vegetation, there are ecotonal areas which intergrade with forest and grassland. Management and natural disturbance temporarily drive vegetation in these areas from shrubland/woodland to forest or grassland and back again.

Vegetation science helps land managers understand the ecological factors which make shrubland/woodland different from forests and grasslands. This knowledge allows range managers to optimize the desired returns from rangelands using ecological principles as tools. The remainder of this chapter will give examples of ecological considerations effecting land use in the context of shrubland/woodlands.

12.3 Uses of shrubland/woodlands

The primary use made of all shrubland/woodland worldwide by man is the grazing of domestic livestock. Of the 500—600 million people estimated to inhabit arid or semiarid tropical and subtropical regions of the world approximately 8% are believed to possess "livestock-dependent" economies (Child *et al.* 1984). Many millions of domestic livestock are supported on shrub or tree dominated rangelands throughout the world. Domestic animal numbers in Africa alone according to the 1979 FAO production Yearbook include:

170,110,000 cattle
171,846,000 sheep
144,684,000 goats
11,983,000 camels
11,734,000 donkeys

This is in part true because net primary productivity is high, especially when compared as a ratio with standing crop. More of shrubland/woodland net primary productivity is immediately available to grazers than is forest biomass. Organic matter turnover is high due not only to large grazing animals, but also to invertebrates (Child *et al.* 1984).

Animal groupings and range use differ between grassland, forest with grass understory and desert scrub vegetation.

Savannas with grasses:

humped cattle
hairy, thin-tailed sheep
Fat-tailed sheep
fat-rumped sheep
goats
donkeys

Deciduous forest with high grass:

humped cattle
hairy, thin-tailed sheep

Desert shrub and grassland:

wooled, thin-tailed sheep
fat-tailed sheep
fat-rumped sheep
goats
camels
donkey

Winter rainfall vegetation:

humpless cattle
wooled, thin-tailed sheep (Child *et al.* 1984).

Primary production of forage is often not sufficient to allow for permanent pasture so some form of nomadic herding is practiced. In Africa and Asia and parts of Australia huge areas of land are grazed by one herd or band throughout the year. In the intermountain western United States, shrublands are used often only as transitional range between summer grazing on mountain meadows and winter grazing or feeding from irrigated haylands. Some shrublands of the American desert southwest are used as winter grazing lands where seeds of shrubs such as fourwing saltbush (*Atriplex confertifolia*) form an important winter high energy food source. In California and mediterranean regions with mild, moist winters, grazing takes place during a winter growing season. In tropical regions, woodlands are often flooded during parts of the year, restricting grazing use to the dry period.

Shrublands used as transition range between winter haymeadows and montane grazing are a very important, but often overlooked resource. Ranchers with access to transition shrubland range are able to get off expensive haylands earlier in the spring and are able to utilize these areas in the fall as gathering areas when the animals move out of the high mountains. Maintenance of high quality forage condition on these important ranges should be a high priority.

Woodlands throughout the world support varying populations of wildlife in addition to the domestic animal component. In much of southern Africa poaching and competition with domestic livestock have effectively removed many populations of wildlife from ancestral savannah and low veld woodlands forcing many areas to be set aside as wildlife parks and preserves. In western North America, deer are dependent on foothill transition winter ranges. Poor condition of these ranges due to overuse by livestock is blamed for the decline of some deer herds such as the King's deer herd in the southwestern Sierra Nevada range of California.

A threat to both livestock and wildlife use of transition ranges is increasing urbanization. Western U.S. cities originally built on low agricultural land are rapidly expanding into the surrounding hills taking away grazing land and presenting wildlife and livestock conflicts with roads, fences, and dogs. Largely because of increased urbanization, emergency feeding programs for deer and elk and reimbursement for depredation by wildlife have become part of the activities of several state fish and game agencies. Increased fire hazard in shrub dominated urban rangelands is an added urban/rangeland problem.

In the western U.S. another competing herbivore on these ranges is the feral horse. Protected by law, management of these animals in balance with wildlife and livestock is increasingly difficult. Being year-round residents in the type instead of seasonal grazers like the cattle and deer, the horses exert considerable competitive pressure on the shrubland forage resource.

The foregoing suggests that the worlds shrublands have many uses. Several uses of individual shrub species not previously mentioned include. (1) Revegetation to increase aesthetics and watershed value of roadsides and mine wastes and for forage improvement. (2) Firewood is a major source of fuel in Africa and is in short supply in vast areas of the continent. Over 50% of the woodlands in the tribal areas of Zimbabwe were removed between 1963 and 1978. In the rural areas of Zululand, three times as many man-hours are spent collecting firewood as in growing food (Anderson 1983). (3) Medicinal and pharmaceutical uses — botanochemicals derived from shrubby species as anticarcinogenic agents, vitamin producers, tanning substances, resin producers, oils and pigments and salt producers (Petrov 1971).

12.4 Fire as an ecological factor

For those shrublands and woodlands with enough annual production to carry fire, fire is probably the most important ecological factor. Periodic fire has the effect of opening up dense, decadent shrub canopies and setting back succession to a grass-forb stage temporarily. Fire adapted shrub species sprout back from root crowns. In these types, relatively frequent, light burns prevent the fuel bed of dead wood from building up to levels where a potentially root killing fire might occur. Range managers in California, Australia and Africa, where shrubs are fire adapted, use prescribed fire as a tool to reduce intensity of wildfires and increase forage for wildlife and livestock. Where not adapted to fires the shrubs are replaced by grasses. Where resistant perennial grasses have been grazed out of the stand, annuals replace them after fire. The forage resources on approximately 3.3 million acres of the sagebrush-grass zone in the western U.S. consists of the annual cheatgrass (*Bromus tectorum*) and others (Stewart G. A. C. and Hull 1949).

There are many examples of periodic fires in savannahs preventing encroachment of a shrub understory with little forage value. Fire control, on the other hand, in ecotonal zones has expanded the range of shrubland and woodland into areas formerly occupied by grassland. The invasion of pinyon and juniper downslope into grasslands has been blamed in part on aggressive wildfire control over the last 50—100 years. Soils with evidence of supporting grasslands now grow open woodland canopies of juniper and pinyon pine.

Fire control in urbanized shrubland-woodland areas has been offset by firewood cutting at least in some areas. In other areas firewood cutting threatens to convert woodland to grassland or desert.

12.5 Disease as an ecological factor

Periodic epidemics or insect plagues are characteristic of late seral stages of many plant communities. When these periodic outbreaks take place in shrublands or woodlands they have the potential of substantially altering the character of the vegetation. Because of the long-lived nature of trees and shrubs, the results of such outbreaks are evident for long periods of time and in some cases may result in permanent conversion.

California live oaks are susceptible to the defoliating attacks of the oak leaf moth. Continued attacks on the same trees for several years may kill the stand. With regeneration of such trees dependent on good growing conditions and good mast crop, establishment of a replacement stand may take a long time.

The sagebrush country of the intermountain western U.S. is subject to the effects of the moth (*Aroga westerii*) larvae that feed on sagebrush (*Artemisia*). Extensive stands of the shrub are sometimes killed converting them to grassland or changing the shrub dominants. Between 1955—1958 a scale insect (*Orthezia annae*) infested shadscale plants in Idaho causing widespread death. Annual plants of halogeton, peppergrass and tansy mustard dominated the stands through dry years 1959, 1960 and 1961. Return of shadscale was noticeable in 1963 and squirreltail increased in abundance after this time. With greater than average precipitation from 1962 to 1965 shadscale vegetation returned (Sharp and Sanders 1978). The Looper (*Aracamptodes clivinavia profanata*), a leaf defoliator, killed 46% of the trees in a 5,900 acre (2,400 hectare) stand of curlleaf mountain mahogany over 3 years in southwestern Idaho (Furniss and Barr 1967).

Bitterbrush (*Purshia tridentata*), an important browse species for deer and livestock is also under attack from insect herbivores (Furniss and Krebill 1972). Tent caterpiller infestations may wipe out major feeding grounds for wintering deer herds (Hubbard *et al.* 1959). The pest is not always an insect herbivore. Parasitic plants take their toll too. Dodder weakens many shrubs in wet years in the Mojave Desert.

12.6 Grazing as an ecological factor

Grazing by herbivores often has little effect on shrubland or woodland habitats unless it is at a high level of use such as occurs during drought or during years of high herbivore population levels. When herbivore numbers exceed the carrying capacity of the shrubland or woodland, damage takes place and some of the typical woody species are either removed or replaced.

In Africa where elephants are protected, their populations within the national parks has been very high at times. Especially during drought conditions elephants turn to thorn trees for food and if their populations are high they may push over and kill enough trees to turn savannah to grassland (Child *et al.* 1984).

Where herbivores specialize in grazing grass rather than browsing, overgrazing may have the effect of extending the limits of the shrubland into grassland. Partly, this conversion is due to selective foraging on grasses in contrast to shrubs, but also, to lowering of the watertable, i.e., meadow water tables drop because of erosion and head cutting as a result of overgrazing causing a general invasion of upland shrubs.

12.7 Climate as an ecological factor

Although shrublands and woodlands occur under a wide variety of climatic regimes, they are still susceptible to climatic shifts particularly long term ones. Generally an increase in moisture and temperature at the edge of the range will cause expansion of other vegetative elements namely wetland, grassland, and forest while a decrease in moisture or a contraction of the growing season will favor expansion of the shrubland, woodland ecosystem.

Areas with precipitation between 350 mm—700 mm, wide ranging daily and yearly temperature, low humidity, intense sunlight, irregular rainfall, seasonal growing season (moisture) and common drought have considerable potential to produce woody plants. Savannahs with continuous understory grassland through east and central Africa, North and South America and Australia have been the center of evolution of diverse large herbivore species. Evolution of niche separation in animal food habits, activity patterns, migrations adaptations of digestive processes and low water requirements serve to make savannas excellent woodland habitats.

Paleobotanical work has mapped the advance and retreat of ice sheets and their concomitant effects on shrub-forest boundaries. Localized and global climatic changes associated with man's activities have also affected shrub and woodland plant distribution. On a localized basis runoff has been altered by roadways producing strips of land with more moisture than surrounding areas. On the edges of these roads plants will grow more profusely and new species colonize. A hedge of creosote or flowers along roadsides serve as one example of this.

On a larger scale, man's mining of water from underground aquifers and tapping of surface flows has effected the climate and thus the distribution of shrublands and woodlands. In some areas grassy shrublands have lost their grass component to a reduction in the water table. On an even larger scale, one can consider the effects of global atmospheric CO_2 shifts — the so-called greenhouse effect. The CO_2 effect may possibly be beneficial with an enrichment of CO_2 in the atmosphere causing an increase in the production of many terrestrial plant species (Kimball 1983).

Climate induced short growing seasons dictate either fast growth followed by seed production and death (annual habit) or periodic dormant seasons and long-lived plants. Often shrublands and woodlands have plants with both growth habits. Annuals are prevalent where seasonal moisture is unpredictable in timing. Soil moisture triggers a flush of growth where soil moisture is recharged deep in the soil profile during the winter by snows. Deeply rooted plants tap that moisture once growing temperatures are reached and continue to grow later into the dry season than can shallow rooted annuals.

Dry season dormancy is common among more shallowly rooted species as well as some of the shrubs. Other adaptations to droughty conditions include small leaf surfaces, pubescence to reduce transpiration, modification of leaves to thorns, waxy cuticles, grasses with narrow, rolled and folded leaves and other adaptations. The climatic regimes also favor positioning of the meristematic tissue at or just above the ground surface. Arid and semiarid lands are characterized by large compliments of hemicryptophytes and chamaephytes.

The cyclical nature of grazing, fire, and climate is necessary to maintain diversity and productivity in diverse shrubland/woodland ecosystems. While vegetation in these regions may be subjected to more fluctuation in food production than wetter rangelands they also have greater resiliency and responds quickly to moisture. For example, browsing supplements the quality of animal diets during dry periods, especially when the browse is from legumes. Tree fodder is useful for drought.

Desert shrub and grasslands with 100 mm—350 mm rainfall have wide diurnal temperature fluctuations and soils high in minerals and salts because of low rainfall, some with high clay. Such arid climates require flexibility in grazing schedules and in movements of livestock for maintenance of both animals and forage resources. Dry area forages are of higher nutritional value even when mature than those of wetter climates. The forage has less lignin and fiber and is less leached by rain. Fewer herbivores are found in these desert areas thus providing low carrying capacities. They are less adapted to fire. Nomadic grazing is often practiced.

Seasonally flooded woodlands and wetlands occur mixed in many shrubland/woodland rangeland types. Livestock grazing is important in these areas because of the high forage value of these woody range types used for dry season grazing.

Mediterranean shrubland types are areas of winter rainfall. They tend to be fire types and may be so thick as to preclude foraging.

12.8 Man is an ecological factor

Man's effects on shrublands are probably the most significant. They bear some important ecological messages as well.

One way man has effected some of the shrublands of the world is by introductions of exotic species from other continents into natural systems. Because one of the primary uses of shrubland and woodland is grazing, man is interested in improving or maintaining grazing potential. Often improvement of grazing land includes seeding of exotic forage grasses. In the North American intermountain west crested wheatgrass has been seeded into 1.6 million acres of sagebrush and pinyon/juniper woodland (Sharp and Sanders 1978).

In California, annual ryegrass is sown on burned chaparral as an erosion control measure. Seeding large acreages has led to fears by some scientists that rare endemic species may not be able to compete with the seeded species.

Weedy species not intentionally seeded and encouraged by increased grazing also are feared to offer a threat to native rare species. In the Fynbos (sclerophyllos shrublands of South Africa) research into the spread, control and requirements of weedy species has been accomplished to improve management for the preservation of the region's large number of endemic species.

For some shrubland and woodland ranges, domestic livestock can be considered to be an exotic species. Australian ecosystems evolved without large ungulate herbivores. The introduction of sheep and cattle has changed the nature of the vegetation in many parts of the world. In many cases this has been shown to result in a greatly expanding shrubland, decreasing woodland and grassland and a threat to the existence of native endemic plant species. The herbaceous species growing with the shrubs have changed considerably since the introduction of livestock 100 years ago. Annual species have replaced many perennials and poisonous plants have become more frequent. Spiny plants are more often prevalent.

12.9 The future for shrubland/woodland

As increased pressure is placed on shrublands/woodlands by all types of users, more intensive management will and should be expected. More products will be sought from a shrinking landbase. Regulation of uses will become more and more stringent. Now in California, the State Board of Forestry is considering regulation of hardwood cutting and management practices on all ownerships in recognition of the dwindling acreage of oak woodland.

In many areas of the world energy demands will increase. More and more thought must be directed toward biomass conversion of shrubland fuels. Rather than control burn decadent chaparral, these areas may be harvested for fuel and burned to produce electricity and heat. In Kenya, small locally made kilns are used to convert stumps and wood from brushland clearings to charcoal. Typically 100 kg of charcoal results from 1m3 of wood. Charcoal production defrays costs of brushland conversion (Little 1972). There has been some effort to use shrubland/woodland botanochemicals as pharmaceuticals and so at some future date the woody biomass may be harvested for these and other products.

Because of the high cost of most forms of manipulative management the use of prescribed burning will increase as a tool to improve forage for domestic livestock as well as cover and food for populations of wildlife. This will await further research into fire behavior and supression and management of prescribed burns.

Decision making for management is requiring more and more public involvement. Public disclosure of past, present and future use and environmental effects are now mandatory for larger projects at least in the United States. Such restrictions are now a reality in many other countries. Decisions will increasingly transferred to the political arena and it will be more important for vegetation scientists to make others aware of their knowledge.

12.10 References

Anderson, J. L. 1983. Sport Hunting in National Parks: Sacrilege or Salvation? Information series: 2. Bophuthatswana National Parks.

Child, R. D., H. F. Heady, W. Hickey, R. A. Peterson, and R. D. Pieper. 1984. Arid and semiarid lands: Sustainable use and management in developing countries prepared for: AID/NPS natural Resources expanded information Base Project, Division of International Affairs, National park Service, Wash, D.C. 20240 pub. by Winrock International, Morrillon, Arkansas 72110, Oct, 1984.

Furniss, M. M., and W. F. Barr. Bionomics of *Anacamptodes clivinaria profanata* (Lepidoptera: Geometridae) on mountain mahogany in Idaho. Res. Bull. 73 Moscow, ID: Univ. of Idaho, Idaho Ag. Exp. Stat. 1967.

Furniss, M. M., and R. G. Krebill. 1972. Insects and diseases of shrubs on western big game. *In*: Wildland shrubs — Their biology and utilization. USDA Intermountain Forest and Range Experiment Station. Forest Service General Technical Report INY-1.

Hubbard, R. L., E. C. Nord, and L. L. Brown. 1959. Bitterbrush reseeding — a tool for the game range manager. Misc. Pap no. 39. Berkeley, Ca: U.S. Dept. of Ag., Forest, and Range Expt. Station.

Kimball, B. A. 1983. Carbon dioxide and agricultural yield: An assemblage and analysis of 770 prior observations. United States Water Conservation Laboratory Report #14. Agricultural Research Service, Phoenix, AZ.

Little, E. C. S. 1972. Kenya brush control. United Nations Development Program/Food and Agriculture Organization (UNDP/FAO) Nairobi, Kenya.

McKell, C. M., J. P. Blaisdell, and J. R. Goodin. 1972. Wildland shrubs — Their biology and utilization. USDA Intermountain forest and Range Experiment Station. USDA Forest Service General Technical Report INT-1.

Naveh, Z., and R. H. Whittaker. 1980. Structural and floristic diversity of shrublands and woodlands in northern Israel and other Mediterranean areas. Vegetatio 41: 171—190.

Petrov, M. P. 1972. Natural features of arid areas of Asia. p. 37—50. *In*: C. M. Mckell, J. P. Blaisdell, and J. R. Goodin. Wildland shrubs — Their biology and utilization. USDA Forest Service General Technical Report INT-1.

Raunkiaer, C. 1934. The life forms of plants and statistical plant geography. Oxford: Clarendon Press.

Sharp, L. A., and K. Sanders. 1978. Rangeland resources of Idaho, a basis for development and improvement. Univ. of Idaho, Forest, Wildlife and Range Expt. Station Contribution no 141, Sept. 1978.

Shmida, Avi. 1985. Biogeography of the desert flora. p. 23—77. *In*: M. Evenari, I. Noy-Meir, and D. W. Goodall (eds.) Ecosystems of the world 12A, Hot deserts and arid shrublands, A.

Stewart, G. A. C., and Hull. 1949. Cheatgrass (*Bromus tectorum*) — An ecologic intruder in southern Idaho. Ecology 30: 58—74.

13. Tundra vegetation as a rangeland resource

PETER C. LENT AND DAVID R. KLEIN

13.1 Abstract

Tundra rangelands comprise about 10% of the earth's land surface. They provide forage all or part of the year for millions of wild and domestic reindeer and for muskoxen as well as summer range for other domestic ungulates. Arctic tundra habitats are of international significance because of their importance to waterfowl and other migratory bird species.

Tundra plant growth is generally nutrient limited, especially N limited. Wildlife and domestic species exploit tundra to take advantage of short periods of vigorous growth and very high quality forage. Nutrient dynamics within individual plants are rapid and complex and herbivores must respond accordingly for optimal forage intakes. Snowcover is an especially important ecological factor in tundra rangelands, affecting soil-moisture, plant survival, plant community composition, phenology of growth, and access to winter forage by large herbivores moving above and through the snowcover and by small herbivores moving below the snow surface.

Heavy grazing has frequently reduced or eliminated deciduous shrubs and preferred lichen species. Graminoids may be either increasers or decreasers depending upon the specific circumstances. Repeated grazing on summer ranges by species such as reindeer and geese may lead to formation of highly productive lawn-like graminoid communities. Numerous secondary metabolites have now been identified in tundra plants, especially evergreen shrubs. Their effectiveness as defenses against mammalian herbivory and their ecological roles, as in influencing microtine population cycles, are still poorly understood.

Tundra vegetation is sensitive to disturbance and stress, especially where permafrost is present but also in alpine environments where effects of recreational use are of concern. Remote sensing techniques for classifying and monitoring tundra rangelands are relatively well developed. Rehabilitation techniques are also being developed and some are now available to the vegetation manager.

Tueller, P. T. (ed.), Vegetation science applications for rangeland analysis and management.
ISBN 90-6193-195-9.
© 1988, *Kluwer Academic Publishers, Dordrecht, Boston, London.*

13.2 Introduction

The word "tundra" is said to come from the Finnish "tunturi" and subsequently to English from the Russian "tundra", meaning a marshy, treeless plain. In a general sense the term refers to any cold climate landscape having vegetation without trees. Such landscapes occur predominantly in the northern hemisphere north of the tree line (arctic tundra) and in mountainous areas of both hemispheres above the tree line (alpine tundra). Tundra plant communities are characterized by short-stemmed perennial herbs, low growing or prostrate shrubs, lichens and mosses. Similar vegetation, where trees are absent due to low temperatures during the growing season, also occurs in mid- to high latitude maritime environments such as the Aleutian Islands, and certain islands in the southern hemisphere, such as South Georgia and the Macquaries (Jenkins and Ashton 1970).

The various types of arctic and alpine tundra occupy 10 to 15% of the earth's land surface, according to data compiled by Bliss (1971). This includes 3 million km^2 in the U.S.S.R., 2.5 million km^2 in Canada and 0.74 million km^2 in Alaska. The proportion in alpine tundra is difficult to ascertain but probably does not exceed 750,000 km^2 including 320,000 km^2 in the southern hemisphere and 30,000 km^2 in the United States, excluding Alaska. In areas such as Alaska and the Yukon Territory of Canada where mountain ranges extend into the Arctic, the delineation between arctic and alpine tundra is not easily made. However, Brown *et al.* (1978) estimated that 48% of the tundra area in Alaska was alpine.

All tundra areas are characterized by short growing seasons, usually fewer than 100 days, and low summer temperatures. In extreme cases the daily summer temperature may average only 2° to 5°C. Although all share this characteristic of being "cold-dominated," the vegetation types encompassed by the term "tundra" are exceptionally diverse. At one extreme in this spectrum of diversity is the high arctic polar desert, where plant cover may be less than 10%, comprised primarily of cushion plant-forms and mosses with a few annuals scattered sporadically. At the other end of the spectrum, tundra grades into forest in a forest-tundra ecotone. Here vegetative biomass can be relatively high with well-developed canopy layers if sufficient soil moisture is available during the growing season. This transition zone has qualities in common with tundra, such as prevalence and importance of lichens, sedges and mosses, and potential presence of permafrost. Important migratory mammals, such as elk (*Cervus elaphus*) and caribou (reindeer) (*Rangifer tarandus*), often depend upon a single system comprised of tundra, forest-tundra, and forest vegetation.

Some writers reserve the term "tundra" to refer to areas with continuous or discontinuous permafrost in the soils. Billings (1973) prefers the use of the terms "meadow" and "fell-field" for the mesic and xeric alpine equivalents of tundra, respectively. Important differences between alpine and arctic tundra do exist; such as the more intense solar radiation in alpine environments coupled with extreme diurnal temperature fluctuations, as opposed to the continuous summer daylight in high latitude tundra environments.

Aside from the similarities in low growth forms, which suggest that similar vegetation study techniques are applicable in alpine and arctic situations, there are some basic similarities arising from the cold-dominated environment of both, such as low primary productivity, low decomposition rates, and high susceptibility to physical disturbance — even where permafrost is not present. The complex spatial heterogeneity of arctic and alpine vegetation, a "patchiness" that is based on responses to micro-relief features, also dictates that similar problems are to be encountered in management of arctic and alpine habitats.

The presence of a perennially frozen layer in the substrate is characteristic of most arctic tundra areas. Excluding Antarctica, about 6% of the earth's surface falls in the continuous permafrost zone. Discontinuous or sporadic permafrost areas comprise an additional 5% including alpine areas in Asia and western North America (Black 1954, Pewé 1983). However, permafrost is not universally present in tundra nor does the presence of permafrost necessarily preclude tree growth. Furthermore, even in the absence of permafrost, *per se*, cryopedogenic forces play a significant ecological role in both arctic and alpine habitats. Thus the presence, absence and the characteristics of the permafrost layer are among the major influences on the nature of a given tundra area and are major considerations in management but are not decisive in determining whether or not an area is considered to be "tundra."

There is considerable overlap of plant species between alpine and arctic tundra. For example, 50% of the vascular plant species in the alpine zone of the Beartooth Plateau of Montana are also present in arctic tundra (Billings 1973). The overlap between arctic and alpine faunas is not as large (Hoffman 1974, 1984). With regard to the Eurasian tundra Chernov (1980) notes that the ancestral forms of the present tundra fauna are primarily steppe and northern forest and that very few alpine-adapted species have established themselves in the tundra. This may be in part due to the more uniform elevation gradations and south to north zonation from forest tundra to the high arctic subzone in Eurasia. In North America, particularly west of the Mackenzie River, mountain ranges extend into the Arctic, providing opportunities for alpine species such as the marmot (*Marmota caligata*), snow sheep (*Ovis dalli*) and brown bear (*Ursus arctos*) to become established in drier arctic tundra habitats.

13.3 Tundra rangeland as wildlife habitat

Various wildlife habitat classification systems have been employed in both arctic and alpine areas (Lent 1987a). Most of these habitat classes represent various subdivisions of the soil moisture gradient from flooded and saturated through desiccated and rocky (Pruitt 1966, Hoffman 1974). This reflects the dominant role of soil moisture as a niche variable both directly and indirectly through its interrelations with the permafrost regime and vegetation. Small mammal winter habitats also reflect similar snowcover gradients (Pruitt 1966). These general habitat classes and other habitat features and their relations to one another are illustrated in Figures 1 and 2. More specialized tundra habitat classifications have been employed for wetlands used by migratory waterfowl and shorebirds (Bergman *et al.* 1977, Walker *et al.* 1978, Derksen *et al.* 1982).

Vertebrates using arctic tundra habitats may be considered in three categories: (1) resident species that remain active year-round, (2) resident species hibernating in winter and (3) migratory species present for only a portion of the year. Tundra rangelands tend to have a larger proportion of migratory and otherwise highly mobile species than most other broad habitat classes.

Resident arctic mammal species include some medium-sized hibernators, such as the ground squirrel (*Spermophilus undulatus*) and marmot and large ones, such as brown bears, plus a greater variety of species that remain active year-round.

The small mammals, rodents and shrews primarily, tend to be "chionophiles" (Pruitt 1978), that is, they require snow-cover to ameliorate an otherwise unsurvivable winter environment. Others are "chionophobes," such as the muskox (*Ovibos moschatus*) and arctic hare (*Lepus arcticus*), requiring habitats of restricted snowcover for winter survival. The role and importance of snow phenomena in the tundra environment is treated later in this chapter.

Resident arctic bird species are extremely few in numbers, but may include ptarmigan (*Lagopus lagopus and L. mutus*), ravens (*Corvus corax*), snowy owls (*Nyctea scandiniaca*) and gyrfalcons (*Falco rusticolus*). The great majority of the 97 or so bird species using the northern Alaska tundra in summer, for example, are migratory (Pitelka 1979). Their dependency upon tundra habitats, however critical, is a brief one and timed precisely according to the adaptations of each species.

Waterfowl and shorebirds are obviously almost entirely migratory except for a few species that may associate in winter with polynyas and other open water marine environments in some coastal arctic areas. Most waterfowl and shorebirds are tied to wet tundra habitats for breeding.

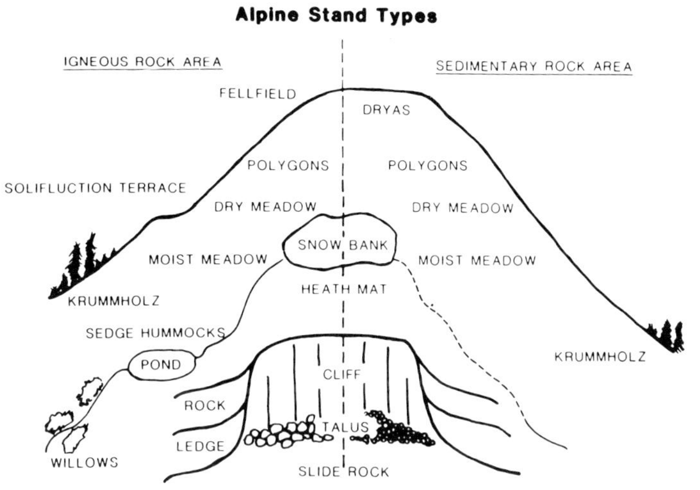

Figure 1. Alpine habitat types.

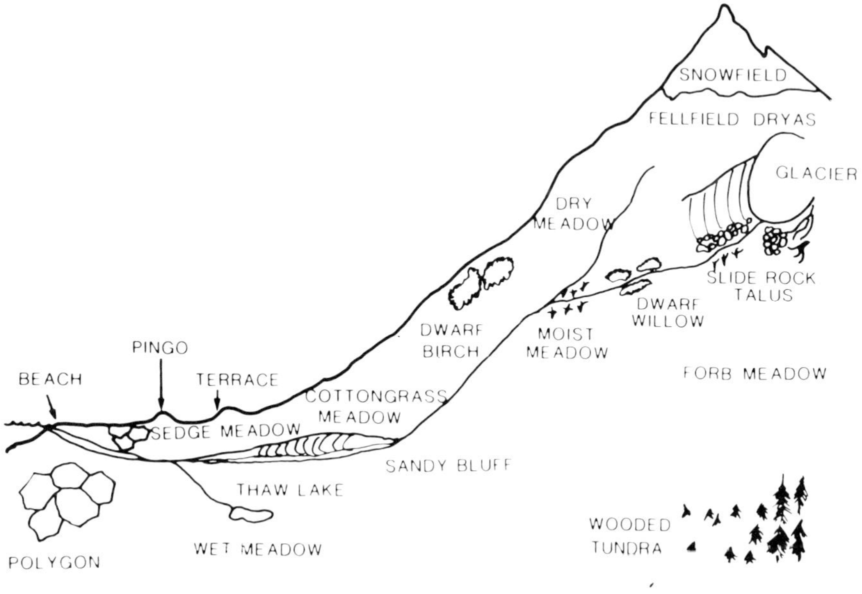

Figure 2. Arctic habitat types and features. Adapted from Hoffman 1974.

However, there are some shorebird species clearly associated with dry, upland or even alpine arctic habitats. The arctic alpine tundra supports a surprisingly diverse array of summer breeding birds, estimated by Pitelka (1979) to be 50 species, of which nearly half are passerines.

In some cases the same species may have both migratory and resident populations, even within the same general area, as has been described for caribou. In other cases, such as the snowy owl and gyrfalcon, a population may shift from resident to migratory status depending upon certain prey availability or other environmental conditions.

The Eurasian tundra supports about 25 herbivorous vertebrate species, many being the same as those in North America or closely related species (i.e., *Anser canadensis* in North America, *Anser fabalis* in the U.S.S.R.).

Phytophagous insects are well represented in tundra ecosystems but species diversity drops off rapidly in true or high arctic areas. In general these invertebrates play a lesser role as secondary producers and in ecosystem dynamics than in temperate or tropical systems, not only because of lower numbers but because of reduced metabolic rates and slower life cycles (Chernov 1980, MacLean and Jensen 1985). Chernov (1980) considers rhizophages, particularly cranefly (*Tipulidae*) larvae, to be more abundant in the High Arctic where bare soils are warmed rapidly in spring. These may number hundreds per m^2.

The importance of small mammal herbivores in tundra communities is highly variable in space and in time. *Lemmus* numbers may vary from fewer than 1/ha to nearly 200/ha during the course of a 3 to 6 year cycle (Bunnell *et al.* 1975)

Of the large herbivore species, one is overwhelmingly the numerical dominant in arctic tundra ecosystems. *Rangifer tarandus*, reindeer or caribou, numbers roughly 3 million in domestication and 3.3—3.9 million in wild or feral populations (Klein and Kuzyakin 1982, Williams and Heard 1986). Of these, 2.5 million domestic and 1 million wild reindeer are in the U.S.S.R. with over half of the latter being in the Taimyr Peninsula population (N. Naurzbaeva in corresp.). In Scandinavia there are over 350,000 domestic reindeer. The remainder are mostly in Alaska (30,000), Canada (15,000) and Greenland (3,000). All of the large herds of caribou and wild reindeer, in both North America and the Soviet Union, have been increasing in recent years. In contrast, the muskox (*Ovibos moschatus*), the other major tundra-adapted ungulate, numbers only about 84,000 world-wide but has also been increasing significantly (Klein in press, Lent 1987b).

Moose (*Alces alces*) use tundra rangelands year-round in both North America and Siberia. Such use has generally increased in this century (Bee and Hall 1956, Kistchinsky 1974). Their distribution on both continents is strongly associated with riparian habitats (Mould 1979), ribbons of high shrubs (willows primarily) that penetrate into the tundra and are also important forage sources for muskoxen (Klein 1986, Lent 1987b). In the U.S.S.R. stands of larch (*Larix* spp.) in the subarctic tundra are also important for moose. Densities of moose in tundra habitats are low, generally 2—13/100 km^2.

As with the Arctic, alpine tundra areas show low species diversity. A major difference is that, especially with resident species, alpine populations often exist in small "islands" of habitat. Thus, extirpated populations do not quickly recolonize and genetic conservation is more likely to be a management concern in alpine habitats.

Thirteen species of herbivorous mammals inhabit the alpine area of the Beartooth Montains, Montana and only six species of herbivorous or insectivorous birds regularly breed above timber line. Eight carnivorous mammal species and seven to eight raptor species spend at least part of the year in this alpine environment (Hoffman 1974). Carnivorous mammals and predatory hawks and owls seem to comprise a relatively greater proportion of the species array compared with temperate or tropical biomes.

In alpine environments the large herbivores, such as bighorn sheep (*Ovis canadensis*), elk, ibex (*Capra ibex*), snow sheep and mountain goats (*Oreamnos americanus*), generally perform seasonal altitudinal movements, descending to more wooded habitats in winter, but at least the latter two may on occasion use tundra ranges in winter.

Except for the reindeer, other domestic species in arctic or alpine tundra rangelands are introduced from temperate areas. Domestic sheep (*Ovis aries*) are the most widespread and numerous, particularly in alpine habitats, but cattle are also grazed throughout temperate alpine areas during summer worldwide. In certain areas (Iceland, Siberia) significant numbers of horses are grazed and yaks (*Bos mutus*) serve as beasts of burden and important source of milk and meat in alpine areas of central Asia.

13.4 Biological characteristics

The biology and primary production characteristics of tundra plant communities have become better understood as a result of the IBP Tundra Biome investigations and related studies over the past two decades (Bliss 1971, 1981, Batzli 1980, Wielgolaski 1975a, Rosswall and Heal 1975). A small number of vascular species (often only two or three) usually dominate primary production in a given plant community. In the High Arctic and increasingly as one moves into polar desert communities, mosses become major primary producers. Because of their high lignin

Table 1. Phytomass and annual production characteristics for selected tundra sites.

	Phytomass g/m² Above/Below Ground	Net Annual Production g/m²					Source
		Total Vascular	*Salix*	Monocot/ Graminoid	Lichen	Moss	
Hardangervidda Plateau							Ostbye 1975
Norway (60—61°N)							
Wet meadow	322/1316	147.0				0	173.0
Dry meadow	211/545	235.0			4.0	48.0	
Lichen heath	439[a]/191	75.0			78.0	10.0	
Salix shrub	1111/1305	285.0			0	210.0	
Putorana Plateau							Deyeva 1986
U.S.S.R. (67°30′N)							
Dryas tundra	164/804	24.5			1.0	0.5	
Wet sedge tundra	158/5555	37.5			0	4.0	
Atkasook (70°N)		179.2	107.5	71.3			Trudell and White 1981
Barrow (71°N)	197/4366	53.0	2.0		T	22.0	Bliss 1986
Devon Island (76°N)							Bliss 1977
Sedge-moss	1191[b]/2023	47.0		44.0	0	102.0	
Dry cushion	451/57	15.0		1.0	2.0	2.0	
Axel Heiberg Isl. (80°N)		19.1	8.5	9.0			Parker and Ross 1976

[a] Incl. 370 g/m² lichen.
[b] Incl. 908 g/m² bryophytes.

content, most of this production is unused by herbivores and goes directly (but slowly) to decomposition. During the short growing season, daily primary production rates may be surprisingly high (3—6 g/m^2) but annual production is low (Table 1).

The generally dominant role of nutrients, especially N, as the limiting factor in tundra plant growth (Ulrich and Gersper 1978) has been further confirmed by recent experimental work (Tissue and Oechel 1987). However, little work has been done in extreme high arctic conditions. Evergreen shrubs, such as *Vaccinium vitis-idaea* and *Ledum* spp., are common in tundra communities and tend to dominate in the most nutrient-poor sites. Leaf retention in these species is an adaptation for nutrient conservation.

Chapin and Shaver (1985) experimentally manipulated environmental factors (air temperature, light regime and nutrient availability) of several key plant species from tussock tundra and wet meadow communities. They found a diversity of growth responses and no single limiting factor. The implication is that overall production in these communities is surprisingly stable from year to year, in spite of the relatively small number of species.

13.4.1 Snowcover and tundra rangelands

Snowcover as an ecological factor has several important influences on tundra rangeland systems. First, its distribution, depth and duration are principal determinants of the underlying plant community. Snowfall contributes 70—80% of annual precipitation. Second, it can directly influence the population structure and the chemical composition of plant species on specific sites. For example, Callaghan and Emanuelsson (1985) report disjunct populations with entire age classes killed as a result of wind action and desiccation during winters of low snowfall. Deep snowcover, however, also provides the sheltered environment in which small rodents (microtines especially) can carry out subnivean herbivory and reproduce successfully (Kaikusalo and Tast 1984).

With regard to the large herbivores snowcover depth, distribution and physical characteristics influence the location, species selection and energy expenditures for foraging. Finally, late melting snow accumulations delay phenology of underlying vegetation, providing areas where green-up occurs late in the growing season. Such patches are particularly important for alpine ungulates (Klein 1970, Skogland 1980, 1984).

Any survey or inventory of tundra rangelands and habitats important to domestic livestock or wildlife must encompass the collection and analysis of information on snowcover conditions if it is to provide a proper understanding of carrying capacity, relative importance of various subareas, and temporal patterns of use. Surveys of snow cover may be limited only to recording the presence or absence and the timing of accumulation and ablation. Data on depth or thickness greatly increase the value of a survey and for many purposes data on density, hardness, ice and crust layers and patterns of drifting or distribution in relation to topography are desirable.

Pruitt (1959) was the first in North America to employ extensive aerial surveys and snow measurements to relate snowcover characteristics to caribou movements and distribution, working primarily in the taiga. He used National Research Council of Canada snow instruments (Klein *et al.* 1950) with which various parameters of snowcover may be measured. Richens and Madden (1973) report on modifications to these instruments to make them more suitable for biologists.

Other workers (LaPerriere and Lent 1977, Lent and Knutson 1971, Skogland 1978, Leader-Williams *et al.* 1981, Brooks and Collins 1985) carried out snow surveys using the Ramsonde penetrometer. The penetrometer was developed in the European Alps for avalanche studies and is most suitable for use with dense, wind-blown snow with complex thin crusts and ice layers, such as is typical of alpine and some tundra environments. It measures the force required to penetrate a given layer from above.

Pruitt's studies and those of subsequent investigators working with barren-ground caribou (LaPerriere and Lent 1977) and reindeer (Skogland 1978, Leader-Williams *et al.* 1981) have confirmed selection of winter ranges with relatively low snow depths, densities and hardnesses. Muskoxen have shown even stronger selection for such winter ranges (Lent and Knutson 1971). In the case of muskoxen it has been shown repeatedly that an understanding and knowledge of the prevalence of crusts, ice layers and groundfast ice on winter ranges is critical to proper habitat evaluation (Vibe 1967, Lent and Knutson 1971, Thomas *et al.* 1981).

Miller *et al.* (1982) combined surveys of caribou winter trails and feeding sites, and surveys of snowcover and ice conditions with subsequent vegetation sampling in summer. Several range types with desirable and nutritious forage species were shown to be underutilized because of difficulty of access. Winter feeding was directed primarily at poor forage species that were exposed and directly available. Similarly, Gaare and Skogland (1975) concluded that only 25% of the potential winter forage on Norwegian alpine reindeer ranges was actually available for consumption. Thus summer-only inventories of winter ranges can greatly exaggerate the winter range carrying capacities. Snow surveys are essential for assessing winter forage availability (Brooks and Collins 1985).

The best lichen growths on reindeer ranges are associated with moderately deep snowcover because of the protection against wind action and the spring moisture the snow provides. Snow accumulations in streambeds and other depressions provide specialized, critical habitats for numerous species by protecting tundra shrubs from winter damage. The plant associations of these snowbeds in turn support some specialized wildlife species such as the singing vole (*Microtus miurus*) in arctic tundra (Pruitt 1966) and the pocket gopher (*Thamomys talpoides*) in alpine communities (Thorn 1982).

13.4.2 Fire in tundra rangelands

Fires in tundra rangelands occur occasionally in low arctic or sub-arctic environments but not in the High Arctic with discontinuous vegetation and extremely low fuel levels (Wein 1976). A shrub tundra area of over 2,124 km^2 with interspersed small islands of black spruce (*Picea mariana*) east of the Mackenzie River Delta (Canada) burned in 1954. Following the fire this winter range for reindeer showed thawing occurred to 41 cm and karst hummocks resulted. Nine years later there was still no significant lichen forage. *Ledum palustre*, dwarf birch and willows to 3 m in height had returned and *Calamagrostis* spp. was among the significant invaders (Cody 1964).

On the Seward Peninsula of Alaska the effects of a 1977 fire were found to be most marked on well drained slopes with birch and ericaceous shrubs. In poorly drained sedge-shrub communities post-fire thaw depths were only 10—15 cm. Sphagnum mats retarded the fire but were killed by it (Racine 1981). *Eriophorum* tussocks recover rapidly from fires in general. Tussocks were resprouting one month after fire had destroyed 80—90% of the shrub and graminoid biomass (Hall *et al.* 1978). Domestic reindeer grazed in the burned areas the following spring, selecting new growth of *Eriophorum* spp., *Rubus chamaemorus* and *Salix* spp. (Klein, unpublished).

13.4.3 Nutrient dynamics and herbivory

The migratory movements and seasonal shifts in range use of continental caribou populations have been known in detail for many years (Kelsall 1965, Lent 1966). However, the dynamics and seasonal changes in chemical constituents of tundra forage species have only recently become better clarified (Skogland 1980), so that a fuller interpretation of their relationship to these movements is possible (Klein 1970).

First, some tundra plants are adapted to initiate growth immediately upon or even slightly before the period of spring snow melt. This can occur through the effect of so-called "snow hot-houses" where low-growing forms, for example of *Potentilla*, may initiate growth under a thin crystalline snow crust (Chernov 1980). Other species, notably *Eriophorum vaginatum*, have a tussock growth form that creates a favorable micro-environment. Wind action causes snow to accumulate in the between-tussock troughs and exposes the dark tussocks to insolation and rapid warming. Chapin *et al.* (1979) reported thaw to a depth of 16.3 cm from the top of a 20 cm high tussock only 10 days after snowmelt. The flowering parts of upland forbs in areas where dry soils are warmed very rapidly are also available early in summer (Lent 1966, Kuropat and Bryant 1980). Chapin *et al.* (1980) report distinct peaks in N content of all parts (flowers, leaves and stems) of the forb *Rubus chamaemorus* around mid-July at Atkasook on the Alaskan Arctic Coastal Plain. Although these reproductive parts are high in N, they represent relatively small commitment of carbon, only 4% in *Eriophorum*, and 5% in the shrub species (Johnson and Tieszen 1976).

The initiation of leaf growth in deciduous shrubs (*Salix* spp., *Betula* spp.) comes almost as soon as the floral parts. Chapin *et al.* (1980) found that on the Arctic Coastal Plain of Alaska (growing season about 75 days) maximum N concentrations (4.5%) in leaves of these shrubs were reached 3—4 weeks after snowmelt but leaf biomass production continued to increase rapidly for about another 3 weeks.

A distinct and important feature of plant nutrient dynamics for tundra herbivores is the relative high importance of translocation processes among plant parts. The nutrient pool for tundra plants is particularly limited at the time of growth initiation when seasonal thaw of soils has only barely commenced. Thus, deciduous shrubs store N, P and K in both stem and root reserves for initiation of leaf growth. Up to 14% of the total plant N and K pool may be moved into leaves from stems and roots during the 3 week initial growth period (Chapin *et al.* 1980). As soils thaw and root growth is possible, there is then a rapid translocation of N back below ground. This process is even more pronounced in *Eriophorum vaginatum* because all roots are replaced annually.

It has long been known that far greater proportions of plant biomass are below ground in tundra communities (Bliss 1971, Wielgolaski 1975a). What is now becoming apparent is that the magnitude and rapidity of these nutrient translocations makes timing and selection all the more critical for tundra herbivores.

In the heaths and steppe-like tundra of West Greenland, caribou in winter seek out species such as *Hippuris vulgaris* and *Equisetum scirpoides* that remain green (Thing 1984). In the taiga as well, although the principal winter dietary component is lichens, caribou will seek out "winter-green" forage species with high nitrogen content (Klein 1982b), for example horsetails (*Equisetum* spp.) (Lent 1966) although such species tend to be only locally abundant. *Equisetum*, like lichens, shows relatively high levels of *in vitro* digestibility in winter (Thomas and Kroeger 1981).

Other winter-green species (the evergreen shrubs) show high nitrogen content for the season but low digestibility (Bryant *et al.* 1983, Russell and Martell 1984). Nevertheless significant use is made of these species by both forest and tundra-wintering populations of *Rangifer* (Lent 1966, Scotter 1967, Parker *et al.* 1975, Parker 1978). It has been hypothesized that these shrubs provide small but important supplements of minerals and protein for pregnant cows in late winter (Russell and Martell 1984).

Thus, while tundra herbivores obviously respond to a variety of factors influencing forage selection (Bryant *et al.* 1983, Chapin *et al.* 1980, 1986, White and Trudell 1980) there is now strong evidence that the influencing role of nutrients is a dominant one. Taking the caribou population of northwestern Alaska as an example, the movements and habitat use patterns over an area of approximately 360,000 km^2 fit the described vegetation phenology and nutrient dynamics very well (Figure 3).

Pregnant females arrive on the *Eriophorum* tussock community of the tundra foothills in close synchrony with the start of *Eriophorum* flowering and move out after anthesis and their own calving (Lent 1966, Kuropat and Bryant 1980). Immediately thereafter there are enormous post-calving aggregations primarily in higher elevation dry tundra sites but also in riparian areas. There, both early flowering forbs, such as *Lupinus* and the first green leaves of shrubs are exploited.

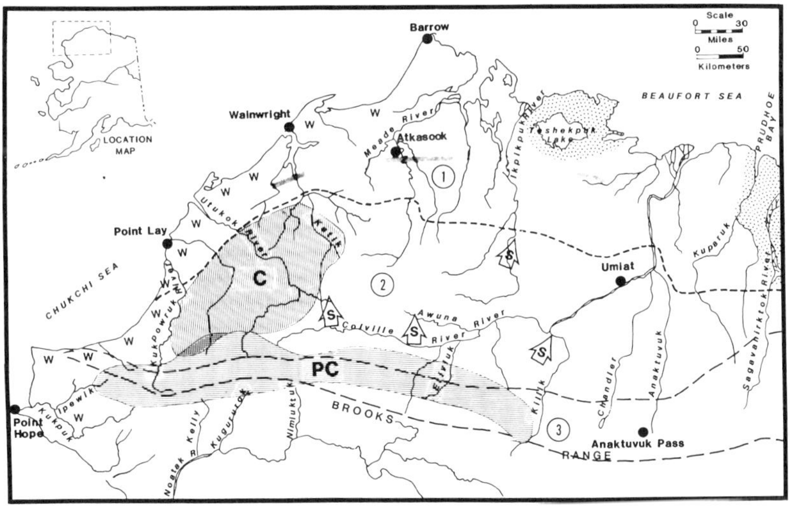

Figure 3. Range of the migratory caribou population of Northwest Alaska. Physiographic provinces = 1. Coastal plain/wet tundra, 2. Foothills/coastal tundra, 3. Brooks range/arctic-alpine tundra. C = Calving area. PC = Post-calving area. W = Winter range. Stippling shows areas also used by resident caribou herds.

In July, there is a general shift back north and to wetter habitats of the cooler coastal tundra. This coincides with the delayed phenology and later peaks in N in new leaves of deciduous shrubs (Chapin *et al.* 1980, White and Trudell 1980). Late summer and fall feeding is more generalized but takes advantage of aquatics, fungi and late snowbed communities. In winters when large numbers of these caribou remain on the tundra, they forage predominantly in *Dryas* heath communities where the over-wintering green material is relatively abundant (Lent 1966, White and Trudell 1980) foliose lichens, although not abundant, are available in some areas and remain an important part of their winter diet (Klein 1982b). In northwest Alaska tundra-wintering caribou are most frequent in a coastal belt of relatively dry and wind-blown uplands (Figure 3).

Migratory willow ptarmigan move onto the tundra rangelands of northwest Alaska, in spring, in a pattern very similar to that of the caribou. Approximately 50,000 move north annually through Anaktuvuk Pass (Figure 3) alone (Irving *et al.* 1967). They arrive on the *Eriophorum* tussock communities, often in flocks of thousands, and then begin to exploit deciduous shrubs only slightly earlier than the caribou (Lent, unpublished).

Similar patterns of movement and habitat use, precisely exploiting the "green wave", but on a much smaller spatial scale, are described for alpine reindeer in Norway (Skogland 1980, 1984) and feral reindeer in Alaska (Klein 1982a). Where they occur together in Norway introduced sheep and native reindeer show dietary overlap and similarity in habitat use patterns, concentrating in summer on snowbed meadows. However, reindeer exploit a greater variety of habitats. With present herding practices and numbers there is relatively little spatial association or direct competition (Skogland 1984).

Because they are less selective bulk feeders with a much larger rumen capacity muskoxen tend to be more sedentary than caribou (Klein 1986). In general muskoxen make more use of graminoids in summer, especially on high arctic ranges. They also make greater use of deciduous and evergreen shrubs in winter (Klein 1986, Lent 1987b).

13.5 Plant-herbivore interactions

13.5.1 Plant defenses against herbivory

Because resources, particularly nutrients, are so limiting in tundra environments, we could expect that plant defenses against tissue loss to herbivory would be well developed (Bryant *et al.* 1983). In tundra environments as elsewhere, non-deciduous dwarf shrubs are strongly associated with the most nutrient-poor sites. These species are long-lived and slow-growing with a low turn-over of plant parts that is clearly advantageous in nutrient-deficient habitats. Because the leaf parts are long-lived and are major nutrient depots, there should be particularly strong evolutionary pressure to protect these parts.

The idea has been advanced that secondary plant metabolites have evolved primarily in response to insect herbivory and only secondarily as defenses against mammalian herbivores (Janzen 1979). In the Arctic, insect herbivory should be of less importance. Others have hypothesized that such defenses are critical at peaks of microtine cycles when rodent biomasses and herbivory rates are extremely high. In turn such defense responses could lead to microtine population declines (Haukioja 1980, Laine and Henttonen 1983, Rhoades 1985). However, a recent examination of the data showed no correlation between stage in microtine population cycle and levels of phenolics in forage species (Jonasson *et al.* 1986). The only significant positive correlation was between levels of tannins and warmth of growing season.

Bryant *et al.* (1983) logically suggest that carbon-based defenses would be most common in nutrient-limited plants, and nitrogen-based in those of shaded sites and carbon-limited situations. However, a survey by Jung *et al.* (1979) found alkaloids to be more frequent than carbon-based tannins, at least in the year and season of sampling. Clearly, our understanding of the roles and relative importance of the various secondary metabolites in tundra range species is still unfolding.

Graminoids depend primarily on large root and rhizome reserves as well as growth form and silica content for resistance to herbivory. Deciduous shrubs also depend upon regrowth and replacement transfers from stems and roots but are less able to cope with repeated defoliation. Tussock graminoids retain unpalatable leaf material over winter which deters herbivory on new leaf growth the following year.

13.5.2 Plant community responses to grazing

Evidence for depletion of lichens from tundra winter ranges, as a result of grazing, comes primarily from studies of domestic reindeer or from islands having introduced *Rangifer* populations. The virtual elimination of lichens on sub-Antarctic South Georgia Island is a well documented example (Bonner 1958, Leader-Williams *et al.* 1981). The depletion of lichens on Nunivak Island, Alaska, also occurred as a result of heavy grazing by introduced reindeer (Palmer and Rouse 1945) as did increases of *Eriophorum* on wetter sites and shrubs on drier sites. In dry tundra communities, lichens comprised 80% of total plant cover in the 1920's, only 12% in the 60's. Reduction of lichens on wet tundra was 30% to under 1% (Bos 1967). On St. Matthew Island, Alaska, introduced reindeer greatly reduced lichens as the population built to high density (Klein 1987). Mosses, sedges and grasses increased in response to the removal of the lichens. Twenty-two years after the crash die-off of the herd from starvation the lichen regrowth was dominated by early pioneering species with low palatability. A lichen biomass of 35 g/m^2 in the best sites was only 10% of the lichen standing crop on similar sites on an ungrazed adjacent island (Figures 4 and 5). Under intense grazing pressure, reindeer also reduce highly palatable forbs such as *Pedicularis* spp. (Klein 1987).

In a study of the effects of grazing by confined muskoxen on tussock and mat cushion tundra McKendrick (1981) measured changes in basal cover and species composition. There was a significant reduction in lichen and shrub standing crops due to heavy use by muskoxen of these tundra types. In West Greenland foliose lichens have apparently been greatly depleted over the past decades as caribou numbers reached a peak of perhaps 100,000 in the 1960's and then declined to under 10,000 (Thing 1984). Here, the high population density of caribou has also led to the reduction of shrubs, *Betula nana* and *Salix glauca* through repeated defoliation and removal of terminal twigs, leading to increased graminoids, notably *Poa* (Thing 1984).

In the sub-antarctic tundra of South Georgia, the deciduous dwarf shrub

Figure 4. Overgrazed tundra rangeland on St. Matthew Island, Alaska, formerly lichen-dominated. Remaining vascular plants are *Salix crassijulis* x *ovalifolia* and *Carex nesophila.*

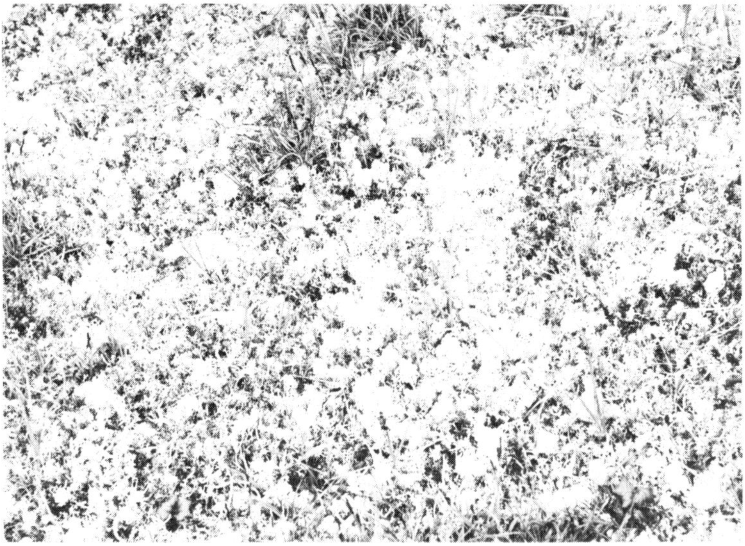

Figure 5. Lichen-dominated tundra on ungrazed Hall Island, adjacent to St. Matthew Island.

Acaena magellanica has been greatly reduced in distribution and abundance by the introduced reindeer (Leader-Williams *et al.* 1981). A *Poa* with high levels of carbohydrates during winter has now also been greatly reduced, suggesting it is a nutritional analog to lichens for the reindeer. On this range the principal increasers have been mosses.

Three years of experimental grazing in Norway by sheep greatly reduced shrub cover (*Salix herbacea*) and herb cover (especially *Astragalus alpinus*) and increased the cover of monocots relative to ungrazed sites in both wet and dry alpine tundra meadows (Wielgolaski 1975b). The increase in monocots was most pronounced on dry sites, less on wet. Therefore, overall vegetated cover decreased only on wet meadows.

Over centuries in Iceland grazing by sheep, horses and cattle has also decreased woodlands and forbs. Monocots have generally increased but in areas of most intensive grazing palatable grasses have also been decreasers and less palatable sedges and rushes have been increasers. These historical trends have been simulated and duplicated in sheep grazing experiments (Thorsteinsson 1986).

Thus, in a variety of circumstances in arctic, subarctic, antarctic and alpine heaths and tundra intensive grazing by large herbivores has led to marked reduction in foliose lichens, loss of deciduous shrubs and related increases in graminoids.

13.5.3 Plant-herbivore co-evolution

McNaughton (1984) implied that caribou may provide an example of his general hypothesis that among grazing animals dense aggregations are advantageous because they alter vegetation structure so that the food intake per bite to the individual grazer is increased. Holt (1980) and Thing (1984) report that concentrations of reindeer intensively graze meadows of *Poa pratensis* in West Greenland, maintaining them in a prostrate, vegetative form and presumably stimulating new, green growth throughout the summer. In the process fecal fertilization of these meadows occurs with an estimated accumulation of 4.5×10^3Kg/ha. Geese also undertake intensive herbivory of arctic graminoids with similar response of the vegetation (Cargill and Jefferies 1984a).

As described earlier, the most pronounced and spectacular aggregations of caribou are associated with extremely short-term herbivory on floral parts and new leaves of deciduous shrubs where growth stimulation and alteration of vegetation structure must be insignificant. It appears that relatively sedentary *Rangifer* populations depend more on repeated grazing whereas the migratory populations emphasize one-time annual convergence on vegetation patches in peak nutritive state and lower intensity grazing (Andreev 1977). Thus, the latter populations do not provide good examples of McNaughton's hypothesis.

Chapin (1980) performed defoliation (clipping) experiments. Initial defoliation of *Eriophorum* removed old leaf growth and leaf nitrogen and phosphorus content. Repeated clipping, simulating severe grazing, led to carbohydrate depletion. *Carex* spp. were most resistant to repeated clipping apparently because of their large underground reserves.

The study by Cargill and Jefferies (1984a) in a subarctic salt marsh is the only one in a boreal regime that demonstrates experimentally that grazing increased net above-ground primary production and increased total nitrogen concentration. This occurred in two graminoid species *Puccinella phryanodes* and *Carex subspathacea*. In one study year production on grazed plots by the lesser snow goose (*Anser caerulescens*) was 135 $g/m^2/year^1$ versus ungrazed with 100 $g/m^2/year^1$. The mechanisms for these effects are not understood because inorganic N was clearly limiting primary production (Cargill and Jefferies 1984b) and in spite of heavy fertilization by geese droppings there was a net loss of 2.2 g/m^2 of nitrogen annually from the marsh. The probable mechanisms for increased plant production are elimination of the litter layer and increase in the N recycling rate. Up to 80% of the above ground graminoid biomass was removed by the geese.

In summary, the evidence for herbivore-plant coevolution in tundra ecosystems is not strong. Gregariousness in *Rangifer* populations is expressed in response to several factors including predation, insect harassment and plant phenology (Bergerud 1974). In some cases, it is documented that populations exert significant influence on plant growth forms and individual grazers apparently benefit from that effect.

13.6 Vegetation management

13.6.1 Range surveys

Special techniques are required for tundra areas where lichens are a significant component or likely to be important forage plants. Pegau (1970) used a one-square-foot frame and visually estimated lichen cover based on random throws. Using this simple technique he was able to assess the damage to lichens caused by holding or herding reindeer on lichen ranges during dry periods. While the randomness of "random throws" has been legitimately criticized, the technique is valid when modified to improve randomization. Racine (1981) used simple m^2 plots, recording stem density and percent cover to measure significant differences before and after a tundra fire.

Thilenius (1979) discusses the problems, concepts and inexact state-of-the-art for assessing and monitoring North American alpine range communities. He notes that there is confusion in alpine range management regarding condition classes and site climax. For example, cushion plant communities dominated by *Geum rossi* have been treated by range managers as "poor condition" communities when in fact such forb-dominated communities are common on xeric sites and may represent excellent range conditions.

13.6.2 Remote sensing applications

Because of the vast areas involved, the difficulty of on-site access and the short field season, there is a particularly strong rationale for the use of remote sensing systems for mapping, inventory and monitoring arctic tundra vegetation. Several efforts have been directed in recent years toward the use of satellite sensors, notably Landsat, for that purpose (Anderson and Belon 1973, George *et al.* 1977, Nodler *et al.* 1978, Lent and LaPerriere 1974, Hall *et al.* 1980, Acevedo *et al.* 1982). This work establishes Landsat as a useful tool to map tundra plant communities and monitor changes in rangelands due to wildfire and subsequent recovery of tundra vegetation. Lent (1980) also used Landsat imagery to monitor gross changes in extent of snowcover due to patterns of accumulation and melt over large areas. Use of satellite imagery for remote sensing of tundra vegetation is best adapted for broad scale inventories and mapping. Where detailed vegetation information is required aerial photography remains a useful tool (Sims 1983).

Satellite imagery from Landsat and weather satellites has also been used to assess the snow and ice conditions encountered by nesting geese in the Canadian Arctic. By assessing the year to year timing of snow and ice melt on a broad scale this nesting habitat information has been used in conjunction with other data to successfully predict productivity of species such as brant (*Branta bernicula*), snow geese (*Chen caerulescens*) and Arctic-breeding Canada (*B. canadensis*) and white-fronted (*Anser albifrons*) geese (Reeves *et al.* 1976).

A major limitation to the use of satellite imagery over much of the Arctic is the high frequency of overcast days. The large number of days in which cloud cover precludes usable imagery is partially compensated for by the polar convergence of Landsat orbits, that results in bursts of imagery on successive days at high latitudes. However, because the phenology changes so rapidly and the windows of good weather are frequently so short, it is difficult to obtain comparable imagery in successive years to document long-term trends.

In a multiple stage sampling program to discriminate and map lichen community types Sims (1983) determined reindeer winter range types and carrying capacities on the Tuktoyaktuk Peninsula, Canada. Microdensitometric readings from aerial photos (70 mm color infrared) were diagnostic for areas with over 5% lichen cover. About 81% of sample sites were correctly classified.

The use of satellite imagery for mapping and monitoring alpine vegetation and habitats is much more difficult because of the topographic complexity and the effects of shadows on surface albedo. Craighead *et al.* (1982) have classified and mapped alpine and subalpine vegetation in western Montana; using Landsat, a computer-assisted multi-spectral analysis system and multiple-level ground truthing to inventory grizzly bear habitat. Landsat imagery has also been used by Wiersema and Schroder (1985) to map snow-free winter range and areas of early green-up for management of European ibex.

Frank and Isard (1986) used CIR aerial photos and indices of topoclimatic conditions to produce a map of alpine tundra types in Colorado. Slope and aspect in relation to prevailing wind and the daily total insolation were used as the principal determinants.

13.6.3 Impact assessment

Tundra rangeland systems are increasingly involved in issues and decisions that relate to potentially conflicting human uses of these lands. In the

Arctic, exploration, development, transportation, and production of oil and gas and mineral ores are prime examples of these land use problems (Bliss and Klein 1981). In alpine environments issues are frequently those relating to intensive recreational use, livestock use or mineral development. Thus there has been a corresponding increase in emphasis on inventory and monitoring techniques that have had demonstrable applications in such management situations.

Surface disturbance and alteration of permafrost regime have clearly had undesirable esthetic effects in arctic areas, as well as causing localized erosion problems. The long-term ecological consequences of such changes have been less clear-cut.

Nevertheless, for purposes of planning and inventory prior to proposed developments in the Arctic, mapping of permafrost temperature, depth and moisture content and protective vegetative cover, have received paramount attention to date (Everett 1982, Linell and Tedrow 1981). The fragility or sensitivity to damage in tundra area is often directly proportional to the ice content of the permafrost layer and to the mean annual ground temperature (Ives 1970) as well as to the insulative qualities of the surface cover. Our ability to predict such sensitivity is now good. Monitoring can identify changes in permafrost status and the overlying insulating vegetation and track the success of stipulations relating to surface disturbance and recovery programs (van Cleve 1977, Brown *et al.* 1978, Brown and Berg 1980).

Because of the rapid succession of seasonal changes and the importance of these physical events to phenology, animal abundance, and sensitivity, it is important that the biologist understand this annual sequence, based as necessary on the literature or on information from individuals familiar with the area. Events such as freezing-over of lakes, rivers and coastal lagoons, reaching of critical ice depths, establishment and amount of permanent winter snowcover, rate of spring ablation are all events critical to planning and environmental assessments. These are also essential information for any environmental impact analysis (Lent 1985). For example, in consideration of permits where surface travel is limited to certain periods with adequate snowcover, it is essential to know whether the applicant's plans are realistic, based on the likely duration of that period.

Walker *et al.* (1980) produced an atlas of the Prudhoe Bay region, Alaska. Among the many maps are ones showing relative sensitivity of areas to oil spills, sensitivity to damage by over-terrain vehicles, lichen cover, breeding bird density and the thickness of the active layer. The oil spill sensitivity map and the procedures for its production are described in detail in Walker *et al.* (1978). They demonstrated great differences in the sensitivity of various plant communities to either diesel or crude oils by measuring live cover on m^2 quadrats before and one year after experimental "spills." Later time-series and cumulative impact analyses (Walker *et al.* 1986) demonstrate that specific types of disturbances tend to be associated with specific moisture-regime categories.

Sensitivities of alpine tundra areas to human impacts are also well documented. Using simple species cover measures Willard and Marr (1971) monitored the disturbance and recovery of alpine vegetation. In general they reported extremely slow recovery rates. In some cases no recovery had occurred after four years of protection, based on both total cover and species composition measures. In the case of *Kobresia* meadows damage to the turf caused such severe erosion and loss of the thin humus-enriched horizons that hundreds of years may be necessary for restoration.

Bell and Bliss (1973) used 10 × 50 cm clipping plots and line intercept sampling of plant cover to measure recovery rates of alpine plant communities after human trampling. They also measured rates of down-slope movement as indicators of disturbance. Dry tundra and scree slopes were most fragile and slowest to recover. Meadows and snow bed communities recovered most rapidly.

Fallout of atmospheric pollutants poses special problems for arctic and alpine systems. The cool temperatures result in slow degradation of contaminants, thus long exposures of plants to them. Lichens are particularly sensitive to sulfur components of atmospheric fallout, and certain species (particularly the foliose forms of importance to caribou and reindeer) are eliminated or show reduced growth close to heavy pollution sources (Puckett *et al.* 1973).

13.6.4 Radioactive fallout contamination

In the late 1950's and early 60's the utilization of tundra rangelands by man was threatened by accumulation of radioisotopes from above-ground nuclear testing by the U.S.A. and U.S.S.R. (Liden 1961, Watson *et al.* 1964). Lichens, because of their long-lived nature and non-rooted, high surface-to-mass growth form, accumulate atmospheric radioisotopes to a greater degree than vascular plants (Gorham 1959) and these isotopes are further concentrated in the flesh and organs of reindeer/caribou and in Eskimos, Lapps and other consumers of these parts (Liden 1961, Hanson *et al.* 1964).

Following the Chernobyl accident in 1986 high levels of radioisotopes were once again apparent in alpine ranges in Scandinavia and mean summer radio-cesium contamination of wild reindeer meat in Norway was 20 times higher than values from the Kola Peninsula during the nuclear arms testing of the 1960's (Skogland 1986). Guidelines recommending maximum safe limits for consumption of reindeer meat have been issued by Scandinavian governments.

13.6.5 Fertilization and rehabilitation of tundra rangelands

In dry hill pastures of northern Iceland, the effects of N and P fertilization over several years have been great. Annual herbage production increased from under 20 kg/ha up to 1800 kg/ha. In general evergreen shrubs, lichens and moss declined, grasses, and sometimes sedges, increased (Garmo 1986). Fertilization in alpine pastures of Norway has also yielded significant increases (Gudmundssen and Bennet 1986). In areas of permafrost, however, it is possible that fertilization could have a negative long-term effect on net annual growth if the increased standing crop/cover density decreases the soil thaw zone and thus further locks up nutrients (Ulrich and Gersper 1978). Conversely, cropping to lawn-like growth form, as described earlier, would presumably increase the summer thaw zone.

Fertilization, at rates of up to 111 kg/ha, is a necessary part of tundra rehabilitation efforts for establishment of plants on disturbed tundra sites. The exact requirements for reseeding and fertilization differ greatly from site to site, however (Brown *et al.* 1978, 1980).

Bulldozers have been frequently used to scrape away vegetation and make firebreaks to stop the spread of tundra fires. However, such techniques have led to deep soil thaw and severe erosion and gullying along such breaks. Patterson and Dennis (1981) report on a technique of replacement of tussocks and sod that allows rapid stabilization and rehabilitation with 75—80% of the tussocks and mat vegetation alive two years later.

Native plant species are often the most suitable for revegetating disturbed ground in arctic and alpine areas but commercial sources for the seed are often lacking. A successful technique employed in northern Alaska has been seeding with available agricultural varieties of grasses that are annuals or biennials. This binds the soil, reduces erosion and allows for the gradual invasion of native species. The Agricultural Experiment Station of the University of Alaska at Palmer has released a few varieties of grasses indigenous to Alaska that have been found suitable for revegetation in the Arctic (Mitchell 1981). These include *Poa glauca*, *Arctagrostis latifolia*, *Calamagrostis canadensis* and *Deschampsia beringensis*.

13.6.6 Grazing systems

In almost all cases of arctic grazing systems involving domestic species tundra ranges are used primarily in the summer months. Winter forage is from forested ranges (primarily lichens), in the case of reindeer, or from hays and supplemental feeds for sheep and horses. Thus, attention has been focused on winter range lichen supplies and little effort has been devoted to tundra grazing systems. With sheep, attention has been given to increasing productivity by grazing to the point where the shrub component is reduced and graminoid annual production is increased (Skogland 1984, Thorsteinsson 1986). This technique, although successfully employed in Iceland by the Viking settlers, ultimately led to severe soil erosion through over-grazing. Thus, what is still lacking is a good understanding of range carrying capacity in tundra systems. Recent estimates of carrying capacity by Bliss (1986) only add confusion by failing to consider the high degree of annual variability in both primary production and winter availability of forage and by interpreting factors such as herbivore disease and predation.

Cultural and technological changes, such as the shift to herderless grazing with sheep (Thilenius 1979) and the mechanization of the reindeer industry (Collins 1986), also present new challenges to successful grazing systems.

13.7 Conclusions

Understanding of tundra rangelands has greatly increased in recent years. Notable improvements have been made in understanding of plant nutrient dynamics, plant defenses and secondary metabolites, primary production estimates, and plant-herbivore interactions. Advances have also been made in the study of snow as an ecological factor and in the use of remote sensing techniques.

The above advances have been applied in impact assessment, mitigation and rehabilitation. It is hoped that vegetation science can contribute further in the future to an improved understanding of carrying capacity in tundra ecosystems. The greatest challenge is the integration of vegetation science with other biological advances and with the social sciences in the development of holistic tundra rangeland management systems.

13.8 References

Acevedo, W., D. Walker, L. Gaydos, and J. Wray. 1982. Vegetation and land cover, Arctic National Wildlife Refuge, Coastal Plain, Alaska. U.S. Geological Survey Map 1—1443. Misc. Investigations Ser. 1 sheet.

Anderson, J. H., and A. E. Belon. 1973. A new vegetation map of the western Seward Peninsula, Alaska, based on ERTS-1 imagery. No. E73-10305. NTIS, Springfield VA.

Andreev, V. N. 1977. Reindeer pastures in the subarctic territories of the U.S.S.R. Handbook of Vegetation Science, Part XIII. W. Krause (ed.) W. Junk Publ., Dordrecht.

Batzli, G. O. (ed.) 1980. Patterns of vegetation and herbivory in Arctic Tundra. Results from the research on arctic tundra (RATE) program. Arctic and Alpine Res. 12: 401—588.

Bee, J. W., and E. R. Hall. 1956. Mammals of northern Alaska on the Arctic Slope. Misc. Publ. Mus. Nat. Hist. No. 8, Univ. Kansas.

Bell, K. L., and L. C. Bliss. 1973. Alpine disturbance studies: Olympic National Park. Biol. Cons. 5: 25—32.

Bergerud, A. T. 1974. The role of the environment in the aggregation, movement and disturbance behavior of caribou. Vol. 2: 552—584. *In*: V. Geist, and F. Walther (eds.) The Behavior of ungulates and its relation to management. IUCN New Ser. Publ. 24, Morges, Switzerland.

Bergman, R. D., R. L. Howard, K. F. Abraham, and M. W. Weller. 1977. Water birds and their wetland resources in relation to oil development at Storkerson Point, Alaska. Resource Publ. No. 129. U.S. Fish and Wildlife Service, Washington, D.C.

Billings, W. D. 1973. Arctic and alpine vegetations: similarities, differences and susceptibility to disturbance. Bioscience 23: 679—704.

Black, R. F. 1954. Permafrost — a review. Geol. Soc. Amer. Bull. 65: 839—856.

Bliss, L. C. 1971. Arctic and alpine plant life cycles. Ann. Rev. Ecol. Syst. 2: 405—438.

Bliss, L. C. 1981. Tundra biome: past and present. p. 3—24. *In*: L. C. Bliss, O. W. Heal, and J. J. Moore (eds.) Tundra ecosystems: A comparative analysis. Cambridge Univ. Press, Cambridge.

Bliss, L. C. 1986. Arctic ecosystems their structure, function and herbivore carrying capacity. p. 5—35. *In*: O. Gudmundsson (ed.) Grazing Research at Northern Latitudes. Plenum Press, New York.

Bliss, L. C., and D. R. Klein. 1981. Current extractive industrial development, North America. *In*: L. C. Bliss, J. B. Cragg, D. W. Heal, and J. J. Moore (eds.) Tundra Ecosystems: A comparative analysis. Cambridge Univ. Press.

Bonner, W. N. 1958. The introduced reindeer of South Georgia. Falkland Islands Dependencies Survey Sci. Rept. 22: 1—8.

Bos, G. N. 1967. Range types and their utilization by muskox on Nunivak Island — A reconnaissance study. Univ. of Alaska, M.S. Thesis.

Brooks, J., and W. Collins. 1985. Snow cover and interpretation of vegetation habitat inventories. p. 203—210. *In*: V. J. LaBau, and C. L. Kerr (eds.) Inventorying forest and other vegetation of the high latitude and high altitude regions. Proc. Intern. Symp. Amer. For. Reg. Conf. Soc. Amer. For. Washington, D.C.

Brown, J., and R. L. Berg (eds.) 1980. Environmental engineering and ecological baseline investigations along the Yukon River — Prudhoe Bay Haul Road. CRREL Report 80-19. Hanover NH.

Brown, R. W., R. S. Johnston, and K. Van Cleve. 1978. Rehabilitation problems in Alpine and Arctic regions. p. 23—44. *In*: Reclamation of drastically disturbed land. Soil Sci. Soc. Amer., Madison, Wisc.

Bryant, J. P., F. S. Chapin III, and D. R. Klein. 1983. Carbon/nutrient balance of boreal plants in relation to vertebrate herbivory. Oikos 40: 357—368.

Bunnell, F. L., S. F. Maclean, and J. Brown. 1975. Barrow, Alaska, USA. p. 73—124. *In*: T. Rosswall, and O. W. Heal (eds.) Structure and functions of tundra ecosystems. Stockholm. Swedish Natural Science Research Council, Ecol. Bul. 20.

Callaghan, T. V., and U. Emanuelsson. 1985. Population structure and processes of tundra plants and vegetation. p. 399—349. *In*: J. White, (ed.) The population structure of vegetation. Dr. W. Junk, Dordrecht.

Cargill, S. M., and R. L. Jefferies. 1984a. The effects of grazing by lesser snow geese on the vegetation of a sub-Arctic salt marsh. J. Applied Ecol. 21: 669—686.

Cargill, S. M., and R. L. Jefferies. 1984b. Nutrient limitation of primary production in a sub-Arctic salt marsh. J. Applied Ecol. 21: 657—668.

Chapin, F. S. III. 1980. Effect of clipping upon nutrient status and forage value of tundra plants in Arctic Alaska. p. 19—25. *In*: E. Reimers, E. Gaare, and S. Skjenneberg (eds.) Proc. 2nd Intern. Reindeer/Caribou Symp, Direktoratet for vilt og ferskvannsfisk, Trondheim, Norway.

Chapin, F. S., III, D. A. Johnson, and J. D. McKendrick. 1980. Seasonal movement of nutrients in plants of differing growth form in an Alaskan tundra ecosystem: Implications for herbivory. J. Ecol. 68: 189—209.

Chapin, F. S., III, J. D. McKendrick, and D. A. Johnson. 1986. Seasonal changes in carbon fractions in Alaskan tundra plants of differing growth form: implications for herbivory. J. Ecol. 74: 707—731.

Chapin, F. S., and G. R. Shaver. 1985. Individualistic growth response of tundra plant species to environmental manipulations in the field. Ecology 66: 564—576.

Chapin, F. S., III, K. Van Cleve, and M. C. Chapin. 1979. Soil temperature and nutrient cycling in the tussock growth form of *Eriophorum vaginatum*. J. Ecol. 67: 169—189.

Chernov, Y. I. 1980. The living tundra. English translation from the Russian 1985. Cambridge Univ. Press.

Cody, W. J. 1964. Reindeer range survey, 1957—1963. Can. Depart. of Agr., Plant Research Inst., Central Experimental Farm, Ottawa.

Collins, W. B. 1986. Control of herd movements and distribution: The foundation of reindeer husbandry. p. 111—120. *In*: O. Gudmundsson (ed.) Grazing Research at Northern Latitudes. Plenum Press, New York.

Craighead, J. J., J. S. Sumner, and G. B. Scaggs. 1982. A definitive system for analysis of grizzly bear habitat and other wilderness resources. Wildlife-Wilderness Inst. Monogr. No. 1, University of Montana, Missoula.

Derksen, D. V., W. D. Eldridge, and M. W. Weller. 1982. Habitat ecology of Pacific Black Brant and other geese moulting near Teshekpuk Lake, Alaska. Wildfowl 33: 39—57.

Deyeva, N. M. 1986. (Phytomass quantity and structure in the mountain tundras of the north-west of the Putorana Plateau.) Bot. Zhurnal 71: 789—794. (In Russian)

Everett, K. R. 1982. Some recent trends in the physical and chemical characterization and mapping of tundra soils, Arctic Slope of Alaska. Soil Science 133: 264—280.

Frank, T. D., and S. A. Isard. 1986. Alpine vegetation classification using high resolution aerial imagery and topoclimatic index values. Photogram. Eng. and Rem. Sens. 52: 381—388.

Gaare, E., and T. Skogland. 1975. Wild reindeer food habits and range use at Hardangervidda. p. 195—205. *In*: F. E. Wielgolaski (ed.) Fennoscandian Tundra ecosystems, Part 2. Springer Verlag, New York.

Garmo, T. H. 1986. Mountain pasture improvement by shrub clearing and fertilization. p. 87—98. *In*: O. Gudmundsson (ed.) Grazing research at northern latitudes. Plenum Press, New York.

George, T. H., W. J. Stringer, J. E. Preston, W. R. Fibiel, and P. C. Scorup. 1977. Reindeer range inventory in western Alaska from computer-aided digital classification of LANDSAT data. 28th AK Science Conf., Anchorage, Alaska.

Gorham, E. 1959. A comparison of lower and higher plants as accumulators of radioactive fallout. Can. J. Bot. 37: 327—329.

Gudmundsson, O., and R. E. Bement. 1986. Grazing intensity and balancing animal numbers with forage resources: sheep responses under subarctic conditions. p. 311—322.

In: O. Gudmundsson (ed.) Grazing research at northern latitudes. Plenum Press, New York.

Hall, D. K., J. Brown, and L. Johnson. 1978. The 1977 tundra fire in the Kokolik River area of Alaska. Arctic 31: 54—58.

Hall, D. K., J. P. Ormsby, L. Johnson, and J. Brown. 1980. Landsat digital analysis of the initial recovery of burned tundra at Kokolik River, Alaska. Remote Sensing of Environment 10: 263—272.

Hanson, W. C., H. E. Palmer, and B. I. Griffin. 1964. Radioactivity in northern Alaskan Eskimos and their foods, summer 1962. Health Physics 10: 421—429.

Haukioja, E. 1980. On the role of plant defences in the fluctuation of herbivore populations. Oikos 35: 202—213.

Hoffmann, R. S. 1974. Terrestrial vertebrates. p. 475—570. *In*: J. D. Ives, and R. G. Barry (eds.) Arctic and Alpine environments. Methuen, London.

Hoffmann, R. S. 1984. Small mammals in winter. The effects of altitude, latitude and geographic history. p. 9—23. *In*: J. E. Merritt (ed.) Winter ecology of small mammals. Carnegie Mus. Spec. Publ. No. 10.

Holts, S. 1980. Vegetation patterns and effects of grazing on caribou ranges in the sondre Stromfjord area, West Greenland. p. 57—63. *In*: E. Reimers, E. Gaare, and S. Skjenneberg (eds.) Proc. 2nd Int. Reindeer/Caribou Symp. Direktoratet for Vilt og Ferskvannsfisk, Trondheim.

Irving, L., G. C. West, L. J. Peyton, and S. Paneak. 1967. Migration of willow ptarmigan in Arctic Alaska. Arctic 20: 77—85.

Ives, J. D. 1970. Arctic tundra: how fragile? A geomorphologist's point of view. p. 39—42. *In*: E. K. Hare (ed.) The Tundra environment. Royal Soc. Canada. Univ. Toronto Press, Toronto.

Janzen, D. H. 1979. New horizons in the biology of plant defenses. p. 331—348. *In*: G. A. Rosenthal, and D. H. Janzen (eds.) Herbivories: Their interaction with secondary plant metabolities. Academic Press, New York.

Jenkins, J. F., and D. H. Ashton. 1970. Productivity studies on Macquarie Island vegetation. p. 951—863. *In*: M. W. Holdgate (ed.) Antarctic ecol. vol. 2. Academic Press, London.

Johnson, D. A., and L. L. Tieszen. 1976. Aboveground biomass allocation, leaf growth, and photosynthesis patterns in tundra plant forms in Arctic Alaska. Oecologia (Berl.) 24: 159—173.

Jonasson, S., J. P. Bryant, F. S. Chapin, and M. Andersson. 1986. Plant phenols and nutrients in relation to variations in climate and rodent grazing. Am. Nat. 128: 394—408.

Jung, H. G., G. O. Batzli, and D. S. Seigler. 1979. Patterns in the phytochemistry of Arctic plants. Biochem. System. Ecol. 7: 203—209.

Kaikusalo, A., and J. Tast. 1984. Winter breeding of microtine rodents at Kilpisjarvi, Finnish Lapland. p. 243—252. *In*: J. Merritt (ed.) Winter ecology of small mammals. Carnegie Mus. Spec. Publ. 10.

Kelsall, J. P. 1968. The migratory barren ground caribou of Canada. Ottawa. Can. Wild. Serv.

Kistchinsky, A. A. 1974. The moose in north-east Siberia. Nat. Can. 101: 179—184.

Klein, G. J., D. C. Pearce, and L. W. Gold. 1950. Method of measuring the significant characteristics of a snowcover. Nat. Res. Coun. Canad. Tech. Mem. 18.

Klein, D. R. 1970. Tundra ranges north of the boreal forest. J. Range Manage. 23: 8—14.

Klein, D. R. 1982a. Factors influencing forage quality for reindeer and caribou. p. 383—393. Wildlife-Livestock Symposium Proc. Univ. Idaho, Forest, Wildlife & Range Exp. Sta.

Klein, D. R. 1982b. Fire, lichens, and caribou. J. Range Manage. 35: 390—395.
Klein, D. R. 1986. Latitudinal variation in foraging strategies. p. 237—246. *In*: O. Gudmundsson (ed.) Grazing research at northern latitudes. Plenum. New York.
Klein, D. R. 1987. Vegetation recovery patterns following overgrazing by reindeer on St. Matthew Island. J. Range Manage. 40: 336—338.
Klein, D. R. (*In Press*) Establishment of muskox populations by translocation. Ceasar Kleberg Wildlife Research Institute and Wisconsin Humane Soc.
Klein, D. R., and V. Kuzyakin. 1982. Distribution and status of wild reindeer in the Soviet Union. J. Wildl. Manage. 46: 728—733.
Kuropat, P., and J. P. Bryant. 1980. Foraging behavior of cow caribou on the Utukok calvong grounds in northwestern Alaska. p. 64—70. *In*: E. Reimers, E. Gaare, and S. Skjenneberg (eds.) Proc. 2nd. Intern. Reindeer/caribou Symp. Direktoratet for vilt og Ferskvannsfisk, Trondheim.
Laine, K., and H. Henttonen. 1983. The role of plant production in microtine cycles in northern Fennoscandia. Oikos 40: 407—418.
LaPerriere, A. J., and P. C. Lent. 1977. Caribou feeding sites in relation to snow characteristics in northeastern Alaska. Arctic 30: 101—108.
Leader-Williams, N., T. A. Scott, and R. M. Pratt. 1981. Forage selection by introduced reindeer on South Georgia, and its consequences for the flora. J. Appl. Ecol. 18: 83—106.
Lent, P. C. 1966. The caribou of northwestern Alaska. Ch. 19. *In*: N. J. Willimovsky, and J. N. Wolfe (eds.) Environment of the Cape Thompson Region, Alaska, U.S. Atomic Energy Comm., Washington, D.C.
Lent, P. C. 1980. Synoptic snowmelt patterns in Arctic Alaska in relation to caribou habitat use. p. 71—76. *In*: E. Reimers, E. Gaare, and S. Skjenneberg (eds.) Proc. Second Intern. Reindeer/Caribou Symp. Direktoratet for Vilt og Verskvannsfisk, Trondheim.
Lent, P. C. 1985. Cold region vegetation information needs from the perspective of wildlife and fisheries. p. 20—27. *In*: V. J. LaBau, and C. L. Kerr (eds.) Inventorying forest and other vegetation of the High Latitude and High Altitude Regions. Proc. Intern. Symp. Amer. For. Reg. Conf. Soc. Amer. For., Washington, D.C.
Lent, P. C. 1987a. Arctic and alpine habitats. p. 149—168. *In*: A. Cooperrider, and R. Boyd (eds.) Inventory and monitoring of wildlife habitats. U.S. Gov't Printing Office, Wash. D.C.
Lent, P. C. 1987b. *Ovibos moschatus*. Mammalian Species. 302: 1—9.
Lent, P. C., and D. Knutsen. 1981. Muskox and snow cover on Nunivak Island, Alaska. p. 50—62. *In*: A. Haugen (ed.) Proc. Snow and Ice in Relation to Wildlife and Recreation Symp. Iowa State Univ., Ames.
Lent, P. C., and A. J. LaPerriere. 1974. Applications of ERTS imagery to the study of caribou movements and winter habitat. Final Rept. NASA Contract NAS521833. NTIS, Springfield, VA.
Liden, Kurt. 1961. Cesium 137 burdens in Swedish Laplanders and reindeer. Acta Radiol. 56: 237—240.
Linell, K. A., and J. C. F. Tedrow. 1981. Soil and permafrost surveys in the Arctic. Clarendon Press, Oxford.
McKendrick, J. D. 1981. Responses of Arctic tundra to intensive muskox grazing. Agroborealis 13: 49—55.
McNaughton, S. J. 1984. Grazing lawns: Animals in herds, plant form, and coevolution. Amer. Nat. 124: 863—886.
MacLean, S. F., Jr., and T. S. Jensen. 1985. Food plant selection by insect herbivores in Alaskan arctic tundra: the role of plant life form. Oikos 44: 211—221.
Miller, F. L., E. J. Edmonds, and A. Gunn. 1982. Foraging behaviour of Peary caribou in response to springtime snow and ice conditions. Canad. Wild. Serv. Occas. Pap. No. 48.

Mitchell, W. W. 1981. Alaska-developed grass varieties coming into use. Agroborealis 13: 56—58.

Mould, E. 1979. Seasonal movements related to habitat of moose along the Colville River, Alaska. Murrelet 60: 6—11.

Nodler, F. A., A. J. LaPerriere, and D. R. Klein. 1978. Vegetation type mapping in northwestern Alaska in relation to caribou and reindeer range potentials. AK Coop. Wildlife Research Unit Spec. Rept. No. 2. Univ. of Alaska Fairbanks AK.

Ostbye, E. (ed.) 1975. Hardangervidda, Norway. p. 225—264. *In*: T. Rosswall, and O. W. Heal (eds.) Structure and function of tundra ecosystems. Stockholm.

Palmer, L. J., and C. H. Rouse. 1945. Study of the Alaska tundra with reference to its reactions to reindeer and other grazing. U.S. Fish and Wild. Serv. Res. Rept. 10: 1—48.

Parker, G. R. 1978. The diets of muskoxen and Peary caribou on some islands in the Canadian High Arctic. Can. Wildl. Serv., Occ. Pap. No. 35.

Parker, G. R., D. C. Thomas, E. Broughton, and D. R. Gray. 1975. Crashes of muskox and Peary caribou populations in 1973—1974 on the Parry Islands, Arctic Canada. Can. Wildl. Serv., Prog. Notes No. 56.

Parker, G. R., and R. K. Ross. 1976. Summer habitat use by muskoxen *Ovibos moschatus* and Peary caribou (*Rangifer tarandus pearyi*) in the Canadian High Arctic. Polarforschung 46: 12—25.

Patterson, W. A., III, and J. G. Dennis. 1981. Tussock replacement as a means of stabilizing fire breaks in tundra vegetation. Arctic 34: 188—189.

Pegau, R. E. 1970. Effect of reindeer trampling and grazing on lichens. J. Range Manage. 23: 95—97.

Pewé, T. 1983. Alpine permafrost in the contiguous United States: A review. Arctic and Alpine Research 15: 145—156.

Pitelka, F. A. 1979. An avifaunal review for the Barrow region and North Slope of Arctic Alaska. Arctic and Alpine Research 6: 161—184.

Pruitt, W. O., Jr. 1959. Snow as a factor in the winter ecology of barren-ground caribou. Arctic 12: 159—179.

Pruitt, W. O., Jr. 1966. Ecology of terrestrial mammals. p. 519— 564. *In*: Environment of the Cape Thompson Region, Northwest Alaska. U.S. Atomic Energy Commission. U.S. Gov't. Printing Office.

Pruitt, W. O., Jr. 1978. Boreal Ecology. Edward Arnold Publ. Ltd., London.

Puckett, K. J., E. Nieboer, W. P. Flora, and D. H. S. Richardson. 1973. Sulfur dioxide: Its effects on photosynthetic ^{14}C fixation in lichens and suggested mechanisms of phytotoxicity. New Phytol. 72: 141—154.

Racine, C. H. 1981. Tundra fire effects on soils and three plant communities along a hill-slope gradient in the Seward Peninsula, Alaska. Arctic 34: 71—84.

Reeves, H. M., F. G. Coach, and R. E. Munro. 1976. Monitoring arctic habitat and goose production by satellite imagery. J. Wildl. Manage. 40: 532—541.

Rhoades, D. F. 1985. Offensive-defensive interactions between herbivores and plants: Their relevance in herbivore population dynamics and ecological theory. Amer. Natural. 125: 205—235.

Richens, V. B., and C. G. Madden. 1973. An improved snow study kit. J. Wildl. Manage. 37: 109—113.

Rosswall, T., and O. W. Heal. (eds.) 1975. Structure and Function of Tundra Ecosystems. Stockholm.

Russell, D. E., and A. M. Martell. 1984. Winter range ecology of caribou (*Rangifer tarandus*). p. 117—144. *In*: R. Olson *et al.* (eds.) Northern ecology and resource management. Univ. Alberta Press, Edmonton.

Scotter, G. W. 1967. The winter diet of barren-ground caribou in northern Canada. Can. Field-Natur. 81: 33—39.

Sims, R. A. 1983. Ground-truth and large scale 70 mm aerial phtographs in the study of reindeer winter rangeland, Tuktoyaktuk Peninsula area, NWT. PhD. Thesis. Univ. of British Columbia, Vancouver.

Skogland, T. 1978. Characteristics of the snow cover and its relationship to wild mountain reindeer (*Rangifer tarandus tarandus* L.) feeding strategies. Arctic and Alpine Research 10: 569—580.

Skogland, T. 1980. Comparative summer feeding strategies of Arctic and alpine *Rangifer*. J. Animal Ecol. 49: 81—98.

Skogland, T. 1984. Wild reindeer foraging-niche organization. Holarctic Ecol. 7: 345—379.

Skogland, T. 1986. High radio-cesium contamination of wild reindeer from southern Norway following the Chernobyl accident. Appendix. p. 72. Rangifer No. 1.

Thilenius, J. F. 1979. Range management in the alpine zone. p. 43—64. *In*: D. A. Johnson (ed.) Special management needs of Alpine ecosystems. Range Science Ser. Society for Range Management, Denver.

Thing, H. 1984. Feeding ecology of the west Greenland caribou (*Rangifer tarandus groenlandicus* in the Sisimiut-Kangerlussuaq Region. Dan. Rev. Game Biol. 12: 1—53.

Thomas, D. C., and P. Kroeger. 1981. Digestibility of plants in ruminal fluids of barren-ground caribou. Arctic 34: 321—324.

Thomas, D. C., F. L. Miller, R. H. Russell, and G. R. Parker. 1981. The bailey point region and other muskox refugia in the Canadian Arctic: A short review. Arctic 34: 34—36.

Thorn, C. E. 1982. Gopher disturbance: Its variability by Braun-Blanquet vegetation units in the Niwot Ridge alpine tundra zone, Colorado Front Range, U.S.A. Arctic and Alpine Research 14: 45—51.

Thorsteinsson, I. 1986. The effect of grazing on stability and development of northern rangelands: A case study of Iceland. p. 37—44. *In*: O. Gudmundsson (ed.) Grazing research at northern latitudes. Plenum Press, New York.

Tissue, D. T., and W. C. Oechel. 1987. Response of *Eriophorum vaginatum* to elevated CO_2 and temperature in the Alaskan tussock tundra. Ecology 68: 401—410.

Ulrich, A., and P. L. Gersper. 1978. Plant nutrient limitations of tundra plant growth . p. 457—481. *In*: L. L. Tieszen (ed.) Vegetation and production ecology of an Alaskan Arctic tundra. Springer Verlag, New York.

Van Cleve, K. 1977. Recovery of disturbed tundra and taiga surfaces in Alaska. p. 422—455. *In*: J. Cairns, K. L. Dickson, and E. Herricks (eds.) Proc. of an International Symp. on the Recovery of Damaged Ecosystems. Virginia Polytechnic Inst. and Univ., Blacksburg, VA.

Vibe, C. 1967. Arctic animals in relation to climatic fluctuations. Meddel. om Grønland 170(5): 1—227.

Walker, D. A., P. J. Webber, K. R. Everett, and J. Brown. 1978. Effects of crude and diesel oil spills on plant communities at Prudhoe Bay, Alaska, and the derivation of oil spill sensitivity maps. Arctic 31: 242—258.

Walker, D. A., K. R. Everett, P. J. Webber, and J. Brown. 1980. Geobotanical Atlas of the Prudhoe Bay Region, Alaska. CRREL Rept. 80—14: 1—69. U.S. Army Corps of Engineers.

Walker, D. A., P. J. Webber, M. D. Walker, N. D. Lederer., R. H. Meehan, and E. A. Nordstrand. 1986. Use of geobotanical maps and automated mapping techniques to examine cumulative impacts in the Prudhoe Bay oilfield, Alaska. Environ. Conserv. 13: 149—160.

Watson, D. G., W. C. Hanson, and J. J. Davis. 1964. Strontium-90 in plants and animals of Arctic Alaska, 1959—1961. Science 144: 1005—1009.

Wein, R. W. 1976. Frequency and characteristics of Arctic tundra fires. Arctic 29: 213—222.

White, R. G., and J. Trudell. 1980. Habitat preference and forage consumption by reindeer and caribou near Atkasook, Alaska. Arctic and Alpine Res. 12: 511—529.

Wielgolaski, F. E. 1975a. Productivity of tundra ecosystems. p. 1—12. *In*: Productivity of world ecosystems. Proc. of a Symposium. National Academy of Sciences. Washington, D.C.

Wielgolaski, F. E. 1975b. Comparison of plant structure on grazed and ungrazed tundra meadows. p. 86—93. *In*: F. E. Wielgolaski (ed.) Fennoscandian Tundra Ecosystems. Part 1: Plants and microorganisms. Springer-Verlag, New York, Berlin.

Wiersema, G., and W. Schroder. 1985. How to find suitable ibex (*Capra ibex L.*) habitat using Landsat imagery. p. 226—230. *In*: S. Lovari (ed.) Biology and management of mountain ungulates.

Willard, B. E., and W. Marr. 1971. Recovery of alpine tundra under protection after damage by human activities in the Rocky Mountains of Colorado. Biol. Cons. 3: 181—190.

Williams, T. M., and D. C. Heard. 1986. World status of wild *Rangifer tarandus* populations. p. 19—28. *In*: A. Gunn, F. Miller, and S. Skjenneberg (eds.) Proc. 4th Intern. Reindeer/Caribou Symp. Rangifer, Spec. Issue 1.

14. Forest rangeland relationships

BARBARA H. ALLEN

14.1 Abstract

Forest rangeland relationships are complex with different forest ecosystems providing an array of potential understory forage species for wild and domestic herbivore utilization. Specific examples from forest ecosystems around the world are used to illustrate principle overstory-understory relationships and management principles. Generally, forest ecosystems can be categorized into two major types, that are structurally similar, even though specific tree and forage species change. In forest ecosystems understory species cover generally declines with increasing tree crown cover, and in many forest ecosystems there is potential to use livestock grazing as a tool in forest management. Field application of the principles described in this chapter must be at the site-specific level, however. Although, livestock grazing on forest and transitory rangelands is a history clouded by general mismanagement of grazing animals, livestock can utilize available forage on these lands, without damage to tree regeneration, and may in the future become more important as a tool in forest management.

14.2 Description of the forest rangeland resource

Forests cover about 39 percent of the earth's surface, with northern coniferous forests occupying some 23.2 million square km or about 17 percent of the earth's land surface (Walter 1979). Coniferous forests are diverse, some comprised of multiple tree species at maturity, while others exhibit single tree species dominance. Many of these forests include high quality, palatable, diverse understory forage species which are consumed by large and small herbivores including domestic livestock.

Coniferous forests form a continuous belt in the northern hemisphere

Tueller, P. T. (ed.), Vegetation science applications for rangeland analysis and management.
ISBN 90-6193-195-9.
© 1988, *Kluwer Academic Publishers, Dordrecht, Boston, London.*

across North America and Eurasia, and also occur in high mountain areas in the tropics (Odum 1959). This forest region is characterized by a climate generally unfavorable to deciduous trees, with short summers and long, cold winters. In North America and Asia, many species of conifers are present, in the genera *Picea, Pinus, Abies, Larix, Tsuga* and *Thuja*. A distinctive moist type of northern coniferous forest can be found along the coast of North America from Alaska to California. In this region, temperatures are higher, and seasonal temperature fluctuations are lower, allowing for development of the magnificent redwood (*Seqouia sempervirens*) forest in the south, to Douglas-fir (*Psuedotsuga menziesii*) and western hemlock (*Tsuga mertensiana*) forests along the Pacific coast of Washington, Oregon and British Columbia, to spruce (*Picea*) forests in Alaska. In Europe, however, spruce and pine are the only conifer genera, and spruce drops out to be replaced by larch in Siberia.

Conifers are generally considered successional species in the temperate deciduous forest regions of the world. The temperate deciduous forest region occupies 7.4 million square km or about 6 percent of the earth's land surface, and has been highly modified by human settlement and manipulation, with many areas converted to agriculture or to coniferous forest through periodic fire and/or intensive forest management (Walter 1979, Odum 1959). These conifer forests in eastern North America, Europe, Japan, Australia and the tip of South America, are long-lived and maintained by periodic fire and human management. For example, the spruce forests of Germany are maintained through forest management in an area once occupied by oak forests (Walter 1979), while much of the southern pine forests in the southeastern United States are maintained by fire and intensive forest management, and are considered successional to the oak-hickory-sweetgum forests (Smith 1966, Odum 1959).

The conifer forests of the world can be labelled *forest rangeland* or *transitory rangeland*. Forests that produce forage throughout their life cycle are forest rangeland, while forests that produce forage only during the early stages of succession (generally less than 15 yrs) are transitory rangeland. Forest rangelands are usually drier forest types which may provide spring/fall and summer forage. Their characteristic open stand structure generally allows for continuous understory forage production and utilization. Transitory rangelands are generally higher elevation and/or more mesic forests, often grazed primarily in the summer. Grazing occurs in openings, streamsides, meadows, along roads and on cutover and burned over land early in the successional life-cycle of the forest.

Wood production usually dominates as a management goal on both forest and transitory rangeland types. Grazing use must be combined with timber, fuel, recreation, watershed and wildlife habitat objectives in the forest. Sometimes, timber and livestock production goals are much more integrated in an intensive agronomic system, unlike the wildland forest production system where livestock grazing has traditionally been incidental or unmanaged. For example, in New Zealand, monterey pine (*Pinus radiata*) forests are managed for both timber production and grazing in an agroforestry production system (Reid and Wilson 1985). Wildland forest grazing is the focus of this chapter, rather than agroforestry systems.

This chapter describes the major coniferous forest and transitory rangeland ecosystems by region, and concludes with a discussion of current management concerns and suggestions for compatible grazing/timber management. Although all temperate regions throughout the world have grazed forests, most examples of conifer forest grazing practice come from the United States and Canada. Exclusion of grazing has been the general rule in European coniferous forests managed for timber production, and other regions have produced little published research. The principles of forest grazing and the forest rangeland relationships apply to ecologically similar forest sites throughout the world.

14.3 Forest rangeland types

14.3.1 Description of types

Forest rangeland types are forest ecosystems that generally retain some quantity of useable forage throughout their lifecycle, although forage quantity and sometimes quality declines with increasing forest cover. The widespread, ecologically diverse pine forests of North America are the primary forest rangeland type because of their characteristic open stand structure at maturity. Two major types of pine forests can be identified: the interior ponderosa pine (*Pinus ponderosa*) forests of the western U.S. and Canada and longleaf-slash pine (*P. palustris-P. elliotii*) forests of the southern United States.

14.3.1.1 Interior ponderosa pine forests. The interior ponderosa pine forest is commonly classified as SAF (Society of American Foresters) Type 237 (Eyre 1980). Ponderosa pine is the dominant tree species occurring in single or mixed species stands. The type occupies 11 million hectares in extensive, non-contiguous areas in western North America, ranging from British Columbia to Mexico, and the Plains states to the Sierra Nevada and Cascade mountains.

Because of its wide geographic and elevational distribution, associated species are diverse (Barrett *et al.* 1980). Douglas-fir, lodgepole pine (*P. contorta*), grand fir (*Abies grandiflora*) and western larch (*Larix* spp.) are main associates in Oregon, Washington, Idaho, Montana, and British Columbia while white fir (*A. concolor*), incense-cedar (*Libocedrus decurrens*), and Jeffrey pine (*P. jeffreyii*) are common associated species in California. In the southwestern United States, *Abies*, western white pine (*P. monticola*), gambel oak (*Quercus gambeli*) and aspen (*Populus tremuloides*) are locally abundant. At high elevations in the Rocky Mountains and Utah, Douglas-fir, blue spruce (*Picea pungens*), lodgepole pine, and aspen are frequent associated species, while at lower elevations ponderosa pine is most often found in pure stands or associated with pinyon pine (*P. edulis*) and juniper (*Juniperus* sp.) at the drier margins. In the Black Hills of South Dakota, aspen, paper birch (*Betula* sp.), bur oak (*Q. macrocarpa*) and white spruce (*P. glauca*) are common associates.

14.3.1.2 Longleaf-slash pine forests. The longleaf-slash pine (SAF Type 83) forest cover type is characterized by stands comprised primarily of longleaf pine and slash pine. For purposes of examining forest range relationships, the longleaf pine (SAF Type 70), slash pine (SAF Type 84), and South Florida slash pine (SAF Type 111) types are included in discussions of this type. These types occur on similar sites varying primarily in the relative dominance of slash or longleaf pine. This forest cover type is limited by the natural range of longleaf and slash pines to the Gulf and Atlantic coastal plains of the United States (Grelen 1980), occupying about 11 million hectares.

As with the interior ponderosa pine forest, the longleaf-slash pine forest occurs on a variety of soil types, moisture conditions, and topographic locations. Associated species vary by site, but commonly include a variety of hardwood and brush species such as gallberry (*Ilex* spp.), saw-palmetto (*Seronoa repens*), blackgum (*Nyssa sylvatica*), blueberries (*Vaccinium* spp.), blackberries (*Rhus* spp.), dogwood (*Cornus* spp.), post oak (*Q. stellata*) and blackjack oak (*Q. marilandica*). Turkeyoak (*Q. laevis*) is especially abundant in the Sandhills of Georgia and North and South Carolina. Herbaceous species include bluestem grasses (*Andropogon* spp.), *Panicum* spp., Curtis dropseed (*Sporobolus curtissii*), *Paspalum* spp., asters and tickclovers (Boyer 1980, Grelen 1980, Hodges 1980).

14.3.2 Forage production

In the 3.5 million ha of ponderosa pine forests of the southwestern United States, ponderosa pine in small even-aged groups may be virtually the only timber species present (Ffolliott 1983). Shrubs, such as mountain mahogany (*Cercocarpus betuloides*), gambel oak, and buckbrush (*Ceanothus fendleri*) play important roles in forage production. These and other shrubs, such as sagebrush (*Artemesia* spp.) and manzanita (*Arctostaphylos* spp.), may interfere with tree regeneration. Grasses predominate in the understory, with mountain muhly (*Muhlenbergia montana*), screwleaf muhly (*M. virescens*), junegrass (*Koeleria cristata*), and Arizona fescue (*Festuca arizonica*) bunchgrasses dominant in late successional stages and pine dropseed (*Sporobolus* spp.), black dropseed (*S. interruptus*), grama grasses (*Bouteloua* spp.) and mountain brome (*Bromus carinatus*) common in early successional stages (Currie 1975).

Open parklike ponderosa pine stands in the Oregon, Washington, and British Columbia region range from 1 to 20 ha in size with common understory forage species including bluebunch wheatgrass (*Agropyron spicatum*), Idaho fescue (*F. idahoensis*), junegrass, sandberg bluegrass (*Poa sandbergii*) and onespike danthonia (*Danthonia unispicata*).

Open pine stands of this region contain elk sedge (*Carex geyeri*) which declines with increased grazing pressure, pinegrass (*Calamagrostis rubens*), Idaho fescue and junegrass. These forest ranges also support bitterbrush (*Purshia tridentata*), a highly palatable browse, as well as sagebrushes, mountain mahogany and manzanitas (Skovlin 1976). In wetter areas, Douglas-fir is codominant with ponderosa pine, with elk sedge and the more shade tolerant grasses in the understory.

Ponderosa pine-bunchgrass range averages 1000 kg/ha forage production, while the pine-elk sedge areas average 800 kg/ha. Dense stands in Arizona and Colorado may provide 50 to 75 kg/ha, while open areas between stands may produce 1400 kg/ha (Clary *et al.* 1974). Patton (1974) found open pine stands producing 600 kg/ha while dense stands produced only 80 kg/ha. Similarly Arnold (1950) found open canopy stands produced an average of 900 kg/ha. McConnell and Smith (1970) thinned ponderosa pine stands in eastern Washington and increased production from 70 kg/ha to 417 kg/ha (Fig. 2). In the open pine stands of the Black Hills of South Dakota, production averaged 1300 kg/ha while in stands with 65% tree cover, production fell to 60 kg/ha (Pase 1958).

In the longleaf-slash pine type, there are two main understory forage types. The pine-threeawn or wiregrass (*Aristida stricta*) type is most common in the states of Georgia and Florida. Wiregrass is a wiry, fire tolerant bunchgrass which provides good forage for grazing cattle from March through June. Associated species include sawpalmetto, curtis dropseed, bluestems, *Panicum* spp., and carpetgrass (*Axonopus affinis*). Stocking rates are commonly 0.6 to 2 ha per AUM (animal unit month), and grazing has traditionally been open, and year round (Hall *et al.* 1964).

The longleaf-slash pine-bluestem type has even higher grazing capacity at 0.4 to 0.8 ha per AUM, but retains high forage quality only during the early spring (March through May) and provides poor quality forage from June through March. The pine-bluestem type is most common in Louisiana, Mississippi, and Alabama (Campbell 1947). Primary forage species include pinehill bluestem (*A. divergens*) and slender bluestem (*A. tener*).

14.3.3 Changes in forage production over time

Forage production varies by site but is generally inversely proportional to tree cover in the southwestern ponderosa pine forest (Fig. 1). Although overall forage production declines with increasing canopy cover, species composition does not change with increasing tree cover. However, species composition does change due to grazing (Arnold 1950). Typically, grazing capacity declines with range deterioration as plant composition changes from bunchgrasses to annuals and weeds (Pearson 1950).

Timber, forage and fire interactions are poorly understood in the Arizona/Colorado ponderosa pine type. Timber production tends to be the primary goal, but whether timber regeneration and grazing are compatible or not is hotly debated. The value of fire is also widely debated. Fire is considered bad because it kills trees (Pearson 1931), however removal of fire has resulted in thickets of poor quality timber. Removal of fire has also resulted in a reduction in grazing capacity. Grazing, on the other hand, may improve tree regeneration by relieving grass competition, and damage to regenerating pine seedlings may be negligible with light to moderate grazing and adequate water. Clary *et al.* (1974) has found the maximum combined production value for beef and sawtimber on the ponderosa pine-bunchgrass range to be 45 to 60 square feet of basal area.

Forage production is also inversely proportional to overstory canopy cover (Fig. 2) in the northern and eastern interior ponderosa pine. Fire is also an important management tool in this region. With removal of fire, ponderosa pine is replaced by Douglas and white fir. Canopy cover then increases, elk sedge and pinegrass drop out and AUMs decline (Hall 1983). Current Forest Service stocking guides in Region 6, the Pacific Northwest Region, recommend thinning pine stands to levels which increase the growth rate of the trees, while at the same time increasing forage. Although, timber production and forage use appear to be compatible in this region the interactions are not simple and again managers must pay close attention to type of vegetation competition, soil moisture regimes, planting stock and timing of grazing.

Forage production in the southeast is also inversely related to tree cover, commonly declining 14 kg/ha for each square foot increase in tree canopy cover (Pearson 1974), though grasses are rarely eliminated.

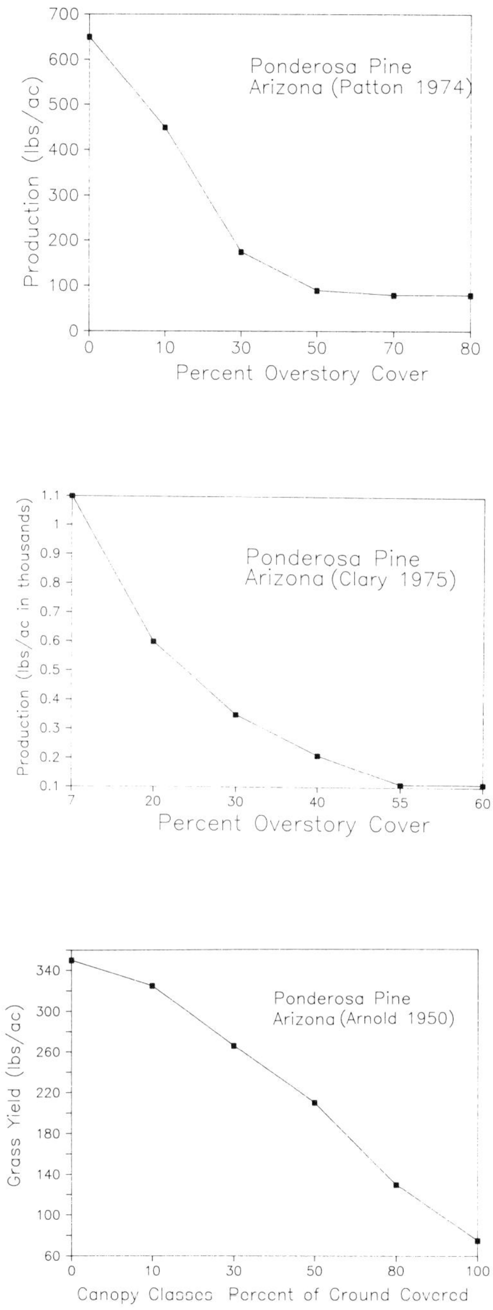

Figure 1. Inverse relationships between ponderosa pine overstory and understory production from three areas in Arizona. Some forage is available throughout the forest rotation. These data have been taken directly from Clary (1975), Patton (1974) and Arnold (1950). Note differences in scale and parameters measured.

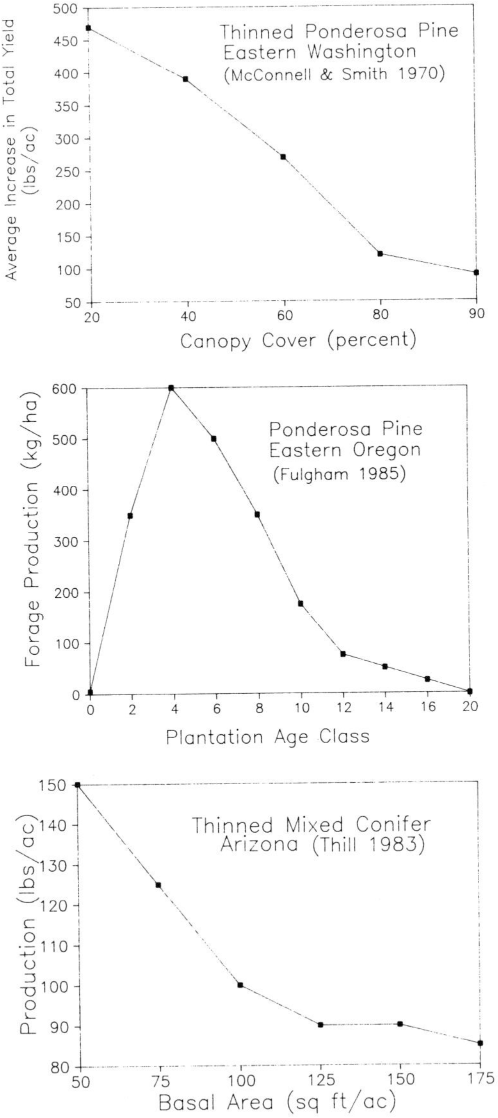

Figure 2. Inverse relationship between northern interior ponderosa pine and crown cover or plantation age class. Forage becomes unavailable only in the plantation situation before thinning. These data are taken directly from Pase (1958), Fulgham (1985) and McConnell and Smith (1970). Note differences in scale and parameters measured.

14.3.4 Historical/current use

The ponderosa pine forests of the interior were part of the traditional large ranching industry that developed in the western United States in the

1800s. Forests and rangelands were grazed in common, with large areas of open public land controlled by ranchers who privately owned and controlled scarce water sources. The result was overuse of large areas of forest rangeland by cattle and itinerant sheep herds. This kind of use continued until about 1906 when large areas of forest came under regulation by the United States government as forest reserves (Kosco and Bartolome 1981, Dana and Fairfax 1980). Today ponderosa pine forests of the U.S. contain a net standing volume of approximately 192 billion board feet of sawtimber (USDA 1986), with approximately 2.4 million AUMs of forage utilized during a spring/fall and summer grazing season (USDA 1984).

The historic overuse in the Arizona/Colorado region, resulted in a reduction in bunchgrasses and increases in grasses less susceptable to grazing (Currie *et al.* 1978). Today grazing continues to occur during the spring/fall seasons at lower elevations and summer at higher elevations. English breeds such as Herford, Angus and Shorthorns continue to be the traditional breeds. The region is characterized by two periods of precipitation, one summer and one winter, unlike the ponderosa pine forests farther to the north.

These northern ponderosa pine forests, in eastern Oregon, California, Washington and British Columbia, like their counterpart to the south, suffered historic grazing abuse. Hereford, Angus, Brangus and other English breeds graze this region as spring/fall and summer range. The open pine and parklike grassland areas were historically the most heavily used and currently stocking rates are keyed toward improving these areas.

The southeastern United States produces 25% of all the cattle and calves in the United States, 33% of all beef cows and heifer replacements and 3% of the sheep (USDA 1976). Timber production is currently a major use of the pinelands, producing half of the nation's wood products (Byington *et al.* 1983). Grazing and crop production has varied in its importance, but has occurred since colonial times. The grazing history has been one of unmanaged grazing of poor stock on cutover conifer and hardwood forest lands. Initially herds of cattle were maintained with minimum investment in capital and labor resulting in minimum profit. Farmers grazed warm weather breeds such as Brahman, as well as some purebred English breeds. Livestock management continues to be extensive (Byington *et al.* 1983), while forage production remains very heterogeneous both in distribution and kinds of vegetation.

Providing quality forage during the winter months is the greatest concern of cattle grazers in the southern pine types. Phosphorus is deficient yearlong, while protein is deficient in the winter. Good management calls for prescribed winter rotational burning to remove old, unpalatable material, improve quality, and maintain a diverse species composition (Halls *et al.* 1964). Native species appear to be adapted to such burning.

As timber production became increasingly economically important, protection of plantations from grazing damage required more intensive grazing management. Livestock owners strategically place supplemental feed to reduce and/or minimized plantation damage. Tree reproduction is also protected by controlling grazing intensity. Most damage occurs during the winter months when forage quality is poor and in the early spring before the grasses green-up. Although, Lewis (1980) reported that slash pine, loblolly and shortleaf pines are highly resistant to grazing damage, managers must continue to properly manage grazing animals as tree seedlings are the main green vegetation during the winter. Historically poor grazing management, and fluctuations in emphasis between timber and agriculture production have promoted the attitude by many that livestock grazing and timber production are incompatible. However, current management practices seek to integrate timber, livestock grazing and enhancement of wildlife habitat (Byington *et al.* 1983).

14.4 Transitory rangeland types

14.4.1 Description of types

Transitory rangeland types in coniferous forest ecosystems generally provide forage only early in forest succession, after fire, clearcutting or other forest disturbance. Timber production is commonly the primary objective in these ecosystems. The forests are diverse and this chapter will focus on 6 major transitory range types which provide summer forage for grazing and browsing animals.

14.4.1.1 Interior and Pacific Douglas-fir forests. The Douglas-fir forests of the Pacific slope (SAF Type 229) and the interior Rocky Mountain region (SAF Type 210) occupy approximately 35 million hectares, with dominance of Douglas-fir, often 80% of the stand, the primary common attribute (Walstad *et al.* 1987). The Pacific Douglas-fir type is one of the world's most productive forest types in volume of wood (Williamson 1980). Douglas-fir occupies a variety of sites, and may be climax or seral depending on moisture regimes and disturbance (Ryker and Steele 1980). It is relatively shade intolerant and requires fire or other disturbance to provide openings for regeneration. Where canopies exceed about 70 percent, other more shade tolerant species, such as western hemlock

(*Tsuga mertensiana*), western redcedar (*Thuja plicata*), and Pacific silver fir (*A. amabilis*) become established (Williamson 1980). Common associated tree species may include western hemlock, grand fir, western white pine and lodgepole pine in the interior region. White fir, Engelmann spruce (*P. engelmannii*) and ponderosa pine are common associates in the southwestern United States, while western hemlock, grand fir, silver fir and other fir species occur with Douglas-fir along the coast.

14.4.1.2 Spruce-fir forests. The spruce-fir forests occupy 9.7 million hectares in the interior and northern interior regions of the United States and Canada (Clary 1983). These forests are highly varied including 6 SAF forest cover types in the northern interior (boreal) region, 4 high elevation SAF types, and 3 middle elevation interior types (Eyre 1980). This type is characterized by several species of spruce and fir, which are locally dominant. The 3 middle elevation interior types, white fir (SAF Type 211), grand fir (SAF Type 213) and blue spruce (SAF Type 216) contain the major forage resource for this type.

14.4.1.3 Lodgepole pine forests. Lodgepole pine forests are one of the most extensive cover types in western North America, occupying 10 million ha (Bartolome 1983). Defined as SAF Type 218 (Eyre 1980), this type includes recognized varieties of *Pinus contorta*, particularly Rocky Mountain lodgepole pine, variety *latifolia*, Sierra lodgepole, variety *murrayana*, and short pine, variety *contorta*. Common associates include subalpine fir (*A. lasiocarpa*), Engelmann spruce, white spruce (*P. glauca*) and Douglas-fir. Other associates vary by location (Pfister and McDonald 1980). Understory species vary, but pinegrass, huckleberry and alder (*Alnus* spp.) provide forage for grazing animals.

14.4.1.4 Pacific ponderosa pine forests. Pacific ponderosa pine (SAF Type 245) is defined by nearly pure stands of ponderosa pine in a narrow elevational band that varies with latitude. The type is most extensive and best developed in the Sierra Nevada of California ranging from 300–600 m in elevation in the north to 1,065 m in the central Sierra Nevada. Douglas-fir, Oregon white oak (*Q. garreya*), bigleaf maple (*Acer macrophylla*), California black oak (*Q. kellogii*) and Pacific madrone (*Arbutus menziesii*) are common associates. Perennial grasses and forbs, annuals and shrubs, especially deerbrush (*Ceanothus integerrimus*), provide forage in clearcuts, along roads and in openings. Forage species are virtually absent in the climax Pacific ponderosa pine forest.

14.4.1.5 Sierran mixed conifer forests. The Sierra Nevada Mixed Conifer forest (SAF Type 243) is characterized by the presence of 5 conifers (ponderosa pine, sugar pine, Douglas-fir, white fir and incense cedar) and one hardwood species (California black oak), either together or in small interspersed groups of pure species. This 6 species mixture is common to the west side of the Sierra Nevada from about 915 to 1830 m elevation. The type occupies approximately 7 million ha (Larson and Wolters 1983), including areas of the Cascade mountains of Oregon, which are not strictly included in the SAF classification. Numerous shrubs and as many as 100 herbaceous species are found in this type. Forage species include deer-brush, bitter cherry (*Prunus emarginata*), *Stipa* spp., and other palatable grasses and forbs. Another 1 million ha of mixed conifer forest also occur in southwestern Colorado, western New Mexico, and eastern Arizona (Larson and Wolters 1983). The mixed conifer forest is a complex association with extreme variation in mixture of species.

14.4.1.6 Loblolly-shortleaf pine forests. The loblolly-shortleaf pine (*P. taeda-P. echinata*) type occupies 22 million ha along the Piedmont and upper coastal gulf plain (Thill 1983). It is the most important for softwood timber production in the region, and may occur in dense stands at maturity with sparse understory forage. Loblolly pine predominates on wetter soils and higher elevations, while shortleaf pine predominates on thinner, drier soils (Mann 1980). The type occurs at elevations from sealevel to 760 m. Associated species may include sweetgum (*Liquidamber styraciflua*), blackgum, southern red oak (*Q. rubra*), post oak, white oak (*Q. alba*) and mockernut hickory (*Carya tomentosa*) on well-drained soils; and sweetgum, red maple (*A. rubrum*), water oak (*Q. nigra*) and willow oak (*Q. phellos*) on moister soils. In unmanaged stands, loblolly-shortleaf forests gradually revert to an upland oak-hickory climax (Mann 1980).

14.4.2 Forage production

Forage production in the mature Douglas-fir forest is almost non-existent. Only shade-tolerant species, primarily composed of shrubs largely unpalatable to livestock, persist in the mature forest (Ingram 1931). However, within two years after fire or logging, palatable species such as fireweed, blackberry, maple, dogwood and grasses and forbs may appear (Ingram 1931). Common graminoid species include *Dactylis glomerata, Holcus lanatus*, and *Carex hendersonii* (Sharrow and Leininger 1982). Palatable shrubs include *Prunus emarginata.* In interior British Columbia, the

Douglas-fir forests may persist in more open stands with lodgepole, aspen and willow present with repetitive fire (Tisdale 1950). The aspen and willow are important in early successional stages. The principle herbaceous species is pinegrass which is dominant over large areas, commonly comprising 40—50% of the ground cover and providing 50—65% of the forage (Tisdale 1950). Other forage species include *Arnica cordifolia, Vaccinium* spp., brome grasses and *Stipa* spp.

Mature spruce-fir forests commonly have an associated understory comprised largely of *Vaccinium* spp., mosses and lichens which are of little importance to livestock (Tisdale 1950). However, the principal grazing value in the spruce-fir region lies in the treeless parks and aspen groves, as well as burned-over areas. Forage yield and grazing capacity vary greatly; however with favorable moisture, excellent forage supporting 1.5—2 ha/AUM is available. Common forage species in the interior Central Rockies into British Columbia may include *Stipa lettermanii, Poa reflexa, Phleum alpinum*, and *Trisetum spicatum* (Stoddart, Smith and Box 1975). The grazing season is short in this region, confined to about two months in mid summer.

Transitory vegetation of clearcut lodgepole pine-mixed conifer forest of Oregon includes *Sitanion hystrix, Carex rossii, Stipa occidentalis, Bromus marginatus*, and *Poa pratensis* (Monfore 1986). Shrub species include *Ceanothus velutinus, Arctostaphylos patula*, and *Purshia tridentata*. Forage production on these plantations, planted to ponderosa pine, range from 140 kg/ha on one-year old plantations to 647 kg/ha on 4 yr old plantations (Fulgham 1985).

McLean and Clark (1980) studied seeded and unseeded lodgepole pine-Engelmann spruce clearcuts in interior British Columbia. They found clearcuts seeded to orchard grass (*Dactylis glomerata*) and timothy (*Phleum pratense*) produced two to four times more herbage than the native pinegrass. Yields averaged between 590 kg/ha and 1540 kg/ha and continued to increase during the first five years following clearcutting. The authors estimate that benefits from seeding grasses may persist up to 20 years before the tree canopy closure (McLean and Clark 1980).

The Pacific ponderosa pine forest lies in a narrow band between the Douglas-fir and Sierra Nevada mixed conifer forests above and oak types below. Forage species include deerbrush, poison-oak (*Rhus diversiloba*), grasses and forbs in young forest stands. Grasses and annual species are virtually absent in mature stands, though palatable shrub species, such as deerbrush may persist for 15—20 years (Roy 1981). Production ranges from 5 ha/AUM in young stands to 50 ha/AUM in older or more dense stands.

The Sierra Nevada Mixed Conifer forest is characterized by a shrub understory interspersed with narrow stringer meadows. Browsed species include deerbrush, California black oak, curleaf mountain mahogany, hazelnut (*Corylus* spp.), willow (*Salix* spp.) and bitter cherry. Herbaceous species include orchardgrass, needlegrasses, and brome grasses, while diverse perennial grasses and forbs occupy the stringer meadows (Kosco and Bartolome 1983). Young clearcuts, occupied primarily by deerbrush may support up to 2.5 ha/AUM (Cronemiller 1953), while production may average 16 ha/AUM over a typical area interspersed with clearcuts of varying ages, mature forest and stringer meadows (Kosco and Bartolome 1983). Larson and Wolters (1983) found that herbaceous production in young open mixed conifer stands may reach 2240 kg/ha per year. Mixed conifer forests in eastern Oregon produce an average of 95 kg/ha current annual growth of shrub forage (Hedrick *et al.* 1968).

The loblolly-shortleaf pine forests of the southeastern United States contain bluestems as the primary forage species. Production varies with stand age and canopy density, and ranges between 12 and about 40 ha per AUM (Halls *et al.* 1964).

14.4.3 Changes in forage production over time

Although very little specific data on overstory-understory relationships have been published for the transitory range types, the primary commonality between all these forests is their lack of understory forage in mature stands. Thill *et al.* (1983) found in mixed conifer stands in Arizona, for example, that clearcuts produce as much as 3300 kg/ha while stands with 27 m2/ha basal area produce only 350 kg/ha. Mitchell (1983) reports that even deer cannot use mature stands of Douglas-fir except for cover near the edges of more open areas. Production in lodgepole pine varied by site, with site tree production potential inversely related to understory production. Mature lodgepole pine forests were found to produce as little as 45 kg/ha in dense stands and as much as 1250 kg/ha in open stands (Bartolome 1983). In transitory range types, understory species are found primarily along roads, and in openings where clearcutting and/or fire has opened up the forest. Forage in the openings may persist 10 to 20 years depending the forest type and site before the forest canopy shades it out (Fig. 3). In addition, several authors have forage quality declines as the forest canopy closes in Douglas-fir and ponderosa pine/lodgepole pine forests (Sharrow and Leininger 1982, Fulgham *et al.* 1985).

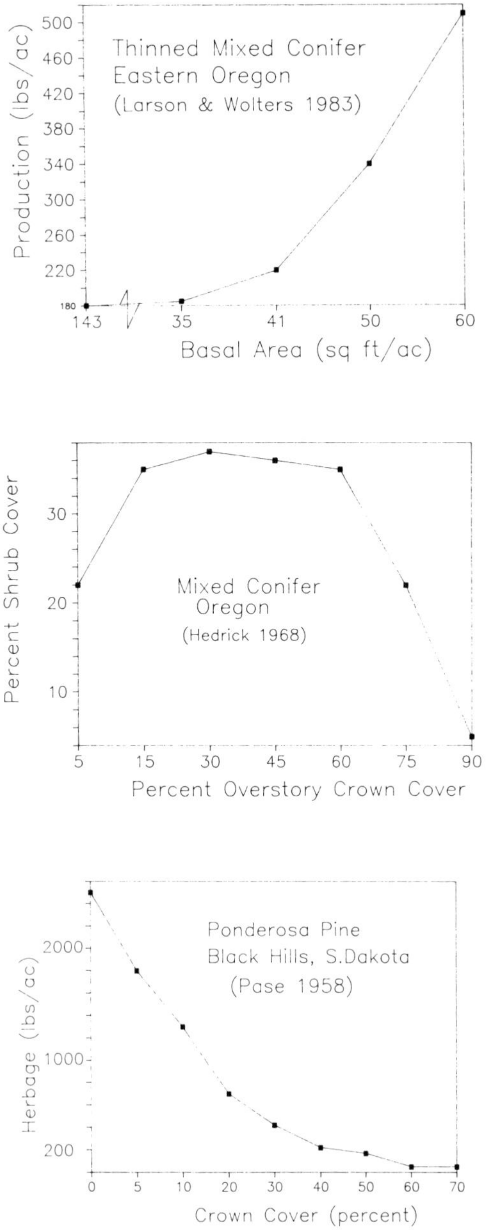

Figure 3. Relationship between understory production or cover and overstory cover or basal area on three mixed conifer sites classified as transitory range. Note break in the scale in Larson and Wolters (1983) as the stand was thinned, understory production increases until basal area increases again. These data are take directly from Hedrick (1968), Thill (1983) and Larson and Wolters (1983). Note differences in scale and parameters measured.

14.4.4 Historical/current use

Timber production is generally the number one use of transitory forest types. Water production, recreation and wildlife habitat are additional important uses and concerns of managers. Grazing by livestock was important historically (Kosco and Bartolome 1981), although conflicts with other uses now require even greater coordination.

Transitory range types continue to provide a high quality, summer forage resource which is often unavailable to ranchers at lower elevations. Water is commonly plentiful, and relative freedom from insects and disease makes the coniferous forest types highly desirable, despite costs associated with trucking animals and collecting animals from the range.

These forests produce 83 percent of the sawtimber, over 200 million recreational visitor days, and approximately 6.5 million AUMs in the United States alone (USDA 1985). Income from timber production was $544 million in 1984, while income from livestock grazing was approximately $9.6 million from National Forest System lands in 1984 (USDA 1985).

14.5 Management concerns

14.5.1 Grazing damage to tree regeneration

Foresters have expressed concern over livestock damage to young conifer plantations since the turn of the century in North America. Examples of conflicts between livestock grazing and intensive timber management are numerous. Damage figures vary between 25% browsing damage (Kosco and Bartolome 1983) to severe damage to tree regeneration (Arvola 1978).

Yet, shrub and herbaceous species growth on plantations following timber harvesting is estimated to reduce timber volumes by up to 30% (USDA Forest Service 1986). Brush densities of 10,000 cubic feet (Bentley 1971) or cover over 30% (Fiske 1981) appear to be sufficient to reduce conifer volume growth. Although some evidence exists that brush may aid initial conifer survival without impairment of early height growth (Roy 1981), subsequent loss of volume growth from competing brush species is significant on all forest types (Walstad and Kuch 1987). Grass competition with young conifers appears to be even more detrimental to conifer volume production than brush on many sites (USDA Forest Service 1986).

Early research emphasized that browsing and trampling damage to plantations resulted from faulty livestock management or wildlife, not inherent incompatibility of grazing and reforestation (Jardine and Anderson 1919, Cassidy 1937, Heerwagen 1954). For example, research on Weyerhauser Corp. lands in southeast Oregon suggests that proper timing of cattle grazing is essential for successful plantation management (Fulgham *et al.* 1985), and tree browsing damage was directly related to starting date of grazing and grazing intensity (Kingery *et al.* 1987). Knowledge of palatability and season has resulted in better management of livestock to meet timber management objectives and protect sensitive riparian areas too (Monfore 1983). Sharrow and Leininger (1982) similarly found on their study of sheep grazing on Douglas-fir plantations in southwest Oregon that tree damage was confined to early spring and that key browse species could be used to define the proper grazing season for effective plantation management. Now even forest management texts suggest that livestock represent one promising tool for suppressing competing vegetation in young conifer plantations (Walstad *et al.* 1987), although livestock grazing is not a panacea for the vast array of vegetative problems in forest management.

Experience indicates that managers must identify the potential causes of damage to conifer seedlings and develop a plan to verify damage source(s) if it occurs. This planning should also initiate a discussion of what percent mortality is acceptable. Some tree damage, browsing and/or trampling will occur even under the best livestock management, yet forest managers have determined, in many cases, that browsing damage from livestock grazing has negligible economic impact (Thomas 1983, Heald 1986). Grazing plantations when forage is more palatable than tree seedlings results in little damage to trees (Fulgham 1985, Sharrow and Leininger 1982, Monfore 1983). And many have shown, deer often cause most of the browsing damage to young plantation conifers (Kosco and Bartolome 1983).

14.5.2 Livestock/wildlife interactions

The impacts of domestic livestock grazers on forest habitats include alteration of forage quantity, quality and species composition; and removal of cover and alteration of structure. These impacts on wildlife, whether they are good or bad, must be considered in terms of individual wildlife species habitat requirements and management objectives.

Sharrow and Rhodes (1982) found sheep reduced brush cover on Oregon Douglas-fir transitory range clearcuts and improved big game habitat by increasing quality of fall forage and quantity of quality spring forage, even though there was an overall reduction in total standing crop. Baron (1962) found that cattle increased grass production on clearcuts with subsequent heavy use by deer, rabbits and mice. Moore and Terry (1978) used a short-duration grazing system with cattle to significantly reduce pineland threeawn and saw-palmetto resulting in increases in herbs desirable to certain wildlife species in longleaf-slash pine forests.

Radosevich and Conard (1982) studied a Sierra mixed conifer forest and found that where cattle removed the rapidly sprouting and seeding brush species on clearcuts, better wildlife habitat resulted from both a forage and ease of movement standpoint. Krueger (1983) determined plant community structure was most influenced by logging, big game, and cattle in that order on Oregon clearcuts. Brown (1983) suggested using livestock as a tool to manipulate wildlife habitat to meet specific management objectives. He indicated that livestock have been successfully used to manipulate wildlife habitat through Coordinated Resource Management Plans in eastern Washington.

Page *et al.* (1978) reported, however, that under heavy stocking rates, cattle changed vegetation structure to the detriment of some wildlife species. The response varied by habitat. Species, such as California and antelope jackrabbits, ground squirrels and pocket gophers, increased as these animals tend to favor areas of lower cover. Meadow mice, on the other hand, declined because they prefer heavy cover. In a heavily grazed aspen community, the removal of a willow midstory also caused profound changes in wildlife species, depending on a species' particular preferences (Page *et al.* 1978).

The impact of domestic livestock grazing on wildlife movements and use of habitats has been studied by several authors. Lonner and Mackie (1983) found elk avoided areas grazed by cattle and thus might be excluded from preferred food or resting grounds. Also, cattle can disturb elk calving, if allowed in calving areas in the spring. This may have long term detrimental effects on herd health. On the other hand, Anderson and Scherzinger (1975) reported a 3-fold increase in elk numbers following planned livestock grazing. Loft *et al.* (1986) found heavy cattle grazing in a Sierra forest range caused mule deer to shift fawning and cover habitat. They also found that moderate grazing minimized effects on deer habitat.

Managers can mitigate livestock impacts by recognizing that the impacts are tied to specific habitats and that the impact itself can be negative or positive depending on wildlife species of interest and management objectives. Specific management options for grazing include changing the kind of livestock or even the particular herd (Monfore 1983), altering the grazing season (Kaufmann *et al.* 1982), changing stocking rates and ensuring animal distribution. Animals of different age classes can also cause different impacts (Thomas 1983).

14.5.3 Livestock grazing as a tool in vegetation management

Forest grazing interactions are complex. For successful use of livestock as a tool in vegetation management, managers must consider a number of interacting factors. These include the browse and herbaceous species composition; the type of grazing animal; age of animals; season of grazing; affect of browse or herbaceous species on tree growth; tree species, age, vigor and planting methods; site prep methods; and site quality differences, to name a few. Timing of grazing initiation is key to minimizing browsing damage to tree seedlings.

The potential effects of livestock grazing on tree seedlings is well documented. Unmanaged domestic livestock grazing on plantations can have negative results (Thomas 1983, Arvola 1978). However, livestock can be used as a tool to reduce unwanted vegetation on plantations; research has documented that livestock grazing and intensive forest management need not conflict, and indeed may be mutually beneficial by using grazing animals to reduce competing vegetation (Kosco and Bartolome 1983, Sharrow and Leininger 1982, Monfore 1983, Krueger 1983, Thomas 1983).

These studies have shown on a number of forest and transitory rangeland types that livestock can reduce shrub or herbaceous cover to levels acceptable to forest managers. For example in Figure 4, in the mixed conifer forest in California, cattle grazing held brush cover to 30% on grazed sites compared to 75% cover of shrubs on ungrazed sites (Kosco and Bartolome 1983). Another study has also shown a 8—14% increase in tree diameter growth on grazed Douglas-fir sites in Oregon (Sharrow and Leininger 1983).

The successful use of livestock as a tool in forest management requires greater coordination between forest and range managers. Managers must determine if permittees and livestock are available, and if incentives will be necessary to attract permittees to implement a forest grazing program that promotes good forest and forage management. If animals are already grazing in the area, what kind? Are they already familiar with forest rangeland grazing? Is the plantation already part of an allotment? Will the plantation fit into a season long grazing plan for the livestock operator? Is the operator reliable? Is grazing a herded or open range situation? Which will be required to meet forest management objectives? Communication and coordination between the forest manager, range manager and permittee are essential to success. The operator must understand the forest manager's goals and conversely, the forest manager must understand that animals will need to move somewhere when the forage on the plantation

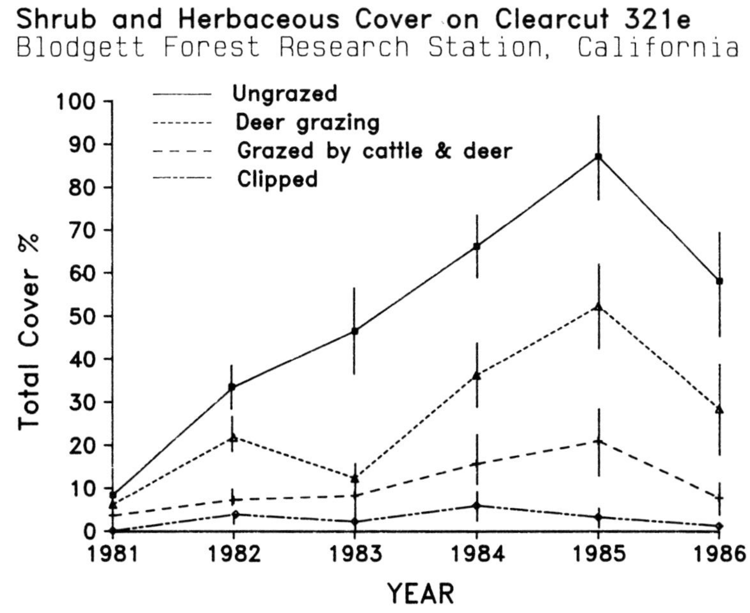

Figure 4. Cattle and deer grazing has held brush cover to 30% on a mixed conifer clearcut in the central Sierra Nevada, while ungrazed sites average 75% brush cover. These data and data from other forest sites indicate that livestock may be used as a tool in forest management to reduce unwanted vegetation on regeneration sites. Data through 1983, are from Kosco and Bartolome (1983). Recent data is supplied by this chapter's author to be published in Allen and Bartolome (1988 in press).

is gone; animals require water, bedding grounds, and protection from predators.

Continued use of forest and transitory rangelands, for domestic livestock production, depends on continued and improved coordination between often conflicting demands for the use of these resources. Successful forest and transitory rangeland grazing depends on managers who understand something about animal grazing behaviour as well as the complexitics of the forest ecosystems in their regions.

14.6 References

Allen, B. H., and J. W. Bartolome. 1988. Grazing in two mixed conifer clearcuts: changes in understory cover and tree growth. W. J. Applied Forestry (in press).

Anderson, E. W., and R. J. Scherzinger. 1975. Improving quality of winter forage for elk by cattle grazing. J. Range Manage. 28(2): 120—125.

Arnold, J. F. 1950. Changes in ponderosa pine bunchgrass ranges in northern Arizona resulting from pine regeneration and grazing. J. Forestry 48: 118—126.

Arvola, T. F. 1978. California Forestry Handbook. California Dept. of Forestry. Sacramento, CA.

Baron, F. J. 1962. Effects of different grasses on ponderosa pine seedling establishment. PSW Res. Note #199 USDA Forest Service.

Barrett, J. W., P. M. McDonald, F. Ronco Jr., and R. A. Ryker. 1980. Interior ponderosa pine. p. 114—115. *In*: F. H. Eyre (ed.) Forest cover types of the United States and Canada. Society of American Foresters. Washington, D.C.

Bartolome, J. W. 1983. Overstory-understory relationships: lodgepole pine forests. p. 1—4. *In*: E. T. Bartlett, and D. R. Betters (eds.) Overstory-understory relationships in western forests. Colorado State Univ. Expt. Stat. Ft. Collins, CO.

Bentley, J. R., S. B. Carpenter, and D. A. Blakeman. 1971. Early brush control promotes growth of ponderosa pine planted on a bulldozed site. USDA Forest Service Research Note PSW-238.

Boyer, W. D. 1980. Longleaf pine. p. 51—52. *In*: F. H. Eyre (ed.) Forest cover types of the United States and Canada. Society of American Foresters. Washington, D.C.

Brown, E. R. 1983. The challenge of managing big game species. p. 79—84. *In*: B. F. Roche, and D. M. Baumgartner (eds.) Forestland grazing. Washington State University. Pullman, WA.

Byington, E., D. Child, N. Byrd, H. Dietz, S. Halverson, H. Pearson, and F. Horn. 1983. Management of Southern U.S. Farms for Livestock Grazing and Timber Production on Forested Farmlands and Associated Pasture and Range Lands. Winrock International. Morrilton, AS.

Campbell, R. S. 1947. Forest grazing in the southern coastal plain. Proc. Society American Foresters. pp. 262—270.

Cassidy, H. O. 1937. How cattle use cut-over ponderosa pine-bunchgrass ranges with minimum injury to reproduction. SW Forest & Range Expt. Stat. Res. Note #15, USDA FS. Tucson, AR.

Clary, W. P., W. P. Kruse, and F. R. Larson. 1974. Cattle grazing and wood production with different basal areas of ponderosa pine. J. Range Manage. 28(6): 434—437.

Clary, W. P. 1983. Overstory-understory relationships: spruce-fir forests. p. 9—13. *In*: E. T. Bartlett, and D. R. Betters (eds.) Overstory-understory relationships in western forests. Colorado State Univ. Expt. Stat. Ft. Collins, CO.

Cronemiller, F. P. 1953. Deerbrush (sweet birch) in California. U.S. Forest Service R5 Unnumbered Publ. 16 p.

Currie, P. O. 1975. Grazing management of ponderosa pine-bunchgrass ranges of the central Rocky Mountains: the status of our knowledge. Res. Paper RM-159, USDA Forest Service. Ft. Collins, CO.

Currie, P. O., C. B. Edminster, and F. W. Knott. 1978. Effects of cattle grazing on ponderosa pine regeneration in central Colorado. USDA Forest Service Res. Paper RM-201. Ft. Collins, CO. 7.

Dana, S. T., and S. K. Fairfax. 1980. Forest and Range Policy. McGraw-Hill Book Company. New York.

Eyre, F. H. 1980. Forest cover types of the United States and Canada. Society of American Foresters. Washington, D.C.

Ffolliot, P. F. 1983. Overstory-understory relationships: ponderosa pine forests. p. 13—18. *In*: E. T. Bartlett, and D. R. Betters (eds.) Overstory-understory relationships in western forests. Colorado State Univ. Expt. Stat. Ft. Collins, CO.

Fiske, J. N. 1981. Evaluating the need for release from competition from woody plants to improve conifer growth rates. *In*: Proc. 3rd Ann. Veg. Mgmt. Conf. Soc. American Foresters. Redding, CA.

Fulgham, K. O. 1985. Plantation grazing in southcentral Oregon. *In*: 7th Annual Forest Vegetation Management Conference. Society of American Foresters. Eureka, CA.

Grelen, H. F. 1980. Longleaf pine-slash pine. p. 52—53. *In*: F. H. Eyre (ed.) Forest cover types of the United States and Canada. Society of American Foresters. Washington, D.C.

Hall, F. C. 1983. Application and interpretation of forest ecosystems classification. p. 7—14. *In*: B. F. Roche, and D. M. Baumgartner (eds.) Forestland grazing. Washington State University. Pullman, WA.

Halls, L. K., R. H. Hughes, and R. S. Rummell. 1964. Forage and cattle management in longleaf-slash pine forests. Farmers Bull. #2199 Washington, D.C.

Heald, R. C. 1987. Management of giant sequoia at Blodgett Forest. In: Giant Sequoia Management Conf. USDA Forest Service, Pacific Southwest Station. Berkeley, CA. (in press).

Hedrick, D. W., J. A. Young, J. A. McArthur, and R. F. Keniston. 1968. Effects of forest and grazing practices on mixed coniferous forests of Northeastern Oregon. Oregon Agric. Experiment Station Bull. 103.

Heerwagen, A. 1954. The effect of grazing use upon ponderosa pine reproduction in the Rocky Mountain area. Soc. American Foresters. Milwaukee, WI. 206—207.

Hodges, John. D. 1980. Slash pine. p. 56—57. *In*: F. H. Eyre (ed.) Forest cover types of the United States and Canada. Society of American Foresters. Washington, D.C.

Ingram, D. C. 1931. Vegetative changes and grazing use on Douglas-fir cut-over land. J. Agric. Research 43(5): 387—417.

Jardine, J. T., and M. Anderson. 1919. Range management on the national forests. USDA Forest Service Bull. #790. Washington, D.C.

Kaufmann, J. B., W. C. Krueger, and M. Vavra. 1982. Impacts of a late season grazing scheme on non-game wildlife in a Wallowa Mountain riparian ecosystem. In: Wildlife-livestock relationships symposium. Peek, J. M., and P. D. Dalke (eds). Univ. of Idaho, Moscow, ID. pp. 208—220.

Kingery, J. L., R. T. Graham, and J. S. White. 1987. Damage to first-year conifers under three livestock grazing intensities in Idaho. USDA Forest Service Research Paper INT-376. Ogden, UT. 8 p.

Kosco, B. H., and J. W. Bartolome. 1981. Forest grazing: past and future. J. Range Manage. 34(3): 248—251.

Kosco, B. H., and J. W. Bartolome. 1983. Effects of cattle and deer on regenerating mixed conifer clearcuts. J. Range Manage. 36(2): 265—268.

Krueger, W. C. 1983. Cattle grazing in managed forests. p. 29—42. *In*: B. F. Roche, and D. M. Baumgartner (eds.) Forestland grazing. Washington State Univ. Pullman, WA.

Kruse, W. H., R. P. Balda, M. J. Simoro, A. M. Macrander, and C. D. Johnson. 1979. Community development in 2 adjacent pinyon-juniper eradication areas 25 years after treatment. J. Envir. Mgmt. 8: 237—247.

Larson, F. R., and G. L. Wolters. 1983. Overstory-understory relationships: mixed conifer forests. *In*: Overstory-understory relationships in western forests. Colorado State Univ. Expt. Stat. Fort Collins, CO.

Lewis, C. E. 1980. Simulated cattle injury to planted slash pine defoliation. J. Range Management 33: 345—348.

Loft, E. R., J. W. Menke, and J. G. Kie. 1986. Interaction of cattle and deer on mountain rangeland. Calif. Agric. 1: 6—9.

Lonner, T. N., and R. J. Mackie. 1983. On the nature of competition between big game and livestock. p. 53—58. *In*: B. F. Roche, and D. M. Baumgartner (eds.) Forestland grazing. Washington State Univ. Pullman, WA.

Mann, W. F. 1980. Loblolly pine-shortleaf pine. p. 56. *In*: F. H. Eyre (ed.) Forest cover types of the United States and Canada. Society of American Foresters. Washington, D.C.

McConnell, B. R., and J. G. Smith. 1970. Response of understory vegetation to ponderosa pine thinning in eastern Washington. J. Range Manage. 23(3): 208—212.

McLean, A., and M. B. Clark. 1980. Grass, trees, and cattle on clearcut-logged areas. J. Range Manage. 33(3): 213—217.

Mitchell, J. E. 1983. Overstory-understory relationships: Douglas-fir forests. p. 27—34. *In*: E. T. Bartlett, and D. R. Betters (eds.) Overstory-understory relationships in western forests. Colorado State Univ. Expt. Stat. Ft. Collins, CO.

Monfore, J. D. 1983. Livestock — a useful tool for vegetation control on ponderosa pine and lodgepole pine plantations. p. 105—108. *In*: B. F. Roche, and D. M. Baumgartner (eds.) Forestland grazing. Washington State Univ. Pullman, WA.

Moore, W. H., and W. S. Terry. 1978. Short-duration grazing may improve wildlife habitat in southeast pinelands. Proc. Ann. Conf. Southeastern Fish and Wildlife Agencies 33: 279—287.

Odum, E. P. 1959. Fundamentals of Ecology. W. B. Saunder Co., Philadelphia, PA.

Page, J. L., N. Dodd, and T. O. Osborne. 1978. The influence of livestock grazing on non-game wildlife. CAL-NEVA Wildlife Transactions 159—173.

Pase, C. P. 1958. Herbage production and composition under immature ponderosa pine stands in the Black Hills. J. Range Manage. 11: 238—243.

Patton, D. R. 1974. Patch cutting increases deer and elk use of a pine forest in Arizona. J. Forestry 72(12): 764—766.

Pearson, D. 1950. Management of ponderosa pine in the Southwest. USDA Monograph #6.

Pearson, G. A. 1931. Recovery of western yellow pine seedlings from injury by grazing animals. J. Forestry 29: 876—895.

Pearson, G. A. 1974. Range and wildlife opportunities. p. 19—27. *In*: Proc. Symp. Mgmt. of Young Pines. USDA Forest Service, S. & SE. Forest Expt. Station.

Pfister, R. D., and P. M. McDonald. 1980. Lodgepole pine. p. 97—98. *In*: F. H. Eyre (ed.) Forest cover types of the United States and Canada. Society of American Foresters. Washington, D.C.

Radosevich, S. R., and S. G. Conard. 1982. Interactions among weeds, other pests and conifers in forest regeneration. Biometeorology in Integrated Pest Management 463—486.

Reid, R., and G. Wilson. 1985. Agroforestry in Australia and New Zealand. Goddard and Dobson Co., Box Hill, Victoria, Australia.

Roy, D. F. 1981. Effects of competing vegetation on conifer performance. In: Proc. Forest Vegetation Workshop. Oregon State Univ. Corvallis, OR.

Ryker, R. A., and R. W. Steele. 1980. Interior Douglas-fir. p. 91—92. *In*: F. H. Eyre (ed.) Forest cover types of the United States and Canada. Society of American Foresters. Washington, D.C.

Sharrow, S. H., and W. C. Leininger. 1982. Forage preferences of herded sheep as related to brush control and seasonal browsing damage to Douglas-fir regeneration. Oregon State Univ. Agric. Expt. Stat. Corvallis, OR. #56.

Sharrow, S. H., and B. Rhodes. 1982. The effect of grazing sheep on clearcuts in Oregon's coast range and its impact on big game habitat. Oregon Agr. Expt. Stat. Project Progress Report #0066. Corvallis, OR.

Skovlin, J. M., R. W. Harris, G. S. Stricler, and G. A. Garrison. 1976. Effects of cattle grazing methods on ponderosa pine-bunchgrass range in the pacific northwest. Tech. Bull. 1531 USDA Forest Service: 40.

Smith, R. L. 1966. Ecology and Field Biology. Harper & Row, New York.

Stoddart, A., A. Smith, and T. Box. 1975. Range Management. John Wiley & Sons. New York.

Thill, R. E. 1983. Deer and cattle forage selection on Louisiana pine-hardwood sites. Res.

Paper SO-196 5.

Thill, R. E., P. F. Ffolliott, and D. R. Patton. 1983. Deer and elk forage production in Arizona mixed conifer forests. Res. Paper RM-248, Rocky Mtn. For. & Range Expt. Station.

Thomas, D. F. 1983. The use of sheep to control competing vegetation in conifer plantations. *In*: Fifth Annual Forest Vegetation Mgmt. Conf. Society American Foresters. Sacramento, CA.

Tisdale, E. W. 1950. Grazing of forest lands in interior British Columbia. J. Forestry 48(12): 856—860.

USDA. 1985. Agriculture statistics. Govt. Printing Office, Washington, D.C. 551.

USDA. 1986. Vegetation management for reforestation. Pacific Southwest Region, Forest Service. San Francisco, CA. 500.

USDA, Forest Service. 1976. RPA Assessment. US Department of Agriculture, Forest Service. Washington, D.C.

Wallmo, O. C., and J. W. Schoen. 1980. Response of deer to secondary forest succession in southeast Alaska. Forest Science 26(3): 448—462.

Walstad, J. D., and P. J. Kuch. 1987. Introduction to forest vegetation management. p. 3—14. *In*: J. D. Walstad, and P. J. Kuch (eds.) Forest Vegetation Management for Conifer Production. John Wiley & Sons. New York.

Walstad, J. D., M. Newton, and R. J. Boyd. 1987. Forest vegetation problems in the northwest. p. 15—54. *In*: J. D. Walstad, and P. J. Kuch (eds.) Forest Vegetation Management for Conifer Production. John Wiley & Sons. New York.

Walstad, J. D., M. Newton, and R. J. Boyd. 1987. Forest vegetation problems in the northwest. p. 15—54. *In*: J. D. Walstad, and P. J. Kuch (eds.) Forest Vegetation Management for Conifer Production. John Wiley & Sons. New York.

Walter, Heinrich. 1979. Vegetation of the Earth. Springer-Verlag. New York.

Wheeler, W. P., W. C. Krueger, and M. Vavra. 1980. The effects of grazing on survival and growth of trees planted in a northeast Oregon clearcut. Oregon State Agric. Expt. Stat. Spec. Rep. # 585. Corvallis, OR.

Williamson, R. L. 1980. Pacific Douglas-fir. p. 106—107. *In* F. H. Eyre (ed.) Forest cover types of the United States and Canada. Society of American Foresters. Washington, D.C.

15. Ecological principles and their application to rangeland management practice in South Africa

O. J. H. BOSCH AND N. M. TAINTON

15.1 Abstract

The backbone of livestock production systems in South Africa is formed by the Karoo, Grassland and Savanna biomes. These biomes cover an estimated 85 percent of the land surface.

The nature of the livestock systems varies according to the nature of the vegetation itself. The natural vegetation of all the biomes is very species rich and is sufficiently productive to have continued to serve as the basis for a well developed livestock industry.

Ecological studies over the last six decades provided useful principles with regards to the inherent limitations to rangeland production, e.g., waterbalance and nutrient availability. These, together with the principles related to the ecological relationships between herbivores and the vegetation, are at present being used in defining sound management strategies for optimal use of the rangelands of South Africa.

15.2 Introduction

Five main biomes (Fig. 1) are distinguished in South Africa which provide the boundaries within which specific ecological principles and management strategies are applicable.

Each of these biomes represents a broad ecological set of life forms (Odum 1971) and is characterized by a distinct physiognomic type and composition. Of these the Grassland, Karoo and Savanna biomes cover an estimated 85 percent of the land surface and form the backbone of livestock production systems in South Africa. Due to structure and floristic composition the Fynbos (Macchia) and Forest biomes are not extensively used for grazing purposes.

The vegetation of South Africa is very species rich (more than 23000

Tueller, P. T. (ed.), Vegetation science applications for rangeland analysis and management.
ISBN 90-6193-195-9.
© 1988, *Kluwer Academic Publishers, Dordrecht, Boston, London.*

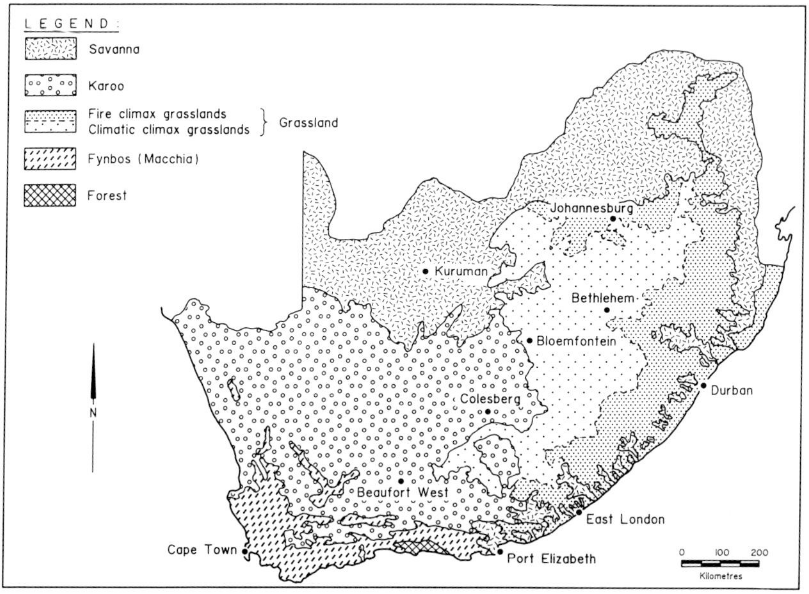

Figure 1. The main biomes of South Africa (adapted from Acocks 1975).

taxa) and evolved in the presence of a great variety of indigenous herbivores, including grazers, browsers and mixed feeders. What evolved was a vegetation well adapted to herbivore use, but more specifically to the seasonal pattern and intensity of use typical of a largely migratory population of these herbivores, and a vegetation which, by international standards, was extremely productive. These factors have had two important consequences. Firstly, the natural vegetation is sufficiently productive to have continued to serve as the basis for a well developed livestock industry. Secondly, researchers have concentrated much of their efforts into developing an understanding of the ecological relationships between herbivores and the vegetation, a relationship which had apparently been successful in allowing the vegetation to support large populations of such herbivores, while itself remaining both stable and productive.

The nature of the livestock systems which currently operate within the natural grazing lands of South Africa varies from place to place according to the nature of the vegetation itself. In much of the sweet grassland and karoo, small stock (and particularly woolled sheep) dominate the livestock system. In sub-humid areas, and particularly in areas subject to the invasion of xerophytic shrubs, beef production is widely practised, with goats increasingly being used to utilize the available browse. Where rainfall is high, both beef and mutton sheep are farmed extensively. In these areas where conditions permit the production of high quality forage from cropping land and sown pastures allow dairy production to become extremely important. Dairy production is also particularly important where irrigation can be widely practised.

However, in spite of considerable impetus given to intensification of land use practices during recent years, the natural vegetation continues to form the most important and economical food source for livestock in South Africa. It is therefore not surprising that considerable attention has been given to this most valuable resource. Ecological studies, especially over the last six decades provided useful principles which are at present being used in defining sound management strategies for optimal use of the rangelands of South Africa.

15.3 The Karoo biome

15.3.1 Vegetation structure and variability

The vegetation of the Karoo biome is dominated by chamaephytes and hemicryptophytes and can be described as a grassy, dwarf shrubland (Edwards 1983). The hemicryptophyte component (mainly C4 -graminoids) makes a low contribution in many parts (Rutherford and Westfall 1986). However, towards the eastern relatively higher rainfall areas, grasses can become more important under better management conditions. They include mainly *Aristida diffusa, A. congesta, Eragrostis lehmanniana* and *E. bicolor.*

In the far western parts, arid grasslands occur along the desert margin in which *Stipagrostis obtusa* and *S. ciliata* are the dominant species. These grasslands are closely associated with sandy substrates and with removal of the sand through erosion the chamaephytes also increase appreciably (Rutherford and Westfall 1986).

Chamaephyte cover in the Karoo biome varies considerably depending on climate (east or west) and soils (under similar climatic conditions). The dominant dwarf shrubs are species of the genera *Pentzia, Salsola, Eriocephalus, Rosenia, Zygophyllum, Ruschia* and *Galenia* (Vorster and Roux 1983). *Chrysocoma tenuifolia* is an important invader in overutilized vegetation of the eastern parts of the Karoo biome. Taller shrubs such as *Rhigozum* spp. occur in varying quantities throughout the flatter country. In the drier areas ephemerals also make a large contribution to the composition of the vegetation.

The mountainous areas are characterized by woody shrubs, (e.g., *Elytropappus rhinocerotis, Euryops* spp.) grasses (e.g., *Merxmuellera disticha*) and dwarf shrubs (mainly *Eriocephalus ericoides, Chrysocoma tenuifolia, Pteronia* spp., *Pentzia* spp. and *Ruschia* spp.) (Vorster and Roux 1983).

Apart from the great variation along rainfall and temperature gradients in the biome, the vegetation composition also varies considerably over short distances. The topographical variations on individual farms (and resultant variations in especially soil moisture conditions) give rise to great diversity in the general structure, composition and palatability of the vegetation on individual farms (Bosch 1974). Vegetation of different topographic positions differ in its reaction to injudicious management, which is also an important reason for the variability found on individual farms.

15.3.2 Factors determining management philosophy

15.3.2.1 Inherent limitations to rangeland production. Drought is a natural and characteristic phenomenon in the Karoo biome and an event that occurs frequently (Booysen and Rowswell 1983). Numerous reasons for the high frequency of drought exist, e.g., the high variability of seasonal rainfall, aridity of the area, low moisture content of the atmosphere and the unfavourable geographical position relative to the general air circulation patterns passing from west to east over South Africa (Booysen and Rowswell 1983; Roux 1979; Schulze 1965).

Waterholding capacities of soils can vary between 70 and 160 mm per meter (Booysen and Rowswell 1983). Low average soil depths of 0.2 to 0.5 m leave the soils with extremely low waterholding capacities of between 14 and 80 mm. Average evaporation rates are more than 10 mm per day, making it clear that the low waterholding capabilities of most soils in the Karoo biome are unable to meet the high demands set by evaporation and evapotranspiration rates under the temperature regime of these parts of the country.

Roux (1979) also points out the existence of pseudo-drought conditions. In this a scarcity of grazing is created by overgrazing, in spite of reasonable climatic circumstances. Injudicious management not only creates pseudo-droughts, but also intensifies the impact and effects of normal seasonal or periodic droughts.

The growth activities of the plants are clearly influenced by general climatic conditions prevailing during the various seasons. Due to inherent characteristics of the plants, the various components of Karoo vegetation have different general growth cycles. Maximum growth activities and development are achieved during different seasons of the year (Roux 1968). Figure 2 provides a general picture of these growth cycles in the eastern parts of the Karoo. The grasses reach their highest growth activity during summer/early autumn, dwarf shrubs show two growth periods (i.e., spring and autumn) while larger shrubs and trees reach their highest growth activity during late summer. The herbaceous annuals and biennials

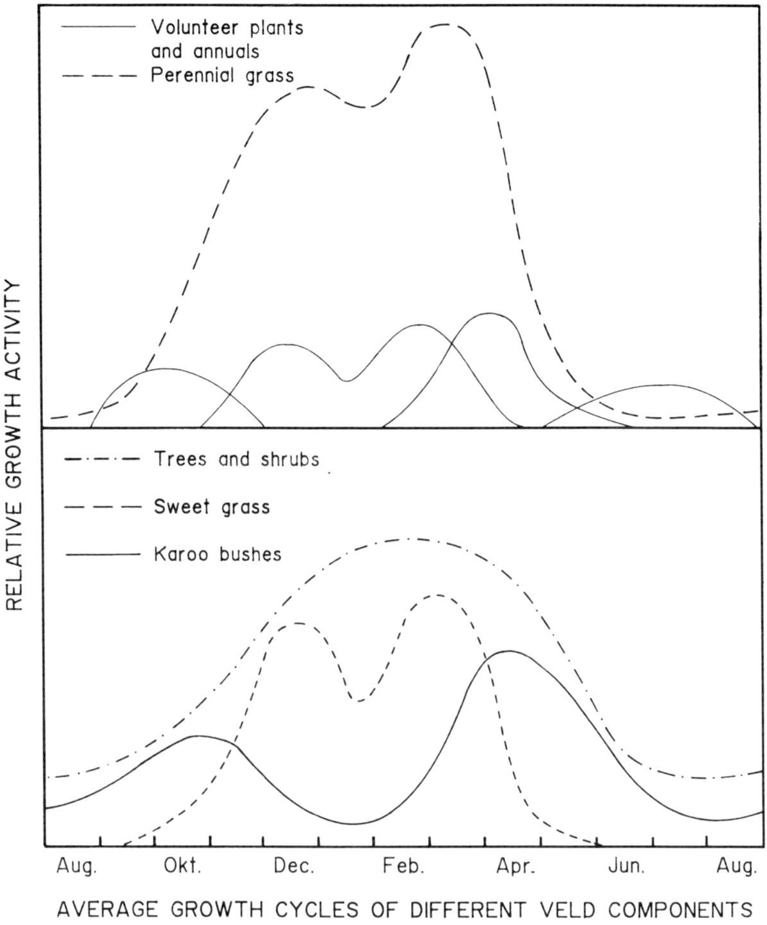

Figure 2. Cycles of growth of different components of the Karoo vegetation (Roux and Vorster 1983).

have short growth cycles. Depending on the species and climate this component can be active throughout the year.

A further important characteristic of Karoo vegetation is the fact that *production of reproductive material* also shows certain seasonal differences (Fig. 3). Desirable dwarf shrubs flower mainly in autumn or spring, while the undesirable and invader species flower mainly in winter. The flowering time of the grass component is in the late summer and early autumn (Vorster and Roux 1983).

15.3.2.2 Forage supply and utilization principles. Effectively available forage: Not all phytomass produced is available to the animal, mainly due to differences in palatability of plant species and selective grazing patterns

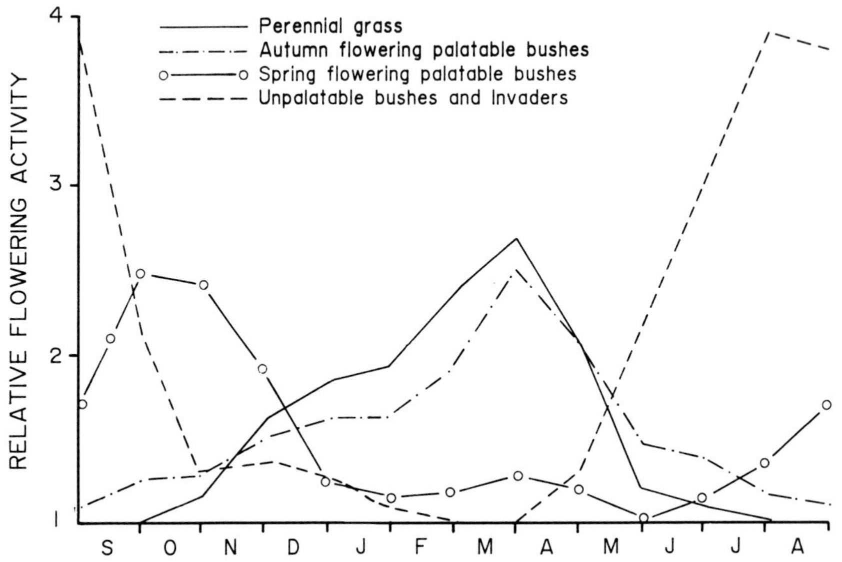

Figure 3. Average flowering activity of some Karoo veld components (Roux and Vorster 1983).

of the various consumer components in the Karoo vegetation (Fig. 4). Cattle concentrate mainly on the grass component as a source of food (Botha 1981). All animals show a preference for the palatable dwarf shrub component during autumn and winter and for the unpalatable component during spring (i.e., mainly during their active growing periods). In the case of small stock the grass component is mainly selected during the summer. The ephemeral component is found in the diet of animals whenever available.

Botha *et al.* (1983) also found an overlap in the diets of different animal types grazing the same vegetation. An overlap of approximately 35 percent exists between large and small stock, while that between the various small stock breeds varies from 36 to 75 percent, depending on the season of use. From above it is clear that the phytomass which is preferred by and therefore effectively available to the animals varies throughout the year. The factors to be considered in defining selection patterns and therefore modelling of the availability of forage are summarized by Botha *et al.* (1983) as follows:

— palatability of plants, as determined by chemical composition, phenological stage, environmental conditions and morphology of the plants;
— associated species;
— climate, soil and topography;
— type of animal; and
— physiological status of the consumers.

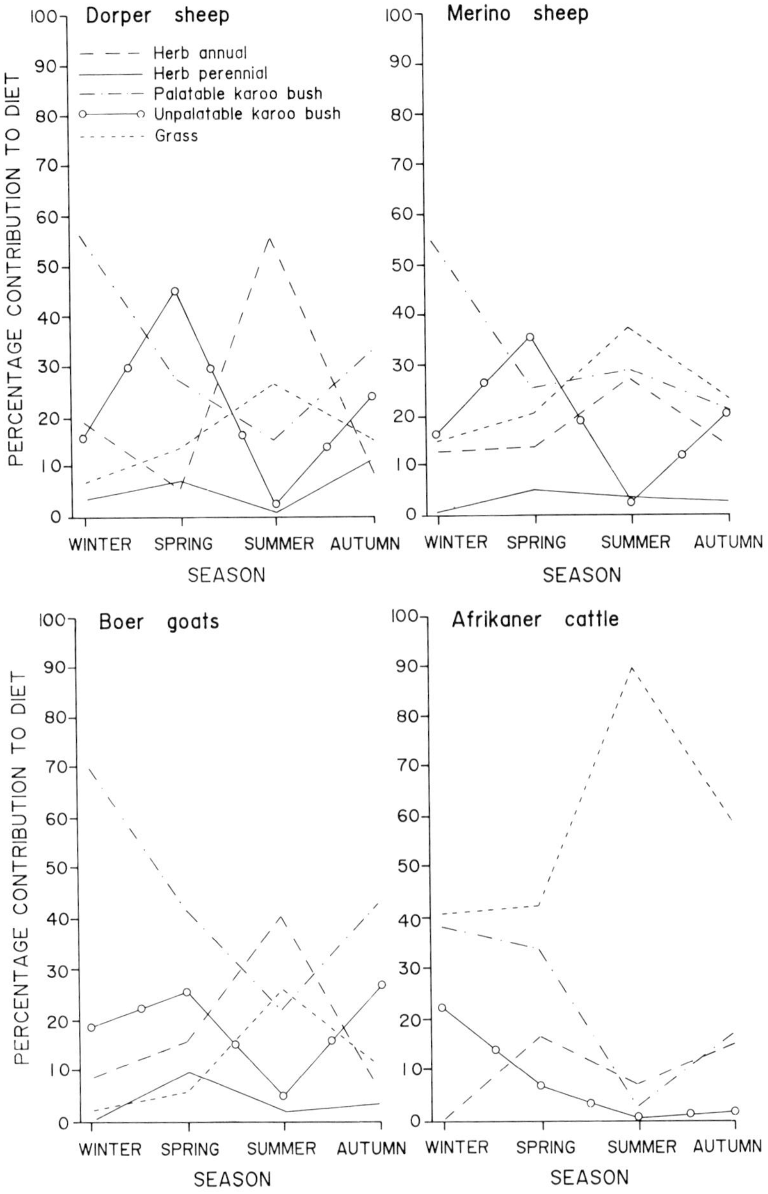

Figure 4. The average contribution of the different components of Karoo vegetation to the diet of live stock (Botha *et al.* 1983).

Grazing pressure is one of the main factors determining the patterns of defoliation, in that the frequency and intensity of defoliation increase with increasing animal numbers. The grazing pressure on a specific species or physiognomic component in one area is not homogeneously spread over the entire area. Defoliation patterns are influenced by factors such as positioning of watering points, size and shape of camps (paddocks) and the presence and spatial distribution of other palatable species (Bosch and Dudzinski 1984).

The sensitivity of the various species towards grazing is related to the season and their growth stage. The month during which physiological damage is most likely to occur can be deduced from Fig. 2, usually within the period from the start of growth to when growth begins to decline (Roux 1968). One of the earliest reports of a defoliation study in the Karoo is that of Henrici (1951) on the effect of cutting and grazing on *Pentzia incana* under different systems of veld management. The study was carried out under drought conditions and illustrated in general a negative effect of grazing on the productivity of the plants. Hobson and Sykes (1980) investigated the effects of defoliation frequency on three key Karoo shrub species, *Felicia muricata, Pentzia incana* and *Eriocephalus ericoides.* Increased production was found with decreasing frequencies of defoliation. It was further illustrated that species differ highly in their recovery potential. At low frequencies of defoliation *E. ericoides* had the highest total production, indicating that it would be able to utilize longer rest periods better than the other two. Root studies have also shown that defoliation frequencies of more than six times per year reduced the root mass and consequent vigour and recovery of the plants.

Van der Westhuizen (1980) studied the influence of pruning on the growth habit, dry matter production, photosynthesis, carbohydrate and nitrogen content of a winter growing grass (*Ehrarta calycina*) and a Karoo bush (*Osteospermum sinuatum*). The reaction that pruning had on the plants, whether a change in dry matter production, nitrogen or carbohydrate content, was in all cases related to the extent at which the photosynthetic capacity of all the plants was affected. Similar studies in Bushman grasslands have shown that summer grazing (November to January) depressed the basal cover of *Stipagrostis ciliata* as well as that of *S. obtusa*, whilst the effect of winter grazing (May to July) was to increase the cover of *S. ciliata* (Van der Walt 1971).

Various ecological studies have been carried out on Karoo vegetation, describing retrogression and succession under grazing conditions (Tidmarsh 1947 and 1948; Skinner 1964; Roux 1966 and 1969; Van den Berg 1972; Van der Walt 1972; Vorster 1980).

The main seral stages of plant retrogression in the semi-arid Karoo have been described as a change from the *Themeda triandra* stage to the *Eragrostis lehmanniana-, E. curvula-* and *Sporobolus fimbriatus*-stage and eventually to communities in which *Aristida* spp., *Tragus racemosus* and *Eragrostis bicolor* became the dominant grasses. The intermediate and pioneer stages are also accompanied by invasion of Karoo shrubs such as

Pentzia spp. and *Chrysocoma tenuifolia*. The invasion of Karoo bushes in the breaking down of communities dominated by grass becomes increasingly important towards the western lower rainfall areas of the False Upper Karoo and central Karoo.

In the arid Karoo areas *Stipagrostis ciliata* and *S. obtusa* are the dominant grasses of both climax and pioneer stages (Skinner 1964). Degradation is accompanied with reduction in the cover of these grasses, proliferation of a complex of ephemerals and subsequently the invasion of undesirable Karoo bushes such as *Pentzia spinescens, Zygophyllum flexiosum, Galenia africana* and *Rhigozum* species (Acocks 1975).

15.3.3 Ecological principles and management practice

Vegetation management strategies have to satisfy two main requirements: they have to maintain cover, quality and production at a high level, and they have to provide for stock-feed requirements (Roux and Skinner 1970). Defining such strategies for management of Karoo vegetation is based on the fundamental principles related to vegetation structure, climate, growth and utilization processes.

(1) *Slope and topographic differences are basic to vegetation structure, composition and palatability differences.* The identification and separation of plant communities on different topographical positions is the first step to create a basic framework for the further planning and management of an individual farm. Fencing off different communities is of particular importance to prevent area-selective grazing and to enable the manager to apply a scientifically sound grazing system. It also provides a means for special treatement of highly eroded or erodible sites on the farm.

(2) *The persistence and continued productivity of plant populations depend on how successful plants are in completing their life cycles.*

(3) *Defoliation patterns determine the successful completion of life cycles.*

(4) *Karoo vegetation consists of various components, which have different responses to grazing and resting during different seasons of the year.*

(5) *Most physiological harm occurs when overutilization takes place during the active growing and flowering/seeding periods of the different components.* The different requirements are satisfied by allowing seasonal periods of rest that succeed each other in a predetermined order. The three-camp (paddock) rotational grazing system, set out in Table 1, is one of the more important management strategies used for these purposes (Roux and Skinner 1970). During a three year cycle, each vegetation unit or camp is rested for at least two out of the three summer, autumn, winter

Table 1. A three-camp rotational grazing system for Karoo and dry sweet grassveld with an average rainfall of more than about 10 inches (Roux and Skinner 1970).

Camp	Grazing year and season								
	First year				Second year		Third year		
A	1			4		3		2	
B		2			1		4		3
C			3		2		1		4

and spring seasons. In this way all the different vegetational components are allowed to produce flowers and seed during their different periods of growth and reproduction activities.

Flexible application of grazing systems is generally achieved by means of at least 15 camps, each between 100 and 200 hectares in extent. Between one and five separate herds can be rotated in the camps according to a prescribed grazing system (Roux and Skinner 1970).

Camp grouping is generally recommended to enable the manager to apply a rotational grazing system in a practical and extremely flexible way. Camps are allocated to a number of groups appropriate to the grazing system envisaged. If a three-camp rotational system is to be applied, the camps are allocated to three groups. Each group contains camps on different topographical positions as well as camps for specific purposes such as lambing. Surveys in each camp prior to grouping are used to ensure that each group has approximately the same yearly grazing capacity.

When the grazing system (e.g., three-camp system) is applied, only the camps of one group are grazed at any one time, allowing at least two-thirds of the camps to rest. Within the camps of each group, the animals are allowed to graze continuously for the prescribed grazing period, or to rotate between the camps of each group (Roberts 1981a).

Provided there are enough camps in each group, certain camps can be withdrawn from the system for special treatment (such as avoiding grazing camps containing poisonous plants at certain times of the year, or allowing forage accumulation for lambing or weaning (Roberts 1981a).

(6) *Drought is a natural phenomenon in the Karoo biome.*

(7) *Growth cycles are of short duration due to low waterholding capabilities of the soils.* As rainfall is low and extremely variable, resting of vegetation for short periods has almost no effect. In three- or group camp rotational systems the periods of absence range from five to six months. These relatively long resting periods improve the chances for the occurrence of rainfall events and therefore the full recovery of the vegetation after grazing.

When sufficient moisture becomes available through rainfall events that stimulate growth, animals concentrate their grazing on the new growth. This may have a damaging effect on the new seedlings as well as on mature plants if the new growth is repeatedly utilized. Ensuring the completion of life cycles and maximising dry matter production of especially the ephemeral component after rainfall events are therefore important management objectives in the Karoo areas. This is achieved by the withdrawal of animals after rainfall events from the various camps, and concentrating them in the least possible number of camps for periods of 14 to 21 days. The highly flexible group-camp system also simplifies the application of this management practice. In fact, if there is only one herd for the three groups of camps, at least two-thirds of the farm derives the full benefit of rainfall events, which makes the withdrawal of animals after rainfall unnecessary.

During good rainy seasons fewer camps per group are grazed, the others being rested to maximize production (accumulation of feed reserves) and ensure continual productivity through the building up of seed resources.

The impact and effects of droughts are ameliorated by means of special precautionary measures (Booysen and Rowswell 1983; Anon 1979). It is a general practice to give supplementary feeding (via licks) in the initial stages of drought while sufficient forage is still available (Hayward 1979). This measure stimulates the stock to utilize the complete source of edible material. However, when drought conditions become more severe, one of the most important measures is to reduce stock numbers on the farm to well below the grazing capacity of the vegetation at that stage.

Many group-camp systems include special camps in which drought fodder crops are produced for periods of feed shortages. Old man saltbush (*Atriplex nummularia*), spineless cactus (*Opuntia* spp.) and in extreme cases Mexican aloe (*Agave mexicana*) are the most common species used as maintenance feed (Anon 1979).

(8) *Grazing pressure is the main determinant of intensity and frequency of defoliation and therefore persistence of plant populations and continual productivity.* As camps are normally large, more than one watering point is provided in each camp to ensure a more homogeneously spread of grazing pressure.

Elongated camps are also avoided and to ensure that the accessibility of all parts of the vegetation in each camp is the same (e.g., by fencing in such a way that erosion gullies or ridges would not divide a camp).

The number of animals or stocking rate is the most important factor determining the grazing pressure on vegetation. Maintaining stock numbers at a level which is compatible with the grazing capacity of the vegetation is therefore one of the most important aspects of management. Determining actual grazing capacities from which optimum stocking rates can be calculated, presents a problem. Vorster *et al.* (1983) ascribe this to the numerous factors that are influencing the determination of grazing capacity, e.g., the edaphic factors, unreliable climate and differences in growth activities of various vegetation components.

Because of the difficulties in devising scientific methods for determining grazing capacities, an alternative, more subjective method is at present being used in management programmes. The whole of the Karoo biome has been demarcated into relatively accurate long term grazing capacity zones (Vorster and Meyer 1981). This zonation was defined on the basis of vegetation composition and structure, climate and growth activities, as well as by using the experience of field officers. Although these zones are relatively subjective, they are highly compatible to the views of leading farmers and agriculturalists.

These long term grazing capacities serve as general guidelines for the stocking rates to be applied on individual farms. Stocking rate adjustments during the different seasons to current grazing capacities are continuously carried out by the manager of a unit. However, economical and other managerial constraints often restrict these adjustments, which could lead to high levels of defoliation during periodic droughts. Vorster *et al.* (1983) recommended that the management plan of each farm should include an infra-structure which could allow for adjustments to stocking rate, without necessarily having to sell excess stock, e.g., withdrawing excess number of stock to feeding lots, irrigated pastures or specially established fodder crops.

Maintaining the stocking rate at 50 to 60 percent of the long term grazing capacity is another alternative for management of Karoo vegetation. By this a buffer of food reserves is left for use during dry seasons. Roux (1964) has also shown that low stocking rates would not necessarily be less economical. An experiment with two stocking rates applied under continuous grazing in the eastern Karoo over a period of 10 years, has shown that a moderate stocking rate resulted in a higher production per animal than the heavy stocking rate (Table 2).

(9) *Animal types show different preferences for different vegetation components.* Due to the absence of a well-developed grass component, the larger part of the Karoo is primarily a small-stock producing region. Wool, mutton, mohair and Karakul pelts are the main products (Roberts 1981a).

The southern and eastern mountain areas are more grassy and therefore form the main beef production areas of the Karoo.

In the central to eastern mixed grass and dwarf shrub vegetation,

Table 2. Average body mass (kg) and production of Merino wethers under two stocking rates grazing continuously for 10 years on vegetation of the eastern Karoo biome (Roux 1964).

Production parameter	Stocking rate	
	Moderate (1.3 ha/sheep)	Heavy (0.86 ha/sheep)
Average body mass/sheep (kg)	38.3	33.0
Average carcass mass/sheep (kg)	15.0	13.1
Average greasy wool/sheep (kg)	4.2	3.5
Average body mass/ha (kg)	29.5	38.4
Average carcass mass/ha (kg)	11.5	15.2
Average greasy wool mass/ha (kg)	3.2	4.1

different animal species and types are used. According to Vorster *et al.* (1983) more complete utilization is obtained when mixed vegetation is grazed by different animal species and types, than by only one. This can be attributed to the variation in grazing habits and preferences of different animal species and types (Skinner 1963; Botha *et al.* 1983). Skinner (1972) also found that mixed flocks of merino sheep and angora goats resulted in higher meat production per hectare than single breed flocks.

It is essential to bring dietary overlap into consideration when the stocking rate of each animal type is calculated for multiple utilization purposes. Botha *et al.* (1983) gave an example for an area in which the grazing capacity is estimated at 1.71 ha. SSU (0.58 SSU/ha or 0.04 AU). Botha (1983) defines 1 AU as a 450 kg steer, which he found to be equal to 14.6 SSU (1 SSU = 1 dry merino ewe of 45 kg). The dietary overlap between Merino sheep and Afrikaner cattle is 37.3%. In calculating the grazing capacity for the combination, the number of each type is reduced by 50 percent of the dietary overlap, i.e., 18.65 percent. For multiple use of mixed Karoo vegetation the grazing capacities for sheep and cattle will therefore be 81.35% of the normal capacity for each type (Merino: 81.35% of 0.58 = 0.47 SSU/ha and cattle: 81.35% of 0.04 = 0.03 AU/ha).

15.4 The grasslands

15.4.1 Vegetation structure and variability

The grasslands are physiognomically monolithic and are characterised by strong dominance of hemicryptophytes of the Poaceae (Rutherford and Westfall 1986). These grasslands range from areas dominated by "sweet" grasses in the more arid western areas adjacent to the Karoo and in semi-arid valley systems (400—500 mm per annum), to a dominance of essentially sour grasses in the more humid regions (greater than 1000 mm per annum) on the eastern fringe and elsewhere at high altitude (Acocks 1975). The sweet grasslands provide nutritious forage throughout the year. The grasses effectively retain their nutritive value and remain acceptable to livestock even when dormant during the dry winter. In contrast, the sour grasslands are acceptable to livestock only during the summer growing season. In the winter both the nitrogen status and the digestibility of the foliage declines to levels too low to provide even for the maintenance requirements of animals (Tainton 1981a).

The sweet to sour gradient in South African grasslands broadly follows a eutrophic-dystrophic soil gradient from the more arid south-western regions (sweet) to regions of high rainfall (sour) and is apparently associated with the degree to which soils are leached. However, the relationship between the environment and the degree of sweetness or sourness of the common species in the climax grassland (whether a "climatic" climax, as in the arid regions, or a biotic (fire/grazing) climax in the more humid regions), is a complex one which continues to receive research attention.

The grasslands of Southern Africa are dominated by C4 grasses, except in the high altitude mountainous areas where C3 grasses become common and may dominate at altitudes above 900 m (Mentis and Huntley 1982). Except at such high altitudes, *Themeda triandra* dominates the grasslands. It often comprises more than 50% of the constituent species where the grasslands are perceived to be in a productive and stable condition, and may comprise as much as 80% of the grassland in certain ecological regions (Acocks 1975). In the sweet grasslands, *T. triandra* is generally associated with various species of the genera *Setaria, Eragrostis, Digitaria, Sporobolus* and *Aristida*, many of which replace the *Themeda triandra* when the grassland degrades through excessively high stocking pressures or generally poor management. In the more humid regions various (but different) species of the same genera listed above may similarly replace the *T. triandra* under excessive grazing pressure. Here, however, infrequent and lax grazing of the grassland or infrequent grassland fires, leads to the dominance of such species as *Tristachya leucothrix, Trachypogon spicatus* and *Cymbopogon validus*, while poor grazing management practice, and particularly selective grazing, leads to the dominance of such "wire" grasses as *Elionurus muticus, Diheteropogon filifolius, Rendlia altera* and *Aristida junciformis.* At high altitudes the C3 grasses *Festuca costata* and *Merxmuellera disticha* dominate under poor management, often in association with a heath component (Tainton 1981a).

The grasslands of the semi-arid regions (400 to 500 mm per annum) of

South Africa are typically relatively tall (0.75 to 2 m). Sward height declines as rainfall increases, so that those of regions receiving 800 to 1000 mm per annum are normally less than 0.5 m tall. However, since altitude normally increases with rainfall, increasing rainfall is generally associated with lower temperatures, so that rainfall and temperature are normally confounded. It is of interest to note, also, that these height changes often occur within the same species, so that *Themeda triandra*, for example, may typically vary in height from around 2 m at low altitude to only about 0.3 m at high altitude.

Small scale variations in the nutritive value and acceptability of the grasslands, such as may often be found on individual farms, is apparently largely a function of soil moisture regime and soil nutrient status. Hence, topography normally has a major influence on these small scale variations (Bosch 1977), apparently through its influence on both slope (and so also soil depth) and aspect. Soil texture also plays a major role, presumably through its influence on the ability of soils to retain moisture and nutrients. The outcome of topographic and textural variation may often be manifested in changes in floristic composition of the grassland communities on a micro-scale, but whether or not this happens, small-scale variations in the nutritive value and acceptability of the grasslands are common. As extreme examples, northern aspects or heavy soils typically produce more nutritious and acceptable forage than southern slopes or grasslands of sandy soils, respectively.

15.4.2 Factors determining management philosophy

15.4.2.1 Inherent limitations to grassland production. Not surprisingly, moisture deficiency imposes a major limitation to grassland production in the semi-arid grassland region, whereas soil nutrient deficiency imposes the greatest limitation to production in the humid regions (Edwards *et al.* 1980). It is consequently these two aspects viz. water-balance and soil nutrient studies, which have received a great deal of attention in the more arid and humid grasslands, respectively.

Water-balance studies in semi-arid grasslands. In these regions the summer rainfall of 400 to 650 mm is generally unreliable. What rain does fall often comes in the form of heavy thunderstorms so that runoff rates may be high. As a consequence soils are generally too dry to support active growth for much of the summer season, and for the entire winter season (Roberts 1981b). Periods of feed shortage are therefore generally common, even during summer.

The sweet grasslands have suffered relatively greater denudation than most other vegetation types in South Africa, primarily because of their greater year-round acceptability to all livestock. Roberts (1981b) has suggested that, unlike the sour grasslands, the species which occupy these areas do not possess the inherent protective mechanism of low palatability, and that they are consequently extremely vulnerable to grazing misuse. Heavy yearlong grazing by sheep, for more than a century has understandably resulted in severe degradation throughout this vegetation type. Such degradation, as Snyman *et al.* (1985) have shown, has served to accentuate the problems of drought stress. They showed, for example, that runoff loss from a degraded pioneer community was 180 percent higher over a 7-year period than the loss from a well preserved "climax" community. During high intensity rainfall events (more than 25 mm per hour), as much as 48 percent of the rain was lost from degraded grassland as runoff. This high percentage contrasted with a loss of only 15 percent from well preserved grassland. Not unexpectedly, soil loss was also substantially different in the two situations. A degraded grassland with a 2 percent cover on a 4 percent slope lost an average of 3.86 tons of soil per hectare per year, whereas grassland in good condition, on the same slope but with a cover of 6 percent, lost an average of only 0.6 tons of soil per hectare per year. Snyman *et al.* (1985) therefore clearly showed that degradation leads to a decrease in the effectivity of rainfall, and it is this increased aridity which is commonly considered to be one of the main reasons for the early movement of the more xerophytic Karoo (Acocks 1975). Roux and Vorster (1983) see this process as among the first phases of desertification which they describe as follows: The initial phase is characterised by a rapid breakdown of the pristine grassland, followed by the elimination of acceptable and nutritious plant species which are in time replaced by xerophytic Karoo shrublets. Rainfall effectivity declines throughout this process, and with continued heavy stocking pressure continued degradation may eventually lead to complete desertification.

In the more humid grasslands of intermediate rainfall between the truly sweet and sour grasslands (500—800 mm per annum) overgrazing frequently leads to the encroachment of shrubs and trees, particularly of the genus *Acacia* (Acocks 1975). Friedel (1987) has, for example, found a significant relationship between the condition of the herbaceous community and the density of encroaching *Acacia karroo*. She established two density thresholds as the condition of the grassland declines (Fig. 5).

The first is at a moderate condition (condition index of 740), beyond which a small decline in the condition of the grassland is accompanied by a large increase in tree density. A second threshold occurs when the grassland is in a moderately poor to poor condition (condition index of

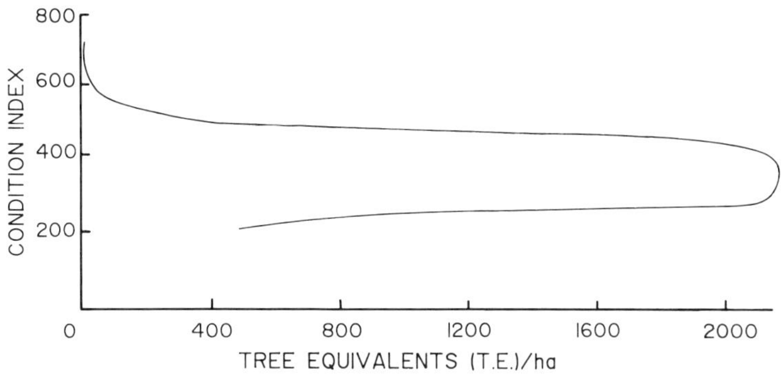

Figure 5. The relationship between condition index and tree equivalent (TE) density. (Friedel 1987).

650—450). As grassland degrades below this level, tree density declines. The first threshold is thought to relate to a change in the competitive balance between the grass and the shrubs for soil moisture (Walker and Noy-Meir 1982; Trollope 1984b; Friedel 1987). As Stuart-Hill *et al.* (1984) have shown, a decline in grass density, and with it a substantial reduction in the rate of soil moisture depletion in the upper 60 cm of soil after rain, results in an increased depth of moisture penetration down the soil profile. This increased availability of moisture at depth in the profile is to the competitive advantage of shrub and tree species. However, as grassland cover declines to very low levels, soil surface conditions become unsuitable for seed germination and seedling survival, so that all plants find it difficult to establish.

Tree invasion into grassland may also, however, be associated with episodic periods of unusually high rainfall (referred to as "pulse" disturbances (Bender *et al.* 1984)), when tree seedlings can survive the competition afforded by even dense grass swards, and also with the elimination, by man, of intense grassland fires that would normally have followed such periods of high rainfall. Fuel loads would have been expected to be high, and fires caused by either lightning (Edwards 1984) or by early man (e.g., by bushmen, who made extensive use of fire (Hall 1984)) are likely to have been sufficiently intense to kill all emerging tree seedlings. Today fire plays only a minor role in the management of these sub-humid grasslands, and when these grasslands are burnt, the requirement for "safe" fires are such that their intensities are too low to do any appreciable damage to tree seedlings.

The influence of climate, and particularly of moisture, on growth processes in the important forage producing species of the semi-arid to sub-humid grasslands, is well illustrated by the work of Danckwerts and Aucamp (1985). Unstressed vegetative tillers of *Themeda triandra* and

Sporobolus fimbriatus accumulate between four and six living leaves per tiller. New leaves emerge in such tillers at a rate of one every 2 to 3 weeks. New tillers therefore grow for at least 80 to 90 days before the rate of leaf senescence balances the rate of leaf emergence. Having senesced approximately 3 months after emergence, the leaves remain attached to the tillers and finally abscise 2 to 3 months later (5 to 6 months after emergence). However, under conditions of moisture stress new leaves appear only every 3 to 4 weeks, and leaf longevity drops in both species. The leaves abscise about 4 months after emergence. Because of these effects on both tiller growth and leaf retention on the tiller, moisture stress clearly affects standing forage yield, and provides an explanation for the findings of Snyman *et al.* (1985) that more than 88 percent of variation in above-ground standing forage in the sweet grasslands of the central Orange Free State is attributable to the variation in annual rainfall.

Roux (1966) and Snyman *et al.* (1985) have reported that short-term fluctuations in seasonal rainfall also effect changes in both the cover and composition of the sweet grasslands. Snyman and Opperman (1983) found statistically significant differences in cover and composition in these grasslands after only three seasons of differential moisture supply. Where normal rainfall was experimentally reduced by 20 percent, basal cover declined from 8.4 percent in 1978 to 5.0 percent in 1981. Also, the species composition changed from one dominated by *Themeda triandra* in 1978 to a sub-climax/pioneer grassland dominated by *Tragus* spp. and *Aristida* spp. in 1981. This change not only influenced infiltration, as discussed earlier, but also had a substantial influence on the efficiency of use of the water which did infiltrate into the profile (Snyman *et al.* 1985). Climax, sub-climax and pioneer communities produced, respectively, 0.21, 0.12 and 0.06 g above-ground phytomass per litre of water evapotranspired. The combined effect of grassland degradation on both infiltration and water use efficiency, therefore, provides a very clear explanation for the much greater reported incidence of drought in degraded vegetation than in climax grassland, and explains why drought conditions are now often reported in areas of degraded grassland even in years of near normal rainfall.

Soil nutrient studies in humid grasslands. Under sub-humid to humid conditions, where nutrient leaching is likely to be intense, it is not unreasonable to suppose that soil nutrient deficiency will restrict the ability of natural grasslands to make use of the abundant moisture provided by the natural rainfall. Under such conditions, nutrient deficiencies may be seen as providing the major limitation to increased productivity. Not surprisingly, therefore, the influence of nutrient addition has received widespread attention in the sub-humid and humid grasslands of South Africa.

The addition of nutrients, and particularly of nitrogen, to natural grasslands in sub-humid and humid regions has indeed increased annual herbage yields substantially (Fig. 6).

The response is closely related to local soil moisture regimes. The higher the rainfall in any season, the better the response, although the response may be surprisingly good even in dry years. As Haylett (1969) reported, "better yields were obtained in dry years with adequate fertilization than in wet years without fertilization".

As the data of Fig. 7 show, however, response to nitrogen declines progressively as increasing levels are applied.

In general phosphorous has had little influence on herbage yields when applied alone, but has in general substantially increased responses to nitrogen. Potash has induced positive responses only in isolated areas of

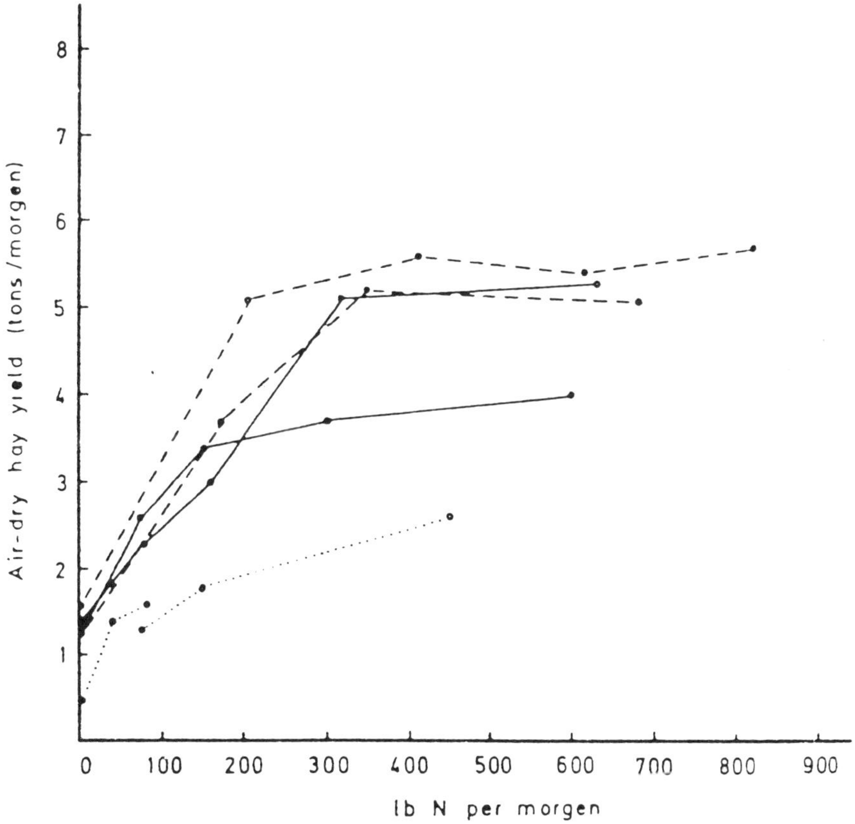

Figure 6. Yield response of grassland to increasing amounts of nitrogen at a number of locations (after Grunow *et al.* 1970).

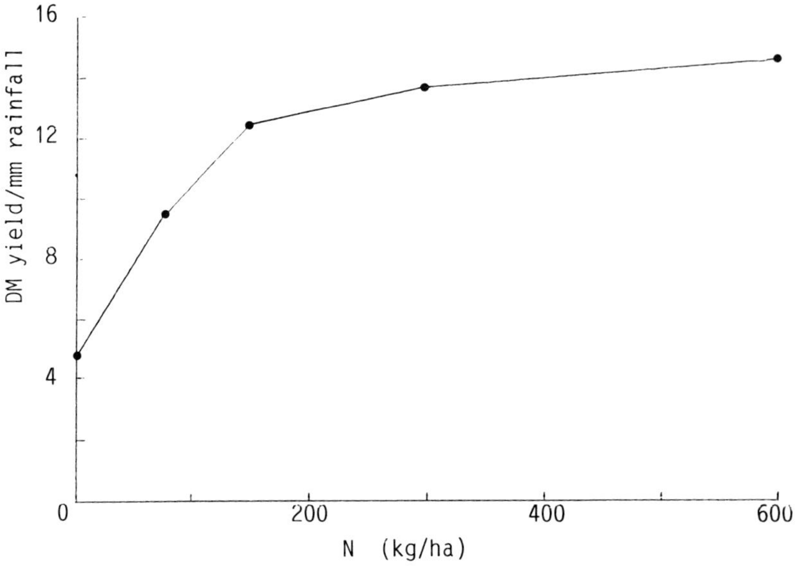

Figure 7. Air dry herbage yields per 1 mm of rainfall.

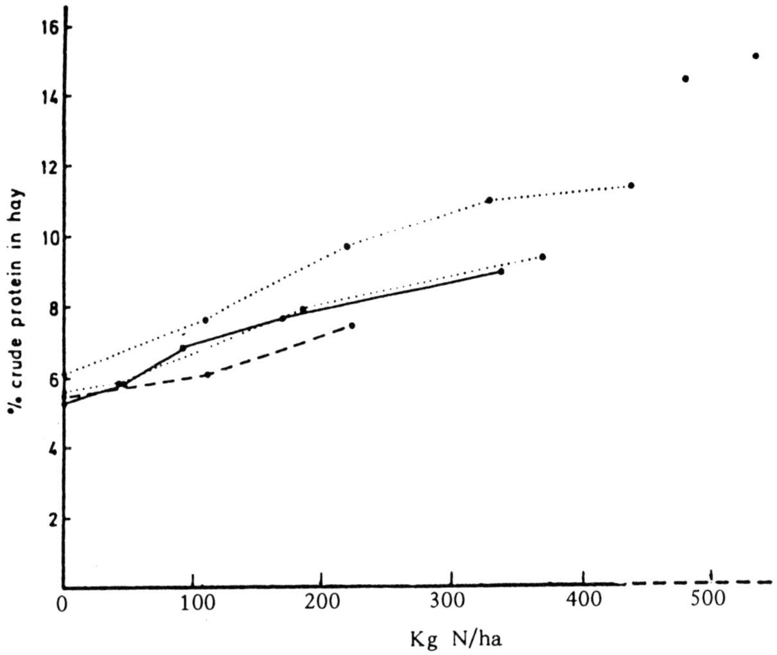

Figure 8. Influence of increasing amounts of nitrogen fertilizer on the crude protein content of harvested material at a number of locations (after Grunow *et al.* 1970).

known low soil potash status, while the application of lime has generally produced little response from the grassland.

Added to the sometimes dramatic influence of the addition of nitrogen on herbage yield, particularly when it is applied together with phosphorus, is a general improvement in the nitrogen status of the herbage produced. This is illustrated in Fig. 8 where the crude protein content of material harvested from nitrogen trials undertaken at a number of sites is shown. The result is an even more impressive response of total crude protein yield as fertilizer nitrogen levels increase (Fig. 9).

However, the above response of the grassland to fertilizer application has in all cases been accompanied by an often dramatic change in its species composition. The species of the fire/grazing climax community are displaced by lower seral species (Grunow *et al.* 1970). In most ecological zones within the sour grassland region such a change in species composition is considered to be undesirable because the lower seral species are often much less acceptable to livestock and are less digestible than the original species which they replace. In some areas, however, the particular species which invade fertilized grassland may be relatively well utilized by livestock, and in such areas the use of nitrogen and phosphatic fertilizers may be an economically viable practice. Where this is so, the rates of application should be governed by seasonal rainfall. Indeed, Grunow *et al.*

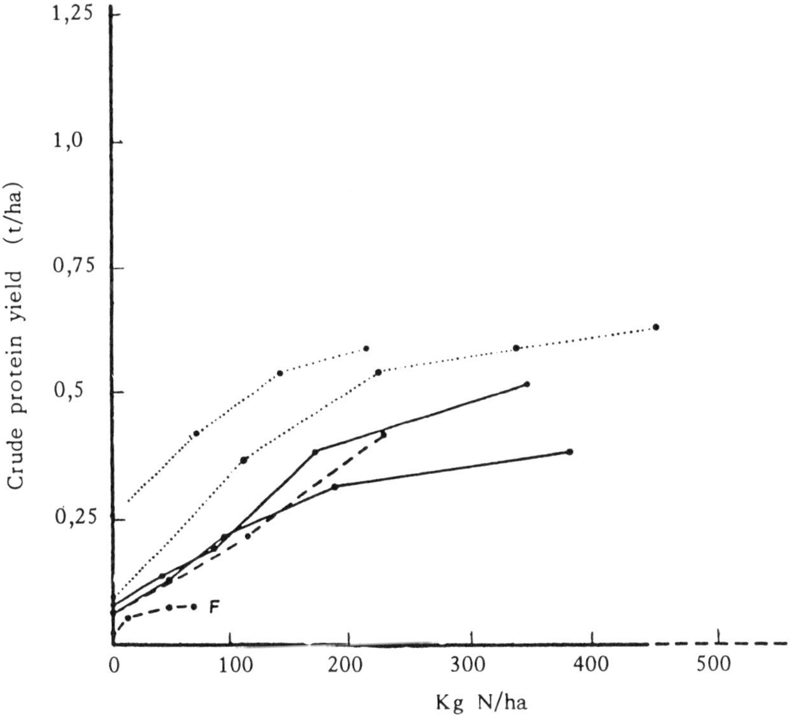

Figure 9. Influence of nitrogen fertiliser on the crude protein yields produced by grasslands at a number of sites (after Grunow *et al.* 1970).

(1970) have suggested that early application of about 22 kg N per hectare should be followed by continued applications of the above amount for each 125 mm of rainfall received during the summer season. This, they suggest, should be accompanied by an early and mid-season application of phosphate.

In general, however, the application of expensive fertilizers to natural grasslands in South Africa is not a common practice. What fertilizer is applied to grassland, is normally reserved for highly modified areas planted to selected pasture species.

15.4.2.2 Acceptability. South African grasslands are typically species rich (Acocks 1975), although in most situations only a few species will normally dominate any particular community. Typically, also, the species in any community vary widely in their acceptability to livestock, so that grazing is normally extremely selective. The result of this is that, where grazing is poorly controlled, the major forage producing species in any community are normally heavily utilized, while unacceptable species may be only lightly used or, in many situations, not used at all. Such a utilization pattern will have an extremely important influence on the competitive balance between acceptable and unacceptable species, which may, depending on circumstances, be used either to the advantage or detriment of the main forage producing species. Used wisely, therefore, differences in acceptability among species can be used to improve long term livestock production off these grasslands by improving the composition of the grasslands themselves.

Also, grazing is normally patchy as a result of spatial differences in acceptability of even the same species. This tendency for area selection by livestock poses serious management problems and if allowed to continue for any length of time results in severe grassland degradation (Tainton 1972).

As a result of these species and spatial differences in the acceptability of the material available to livestock in the grassland areas, not all of the dry matter produced can normally be classed as usable forage. Adjustments are therefore necessary in the determination of permissable stocking rates, as well as in management programmes if the grassland is to be maintained in a productive condition.

15.4.2.3 The ontogeny of tiller development. Early work in this field centered on an application of the need to understand the patterns of tiller development among the important species in any grassland community. A number of species were examined in the semi-humid grasslands of Natal (Tainton 1964; Tainton and Booysen 1963, 1965a and 1965b; Rethman and Booysen 1967 and 1968.) Rethman (1971) and Danckwerts *et al.* (1984) have continued this work in other grassland communities.

It is with respect to the work on the generally abundant *Themeda triandra* (in grasslands in good condition) that such work has done much to shape grazing management philosophy. In the Natal sub-humid grasslands a new population of tillers emerges in this species in the early spring from node positions on the parent tiller population. In undisturbed grassland, the stem itself remains compact until mid-summer as it continues to produce new leaves. However, stem elongation commences in mid-summer (through the extension of internodes at intermediate positions on the stem), and the still vegetative stem apex is gradually elevated above the soil surface throughout the remainder of the growing season (Tainton 1981b). Stem growth ceases temporarily with the onset of unsuitable growth conditions in early winter but the apex itself remains active, and becomes reproductive in mid to late winter. With the onset of good conditions in the spring, stem elongation is rapid and the tillers flower soon thereafter. At this stage tillers of this population will have completed their life cycle, and will begin to senesce from the terminal floral branches towards the tiller base. However, in the continued absence of defoliation, tiller initials at elevated node positions will commence their development to give rise to "aerial" tillers, while those at basal node positions will remain inactive. Since these aerial tillers are unable to develop effective root systems, and since they are in no position to produce an effective generation of new tillers when they mature, such treatments lead finally to the death of the plant.

In contrast, when such grasslands are defoliated in the spring or early summer, the stems of the new generation of *T. triandra* tillers remain short through to the following spring, when rapid elongation takes place immediately before flowering. The apices therefore remain near the soil surface where they are protected against removal during grazing and from possible winter frost damage. However, if defoliation is delayed until the late summer (by which time the stem apices are likely to have been elevated above the defoliation height), the growth of the decapitated tiller is terminated and the next generation of tillers is immediately stimulated to develop from basal node positions.

Because of these patterns of morphological development, plants of *T. triandra* are extremely responsive to the season in which defoliation treatments are applied. In particular, however, this species requires regular defoliation if it is to persist in a grassland community. However, if such defoliation is too frequent, root systems become small and the initiation of new tillers is impaired, presumably because of the lack of sufficient energy resources to permit continued vigorous growth.

In the conditions substantially colder than those experienced by the ecotype examined in Natal, Rethman (1971) reported two behavioural patterns among *T. triandra* plants. The one exhibited a pattern similar to the Natal type already described, but in a second type, stem elongation was delayed until just before flowering. Danckwerts *et al.* (1984) also reported this pattern of behaviour in the *T. triandra* of semi-arid grasslands but when additional irrigation water was provided, the plants exhibited a pattern of development similar to the Natal plants. He concluded from his work that this species is well adapted to infrequent but severe defoliation, suggesting that fire may have been the most important defoliating agent during the evolutionary development of the ecotype.

These investigations have made an important input into management philosophy in South Africa. In particular they have permitted a precise definition of the treatment needed to promote seed production, and they have provided information on the timing of grazing and recovery rest periods for grassland dominated by this species. Also, the tendency for plants to smother themselves when they remain undefoliated for any length of time, has pointed to a grazing mechanism whereby plants of unacceptable species, assuming they react in the same way as *T. triandra*, are deliberately left ungrazed in order to reduce their ability to continue to produce effective populations of new tillers. Such treatments should improve the relative competitiveness of acceptable plants provided they are grazed at an intensity and frequency which will ensure their vigorous growth.

15.4.2.4 Carbohydrate production and utilization. The physiological work most pertinent to the development of management principles in South African grasslands has centred around carbohydrate storage and its re-use during regrowth. The first worker to investigate these aspects was Weinmann (1955), who concluded that "during the latter part of the growing season (autumn) sugars and starch are photosynthesized by the leaves at a rate exceeding the requirement of aerial growth, and the excess is translocated to the roots and other underground organs for storage". He showed a distinct rise in carbohydrate content of the "storage" organs during autumn and winter, and a retranslocation of these substances to the shoots in spring and summer. He further maintained that grasses draw on these reserves whenever new growth is produced after cutting or grazing in the summer, but excessive defoliation will result in a depletion of underground reserves and a reduction in the size of the root system. This would, he claimed, lead to reduced vigour and in extreme cases to the death of the plant. The final outcome of excessive defoliation would therefore be to reduce herbage yield and in extreme cases would lead to a change in species composition of the sward.

The results of this and later work (Steinke and Booysen 1968; Bartholomew and Booysen 1969; Steinke 1969, 1975; Opperman and Human 1977; Snyman and Opperman 1983) has largely supported the conclusions drawn by Weinmann (1955), although the site of storage has been widened to include stem bases as well as roots and rhizomes. In recent years, however, the role of "storage" reserves in promoting regrowth has been severely criticised, at least in the international literature (Ryle and Powell 1974; Atkinson and Ferrar 1983; Richards and Caldwell 1985). But as Danckwerts and Gordon (1986) have shown, labelled "storage" carbohydrates are themselves not found in any appreciable amounts in regrowth. Here they must surely play a vital role in recovery growth even if this role is none other than to provide the necessary energy for the development of the first one or two leaves following defoliation. After all, such leaves cannot develop without an energy source, and what other possible energy source can there be? There remains, however, the question of whether the level of carbohydrates in storage plays any role in determining the rate of recovery growth. Danckwerts and Aucamp (1985) has indeed shown this to be so. Any advantage gained by a plant in the rate of extension of the first leaves to develop after defoliation translates into a substantial advantage at later growth stages.

In spite of these uncertainties, the principles outlined by Weinmann (1955) remain central to recommendations on both stocking rate and grazing systems in South Africa: Heavy stocking rates are condemned because they do not permit a sufficient accumulation of leaf material to permit the synthesis of sufficient energy substrates to support vigorous growth; continuous grazing is condemned because the acceptable species are over-utilized, while the unacceptable species are avoided by the animals, so influencing the competitive balance between the species to the disadvantage of the forage producing species; rotational grazing systems should be such that the acceptable forage producing species alone determine the grazing practice which should be adopted. During the period of animal occupation in a paddock the acceptable species should not be overgrazed, animals should not be permitted to stay in a camp long enough to permit them to graze recently produced regrowth (e.g., the period of occupation should be relatively short) and the interval between successive periods of occupation in a paddock should be sufficiently long to permit an adequate level of recovery of the important forage species.

These principles are embodied in recommended programmes in which between 6 and 8 paddocks are normally recommended per animal group.

Central to these programmes, is the requirement to divide the vegetation into homogeneous units, at least in so far as it is possible to separate out different categories of sweet and sour grassland into different paddocks.

15.4.2.5 The determination of grazing capacity. A requirement for the introduction of legislative control to stem the rate of degradation of South African grasslands has been the need to estimate permissible stocking rates which may be applied to different grassland types. To do this, two conceptually different procedures have been developed. Barnes *et al.* (1984) have concentrated their efforts on an essentially agronomic method in which the potential dry matter production of each species in the sward is estimated, together with the extent to which each species would be utilized when the stocking rate is such that it would not lead to degradation of the grassland. The alternative procedure which is now widely used is essentially ecologically based. It is based on the premise that the species composition of a community responds to grazing pressure, and that there is some community along the composition gradient (from that characteristic of very heavily grazed sites to that characteristic of lightly grazed sites) which can be defined as optimum for the purposes for which the grassland is used. In practice such gradients are defined using quantitative ecological techniques (Bosch *et al.* 1987) and any sample of grassland may be indexed according to its position along this gradient. Also, the stocking rate appropriate to a community at any position along a gradient can be established experimentally, allowing also for natural ecological processes (secondary succession) to react on any community which is not in optimum condition. Here the recommended grazing pressure for communities whose composition is characteristic of overutilized sites, is deliberately set below the short term agronomic value for the site, while in underutilized sites either the recommended grazing pressure is higher than that which would be indicated by agronomic methods, or more frequent fires are recommended in an effort to increase defoliation frequency (Foran *et al.* 1978; Edwards 1981a). In an attempt to do this, a number of procedures, based primarily on some assessment of the condition of a particular site relative to the potential for the ecological region to which it belongs, have been developed and are currently in general use (Tainton *et al.* 1980; Danckwerts 1982). However all the procedures need further testing.

15.4.2.6 The role of fire. Fire is seen as a natural factor which played a major role in the evolution of South African grasslands. It is generally conceded that it is, above all, the regular occurrence of fire which has maintained a sub-climax grassland in sub-humid and humid regions known to have an environment capable of supporting mesophytic shrub or forest. Similarly, fire is believed to have controlled the invasion of xerophytic scrub (primarily *Acacia* spp.) and Karoo vegetation (shrublets) into arid grasslands.

Fire still plays a major role in present day management systems. In the humid regions this is seen as an acceptable practice if its purpose is to remove an accumulation of unpalatable senescent material of low quality, whereas in the arid regions it is considered acceptable when it is designed to control encroaching scrub or Karoo shrublets. In both situations, however, the use of fire is legislatively controlled. In the humid regions, for example, burning is permitted only in late winter or early spring, and where the grassland is burnt it may be grazed only after a defined level of recovery growth. In the more arid regions, burning is permitted only in areas in which fuel loads are sufficient to achieve the primary objective of controlling encroaching xerophytic scrub. Here the recommendation is to burn when the air temperature is high ($>25°C$) and the humidity is low ($<30\%$) so that fires are sufficiently intense to damage scrub species (Trollope 1984a). As in the humid regions, grazing in the post-fire season is strictly controlled, and land users are normally strongly advised to introduce browsing animals onto the recovery growth where scrub species coppice after fire (Trollope 1984b).

In practice, the objectives of using fire are readily achieved provided post-fire management permits adequate recovery of the grassland. In the more arid sweet grassland areas, however, land-users are loathe to burn large quantities of potentially useful forage. Fires are therefore seldom intensive enough to severely damage the xerophytic scrub, and the stigma attached to the only available domestic browser (the goat) inhibits their use during the post-fire period. Also, spring rains are normally unreliable so that recovery may be severely inhibited. The result of this, and the almost complete opposition to the use of fire in the very sweet grasslands subject to Karoo invasion, has been an extensive replacement of these valuable grasslands by xerophytic scrub.

15.4.2.7 Selective grazing. Selective grazing is widely accepted as having contributed significantly to the degradation of South African grasslands (Acocks 1975; Tainton 1972). It is not surprising, therefore, that a great deal of attention has been given to the need to reduce the detrimental effects of selective grazing programmes.

The main principle which emerges from a recognition of acceptability differences among the species making up a community is that continuous grazing must inevitably lead to the degradation of the species composition of any community. Such grazing practice leads to the season-long grazing of the most acceptable species in the community, irrespective of the stocking rate applied. Any reduction in stocking rate designed to reduce the grazing pressure on acceptable species merely reduces the number of species which are heavily grazed to only the highly acceptable ones. Area selection, whereby animals concentrate their grazing on specific areas within any paddock, is similarly merely confined to smaller areas.

For the above reasons, rotational grazing procedures are widely, and unanimously, recommended in South Africa. Controlled rotational use permits a degree of control of the intensity at which any particular species or area is grazed and, in particular, it provides a mechanism whereby grazed species or areas can be provided with the opportunity to recover from such treatment.

Two contrasting procedures are used in the design of rotational programmes for South African grasslands. In one, the objective is to minimise the degree of selection by concentrating large numbers of animals in relatively small paddocks for short periods, and following such treatment with a relatively long recovery period before the paddocks are again grazed. The intention here is to force the utilization of even the most unacceptable species so that these species are not afforded a competitive advantage during the subsequent recovery period. It is also often supposed that the unacceptable species are generally less tolerant of heavy grazing than the acceptable species, and that they are therefore disadvantaged relative to the acceptable species by such treatment (Booysen 1969). This procedure is commonly referred to as non-selective grazing (NSG).

In practice, however, such uniform utilization is often difficult to achieve and preferred species or grazing areas often need to be over-utilized in an effort to achieve a reasonable level of utilization of unpreferred species or grazing areas. Indeed, such procedures are currently recommended only where unacceptable species predominate in the grassland, and where "forced" grazing is required to achieve a reasonable level of utilization of the grassland (Tainton 1985).

The second, and now more widely recommended rotation procedure for South African grasslands is based on the deliberate control of the degree of use of the different species in the community (Tainton 1985) in a procedure known as controlled selective grazing (CSG). The main forage producing species in any community form the basis of the management procedure used. When a paddock is grazed, only the welfare of these key species is taken into account. Unacceptable species are normally left ungrazed. Such treatment, if correctly applied, can promote the vigour and improve the competitive ability of the key forage species, while the non-grazed unacceptable species become increasingly non-competitive through a decline in their tiller population resulting from extended periods of self shading. However, not all unacceptable species behave in this way so that this perceived advantage may not always apply.

15.4.2.8 Animal type. The grassland areas of South Africa are farmed almost exclusively with sheep and cattle. Because cattle are less selective than sheep, they are acknowledged to be much less severe on the grasslands than are sheep. The latter do, however, play an extremely important role in the economy of the livestock system of these areas, and so they are generally accepted as a necessary evil which must be accommodated in management programmes. To do this, a compromise is normally recommended whereby sheep farmers are encouraged to introduce cattle into their livestock systems. A maximum ratio of 6 sheep to each cattle unit is widely recommended. This invariably leads to a more uniform utilization of grasslands, and in so doing facilitates management and generally promotes the maintenance of grassland condition.

15.4.3 Ecological principles and management practice

Management practice in the grasslands of South Africa depends on the principles discussed in the preceding section. Rotational methods are recommended for all vegetation types, and particular emphasis is given to the following principles:

(1) grazing periods should be short to ensure that forage plants are grazed only once during any period of occupation of a paddock;

(2) recovery periods should be sufficiently long to ensure the recovery of the plant, but not so long that forage quality declines to unacceptable levels;

(3) specific resting periods should be provided on a regular basis during spring and autumn (to promote plant vigour) and during summer (to promote seed production);

(4) animals should be removed from a paddock before the key forage species are grazed too short;

(5) fire should be introduced towards the end of the seasonal dormant period to remove accumulated low quality material where this becomes excessive, and to even out the sward so that subsequent grazing is less selective. Fire may also be used to control invading scrub, particularly if browsing animals can then be introduced to apply pressure to the new emerging recovery growth;

(6) grassland should be fenced along vegetation boundaries to ensure that, as far as is practically possible, the vegetation within any paddock is homogeneous;

(7) grazing must be within specific grazing capacity limits which are based on the production capacity of any particular community, and must be such that the long term stability and productivity of the grassland community is guaranteed.

These principles are embodied in management programmes which normally require as many as 8 paddocks for each separate animal group maintained on a property. Various layout procedures have been designed for such rotational systems. These are designed to facilitate day-to-day management and to reduce the costs of developing the infrastructure necessary for the application of appropriate management procedures (e.g., the wagon wheel or cart wheel layout) (Edwards 1981b). A variety of these are acceptable, provided they permit the application of the basic principles of grassland management outlined above.

15.5 The savannas

The herbaceous component of South African savannas varies from a grassland in the sub-humid regions to a xerophytic Karoo shrublet in the more arid regions (Acocks 1975). This herbaceous layer is managed in the same way as the grassland and Karoo communities of non-savanna areas. In the savannas, however, the shrub component is seen as requiring its own specialised treatment. Here browsing animals (wildlife and goats) are normally incorporated in the livestock system together with cattle and/or sheep. In many areas indigenous game species, which may be either obligated browsers or mixed feeders, are widely used on a commercial basis. In other areas, either the boer goat is used for meat production, or the angora goat for the production of mohair. Goats will include appreciable quantities of browse in their diets, provided it is of reasonable quality (du Toit 1972). They serve, therefore, to use a component of the vegetation which is not effectively used by cattle and sheep.

Depending on the circumstances, goats may be used to achieve one of two main objectives in savanna areas. Where the intention is to reduce the density of the shrub component, goats may be used to exert heavy pressure on the young regrowth of these species following either a fire or a slash treatment. Here the goats normally remain continuously in the burnt or slashed paddocks to force repeated browsing of new growth for a period of about two years. This is normally sufficiently long enough to kill the shrub species and leave a grassland suited to sheep and cattle grazing (Trollope 1980). In areas which are not natural savannas, but where shrubs have invaded grasslands as a result of poor management practice, such treatment may be relatively effective, provided the grasslands are then well managed. However, in natural savanna areas there is a strong tendency for shrubs to re-invade, and it is considered more appropriate to maintain goats as a permanent component of the livestock system. They have shown themselves to be highly profitable, and so researchers have been encouraged to examine management strategies designed to ensure the continued productivity of the key shrub species. These strategies have been shown to be different from those required by the herbaceous species, particularly in so far as the required duration of recovery periods following browsing. Shrubs have been shown to require periods of recovery ranging from about 6 months to 18 months, depending on the intensity of browsing and the prevailing environmental conditions. It is therefore necessary to design specialised procedures for shrub management (Aucamp 1976).

In practice, grazing animals and browsing animals are operated on separate rotation systems. Where, for example, a recovery period is designed for the herbaceous component, goats may be introduced into a paddock, but sheep or cattle are excluded. The reverse applies where a recovery period is specifically provided for the browse component. In this way effective economic use can be made of both herbaceous and shrub species. The combination of cattle and goats has led to substantially higher profits than systems using only cattle (Aucamp *et al.* 1984), with estimated bush densities for maximum profit being derived from tree equivalents per hectare (Stuart-Hill 1985).

15.6 References

Acocks, J. P. H. 1975. Veld types of South Africa. Mem. Bot. Surv. Sth. Afr. 40. Govt. Printer, Pretoria.

Anon (*Unpublished*). 1979. Feeding and management of small stock in times of drought. Dept. Agriculture Middelburg KP.

Atkinson, C. J., and J. F. Ferrar. 1983. Allocation of photosynthetically fixed carbon in Festuca ovina L and Nardus strutah. New Phytologist 95: 519—531.

Aucamp, A. J. 1976. The role of the browser in the bushveld of the eastern Cape. Proc. Grassl. Soc. Sth. Afr. 11: 135—138.

Aucamp, A. J., J. E. Danckwerts, and J. J. Venter. 1984. The production potential of an Acacia karoo community utilised by cattle and goats. J. Grassl. Soc. Sth. Afr. 1: 29—32.

Barnes, D. L., N. F. G. Rethman, B. H. Beukes, and G. D. Kotze. 1984. Veld composition in relation to grazing capacity. J. Grassl. Soc. Sth. Afr. 1(1): 16—19.

Bartholomew, P. E., and P. de V. Booysen. 1969. The influence of clipping frequency on reserve carbohydrates and regrowth of Eragrostis curvula. Proc. Grassl. Soc. Sth. Afr. 4: 35—43.

Bender, E. A., T. J. Case, and M. E. Gilpin. 1984. Perturbation experiments in community ecology: Theory & Practice. Ecology 65: 1—13.

Booysen, P. de V. 1969. An analysis of the fundamentals of grazing management systems. Proc. Grassl. Soc. Sth. Afr. 4: 84—91.

Booysen, J., and D. I. Rowswell. 1983. The drought problem in the Karoo areas. Proc. Grassl. Soc. Sth. Afr. 18: 40—45.

Bosch, O. J. H. (*Unpublished*) 1974. Die wisselwerking tussen die habitat en 'n aantal grasgemeenskappe in die suidoostelike Oranje-Vrystaat. D.Sc.-thesis, Potchefstroom University for CHE.

Bosch, O. J. H. 1977. Ordination of a number of soil profiles of the Sterkspruit form — A basis for the explanation of floristic differences between some veld types occurring on soils of the Sterkspruit form. Agrochemophysica 9: 21—52.

Bosch, O. J. H., and M. L. Dudzinski. 1984. Defoliation and its effects on Enneapogon avenaceous and Cenchrus ciliaris populations during two summer growth periods in central Australian rangelands. Aust. Rangel. J. 6(1): 17—25.

Bosch, O. J. H., F. P. Janse van Rensburg, and S. du T. Truter. 1987. Identification and selection of benchmark sites on the litholitic soils of the western grassland biome of South Africa. J. Grassl. Soc. Sth. Afr. 4: 59—62.

Botha, P. (*Unpublished*) 1981. Die invloed van spesieseleksie deur skape, beeste en bokke op die floristiese samestelling van die gemengde Karooveld. D.Sc.-thesis, Potchefstroom University for CHE.

Botha, P., C. D. Blom, E. Sykes, and A. S. J. Barnhoorn. 1983. A comparison between the diets of small and large stock on mixed Karooveld. Proc. Grassl. Soc. Sth. Afr. 18: 101—105.

Danckwerts, J. E. 1982. The grazing of sweetveld. 2. A model to estimate grazing capacity in the false thornveld of the eastern Cape. Proc. Grassl. Soc. Sth. Afr. 17: 94—98.

Danckwerts, J. E., and A. J. Aucamp. 1985. The rate of leaf emergence and decay as criteria for optimising the grazing rotation in semi-arid grassveld. J. Grassl. Soc. Sth. Afr. 2(1): 28—34.

Danckwerts, J. E., A. J. Aucamp, and L. P. du Toit. 1984. Ontogeny of Themeda triandra tillers in the false thornveld of the eastern Cape. J. Grassl. Soc. Sth. Afr. 1(3): 9—14.

Danckwerts, J. E., and A. J. Gordon. 1986. Long-term partitioning, storage and remobilization of 14C assimilated by Lolium perenne (cv. Melle). Annals of Botany 59: 55—66.

Du Toit, P. F. 1972. The goat in the bush-grass community. Proc. Grassl. Soc. Sth. Afr. 7: 44—50.

Edwards, D. 1983. A broad-scale structural classification of vegetation for practical purposes. Bothalia 14: 705—712.

Edwards, D. 1984. Fire regimes in the biomes of South Africa. p. 19—37. *In*: P. de V. Booysen, and N. M. Tainton (eds.) Ecological effects of fire in South African ecosystems. Ecol. Studies 48, Springer-Verlag, Berlin.

Edwards, P. J. 1981a. Grazing management: stocking rate. p. 334—339. *In*: N. M. Tainton (ed.) Veld and pasture management in South Africa. Shuter and Shooter in association with Natal University Press.

Edwards, P. J. 1981b. Terms describing aspects of vegetation and its management: the management unit. p. 316. *In*: N. M. Tainton (ed.) Veld and pasture management in South Africa. Shuter and Shooter in association with Natal University Press.

Edwards, P. J., D. M. Scotney, P. E. Bartholomew, and N. M. Tainton. 1980. Environmental modification for improved grassland production. Proc. Grassl. Soc. Sth. Afr. 15: 19—30.

Foran, D. B., N. M. Tainton, and P. de V. Booysen. 1978. The development of a method for assessing veld condition in three grassveld types in Natal. Proc. Grassl. Soc. Sth. Afr. 13: 27—33.

Friedel, M. H. 1987. A preliminary investigation of woody plant increase in the western Transvaal and implications for veld assessment. J. Grassl. Soc. Sth. Afr. 4: 25—30.

Grunow, J. O., A. J. Pienaar, and C. Breytenbach. 1970. Long term nitrogen application to veld in South Africa. Proc. Grassl. Soc. Sth. Afr. 5: 75—90.

Hall, M. 1984. Man's historical and traditional use of fire in South Africa. p. 39—52. *In*: P. de V. Booysen, and N. M. Tainton (eds.) Ecological effects of fire in South African ecosystems. Ecol. Studies 48, Springer-Verlag, Berlin.

Haylett, D. G. 1969. Fertilization of Pretoria veld. Tech. Bull., Dept. Agric. Tech. Serv., Pretoria.

Hayward, F. C. 1979. Supplementary feeding. *In*: Feeding and management of small stock in times of drought. Dept. Agriculture (Karoo Region), Middelburg. K.P.

Henrici, M. 1951. The effect of cutting and grazing in Pentzia incana under different systems of veld management. Science Bulletin 282. Dept. Agriculture, Pretoria.

Hobson, F. O., and E. Sykes. 1980. Defoliation frequency with respect to three Karoo bush species. Karoo Agric. 1(5): 9—11.

Mentis, M. T., and B. J. Huntley. 1982. A description of the Grassland Biome Project. CSIR, South African National Scientific Programmes Report 62: 1—25.

Odum, E. P. 1971. Fundamentals of Ecology. W. B. Saunders Co. London.

Opperman, D. P. J., and J. J. Human. 1977. The effect of defoliation and moisture stress on the productivity of Themeda triandra Forsk under controlled conditions. Proc. Grassl. Soc. Sth. Afr. 12: 65—69.

Rethman, N. F. G. 1971. Elevation of shoot apices in two ecotypes of Themeda triandra on the Transvaal Highveld. Proc. Grassl. Soc. Sth. Afr. 6: 86—92.

Rethman, N. F. G., and P. de V. Booysen. 1967. Growth and development of Cymbopogon excavatus tillers. S. Afr. J. Agric. Sci. 10: 811—822.

Rethman, N. F. G., and P. de V. Booysen. 1968. Growth and development of Heteropogon contortus tillers. S. Afr. J. Agric. Sci. 11: 259—272.

Richards, J. H., and M. M. Caldwell. 1985. Soluble carbohydrates, concurrent photosynthesis and efficiency in regrowth following defoliation: a field study with Agropyron species. J. Appl. Ecol. 22: 907—920.

Roberts, B. R. 1981a. Application of grassland management principles: Karoo. p. 417—425. *In*: N. M. Tainton (ed.) Veld and pasture management in South Africa. Shuter and Shooter in association with Natal University Press.

Roberts, B. R. 1981b. Application of grassland management principles: Sweet and mixed grassveld. p. 383—391. *In*: N. M. Tainton (ed.) Veld and pasture management in South Africa. Shuter and Shooter in association with Natal University Press.

Roux, P. W. 1964. Progress report K-Gf 1. Veld grazing experiment. 1. Camp No. 6 (Apron veld). Agricultural research in the Karoo Region. Section C: Pastures. Dept. Agriculture, Middelburg, (CP).

Roux, P. W. 1966. Die uitwerking van seisoensreënval en beweiding op gemengde Karooveld (The influence of seasonal rainfall and grazing on mixed Karoo veld). Proc. Grassl. Soc. Sth. Afr. 1: 103—110.

Roux, P. W. 1968. Veld management in the Karoo and adjacent sweetgrass veld regions. Ch. 19. *In*: W. J. Hugo (ed.) Small stock industry in South Africa. Government Printer, Pretoria.

Roux, P. W. (*Unpublished*) 1969. The autecology of Tetrachne dregei Nees Ph.D-thesis. University of Natal.

Roux, P. W. 1979. Die veld as voedingsbron vir die Kleinveebedryf. Small stock symposium 26—28 Sept. 1979. University of the Orange Free State.

Roux, P. W., and T. E. Skinner. 1970. The group camp system. Fmg S. Afr. Jan: 25—28.

Roux, P. W., and M. Vorster. 1983. Development of veld management research in the Karoo region. Proc. Grassl. Soc. Sth. Afr. 18: 30—34.

Rutherford, M. C., and R. H. Westfall. 1986. Biomes of Southern Africa — an objective categorization. Mem. Bot. Surv. of S. Afr. 54 Government Printer, Pretoria.

Ryle, G. J. A., and C. E. Powell. 1974. The utilization of recently assimilated carbon in graminaceous plants. Ann. Appl. Biol. 77: 145—158.

Schulze, B. R. 1965. Climate of South Africa. Part 8. General survey. WB. 28 Dept. Transport, Pretoria.

Skinner, T. E. 1963. Progress report K-Gf 84. Dept. Agriculture, Middelburg (CP).

Skinner, T. E. (*Unpublished*) 1964. 'n Fisiologies-ekologiese studie van Stipagrostis ciliata en S. obtusa. M.Sc. Agric.-thesis. University of Pretoria.

Skinner, T. E. 1972. Effects and production of Merino- and Angora goat ewes on Central Lower Karoo veld. Final report K-Gf 84. Dept. Agriculture, Middelburg (CP).

Snyman, H. A., and D. P. J. Opperman. 1983. (The effects of moisture and defoliation treatments in hydrological units on natural veld of the central Orange Free State). Proc. Grassl. Soc. Sth. Afr. 18: 124—130.

Snyman, H. A., W. L. J. van Rensburg, and D. P. J. Opperman. 1985. Soil loss and run-off measurements from natural veld with a rainfall simulator. J. Grassl. Soc. Sth. Afr. 2(4): 35—40.

Steinke, T. D. 1969. The translocation of C14 assimilates in Eragrostis curvula: an autoradiographic survey. Proc. Grassl. Soc. Sth. Afr. 4: 19—34.

Steinke, T. D. 1975. Effects of height of cut on translocation of C14 labelled assimilates in Eragrostis curvula (Schradn) Nees. Proc. Grassl. Soc. Sth. Afr. 10: 41—47.

Steinke, T. D., and P. de V. Booysen.1968. The regrowth and utilization of carbohydrate reserves of Eragrostis curvula after different frequencies of defoliation. Proc. Grassl. Soc. Sth. Afr. 3: 105—110.

Stuart-Hill, G. C. (*Unpublished*) 1985. Competitive interaction between Acacia karoo and the grass sward in the false thornveld of the eastern Cape. M.Sc. Agric.-thesis. University of Natal.

Stuart-Hill, G. C., N. M. Tainton, A. J. Aucamp, and J. E. Danckwerts. 1984. Infiltration and water use patterns in an arid South African savanna. Proc. 2nd Rangeland congress, Adelaide, Australia.

Tainton, N. M. 1964. Developmental morphology of the apical meristem of Themeda traindra Forsk. S. Afr. J. Agric. Sci. 7: 93—100.

Tainton, N. M. 1972. The relative contribution of overstocking and selective grazing to the degradation of tall grassveld in Natal. Proc. Grassl. Soc. Sth. Afr. 6: 39—43.

Tainton, N. M. 1981a. The ecology of the main grazing lands of South Africa: Production characteristics of the main grazing lands. p. 41—46. *In*: N. M. Tainton (ed.) Veld and pasture management in South Africa. Shuter and Shooter in association with Natal University Press.

Tainton, N. M. 1981b. The grass plant and its reaction to treatment: the morphology of tiller growth. p. 217—224. *In*: N. M. Tainton (ed.) Veld and pasture management in South Africa. Shuter and Shooter in association with Natal University Press.

Tainton, N. M. 1985. Recent trends in grazing management philosophy in South Africa. J. Grassl. Soc. Sth. Afr. 2(4): 4—6.

Tainton, N. M., and P. de V. Booysen. 1963. The effects of management on apical level development and seeding in Themeda triandra and Tristachya hispida. S. Afr. J. Agric. Sci. 6: 21—30.

Tainton, N. M., and P. de V. Booysen. 1965a. Growth and development in perennial veld grasses. 1. Themeda triandra tillers under various systems of defoliation. S. Afr. J. Agric. Sci. 8: 93—110.

Tainton, N. M., and P. de V. Booysen. 1965b. Growth and development in perennial veld grasses. 2. Hyparrhenia hirta tillers under various systems of defoliation. S. Afr. J. Agric. Sci. 8: 745—760.

Tainton, N. M., P. J. Edwards, and M. T. Mentis. 1980. A revised method for assessing veld condition. Proc. Grassl. Soc. Sth. Afr. 15: 37—42.

Tidmarsh, C. E. M. 1947. Veld management in the marginal grassveld areas. Fmg. S. Afr. 22: 13—16.

Tidmarsh, C. E. M. 1948. Conservation problems of the Karoo. Fmg. S. Afr. 23: 519—530.

Trollope, W. S. W. 1980. Controlling bush encroachment with fire in the savanna areas of South Africa. Proc. Grassl. Soc. Sth. Afr. 15: 173—177.

Trollope, W. S. W. 1984a. Fire behaviour. p. 199—217. *In*: P. de V. Booysen, and N. M. Tainton (eds.) Ecological effects of fire in South African ecosystems. Ecol. Studies 48, Springer-Verlag, Berlin.

Trollope, W. S. W. 1984b. Fire in savanna. p. 149—175. *In*: P. de V. Booysen, and N. M. Tainton (eds.) Ecological effects of fire in South African ecosystems. Ecol. Studies 48, Springer-Verlag, Berlin.

Van den Berg, J. A. (*Unpublished*) 1972. Invloed van seisoensbeweiding op Cymbopogon-Themeda-veld. M.Sc. Agric.-thesis, University of the Orange Free State.

Van der Walt, J. L. 1971. Voorlopige verslag oor die reaksie van Dorre Karooveld op beweiding gedurende spesifieke jaargetye: Boesmangrasveld. Proc. Grassl. Soc. Sth. Afr. 6: 82—85.

Van der Walt, P. T. (*Unpublished*) 1972. 'n Plantekologiese opname van die Boschberg en sy omringende gebiede met spesiale verwysing na die weidingsfaktor. Ph.D-thesis, University of Natal.

Van der Westhuizen, F. G. J. (*Unpublished*) 1980. Invloed van snoei op fotosintese, reserwestatus en droëmateriaalproduksie van Ehrharta calycina en Osteospermum sinuatum. Ph.D-thesis, University of Stellenbosch.

Vorster, M. (*Unpublished*) 1980. Eienskappe wat 'n rol speel by veldevaluering in die oostelike Karoo. M.Sc. Agric.-thesis, University of the Orange Free State.

Vorster, M., P. Botha, and F. O. Hobson. 1983. Utilization of Karoo veld by livestock. Proc. Grassl. Soc. Sth. Afr. 18: 35—39.

Vorster, M., and E. M. Meyer. 1981. Grazing capacity zones in the Karoo Agricultural Region for the Drought Relief plan. Unpublished map. Dept. Agriculture, Middelburg (CP).

Vorster, M., and P. W. Roux. 1983. Veld of the Karoo areas. Proc. Grassl. Soc. Sth. Afr. 18: 18—24.

Walker, B. H., and I. Noy-Meir. 1982. Aspects of the stability and resilience of savanna ecosystems. p. 556—590. *In*: B. J. Huntley, and B. H. Walker (eds.) Ecology of Tropical savannas. Ecol. Studies 42, Springer-Verlag, Berlin.

Weinmann, N. 1955. The chemistry and physiology of grasses. p. 571—600. *In*: Meredith (ed.) Grasses and pastures of South Africa. Central News Agency Ltd. Johannesburg.

16. Range management from grassland ecology

R. NICHOLSON

16.1 Abstract

The frequency of activity in institutional research in grassland ecology has had a bimodal temporal distribution. The first peak occurred about 1940 and was derived from the cadre of ecologists that emanated from the Department of Botany at the University of Nebraska. Range management as a profession partially distinct from grassland ecology, evolved from the Nebraskan network. The first era lasted about 30 years. The inception of the Grassland Biome component of the International Biological Program (IBP) marked the inception of the second era of intense activity by grassland ecologists, the duration of which was about 10 years, peaking in the mid-1970's. The program produced prodigious detail on grassland ecology, but no new paradigms for the practices of range management. The existence and character of grassland is determined on the basis of prevailing climate, age and human intervention. Community composition, seasonal, trophic and nutrient dynamics, edaphic traits and adaptive fitness are a few of the many ecological concepts that are applied in range management practices. Knowledge of ecological traits of grassland can be used in harmony with knowledge of external forces such as marketability to formulate management plans that might include when, where, with what, how many and how often to graze livestock. From the domain of grassland ecology, the practices of describing range sites and evaluating range condition were brought forth from the concepts of community classification and ecological succession, respectively. Governmental agencies generously utilize these applications, although change and academic controversy are enmeshed. The ecosystem concept, although not directly derived from grassland ecology, is widely vocalized in range management, primarily as a heuristic device rather than one of practicability. The hierarchy is a different approach that recently has been advocated by some ecologists. The relatively recently derived sphere of research in the

Tueller, P. T. (ed.), Vegetation science applications for rangeland analysis and management.
ISBN 90-6193-195-9.
© 1988, *Kluwer Academic Publishers, Dordrecht, Boston, London.*

population biology of plants holds promise for applications in range management, such as revegetation and pest control. Contemporary research in natural plant/herbivore relationships in grasslands has revealed detail on the interdependence of these 2 conspicuous trophic levels, and has provided abundant theorizations for experimentation with domestic animals on rangeland.

16.2 Introduction

The first agricultural experiment station was established in 1843 at Rothamsted, England, which is where the modern era of research on grasslands and pastures had its origin (Barnard and Frankel 1964). By 1880, the results of 20-year experiments on grasslands were being reported, which included plot measurements of standing crop, species composition and chemical analyses. The roots of academic ecology are interred in the Germanic region of Europe of the early 19th century and this history has been previously summarized (see Edgerton 1976). In the United States, C. E. Bessey reported original observations on the grasses and forage plants of Nebraska in 1887 (Tobey 1981). J. G. Smith in 1895 wrote of forage conditions he observed in the prairie region (Dyksterhuis 1980). The details of the history of range management have been recounted elsewhere (see Chapline 1980, Dyksterhuis 1980, Holechek 1981).

Deterioration of grasslands in the arid and semiarid regions of America and Australia in the latter part of the 19th century became an important stimulus for research in grassland ecology. Excessive grazing of these lands was resulting in regressive changes in botanical composition, and it was becoming evident that by controlling the numbers of livestock would it be possible to stabilize these grasslands, but only with a thorough knowledge of grassland ecology. So that management could occur on an ecological basis has been a major thrust of research in grassland sciences for the past 40+ years (Barnard and Frankel 1964).

This chapter intends to describe selected fundamentals of grassland ecology that are relevant to the work of range managers. Emphasis leans more toward the origin of the information rather than its substance. No single volume, let alone a single chapter, could provide a complete description of grassland ecology that has utility for range management. Many relevant papers could not be cited and no doubt many were overlooked; the method does not invalidate the attempt, but it does condition it (Spedding 1971). Ecology of most non-American grasslands was neglected, primarily because it was incognate to the development of range management and literature was obscure. Not covered herein or only indirectly so was the ecology of common grassland components such as poisonous plants, pathogens, rodents, rabbits, predators, insects, birds, and big game and feral animals, as well as individual soil/plant relationships. Methodology of ecological sampling was treated very lighly and not independently. At least some of the above topics are important to all range managers, but all hereafter ought to be important to all range managers. Three major topic areas will be highlighted: (1) the birth of range management from grassland ecology, (2) the characteristics of the ecology of grazed grassland that distinguishes if from other land types and (3) fundamentals that have emanated from grassland ecology, which have utility in rangeland management, both old and new.

The ancestral roots of range management are embedded in grassland ecology, but today the scope of range management is broad, drawing fundamental principles from physiology, soil science, animal husbandry, plant breeding, ecology and other related sciences. It is assumed that the reader possesses a basic literacy of ecology and range management. Many of the references cited hereafter are superb, and can provide readers with detail and additional sources unaffordable in this format. A selected list of briefly annotated volumes on grassland ecology is appended.

16.3 Range management from grassland ecology

Grassland ecology appears to have had 2 heydays, both of which occurred primarily in the central plains of North America. The first was the era of intense research spanning roughly the first 1/2 of the 20th century, reported by J. E. Weaver of the University of Nebraska and his successors. Much of that research dealt with the effects of the drought of the 1930's. The second era was in the late 1960's and 1970's, an intensive and extensive study of grassland ecosystems under the auspices of the Grassland Biome project of the International Biological Program (IBP). Both eras resulted in productive applications of grassland ecology for management. Today, vigorous research in grassland ecology continues on the plains of North America, conducted mostly by investigators at colleges and universities therein, such as Augustana (SD), Macalester (MN), Colorado and Kansas State. Recent acquisitions of large tracts of grassland by The Nature Conservancy in South Dakota, Nebraska and Kansas have been beneficial to current research programs in grassland ecology. On east African savannahs, programs such as those at Syracuse University are actively pursuing research in grassland ecology and on the veld, scientists at major universities in South Africa are doing likewise.

16.3.1 The Nebraskan era

To appreciate the work of the first era of grassland ecology, it is important to understand its conception, which occurred in the Botany Department at the University of Nebraska before the end of the 19th century. In his treatment of the founding school of American plant ecology, historian Tobey (1981) details how range management as a distinct discipline emerged almost entirely from American plant ecology, more specifically American grassland ecology. The history of range management originated with the development of concepts at the University of Nebraska by C. E. Bessey concerning the pragmatic botany of developing agriculture in the Great Plains. Of profound and lasting influence to range management have been the ideas of R. Pound, F. E. Clements, J. G. Smith, J. E. Weaver, A. W. Sampson, F. W. Albertson, L. A. Stoddard and others who as intellectual cohorts and students received their early guidance in the Bessey system. Of Clements contemporaries, Roscoe Pound was most influential in the development of rigor in ecological methods and community succession (Clements 1905, 1916). The earliest practice of ecology by his students found no favor with Bessey, however the tension resulting thereof motivated the rigor and quantification of grassland ecology as later espoused by F. E. Clements.

Based on an analysis of 535 publication records of grassland ecologists between 1895—1955, *Ecological Succession* by Clements in 1916 was considered by Tobey (1981) to be the "paradigm-setting work for [grassland ecology]". He reminded readers that the Ecological Society of America was formed in 1914 and began publishing the periodical *Ecology* in 1920, presumably additional motivating forces to the inception of an era of intense publication of research in grassland ecology. This era culminated in 1939—1941, with approximately 75 papers published in those 3 years. Tobey (1981) performed citation analysis on a sample of 210 papers of 58 grassland scientists, revealing that J. E. Weaver (Univ. Nebraska), F. E. Clements (Carnegie Institution) and F. W. Albertson (Fort Hays KS St. Col.), in that order, accounted for 1109 of 5717 citations. Weaver alone was cited 679 times, while Clements contemporary at the University of Chicago, H. C. Cowles, ranked 16th in the list of grassland ecologists with 32. One-third of all citations were to the 58 scientists themselves. Typically citation counts are considered to be a mechanism for ranking prestige, but Tobey (1981) thought that the statistics "reinforce the image, . . . , of a tighly interwoven, inward-looking, conservative specialty".

No other individual is more likely to be thought of as the father of American plant ecology that F. E. Clements. Although born, raised and educated in the grasslands of Nebraska, it was in the forests of the Colorado Rockies while on vacation with his wife in the summer of 1898, that Clements conceived significant portions of his philosophy of vegetation structure and plant succession. Although Clements left Nebraska before his most influential works were published, his ties with the grasslands of the Great Plains were never severed (Tobey 1981).

Nebraska was not the sole source of original thinking during the embryonic stages of grassland ecology. H. C. Cowles at the University of Chicago (who was preceded and influenced by J. M. Coulter) researched and wrote (Cowles 1899, 1901) of the sand dune communities of Lake Michigan and of ecological succession (Cowles 1911). But in contrast to the Nebraskans, those of the Chicago school of ecology (Cowles and successors) were not prone to research of practical problems, thus it was Clements successors who came to dominate American grassland ecology and consequently range management. McIntosh (1976) wrote comprehensively about the history of ecology in the 20th century. In a discussion of the popular emergence of ecology in the 1960's, he noted that the emergence of grassland ecology in the Great Plains was also partially the result of an environmental crisis (drought). Naturally the affliction of the drought on grassland communities was of direct practical application to range management.

The succession and accomplishment of J. E. Weaver's students was utlimately responsible for the origin of range management as a profession. Many of his successors worked in governmental agencies, whereas others established themselves in experiment stations and universities with emerging programs of research and education in range ecology. Much of the research was pure grassland ecology, but little insight is required to perceive that nearly all genre of basic ecological research in grasslands can be applied to range management. Most contemporary American range managers and many others can trace their educational and/or experiential roots to Charles Edwin Bessey.

16.3.2 The IBP era

The Grassland Biome was the centerpiece of the International Biological Program and evolved over a period of time from vague conception in 1960, until its formal designation and subvention by the National Science Foundation and U.S. Atomic Energy Commission in 1968 (Risser *et al.* 1981). It was a substantial project to integrate the study of structure, function and utilization of grassland ecosystems. Studies were conducted at 11 North American sites. At one site, Pawnee (north-central Colorado), data were collected from 1969—1974 and at most of the other sites from

1970–1972. All sites had both grazed and ungrazed treatments and the Pawnee site included large plot studies of water and nitrogen amendments. Early phases emphasized measuring biomass and nutrient dynamics to be compared with climatic and physical/chemical data, but later the program emphasized measuring processes to explain dynamics. The final year of support by the National Science Foundation was 1977 (Laycock 1979). Cragg (1979) said the synthesis phase was hampered by lack of funding, but the IBP effort put grassland management on a more rational basis than before.

Voluminous preliminary reviews and technical reports of works in progress of the Grassland Biome were written (e.g., Dix and Beidleman 1969, Sparrow and Doxtader 1973) as well as numerous publications in the open literature (e.g., French *et al.* 1976, Clark 1975, Woodmansee 1978a). Kumar *et al.* (1976) identified over 1600 species of invertebrates inhabiting the 25 (approx.) square miles of the Central Plains Experimental Range of the Pawnee National Grassland in north-central Colorado. In a series of 4 reports, Sims *et al.* (1978) and Sims and Singh (1978a, 1978b, 1978c) described in detail the results of data collected from 10 grassland sites. These articles covered structural attributes of the vegetation and abiotic environment, and described functional aspects of the vegetation such as seasonal dynamics and efficiency of primary production, compartmental energy flow, efficiency of water use and the effects of livestock grazing on such.

Mathematical simulation was a substantial part of the Grassland Biome study. Innis (1978) summarized much of the work, including models of vegetation, consumers and abiotic components. Dynamic, whole system, deterministic models were constructed to simulate biomass dynamics and circulation of nutrients, primarily nitrogen. "The goals of the modeling effort were to create a model that would serve as a communications device and organizer of information, be useful as a research tool, and yield results that could help in elucidating biological phenomena in grassland ecosystems" (Woodmansee 1978b). ELM, a total ecosystem simulation was constructed of 9 major sub-models: plant phenology, producer, mammalian consumer, insect consumer, decomposer, water, temperature, nitrogen and phosphorus. Abiotic driving variables were wind speed, cloud cover, relative humidity and short-wave radition, using both observed and simulated weather data. Woodmansee (1978), in a critique of the model, stated that the utility of the model as a predictive aid to management, such as livestock grazing practices, could only be judged by the individual(s) attempting to use the model. He said the best use of the model was "as an aid to, rather than the basis of, decision making." A more international perspective of efforts and results of grassland ecosystem modeling was edited by Breymeyer and Van Dyne (1978).

French (1979) edited a volume entitled *Perspectives in Grassland Ecology*, to which 13 scientists of the Grassland Biome project contributed articles. Simulation models of blue grama (*Bouteloua gracilis*) indicated that large amounts of carbon were translocated (85%) to belowground and high proportions of total respiration (75%) occurred belowground (Detling 1979). Simulated irrigation increased net primary productivity by a factor of 5, whereas both irrigation and fertilization produced increases of 10 times. In a field experiment, Dodd and Lauenroth (1979) reported actual increases from irrigation, nitrogen fertilizer and both to range from 2.0 to 4.7, 1.3 to 2.7 and 5.0 to 13.0 times, respectively. The latter treatment also resulted in large increases and dominance of prairie vole (*Microtus ochrogaster*), a species of rodent rarely found in the area. A survey of literature yielded estimates of natural gains and losses of nitrogen in grasslands to range from 0.2 to 1.0 g m^{-2} $yr.^{-1}$ and 0.4 to 2.0 g m^{-2} $yr.^{-1}$, respectively (Woodmansee 1979). Gains were from wet and dry deposition and fixation, whereas losses were mostly by volatilization of ammonia from animal excrement. Coupland (1979a) summarized major international efforts of the Grassland Biome project in a volume describing structure, development and utilization of the grassland's of the world. He said that early in the development of IBP research, 30 nations began studies of grassland (Coupland 1979b). In 1978, in order to integrate focus on grassland research, several organizations in South Africa coalesced to form the Grassland Biome Project within the National Programme for Environmental Sciences. For over 60 years, ". . . much of the work has been directed to isolated topics, often only remotely integrated with the findings or efforts of other studies" (Tainton 1984). Over 600 references were included in that list of reports.

ELM modeling in the true prairie was reported by Risser *et al.* (1981). Their book (*The True Prairie Ecosystem*) comprehensively summarized most of the IBP work at the tallgrass prairie site (Osage) in Oklahoma, including reports on abiotic properties, producers, consumers, decomposers, whole system properties and successional dynamics. Chapters included studies of responses to stresses induced by livestock grazing, fire, nutrients, irrigation and pesticides; a superb effort, the book integrates and complements major studies of the ecology of grasslands, particularly the tallgrass prairie.

16.4 Ecological traits of grassland

Grassland is a type of land dominated by vegetation in which the major species of plants are members of the grass family (*Poaceae*). A few grasslands are predominately composed of annual species, whereas perennials dominate the majority of grasslands. Only about 3% of North American grasslands are the annual type (Lauenroth 1979). Frequently grasslands contain moderate to large amounts of herbaceous plants, known as forbs (Bazzaz and Parish 1982) and many of the drier grasslands (savannahs) have scattered trees and shrubs. The majority of biomass and biological activity occurs underground (Clark 1975). Availability of nitrogen is the most universal limiting factor (Barbour *et al.* 1980). Spedding (1971) allowed only 2 types of grassland, natural and artificial, depending largely on the degree of influence by man. Coupland (1979b) recognized 6 types of grassland, based on climate, successional status and land use. Lauenroth (1979) stated that there are 3 occasionally interrelated mechanisms that determine the occurrence of grasslands: climate, succession, and agriculture. Of the 3, the former determinant accounts for about 25% of the surface area of the land of the earth. These areas, such as the Great Plains of North America, usually lie in the rain shadow of a range of mountains, whereas others such as the African grasslands, represent a transition between deserts and forest. Although grasslands cover a large amount of the earth's surface (46×10^6 km^2) (Shantz 1954), the temperature/precipitation relationships are more restrictive than those of forest. Precipitation is seasonal and usually insufficient in amount to support forest vegetation with the annual averages between 0.25 and 1.0 m. Many attempts to construct meaningful functional relationships between annual average net primary productivity (AANPP) and climatic factors, principally average annual precipitation, were analyzed by Lauenroth (1979). None of those attempts resulted in prediction efficiencies of greater than about 50%. Grassland temperatures vary from warm year-around to hot summers with cold winters, averaging between 15 and 25 C on an annual basis.

Successional grasslands are those intentionally determined by the activities of man such as mowing, livestock grazing or burning, although some are naturally maintained by the effects of defoliation and trampling by wild herbivores (Harper 1977). Much of this category of land would otherwise be forest, thus productivity is usually greater than most of the climatically determined grasslands. Harper (1977) described European successional grasslands determined by grazing that have a recorded history of over 900 years. Grasslands of the third category (agricultural) are determined by an intensification of man's activities of the successional category, whereby agronomic practices such as seeding, irrigation, and or fertilization are used to boost productivity. Wright and Bailey (1982) described the historical occurrence of fire in North American grasslands, and opined that climate has been the dominant factor in determining their distribution. However, they added that wide fluctuation in the areal extent of woody vegetation would occur were it not for other determining factors such as fire.

Coupland (1979b) concisely described the traits of grassland, both biotic and abiotic. Risser *et al.* (1981) presented a succinct, but thorough description of the ecology of grasslands, particularly those of North America. Spedding's (1971) *Grassland Ecology* is an excellent treatment, especially of Continental grasslands and although somewhat dated, it could serve as a good text for introductory range management.

North American grasslands are 12 to 16% of the grasslands of the earth and produce 9 to 16% of the AANPP or 9.12×10^8 tons $yr.^{-1}$ (Risser *et al.* 1981). All grassland types are utilized to some degree for production of forages for domestic livestock and/or wild herbivores. Despite recent trends in the use of grain-based concentrates for feeding livestock, most of the animal products derived from ungulates come from grazed forages, particularly by beef cows and their progeny to about 350 kg (Ward *et al.* 1975). Testimonial to the productivity for meat production by the major remaining North American grasslands, is the fact that 42% of all the beef cows that produced a calf in the U.S. in 1985 were in the 7 plains states of Montana and North Dakota through Texas. All these states except North Dakota ranked in the top 7 by the above criterion (USDA 1986).

In *Poaceae*, the evolution of 2 different chemical pathways of photosynthesis has resulted in distinct patterns of adaptation (Waller and Lewis 1979). Grasses having the C-3 pathway are adapted to conditions that are more cool than others in which the pathway is C-4. Likewise C-4 grasses are most efficient in areas that are drier and warmer. For example in North America, the abundance of C-3 grasses decreases from north to south and higher to lower altitudes, whereas the distribution of C-4 species is the inverse. In temperate regions with distinct seasons, the rate of growth of C-3 grasses is greatest in spring and secondarily in the fall, whereas summer if the principal season of growth for C-4 grasses. Seasonal grazing is a common application of the knowledge of C-3, C-4 behavioral traits, because often it is possible to obtain profitable rates of animal growth only during the season of most active growth of the forage. For 20 million years grazing animals evolved to survive by consuming grass, not to produce marketable quantities of meat, milk or wool. To obtain the highest possible rates of conversion of forage to livestock product, it is advantageous to match the spatial and temporal attributes of grasses with seasons and intensity of grazing. Management schemes such as intensive early stocking (Klipple 1964, Smith and Owensby 1978) and complementary forages (Lorenz 1977, Launchbaugh 1982) have been demonstrated to greatly increase secondary productivity, while maintaining favorable ecological conditions.

Major environmental factors such as climate determine the distribution, composition and physiognomy of grasslands, but what determines the functional uniqueness of grassland ecology? An important, but intangible trait of grasslands is stability. McNaughton (1968) applied Margalef's (1965) widely accepted model of community stability to California grasslands and pronounced it generally applicable, but with some differences. He discovered "no relationship between stability and productivity, dominance, or diversity.", concluding, unlike Margalef, that no tradeoff relationship between productivity and stability existed in those grasslands. In the Serengeti, 17 years later, McNaughton (1985) said functional properties of vegetation were positively related to community species diversity, in 5 of 7 tests. Harper (1977), in discussing grazing exclusion experiments by Tansley and Adamson (1925), described how selective grazing behavior by rabbits was responsible for maintaining a high level of floristic diversity in a grassland that would otherwise be composed of a few dominant species. Great variability exists in estimates related to trophic dynamics in grassland.

Redmann (1975) reported annual solar efficiencies from estimates of his and others research in grasslands to range from 0.09 to 0.33% capture. MacFadyen (1964) [see Lewis 1973], estimated that livestock removed about 14% of the net annual primary production of an intensively managed grassland in Great Britain. In European pasture ecosystems, cattle and sheep were estimated to have removed 85% and 86% respectively, of aboveground net primary productivity (Haas 1979). In 1970, on a grazed South Dakota grassland in "high range condition", peak standing crop was estimated to be 200 g m^{-2}, whereas the estimate for cumulative net primary production was 500 to 1000 g m^{-2} (Lewis 1973). Stocked at an appropriate rate (1.0 AUM $acre^{-1}$), livestock on this area would have consumed about 37 g m^{-2} of dry matter, (conservatively using the lower estimate, less than 8% of the net primary production would be routed through the livestock). In this scenario, at the end of a grazing season of 6 months, aboveground standing crop would be about 130 g m^{-2} inside exclosures, whereas outside areas would have about 65 g m^{-2}, for an apparent usage rate of 50% (Lewis 1973). McNaughton (1985) said the inside vs: outside procedure in the Serengeti would produce apparent consumption rates of 92%, but that actual consumption of primary production averaged 66%. Furthermore, grazing effectively stimulated net productivity at most study sites. Doubling the rate of stocking in South Dakota would have resulted in a rate of consumption of about 74 g m^{-2}, or $<15\%$ of the cumulative net primary productivity, but that such an increase would not be "possible without injuring the resource with our present understanding" (Lewis 1973). Although European, South Dakota and Serengeti rangelands are characteristically different, one necessarily ponders the magnitude of difference in consumption/production rates.

Most grasslands are inherently stable systems, resistant to perturbations such as hail, fire or drought. Consider how devastating these naturally occurring phenomena are to cropland (naturally unstable systems) compared to grassland (naturally stable). Energy costs for system maintenance are substantial, whether the system is a native grassland or an adjacent field of wheat, differing only in the sources and routing of energy. On cropland, large amounts of phytomass are removed by man, maintenance comes in the form of petroleum energy subsidies, whereas rangeland is managed with little such subsidy (Ward *et al.* 1975). In grassland, an adequate amount of phytomass (photosynthetic and perennating tissues) is the necessary ingredient for maintenance and some amount is a surplus available for harvest by herbivores. Clark (1975) said the average root $shoot^{-1}$ ratio of mass was about 6, in the 9 North American IBP grassland sites and Redmann (1975) reported ratios between 8 and 14 in different grassland community types in the northern Great Plains. Clark (1975) estimated average microbial mass at 100—200 g m^{-2} and about 60% of the total annual net primary productivity was routed belowground. Alternatively or additionally, because such large amounts of matter and energy in grasslands are routed to the subsurface, large populations (e.g., >5 million phytophagous nematodes m^{-2}) of belowground herbivores can be supported (Smolik and Lewis 1982). Nematodes probably consume more than livestock because of higher rates of metabolism, thus competing with aboveground herbivores for harvestable surpluses available from the grasses. The preponderance of invisible activity is a unique trait of grassland ecology, the mechanisms of which are less well understood than the aboveground counterparts. The belowground does not lend itself well to direct manipulations, but because management practices affect the flow of energy and matter (e.g., grazing, burning, fertilizing), its components are indirectly affected. The maintenance of the dominant belowground portion of the grassland ecosystem is requisite to maintaining reasonably stable rates of productivity for aboveground grazing.

Despite the foregoing, grasslands are capable of supporting the largest populations of animals of any biome (Wallace *et al.* 1986). Additional understanding of the dynamics of energy, allocation of carbon and other matter in grassland ecosystems can potentially result in refinement of range management practices that will produce improvement in trophic efficiencies.

16.5 Application of fundamentals

In discussing the role of plant ecology in range management, Parker (1952) grouped problems in range management under 3 major headings: (1) grazing management, (2) artificial revegetation, and (3) control of undesirable range plants. The concepts that he listed and described as most useful in range management and research were: (1) plant and animal competition, (2) indicators, (3) vegetation as a reflection of the major environmental influences of climate, soil and animals, and (4) primary and secondary plant succession. Five years later, Costello (1957) outlined 5 developments in ecology that he considered important to range management: energy relations, ecologic balance, reproduction, competition and ecologic patterns. He expounded upon "ecological blindspots": oversimplification of nature, impatience or desire for quick fixes, basing management on physical or political boundaries rather than ecosystems and ecological leadership. Furthermore he questioned the validity and concreteness of the concept of climax, increasers, decreasers, invaders and the scorecard system of range condition evaluation. However he allowed that "With appropriate application of the principles of their science, ecologists can give unmeasured assistance to range managers". Since 1957, some of the problems discussed by Costello have been ameliorated, others not. Following are 5 selected topic areas within grassland ecology with discussion of fundamentals having applicability in range management: community classification, ecological succession, ecological systems, population biology and plant/herbivore relationships. The first 2 were chosen because of universal applicability, their origin was from the school of grassland ecology and they represent underpinnings in range management. The systems approach represents a more recent paradigm in range management although its origin was not traceable to grassland ecology. The latter 2 topic areas are relatively newer areas of research in grassland ecology with encouraging contributions for range management.

16.5.1 Community classification

The plant community according to Clements was a conceptually concrete entity, having discrete boundaries and the singular characteristic that the whole (community) possessed attributes nonexistent in a mere aggregation of its component parts. These attributes were generally considered to be mechanisms of self-regulating control of metabolism within the community, thereby perceptually analogous to an organism (Clements 1916). From this infrastructure emerged an elaborate system of classifying vegetation into discrete units. Similar floristic assemblages having similar environments (primarily climate) was the basis of Clements scheme of community classification. Over the years many of the details of Clementsian ecology have been criticized, modified, amended and extended, in works of number far beyond the scope of this paper. However in the next section, additional details on the utility of Clementsian ecology are discussed.

As range management was evolving into a unique discipline, grouping and categorizing types of range vegetation was a natural derivation of the concept of plant community, resulting in the classical construct range site (see Shiflet 1973). Formal designation and description of range resources for utilitarian purposes apparently first occurred in 1911 when Federal agencies (USDA Forest Service) became involved in the conservation and management of rangeland (Dyksterhuis 1958). By 1949, the USDA Soil Conservation Service had developed, in the grasslands of the Great Plains, a systematic procedure for classifying range sites based primarily on floristics, but also recognizing abiotic traits of sites (Dyksterhuis 1949). Using these criteria it became possible to inventory and map the areal extent of range sites, which were homologous to grassland community types. Interfusing the community concept with ecological succession resulted in a system to assess the condition of rangeland, which was further extrapolated to produce estimates of the carrying capacity of herbivores (Soil Conservation Service 1985). Bonham (1983) stated that until a better system is derived, "management of rangelands will continue to be applied to vegetation as groups of species".

16.5.2 Ecological succession

Although Sampson (1919) was considered the first to have written about plant succession in direct relation to range management, Dyksterhuis (1949) was most responsible for initiating the widespread use Clementsian principles in evaluation of the condition of rangeland, another classical construct of range management derived from grassland ecology. Whereas spatial changes in floristic composition were used to delineate range sites, the perceived product of temporal change was used to assess the condition of rangeland. Thus range sites and range condition were interdependent. Assessment of condition in this sense referred to an ability of the range technician to judge the degree of floristic departure from the Clementsian climax of a given range site. Condition at any point in time was simply a state (Clementsian *sere*) representing some extent of regressive floristic change that might have occurred. Furthermore, this state was assumed to be correlated with the grazing value of the site, wherein sites that were floristically similar to climax had highest value for grazing and vice versa. Similar to the classification of range sites, quantifiable floristic difference was the dominant criterion. Each species of plant was categorized increaser, decreaser or invader depending upon the species perceived response of its population to the effects of grazing. The floristic mix quantitatively determined the outcome of range condition assessment, on a gradient scale beginning with poor condition composed of mostly increasers and/or invaders, through excellent condition composed of mostly decreaser species. The highest attainable range condition and the Clementsian climax grassland community were assumed to have the same floristic structure (Soil Conservation Service 1985). Understanding the temporal dynamics of plant populations and factors controlling trophic efficiencies in grassland ecosystems would subsequently reveal inadequacies of this approach. For example, McNaughton (1984) described "grazing lawns" in the Serengeti created by gregarious ungulates. These grasslands supported dense, but low growing swards of perennial grasses composed of what would necessarily be defined as decreasers in the range condition concept of Dyksterhuis (1949). McNaughton (1984) concluded that these areas support more grazers, proximally because of the intense grazing. Westoby (1985) subsequently disagreed, saying that much literature in range science contradicts McNaughton's (1984) conclusions about intense grazing. Range condition evaluation based on climax (see Dyksterhuis 1949) suffered from misplaced concreteness. Smith (1978) proposed a system of range site evaluation based on soil characteristics and "productivity of present vegetative cover relative to potential for any particular land use". He criticized the climax approach because climax is not always desirable, pristine conditions might not be climax, no allowance made for annual forages, and the system lacked universal applicability. Range condition concepts continue to be evaluated and criticized by range professionals.

16.5.3 Ecosystem concepts

In an article intensely critical of Clementsian concepts, the term ecosystem apparently first appeared in the open literature (Tansley 1935). Tansley said the ecosystem was a unit of nature in which component organisms and inorganic factors occur in a relatively stable dynamic equilibrium. Earlier in the century, A. G. Tansley had been Britain's leading Clementsian ecologist, but his article of 1935 "signaled the philosophical shift that was isolating Clementsian ecology." (Tobey 1981). For most ecologists it was not until the 1940's and 1950's that sufficient empirical evidence accumulated to reject the Clementsian concepts of the biological community as a superorganism and the climatically determined climax (McIntosh 1985). Lindeman (1942) contributed immensely to the chronology by crystallizing concepts of trophic dynamics, but widespread use of the ecosystem concept did not occur until after the publication of *Fundamentals of Ecology* (Odum 1953), over 10 years after the introduction of trophic dynamic ecology by Lindeman. Dyksterhuis (1958) emphasized the importance of recognizing that "range is an ecosystem".

Several of Clements contemporary ecologists doubted the validity of his premises (McIntosh 1985). McIntosh (1985) commented that in 1947 in separate articles in a single volume of *Ecological Monographs*, 3 prominent ecologists rejected the Clementsian concepts of community in favor of the individualistic approach of Gleason (1926, 1939). Ecology became increasingly quantitative in the 1950's and 1960's, with emphasis on the continuum and gradient theory (see Whittaker 1962, 1967). Major (1969) described the historical development of the ecosystem concept, citing over 60 references. This new paradigm in ecology, which did not originate in the grasslands, apparently had little influence in range management in governmental agencies wherein the discrete community in the form of range sites is in current use, and Federal agencies use of climax concepts persists in their evaluation of range condition (Soil Conservation Service 1985).

Blended in the effort of pedogenecist Jenny (1961) were the basic components and fundamental processes of ecosystems including succession. He described the ecosystem in a series of state factor equations, considering the system's initial state, its flows of energy and matter, and time. Collectively these equations expressed the functional relationship between the independent or controlling factors of the environment (climate, geological materials and available organisms) and the interrelated dependent factors of the ecosystem (vegetation, soils, consumers, decomposers, and the microclimate). Jennys heuristic, orderly and functional descripton of nature was further utilized as the ecosystem framework for range management by Lewis (1969). All practices of range management can be reduced to the manipulation of 1 or more of the controlling and dependent factors; the former generally more unusual and difficult (e.g., weather modification) and the latter conventional and simple (e.g., adjusting the stocking rate of livestock). Furthermore, ecological succession is axiomatic when ecosystem components are manipulated by nature or man. The stated goal of range management was to attain a stable dynamic equilibrium through manipulation of 1 or more of the dependent or controlling factors of the grassland ecosystem.

Like others, grassland ecology and rangeland management adopted the ecosystem concept (e.g., Watt 1968, Van Dyne 1969, Spedding 1971). Jameson (1970) said successional ecology was being replaced on "the theoretical front" by a collection of concepts and analytical techniques known as systems ecology. By the early 1970's even range management curricula in universities adopted the ecosystem approach, offering courses entitled "producers", "consumers", etc. Although the ecosystem has become widely accepted as the fundamental unit of study in ecology and management of natural resources (see Van Dyne 1969), some criticize that it "remains the least coherent of the organizational levels of ecology. It lacks a logical system of interconnected principles and a well-understood and widely accepted focus." (Reiners 1986). Engelberg and Boyarsky (1979) questioned the seemingly dogmatic existence of cybernetic control in ecosystems (see Margalef 1968), and that to presume such existence was to analogize the ecosystem with an organism, a tenet of Clements that most ecologists no longer consider valid. Jordan (1981) and others countered that to deny self regulation in ecosystems was to deny that ecosystems exist and that real-world resource management practices demonstrate the existence of informational networks in ecosystems. Allen and Star (1982) stated that the approach to studying ecosystems has been one of massive reductionism (e.g., International Biological Program), and that the approach has reached its limits. O'Neil, *et al.* (1986) suggested an hierarchical approach in order to assure that appropriate complexity as a definable ecosystem property is included in scientific models. Snaydon (1984) stated that grazed pastures demand management as ecosystems, not populations or communities, thus most of the information needed to define management practices must be obtained at the ecosystem level. Furthermore, he stated that the present management practices used in pasture ecosystems are based almost entirely on the results of empirical studies (e.g., stocking rate — weight gain studies). Such studies represent a bioassay approach that accounts, in end results, for complexity as an entity that concerned Allen, Star and O'Neil.

16.5.4 Population ecology

Whereas most paradigms of range management have emanated from the theory of plant community and ecosystem ecology, relatively fewer applications of the more recently derived plant population ecology have occurred. However, numerous basic studies of the population ecology of grassland species are of potential and immediate relevance to range management (e.g., Kays and Harper 1974, McCanny and Cavers 1987). West (1968) stated that research on the autecology of range plants was "fragmentary". He presented a comprehensive outline for the study of the autecology of grasses, with 6 major components: (1) taxonomy, (2) genecology, (3) developmental history, (4) ecological relationships, (5) physiological relationships, and (6) economic considerations. The paucity of autecological information on range plants was a major problem. He suggested that no one individual is capable of a complete autecological analysis of a species, thus coordination among several researchers is needed to handle the task.

The significance of population biology to range management has become transparent, especially from a standard such as Harper (1977). Among many relevant concepts, he wrote that the population biology of higher plants occurs on 2 interrelated levels — the number of genets (the product of a zygote) and the number of ramets or repetitive units of structure that occur on each genet (e.g., tillers in grasses). Grassland plants that reproduce exclusively by seed (annuals) require different management than perennials in which reproduction is primarily vegetative (rhizomes, stolons, tillers). The maintenance of a seed bank of desirable species is essential in an annual grassland, in order to perpetuate the species supplying forage for livestock. For successful reproduction by perennials, adequate quantities of non-structural carbohydrates, proteins and minerals must be stored in order to endure periods of drought. Because defoliation can affect the efficiency of metabolism in plants, considerable research has been devoted to the study of reserve substances, mostly photosynthates (see White 1973). Harper (1977) said this well illustrates the domain of the plant population biologist, i.e., more concern with what is left after herbivory than that which is grazed. However, grazing can result in the consumption of seeds and the subsequent dissemination thereof, a matter of concern to the range manager if the plant is considered to be a weed (e.g., *Prosopis*). Conversely, fertile tillering in perennial grasses produces stalks of low nutrient density for grazers, thus wasting photosynthate that might otherwise be converted to leaf, which has high nutrient density. Furthermore, Hyder *et al.* (1975) stated that seed production by established perennial grasses is of little importance in their reproductive success. It has been frequently claimed that trampling of seed helps plant seeds of forage species, but Ellison (1960) said no experiments had been performed to substantiate such. Hyder *et al.* (1975) were unsuccessful in attempting to establish perennial grasses by broadcast seeding, even when followed by trampling by livestock.

Dispersed seed forms a living, but quiescent population of plants (seed bank) in the soil (Harper 1977). Seed banks of undesirable species can result in population pulses of worthless or poisonous forage or in increased competition with valuable forage. The relative rates of production (seed rain), dispersal, longevity, dormancy and recruitment of seeds determine the size of the seed bank. Competition, predation, parasitism and physical components of the environment also act to reduce reproductive potential. Collectively these processes act as an environmental sieve to determine the success of a seed producing more seed. The size and composition of the seed bank is of only proximate importance to the range manager; ultimately it is the products of the seed bank, such as forage or weeds that are of concern. Management goals are oriented toward manipulating the seed bank of such plants through grazing management, revegetation and pest control practices.

Other applications of population ecology include notable instances of deliberate introductions of predatory organisms (insects) to control undesirable species of plants, an example of a practice loosely referred to as biological control. The most efficacious applications of these principles have all been in grasslands: *Opuntia* controlled by *Cactoblastis* in Australia, *Hypericum* controlled by *Chrysolina* in California, and *Senecio* controlled by *Tyria* in Australia (Harper 1977). Plants controlling the success of other plants through the production of chemical substances (allelopathy) has been the subject of interest in studies of oldfield succession. The standard of Rice (1982), *Allelopathy*, details the role of chemical inhibitors produced by plants in ecological succession in grassland. Attempts at revegetation in grasslands must account for the potential of allelopathic effects.

16.5.5 Plant/herbivore relationships

Nearly all of the grassland of the world is used to some degree for livestock production, regardless of land tenure. Thus perhaps no relationship in grassland ecology is of more significance to range management than that which exists between grasses and herbivores, particularly large mammalian grazers. Extensive areas of grasslands not used by livestock are those found primarily in Africa, where large herds of wild ungulates graze (Sinclair and Norton-Griffiths 1979). Whereas the grassland ecologist studying plant/herbivore relationships is concerned with discovery of the natural relationships that exist, the range ecologist in the same situation would want to apply that knowledge to improving the efficiency rate of grass to livestock energy conversion. Noy-Meir (1975) lamented that there was too little interchange between theoretical and practical ecologists who study grazing, and it was limiting to both phases of the discipline.

Grazing is an ecological process involving the transfer of energy and matter from plant to grazing animal, which is usually considered to be a variety of predation. As in any predator/prey relationship, coevolution by natural selection has resulted in adaptations of the predator to be as efficient as necessary in order that enough individuals survive to perpetuate the species (Darwinian fitness), and for the prey species to evolve characteristics to avoid enough predation to do likewise. Most of the large mammalian grazers have evolved to be generalists, i.e., users of multiple species of prey, although interspecific grazing preference is common. Grasses have evolved mechanisms to avoid grazing such as low apical meristems and siliceous epidermal cells.

Other types of plant/herbivore relationships exist in grasslands, such as mutualism and parasitism, but gain somewhat less attention in range management circles, (e.g., ground squirrels that cache seeds or insects that suck juices). The implication of a mutualistic relationship between livestock and grasses is that some amount of defoliation will result in the maximum rate of primary productivity. The exact ecological status of the plant/herbivore relationship in grasslands is controversial. Ellison (1960) stated that evidence of the dependence of the grazed plant upon animals grazing them was negative, referring to the relationship as apparent "parasitism by the animals". However, he allowed that the nature of dependence of plants on animals might not yet be discovered. Chew (1974) said that grazing increases heterogeneity in grassland vegetation by affecting both plant populations and their milieu, deducing that grazers benefit grassland ecosystems (increase productivity, complexity or homeostasis).

It has been hypothesized that partial defoliation by grazing might be beneficial to the defoliated plant (Owen and Wiegert 1976). McNaughton (1979) listed several mechanisms that plants might utilize to compensate for tissue lost to grazing, both biochemically and biomechanically. Dyer *et al.* (1982), reviewed the hypothesized salutory mechanisms: (1) biomechanical (e.g., increased tillering) and (2) biochemical (e.g., growth stimulation by saliva). Crawley (1983) stated that in plants subject to continuous feeding by food-limited herbivores one would not expect strong compensation to occur. In a review of more than 40 papers supporting the hypothesis of herbivore benefit, Belsky (1986) concluded that except for a few experiments in controlled environments, there was insufficient rigor in the evidence to refute the concept in the original statement of Ellison (1960). Citing research of his own and others, McNaughton (1986) disagreed, saying that Belsky (1986) among several particulars, failed in her attempt by posing a single question ("Does herbivory benefit plants?"), because the plant/herbivore relationship is too complex to apply this single question. He quoted himself in an earlier study (McNaughton 1983) that supported the hypothesis of compensatory growth: "Many effects of herbivores upon their food plants that are presently perceived to be deleterious may be less deleterious than expected due to compensatory growth of the plant ... I do not contend that herbivory maximizes plant fitness, but that plants have the capacity to compensate for herbivory and may, at low levels of herbivory, overcompensate for damage so that fitness may be increased". Disputes such as this often arise because the object of interest is viewed at differing levels of ecological scale (Spedding 1971, Allen and Star 1982, McIntosh 1985). Belsky (1987) said there was much confusion in the literature due to arguments made on differing hierarchical levels namely, organism, population and ecosystem. She said it is unlikely that overcompensation in total productivity will ever be found in natural communities, but if it is, it will not be in grassland. Plants are benefited by herbivores, but not directly because of being defoliated, rather indirectly because of reduced growth of defoliated competitors, altered trends of succession, litter removal and amelioration of soil fertility. Thus benefits accrue to plants at the community and ecosystem level, not at the organismal level. She concluded that results of research that are more convincing need to be presented to resolve the question of whether or not individual plants benefit from herbivory.

16.6 References

Allen, T. F. H., and T. B. Starr. 1982. Hierarchy: perspectives for ecological complexity. Univ. of Chicago Press, Chicago.

Barnard, C., and O. H. Frankel. 1964. Grass, grazing animals, and man in historic perspective. p. 1—12. *In*: C. Barnard (ed.) Grasses and grasslands. Macmillan and Co. Ltd., London.

Barbour, M. G., J. H. Burk, and W. D. Pitts. 1980. Terrestrial plant ecology. The Benjamin/Cummings Pub. Co., Inc., Menlo Park, CA.

Bazzaz, F. A., and J. A. D. Parrish. 1982. Organization and grassland communities. p. 233—254. *In*: J. R. Estes, R. J. Tyrl, and J. N. Brunken (eds.) Grasses and grasslands: systematics and ecology. Univ. of Oklahoma Press, Norman.

Belsky, A. J. 1986. Does herbivory benefit plants? A review of the evidence. Am. Nat. 127: 870—892.

Belsky, A. J. 1987. The effects of grazing: confounding of ecosystem, community, and organism scales. Am. Nat. 129: 777—783.

Bonham, C. D. 1983. Range vegetation classification. Rangelands 5: 19—21.

Breymeyer, A. I., and G. M. Van Dyne (eds.) 1978. Grasslands, systems analysis and man. IBP synthesis series 18. Cambridge Univ. Press, London.

Chapline, W. R. 1980. First 10 years of the office of grazing studies. Rangelands 2: 223—227.

Chew, R. M. 1974. Consumers as regulators of ecosystems: an alternative to energetics. Ohio J. Sci. 74: 359—369.

Clark, F. E. 1975. Viewing the invisible prairie. p. 181—197. *In*: M. K. Wali (ed.) Prairie: A multiple view. Univ. of North Dakota Press, Grand Forks.

Clements, F. E. 1905. Research methods in ecology. Lincoln: Univ. Publ. Co. Reprinted New York: Arno Press (1977).

Clements, F. E. 1916. Plant succession: an analysis of the development of vegetation. Publication No. 242. Washington, D.C.: Carnegie Inst. of Washington.

Clements, F. E. 1936. Nature and structure of the climax. J. Ecol. 24: 552—584.

Costello, D. F. 1957. Application of ecology to range management. Ecology 38: 49—53.
Coupland, R. T. (ed.) 1979a. Grassland ecosystems of the world: analysis of grasslands and their uses. IBP synthesis series 18. Cambridge Univ. Press, London.
Coupland, R. T. 1979b. Introduction. p. 1—37. *In*: R. T. Coupland (ed.) Grassland ecosystems of the world: analysis of grasslands and their uses. IBP synthesis series 18. Cambridge Univ. Press, London.
Cowles, H. C. 1899. The ecological relations of the vegetation of the sand dunes of Lake Michigan. Bot. Gaz. 27: 95—117, 167—202, 281—308, 361—391.
Cowles, H. C. 1901. The physiographic ecology of Chicago and vicinity: a study of the origin, development, and classification of plant societies. Bot. Gaz. 31: 73—108, 145—181.
Cowles, H. C. 1911. The causes of vegetative cycles. Bot. Gaz. 51: 161—183.
Cragg, J. B. 1979. Foreword. *In*: R. T. Coupland (ed.) Grassland ecosystems of the world: analysis of grasslands and their uses. Cambridge Univ. Press, London.
Crawley, M. J. 1983. Herbivory: the dynamics of animal-plant interaction. Univ. of California Press, Berkeley.
Detling, J. K. 1979. Processes controlling blue grama production on the shortgrass prairie. p. 25—42. *In*: N. R. French (ed.) Perspectives in grassland ecology. Springer-Verlag, New York.
Dix, R. L., and R. G. Beidleman (eds.) 1969. The grassland ecosystem: a preliminary synthesis. Range Sci. Dep. Sci. Ser. No. 3. Colorado State Univ., Fort Collins.
Dodd, J. L., and W. K. Lauenroth. 1979. Analysis of the response of a grassland ecosystem to stress. p. 43—58. *In*: N. R. French (ed.) Perspectives in grassland ecology. Springer-Verlag, New York.
Dyer, M. I., J. K. Detling, D. C. Coleman, and D. W. Hilbert. 1982. The role of herbivores in grasslands. p. 255—295. *In*: J. R. Estes, R. J. Tyrl, and J. N. Brunken (eds.) Grasses and grasslands: systematics and ecology. Univ. of Oklahoma Press, Norman.
Dyksterhuis, E. J. 1949. Condition and management of range land based on quantitative ecology. J. Range manage. 2: 104—115.
Dyksterhuis, E. J. 1958. Ecological principles in range evaluation. Bot. Rev. 24: 253—272.
Dyksterhuis, E. J. 1980. Notes on the origin of range science. Rangelands 2: 228—229.
Egerton, F. N. 1976. Ecological studies and observations before 1900. p. 311—351. *In*: B. J. Taylor, and T. J. White (eds.) Issues and Ideas in America. Univ. of Oklahoma Press, Norman.
Ellison, L. 1960. Influence of grazing on plant succession of rangelands. Bot. Rev. 26: 1—78.
Engelberg, J., and L. L. Boyarsky. 1979. The noncybernetic nature of ecosystems. Am. Nat. 114: 317—324.
French, N. R., W. E. Grant, W. Grodzinski, and D. M. Swift. 1974. Small mammal energetics in grassland ecosystems. Ecol. Monogr. 46: 201—220.
French, N. R. (ed.) 1979. Perspectives in grassland ecology. Springer-Verlag, New York.
Gleason, H. A. 1926. The Individualistic concept of the plant association. Bull. of the torrey Bot. Club 53: 1—20.
Gleason, H. A. 1939. The individualistic concept of the plant association. Am. Midland Nat. 21: 92—110.
Harper, J. L. 1977. Population biology of plants. Acad. Press, London.
Holechek, J. L. 1981. A brief history of range management in the United States. Rangelands 3: 16—18.
Hyder, D. N., R. E. Bement, E. E. Remmenga, and D. F. Hervey. 1975. Ecological responses of native plants and guidelines for management of shortgrass range. USDA Tech. Bull. No. 1503.

Innis, G. S. 1978. Objectives and structure for a grassland simulation model. p. 1—30. *In*: G. S. Innis (ed.) Grass simulation model. Springer-Verlag, New York.

Jameson, D. A. 1970. Land management policy and development of ecological concepts. J. Range Manage. 23: 316—321.

Jenny, H. 1961. Derivation of state equations of soils and ecosystems. Soil Sci. Soc. Amer. Proc. 25: 385—388.

Jordan, C. F. 1981. Do ecosystems exist? Am. Nat. 118: 284—287.

Kays, S., and J. L. Harper. 1974. The regulation of plant and tiller density in a grass sward. J. Ecol. 62: 97—105.

Klipple, G. E. 1964. Early- and late-season grazing versus season-long grazing of short-grass vegetation on the Central Great Plains. USDA, Forest Ser. Res. Paper RM-11.

Kumar, *et al.* 1976. Insects of the Central Plains Experimental Range, Pawnee National Grassland. Univ. of Wyoming Sci. Monogr.

Lauenroth, W. K. 1979. Grassland primary production: North American grasslands in perspective. p. 3—24. *In*: N. R. French (ed.) Perspectives in grassland ecology. Springer-Verlag, New York.

Launchbaugh, J. L. 1982. The use of complementary forages in a reproductive beef cattle operation. Dept. Rep. Fort Hays Branch Exp. Sta. Contrib. 82—165—D, Kansas State Univ.

Laycock, W. A. 1979. Introduction. p. 1—2. *In*: N. R. French (ed.) Perspectives in grassland ecology. Springer-Verlag, New York.

Lewis, J. K. 1969. Range management viewed in the ecosystem framework. p. 97—187. *In*: G. M. Van Dyne (ed.) The ecosystem concept in natural resource management. Academic Press, New York.

Lewis, J. K. 1973. Improving the efficiency of calf production in the 80's using resources of the great plains: an educator's view. p. B-1—B-29. *In*: F. H. Baker (ed.) Great Plains Beef Symposium. Great Plains Agr. Counc. Pub. No. 63, Univ. of Nebraska, Lincoln.

Lind, E. M., and E. S. Morrison. 1974. East African vegetation. Longman Group Limited, London.

Lindeman, R. L. 1942. The trophic-dynamic aspect of ecology. Ecology 23: 399—418.

Lorenz, R. J. 1977. Complementary grazing systems for the Northern Great Plains. Proc. The Range Beef Cow Symp. V, Chadron, NB.

McFayden, A. 1964. Energy flow in ecosystems and its exploitation by grazing. *In*: D. J. Crisp (ed.) Grazing in terrestrial and marine environments. Blackwell Sci. Publ., Oxford.

Major, J. 1969. Historical development of the ecosystem concept. p. 9—22. *In*: G. M. Van Dyne (ed.) The ecosystem concept in natural resource management. Academic Press, New York.

Margalef, R. 1965. Ecological correlations and the relationship between primary productivity and community structure. *In*: C. R. Goldman (ed.) Primary productivity in aquatic environments. Mdm. Inst. Ital. Idrobiol., 18 Suppl., Univ. of California Press, Berkeley.

Margalef, R. 1968. Perspectives in ecological theory. Univ. of Chicago Press, Chicago.

McCanny S. J., and P. B. Cavers. 1987. The escape hypothesis: a test involving a temperate, annual grass. Oikos 49: 67—76.

McIntosh, R. P. 1976. Ecology since 1900. *In*: B. J. Taylor, and T. J. White (eds.) Issues and ideas in America. Univ. of Oklahoma Press, Norman.

McIntosh, R. P. 1985. The background of ecology: concept and theory. Cambridge Univ. Press.

McNaughton, S. J. 1968. Structure and function in California grasslands. Ecology 49: 962—972.

McNaughton, S. J. 1979. Grassland-herbivore dynamics. p. 82—103. *In*: A. R. E. Sinclair, and M. Norton-Griffiths (eds.) Serengeti: dynamics of an ecosystem. Univ. of Chicago Press, Chicago.

McNaughton, S. J. 1983. Serengeti grassland ecology: the role of composite environmental factors and contingency in community organization. Ecol. Mongr. 53: 291—320.
McNaughton, S. J. 1984. Grazing lawns: animals in herds, plant form, and coevolution. Am. Nat. 124: 863—886.
McNaughton, S. J. 1985. Ecology of a grazing ecosystem: the Serengeti. Ecol. Monogr. 55: 259—294.
McNaughton, S. J. 1986. On plants and herbivores. Am. Nat. 128: 765—770.
Noy-Meir, I. 1975. Stability of grazing systems: an application of predator-prey graphs. J. Ecol. 63: 459—481.
Odum, E. P. 1953. Fundamentals of ecology, 1st ed. Saunders, Philadelphia.
O'Neil, R. V., D. L. DeAngelis, J. B. Waide, and T. F. H. Allen. 1986. A hierarchical concept of ecosystems. Princeton Univ. Press, Princeton, NJ.
Owen, D. F., and R. G. Wiegert. 1976. Do Consumers maximize plant fitness? Oikos 27: 488—492.
Parker, K. W. 1952. The role of plant ecology in range research and range management. Int. Grassland Cong. Proc. 6: 618—624.
Redmann, R. E. 1975. Production ecology of grassland plant communities in western North Dakota. Ecol. Monogr. 45: 83—106.
Reiners, W. A. 1986. Complementary models for ecosystems. Am. Nat. 127: 59—73.
Rice, E. L. 1983. Allelopathy, 2nd ed. Academic Press, Inc., Orlando, FL.
Ricou, G. A. 1979. Pastures. p. 147—153. *In*: R. T. Coupland (ed.) Grassland ecosystems of the world: analysis of grasslands and their uses. Cambridge Univ. Press, London.
Risser, P. G., E. C. Birney, H. D. Blocker, S. W. May, W. J. Parton, and J. A. Wiens. 1981. The true prairie ecosystem. Hutchinson Ross Publ. Co.
Sampson, A. W. 1919. Plant succession in relation to range management. USDA Bull. No. 791.
Shantz, H. L. 1954. The place of grasslands in the earth's cover of vegetation. Ecology 35: 142—145.
Shiflet, T. N. 1973. Range sites and soils in the United States. p. 26—39. *In*: D. N. Hyder (ed.) Arid Shrublands — Proceedings of the Third Workshop of the United States/ Australia Rangelands Panel. Soc. Range Manage., Denver.
Sims, P. L., J. S. Singh, and W. K. Lauenroth. 1978. The structure and function of ten western North American grasslands. I. Abiotic and vegetational characteristics. J. Ecol. 66: 251—285.
Sims, P. L., and J. S. Singh. 1978a. The structure and function of ten western North American grasslands. II. Intra-seasonal dynamics in primary producer compartments. J. Ecol. 66: 547—572.
Sims, P. L., and J. S. Singh. 1978b. The structure and function of ten western North American grasslands. III. Net primary production, turnover and efficiencies of energy capture and water use. J. Ecol. 66: 573—597.
Sims, P. L., and J. S. Singh. 1978c. The structure and function of ten western North American grasslands. IV. Compartmental transfers and energy flow within the ecosystem. J. Ecol. 66: 983—1008.
Sinclair, A. R. E., and M. Norton-Griffiths (eds.) 1979. Serengeti: dynamics of an ecosystem. Univ. of Chicago Press, Chicago.
Smith, E. L. 1978. A critical evaluation of the range condition concept. p. 266—267. *In*: D. N. Hyder (ed.) Proceedings of the First International Rangeland Congress. Soc. Range Manage., Denver.
Smith, E. F., and C. E. Owensby. 1978. Intensive-early stocking and season-long stocking of Kansas Flint Hills Range. J. Range Manage. 31: 14—17.
Smolik, J. D., and J. K. Lewis. 1982. Effect of range condition on density and biomass of nematodes in a mixed prairie ecosystem. J. Range manage. 35: 657—663.

Snaydon, R. W. 1984. Plant demography in an agricultural context. p. 389—407. *In*: R. Dirzo, and J. Sarukhan (eds.) Perspectives on plant poulation ecology. Sinauer Assoc. Inc., Pub., Sunderland, MA.

Soil conservation Service. 1985. National range handbook. 190-IV, Notice-5. USDA, Washington, DC.

Sparrow, E. B., and K. G. Doxtader. 1973. Adenosine triphosphate (ATP) in grassland soil: Its relationship to microbial biomass and activity. US/IBP Grassland Biome Tech. Rep. No. 224. Colorado State Univ.

Spedding, C. R. 1971. Grassland ecology. Clarendon Press, Oxford.

Tainton, M. N. 1984. A guide to the literature on research in the grassland biome of South Africa. South African national Scientific Programmes Rep. No. 96.

Tansley, A. G., and R. S. Adamson. 1925. Studies of the vegetation of the English chalk. III. The chalk grasslands of the Hampshire-Sussex border. J. Ecol. 13: 177—223.

Tansley, A. G. 1935. The use and abuse of vegetational concepts and terms. Ecology 16: 284—307.

Tobey, R. 1981. Saving the prairies: the life cycle of the founding school of American plant ecology, 1895—1955. Univ. of California Press, Berkeley.

USDA 1986. Agricultural statistics. U.S. Gov. Print. Off., Washington, DC.

Van Dyne, G. M. (ed.) 1969. The ecosystem concept in natural resource management. Acad. Press, New York.

Wallace, R. A., J. L. King, and G. P. Sanders. 1986. Biology: the science of life. Scott, Foresman and Co., Glenview, IL.

Waller, S. S., and J. K. Lewis. 1979. Occurrence of C3 and C4 photosynthetic pathways in North American grasses. J. Range. Manage. 32: 12—28.

Ward, G. M., P. L. Knox, and B. W. Hobson. 1975. Beef production options and requirements for fossil fuel. Science 198: 265—271.

Watt, K. E. F. 1968. Ecology and resource management: a quantitative approach. McGraw-Hill Book Co., New York.

Westoby, M. 1985. Does heavy grazing usually improve the food resource for grazers? Am. Nat. 126: 870—871.

White, L. M. 1973. Carbohydrate reserves in grasses: a review. J. Range Manage. 26: 13—18.

Whittaker, R. H. 1962. Classification of natural communities. Bot. Rev. 28: 1—239.

Whittaker, R. H. 1967. Gradient analysis of vegetation. Biol. Rev. 42: 207—264.

Woodmansee, R. G. 1978a. Critique and analyses of the grassland ecosystem model ELM. p. 257—281. *In*: G. S. Innis (ed.) Grass simulation model. Springer-Verlag, New York.

Woodmansee, R. G. 1978b. Additions and losses of nitrogen in grassland ecosystems. Bioscience 28: 448—453.

Woodmansee, R. G. 1979. Factors influencing input and output of nitrogen in grasslands. p. 117—134. *In*: N. R. French (ed.) Perspectives in grassland ecology. Springer-Verlag, New York.

Wright, H. A., and A. W. Bailey. 1982. Fire ecology: United States and southern Canada. John Wiley & Sons, New York.

16.7 Annotated listing of books of grassland ecology

Allred, B. W. 1950. Practical grassland management. Sheep and Goat Raiser Mag., San Angelo, TX.

— A more or less nontechnical review of range management of the era from the perspective of a federal range conservationist.

Barnard, C. (ed.) 1974. Grasses and Grasslands. McMillan and Co. Ltd., London.
— The majority of this volume deals with the biology of grasses, with only 4 of 14 chapters on ecological subject matter.

Breymeyer, A. I., and G. M. Van Dyne (eds.) 1978. Grasslands, systems analysis and man. IBP synthesis series 18. Cambridge Univ. Press, London.
— A systematic synthesis of Grassland Biome studies, emphasizing processes related to productivity.

Castello, D. F. 1969. The prairie world. Thomas Y. Crowell Co., NY.
— A natural history of grassland life including chapters on plants, mammals, birds, insects, reptiles and amphibians, and man.

Coupland, R. T. (ed.) 1979. Grassland ecosystems of the world: analysis of grasslands and their uses. IBP synthesis series 18. Cambridge Univ. Press, London.
— Over 30 international contributors on the structure, development and use of the grasslands of the world.

Duffey, E. 1974. Grassland ecology and wildlife management. Chapman and Hall, London.
— An overview of the ecology of the grasslands of Great Britain including a number of chapters on management.

Estes, J. R., R. J. Tyrl, and J. N. Brunken (eds.) 1982. Grasses and grasslands: systematic and ecology. Univ. of Oklahoma Press, Norman.
— Proceedings of a symposium sponsored by the American Society of Plant Taxonomists, Ecological Society of America and Botanical Society of America. An excellent collection of papers by several prominent grassland ecologists.

French, N. R. (ed.) 1979. Perspectives in grassland ecology. Springer-Verlag, NY.
— A collection of articles synthesizing and summarizing IBP Grassland Biome studies.

Harlan, J. R. 1956. Theory and dynamics of grassland agriculture. D. Van Nostrand Co., Inc., Princetown, NJ.
— A summary of American grassland science including ecology, geography, evolution, distribution, physiology, soils, mensuration, and range management.

Humphrey, R. R. 1962. Range ecology. The Ronald Press Co., NY.
— Although broadly written, includes significant amounts of information on desert grasslands of the American Southwest.

Humphrey, R. R. 1968. The desert grassland. The Univ. of Arizona Press.
— Primarily a history of the ecology of American desert grasslands. Also it was published in 1968 in Bot. Rev. 24(4).

Innis, G. S. (ed.) 1978. Grassland simulation model. Springer-Verlag, NY.
— From the IBP Grassland Biome studies, an overview and detailed description of the 9 submodels of ELM and critique.

Moore, R. M. (ed.) 1970. Australian grasslands. Australian National Univ. Press, Canberra.
— Twenty-seven contributions from 35 authors detailing ecology, soils, vegetation types, and management of Australian grasslands.

Risser, P. G., E. C. Birney, H. D. Blocker, S. W. May, W. J. Parton, and J. A. Wiens. 1981. The true prairie ecosystem. Hutchinson Ross Publ. Co., Stroudsburg, PA.
— A detailed summary of the IBP Grassland Biome studies of the Osage Oklahoma site, including background information and relationships with previous relevant work.

Sinclair, A. R. E., and M. Norton-Griffiths (eds.) 1979. Serengeti: dynamics of an ecosystem. The Univ. of Chicago Press.
— Thirteen chapters by 12 authors on Serengeti ecology, including vegetation, but mostly studies on the large primary and secondary consumers.

Spedding, C. R. W. 1971. Grassland ecology. The Clarendon Press, Oxford.
— A very comprehensive, but succinct treatment of grassland ecology and management, including topics such as individual plants, trophic dynamics and efficiency, animal production and ecosystem modeling.

Sprague, H. B. (ed.) 1959. Grasslands. Publ. No. 53 of the Amer. Assoc. for the Advancement of Sci. Washington, DC.
— Taken from a AAAS symposium in 1956; 37 short papers on grassland agriculture, animal nutrition, climatology, ecology and range management.
Sprague, H. B. (ed.) 1974. Grasslands of the United States: their economic and ecologic importance. The Iowa State Univ. Press, Ames, IA.
— Taken from a symposium of the American Forage and Grassland Council. Topics include a wide variety of subjects such as grassland resources, hydrology, economics, research, wildlife, public lands and recreation.
Staten, H. W. 1952. Grasses and grassland farming. The Devin-Adair Co., NY.
— A nontechnical presentation of grassland management, mostly oriented at production of grazable forages.
Swierenga, R. P. (ed.) 1984. History and ecology: studies of the grassland. Univ. of Nebraska Press, Lincoln.
— A collection of papers by historian James C. Malin on the natural history and settlement of the grassland region of North America.
Tobey, R. C. 1981. Saving the prairies: the life cycle of the founding school of American plant ecology, 1895—1955. Univ. of California Press, Berkeley.
— An outstanding sociohistorical analysis of the rise and fall of a single scientific discipline.
Weaver, J. E. 1954. North American prairie. Johnsen Publ. Co., Lincoln, NB.
— Largely a compilation and synthesis of Weavers previous studies of Great Plains grasslands.
Weaver, J. E., and F. W. Albertson. 1956. Grasslands of the great plains. Johnsen Publ. Co., Lincoln, NE.
— A comprehensive descriptive ecology of the short and mixed grasslands from Texas and New Mexico to Canada.
Weaver, J. E. 1968. Prairie plants and their environment. Univ. of Nebraska Press, Lincoln.
— Published posthumously, this volume includes a complete Weaver bibliography and 14 chapters on the ecology of the true prairie.
Whyte, R. O. 1968. Grasslands of the monsoon. Frederick A. Praeger, Publ., NY.
— A biogeographic treatment of grasslands of Africa, the Middle East, and Asia.
Whyte, R. O. 1974. Tropical grazing lands: communities and constituent species. Dr. W. Junk b.v. — Publ. — The hague.
— Grass resources of tropical Africa, Asia and Australia; their origin, composition and management.
Youngner, V. B., and C. M. McKell (eds.) 1972. The biology and utilization of grasses. Academic Press, NY.
— Resulting from a symposium in 1969, this highly cited 28 chapter volume, covers grass systematics, breeding, ecology and grazing management.

17. Riparian values as a focus for range management and vegetation science

SHERMAN SWANSON

17.1 Abstract

Riparian vegetation is defined by its association with free or unbound water. The productivity and diversity of riparian vegetation and its relationship with aquatic ecosystems makes it extremely valuable to fish, wildlife, livestock, and people. Intense human and livestock use has accentuated the naturally high disturbance in most riparian areas. This has caused significant alteration to riparian vegetation and hydrologic processes, threatening the quality of riparian habitats. Even though riparian areas have long been neglected because of their small area, the values and threats have focused much attention on riparian management and science in recent decades. The long process of developing an integrated interdisciplinary and hierarchical riparian classification is just beginning. A significant challenge facing riparian classifiers, scientists, and managers is the interrelationships among vegetation, hydrology, and fluvial geomorphology. In spite of the challenges, many riparian managers have applied a variety of management strategies and the range profession is beginning to learn which strategies are appropriate in which settings and for which combinations of resource values. The process of managing, and therefore monitoring, specific riparian areas will provide insight and motivation for the challenges ahead.

17.2 Introduction

Riparian is derived from the latin "riparius", which is derived from "ripa", meaning bank in the sense of the bank of a water body. Just as range is difficult to define as a kind of land with clear borders that apply in all areas, so too is the land that provides riparian values. In arid or semiarid landscapes, water abundantly supplied to vegetation or in the form of an

Tueller, P. T. (ed.), Vegetation science applications for rangeland analysis and management.
ISBN 90-6193-195-9.
© 1988, *Kluwer Academic Publishers, Dordrecht, Boston, London.*

aquatic ecosystem is of special significance. Since about 1970, riparian values have led to research, management policies, and, in turn, to definitions.

17.3 Definitions

17.3.1 Riparian

The U.S. Forest Service (1980) defines riparian areas as:

> geographically delineated areas, with distinctive resource values and characteristics, that are comprised of the aquatic and riparian ecosystems, floodplains and wetlands. They include all areas within a horizontal distance of 100 feet from the edge of perennial streams or other water bodies.

This implies that perennial surface water must be part of a riparian area. Riparian ecosystems are defined as:

> a transition between the aquatic ecosystem and the adjacent terrestrial ecosystem . . . identified by soil characteristics and distinctive vegetation communities that require free or unbound water.

The U.S. Bureau of Land Management (1987a) suggests but does not require that permanent water be above ground by defining a riparian area as:

> an area of land directly influenced by permanent water. It has visible vegetation and or physical characteristics reflective of permanent water influence. Lake shores and stream banks are typical riparian areas. Excluded are such sites as ephemeral streams or washes that do not exhibit the presence of vegetation dependent upon free water in the soil.

The last sentence appears to distance the agency from an overzealous or over broad definition that could prescribe intensive management on areas that do not warrant the attention. No definition can, however, ensure that all land included will be more deserving of special treatment than all land excluded. Johnson *et al.* (1984) deal with the gradient of values and the transition to nonriparian lands by suggesting the terms hydroriparian, mesoriparian, and xeroriparian. Fortunately, the impetus for riparian management has not been translated into laws prescribing specific management actions on all riparian areas. This recognizes that range and riparian management is an art based on science that must be locally adapted.

17.3.2 Wetland

Presumably, it takes greater availability of water to support special vegetation in wetter climates. This is perhaps why the issue is generally focused over the term wetlands in the eastern United states. A wetland differs from a riparian area chiefly by its exclusion of land dominated by vegetation that reaches below the depth of soil to tap into free or unbound water. A similar problem of definition confronts those fulfilling the legal mandates to protect wetlands, jointly defined by the U.S. Environmental Protection Agency (Federal Register 1980) and the U.S. Army Corps of Engineers (Federal Register 1982) as:

> Those areas that are inundated or saturated by surface or ground water at a frequency and duration sufficient to support, and that under normal circumstances do support, a prevalence of vegetation typically adapted for life in saturated soil conditions. Wetlands generally include swamps, marshes, bogs, and similar areas.

The subjectivity of this definition is reduced by manuals for wetland identification and delineation. They discuss indicators based on vegetation, soil, and hydrology (Sipple 1987 and U.S. Army Corps of Engineers 1987). Furthermore, voluminous lists of hydrophytes (U.S. Fish and Wildlife Service 1986) and hydric soils (U.S. Soil Conservation Service 1985) have been compiled. However, some plants and some soils indicate somewhat different conditions across their range, and soil wetness certainly varies between wet and dry phases in climate. Many plants have therefore become adaptable, hence the practice of labeling species as obligate, facultative wetland, facultative, facultative upland, nonwetland, and drawdown. The terms refer to the proportion of the regional population that occurs in wetlands (> 99%, 67—99%, 34—66%, 1—33%, and < 1%) or, in the case of drawdown, to those plants typically associated with drier stages of wetlands such as mud flats, or playa lakes (U.S. Fish and Wildlife Service 1986). The greater attention paid to definition and delineation of wetlands as opposed to riparian areas undoubtedly stems from the prohibition against draining wetlands (with penalties to farmers specified in the "swampbuster" part of the 1985 Food Security Act) and filling wetlands (for which permits are required from the U.S. Army Corps of Engineers).

17.4 Riparian values

In the case of both riparian and wetland management, the extremely high values provide the focus. Johnson and Carothers (1982) found that riparian habitats attract more recreation use than any other Southwest habitat. Roath and Krueger (1982) found that although the riparian zone comprised only 1.9% of their study area it provided 81% of the forage consumed by cattle in a partially forested eastern Oregon pasture. Thomas *et al.* (1979) listed several studies demonstrating the disproportionate use of riparian zones by wildlife and reported that "of the 363 terrestrial species known to occur in the Great Basin of southeastern Oregon, 288 are either directly dependent on riparian zones or use them more than other habitats". Johnson (1978) summarized the work of many studies conducted primarily in the desert southwest in a variety of 21 nonriparian and 12 riparian habitats. The number of breeding bird pairs per 100 acres (40.5 ha) in nonriparian habitats ranged from 6 to 380 and averaged 151. Whereas in a variety of riparian habitats the number of pairs per 100 acres ranged from 175 to 1059 and averaged 514. The density of migrating birds associated with riparian areas also appears to be much higher than in surrounding desert uplands (Stevens *et al.* 1977). Johnson *et al.* (1977) noted that 64 endangered wildlife species and 47 more being considered for listing depend on riparian habitats.

17.5 Riparian vegetation

Unique and productive vegetation induced by the abundant water is the most outstanding feature of most riparian areas. The diversity of habitats and the number of edges created by generally narrow corridors of heterogeneous vegetation stands in marked contrast to homogeneous uplands. The structure of woody riparian vegetation is most beneficial to wildlife for providing nest sites and niches where there are many layers, especially where some trees attain great height, and where surrounding upland vegetation is uniformly low in stature. In the lower Colorado River drainage, the dependency of avian communities upon riparian habitat varies with elevation (Johnson 1978) as does precipitation and presumably upland vegetation stature. Some raptors commonly use riparian trees for nesting while they forage widely over broad largely nonriparian landscapes. Standing biomass and annual production in riparian ecosystems can be 10 — 100+ times higher than in nearby nonriparian communities. This provides the structure and the food supply for dense biotic communities.

That riparian vegetation is well supplied with water and remains green after other landscapes desiccate provides habitat features upon which many animals depend. Extension of the green-feed season and high productivity are two of the reasons cattle concentrate in riparian areas. The extended period of weight gain from nutritious forage makes riparian grazing, such as mountain meadow grazing, an essential component of economic livestock production over much of the western United States (Gomm 1979).

17.6 Hydrology

The dependence of riparian vegetation on abundant free water is obvious from the definition; the dependence of the hydrologic system upon that vegetation is a phenomenon gaining increased recognition. Cottam and Stewart (1940) described the phenomenon of a wet meadow gullying after overgrazing. The dropping water table drastically changed water availability in the meadow and the plant community shifted from hydric to mesic plant species. Similar scenarios could be described for many areas in the western United States. Strong vegetation supplied with abundant water has the capacity to withstand high hydraulic energy and can dissipate much of the potential energy of streams. As vegetation stabilizes stream banks or meanders on low-gradient reaches, floodwaters retain access to a wider floodplain where they infiltrate to recharge an aquifer and ameliorate flood flows. Floodplain vegetation maintains improved water quality by decreasing erosion, increasing deposition of sediment from upstream and upslope, and then stabilizing it.

Some stream valleys thus collect sediment to a level above that otherwise set by a downstream base level. As the sediments collect, the floodplain widens. Also, the stream and water table rises, staying within a short distance of the floodplain surface. In some valleys a resulting wet-meadow riparian ecosystem can occupy hundreds of acres. Eventually a hydrologic event triggers an erosion process caused by gradient oversteepening, vegetation weakening, and/or an alteration of watershed hydrology. This can convert the old floodplain into a desiccated terrace. After which a new process of floodplain formation can eventually initiate in the gully bottom. The timing and magnitude of all these events is at least partially determined by characteristics of riparian vegetation. Riparian vegetation management has, therefore, become a primary emphasis of nonpoint pollution control and of watershed management.

Much as the value of riparian areas and riparian vegetation to wildlife became an issue in the desert southwest, so too, its value to fish, especially anadromous fish, became an issue in the Pacific Northwest. Riparian vegetation shades streams and can prevent an increase in water temperature that would deplete dissolved oxygen and kill cold-water fish such as trout and salmon. By reducing fine sediments in streams, riparian vegetation indirectly protects spawning gravels and other stream habitats necessary for aquatic life. Riparian vegetation provides much of the organic energy supplied to small stream ecosystems (Vanote *et al.* 1980). It also shapes stream banks and stream channels to create a diversity of pool, riffle, and overhanging-bank habitat. The ideal stream for salmonids (trout and salmon) in the western United States would have maximum undercut bank and vegetation overhang and minimum altered (exposed) bank, embeddedness (of gravels with fines), and width to depth ratio (U.S. Forest Service 1985).

Many authors point to the soil-holding and stream-bank-stabilizing capacity of riparian vegetation and a few have measured it. Zimmerman *et al.* (1967) found that in a headwater stream, the width depth ratio of streams was much lower in reaches vegetated by herbaceous sod rather than forest. Smith (1976) found a ten-fold decrease in bank erosion rate between sediment samples containing 10 and 20 percent (by volume) sod roots.

17.7 Riparian uses

In spite of, or perhaps because of, the tremendous resource values in riparian areas, they have been the center of human activities and a principle target of human impacts. Throughout history man has built cities and developed agriculture or transportation based civilizations adjacent to rivers and streams. Agriculture, especially irrigated agriculture, has led to removal of the vast majority of riparian vegetation from the banks of many low gradient rivers and streams. Whether by clearing land for crops (Katibah 1984 and Hoar and Erwin 1985), bulldozing floodplain vegetation to save water from consumption by phreatophytes (Horton and Campbell 1974), or spraying out willows to increase the area and productivity of native hay lands (Price and Ridd 1983), the effects have been drastic. The process of water control for irrigation, flood control, drainage, power generation, and sundry other purposes has, in many cases, drowned or otherwise altered the physical environment of much of the remaining riparian vegetation (Johnson and Carothers 1987). Along the small streams of the arid and semiarid regions in the West, cattle grazing has caused significant impacts to riparian and aquatic ecosystems (Behnke and Raleigh 1978 and Meehan and Platts 1978). In some regions, timber harvesting, mining, or road building impacts have also been severe. There was little concern and less understanding of the effects to riparian and aquatic ecosystems of maintaining neither minimum nor maintenance flows (Rosgen 1982). Transportation corridors in riparian areas characteristically impair the floodplain function and upset the hydraulics of flowing water. By altering the infiltration, runoff, evapotranspiration, and sediment-producing attributes of the watershed, impacts to riparian areas can originate in far distant locations (Rosgen 1981). By now few large watersheds and their riparian areas continue to function with a natural regime of high, moderate, and low flows carrying sediment as appropriate.

17.8 Physical environment

In addition to the impacts of man, riparian ecosystems are subjected to the perpetual instability of floods, erosion, and sedimentation. The physical environment of most individual riparian organisms and communities is likely to change dramatically within a period of years, decades, or at most centuries. Some riparian species have adapted to this disturbance and even require it for regeneration (Strahan 1984 and McBride and Strahan 1984). Although upland physical environments change under the influence of climatic variation and soil formation or erosion, the rate of change is far greater in riparian areas.

Water availability is perhaps the environmental parameter of greatest significance to most plants. Although riparian areas are defined by the availability of free or unbound soil water, there remains considerable variation in the depth, duration, and quality of that water, and the effects on vegetation are both direct and indirect (Courtois 1984, Fenner *et al.* 1985, and Kovalchik 1987). The preponderance of soils recently deposited by water creates a highly heterogeneous mixture of soil textures, depths, and arrangement of lenses and layers (Platts *et al.* 1987). A broad meadow of soil formed in place or deposited gradually in a series of similar flood events that consistently deposited small amounts of fine sediment is the rare exception. Even there, continued sediment deposition may gradually separate the land surface and vegetation from the underlying water and thereby change species composition. Moisture availability in riparian soils is likely to change in response to raising and lowering of the water table as streams meander across a floodplain or aggrade and degrade in response to the evolution of fluvial geomorphology (Youngblood *et al.* 1985a) or to the activities of waxing and waning beaver populations (Parker *et al.* 1985).

The heterogeneous nature of riparian soils and the relatively small area they comprise within most large blocks of land caused riparian areas to be lumped with uplands or treated as one generally undifferentiated unit in most inventories or on resource maps. In spite of their high values, managing riparian areas as sacrifice areas was acceptable in range livestock grazing (Stoddart *et al.* 1956) if the sacrificed area was not too large. Prior to the last few years, riparian classification and detailed inventory of riparian areas had been too complex for consideration by agencies charged with the management of extensive range acreages. Consequently, the science of riparian classification is in its infancy.

17.9 Classification

There has been, however, much interest in riparian classification in recent years. A riparian classification system should serve as the basis for communication among researchers and managers trying to understand the domain of ecological relationships and of promising riparian management strategies. The utility of land and vegetation classification systems has led to their widespread use by many agencies in nonriparian ecosystems. The same holds true for riparian systems.

17.9.1 Upland

The process of habitat typing National Forest System Lands began with the work of Daubenmire who taught (Daubenmire 1968) that climax vegetation could be used to simplify the mosaic of a landscape where the stands are components of a few successional sequences and even fewer climax types. The land occupied by each of these climax types or plant associations could then be considered together as one habitat type. Plant associations respond to the effective environment, and because they are more easily observable than soil or climate, the habitat types they occupy can be considered as units of land with specific management response potential (see Leonard *et al.* this volume). The work of DeVelice *et al.* (1986) is but one recent example of ongoing efforts to habitat type National Forests.

The Bureau of Land Management is continuing the long established thrust of the Soil Conservation Service by acquiring a range site or ecological site classification of its vast acreages. The Clementsian base for this classification (and habitat typing) is discussed by Nicholson (this volume). Range sites have been used to identify potential vegetation and to evaluate the condition of present vegetation (Dyksterhuis 1949). Management has assumed that climax vegetation was characteristically desirable and was attainable by simply allowing succession to proceed without undue disturbance. The weaknesses and strengths of this approach are discussed by Smith (this volume).

17.9.2 Riparian

In spite of the desire of many range professionals to extend the concepts of land classification into the riparian area, there remain two looming problems, disturbance and naturally fluctuating water conditions. Disturbed vegetation is not a new problem. However, in a consistent environment there usually remain indicators of the previous and potential vegetation. They can be used in combination with characteristics of the soil to accurately stratify the landscape in most situations. Fluctuating water table and soil moisture conditions, however, create an inconsistency of the physical environment that complicates the use of climax vegetation for stratification. Large or old plants that might be considered indicators of climax in an upland setting are not reliable in riparian areas. This represents a complication of the climax concept that goes beyond the problem caused by lack of soil maturity required by mono-, poly-, and polyclimatic-climax theories, thus affecting site and climax-pattern theories as well (Meeker and Merkel 1984).

17.9.2.1 *Physical environment.* In riparian areas, especially in those parts of riparian areas with active processes of fluvial geomorphic evolution, one must learn the potential for the physical environment of the plant communities before one could know the potential for the vegetation. This has led to attempts to classify the land and/or the aquatic ecosystem associated with riparian vegetation as an integral part of the process of classifying the vegetation itself (Platts 1980, Swanson *et al.* in press, Kovalchik 1987). This is different from soil classification incorporated by most riparian and upland classifiers primarily by its emphasis on physical processes of future development as opposed to emphasis on the physical state to which pedogenesis has currently delivered the soil.

The land-systems-inventory approach of Wertz and Arnold (1972) seeks to use geology and geomorphology to stratify the whole landscape into units with similar sets of management response attributes. Their landtype phase could perhaps find close similarity with vegetation and with stream morphology for which vegetation plays such an important role. This approach was suggested by Platts (1980) in his argument for a hierarchical fishery classification system that could be linked with the land classification systems in use by land ecologists and managers. Morphology of the stream and its floodplain are the basis for the stream classification system developed by Rosgen (1985). It differs from the approach of Wertz and Arnold by its focus on the stream and by its reliance on present conditions. Although his stream types can be linked into successional sequences corresponding to geomorphic evolution and can be used to identify units of similar response, he did not claim to classify streams on the basis of potential.

The linkage through time of stream and riparian developmental states was emphasized by Frissell *et al.* (1986). They discuss the potential capacity of a habitat system. From a given state, a habitat system could develop into any of a number of developmental states. Depending on which of these states a system develops into, it then has an additional set of possible developmental states that could occur under future environmental conditions. This implies that future hydrologic environments are very difficult to predict and that predictions decrease in their accuracy with time away from present. The ambiguity associated with predicting future states of riparian and stream systems has led to the reliance on present conditions for most riparian classifications.

17.9.2.2 *Vegetation.* Even though the progress of succession in riparian areas is fast, (major changes can occur in 10 to 20 years (Winward 1986)), the process of riparian synecology has so far dealt almost entirely with present vegetation. Ecologists have described: (a) riparian dominance types, named for the plant species that dominate the tallest appreciable layers (Cowardin *et al.* 1979 and Hansen *et al.* 1987); (b) riparian communities identified by repeating assemblages with similar species composition within a local area (Ellison 1954, Kuramoto and Bliss 1970, Medina 1986, Kauffman *et al.* 1985); or (c) riparian community types that repeat themselves over broader regions (Hall 1979, Paysen *et al.* 1980, Ratliff 1982, Mutz and Querioz 1983, Youngblood *et al.* 1985a and 1985b and Kovalchik 1987).

Many workers, however, would like to proceed from the level of current efforts to eventually describe plant associations representing climax under a given environment. Hansen *et al.* (1987) portrayed their work describing dominance types as a phase 1 and stated that phase 2 will be to develop a "riparian classification that reflects site potential". Swanson *et al.* (in press) designed a hierarchical and interdisciplinary riparian classification system that utilized (but did not identify) both riparian community types and ecological sites that would be homes for plant associations. Kovalchik (1987) believed that his riparian plant communities generally reflected plant associations in the forested part of his study area but that they could not be relied upon in the grazed and disturbed rangeland riparian settings. He believed that geomorphology could be used to identify potential within these settings and he divided his study area into 9 physiographic areas. Within these he used riparian landform to "reflect uniformity in elevation, valley gradient, fluvial processes, water regime, and soil". He also asserted that "riparian associations occur on predictable fluvial surfaces within riparian landforms".

17.9.2.3 *Integration.* The physical environment, especially water level fluctuation, persistence of flow, scouring, and soil type were used to predict the potential plant community of sites by Crouse and Kindschy (1981). Similar criteria were used to predict the habitats for eventual development of gallery forests along two rivers in the American Southwest by Brady *et al.* (1985). The role and importance of linking vegetation classification with classification of the physical environment and with ecological processes is further discussed by West (this volume).

The reliance on riparian landform is similar to the approach of Swanson *et al.* (in press) who suggested using the type of aquatic ecosystem to identify similar units in riparian areas. For nonstream riparian areas the "types" included ponds, reservoirs, seeps, etc., whereas along streams the stream types of Rosgen (1985) were recommended. This is similar to the approach taken by Cooper (1986) and Cowardin *et al.* (1979) in their wetland classifications. Cowardin *et al.* used five systems: marine, estuarine, riverine, lacustrine, and palustrine. Within each of the first four, they used subsystems such as (for the riverine system) tidal, lower perennial, upper perennial, and intermittent. Each subsystem was in turn broken into classes based on either dominant life form of the vegetation or the physiography and composition of the substrate. Cowardin *et al.* (1979) was considered for use by Swanson *et al.* (in press) but rejected because of its lack of refinement within the riverine and palustrine systems that dominate the riparian resource of the western United States. The stream types of Rosgen (1985) are similar to the segment system level of Frissell *et al.* (1986) with the distinct advantage that the units of the classification were previously delineated and described. This facilitates stratification of any river, stream, or ephemeral channel into units that responded similarly to natural events and management activities.

Winward and his associates in the Intermountain Region of the Forest Service have been exploring the linkages between stream types (Rosgen 1985) and complexes:

> The riparian complex represents a unit of land which supports or may potentially support a similar grouping of community types. It is identified on the basis of its overall geomorphology and substrate, as well as its general vegetation pattern. (Winward and Padgett 1987).

Their preliminary data (Winward personal communication) suggest that there may be one or a few complexes that repeat along any given stream type. Also, a break in stream type normally corresponds with a break in riparian complex. The concept of complexes with repeating clusters of riparian community types is particularly attractive for range management because of its scale. In the fluvially active riparian area, riparian community type complexes might be more temporally stable than the community types that comprise them. A given reach of stream may provide the mix of habitats for a complex of riparian community types, each covering a relatively consistent proportion of the area, even though the specific location of each stand and each community type changes through time.

The form of vegetation (formation class and subclass such as "needle leaved evergreen forest") provides a useful level in riparian classification (Hansen *et al.* 1987 and Swanson *et al.* in press). Others accomplish a similar effect by grouping community types (Hall 1979, Paysen *et al.* 1980, Ratliff 1982, and Youngblood *et al.* 1985a and 1985b) dominance types (Hansen *et al.* 1987), or associations (Kovalchik 1987) into meadow, low shrub, tall shrub, tree dominated, etc. assemblages. This reflects the importance of vegetation structure in riparian values, especially for wildlife and hydrology. It also reflects a logical organization for grouping rather tightly defined vegetation units into broad groups that have some similar function. Cowardin *et al.* (1979), Brown *et al.* (1980), and Driscoll *et al.* (1983) describe the use of vegetation formation over broad regions. Brown *et al.* (1980) use formation as the third level in a hierarchical ecosystem classification based primarily on vegetation as it reflects environmental conditions. Levels in the hierarchy are: (1) biogeographic (continental) realm; (2) vegetation (upland, wetland, or cultivated); (3) formation-type; (4) climatic (thermal) zone; (5) regional formation (biome); (5) series (community of generic dominants); (6) association (of specific dominants); and (7) composition-structure-phase. This system, suggested for riparian classification by Dick-Peddie and Hubbard (1977) and by Brown (1978 and 1984), is an extension of work that began in the American Southwest by Lowe (1961).

17.9.3 Management implications

The riparian classification systems developed by land managers or by land management agencies stand out by virtue of the management information they convey along with descriptions of the vegetation units. Youngblood (1985a) included a succession/management section with each riparian community type description. Relationships to other community types and normal shifts in species composition with disturbance are discussed along with values to livestock and wildlife and values to stream stability and erosion control. Hansen *et al.* (1987) list for each dominance type the: (1) timber productivity; (2 and 3) forage value to livestock and wildlife of the community and of the individual dominance type species; (4) fishery value; (5) hydrologic characteristics (including compaction by heavy equipment or livestock, erosion, flooding, bank stability, sedimentation, and road placement); (6) fire susceptibility and potential for a fire break; and (7) considerations for recreational use. Likewise, Kovalchik (1987) included a management and rehabilitation section with each plant association description. He described: (1) "livestock relationships" by discussing forage productivity and palatability as well as vegetation response to grazing; (2) "wildlife and fisheries values" by describing use of the vegetation for nesting cover, hiding cover, or forage, by various groups of wildlife and by describing the potential influence of the plants on stream bank stability; (3) "fire" susceptibility and flammability; (4) "recognizing or estimating potential on disturbed sites" by listing indicators of the site and potential for rehabilitation; and (5) "rehabilitation pathways" by discussing the need for, and potential of, management actions to rehabilitate disturbed sites.

17.10 Grazing management

Where managers do not have the benefit of a complete riparian classification system, and even where they do, a great deal of progress toward good management can be achieved by observing the effects of current management and seeking alternatives that would reduce adverse impacts. Ratliff (1985) reviews the effects on meadows of defoliation, preferential grazing, trampling, mineral redistribution, and burrowing by rodents. Where grazing is too severe, stream-side vegetation may not be able to fulfill its function concerning the "four habitat components that make up the aquatic environment: stream-side vegetation (riparian zone), the stream channel, the water column, and the stream banks". These four components determine the quality of aquatic habitat and the productivity of the fishery (Platts 1983). The significant effects of livestock grazing on wildlife habitat have been reviewed by Kauffman and Krueger (1984), Skovlin (1984), Platts and Raleigh (1984), and Ohmart and Anderson (1986). Land management agencies could be forced through legal action to improve riparian management if current practices do not meet state water quality standards mandated by the Clean Water Law (Braun 1986). Such water quality standards may include sediment, temperature, bacteria, and salinity, each of which interrelate with riparian vegetation and with riparian water storage and flow regime.

Uneven distribution of grazing is typically the number one problem with livestock, especially cattle, in arid or semi-arid pastures containing little riparian vegetation. This problem is more pronounced if: (1) water is not well distributed; (2) the land near riparian water is steep or rocky (especially if all water is riparian); (3) salt is placed in or near riparian areas; (4) the weather is hot and riparian shade is available; (5) nonriparian forage is less palatable than riparian forage; (6) the herd is composed of cows with calves as opposed to yearlings; (7) Individual animals develop behavior patterns that favor riparian areas; (8) animal distribution is not maintained by herding; and/or (9) the grazing season is long (Swanson 1987).

Except for the intense utilization of riparian vegetation, it might withstand grazing much better than upland vegetation. Typically, due to a lack of soil moisture, herbaceous arid and semiarid range plants mature shortly after their early peak growing period. Riparian plants, however, generally have more soil moisture and, therefore, have longer to recover from defoliation. Winward (1986) draws the distinction between riparian vegetation, which is not fragile so long as the water remains abundantly available, and riparian soils and stream banks, which may be fragile and upon which water relationships depend. Compaction is most commonly a problem on moist as opposed to either wet or dry soil.

Exclusion of livestock grazing in riparian areas is an option, but it is by no means the only option. It is common for other approaches to be cheaper (Platts and Wagstaff 1984) and to provide a greater total mix of resource values. An array of grazing management options such as grazing systems, season of use, intensive management in riparian pastures, kind and class of livestock, selection of individual animals, and range improvements can each be used in selected settings to enhance riparian areas or optimize management for hydrology or fish, wildlife, and livestock habitat (American Fisheries Society, Western Division 1982, Skovlin 1984, Platts 1984, Platts and Raleigh 1984, Ohmart and Anderson 1986, and Swanson 1986). Gomm (1979) discussed mountain meadow grazing under the headings "grazing management on wet meadows" and "grazing management on improved meadows".

There is a temptation to speed up the process of stream habitat enhancement by using in-stream structures (DeBano and Heede 1987). Rosgen and Fittante (1986) explain that fish habitat structures often meet with great success on certain streams and are total disasters on others. They use the stream classification system of Rosgen (1985) to explain and predict the variability of success by presenting a table of compatibility between certain structures and each of the stream types. Elmore and Beschta (1987) argue against the use of instream structures, especially before land management has been adjusted to achieve vegetation recovery.

17.11 Objectives and monitoring

Riparian management can benefit from critical thought devoted to the use of land and vegetation classification in the "quantitative ecology" manner proposed by Dyksterhuis (1949) and applied by the Soil Conservation Service and Bureau of Land Management (See Smith this volume). Management focused on climax or potential natural vegetation would be especially problematical in riparian areas if it automatically assumed that this state or a percent approximation of this state is a clear objective for management. The natural tendency for the physical environment to vacillate due to the disturbance of fluvial processes would frequently preclude the opportunity for potential natural communities to express themselves. Describing vegetation management objectives in terms of percent approximation to climax would obscure the aim of management on sites having alternate routes to climax. This is especially true if these routes progress through seral stages having vastly different characteristics such as structure or palatability. The success of management would also remain obscure.

An approach whereby a community that could occupy a site under different management scenarios is selected and described as the vegetation objective has some appeal (Leonard 1987). Managers and users consider the differing resource values of each possible community, and the management required to achieve each community, before choosing the one most capable of meeting the multiple use resource management objectives within budgetary constraints and other considerations. Once management strategies have been implemented to reach the "desired plant community", success can be measured in a manner similar to the quantitative ecology approach of Dyksterhuis (1949). However, the basis of rating success would be proximity to the "desired plant community" as opposed to the climax community (although the desired plant community could be the climax community).

A similar approach could use the relative composition of riparian community types, as opposed to species within selected community types, to identify the riparian complex desired in heterogeneous areas. Winward and Padgett (1987) suggested using this approach to monitor riparian management. They recommended line intercepts across riparian areas and along the green line of perennial vegetation adjacent to the aquatic ecosystem. The objective of management could relate to changes in water relations, for example increased water storage in a higher water table. These changes would likely be reflected by vegetation, but could also be measured directly.

The fundamental problem confronting riparian managers is the setting of objectives. The more future oriented and tightly defined the objectives are, the more important it will be to use insights from classification to understand the realm of possibilities; also, the more tenuous will be the assurance that they are both achievable and desirable. Management and monitoring must follow objective setting. "Monitoring is management" (Tidwell 1987), and numerous techniques are available for rangeland monitoring (Bureau of Land Management 1987b), for rangeland watershed monitoring (Jackson *et al.* 1985), and for riparian monitoring (Platts *et al.* 1987). Any techniques chosen for a specific setting must be capable of answering the question "is management working to meet objectives".

17.12 Conclusion

The field of riparian management science has borrowed heavily from knowledge learned on uplands. Only some of it applies in riparian settings. Many commonly held precepts must be adapted for understanding this valuable resource that changes so abruptly both spatially and temporally. Long neglected, the process of riparian classification will require concentrated interdisciplinary effort before the appropriate diversity of scientists and managers will have a common language. Along with the process of delineating similar units of vegetation and physical environment, we must study the ecological relationships that may eventually enable us to describe successional potentials and pathways under differing management approaches. A principal component of this task is linking hydrology with vegetation science and vice versa. Little is quantitively known of the role hydrology plays in vegetation science and even less is known of the role vegetation plays in fluvial geomorphology. In an environment so defined and dominated by relationships with water, the motivation is becoming clear.

Ultimately the insight from classification and ecological study must also be made concise. Although we have no right to expect nature to be simple, we must appreciate the fact that it must appear simple if many people are to think they understand it well enough to apply management. It is now apparent that society will no longer allow us to ignore riparian areas. The challenge facing riparian managers is to apply appropriate strategies to optimize multiple use management.

17.13 References

American Fisheries Society, Western Section. 1982. Best management practices for the management and protection of western riparian stream ecosystems. Paper prepared by the Riparian Habitat Committee and Its Best Management Practices Subcommittee, Haugen, G. and D. Duff Chairmen.

Behnke, R. J., and R. F. Raleigh. 1978. Grazing and the riparian zone: Impact and management perspectives. p. 263—267. *In*: Strategies for Protection and Management of Floodplain Wetlands and other Riparian Ecosystems. Proc. Symp. Dec. 11—13 1978, Callaway Gardens, GA. Gen. Tech. Rep. WO-12, USDA For. Serv., Wash. D.C.

Brady, W., D. R. Patton, and J. Paxson. 1985. The development of southwestern riparian gallery forests. p. 39—43. *In*: Riparian Ecosystems and their management: Reconciling conflicting uses. First North American Riparian Conference, April 16—18 1985. USDA For. Serv. Gen. Tech. Rep. RM-120, Rocky Mountain For. and Rng. Exp. Sta., Fort Collins, Colo.

Braun, R. H. 1986. Emerging limits on federal land management discretion: Livestock, riparian ecosystems, and clean water law. Environmental Law, 17: 43—79.

Brown, D. E. 1978. Southwestern wetlands — Their classification and characteristics. p. 269—282. *In*: Strategies for Protection and Management of Floodplain Wetlands and other Riparian Ecosystems. Proc. Symp. Dec 11—13 1978, Callaway Gardens, GA. Gen. Tech. Rep. W0-12, USDA For. Serv. Wash. D.C.

Brown, D. E. 1984. Evolution and riparian systematics. p. 288—306. *In*: R. E. Warner, and K. M. Hendrix (eds.) California Riparian Systems: Ecology, Conservation, and Productive Management. Proc of the California Riparian Systems Conf., 1981. Univ. Cal. Berkeley Press.

Brown, D. E., C. H. Lowe, and C. P. Pase. 1980. A Digitized Classification for Ecosystems with and Illustrated Summary of the Natural Vegetation of North America. USDA For. Serv. Gen. Tech. Rep. RM-73, Rocky Mountain For. and Rng. Exp. Sta., Fort Collins, Colo.

Cooper, D. J. 1986. Community structure and wetland classification of Rocky Mountain wetland ecosystems. Chapter 3. p. 66—147. *In*: An Ecological Characterization of Rocky Mountain Montane and Subalpine Wetlands. USDI Fish and Wildlife Serv. Biol. Rep.

Cottam, W. P., and G. Stewart 1940. Plant succession as a result of grazing and of meadow desiccation by erosion since settlement in 1862. J. Forestry, 38(8): 613—626.

Cowardin, L. M., V. Carter, F. C. Golet, and E. T. LaRoe. 1979. Classification of Wetlands and Deepwater Habitats of the United States. US Fish and Wildl. Serv./Off. of Biol. Serv. — 79/31.

Crouse, M. R., and R. R. Kindschy. 1981. A method for predicting riparian vegetation potential of semiarid rangelands. p. 110—116. *In*: proc of the Symposium on acquisition and utilization of aquatic habitat inventory information, Portland, Ore. Oct. 28—30, 1981.

Courtois, L. A. 1984. Temporal desert riparian systems — the Mojave River as an example. p. 688—693. *In*: E. Warner, and K. M. Hendrix (eds.) California Riparian Systems: Ecology, Conservation, and Productive Management. Proc of the California Riparian Systems Conf., 1981. Univ. Cal. Berkeley Press.

Daubenmire, R. 1968. Plant Communities — a textbook of plant synecology. Harper and Row, New york.

DeBano, L. F., and B. H. Heede. 1987. Enhancement of riparian ecosystems with channel structures. Water Resources Bull., 23(3): 463—470.

DeVelice, R. L., J. A. Ludwig, W. H. Moir, and F. Ronco Jr. 1986. A classification of Forest Habitat Types of Northern New Mexico and Southern Colorado. USDA For. Serv. Gen. Tech. Rep. RM-131, Rocky Mountain For. and Rng. Exp. Sta., Fort Collins, Colo.

Dick-Peddie, W. A., and J. P. Hubbard. 1977. Classification of riparian vegetation. p. 85—90. *In*: R. R. Johnson, and D. A. Jones (coordinators) on the Importance, Preservation and Management of the Riparian Habitat: A Symposium. USDA For. Serv. Gen. Tech. Rep. RM-43, July 9, 1977, Tucson, Ariz. Rocky Mountain For. and Rng. Experiment Station, Fort Collins, Colo.

Driscoll, R. S., D. L. Merkel, J. S. Hagihara, and D. L. Radloff. 1983. A component Land Classification for the United States: Status Report. USDI Bur. of Land Manage. Tech. Rep. 360.

Dyksterhuis, E. J. 1949. Condition and management of range land based on quantitative ecology. J. Range Manage. 2(3): 104—115.

Ellison, L. 1954. Subalpine vegetation of the Wasatch Plateau, Utah. Ecol. Monogr., 24: 89—184.

Elmore, W., and R. L. Beschta. 1987. Riparian areas: Perceptions in management. Rangelands 9(6): 260—265.

Federal Register. 1980. "40 CFR Part 230: Section 404(b)(1) Guidelines for specification of disposal sites for dredged or fill material," Vol. 45, No. 249, p. 85352—85353, U.S. Govt. Printing Off., Washington, D.C.

Federal Register. 1982. "Title 33: Navigation and Navigable Waters; Chapter II, Regulatory Programs of the Corps of Engineers." Vol. 47, No. 138, p. 31810, U.S. Govt. Printing Off., Washington, D.C.

Fenner, P., W. W. Brady, and D. R. Patton. 1985. Effects of regulated water flows on regeneration of fremont cottonwood. J. Range Manage., 38(2): 135—138.

Frissell, C. A., W. J. Liss, C. E. Warren, and M. D. Hurley. 1986. A hierarchical framework for stream habitat classification: viewing streams in a watershed context. Environmental Manage., 10(2): 199—214.

Gomm, F. B. 1979. Grazing management on mountain meadows. p. 141—157. *In*: R. H. Delaney, and J. Borrelli (eds.) Proc. of the Symposium on Management of Intermountain Meadows, June 7—9, Jackson, Wy. Univ. of Wyo. Col. of Agric. and Colo. St. Univ. Mountain Meadow Res. Cntr.

Hall, F. C. 1979. Codes for Pacific Northwest Ecoclass Vegetation Classification. USDA For. Serv. Pacific Northwest Region R6 ECOL 79-002.

Hansen, P. L., S. W. Chadde, and R. D. Pfister. 1987. Riparian Dominance Types of Montana (Review Draft). Montana Riparian Association, School of Forestry, University of Montana, Missoula, Montana.

Hoar A. R., and M. J. Erwin. 1985. Relationships between the expansion of agriculture and the reduction of natural riparian habitat in the Missouri River Floodplain of northeast Montana, 1938—1982. p. 250—256. *In*: Riparian Ecosystems and their management: Reconciling conflicting uses. First North American Riparian Conference, April 16—18 1985. USDA For. Serv. Gen. Tech. Rep. RM-120, Rocky Mountain For. and Rng. Exp. Sta., Fort Collins, Colo.

Horton, J. S., and C. J. Campbell. 1974. Management of phreatophyte and riparian vegetation for maximum multiple use values. USDA For. Ser. Res. Paper RM-117, Rocky Mountain For. and Rng. Exp. Sta., Fort Collins, Colo.

Jackson, W. L., S. Hudson, and K. Gebhardt. 1985. Considerations in Rangeland Watershed Monitoring. USDI Bur. of Land Manage., Tech. Note No. 369.

Johnson, R. R. 1978. The lower Colorado River: A western system. p. 41—55. *In*: Strategies for Protection and Management of Floodplain Wetlands and other Riparian Ecosystems. Proc. Symp. Dec. 11—13 1978, Callaway Gardens, GA. Gen. Tech. Rep. WO-12, USDA For. Serv. Wash. D.C.

Johnson R. R., and S. W. Carothers. 1987. External threats: the dilemma of resource management on the Colorado River in Grand Canyon National Park, USA. Environmental Manage., 11(1): 99—107.

Johnson R. R., L. T. Haight, and J. M. Simpson. 1977. Endangered species vs. endangered habitats: A concept, p. 68—79. *In*: R. R. Johnson, and D. A. Jones (Coordinators) Importance, Preservation and Management of the Riparian Habitat: A Symposium. July 9, 1977, Tucson, Ariz. USDA For. Serv. Gen. Tech. Rep. RM-43, Rocky Mountain For. and Rng. Exp. Sta., Fort Collins, Colo.

Johnson R. R., and S. W. Carothers. 1982. Southwestern riparian habitats and recreation: interrelationships and impacts in the Southwest and Rocky Mountain Region. Eisenhower Consortium Bull. 12, USDA For. Serv., Rocky Mountain For. and Rng. Exp. Sta., Fort Collins, Colo.

Johnson, R. R., S. W. Carothers, and J. M. Simpson. 1984. A riparian classification system. p. 375—382. *In*: R. E. Warner, and K. M. Hendrix (eds.) California Riparian Systems: Eclogy, Conservation, and Productive Management. Proc of the California Riparian Systems Conf., 1981. Univ. Cal. Berkeley Press.

Katibah, E. F. 1984. A brief history of riparian forests in the Central Valley of California. p. 23—29. *In*: R. E. Warner, and K. M. Hendrix (eds.) California Riparian Systems: Ecology, Conservation, and Productive Management. Proc of the California Riparian Systems Conf., 1981. Univ. Cal. Berkeley Press.

Kauffman, J. B., and W. C. Krueger. 1984. Livestock impacts on riparian ecosystems and streamside management implications . . . A review. J. Range Manage., 37(5): 430—438.

Kauffman, J. B., W. C. Krueger, and M. Vavra. 1985. Ecology and plant communities of the riparian area associated with Catherine Creek in Northeastern Oregon. Oregon State University Agricultural Experiment Station, Tech. Bull. 147.

Kovalchik. B. L. 1987. Riparian zone associations: Deschutes, Ochoco, Fremont, and Winema National Forests. USDA For. Serv. Pacific Northwest Region R6 ECOL TP-279-87.

Kuramoto, R. T., and L. C. Bliss. 1970. Ecology of subalpine meadows in the Olympic Mountains, Washington. Ecol. Monogr., 40: 317—347.

Leonard, S. 1987. Determining and reporting range condition. mimeograph produced by the Soil Range Team of the USDI Bur. of Land Management, and the USDA Soil Conservation Service, University of Nevada, Reno, Nev.

Lowe, C. H. 1961. Biotic communities in the sub-Mogollon region of the inland Southwest. J. Ariz. Acad. Sci. 2: 40—49.

McBride, J. R. and J. Strahan. 1984. Fluvial processes and woodland succession along Dry Creek, Sonoma County, California. p. 110—119. *In*: R. E. Warner, and K. M. Hendrix (eds.) California Riparian Systems: Ecology, Conservation, and Productive Management. Proc. of the California Riparian Systems Conf., 1981. Univ. Cal. Berkeley Press.

Medina, A. L. 1986. Riparian plant communities of the Fort Bayard watershed in southwestern New Mexico. The Southwestern Naturalist 31(3): 345—359.

Meehan, W. R., and W. S. Platts. 1978. Livestock Grazing and the aquatic environment. J. Soil and Water Conservation 33(6): 274—278.

Meeker, D. O., and D. L. Merkel. 1984. Climax theories and a recommendation for vegetation classification — A viewpoint. J. Range Manage. 37(5): 427—430.

Mutz, K. K., and J. Queiroz. Riparian community classification for the Centennial Mountains and South Fork Salmon River, Idaho. Contract No. 53-84M8-2-0048, Meiiji Resource Consultants, Layton, UT.

Ohmart, R. D., and B. W. Anderson. 1986. Riparian Habitat. Ch. 9. *In*: A. Y., Cooperrider, R. J. Boyd, and H. R. Stuart (eds.) Inventory and monitoring of wildlife habitat. USDI Bur. Land Manage. Service Center. Denver, Co. xviii.

Parker, M., F. J. Wood, B. H. Smith, and R. G. Elder. 1985. Erosional downcutting in lower order riparian ecosystems: Have historical changes been caused by removal of beaver? p. 35—38. *In*: Riparian Ecosystems and their management: Reconciling conflicting uses. First North American Riparian Conference, April 16—18 1985. USDA For. Serv. Gen. Tech. Rep. RM-120, Rocky Mountain For. and Range Exp. Sta., Fort Collins, Colo.

Paysen, T. E., J. A. Derby, H. Black Jr., V. C. Bleich, and J. W. Mincks. 1980. A vegetation classification system applied to southern California. USDA For. Serv. Gen. Tech. Rep. PSW-45 Pacific Southwest For. and Range Exp. Sta. Berkeley, Calif.

Platts, W. S. 1980. A plea for fishery habitat classification. Fisheries, 5(1): 2—6.

Platts, W. S. 1983. Vegetation requirements for fisheries habitats. p. 184—188. *In*: Managing Intermountain Rangelands — Improvement of Range and Wildlife Habitats. Proceedings of Symposia: September 15—17, 1981, Twin Falls, ID and June 22—24, 1982, Elko, NV.

Platts, W. S. 1984. Compatibility of livestock grazing strategies with riparian-stream systems. p. 67—74. *In*: Pacific Northwest range management short course: Range watersheds, riparian zones, and economics. Ore. St. Univ. Corvallis, Ore.

Platts, W. S., and R. F. Raleigh. 1984. Impacts of grazing on wetlands and riparian habitat. p. 1105—1118. *In*: Developing strategies for rangeland management: A Report Prepared by the Committee on Developing Strategies for Rangeland Management. National Research Council/National Academy of Sciences. Westview Press, Boulder.

Platts, W. S., and F. J. Wagstaff. 1984. Fencing to control livestock grazing on riparian habitats: Is it a viable alternative? N. Amer. Fish Manage. 4: 266—272.

Platts, W. S., C. Armour, G. D. Booth, M. Bryant, J. L. Burford, P. Cuplin, S. Jensen, G. W. Lienkaemper, G. W. Minshall, S. B. Monson, R. L. Nelson, J. R. Sedell, and J. S. Tuhy. 1987. Methods for Evaluating Riparian Habitats With Applications to Management. USDA For. Serv. Gen. Tech. Rep. INT-221. Intermountain Res. Sta. Ogden, Ut.

Price, K. P., and M. K. Ridd, 1983. Riparian habitat on the Humboldt River, Deeth to Elko, Nevada. Center for Remote Sensing and Cartography Report 83-3, University of Utah Research Institute, Salt Lake City, Utah.

Ratliff, R. D. 1982. A Meadow Site Classification for the Sierra Nevada, California. USDA For. Serv. Gen. Tech. Rep. PSW-60. Pac. SW For. & Rng Exp. Sta., Berkeley, Calif.

Ratliff, R. D. 1985. Meadows in the Sierra Nevada of California: State of Knowledge. USDA For. Serv. General Tech. Rep. PSW-84 Pac. SW For. & Rng. Exp. Sta., Berkeley, Calif.

Roath, L. R., and W. C. Krueger. 1982. Cattle grazing and behavior on a forested range. J. Range Manage., 35(3): 332—338.

Rosgen, D. L. 1981. Total potential sediment., Ch. VI. *In*: Water resources evaluation of nonpoint silvicultural sources. (WRENSS), U.S. Environmental Protection Agency, Washington, D.C.

Rosgen, D. L. 1982. A procedure and Rationale for Securing Favorable Conditions of Water Flow on National Forest System Lands in Northern Wyoming. Dave Rosgen was hydrologist for Arapaho and Roosevelt National Forests, Region 2 USDA For. Serv.

Rosgen, D. L. 1985. A stream classification system. p. 39—43. *In*: Riparian ecosystems and their management: Reconciling conflicting uses. First North American Riparian Conference, April 16—18 1985. USDA For. Serv. Gen. Tech. Rep. RM-120, Rocky Mountain For. and Rng. Exp. Sta., Fort Collins, Colo.

Rosgen, D., and B. L. Fittante. 1986. Fish habitat structures — A selection guide using stream classification. *In*: Fifth Trout Stream Habitat Workshop: Lock Haven University, Lock Haven, Pennsylvania August 1986.

Sipple, W. S. 1987. Wetland Identification and Delineation Manual, Volumes I and II. Office of Wetlands Protection, Office of Water, U. S. Environmental Protection Agency.

Smith, D. G. 1976. Effect of vegetation on lateral channel migration of anastomosed channels of a glacier meltwater river. Geological Society of America Bull., 87: 857—860.

Skovlin, J. M. 1984. Impacts of grazing on wetlands and riparian habitat: A review of our knowledge. p. 1001—1103. *In*: Developing strategies for rangeland management: A Report Prepared by the Committee on Developing Strategies for Rangeland Management. National Research Council/National Academy of Sciences. Westview Press, Boulder.

Stevens, L. E., B. T. Brown, J. M. Simpson, and R. R. Johnson. 1977. The importance of riparian habitat to migrating birds, p. 156—164. *In*: R. R. Johnson, and D. A. Jones (Coordinators) Importance, Preservation and Management of the Riparian Habitat: A Symposium. July 9, 1977, Tucson, Ariz. USDA For. Serv. Gen. Tech. Rep. RM-43, Rocky Mountain For. and Rng. Exp. Sta., Fort Collins, Colo.

Stoddart, L. A. and A. D. Smith. 1955. Range management. 2nd. Ed. McGraw-Hill, New York.

Strahan, J. 1984. Regeneration of riparian forests of the Central Valley p. 58—67. *In*: R. E. Warner, and K. M. Hendrix (eds.) California Riparian Systems: Ecology, Conservation, and Productive Management. Proc of the California Riparian Systems Conf., 1981. Univ. Cal. Berkeley Press.

Swanson, S. 1986. Options for riparian grazing management. Fact Sheet 86-77, Col. of Agric. Univ. Nev. Reno, Nev.

Swanson, S. 1987. Riparian Pastures. Fact Sheet 87-53, Col. of Agric. Univ. Nev. Reno, Nev.

Swanson, S., R. Miles, S. Leonard, and K. Genz. (*In Press*). Classifying rangeland riparian areas: The Nevada task force approach. J. Soil and Water Conservation, Vol. 43.

Thomas, J. W., C. Maser, and J. E. Rodiek. 1979. Riparian zones. *In*: Wildlife habitats in managed rangelands — The great basin of Southeastern Oregon. USDA For. Serv. and USDI Bur. of Land Manage., Gen. Tech. Rep. PNW-80.

Tidwell, D. P. 1987. Monitoring is management. p. 15—17. *In*: G. Secrist, and K. Eschelman (Coordinators). Rangeland Monitoring Workshop Proceedings. U. S. Bureau of Land Management. January 12—16, 1987, Golden, Colo.

U. S. Army Corps of Engineers. 1987. Corps of Engineers Wetland Delineation Manual. Technical Report Y-87-1 Environmental Laboratory, Waterways Experiment Station.

U. S. Bureau of Land Management. 1987a. Riparian Area Management Policy. Signed by R. Burford on January 22, 1987.

U. S. Bureau of Land Management. 1987b. Rangeland Monitoring Workshop Proceedings. Secrist, G. and K. Eschelman Coordinators. January 12—16, 1987, Golden, Colo.

U. S. Fish and Wildlife Service. 1986. Wetland plants of the United States of America 1986. In cooperation with the National Wetland Plant List Review Panel. (continually updated computer file).

U. S. Forest Service. 1980. Forest Service Manual 4/80 Amend 26 section 2526.05, Definitions.

U. S. Forest Service. 1985. COWFISH: A FHR (Fish Habitat Relationships) Habitat Capability Model. Wildlife Management Staff, Intermountain Region.

U. S. Soil Conservation Service. 1985. Hydric Soils of the United States 1985. In cooperation with the National Technical Committee for Hydric Soils. (continually updated computer file).

Vanote, R. L., G. W. Minshall, K. W. Cummins, J. R. Sedell, and C. E. Cushing. 1980. The river continuum concept. Can. J. Fish and Aquat. Sci., 37: 130—137.

Wertz, W. A., and J. F. Arnold. 1972. Land Systems Inventory. USDA For. Serv. — Intermountain Region, Div. Soil & Water Management, Ogden, UT.

Winward, A. H. 1986. Vegetation characteristics of riparian areas. Trans. West. Sect. of The Wildl. Soc., 22: 98—101.

Winward, A. H., and W. G. Padgett. 1987. Special considerations when classifying riparian areas. Paper presented at the symposium "Land Classifications Based on Vegetation: Applications for Resource Management," Moscow, ID. November 17—19, 1987.

Youngblood, A. P., W. G. Padgett, and A. H. Winward. 1985a. Riparian community Type Classification of Eastern Idaho — Western Wyoming. USDA For. Serv. Intermountain Region R4-Ecol-85-01.

Youngblood, A. P., W. G. Padgett, and A. H. Winward. 1985b. Riparian community Type Classification of Northern Utah and adjacent Idaho (Preliminary Draft) USDA For. Serv. Intermountain Region, Ecology and Class. Prog.

Zimmerman, R. C., J. C. Goodlett, and G. H. Comer. 1967. The influence of vegetation on channel form of small streams. p. 255—275. Internat. Assoc. Sci. Hydrology, Symposium on River Morphology, Pub. no. 75.

PART THREE

Vegetation science rangeland applications

18. Rangeland vegetation productivity and biomass

REX D. PIEPER

18.1 Abstract

Herbage weight or biomass is an important characteristic for range vegetation since it supports directly or indirectly all consumer groups. Plant biomass is also important as a measure of dominance in plant communities. Many methods have been devised to estimate plant biomass, but all have some limitations. Estimation techniques, either by plot or plant, often involve a double sampling procedure to improve on the estimates. Direct harvesting is a destructive sampling procedure which is often used when biomass is the response measured in research studies. Precipitation has been used as an independent variable to predict end-of-season biomass. However, considerable data are necessary to develop reliable relationships. Capacitance meters, biomass meters, and β-attenuation techniques have promise for specific situations, but have not received widespread support among range researchers and ecologists. All these techniques require considerable calibration before they can be used for specific situations. Plant biomass data are useful for establishing dominance of individual species, estimating livestock stocking rates, and determining range condition.

18.2 Introduction

Measurement of plant biomass or standing crop has been of interest to range workers for some time because herbivores depend directly upon plant biomass for their food. Interest in production ecology is of more recent emphasis by plant ecologists, particularly during the International Biological Program Studies during the 1970's (Milner and Hughes 1968). These studies illustrate the importance of plant biomass as an indicator of general ecological importance of individual plants and communities.

Tueller, P. T. (ed.), Vegetation science applications for rangeland analysis and management.
ISBN 90-6193-195-9.
© 1988, *Kluwer Academic Publishers, Dordrecht, Boston, London.*

In many phytosociological schemes, plant cover, density, frequency, etc., have been used, but they may not reflect the ecological role of species in a plant community or the plant biomass available for various herbivores. Plant weight or biomass probably comes closer than other characteristics to expressing the three dimensional structural nature of vegetation and dominance of individual species. On the other hand, plant biomass is constantly changing during the year. Herbaceous plant weight generally increases during the growing season, reaches a peak late in the growing season and then declines during the dormant season (Pieper *et al.* 1974). Consequently, estimation of plant biomass or standing crop at any point in time is difficult, and estimating net aboveground primary productivity even more challenging.

The primary objective of this chapter is to review the methods presently available to determine plant biomass and the uses to which the data may be put. Primary emphasis will be on a description of the methods with advantages and limitations considered for different vegetation types.

18.3 Estimation techniques

The weight-estimate method was first described by Pechanec and Pickford (1937). The method was designed to provide a rapid method of determining herbage weight for extensive range surveys by land management agencies (Goebel 1955). However, it has also been used in research projects (Pieper 1978). In the original description of the method, Pechanec and Pickford recommended that the estimator mentally visualize 10, 50 or 100 g units of plant material of each species in the quadrat and add them together to determine total plant weight. Considerable training is necessary and can be easily accomplished by estimating quadrats and then clipping the plant material which has been estimated. The estimators continue training until their estimates are as close to the estimates as possible. Studies in the Panhandle of Oklahoma indicate that the method was reasonably accurate (Shoop and McIlvain 1963). However, variances may be lower for estimated quadrats than for clipped quadrats. Apparently, estimators tend to overestimate quadrats with low biomass and to underestimate biomass of those quadrats with high weight. Increasing sample size can increase the precision of the weight-estimate method, but caution should be exercised because of reduced variances on estimated quadrats.

The weight-estimate method is attractive even for researchers because the estimates can be checked and adjustments made when warranted. Arithmetic adjustments can be made if a person's estimates are either consistently high or low. In most cases, adjustments are made by regression analysis following a double sampling procedure (Wilm *et al.* 1944, Hilmon 1959). Herbage weight is estimated for all quadrats, but only a relatively small portion are clipped (Wilm *et al.* 1944, Hilmon 1959, Cook and Bonham 1977). The regression equation is developed from the data set of estimated and clipped weights and all estimated weights are adjusted by the regression equation.

The critical decision, from a cost standpoint, is the number of clipped quadrats. If an excessive number of quadrats is clipped, the cost of sampling may be unnecessarily high; conversely, if a sufficient number of quadrats is not clipped, the regression may be unreliable (Hilmon 1959). Some authors have used a fairly straight-forward approach utilizing differences in time required to estimate and clip a quadrat, the standard error of the clipped quadrats and the standard error from regression (Schumacher and Chapman 1944). Both ratio and regression techniques have been used to develop optimum ratios of estimated to clipped quadrats. In some cases, the authors have recommended ratio techniques (Blair 1959), while in others, regression techniques appeared to be more appropriate (Francis *et al.* 1979, Ahmed *et al.* 1983). Detailed analysis of several methods of determining ratio of clipped to estimated quadrats is presented by Cochran (1963), Ahmed and Bonham (1980, 1982) and Cook and Stubbendieck (1986).

The technique requires a linear relationship between estimated and clipped weights (Francis *et al.* 1979, Ahmed *et al.* 1983). Many authors have found the relationship to be linear (Francis *et al.* 1979, Ahmed *et al.* 1983), but others have detected curvilinear relationships (Tadmor 1975).

For some individuals, ranking or comparing the weight of herbage in different quadrats is easier than estimating absolute weight (Morris 1962). Hutchings and Schmautz (1969) described a "relative weight-estimate method" whereby a central control plot was clipped and four quadrats estimated in percent herbage weight of the central quadrat. The weight of each of the four estimated quadrats is calculated by multiplying its percentage by the weight of the central clipped quadrat. Accuracy was fairly high for production of all species, but less so for classes (grass, forbs and shrubs) and individual species (Hutchings and Schmautz 1969). The method was both accurate and precise on four alpine areas in northern Utah (Reese *et al.* 1980).

The ranked-sets method involves ranking quadrats based on the absolute weight of all species (McIntyre 1952). As an example, if three quadrats are used in each set, the highest ranking one is clipped in the first set, the intermediate-ranked quadrat in the second set and the lowest one in the third. Different numbers of quadrats can be ranked in each set, but it may not be practical to use more than five (McIntyre 1952). The main advantage of the method is the increased efficiency of sampling (McIntyre 1952 and Morris 1962). Studies in Louisiana (Halls and Dell 1966) and in the California annual grassland (Morris 1962) indicated that less than half the number of quadrats were required using a ranked-sets method compared to clipping randomly-placed quadrats. One limitation of the method is that only total herbage weight is determined. Other procedures are necessary to determine weight of individual species.

18.4 Direct harvest methods

Harvest methods involve clipping herbage or browse of individual plants or that contained in a quadrat and weighing the material (Milner and Hughes 1968, Pieper 1978, Risser 1986). Harvesting is a direct and objective method and is one of the most commonly used methods under research conditions. However, clipping is a slow and relatively expensive method (Blankenship and Smith 1966). In addition, clipping is a destructive sampling procedure which influences vigor and subsequent growth rate of the clipped plants. Consequently, the same plants or quadrats cannot be used for herbage weight determination after the first clipping.

Several sources of variation may lead to inconsistent results with clipping methods. In some case a specific clipping height is designated, while in others clipping is done as close to the soil surface as possible. In both cases, individuals may not use the same standards (Milner and Hughes 1968). This problem is often pronounced since the weight of most grass plants is concentrated near the base of the plant. Plants which are pedestalled or those with large crowns may prove to be particularly troublesome.

For herbaceous plants, several categories may be separated: living, actively growing material; recent dead (produced during the current season, but no longer actively growing); old dead (produced during previous growing seasons); and mulch or litter (detached plant material lying on soil surface). Recent dead and old dead may be lumped as standing dead to distinguish these categories from the litter. However, if productivity (expressed as a rate) is being estimated, then recent dead should be added to living herbage.

For woody plants, leaders and leaves produced during the current growing season are often included in the harvested sample. For some species, distinguishing new growth from old growth is relatively easy, but in others there is no distinct separation. Development of branches from lateral buds on old leaders may also complicate the situation.

Harvests may be conducted on an individual plant basis or on an area basis. If individual plants are used, then some estimate of density is needed to convert data to weight per unit of area (Laycock 1965). Generally, some type of quadrat or plot is used and all plants within the quadrat are clipped and weighed. Use of quadrats with a defined area enables one to determine biomass per unit area directly.

Decisions which have to be made when using quadrats include size, shape and number to be used. One concern with quadrat size and shape is the number of borderline decisions which have to be made. The number of such decisions is directly related to the perimeter of the quadrat (Pieper 1978). The perimeter : area ratio is lowest for circular quadrats and increases as the length of a rectangular quadrat increases (Table 1). Perimeter:area ratios also decrease as quadrat size increases. Van Dyne *et al.* (1963) presented data indicating a positive bias in herbage weight associated with increasing perimeter:area ratios on several bunchgrass ranges in Montana. However, statistical and time efficiency must also be considered in selecting an optimum quadrat. Unfortunately, one quadrat size and shape may not be equally suitable for all species in a plant community (Wiegert 1962, Cook and Stubbendieck 1986). For scattered or clumped species, rectangular quadrats may lower interquadrat variances compared to those from circular or square quadrats (Soplin *et al.* 1975). Studies in salt desert shrub vegetation in southern Wyoming (Wight 1967) (Fig. 1) and desert grassland in southern New Mexico (Pino 1954) showed a reduction of variances with linear quadrats. However, in bunchgrass ranges in Greece, quadrat shape did not change herbage weight determinations (Papanastasis 1977).

Generally, as quadrat size increases, variances among quadrats decrease (Cook and Stubbendieck 1986). Moreover, total time required to clip large quadrats may be less than that for small quadrats assuming the same area is clipped because of reduced travel time. However, distribution of sampling units over the area may favor more small quadrats rather than fewer

Table 1. Perimeter and area relations for quadrats of various shapes and sizes.

Quadrat size and shape	Area	Length of perimeter	P : A ratio
Circle	1 M^2	1.13 m	1.13
Square, 1 M × 1 M	1 M^2	4.00	4.00
Rectangle, 0.5 M × 2.0 M	1 M^2	5.00	5.00
Rectangle, 0.2 × 5 M	1 M^2	11.00	11.00
Square, 0.5 m × 0.5 m	0.25 M^2	2.00	8.00
Square, 0.75 m × 0.75 m	0.56 M^2	3.00	5.36
Square, 1 m × 1 m	1.00 M^2	4.00	4.00
Square, 2 m × 2 m	4.00 M^2	8.00	2.00

large quadrats. Papanastasis (1977) recommended using small quadrats for sampling bunchgrass vegetation in Greece based on both time and statistical efficiency.

Precision of sampling for herbage weight (as indicated by the standard error of the mean) can also be increased by increasing sample size. Sample sizes necessary for a given degree of reliability depends upon the natural variability of the vegetation and how well the sample unit fits the vegetation. Mueggler (1976) calculated that less than ten 4.8 sq. ft. quadrats were necessary to sample within 30% of the population mean for total herbage weight with 80% confidence, but as many as 60 quadrats were required for some habitat types to sample total herbage weight within 10% of the population mean on bunchgrass range in Montana. As many as 730 4.8 ft^2 quadrats were required to sample weight of forbs in bunchgrass vegetation (Mueggler 1976). Over 120 1 × 1 ft. quadrats were necessary to sample total yield of nuttall saltbush within 20% of the population mean and the 90% probability level, while less than ten quadrats were required for quadrats larger than 45 sq. ft. (Fig. 1).

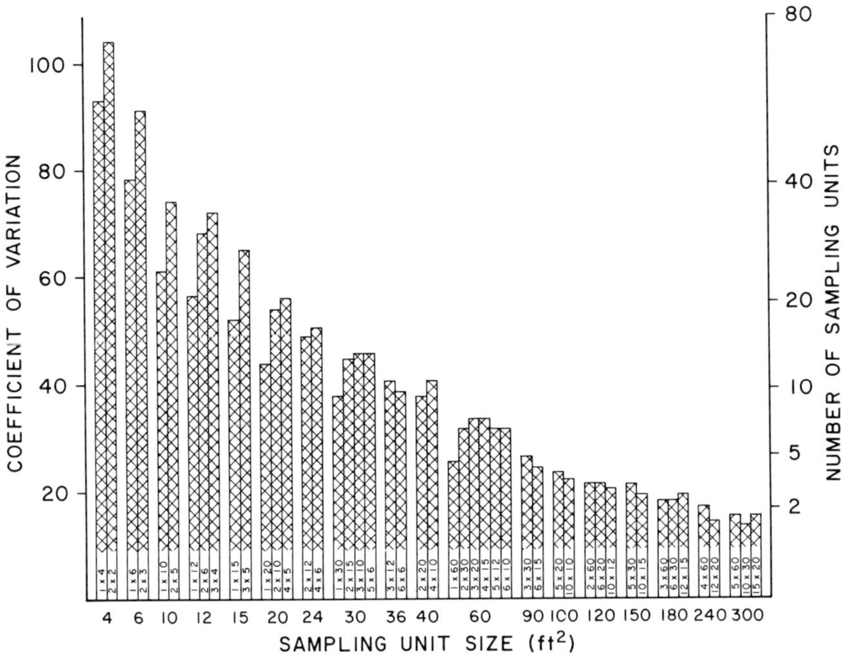

Figure 1. Relationship between sampling unit shape and size and the coefficient of variation and number of sampling units necessary to obtain a yield estimate that is within 20% of the population mean 90% of the time (from Wight 1967).

18.5 Indirect methods

Indirect methods of determining herbage weight usually involve measurements of some variable which is closely related to herbage weight and relatively easy to measure and relating this variable to herbage weight by regression analysis in a calibration procedure. Thereafter, only the related variable would be measured.

18.5.1 Precipitation

On most rangelands, precipitation, and hence soil water, is one environmental factor controlling herbage production. Consequently, precipitation has often been used as the independent variable to predict end-of-season herbage standing crop or biomass (Rogler and Haas 1947). In these studies several years of data are necessary to develop reliable regression equations. It is important to have a wide range in precipitation values to develop a model with wide applicability. In some cases, annual precipitation can be used as the independent variable (e.g., on salt desert shrub vegetation in western Utah [Hutchings and Stewart 1953]) or seasonal precipitation (e.g., short grass vegetation in southern Canada [Smoliak 1956] or New Mexico [Pieper *et al.* 1971], sagebrush-grass vegetation of southeastern Idaho [Passey *et al.* 1964, Blaisdell 1958], or California annual grassland [Murphy 1970]) (Table 2).

There are several limitations involved in using precipitation: biomass relations to predict herbage weight. In many cases, it is difficult to determine how precipitation influences herbage production. Small showers during certain portions of the year may have little influence on subsequent

Table 2. Precipitation: herbage weight relationships for several ranges showing range in coefficients of determinations.

Independent variable	r^2	Vegetation	Author
Annual ppt.	0.89	Salt desert shrub	Hutchings and Stewart 1953
May—June ppt.	0.86	Mixed prairie	Smoliak 1956
June—August ppt.	0.78	Desert grassland	Cable 1975
April ppt.	0.41	California annual grassland	Duncan and Woodmansee 1975
July—August ppt.	0.42	Pinyon juniper	Pieper *et al.* 1971
Annual ppt.	0.79	Sahel annual grassland	Lehouerou and Hoste 1977
Crop year ppt.	0.77	Crested wheatgrass seedings	Sneva and Hyder 1962
November ppt.	0.41	Ponderosa pine forest	Pumphrey 1980

herbage production. Duncan and Woodmansee (1975) found that total herbage biomass was most closely correlated with April precipitation if only one month was considered for California annual grassland in southern California. However, in northern California, November was the most important month (Murphy 1970). Pumphrey (1980) found that precipitation was a relatively poor predictor of herbage production on ponderosa pine ranges in eastern Oregon, but could be used when these same ranges were fertilized. Desert grasslands pose some special problems (Parker 1963, Russell 1973, Cable 1975). Cable (1975) found that August precipitation was the best single predictor of summer herbage production in southern Arizona. However, adding the previous June-September precipitation to the model increased the reliability of the relationship. Hart and Samuel (1985) found a fairly high correlation between spring and summer precipitation and herbage yield ($r^2 = 0.95$) for soils with sandy subsoils but not for those with loamy subsoils. In southern New Mexico, any combination of monthly precipitation was poorly correlated to herbage production (Parker 1963). A precipitation × basal cover index improved the prediction considerably over using precipitation alone. In other cases, more complex models may be necessary to predict range production (Wight *et al.* 1984).

Despite the appeal of using precipitation : biomass relationships, these predictions may not be sensitive enough for all situations. For example, Cable (1975) found that standard errors of estimate varied from about 117 to 80 kg/ha on desert grassland in southern Arizona when pasture production varied from 394 to 587 kg/ha. Hence, one could predict herbage production only within 160 to 234 kg/ha.

18.5.2 Capacitance meters

Considerable interest has been generated in the use of capacitance meters to estimate herbage biomass since the method was introduced by Fletcher and Robinson (1956). The instrument is based on the differences in dielectric constants between air and the vegetation (Fletcher and Robinson 1956, Neal and Neal 1973). The instrument has conductors separated by the air-herbage mixture which serves as a capacitor to measure potential differences between the conductors. The amount of the stored electrical charge is directly related to the amount of herbage present.

Many different capacitance meters have been developed and tested (Van Dyne *et al.* 1968, Neal and Neal 1973). Campbell *et al.* (1962) described a model with several improvements over earlier models and Neal and Neal (1965) described their "Heterodyne Vegetation Meter". Different meters have been described by Alcock (1964), Johns *et al.*

(1965), Alcock and Lovett (1967), Kreil and Matschke (1968), and Van Dyne *et al.* (1968).

Many of the "multiple-probe" meters are similar in appearance, consisting of four supporting legs on the corners and a series of probes nearly as long as the legs arranged in rows between the legs. In some cases, the legs are omitted and the instrument rests on the probes. The "earth-plate" meters operate on a slightly different basis from the probe types with a plate capacitor parallel to the ground surface (Angelone *et al.* 1980a, 1980b and Toledo *et al.* 1980). In actual use readings are taken on several placements of the instrument and the herbage beneath the instrument clipped in a double-sampling procedure. Regression analysis is then used to convert meter readings to herbage weight. Neal *et al.* (1976) reported a 1 : 5 ratio of clipped plus meter-estimated samples to meter-estimated only samples was adequate for perennial vegetation and a 1 : 10 ratio for annual vegetation.

The method was developed primarily for herbaceous vegetation. It has been tested on seeded ranges in Arizona and Colorado (Currie *et al.* 1973), grasslands in New Zealand (Campbell *et al.* 1962), Australia (Johns *et al.* 1965) and Scotland (Alcock 1964), bunchgrass range in California and Colorado, oak-hickory savanna in Louisiana, annual grassland in California (Neal *et al.* 1976) and rangeland in Florida (Terry *et al.* 1981). The method has also been tested for estimating woody biomass in sagebrush-grass range in Colorado and Wyoming (Carpenter 1973, Morris 1976).

In using a capacitance meter, one should be aware of variations in method. Species in different phenological stages result in different readings because of differences in moisture content of the herbage (Neal *et al.* 1976). Sampling when plants are not growing rapidly may require larger sample sizes. Standing dead material and woody stems can also result in highly variable results (Carpenter *et al.* 1973), Morris *et al.* 1976, Neal *et al.* 1976). Terry *et al.* (1981) reported that predictability was improved by converting both meter readings and herbage weights to a logarithmic scale on Florida rangelands.

18.5.3 Beta attenuation technique

The beta attenuation technique was introduced by workers at Purdue University (Mott *et al.* 1965, Teare and Mott 1965, Teare *et al.* 1965) for crops and pastures. Basically, the method involves emitting β-particles on one side of a plot and then counting the particles which emerge on the other side of the plot. Teare *et al.* (1966) found herbage weight was directly related to the number of particles absorbed or deflected by the vegetation. The radioactive counts are related to herbage weight by regression techniques.

Mitchell (1972) evaluated the technique for short-grass vegetation in Colorado. He reported r^2 values between 0.94 and 0.92 for the relationship of activity and wet weight biomass of herbaceous vegetation. Knapp *et al.* (1985) reported that they could estimate biomass of burned tallgrass prairie vegetation within ±5% of clipped values until plants began to senesce late in the growing season. Possible β emiters include: carbon^{14}, chlorine^{36}, krypton^{85}, strontium^{90} and thallium^{204}.

The method apparently has not been widely adopted. Probably, the necessity of working with β-emitting radioactive material and the need for a Geiger counter or other detecting device minimizes use of the technique.

18.5.4 Cover

It appears logical that vegetational cover is related to herbage weight and biomass, but differences in plant height could complicate the relationship. Cover is a single-dimension parameter, while weight appears to be three dimensional and more closely related to volume than cover.

In mountain ranges of southwestern Montana, Payne (1974) found a significant correlation greater than 0.90 between canopy cover and herbage weight for 16 species. Only six species showed a significant correlation coefficient (r) less than 0.7. Goebel *et al.* (1958) found a close relationship between canopy cover and weight of shadescale (*Atriplex confertifolia*) in Utah, but more variable results for winterfat (*Ceratoides lanata*), presumably because of irregular shape of winterfat plants.

18.5.5 Dimension analysis

Dimension analysis involves measurements of some easily measured parameter of a plant and relating these measurements to biomass (Pieper 1984). Often, these measurements result in biomass estimates for individual plants. Consequently, in addition to calibration involving development of appropriate regression models, some estimate of density by species is necessary to express biomass or standing crop on an area basis. Dimension analysis has been used most often for shrubs, but also has application for grasses.

Shrubs. In some cases, shrub biomass has been estimated from crown or basal diameter alone (Brown 1976). Medin (1960) developed regression models relating diameter of mountain mahogany (*Cercocarpus montanus*) to biomass. The relationship was logarithmic with $r = 0.84$. Utah juniper foliage biomass was estimated from crown diameter measurements with r^2 values above 0.90 for 7 out of 12 sites involved in a Utah study (Mason and Hutchings 1967, Peek 1970). Hutchings and Mason (1970) related both foliage and basal area of Gambel oak (*Quercus gambelii*) to weight of leaves and twigs. They found that over 75% of the variation in foliage production could be accounted for by foliage area and basal area. Murray and Jacobson (1982) used 15 independent variables to predict biomass of four shrubby species in eastern Idaho. In other cases, several dimensional measurements have been used and estimates improved by double sampling (Uresk *et al.* 1977).

Volume has also been used as an independent variable to estimate browse biomass (Nord 1965, Lyon 1968, Bentley *et al.* 1970, Ludwig *et al.* 1974, Bryant and Kothmann 1979, Nabado *et al.* 1980). Different volume formulas were used for mesquite (*Prosopis glandulosa*), snakeweed (*Xanthocephalum sarothrae*) and creosotebush (*Larrea tridentata*) (Ludwig *et al.* 1974). Coefficients of determination for relationship between aboveground biomass compartments to canopy volume varied from 0.89 for dead stems to 0.99 for leaves of creosotebush (Ludwig *et al.* 1974).

In several examples, log transformation of dimension and weight measurements has increased the correlation coefficients. Log transformations have been used for big sagebrush (*Artemisia tridentata*) (Rittenhouse and Sneva 1977), mountain mahogany (Medin 1960), Gambel oak (Hutchings and Mason 1970), various shrubby species on the Edwards Plateau of Texas (Bryant and Kothmann 1979), several California shrubs (Bentley *et al.* 1970) and several Great Basin shrubs (Dean *et al.* 1981).

Twig length or diameter has been used to predict shrub biomass (Basile and Hutchings 1966, Telfer 1969, Halls and Harlow 1971, Ferguson and Marsden 1977, Lyon 1980, Dean *et al.* 1981, Provenza and Urness 1981, Bartolome and Kosco 1982, Ruyle *et al.* 1983) for a variety of browse species. In some cases, additional measurements increased the r^2 values (Schuster 1965). However, even if twig weights can be predicted from lengths or diameters, the number of twigs per plant and the density of plants must be known to express biomass on an area basis.

For any of these methods involving measurements of plant dimensions or twig length and diameters, one must be careful to ensure that the relations hold for the specific situation to which they apply. Often such relationships are specific for each site, year, condition class, aspect, etc. (Hutchings and Mason 1970, Pieper 1978). Serious errors can be introduced if equations are applied to situations which do not apply.

Grasses. While dimension analyses are most often applied to shrubs, several studies have included grass species. These studies have developed relationships between plant height and biomass (Heady 1957, Pasto *et al.*

1957, Reppert *et al.* 1962, Carpenter 1980). In the California annual type, plant height alone or an index of plant height × basal cover showed promise in predicting herbage biomass (Evans and Jones 1958). However, Reppert *et al.* (1962) found that this index overestimated plant biomass as determined by clipping.

Some researchers have used a disc meter or rising plate meter to estimate plant biomass (Bransby *et al.* 1977, Mitchell 1982, Michael and Nicholson 1987, Scrivner *et al.* 1986). These techniques require calibration for each specific situation.

18.6 Uses of biomass data

18.6.1 Estimating carrying capacity

Estimating carrying capacity from plant biomass data has gradually replaced the use of cover and composition as used in the range survey and reconnaissance methods (Smith 1984). The basic approach appears to be sound and easily understood. If herbage biomass is known and the amount of this which can safely be utilized is known (two rather important variables), the amount of forage available for livestock and game can be calculated as the product of the two. Previous sections have described limitations in determination of herbage standing crop at one point in time. Additional complications involve variations in the annual standing crop curve among years (Pieper *et al.* 1974). Smith (1984) discussed limitations in the development and use of proper use factors. There is also considerable variation in consumption of available herbage (Cordova *et al.* 1978), depending on species of animal, sex, age, reproductive status, type of forage available, etc. In some early surveys, an annual consumption of 3636 kg (8000 lbs) was used for an animal unit. This represents about 2.2% of the body weight of a 454 kg cow and is in the range of 1 to 2.8% of body weight (on a daily basis) reported by Cordova *et al.* (1978).

Forage supply, also basic for supporting herbivores, is only one factor contributing to carrying capacity. Smith (1984) lists several aspects of both the animal and management factor which must be considered in estimating carrying capacity. Carrying capacity estimates based on forage availability serve only as initial starting points once basic stocking plans have been developed (Pieper 1981) and should be adjusted as needed based on utilization surveys (Bryant *et al.* 1978) or monitoring range trend (Smith 1984). Other studies have considered herbage production and utilization by species to estimate stocking rates (Aldon and Garcia 1971, Reid *et al.* 1963).

18.6.2 Range condition

Herbage biomass estimates have been used extensively as a basis for range condition analysis, regardless of method used for analysis. Herbage biomass by species is often compared to conditions believed to represent the potential or climax vegetation (Smith 1978). In this context, composition of biomass among species may be more critical than absolute amount. Several reviews have shown limitations of older methods of determining range condition (Humphrey 1962, Smith 1978, Wilson and Tupper 1982, Hacker 1983). Several alternative approaches have been used to circumvent problems encountered in some of the earlier approaches (Tueller and Blackburn 1974, Wilson and Tupper 1982, Hacker 1983, Pamo 1983).

18.6.3 Dominance

Dominance is an expression of the relative role or importance of species in a plant community (Kucera 1973). Generally, the dominant species has the highest requirements for requisites in the environment. If more than one species occupies a place of importance in a plant community, then they may be considered codominants.

Often, dominance is determined by an attribute such as basal area (Barbour *et al.* 1987), while in other cases several attributes may be used (e.g., importance values) (Whittaker 1970). However, biomass may also be used as an integrative characteristic to denote dominance (Kucera 1973).

18.7 Conclusions

The choice of method used to estimate herbage biomass should depend upon the objectives of the project. In some cases, speed and convenience are prime considerations, while in others accuracy and reliability are more important. The nature of the vegetation should be a prime consideration in the selection of type of sampling unit. Density and pattern of individual plants are important characteristics in selecting quadrat size and shape to maximize accuracy and minimize sampling variability. For extensive range surveys, some type of estimation techniques alone or in conjunction with appropriate double sampling procedures, appear to be most appropriate. Many researchers will continue to use the harvest method to estimate plant biomass in spite of time disadvantages and the fact that it is a destructive technique. There remains ample opportunity for innovative research into methods of improving our estimates of plant biomass.

18.8 References

Ahmed, J., and C. D. Bonham. 1980. DUBSAM — Alogarithm and computer program for optimum allocation in multivariate double sampling for biomass estimation. Colo. State Univ., Range Sci. Ser. 33.

Ahmed, J., and C. D. Bonham. 1982. Optimum allocation in multivariate double sampling for biomass estimation. J. Range Manage. 35: 777—779.

Ahmed, J., C. D. Bonham, and W. A. Laycock. 1983. Comparison of techniques used for adjusting estimates by double sampling. J. Range Manage. 36: 217—221.

Alcock, M. B. 1964. An improved electronic instrument for estimation of pasture yield. Nature 203: 1309—1310.

Alcock, M. B., and J. V. Lovett. 1967.The electronic measurement of the yield of growing pasture. 1. A statistical assessment. J. Agric. Sci. Camb. 68: 27—38.

Aldon, E. F., and G. Garcia. 1971. Stocking rangelands on the Rio Puerco in New Mexico. J. Range Manage. 24: 344—345.

Angelone, A., J. M. Toledo, and J. C. Burns. 1980a. Herbage measurements *in situ* by electronics. 1. The multiple-probe type capacitance meter: A brief review. Grass and For. Sci. 35: 25—33.

Angelone, A., J. M. Toledo, and J. C. Burns. 1980b. Herbage measurements *in situ* by electronics. 2. Theory and design of an earth-plate capacitance meter for estimating forage dry matter. Grass and For. Sci. 35: 95—103.

Barbour, M. G., J. H. Burk, and W. D. Pitts. 1987. Terrestrial Plant Ecology. 2nd Ed. Benjamin Cummings Publ. Co., Inc., Menlo Park, CA.

Bartolome, J. W., and B. H. Kosco. 1982. Estimating browse production by deerbrush (*Ceanothus integerrimus*). J. Range Manage. 35: 671—672.

Basile, J. V., and S. S. Hutchings. 1966. Twig diameter-length-weight relations for bitterbrush. J. Range Manage. 19: 34—38.

Blair, R. M. 1958. Weight techniques for sampling browse production on deer ranges. *In*: Techniques and methods for measuring understory vegetation. U.S. For. Serv. Proc. Sym., Tifton, GA.

Blaisdell, J. P. 1958. Seasonal development and yield of native plants on the upper Snake River Plains and their relation to certain climatic factors. U.S. Dept. Agr. Bull. 1190.

Blankenship, J. O., and D. R. Smith. 1966. Indirect estimation of standing crop. J. Range Manage. 19: 74—77.

Bransby, D. I., A. G. Matches, and G. F. Krause. 1977. Disc meter for rapid estimation of herbage yield in grazing trials. Agron. J. 69: 393—396.

Brown, J. K. 1976. Estimating shrub biomass from basal stem diameters. Can J. For. Res. 6: 153—158.

Bryant, F. C., and M. M. Kothmann. 1979. Variability in predicting edible browse from crown volume. J. Range Manage. 32: 144—146.

Bryant, D., E. Leviness, P. Ogden, and L. Smith. 1978. Guide to Arizona range analysis. Univ. Ariz., Tucson.

Cable, D. R. 1975. Influence of precipitation on perennial grass production in the semidesert southwest. Ecology 56: 981—986.

Campbell, A. G., D. S. M. Phillips, and E. D. O'Reilly. 1962. An electronic instrument for pasture yield estimation. J. Brit. Grassl. Soc. 17: 89—100.

Carpenter, L. H., O. C. Wallmo, and M. J. Morris. 1973. Effect of woody stems on estimating herbage weights with a capacitance meter. J. Range Manage. 26: 151—152.

Clary, W. P. 1964. A method for predicting potential herbage yield on the Beaver Creek pilot watersheds. p. 244—250. *In*: M. Stelley (ed.) Forage plant physiology and soil-range relationships. Amer. Soc. Agron. Spec. Publ. No. 5.

Cochran, W. G. 1963. Sampling techniques. John Wiley and Sons, Inc., New York.
Cook, C. W., and J. Stubbendieck (eds.) 1986. Range research: Basic problems and techniques. Soc. Range Manage., Denver, CO.
Cordova, F. J., J. D. Wallace, and R. D. Pieper. 1978. Forage intake by grazing livestock: A review. J. Range Manage. 31: 430—438.
Currie, P. O., M. J. Morris, and D. C. Neal. 1973. Uses and capabilities of electronic capacitance instruments for estimating standing herbage. Part 2. Sown ranges. J. Brit. Grassl. Soc. 28: 155—160.
Dean, S. A., J. W. Burkhardt, and R. O. Meeuwig. 1981. Estimating twig and foliage biomass of sagebrush, bitterbrush and rabbitbrush in the Great Basin. J. Range Manage. 34: 224—227.
Duncan, D. A., and R. G. Woodmansee. 1975. Forecasting forage yield from precipitation in California's annual rangeland. J. Range Manage. 28: 327—329.
Evans, R. A., and M. B. Jones. 1958. Plant height times ground cover versus clipped samples for estimating forage production. Agron. J. 50: 504—506.
Ferguson, R. B., and M. A. Marsden. 1977. Estimating overwinter bitterbrush utilization from twig diameter-length-weight relations. J. Range Manage. 30: 231—236.
Fletcher, J. E., and M. E. Robinson. 1956. A capacitance meter for estimating forage weight. J. Range Manage. 9: 96—97.
Francis, R. C., G. M. Van Dyne, and B. K. Williams. 1979. An evaluation of weight estimation double sampling as a method of botanical analysis. J. Environ. Manage. 8: 55—72.
Goebel, C. J. 1955. The weight-estimate method at work in southeastern Oregon. J. Range Manage. 8: 212—213.
Goebel, C. J., L. F. DeBano, and R. D. Lloyd. 1958. A new method of determining forage cover and production on desert shrub vegetation. J. Range Manage. 11: 244—246.
Gould, W. L., V. W. Howard, and K. A. Valentine. 1972. Soil characteristics, biotic composition and vegetative production of areas leased by Western Coal Company for strip mining near Fruitland, New Mexico. New Mex. State Univ. Agric. Exp. Sta. Spec. Rep. No. 20.
Hacker, R. B. 1983. Use of reciprocal averaging ordination for the study of range condition. J. Range Manage. 36: 25—30.
Halls, L. K., and T. R. Dell. 1966. Trial of ranked-set sampling for forage yields. For. Sci. 12: 22—26.
Halls, L. K., and R. F. Harlow. 1971. Weight-length relations in flowering dogwood twigs. J. Range Manage. 24: 236—237.
Hansen, R. M. 1972. Estimation of herbage intake from jackrabbit feces. J. Range Manage. 25: 468—471.
Heady, H. F. 1957. The measurement and value of plant height in the study of herbaceous vegetation. Ecology 38: 313—320.
Hilmon, J. B. 1959. Determination of herbage weight by double-sampling: Weight estimate and actual weight. *In*: Techniques and methods of measuring understory vegetation. USDA For. Serv. Proc. Symp., Tifton, GA.
Hughes, E. E. 1962. Estimating herbage production using inclined point frame. J. Range Manage. 15: 323—324.
Humphrey, R. R. 1962. Range Ecology. Ronald Press Co., New York.
Hutchings, S. S., and L. R. Mason. 1970. Estimating yields of Gambel oak from foliage cover and basal area. J. Range Manage. 23: 430-434.
Hutchings, S. S., and J. E. Schmautz. 1969. A field test of the relative weight-estimate method for determining herbage production. J. Range Manage. 22: 408—411.
Hutchings, S. S., and G. Stewart. 1953. Increasing forage yields and sheep production on intermountain winter ranges. USDA Circ. 925.

Johns, G. G., G. R. Nicol, and R. B. Watkin. 1965. A modified capacitance probe technique for estimating pasture yield. I. Construction and procedure for use in the field. J. Brit. Grassland Soc. 20: 212—217.

Karl, M. G., and A. Nicholson. 1987. Evaluation of the forage-disc method in mixed-grass rangelands of Kansas. J. Range Manage. 40: 467—471.

Knapp, A. K., M. D. Abrams, and L. C. Hurlbert. 1985. An evaluation of beta-attenuation for estimating aboveground biomass in a tallgrass prairie. J. Range Manage. 38: 556—558.

Kreil, W., and J. Matschke. 1968. Studies on the determination of pasture yields with an electronic measuring instrument.Ztschr. Landeskuit. 9: 75—99.

Kucera, C. L. 1973. The Challenge of Ecology. The C.V. Mosby Company, St. Louis, MO.

Laycock, W. A. 1965. Adaptation of distance measurements for range sampling. J. Range Manage. 18: 205—211.

Ludwig, J. A., J. F. Reynolds, and P. D. Whitson. 1975. Size-biomass relationships of several Chihuahuan desert shrubs. Amer. Midl. Nat. 94: 451—461.

Lyon, L. J. 1968. Prediction of herbage yields of serviceberry from crown volumes. J. Wildl. Manage. 32: 115—118.

Lyon, L. J. 1970. Length-and-weight diameter relations of serviceberry twigs. J. Wildl. Manage. 34: 456—460.

Mason, L. R., and S. S. Hutchings. 1967. Estimating foliage yields on Utah juniper from measurements of crown diameter. J. Range Manage. 20: 161—166.

McIntyre, G. A. 1952. A method for unbiased selective sampling, using ranked sets. Austral. J. Agr. Res. 3: 385—390.

Medin, D. E. 1960. Physical site factors influencing annual production of true mountain mahogany. Ecology 41: 454—460.

Milner, C., and R. E. Hughes. 1968. Methods for the measurement of primary production of grasslands. Blackwell Scientific Publications, Oxford, Great Britain.

Mitchell, J. E. 1972. An analysis of the beta-attenuation technique for estimating standing crop of prairie range. J. Range Manage. 25: 300—304.

Mitchell, P. 1982. Value of a rising-plate meter for estimating herbage mass of grazed perennial ryegrass-white clover swards. Grass and Forage Sci. 37: 81—87.

Morris, M. J., K. L. Johnson, and D. L. Neal. 1976. Sampling shrub ranges with an electronic capacitance instrument. J. Range Manage. 29: 78—81.

Morris, M. J. 1962. The use of rating or ranking in vegetation measurement. *In*: Range research methods — a symposium. USDA For. Serv. Misc. Publ. 940.

Mott, G. O., R. F. Barnes, and C. L. Rhykerd. 1965. Estimating pasture yield *in situ* by beta-ray attenuation techniques. Agron. J. 57: 512.

Mueggler, W. F. 1976. Number of plots required for measuring productivity on mountain grasslands in Montana. USDA. For. Serv. Res. Note INT.-207.

Murphy, A. H. 1970. Predicted forage yield based on fall precipitation in California annual grasslands. J. Range Manage. 23: 363—365.

Murray, R. B., and M. Q. Jacobson. 1982. An evaluation of dimension analysis for predicting shrub biomass. J. Range Manage. 35: 451—454.

Nadabo, S., R. D. Pieper, and R. F. Beck. 1980. Growth patterns and biomass relations of *Xanthocephalum sarothrae* on sandy soils in southern New Mexico. J. Range Manage. 33: 394—397.

Neal, D. L., P. O. Currie, and M. J. Morris. 1976. Sampling herbaceous native vegetation with an electronic capacitance instrument. J. Range Manage. 29: 74—77.

Neal, D. L., and R. L. Neal. 1965. A new electronic meter for measuring herbage yield. USDA For. Serv. Res. Note PSW-56.

Neal, D. N., and J. L. Neal. 1973. Uses and capabilities of electronic capacitance instruments

for estimating standing herbage. Part 1. History and development. J. Brit. Grassl. Soc. 28: 81—89.
Nord, E. C. 1965. Autecology of bitterbrush in California. Ecol. Monogr. 35: 307—334.
Pamo, E. T. 1983. Mathematical approach to range condition in comparison to the SCS method. Ph.D. dissertation, New Mex. State Univ., Las Cruces.
Papanastasis, V. P. 1977. Optimum size and shape of quadrat for sampling herbage weight in grasslands of northern Greece. J. Range Manage. 30: 446—448.
Passey, H. B., V. K. Hugie, and E. W. Williams. 1964. Herbage production and composition fluctuations of natural plant communities as related to climate and soil taxonomic units. p. 206—221. *In*: M. Stelley (ed.) Forage plant physiology and soil-range relationships. Amer. Soc. Agron. Spec. Publ. No. 5.
Pasto, J. K., J. R. Allison, and J. R. Washku. 1957. Ground cover and height of sward as a means of estimating pasture production. Agron. J. 407—409.
Payne, G. F. 1974. Cover-weight relationships. J. Range Manage. 27: 403—404.
Pechanec, J. F., and G. D. Pickford. 1937. A weight estimate method for determination of range or pasture production. J. Amer. Soc. Agron. 29: 894—904.
Peek, J. M. 1970. Relation of canopy area and volume to production of three woody species. Ecology 51: 1098—1101.
Pieper, R. D. 1978. Measurement techniques for herbaceous and shrubby vegetation. New Mex. State Univ., Las Cruces, NM.
Pieper, R. D. 1981. The stocking rate decision. *In*: L. D. White, and A. L. Hoermann (eds.) Proc. Inter. Ranchers' Round-up. Texas A & M Univ., College Station, TX.
Pieper, R. D. 1984. A critique of "Methods for Inventory and Monitoring of Vegetation, Litter and Soil Surface Condition". *In*: Developing Strategies for Rangeland Management. Westview Press, Boulder, CO.
Pieper, R. D., J. R. Montoya, and V. L. Groce. 1971. Site characteristics on pinyon-juniper and blue grama ranges in south-central New Mexico. New Mex. State Univ. Agric. Exp. Sta. Bull. 573.
Pieper, R. D., C. H. Herbel, D. D. Dwyer, and R. E. Banner. 1974. Management implications of herbage weight changes on native rangeland. J. Soil and Water Cons. 29: 227—229.
Pino, R. 1954. Comparative efficiency of various lengths of belt transects in estimating forage production on semidesert grassland range. M.S. thesis, New Mex. State Univ., Las Cruces.
Provenza, F. D., and P. J. Urness. 1981. Diameter length-weight relations for blackbrush (*Coleogyne ramosissimus*) branches. J. Range Manage. 34: 215—217.
Pumphrey, F. V. 1980. Precipitation, temperature and herbage relationships for a pine woodland site in northeastern Oregon. J. Range Manage. 33: 307—310.
Reid, E. H., J. L. Kovner, and S. C. Martin. 1963. A proposed method of determining cattle numbers in range experiments. J. Range Manage. 16: 184—187.
Reppert, J. N., M. J. Morris, and C. A. Graham. 1962. Estimation of herbage on California annual-type range. J. Range Manage. 15: 318—323.
Reese, G. A., R. L. Bayn, and N. E. West. 1980. Evaluation of double-sampling estimators of subalpine herbage production. J. Range Manage. 33: 300—306.
Risser, P. G. 1984. Methods for inventory and monitoring vegetation, litter and soil surface conditions. *In*: Developing strategies for rangeland management. Westview Press, Boulder, CO
Rittenhouse, L. R., and F. A. Sneva. 1977. A technique for estimating big sagebrush production. J. Range Manage. 30: 68—70.
Rogler, G. A., and H. J. Haas. 1947. Range production as related to soil moisture and precipitation on the Northern Great Plains. J. Amer. Soc. Agron. 39: 378—389.
Russell, J. W. 1973. Soil-plant-livestock-weather relationships of a range ecosystem: Interac-

tive timeshare regression and simulation model approaches. M.S. thesis, New Mex. State Univ., Las Cruces.

Ruyle, G. B., J. F. Bowns, and A. F. Schlundt. 1983. Estimating snowberry (*Symphoricarpos oreophilus*) utilization by sheep from twig diameter-weight relations. J. Range Manage. 36: 472—474.

Schumacher, F. X., and R. A. Chapman. 1948. Sampling methods in forestry and range management. Duke Univ. School of For. Bull. No. 7.

Schuster, J. L. 1965. Estimating browse from twig and stem measurements. J. Range Manage. 18: 220—223.

Scrivner, J. H., D. M. Centner, and M. B. Jones. 1986. A rising platemeter for estimating production and utilization. J. Range Manage. 39: 475—477.

Shoop, M. C., and E. H. McIlvain. 1963. The micro-unit forage inventory method. J. Range Manage. 16: 172—179.

Smith, E. L. 1978. Evaluation of the range condition concept. *In*: D. N. Hyder (ed.) Proc. First Inter. Rangeland Congr. Soc. Range Manage, Denver, CO.

Smith, E. L. 1984. Use of inventory and monitoring data for range management purposes. *In*: Developing Strategies for Rangeland Management. Westview Press, Boulder, CO.

Smoliak, S. 1956. Influence of climatic conditions on forage production of shortgrass rangeland. J. Range Manage. 9: 89—91.

Soplin, H., H. D. Gross, and J. D. Rawlings. 1975. Optimum size of sampling unit to estimate coastal bermudagrass yield. Agron. J. 67: 533—537.

Sneva, F. A., and D. N. Hyder. 1962. Estimating herbage production on semiarid ranges in the Intermountain Region. J. Range Manage. 15: 88—93.

Tadmor, N. H., A. Brieghet, I. Noy-Mier, R. W. Benjamin, and E. Eyal. 1975. An evaluation of the calibrated weight-estimate method for measuring production in annual grassland vegetation. J. Range Manage. 28: 65—69.

Teare, I. D., and G. O. Mott. 1965. Estimating for yield *in situ*. Crop Sci. 5: 311—313.

Teare, I. D., G. O. Mott, and J. R. Eaton. 1966. Beta attenuation — a technique for estimating forage yield *in situ*. Rad. Bot. 6: 7—11.

Telfer, E. S. 1969. Twig weight-diameter relationships for browse species. J. Wildl. Manage. 33: 917—921.

Terry, W. S., D. H. Hunter, and B. F. Swindel. 1981. Herbage capacitance meter: An evaluation of its accuracy in Florida rangelands. J. Range Manage. 34: 240—241.

Toledo, J. M., J. C. Burns, and A. Angelone. 1980. Herbage measurements *in situ* by electronics. 3. Calibration, characterization and field application of the earth-plate forage capacitance meter: A prototype. Grass and For. Sci. 35: 189—196.

Tueller, P. T., and W. H. Blackburn. 1974. Condition and trend of the big sagebrush/needle-and-thread grass habitat type in Nevada. J. Range Manage. 27: 36—40.

Uresk, D. W., R. O. Gilbert, and W. H. Richard. 1977. Sampling for sagebrush phytomass. J. Range Manage. 30: 311—314.

Van Dyne, G. M., F. M. Glass, and P. A. Opstrup. 1968. Development and use of capacitance meters to measure standing crop of herbaceous vegetation. Oak Ridge Nat. Lab. ORNL-TM02247.

Van Dyne, G. M., W. G. Vogel, and H. G. Fisser. 1963. Influence of small plot size and shape on range herbage production estimates. Ecology 44: 746—759.

Weigert, R. G. 1982. The selection of an optimum quadrat size for sampling the standing crop of grasses and forbs. Ecology 43: 125—129.

Whisenant, S. G., and D. F. Burzlaff. 1978. Predicting green weight of mesquite. J. Range Manage. 31: 396—397.

Whittaker, R. H. 1970. Communities and ecosystems. McMillan Co., London.

Wight, J. R. 1967. The sampling unit and its effect on saltbush yield estimates. J. Range Manage. 20: 323—325.
Wilm, H. G., D. F. Costello, and G. E. Klipple. 1944. Estimating forage yield by the double sampling method. J. Amer. Soc. Agron. 36: 194—203.
Wilson, A. D., and G. J. Tupper. 1982. Concepts and factors applicable to the measurement of range condition. J. Range Manage. 35: 684—689.

19. Rangeland vegetation – hydrologic interactions

M. KARL WOOD

19.1 Abstract

Hydrologic resources are closely tied to an area's climate and rangelands encompass many different climatic zones. Approximately 40% of the world's land surface is classified as rangeland, and more than 80% is within arid and semiarid zones. Hydrologic processes are influenced greatly by the kinds and amounts of plants growing on rangelands. Important plant features include growth forms, phytomass, density, distribution, leaf-area index, height, age, and more. Plants probably have their greatest influence on infiltration rates and these rates influence several other processes in the hydrologic cycle, including runoff, erosion, and permeability. An important key for meeting future water needs in many places throughout the world is water control and efficient use on rangelands.

19.2 Rangeland vegetation and precipitation

The kinds and amounts of plants associated with rangelands are influenced greatly by the forms, types, amounts, and seasonal variations of precipitation. Precipitation, the depositing of rain, snow, sleet, hail, rime, etc., is sometimes influenced by plants. A news release by Field Enterprises, Inc., in 1980 stated that the most hail-prone region of the world is in Kenya, where hailstorms occur on an average of 132 days a year. Kenya's lucrative tea plantations are often seriously damaged. Tea dust particles are churned into the atmosphere by activity of tea pickers and nuclei around which hailstones can easily grow, which induce storms. Thus, the tea produces the seeds of its own destruction.

Particles for condensation nuclei can be added to the atmosphere from rangelands by soil being blown into the air, smoke from combustion processes, and chemicals emitted through plant gas exchange. Landsberg (1938) gave the concentrations of Aitken particles (particles with radii >0.001 microns) over the ocean as $940\ cm^{-3}$. The atmosphere over mountains above 2000 m contained $950\ cm^{-3}$, over mountains from 1000

Tueller, P. T. (ed.), Vegetation science applications for rangeland analysis and management.
ISBN 90-6193-195-9.
© 1988, *Kluwer Academic Publishers, Dordrecht, Boston, London.*

to 2000 m contained 2130 cm^{-3}, and over mountains from 500 to 1000 m contained 6000 particles cm^{-3}. Inland country with crop and rangeland had 9500 particles cm^{-3} in the atmosphere, while there were 34000 and 147000 particles cm^{-3} over towns and cities, respectively. The range for inland country was from 180 to 336000 cm^{-3}. The particles in the air on which cloud droplets grow have radii of 0.05 microns and are found in concentrations of about 100 to 1000 cm^{-3}, so that most particles found in the atmosphere are too small to be condensation nuclei. Particles in the lower atmosphere remain only 1 to 4 weeks. Those in the atmosphere above 10000 m may stay there for a year or two. Particles in the lower atmosphere are the ones that most effect precipitation.

19.3 Rangeland vegetation and interception

Dew and frost are common, highly visible manifestations of condensation on rangelands and have received much biological attention everywhere. A small downward flux of water vapor condenses on ecosystem surfaces, especially those of plants. Much interest has been given to water that condenses on a leaf or stem and its possible absorption into the plant without traversing the tortuous paths through the canopy and soil. Most often, the inputs of water into rangeland ecosystems from condensation are marginal, but interesting. Common ragweed (*Ambrosia artemisiifolia* L.) has morphological characteristics that channel dew into stemflow that moistens its own roots and gives itself some advantage "in early plant succession within old-field ecosystems" (Shure and Lewis 1973). But the amounts of stemflow were only 0.05 mm per night or about 18 mm per year. Therefore, the importance of dew and frost in the total water budget of range ecosystems is minor. Their importance to agriculture in dry climates is found in folklore where stones piled around fruit trees in the Mideast supposedly provided enough condensated water to allow growth with little or no irrigation. However, energy-budget analysts have deflated it to a minor item in the energy budget and a negligible one in the water budget (Miller 1977).

Dew from upper air sources and water evaporated from the soil condenses onto external vegetation surfaces. These moistened surfaces become recepticles for fungus spores that are transported by atmospheric eddies. Blights can then start following long periods of dew on leaves. In a savanna situation, less radiant heat is received in open spaces than underneath trees. This allows dew or frost to form more often and remain longer (Geiger 1965), enhancing fungal growth.

The process of catching and redistributing precipitation by vegetation back to the atmosphere and to the ground is called interception. In some ecosystems, plants actually have net precipitation under their canopies that is greater than gross precipitation between plants. Delivery of precipitation is an aerodynamic process that depends on the sizes of the water particles. A typical liquid-water content of a cloud is 0.5 g m^{-3}, but each droplet is 10^{-6} of the size of a raindrop. These tiny particles do not readily fall, but are easily transported by a strong wind. When clouds or fog blow across a mountain range, such as the chaparral communities of the Sierra Nevada, the water particles are mechanically screened onto the vegetation. They evaporate from the plant or coelesce and drop through the canopy. Most measurements on this phenomena have occurred in forests such as in Germany, Berkeley, California, Portland, Oregon, Hawaii, New Mexico, the Canary Islands, and Margarita off the coast of Venezuela. In Hawaii, this input is 750 mm yr^{-1}. Estimations of fog drip on rangelands are more limited but may be high on some rangelands such as alpine areas.

For most storms, interception results in less water falling through the canopy and reaching the ground. Water intercepted by leaves, twigs, and stems may return to the atmosphere by evaporation as water vapor. It may drip from the foliage, or it may run down the stem. The rainfall reaching the ground as throughfall, drip, or stemflow, is distributed quite differently than when it fell on the canopy.

Many visitors to the range are caught in rainfall away from shelter. To keep dry, the shelter of a tree is often sought. Some raindrops fall through the densest vegetation, but the visitor can stay dry for awhile. If the rainfall event is short, little water falls under the tree, most water is intercepted by the canopy. But when the interception storage capacity of the tree is reached, it begins to drip. Little comfort is then acquired under the tree. The amount of interception storage capacity varies with vegetation type, species, and density.

Any analysis of water delivery to plant canopy requires data on the pertinent characteristics of foliage and branch structure. The impact conditions from rain and snow should be measured as well as the buildup of rain and snow. These kinds of data are rare. Miller (1977) reports that literature was reviewed a few years prior to 1977 on interception of snowfall. Virtually the entire recorded literature only included 110 studies. Only one reported concurrent measurements of the canopy, input of snow, and amounts intercepted. Miller concluded that the plant characteristics that determine the form of storage, strength and stability of their physical support, their thermal surroundings, and conditions under which inputs and outputs of rain and snow take place are generally not known. Data on the rate of snow and rainfall, storm temperature, wind speed, and mass of stored water are scant and sporadic. Interception knowledge is important in studies of the other parts of the hydrologic cycle. More than 100 studies have probably been conducted on infiltration rates by applying simulated rainfall, catching runoff, and ignoring interception.

Most interception studies have been conducted on trees. Generally, these studies have found that coniferous or evergreen trees intercept more precipitation than deciduous trees in full leaf. Deciduous tree losses to interception are two to three times greater in full leaf than after leaf fall. Conifers and deciduous trees without leaves will usually intercept more water in the form of snow than rainfall. Interception has no upper boundary limit due to plants when snow falls on shrubs or herbaceous plants. The intercepted snow merges with the snow that has accumulated on the ground. But the snowpack often contains large voids, as Wilken (1967) showed in a manzanita (*Arctostaphylos* spp. Adans.) community in the Sierra Nevada. Although the snow accumulation on the ground and the intercepted snow merge, the vegetation continues to support snow and voids in the mass of snow and vegetation may total 40% of its volume.

Reliable interception data for grasses, forbs, and other herbaceous plants are difficult to obtain. Their values are probably less in most instances than trees and shrubs. Interception may not be as low as their relative height suggests, however. The leaf surface per area of many herbaceous plants often approaches that of trees and shrubs. A common way to express interception is on an average annual basis. Branson *et al.* (1981) refer to dozens of studies where a percentage of the average annual precipitation is reported as being intercepted. When only the percentage of annual precipitation resulting in interception is reported for a vegetation type or community, the report lacks necessary information. Hull and Klomp (1974) and West and Gifford (1976) found that spring-summer-fall rainfall interception by big sagebrush (*Artemisia tridentata* Nutt.) ranged from 4 to 31% depending on the size and density of plants within the community. One useful approach is that of Tromble (1983a, 1983b). Individual plants were measured for intercepted water throughout a rainfall event and correlated with crown cover, shrub height, shrub green-weight, green-weight of stems, oven-dry weight of stems, green-weight of leaves, oven-dry weight of leaves, number of stems, leaf area, and shrub volume. Predictive models for the species were determined that allowed calculation of interception for individual plants. An individual plant community can be sampled for individual plant characteristics and density to determine an interception value for that particular creosotebush community. But it is erroneous to give an interception value for all creosotebush communities.

Range plants have been credited with changing the distribution and increasing the spatial variability of precipitation within a site and altering raindrop. The spatial variability can be great. Kimmins (1974) found a very large number of gauges were needed to sample within 5% of the mean with a 95% confidence interval: 272 gauges for water flux on 30 × 30-m plots and 2991 gauges for potassium flux.

Young and Evans (1987) studied precipitation under western junipers (*Juniperus occidentalis* Hook.) and compared it to amounts received between plants. The edge of the canopy received 81% of that received in gauges located in 10 ha openings in the woodlands, while 49 and 31% were received at one-half of canopy radius and at the trunk, respectively. Armentrout (1986) found phytomass of herbaceous plants to vary from near the base of the trees to the drip line or outer edge of the canopy to the tree interspaces. Greatest phytomass was at the outer canopy edges and was partially attributed to redistribution of precipitation. This study was conducted beneath pinyon pine (*Pinus edulis* Engelm.) and oneseed juniper (*Juniperus monosperma* [Engelm.] Sarg.). The branches on Utah juniper (*Juniperus osteosperma* [Torr.] Little) extend up and outward from the stem at an angle of about 45°. These branches act as a funnel to direct water to the stem. But the stem is so shaggy that little water reaches the base but is lost to evaporation. In central Australia, mulga (*Acacia aneura* F. Muell.) redistributes precipitation so that as much as one-third of the total rainfall in storms greater than 10 mm results in stemflow (Slayter 1965).

The water flux from foliage can also be altered in chemical composition. Dry particles from atmospheric deposition can be dissolved or washed off the foliage. Other materials such as metal ions may originate in deep soil layers and be transported up through the plant to be deposited on the foliage (Miller 1977). Some substances have nutrient value, some influence the formation of the soil, and others are allelopathic or they make the soil repellent to infiltration (Eckert and Kinsinger 1960).

Water falling from plants may reach the soil in larger drops than the original precipitation particles and hit with greater impact. Their erosivity would be increased. This increase is greatest when snow lands on a plant and melts or agglomerates into a larger, denser body.

Litter covering the soil are aerial parts of the plant that have been shed. Litter forms another intercepting layer to falling precipitation. Water is held on litter surfaces and absorbed within the litter. Much of the precipitation caught on shrub and woodland litter does not penetrate to the soil because of the litter depth. The litter layer can also contain molds, fungi, or other agents of decomposition that cause the litter layer to be hydrophobic (Bond 1964). This was observed by Blackburn and Skau (1974) under pinyon-juniper woodlands in Nevada and by Wood (1979) under mesquite (*Prosopis glandulosa* Torr.) in northcentral Texas. Debano and Rice

(1973) attributed the hydropholic conditions to leachate from litter of chaparral. Clary and Ffolliott (1969) studied water retention of undisturbed forest floor of ponderosa pine (*Pinus ponderosa* Laws.) in Arizona. They found the forest floor weighed an average of 21135 kg ha^{-1} and included three layers with a mean total depth of 4 cm. The freshly fallen litter retained 0.008 cm of water, while the partially decomposed litter layer held 0.03 cm, and the completely decomposed layer held 0.19 cm of water. The top 2.5 cm of forest floor held 0.18 cm of water. Annual interception losses for grass, chaparral, and grass litter on the San Dimas Experimental Forest, California, were quite variable between vegetation types during 32 years of accumulated rainfall data (Table 1) (Corbett and Crouse 1968). For 10 of the years, grass litter intercepted from 4.0 to 4.9% of the annual precipitation, and for 24 of the years, grass litter intercepted between 3.0 and 5.9%. Up to 9.2% was intercepted 1 year.

The hydrologic importance of precipitation interception by plants is that the plants create a pattern of delivery to the soil surface that greatly differs from the initial pattern of precipitation delivery. There are different air motion systems in the atmosphere, different passive states and temperature

Table 1. Distribution of annual interception losses for grass, chaparral, and grass litter, San Dimas Experimental Forest, California (Corbett and Crouse 1968).

Interception (Percent)	Number of Years		
	Grass	Chaparral	Grass Litter
2.0—2.9	—	—	2
3.0—3.9	3	3	7
4.0—4.9	—	—	10
5.0—5.9	5	—	7
6.0—6.9	7	—	5
7.0—7.9	6	—	0
8.0—8.9	1	—	1
9.0—9.92	1	—	1
10.0—10.9	1	4	—
11.0—11.9	2	5	—
12.0—12.9	5	6	—
13.0—13.9	3	4	—
14.0—14.9	—	2	—
15.0—15.9	—	3	—
16.0—16.9	—	1	—
17.0—17.9	—	3	—
18.0—18.9	—	1	—
19.0—19.9	—	2	—

conditions, different chemical contents, and differences in spatial patterns. It can be smaller or larger in magnitude.

19.4 Rangeland vegetation and infiltration

Infiltration refers to the downward movement of water through the soil-air interface. This surface is ever so thin, but ever so important a feature of the hydrologic cycle. Kendall Johnson (1978) once said, "Infiltration is of concern to every range manager." Percolation is the movement of water after it has entered the soil. Percolation rates may or may not be related to infiltration rates depending on subsurface conditions. The rate at which water infiltrates the soil is dependent upon many plant, animal, soil, topographic, and atmospheric factors. The concept of infiltration is relatively new, with literary beginnings in this century. Early soil and agronomic texts (Lyon and Fippin 1909, Harris and Stewart 1930) refer to percolation and soil moisture evaporation but not infiltration. Indeed, the first to recognize the actual role of infiltration in the hydrologic cycle were probably Green and Ampt (1911), chemists at the University of Melbourne, Australia. The Green-Ampt model is interpreted in terms of the analysis of water movement in unsaturated soils based on the concepts of moisture potential and unsaturated hydraulic conductivity. These were embodied in the flow equation by Richards (1931). In the United States, Horton (1933) first recognized the role of infiltration in the hydrologic cycle. Kostiakov (1937). like Horton, developed empirical infiltration equations that contained parameters not necessarily having a physical significance and were required to have a simple form so that parameters could be found using simple curve fitting. Kostiakov's model is most useful for irrigation work. Philip (1957) developed another infiltration equation in Australia. He continued to publish refinements of his equation at least into the 1980s (Philip 1983). Philip's equations are based on physical theory and have been found to be accurate for all but times greater than 24 hours.

A National Conference on Advances in Infiltration was held in Chicago in 1983. Some 36 papers were presented and nearly all related to the models of Green-Ampt, Richards, Horton, Kostiakov, and Philip. It was apparent that infiltration theory has been developed by agronomists and engineers. Range and forest hydrologists were noticeably absent from the conference. As expected, their models need some extensive changes to be applicable and useful on rangelands.

Infiltration rates plotted with time usually form a graph similar to Figure 1. Horton (1940) used the following formula to express this curve:

$$f = f_c + (f_o - f_c)e^{-k_f t}$$

where f = infiltration rate at time t

f_c = terminal infiltration rate or infiltration capacity
f_o = initial infiltration rate
k_f = constants depending on soil characteristics.

The shape of infiltration curves are similar to Figure 1 for several reasons. The curve usually does not intercept the y-axis because infiltration is often found by subtracting runoff rate from precipitation rate and, early in the event, the precipitation rate does not exceed the infiltration rate which would produce runoff. The infiltration rate decreases with time due to several factors. Soil pores are not full of water but fill rapidly at the beginning of rainfall. Raindrops often have enough energy to smooth the soil surface and erode the aggregates which leave the soil surface sealed. Cracks swell shut and frozen layers may thaw. The size and amount of pores decreases with soil depth. Air pressure against the wetting front increases with depth. Finally, a terminal or constant infiltration rate is reached.

Gray *et al.* (1969) attempted to diagram the factors affecting the infiltration rate (Figure 2). This figure represents the known relationships of that time, and the engineering and agronomic influence can be seen in terms

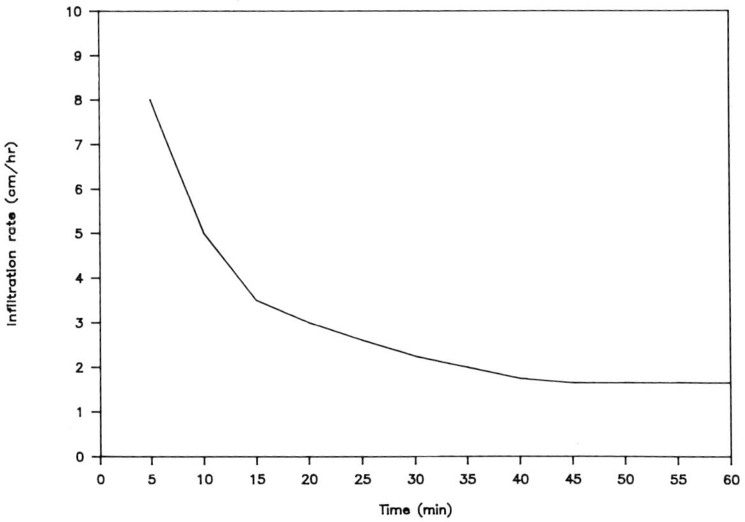

Figure 1. The general shape of most curves expressing infiltration rates on rangelands.

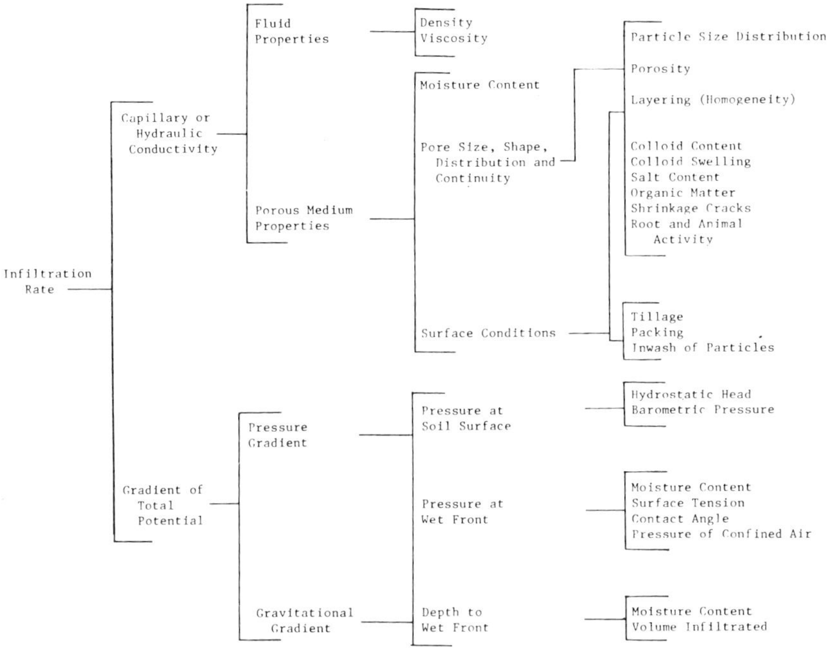

Figure 2. Factors affecting the infiltration rate into unfrozen soil (Gray *et al.* 1969).

such as tillage, packing, and hydraulic conductivity. Variables relating to range plants are quite limited, and this lack of knowledge concerning the influence of non-crop plants is found throughout hydrologic models used on rangelands. These models are extrapolations from croplands and urban areas. A good example of incompleteness is the SCS runoff model that has an infiltration rate base and uses only cover of grass, shrubs, and trees (Wood and Blackburn 1984). The Universal Soil Loss Equation, a model used to estimate erosion, has the same limitations (Blackburn 1980).

An alternative conceptual model is shown in Figure 3 (Wood *et al.* 1987). This model is the result of extensive rangeland watershed research in many locations since 1969. The model is not complete because some relationships are not known such as the influence of soil microorganisms. But research has discovered the tremendous importance of plants as they influence infiltration rates which affect the entire hydrologic cycle.

Early studies found that standing plant and plant litter cover on rangelands have a considerable influence on infiltration (Beutner *et al.* 1940). Importance of cover has been documented by Aldon (1964), Gamougoun *et al.* (1984), Balliette *et al.* (1986), Weltz and Wood (1986a), Wood *et al.* (1986), and Wood *et al.* (1987) in New Mexico. Importance of cover was reported in Texas by McGinty *et al.* (1979), Knight (1980), Wood and Blackburn (1981a, b), McCalla *et al.* (1984), Pluhar (1984). Thurow *et al.*

RANGELAND MANAGEMENT INFLUENCES

ON INFILTRATION RATES

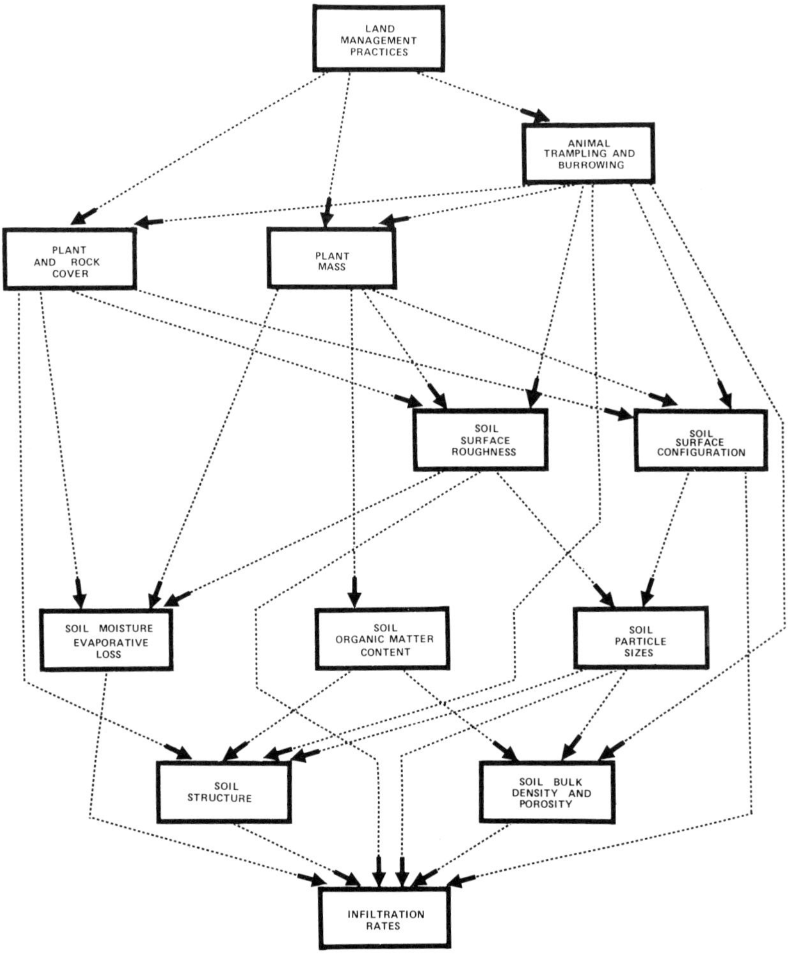

Figure 3. A conceptual model showing relationships between land management practices, plant characteristics, soils, and infiltration rates.

(1986), and Warren *et al.* (1986a, b, c). These represent the most extensive, recent range research related to plant cover and infiltration rates. Studies in other areas with a representative reference include Woodward (1943) in central Utah, Rauzi and Kuhlman (1961) in the northern Great Plains, and Blackburn and Skau (1974) in Nevada. In cropland and forests, infiltration rates have usually increased with increased cover unless the soils were so sandy that no differences were found. Studies in cropland and forests usually examined basal cover. In rangelands, canopy cover (Blackburn 1975) and total foliar cover (Wood and Blackburn 1981b, and others) have been examined. Sanchez-Brito (1984) and Sanchez and Wood (1987) found stronger correlations between total foliar cover and infiltration rates than basal cover and infiltration rates. Gamougoun *et al.* (1984) found infiltration rates decreased as basal cover of blue grama (*Bouteloua gracilis* [H.B.K.] Lag.) increased from overgrazing. Overgrazing caused blue grama to change from a bunchgrass to a sodformer. Therefore, basal cover is not too useful in a predictive model.

Closely associated with plant and litter cover is phytomass. Much confusion exists in the literature comparing infiltration rates between trees, shrubs, grasses, and bare ground, between bunchgrasses and sodgrasses, between grasses, cryptogams, and bare ground, and between about any other imaginable combination. Generally, for a given cover, say 60%, infiltration rates will differ between sites because of differences in plant density and phytomass. When large trees exist in a stand of pinyon-juniper, the cover may be high, but density is low and infiltration rates between trees is often low (Blackburn 1973). If the same site were dominated by small trees or grasses with the same cover but with a much greater density and equal distribution, infiltration rates would be expected to be much higher. Times of concentration of overland flow are increased. Likewise, if a site is dominated by a given cover of the sodforming buffalograss (*Buchloe dactyloides* [Nutt.] Engelm.), infiltration rates will likely be lower than if the site were dominated by sideoats grama (*Bouteloua curtipendula* [Michx.] Torr.) unless both species have been removed by some perturbance. The sideoats grama produces more phytomass which adds to the litter layer. It will also have more root mass than buffalograss. This extensive root system holds soil and adds enough organic matter to the soil for structure stability so that these variables were more important than the type of roots, i.e., rhizomes as found in buffalograss.

Plant cover, phytomass, and density affect the soil surface configuration. This was observed by Wilcox *et al.* (1987) in an extreme situation on steep slopes (up to 70%) in the Guadalupe Mountains in southcentral New Mexico. Dead pinyon and juniper trees laying on the ground had a terracing effect on the slopes. These shortened the slope lengths and increased detention storage. The variance of roughness up and down the slope was greater than the variance of roughness that would be found across the slope because of rilling. Hugie and Passey (1964) observed miniature terraces behind grass plants on 18% slopes.

Increases in soil roughness caused by plants and a subsequent increase in infiltration have been observed on several rangelands. Melton (1940) described coppice dunes and dune interspaces in the southern Great

Plains. He related how mesquite grows vigorously on loose sand and is not readily killed by slow sand burial. The bush causes decreased wind velocity which results in sand falling into the bush and a mound being built. These mounds can approach 3 m height. The windward side of the mound is covered and eroded at the same time that sand accumulates within the bush and on the lee side. In time, the plants on the windward side have their roots exposed and they die. But sand deposited on the lee side furnishes suitable soil for the advancing bush. These dunes migrate very slowly, probably less than 1 m yr^{-1}. Many of these areas now covered by 3 m high coppice dunes were once flat and covered by grasses. The increase in spatial distribution of the dunes has been attributed to periodic droughts, selective grazing of grasses, and seed dispersal by humans, livestock, and wildfire (Herbel *et al.* 1972). On the Jornada Experimental Range, these coppice dunes began in areas that were grazed by livestock and extended through permanent exclosures. Some runoff occasionally occurs. If these mesquite plants were removed and not replaced by grasses, Melton (1940) postulated that a large dune of many hectares in size would result. Stuart *et al.* (1971) described smaller coppice dunes in northern Nevada that contained very fine sands and silts blown off the floodplains and dry playas. Blackburn (1973) found the infiltration rates of these sagebrush (*Artemisia* spp. L.) — dominated coppice dunes were two or three times greater than the dune interspace areas. The entire site had reduced infiltration as a result of understory grass depletion. Many studies have found no relationship between soil surface roughness and infiltration rates, but Sanchez and Wood (1987) suspect that roughness is important but methods to measure roughness are not sensitive enough to reflect high correlations. Soil surface roughness or microtopographical influences were detected by Sanchez (1985) in reclaimed coal mined sites where contour furrowing was compared to undisturbed areas with rills.

Plant cover and mass affect the soil moisture evaporative loss (Figure 3), which affects the soil air content. Different plant species have different evapotranspiration rates which change the soil moisture content or potential to different levels. As an example, Ueckert (1979) studied soil water depletion in a dense broom snakeweed (*Gutierrezia sarothrae* [Pursh] Britton and Rusby) population (387 m^{-2}) and a subsequent blue grama and buffalograss community following snakeweed control. Soil moisture levels were measured for two years. Soil moisture extraction by vigorous shortgrasses was significantly greater than by dense stands of broom snakeweed.

In a stand of crested wheatgrass (*Agropyron cristatum* [L.] Gaertn.) and alfalfa (*Medicago sativa* L.). plants were clipped to 0, 10, 20, and 30 cm stubble heights to simulate extreme, heavy, medium, and light forage removal, respectively, with no clipping on control areas (Buckhouse and Coltharp 1976). The areas not clipped resulted in 51.5 cm of soil moisture depletion through 127 cm of soil. The clipped areas resulted in at least 48.6 cm of depletion except for the extreme treatment (0 cm stubble height), which depleted only 37.8 cm. This was significantly lower than for the other plant heights. Weltz (1983) found a similar situation at Fort Stanton, New Mexico, where the lowest soil moisture contents were found in an exclosure (8.6%) and a rested pasture in a grazing system (6.9%). A moderately stocked, continuously grazed pasture had 8.8%, a heavily stocked, continuously grazed pasture had 9.0%, and a pasture that had just been grazed in a rapid rotation, short duration system had 12.4%. A lot of phytomass uses a lot of water which leaves the soil drier than the soil with few plants so that the soils with the most plants have the highest infiltration rates.

The primary and ultimate source of organic matter in the soil comes from plants. Most of it comes from decomposition of plant litter laying on the soil surface. This results from direct leaf fall, dead plants, stems, etc. or fecal additions from herbivores. Although decomposing animals add some, most of the remainder comes from decadent roots. Organic matter is important to infiltration because it is a building block of soil structure and increases soil porosity. Generally, infiltration rates increase with increases in organic matter. But as previously noted, hydrophobic conditions may form under shrubs and woodland tree species. The hydrophobic layer may be discontinuous or quite widespread as under chaparral vegetation of California and Arizona. Here, infiltration rates are reduced as a result of organic matter additions.

Soil texture may be changed by plants. The particles themselves may not be changed, but texture or the ratio of sand, silt, and clay may be changed by redistribution. The coppice dunes of northern Nevada (Stuart *et al.* 1971, Blackburn 1973) resulted from sagebrush reducing the velocity of winds which allowed the deposition of sands around the base of the plants. Sand particles, being the heaviest, drop from the air. This results in coppice dunes and higher infiltration rates.

Bulk density of soil is usually negatively associated with infiltration rates and is influenced by organic matter additions from plants, soil particle size, and animal trampling and burrowing. The kinds of plant species influence the kinds of animal species through forage preference which determine the kinds and amounts of trampling and burrowing. Porosity is inversely related to bulk density. Bulk and particle density are used to calculate porosity. Particle density fluctuates only a hundredth of a gram per cubic centimeter, so porosity is closely dependent on bulk density. Blackburn

(1973) found porosity to be negatively correlated with infiltration rates when vesicular crusts are present. The vesicular crusts were not interstitial and prevented water entry.

Dixon (1975) developed an air-earth interface concept that relates to both roughness and porosity. Soil surface roughness and openness control infiltration of free surface water by governing the flow of air and water in underlying macropore and micropore systems. For a given volume of water to infiltrate through the soil surface, a nearly equal volume of air must be able to leave the space occupied by the incoming water. The direction air leaves is usually upward against the direction of infiltrating water. Thus, size and shape of pores and their position relative to roughness have a tremendous influence on the infiltration process. Pores left by decayed roots are of tremendous importance as a passage for soil air to escape to the atmosphere and for infiltrated water to percolate deeper into the soil profile.

Not shown in Figure 3 is a box for microorganisms because little is known about their relationships with plants, soils, and water. Whitford (1976) reported that subterranean termites are abundant in rangeland soils of the southwestern United States. Estimates of their biomass range from 3.0 to 5.0 g m^{-2} or approximately ten times the mass of domestic livestock that can be supported on that rangeland. Whitford treated a series of plants with an insecticide that killed the termites and prevented recolonization. Infiltration rates were less on plots without termites and runoff started within 5 minutes of rainfall in comparison to 25 minutes for runoff initiation on plots with termites. Soil without termites had a lower total porosity (24.9%) than with termites (35.8%). The decreased porosity and infiltration rates were attributed to the lack of tunnels.

The infiltration process on rangelands is quite complicated and spatially variable and is influenced greatly by plants and their management. Each of the influences in Figure 3 usually has a positive or negative correlation, but enough exceptions occur with each influence to make modeling difficult. These correlations were calculated for plant and soil variables with terminal infiltration rates or infiltration capacity. Usually, this was after 30 to 60 minutes of rainfall. Further complications were discovered by Wilcox *et al.* (1987) when they compared independent plant and soil variables to infiltration rates at 5 minute intervals after rainfall commenced. Rainfall lasted for 45 minutes on dry antecedent soils and 35 minutes on wet antecedent soils. Infiltration became more correlated with total foliar cover, rock cover, and soil depth as rainfall progressed until a maximum correlation coefficient was reached. Slope, basal cover, and litter cover became less correlated with infiltration from beginning of rainfall until the end. Organic matter content, bulk density, and percentage of bare ground became more strongly correlated up to a maximum level and then rapidly decreased toward the end of rainfall. Therefore, infiltration rates have high spatial variability and different controlling factors through time.

19.5 Rangeland vegetation and runoff and erosion

Plant communities have various depths of runoff and this becomes important in vegetation conversion. Conversion techniques can be well or poorly planned or accidental. Management goals may be to increase or decrease runoff. Often, the amount and duration of runoff achieved is opposite of that desired with accompanied adverse environmental consequences. Generally, the key to runoff management is controlling infiltration rate. The greatest influence of plants on runoff is probably the changes in sinuosity of flow. This is important for erosion, but plants also protect the soil and add organic matter which strengthens the soil structure or lowers its erodibility.

Plant density, cover, phytomass, and distribution affect the paths runoff take in leaving a site (Figure 4). Croplands are much different from rangelands in that fields are often completely bare or have sparce cover between plant emergence and harvest. Croplands often have a uniform slope steepness and slope lengths are often more than several hundred meters long. Runoff can easily concentrate into rills, increase velocity, and be highly erosive as it moves across the landscape.

Rangelands usually have perennial plants or remnants of plants throughout the year. The importance of cover can be seen in examples A through D of Figure 4 which are 60% covered compared to examples E through H which have 10%. Generally, as density, cover, and phytomass of plants increases, runoff and erosion decreases (Figure 4, example C). The various sizes in the examples roughly simulate the comparative sizes of shrubs (example A), grasses (example C), and forbs (example G). Only plant cover has been used in most runoff and erosion models (Amorocho and Hart 1964, Wood and Blackburn 1984a, Renard and Foster 1985). Plant density and distribution have not. Having the same plant cover but great density affords more site protection and less runoff and erosion. Examples A, C, E, and G have equal plant distribution and would prevent concentration of flow, cause longer pathways, and have decreased runoff and erosion compared to the unequal distribution examples B, D, F, and H. Example F has low density, low cover, and unequal distribution of plants which would make this site the most susceptible to runoff and erosion if all other variables were similar.

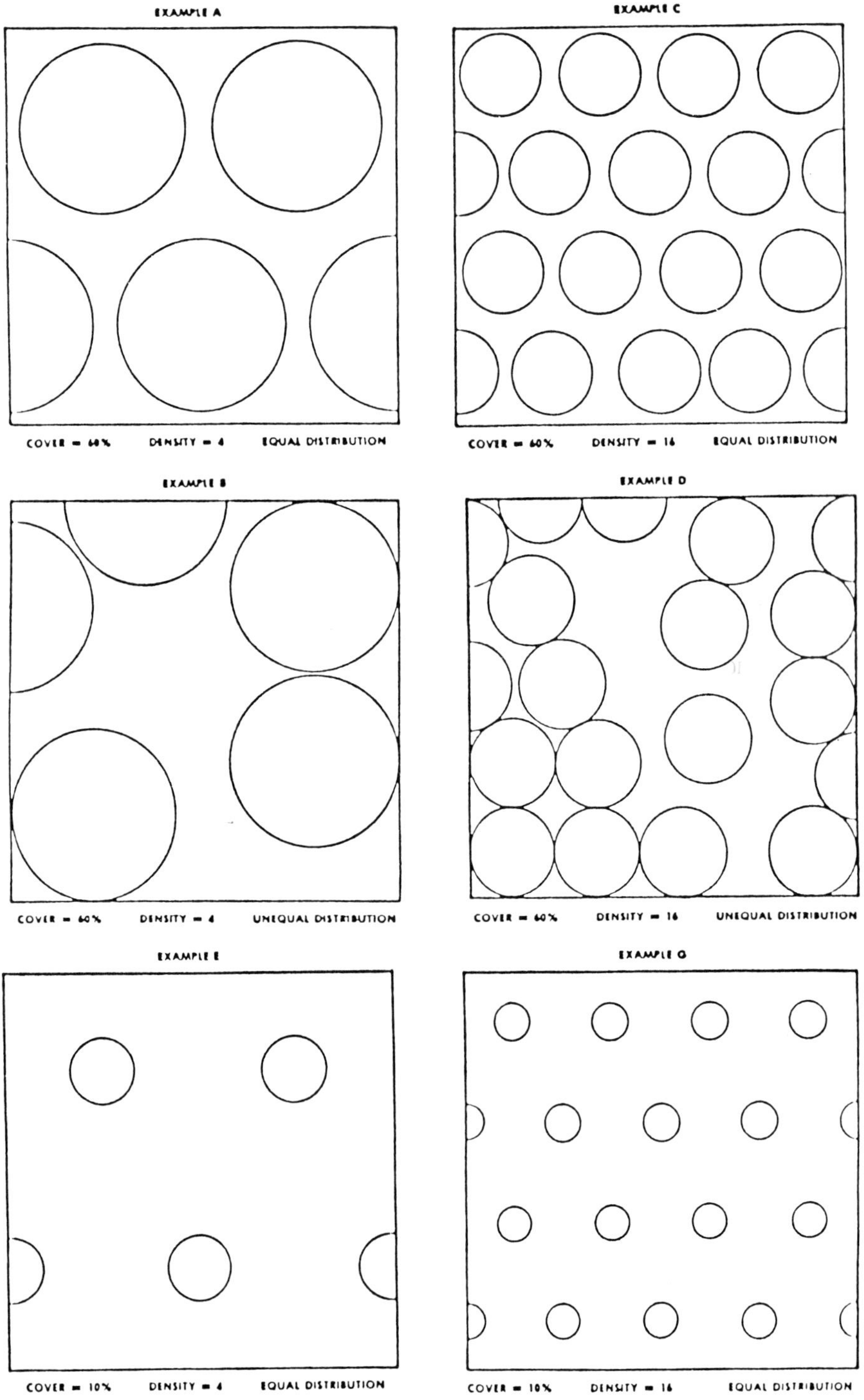
EXAMPLE A
COVER = 60%
DENSITY = 4
EQUAL DISTRIBUTION
EXAMPLE C
COVER = 60%
DENSITY = 16
EQUAL DISTRIBUTION
EXAMPLE B
COVER = 60%
DENSITY = 4
UNEQUAL DISTRIBUTION
EXAMPLE D
COVER = 60%
DENSITY = 16
UNEQUAL DISTRIBUTION
EXAMPLE E
COVER = 10%
DENSITY = 4
EQUAL DISTRIBUTION
EXAMPLE G
COVER = 10%
DENSITY = 16
EQUAL DISTRIBUTION

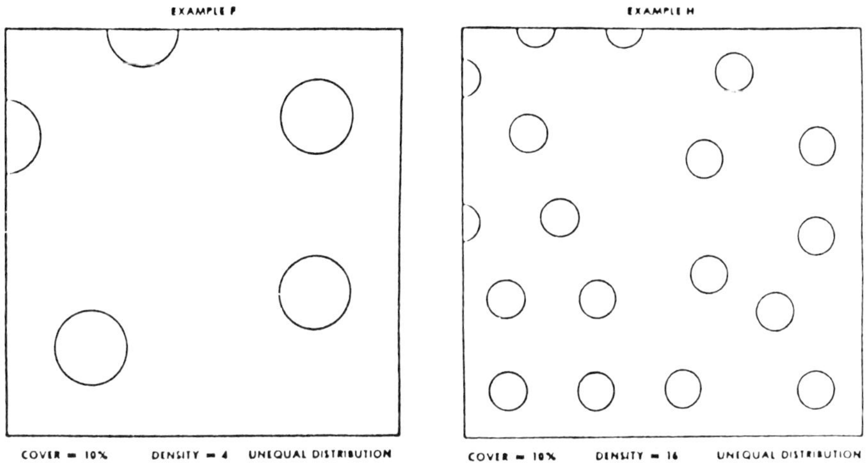

Figure 4. Examples of various combinations of cover, density, and distribution.

19.6 Rangeland succession and the hydrologic cycle

The relationship between natural or geologic erosion and mean annual precipitation was shown by Langbein and Schumm (1958) (Figure 5). At low mean annual precipitation, sediment yield is low because runoff is low since most water infiltrates or evaporates. From 300 to 750 mm of precipitation enough water is present to create runoff and not enough plants are present to prevent erosion. Above about 750 mm, plants protect the site which results in low sediment yields. Many rangelands receive between 300 to 750 mm. Attempts have been made to verify the relationship in Figure 5. It is difficult to find rangeland sites where natural or geologic sediment yields are being produced without the total sediment production also including some accelerated erosion due to activity of humans, their machines, or their animals. When sediment yield data points from geologic and accelerated erosion are added to Figure 5, no correlation can be found. This can be explained or shown by graphing normal erosion and potential erosion or sediment production for different mantle conditions (Figure 6).

The precipitation-erosion relationship (Figure 5) may be closely approached by ecosystems in undisturbed climax or disclimax conditions. This has not been considered by most hydrologists. Climax conditions result from succession or the process of change by which biotic communities are altered and replace each other over time. Change does not continue indefinitely. The final or indefinitely prolonged stage is called the climax community. A disclimax community depends on continuing dis-

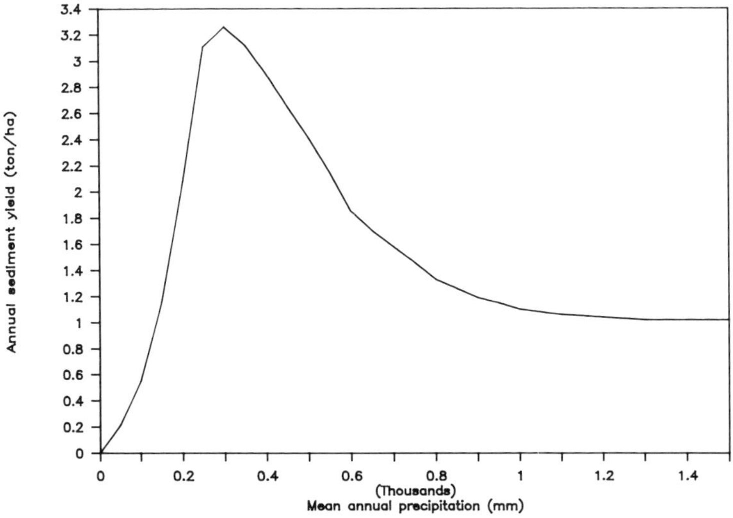

Figure 5. Relationship between natural (geologic) sediment yields and mean annual precipitation (Longbein and Schumm 1958).

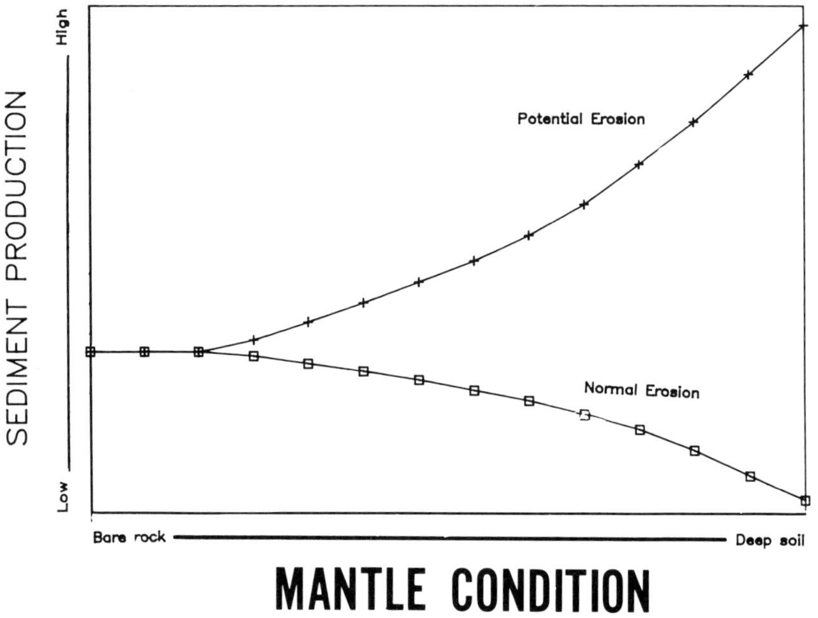

Figure 6. The decrease in normal erosion and the increase in erosion potential that accompany development of a soil mantle under the protective influence of vegetation (Bailey and Copeland 1961).

turbance for its maintenance. Many rangeland ecosystems are maintained by fire or animal disturbance. Over centuries or millenia, a site may progress toward a tree climax but because of fire, a grass or shrub-grass dominated seral stage results. An example is the pinyon (*Pinus* spp. L.) — juniper (*Juniperus* spp. L.) type which tends to be climax on hillsides that are steep enough to have shallow, rocky soils. These shallow, rocky soils do not have enough understory to carry a fire so that the pinyon and juniper trees are unharmed. But their seeds get carried to other areas that are less steep. These less steep areas have deep soils and a shrub-grass or grass community. They burn frequently and any pinyon or juniper seedlings are prevented from dominating the site.

If the fires are stopped, the site with deep soils may become dominated by pinyon and juniper trees (Barney and Frischknect 1974, Rogers 1982). Their lateral root system extends two and a half times their radius. The understory virtually disappears. This can lead to important hydrologic consequences (Carrara and Carroll 1979). The area directly under the trees becames covered with litter which results in high infiltration rates. But in the interspaces, much bare soil is often exposed which leaves the site vulnerable to accelerated erosion. Pinyon and juniper trees do not readily replace themselves with seedlings on site because of their intolerance to shade. The site may have irreversible gully erosion which alters the site for at least centuries. Pinyon and juniper are considered to be brush or weed species. This may be an extreme example of their hydrologic influence on an invaded site. But the unnatural invasion of other brush species on many rangelands probably results in accelerated runoff and erosion also. Southern New Mexico was a desert grassland under pristine conditions. In 1846, a U.S. battalion in the Mexican-American war camped at the foot of Cooke's Peak and one soldier wrote in his diary how they had to search for twigs to build a campfire. Today, one has to search to find a clump of grass. The site is dominated by mesquite and creosotebush.

The hydrologic health of other rangelands may be sustained by disclimax parturbations also. Wood and Blackburn (1984b) reported vegetation and soil responses to livestock grazing and exclusion in the Rolling Plains of Texas on an area with about 10% canopy cover of mesquite. Underneath the mesquite was a dominant understory of annual grasses in the exclosures and perennial grasses in the grazed areas. The low canopy cover (10%) was a result of brush control. Without brush control, the mesquite could dominate the grazed and excluded sites but, with different understories, the grazed areas having the hydrologically most desirable perennial grass understory.

19.7 Conclusions

From one rangeland ecosystem to another, the hydrologic processes are quite different, depending upon climate, topography, elevation, soils, parent materials, and time. Water does not penetrate past the vadose zone on some rangelands, while other areas experience ground water recharge. Parts of eastern New Mexico, for example, have no runoff because of the deep, sandy soil surface. Within a rangeland ecosystem, the hydrologic processes are quite different, depending greatly on vegetation changes. These changes may be from natural succession or land management practices.

Plants and their management or manipulation can affect every part of the hydrologic cycle. Plants can add condensation nuclei for precipitation and change the quality through emission of acids. Interception can change the size and form of precipitation, distribution, quantity, and quality. Plants especially affect water infiltration rates into the soil. There usually is a strong relationship between infiltration rates, runoff, and erosion. The kinds and amounts of plants will determine the rate of soil water depletion. Some rangeland plants reach deep into the soil to remove groundwater. Differences in hydrologic processes within a rangeland ecosystem due to plant changes may be as great as differences between rangeland ecosystems.

Several recommendations have been made to alter the quantity, quality, timing, and placement of waters on rangeland. These recommendations include interbasin transfer of water, weather modification, desalination, and suppression of evapotranspiration. But these techniques are expensive, limited in spatial usefulness, risky, and probably far away in time on a wide-scale basis. Therefore, the key to water control and efficient use on rangelands is best management practices for the plant-soil complex on-site.

19.8 References

Aldon, E. F. 1964. Ground cover changes in relation to runoff and erosion in westcentral New Mexico. U.S. Dept. Agric. For. Serv. Res. Note RM-34.

Amorocho, J., and W. E. Hart. 1964. A critique of current methods in hydrologic systems investigations. Trans. Amer. Geophys. Union 45: 307—321.

Armentrout, S. M. 1986. Vegetational gradients surrounding pinyon-juniper and oneseed juniper in southcentral New Mexico. M.S. Thesis, New Mexico State Univ., Las Cruces.

Bailey, R. W., and O. L. Copeland, Jr. 1961. Low flow discharges and plant cover relations on two mountain watersheds in Utah. Intl. Assoc. Sci. Hydrol. Pub. 51: 267—278.

Balliette, J. F., K. C. McDaniel, and M. K. Wood. 1986. Infiltration and sediment production following chemical control of sagebrush in New Mexico. J. Range Manage. 39: 160—165.

Barney, M. A., and N. C. Frischknecht. 1974. Vegetation changes following fire in the pinyon-juniper type of westcentral Utah. J. Range Manage. 27: 91—96.

Beutner, E. L., R. R. Garb, and R. E. Horton. 1940. Sprinkled-plot-runoff and infiltration experiments on Arizona desert soils. Trans. Amer. Geophys. Union 20: 550—558.

Blackburn, W. H. 1973. Infiltration rate and sediment production of selected plant communities and soil in five rangelands in Nevada. Ph.D. Dissertation, Univ. Nevada, Reno.

Blackburn, W. H. 1975. Factors influencing infiltration and sediment production of semiarid rangelands in Nevada. Water Resources Res. 11: 929—937.

Blackburn, W. H. 1980. Universal soil loss equation and rangelands. Proc. Symp. on Watershed Manage., Amer. Soc. Civil Eng., New York.

Blackburn, W. H., and C. M. Skau. 1974. Infiltration rates and sediment production of selected plant communities in Nevada. J. Range Manage. 27: 476—479.

Bond, R. D. 1964. The influence of the microflora on the physical properties of soils. II. Field studies on water repellent sands. Austral. J. Soil Res. 2: 123—131.

Branson, F. A., G. F. Gifford, K. G. Renard, and R. F. Hadley. 1981. Rangeland Hydrology. Kendall/Hunt Publ. Co., Dubuque, Iowa.

Buckhouse, J. C., and G. B. Coltharp. 1976. Soil moisture response to several levels of foliage removal on two Utah ranges. J. Range Manage. 29: 313—315.

Carrara, P. E., and T. R. Carroll. 1979. The determination of erosion rates from exposed tree roots in Piceance Basin, Colorado. Earth Science Processes 4: 207—217.

Clary, W. P., and P. F. Ffolliott. 1969. Water holding capacity of ponderosa pine forest floor layers. J. Soil Water Conserv. 24: 22—23.

Corbett, E. S., and R. P. Crouse. 1968. Rainfall interception by annual grass and chaparral. U.S. Dept. Agric. For. Serv. Res. Paper PSE-48.

Debano, L. F., and R. M. Rice. 1973. Water-repellent soils: Their implications in forestry. J. For. 71: 220—223.

Dixon, R. M. 1975. Infiltration control through soil surface management. Proc. Watershed Manage. Symp. sponsored by Amer. Soc. Civil Eng., New York.

Eckert, R. E., Jr., and F. E. Kinsinger. 1960. Effects of *Halogeton glomeratus* leachate on chemical and physical characteristics of soils. Ecology 41: 764—772.

Gamougoun, N. D., R. P. Smith, M. K. Wood, and R. D. Pieper. 1984. Soil, vegetation, and hydrologic responses to grazing management at Fort Stanton, New Mexico. J. Range Manage. 37: 538—541.

Geiger, R. 1965. The Climate Near the Ground. Harvard University Press, Cambridge, Massachusetts.

Gray, D. M., D. I. Norum, and J. M. Murry. 1969. Infiltration characteristics of prairie soils. Paper presented at Amer. Soc. Range Manage. Ann. Mtg., Calgary, Alberta, Canada.

Green, W. H., and G. A. Ampt. 1911. Studies in soil physics. I. The flow of air and water through soils. J. Agric. Sci. 4: 1—24.

Harris, F. S. and G. Stewart. 1930. The Principles of Agronomy. MacMillan Co., New York.

Herbel, C. H., F. N. Ares, and R. A. Wright. 1972. Drought effects on a semidesert grassland range. Ecology 53: 1084—1093.

Horton, R. E. 1933. The role of infiltration in the hydrologic cycle. Trans. Amer. Geophys. Union 14: 446—460.

Horton, R. E. 1940. An approach toward physical interpretation of infiltration capacity. Soil Sci. Soc. Amer. Proc. 5: 399—417.

Hugie, V. K., and H. B. Passey. 1964. Soil surface patterns of some semiarid soils in northern Utah, southern Idaho, and northeastern Nevada. Soil Sci. Soc. Amer. Proc. 28: 786—792.

Hull, A. C., Jr., and G. J. Klomp. 1974. Yield of crested wheatgrass under four densities of big sagebrush in southern Idaho. U.S. Dept. Agric. Tech. Bull. 1438.

Johnson, K. L. 1978. Managing livestock grazing in relation to runoff and erosion. Rangeman's J. 5: 147—149.

Kimmins, J. P. 1974. Some statistical aspects of sampling through fall precipitation in nutrient cycling studies in British Columbian coastal forests. Ecology 54: 1008—1019.

Knight, R. W. 1980. Hydrologic response of selected grazing systems on the Edwards Plateau. Ph.D. Dissertation, Texas A&M Univ., College Station.

Kostiakov, A. N. 1937. On the dynamics of the coefficient of water-percolation in soils and on the necessity for studying it from a dynamic point of view for the purposes of amelioration. p. 17—21. Trans. 6th Comm. Intl. Soc. Soil Sci., Part A.

Landsberg, H. 1938. Atmospheric condensation nuclei. Ergebn. Kosm. Phys.

Langbein, W. B., and S. A. Schumm. 1958. Yield of sediment in relation to mean annual precipitation. Trans. Amer. Geophys. Union 39: 1076—1084.

Lyon, T. L., and E. O. Fippin. 1909. The Principles of Soil Management. Macmillan Co., London.

McCalla, G. R., W. H. Blackburn, and L. B. Merrill. 1984. Effects of livestock grazing on infiltration rates, Edwards Plateau of Texas. J. Range Manage. 37: 265—269.

McGinty, W. A., F. E. Smeins, and L. B. Merrill. 1979. Influence of soil, vegetation, and grazing management on infiltration rate and sediment production of Edwards Plateau rangeland. J. Range Manage. 32: 33—37.

Melton, F. A. 1940. A tentative classification of sand dunes, its application to dune history in southern high plains. J. Geol. 48: 113—174.

Miller, D. H. 1977. Water at the Surface of the Earth. Academic Press, New York.

Philip, J. R. 1957. The theory of infiltration: 1. The infiltration equation and its solution. Soil Sci. 83: 345—357.

Philip, J. R. 1983. Infiltration in one, two, and three dimensions. p. 1—13. Proc. Natl. Conf. Advances in Infiltration, Chicago, Amer. Soc. Agric. Eng.

Pluhar, J. J. 1984. Hydrologic impacts of selected grazing systems on the Texas Rolling Plains. M.S. Thesis, Texas A&M Univ., College Station.

Rauzi, F., and A. R. Kuhlman. 1961. Water intake as affected by soil and vegetation on certain western South Dakota rangelands. J. Range Manage. 14: 267—271.

Richards, L. A. 1931. Capillary conduction of liquids through porous mediums. Physics 1: 318—333.

Renard, K. G., and G. R. Foster. 1985. Managing rangeland soil resources: The Universal Loss Loss Equation. Rangelands 7: 118—122.

Rogers, G. F. 1982. Then and Now: A Photographic History of the Central Great Basin Desert. University of Utah Press, Salt Lake City.

Sanchez, C. E. 1985. Investigation of infiltration rates and sediment production associated with reclaimed coal mine spoils in westcentral New Mexico. M.S. Thesis, New Mexico State Univ., Las Cruces.

Sanchez, C. E., and M. K. Wood (*In Press*). 1987. Hydrologic impacts associated with reclaimed coal mine spoils in westcentral New Mexico.

Sanchez-Brito, C. 1984. Effects of livestock grazing and exclusion on infiltration rate and sediment yields for different range sites on El Plateado watershed, Zacatesas, Mexico. Ph.D. Dissertation, New Mexico State Univ., Las Cruces.

Shure, D. J., and A. J. Lewis. 1973. Dew formation and stemflow on common ragweed (*Ambrosia artemisiifolia*). Ecology 54: 1152—1155.

Slayter, R. O. 1965. Measurement of precipitation interception by an arid zone plant community (*Acacia aneura* F. Muell.). Arid Zone Res. 25: 181—192.

Stuart, D. M., G. E. Schuman, and A. S. Dylla. 1971. Chemical characteristics of the coppice dune soils in Paradise Valley, Nevada. Soil Sci. Soc. Amer. Proc. 35: 607—611.

Thurow, T. L., W. H. Blackburn, and C. A. Taylor, Jr. 1986. Hydrologic characteristics of vegetation types as affected by livestock grazing systems, Edwards, Plateau, Texas. J. Range Manage. 39: 505—508.

Tromble, J. M. 1983a. Interception of rainfall by creosotebush (*Larrea tridentata*). p. 373—375. Proc. XIV Intl. Grassl. Congr., Lexington, Kentucky.
Tromble, J. M. 1983b. Interception of rainfall by tarbush. J. Range Manage. 36: 525—526.
Ueckert, D. N. 1979. Broom snakeweed: Effect on shortgrass forage production and soil water depletion. J. Range Manage. 32: 216—220.
Warren, S. D., W. H. Blackburn, and C. A. Taylor, Jr. 1986a. Soil hydrologic response to number of pastures and stocking density under intensive rotation grazing. J. Range Manage. 39: 505—509.
Warren, S. D., W. H. Blackburn, and C. A. Taylor, Jr. 1986b. Effects of season and stage of rotation cycle on hydrologic condition of rangeland under intensive rotation grazing. J. Range Manage. 39: 486—491.
Warren, S. D., T. L. Thurow, W. H. Blackburn, and N. E. Garza. 1986c. The influence of livestock trampling under intensive rotation grazing on soil hydrologic characteristics. J. Range Manage. 39: 391—395.
Weltz, M. 1983. The influence of short duration and continuous grazing on infiltration rate and sediment yield in southcentral New Mexico. M.S. Thesis, New Mexico State Univ., Las Cruces.
Weltz, M., and M. K. Wood. 1986a. Short duration grazing in central New Mexico: Effects on infiltration rates. J. Range Manage. 39: 365—368.
West, N. E., and G. F. Gifford. 1976. Rainfall interception by cool desert shrubs. J. Range Manage. 29: 171—172.
Whitford, W. G. 1986. Termites, water, and nutrient relationships on a semiarid rangeland. Proc. Second Intl. Rangel. Congr. Austral. Acad. Sci., Canberra.
Wilcox, B. P., M. K. Wood, J. M. Tromble, and B. A. Buchanan (*In Press*). 1987. Infiltrability of steep slopes in the Guadalupe Mountains of New Mexico.
Wilken, G. C. 1967. Snow accumulation in a manzanita brush field in the Sierra Nevada. Water Resources Res. 3: 409—422.
Wood, J. C., M. K. Wood, and J. M. Tromble. 1987. Important factors influencing water infiltration and sediment production on arid lands in New Mexico. J. Arid Environ. 12: 111—118.
Wood, M. K. 1979. The influence of grazing systems on infiltration rate and sediment production in the Rolling Plains of Texas. Ph.D. Dissertation, Texas A&M Univ., College Station.
Wood, M. K., and W. H. Blackburn. 1981a. Sediment production as influenced by livestock grazing in the Texas Rolling Plains. J. Range Manage. 34: 228—231.
Wood, M. K., and W. H. Blackburn. 1981b. Grazing systems: Their influence on infiltration rates in the Rolling Plains of Texas. J. Range Manage. 34: 331—335.
Wood, M. K., and W. H. Blackburn. 1984a. An evaluation of the hydrologic soil groups as used in the SCS runoff method on rangelands. Water Res. Bull. 20: 379—389.
Wood, M. K., and W. H. Blackburn. 1984b. Vegetation and soil responses to cattle grazing systems in the Texas Rolling Plains. J. Range Manage. 37: 303—308.
Wood, M. K., G. B. Donart, and M. Weltz. 1986. Comparative infiltration rates and sediment production on fertilized and grazed blue grama rangeland. J. Range Manage. 39: 371—374.
Woodward, L. 1943. Infiltration capacities on some plant-soil complexes on Utah range watershed lands. Trans. Amer. Geophys. Union 24: 468—475.
Young, J. A., and R. A. Evans. 1987. Stemflow on western juniper (*Juniperus occidentalis*) trees. p. 373—381. Proc. Pinyon-juniper Conf. U.S. Dept. Agric. For. Serv., Interm. Res. Sta. Gen. Tech-Rep. INT-215.

20. Grazing management and vegetation response

RICHARD H. HART AND B. E. NORTON

20.1. Abstract

Conventional ecological theory holds that grazing by livestock reduces productivity and alters botanical composition of rangeland plant communities. Changes in composition from that of the "climax" community are considered to be detrimental, but it is usually assumed that these changes will be reversed when grazing animals are removed. Reducing stocking rate and alternating periods of rest and grazing are attempts to mitigate undesirable effects of grazing. Recent research indicates that range ecosystems may retain their productivity and composition when grazed by livestock at seasons and stocking rates which resemble the grazing patterns under which the ecosystems evolved. Weather and fire may have more effect than grazing on productivity and composition, or may interact with grazing. In ecosystems with long-lived woody perennials, changes in composition may be reversible only after centuries or not at all. Finally, communities at early seral stages or communities including introduced plant species may be more productive than climax communities. All these findings indicate that protection of soil and maintenance of stable plant communities at whatever stage of succession may be more reasonable goals of grazing management than efforts to produce and maintain climax communities.

20.2 Introduction

The impacts of grazing on vegetation are best examined from a historical perspective. Current concepts in range management originated in a dream of recovering what was lost, not in the necessity of using and preserving what now exists.

The uninformed often envision the rangelands of the United States,

Tueller, P. T. (ed.), Vegetation science applications for rangeland analysis and management.
ISBN 90-6193-195-9.
© 1988, *Kluwer Academic Publishers, Dordrecht, Boston, London.*

before European settlement, as Elysian fields of "tremendous areas of luxuriant grass" (Fremont 1843), supporting uncounted numbers of bison (usually called "buffalo") and other native herbivores. But this was not uniformly the case; Fremont (1843) also reported that along the Platte River in 1842 near the present town of Douglas, Wyoming, ". . . buffalo were very scarce, and little or no grass (was) to be found." Lewis (1814) found the land on the site of Mandan, North Dakota was "generally covered with a short grass resembling very much the blue grass."

The plains and prairies were grazed by wild animals as fully as the forage supply would permit (Larson 1940). Hornaday (1889) reported "(Buffalo) at times so completely consumed the herbage of the plains that detachments of the United States Army found it difficult to find sufficient grass for their mules and horses." Lewis (1814) observed a herd near present-day Missoula, Montana; from his report a stocking density of nearly 1.25 buffalo per acre (3 buffalo/ha) can be calculated. Morfi (1777) wrote "All the grass on the vast ranges has been consumed by (feral horses), especially around the waterings." Not only large herbivores, but rabbits, prairie dogs and other rodents (Agnew *et al.* 1986), jackrabbits (Anderson and Shumar 1986) and grasshoppers and other insects (Blocker 1977) consume range vegetation. Under such grazing pressure, health and reproduction of the animals is sure to suffer. Charles Goodnight (Haley 1936) described starvation of buffalo by the thousands in the winter of 1876; dead buffalo were "as thick as pumpkins in a field."

In the Great Basin both plant and animal life were sparse. Smith (1827) wrote "After travelling twenty days from the east side of Mount Joseph, I struck the S. W. corner of the Great Salt Lake. The country between the mountain and this Lake is completely barren and entirely destitute of game. We frequently travelled two days . . . where no sign of vegetation was to be seen." Evidence from numerous travellers' journals, written in the early 1800's, indicates that the extent of sagebrush in the Basin was about the same then as now (Vale 1975, Dorn 1986). Total cover of sagebrush may have increased; this would be difficult to determine from brief descriptions. Photographs of Tueller (1975) illustrate the similarity in general appearance of sagebrush-bluebunch wheatgrass range with 2.5% and 18% crown cover of sagebrush.

Large herbivores were an essential component of African grasslands. Rattray (1960) writes "An undisturbed grass cover will . . . deteriorate in a few years if it is not grazed or burned periodically." But Walter (1963) observed in Namibia "Equilibrium of the savannah ecosystem can only be found under natural conditions, with some grazing by game animals and rare burning by lightning. (Under) grazing by cattle, the equilibrium is radically disturbed." Shantz and Turner (1958) concluded that excessive grazing by domestic stock had moved the vegetation of African grasslands away from climax, but that fire was more important than grazing in converting woodland to savannah. In India, where most native herbivores were replaced by domestic stock centuries ago, "Most of the existing grasslands . . . have been so misused that their plant associations are very low in the ecological succession" (Whyte 1957).

In regions with few large native herbivores, the impact of livestock grazing has been greater and more rapid. Roseveare (1948) describes the vegetation of the pampas in the sixteenth century as "tall, plumed bunchgrasses" but continues "In the moister regions the grazing of cattle, and especially of sheep, destroyed the taller elements of the vegetation and produced a lawn-like, low, dense sod. In the more arid west . . . grazing of cattle . . . destroyed the more tender grasses growing among the tall, hard grasses, thus promoting the predominance of the latter."

At the time of the first European settlements in Australia, native herbivores had been greatly reduced below Pleistocene levels by Aboriginal hunters. Widespread fires set by the Aborigines may have "caused considerable changes to the original vegetation over great areas and contributed greatly to the extinction of the giant herbivorous marsupials" (Frith 1970). Vegetation of the arid grass- and shrublands of Australia appears to have evolved without heavy grazing; Perry (1970) comments ". . . the vegetation is remarkably benign, the plants, with a few notable exceptions, lacking spines or thorns." Newman (1974) and Newsome and Corbett (1977) state that many range areas in Australia have suffered irreversible damage from erosion and invasion of unpalatable plant species since the introduction of livestock.

Larson (1940) concluded that the vegetation of the shortgrass plains was created and maintained by grazing pressure of native herbivores, and that the introduction of large domestic herbivores maintained that vegetation, except in cases of gross overgrazing. Unfortunately, gross overgrazing was all too common on the ranges of the western United States a century ago. The range livestock industry was perhaps at its zenith: immense landscapes of apparently limitless forage were occupied by great herds of cattle, and the only regulation was that imposed by forage supply and weather. But as in the case of wild herbivores, the regulation of a herbivore/forage system by forage supply for the herbivores has several predictable ecological consequences. A majority of the herbivore population becomes weakened in the competition for limited forage. Grazing pressure on the vegetation grants successive advantages to the most abundant unpalatable species. Finally, vegetative trend is toward a new combination of less-palatable species generating a more limited food supply. When these forces are combined with the stresses of a capricious climate, deadly in its periodic shortages of warmth or water, then it is not surprising that the carrying capacity of much rangeland declined in the last two decades of the nineteenth century.

Rangeland degradation across the western United States in the early twentieth century promoted the establishment of research programs to set standards to control grazing use; eventually these served as justification for Federal legislation. Because most workers in the field believed that the productivity of nearly all rangelands had declined drastically from pre-settlement levels, the emerging guidelines for rangeland management emphasized conservation and restoration rather than production.

At the same time, an indigenous body of ecological theory of vegetation change emerged. These concepts provided a ready-made framework on which to construct a basis for range management decisions. Pound and Clements (1898, 1900) and Cowles (1899) seem to have originated the concepts of orderly plant succession to a stable climax, and Sampson (1917) applied it to range management. Clements (1928) and other ecologists fathered the range condition concepts specifically applied to central US grasslands (Dyksterhuis 1949), to the semidesert grasslands (Canfield 1948), and to sagebrush-grass ranges (Humphrey 1947). Heady (1975) reviewed the history of these and other successional theories. What made the climax and successional theories of Clements so attractive to range managers was the promise they held of a chance to return to the conditions under which wild and domestic herds exploited the pristine prairies. It is not surprising that this condition was named "excellent." It was suggested that a site's carrying capacity was maximized at climax condition, and that a lower stocking rate on less-than-climax vegetation would lead automatically, even in a semi-arid zone, to "rapid and obvious improvement in range condition" (Dyksterhuis 1949). Such expectations were a natural extension of Clementsian successional theory, and they are embedded in the prevailing philosophy of range management on public lands.

In this review, we propose to examine those elements in our conventional understanding of livestock/vegetation interactions that we believe are most vulnerable to disputation and are most likely to cause errors of judgment in range management practices. We will rely heavily on earlier specialized reviews, but will present a sampling of original articles which are more recent than the corresponding review or which offer new concepts or data.

20.3 Effects of grazing on forage production

It is difficult to separate the effect of grazing on subsequent forage production from the effect of grazing on botanical composition. In the field or greenhouse, it can be demonstrated readily that frequent or severe defoliation will generally depress concurrent as well as subsequent production of individual plants. In a plant community in the field the reduced vigor of the grazed or clipped species is gradually compensated for by more growth of competing species, so that overall change in total community production is slight in spite of less production from the defoliated component.

20.3.1 Effects of grazing on plants

If we confine our attention to forage production in the current or immediately succeeding growing season, and assume that botanical composition is constant, then grazing usually decreases production but may potentially increase it (Jameson 1963, Hyder 1972, Trlica 1977, Sims *et al.* 1982, McNaughton 1983, Caldwell 1984, Kauffman and Krueger 1984). Production may be reduced directly, through loss of photosynthetic tissue, meristems (Arnold 1977, Briske 1987), or propagules (Packard 1977, Williams 1977) or indirectly through reduction in root growth (Crider 1955, Hodgkinson and Baas Becking 1977) and carbohydrate reserves in the roots or stem bases (Cook 1966, Trlica and Cook 1971, White 1973, Menke and Trlica 1981 and 1983).

Grazing may also reduce production through its influences on the microenvironment (Lorenz 1974, Marshall 1974, Whitman 1974, Lee 1977, Klemmedsen 1977). Trampling may decrease the availability of water and nutrients by slowing the rate of water infiltration into the soil (Gifford and Hawkins 1978, Blackburn 1984, Wood *et al.* 1986, Thurow *et al.* 1986), thereby increasing runoff and erosion with the accompanying loss of soil and mineral nutrients. Trampling also destroys cryptogamic crusts formed by lichens, mosses and algae; these crusts speed infiltration of precipitation, reduce runoff and erosion, fix significant amounts of nitrogen, and facilitate seedling establishment (St. Clair *et al.* 1984).

Grazing may enhance forage production if removal of mature tissue stimulates growth of dormant meristems or creates a more favorable environment for their development (Deregibus *et al.* 1985, Sala *et al.* 1986, Paige and Whitham 1987). Removal of some leaf area may increase photosynthetic rate (Detling *et al.* 1979a and b, Caldwell *et al.* 1981, Painter and Detling 1981, Nowak and Caldwell 1984) or reduce water stress (Svejcar and Christiansen 1987) in remaining leaf area. Grazing also may change allocation of assimilates (Ryle and Powell 1975, Trlica and Singh 1979) or accelerate nutrient cycling and redistribution of nutrients (Cowling 1977, Pieper 1977, Ruess and McNaughton 1984). Substances in saliva of grazing animals may directly stimulate regrowth (Reardon and Merrill 1978).

Trampling by grazing animals may bury seeds and enhance seedling establishment (Hormay 1970, Savory 1983), but Hyder *et al.* (1975) noted that seedling establishment plays a very small role in maintaining productivity of perennial grass ranges. Williams (1977) pointed out that successful seedling establishment on range requires two or more favorable years in succession, a rare circumstance. A negligible number of viable seeds were found in range soils in the western US (Hyder *et al.* 1975) and Tanzania (Belsky 1986b).

Much of the published information on the response of plants to grazing derives from clipping experiments and extrapolation of the results to grazed vegetation. But grazing is poorly simulated by clipping; livestock defoliate at non-uniform heights and intervals, and trampling and litter accumulation differ under clipping and grazing (Jameson 1963). The amount of herbage removed by grazing is usually less than that removed in clipping experiments. Grazing animals usually select particular plant parts rather than harvesting entire plants. Finally, when rangeland vegetation is grazed rather than clipped, defoliation is dispersed over a wide area and there is great variability in the fraction of forage harvested from an individual range plant (Norton and Johnson in press, Hart and Balla 1982, Test 1984).

Reliance on data from clipping experiments without field verification may contribute to our state of confusion over contradictory results coming from grazing studies. For example, Cook (1971) concluded from clipping studies that more than 50 percent winter use of salt desert shrub vegetation in Utah was damaging, yet earlier Hutchings and Stewart (1953) had recommended winter stocking rates that would utilize 75 percent of average forage production and the cover of desirable forage species increased at this level of utilization (Norton 1978). Stroud *et al.* (1985) found that 80% use of western wheatgrass caused a much smaller reduction in current and subsequent forage production when the use was dispersed spatially and temporally as in grazing, rather than being synchronized on a few dates as in clipping.

20.3.2 Resistance to and avoidance of grazing

Range plants have evolved a variety of mechanisms to prevent damage from grazing (Sims *et al.* 1982). Sod-forming grasses, which are resistant to grazing, produce many culmless tillers with meristems at or below ground level; these escape damage from grazing and continue to produce new leaves (Hyder 1972). If a culmed tiller with elevated meristem is grazed, the tiller meristem and many of the leaf intercalary meristems will be removed, and any regrowth must be initiated as new tillers from basal or axial buds (Enevoldson and Lewis 1978). For regrowth to occur, these buds must be sufficiently mature to initiate growth, carbohydrate reserves in the roots or stem bases of the plant must be adequate to maintain the initial growth, and environmental conditions must be favorable. Reproductive tillers react the same to grazing as do culmed vegetative tillers. Under grazing, the growth form of whole plant communities may shift to place meristems and most of the aboveground biomass so close to the ground that they cannot be grazed easily (McNaughton 1984). Belsky (1986a) found that *Agropyron greenwayi* in the Serengeti depended on canopy removal by grazing to initiate tillers and on trampling to root tillers and initiate mat formation.

The growing points of forbs and shrubs are located at the ends of branches, and are as susceptible to removal as the meristems of culmed grass tillers. Further growth usually arises from lateral buds of shrubs and basal buds of forbs. Forbs and shrubs which branch freely are more resistant to grazing than less freely branching species (Hyder 1972).

Some plants escape grazing because they are physically or chemically undesirable. Poisonous plants are seldom grazed because grazing animals have evolved the ability to recognize and avoid them (Arnold and Hill 1972, Laycock 1978, Reichardt *et al.* 1987). The silicafication of grasses may be an adaptation to inhibit grazing (McNaughton and Tarrants 1983). Many forbs and woody plants have developed spines or thorns; other forbs are hairy or sticky. Such plants often dominate overgrazed ranges because they escape grazing. Unpalatable and seldom-grazed "wolf plants" may develop if they escape grazing for a season, because of the accumulation of dead stems. Burning, mowing or forced heavy grazing can remove these stems to restore palatability (Willms *et al.* 1980) and reinstate more uniform grazing (Anderson 1982).

Wide dispersal of tillers on a single large plant may allow some tillers to escape grazing. Jeffries (1984) concluded that the probability of a clone's death decreases as the number of stems or tillers increases. Stout and Brook (1985) found more carbohydrate reserves in roots and rhizomes of clipped pinegrass tillers if they remained attached to unclipped tillers, but Archer and Detling (1984) suggested that the positive aspects of resource sharing among interconnected tillers were outweighed by the increased competitive ability of ungrazed vs. grazed tillers.

Individual patches (Morris 1969) or tillers of a sward may be grazed in a de facto rotational pattern under a continuous grazing system. Gammon (1978b) found little diference between continuous and rotation grazing in defoliation patterns of single tillers, and Test (1984) found no difference between season-long and short-duration grazing systems in the frequency and severity of defoliation of tillers of western wheatgrass and blue grama. Hart and Balla (1982) found that repeated defoliation of tillers of crested and western wheatgrasses was rare under grazing, but the number of defoliations was directly proportional to stocking rate.

20.3.3 Weather and forage production

On rangelands the influence of grazing on forage production may be less significant than the effect of weather. The inherent vagaries of climate, expressed in daily to yearly variations in weather, are unpredictable and are responsible for large year-to-year variation in available forage. Such variation creates serious management problems.

Bryson (1974) has stated that the 30—50 percent decrease in summer precipitation over the past century (Wahl and Lawson 1970) has reduced the carrying capacity of the Western U.S. range by 50—75 percent. This estimate seems excessive, but in 1934, when precipitation in the West was about 20 percent below normal, losses of 50—75 percent of the plants on the range were reported (Chapline and Cooperrider 1941; Albertson and Weaver 1942; Weaver and Albertson 1936 and 1944). Donaldson (1967) reported severe losses from the 1964—1966 drought in southern Africa. As average annual precipitation declines, the yearly and shorter-term variability of the precipitation increases (Campbell 1936, Reitan and Green 1968, Noy-Meir 1973, Wallen and Gwynne 1978, Norwine 1978), and considerable spatial variation can be expected (Sharon 1972). Experiments designed to compare different stocking rates or grazing systems often reveal that weather has greater influence on the vegetation than does the grazing animal (Launchbaugh 1967, Burzlaff and Harris 1969, Reed and Peterson 1961, Klipple and Costello 1960, Sarvis 1941).

Hutchings and Stewart (1953), at the Desert Experimental Range in Utah, developed a correlation (r^2) between herbage production of salt desert shrub vegetation and precipitation of 0.88. Hart and Carlson (1975) reported correlations (r^2) of range forage production with precipitation of 0.37 to 0.61 in Colorado, Wyoming and Kansas. Smoliak (1986) found that April—July precipitation accounted for 74% of the variation in annual production of needleandthread-blue grama range in Alberta. Other studies of precipitation-production relationships were reviewed by Wagner (1976), Le Hourerou and Hoste (1977) and Webb *et al.* (1978).

Vorster (1975) suggested that grazing of veld during recovery from drought may decrease forage production more than high grazing pressure during drought. Martin and Cable (1974) noted that forage production was reduced severely in the year after drought, regardless of current precipitation. Approximately 22 cm of precipitation in the summer of 1963 produced only 40% as much perennial grass as 18 cm of precipitation in the summer of 1965; precipitation the preceding summers had been 12 and 30 cm, respectively.

In light of the above, is there an interaction between grazing pressure and weather that affects forage production? In drought years, forage production is reduced by moisture stress, while grazing pressure increases if stocking rate remains static. The combination of these two factors is most likely to occur in a multiple-pasture rotation grazing system with high stocking density, and may cause such a system to abort because of imminent livestock mortality (Gammon 1978a). Under these circumstances forage utilization will approach maximum and grazing is most likely to depress subsequent production. A practical measure adopted by land management agencies to avoid damage from high grazing pressures is to set stocking rates at conservative levels, which may mean that in years of average rainfall or better forage utilization is rather low.

20.3.4 Grazing pressure and forage production

Lacey and Van Poollen (1981) reviewed 14 grazing studies on range in the Western US, and calculated that annual herbage production increased 68 ± 46% on ranges protected from "a moderate level of livestock grazing." However, misleading conclusions may be drawn from comparisons between grazed range and range from which all grazing by large mammals is excluded. Ungrazed ranges are seldom found in nature, occurring only in those rare locations where topographical barriers prevent large grazing animals from reaching them (Madany and West 1984).

A few grazing studies have measured forage production over enough years so that the impact of different levels of grazing pressure on production can be evaluated (Fig. 1). Production in the last year of each study was expressed as percent of production in the first year, to minimize the effects of differences in initial productivity among pastures. Average production of buffalograss-blue grama-western wheatgrass range in the tenth year of grazing was much greater than in the first year but did not increase as much under high as under low grazing pressure (Launchbaugh 1967). However, the decline in production of blue grama-prairie sandreed range over 9 years (Sims *et al.* 1976) and of blue grama-buffalograss range over 15 years of grazing (Klipple and Costello 1960) was very small; production in both cases was about the same in the first and last years. Van Poollen and Lacey (1979) reviewed 18 grazing studies in the western US and calculated that forage production increased 35% and 27%,

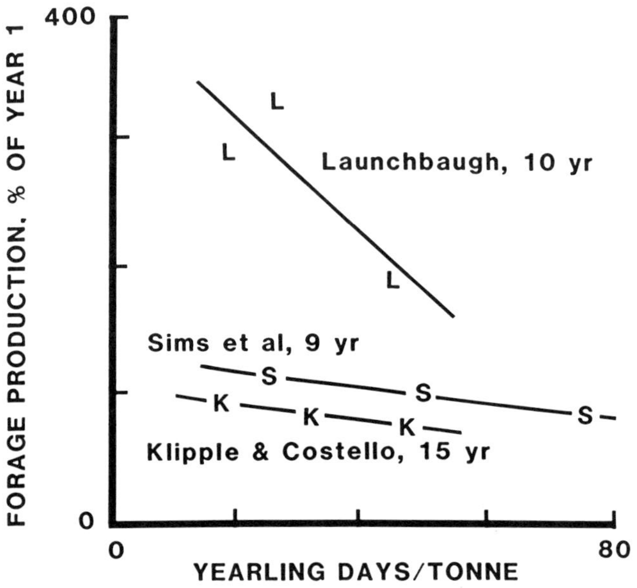

Figure 1. Response of forage production to grazing pressure after 9 to 15 years of grazing (Klipple and Costello 1960, Launchbaugh 1967, Sims *et al.* 1976).

respectively, when stocking rate was decreased from heavy to moderate and moderate to light under continuous grazing.

In Oregon, eleven years of cattle grazing at light, moderate and heavy stocking rates had little effect on production of bluebunch wheatgrass-Sandberg bluegrass grassland, but production of elk sedge and pinegrass under ponderosa pine forest appeared to decline as stocking rate increased (Skovlin *et al.* 1976). After 19 years of grazing on blue grama-western wheatgrass range in Wyoming, production of midgrasses was higher under lighter stocking while production of shortgrasses was the same at all stocking rates (Rauzi and Lang 1967), but without data on initial production of the pastures interpretation is uncertain. Similar difficulty is encountered in interpreting the effects of 25 years of grazing on western wheatgrass-blue grama range in Montana (Houston and Woodward 1966), and is compounded by apparent interactions between range sites, range condition, and stocking rates.

20.3.5 Grazing systems and forage production

Rotation and deferred grazing systems may increase forage production, but production is so confounded with botanical composition and weather that is difficult to draw definite conclusions. Gammon (1978a) found after an extensive literature review that "No rotational system consistently resulted in improved pasture or increased animal production. Where benefits in terms of pasture or animal production have been revealed these have been modest." Van Poollen and Lacey (1979) reported a 13% increase in forage production on rotation and deferred grazing systems over production under continuous grazing; all comparisons were made at a moderate stocking rate. Effects of systems differed considerably among regions, ranging from a 4% decrease in forage production in the northern Great Plains (range −13 to +5%) to a 30% increase in Texas (+24 to +39%). Sampson (1951) and Hickey (1969) concluded rest or deferral might increase forage production in some cases; Herbel (1974) concluded forage production was as high under season-long grazing as under any other system when grazing was less than year-long.

Forage production under season-long grazing was as high as or higher than production under rotation or deferred systems in Oregon (Skovlin *et al.* 1976); Utah (Laycock and Conrad 1981); Wyoming (Hart *et al.* 1986); and Texas (Heitschmidt *et al.* 1985, Heitschmidt 1986, Dahl 1986). In New Mexico, forage production appeared to be higher under 4-pasture rotation than under continuous grazing, but the apparent higher production was not confirmed by higher cattle gains or increased plant cover (Pieper *et al.* 1978, Gamogoun *et al.* 1984).

Comparions of forage and animal production between systems and stocking rates may be confounded by interactions between systems and rates. Production after years of grazing is affected by original and final botanical composition, original productivity of different pastures, and amount and timing of precipitation in the current and preceding years (Clary and Holmgren 1987). The most that can be concluded is that total annual forage production is usually less under grazing than in exclosures, and usually decreases with increased stocking rate or grazing pressure. Grazing systems appear to have little effect on long-term average forage production, other than those effects arising from a shift in botanical composition to less productive plant species.

20.4 Effects of grazing on botanical composition

The history of range management places paramount importance on the botanical composition of the plant community. Plant species are assigned management values, carrying capacity is assessed, and stocking rates are calculated or adjusted according to the accepted model of vegetation response to livestock grazing.

The early degradation of rangelands described in the Introduction provided a convincing demonstration that the grazing animal could substantially influence botanical composition. Selective grazing on range vegetation shifts competitive relationships among species in the community and eventually generates successional or retrogressional changes. Reichman *et al.* (1979), Skiles (1984), and Senft (1987) described selective grazing as a function of the nutrient requirements of herbivores and their physiological, anatomical and behavioral adaptations to meet those requirements, and of the pre- and post-ingestion characteristics of the forage. Ellison (1960) reviewed the subject of grazing and succession in considerable depth; more recent research was reviewed by Newsome and Corbett (1977), Gammon (1978a), McNaughton *et al.* (1982), Bartolome (1984) and Branson (1985). All concluded that vegetation changes can and do occur as a result of selective defoliation pressures.

20.4.1 Models of vegetation change

The most popular model used to depict the process of vegetation change due to grazing is the one originally proposed by Dyksterhuis (1949). A composite representation of the model, as elaborated by Sims and Dwyer (1965) and Schmutz (1978), is shown in Fig. 2A. A conceptual modification (Fig. 2B) was advanced by Poulton (1959) who recognized that some increasers (Type II increasers) are quite unpalatable and increase steadily as ecological condition declines, while others are simply less preferred than the decreaser group and eventually succumb to grazing pressure in the typical fashion (Type I increasers). There are several problems associated with the practical application of this model as a management tool:

(1) The model assumes knowledge of the climax vegetation.

(2) The concept of changing range condition in a successional manner relies on the validity of Clementsian succession, which assumes a kind of "relay floristics" in which successive communities predictably exploit the microenvironment generated by their predecessors.

(3) Early range scientists adopted the view that whenever rangeland is grazed by livestock or other large herbivores, grazing pressure is the cause of any changes observed in botanical composition.

(4) There is an implicit assumption that if retrogression was due to overgrazing, then relaxing the grazing pressure will allow successional processes to generate a natural improvement in range condition. In other words, vegetation change is reversible.

(5) The model (as originally proposed) implies that the climax vegetation is the most desirable and will support the largest number of livestock.

Doubt has been cast on all these assumptions (Smith 1979, Wilson and Tupper 1982). Recent research provides some instructive examples.

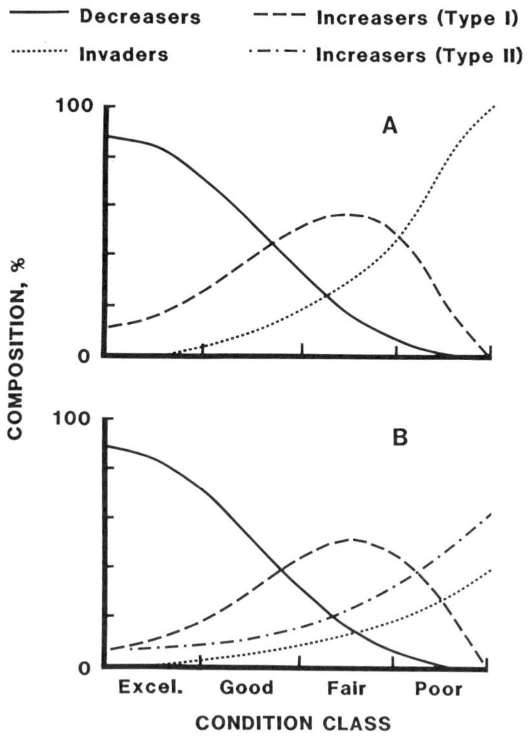

Figure 2. Models of vegetation change in which increasing utilization forces condition class from excellent to poor. 2A. Type I increasers only (Dyksterhuis 1949, Sims and Dwyer 1976, Schmutz 1978). 2B. Both Type I and Type II increasers (Poulton 1959).

(1) Climax condition often is inferred from the vegetation of exclosures or naturally-isolated areas which have been protected from grazing for a number of years (Clements 1934). Such areas "are not completely representative of the natural equilibrium (equivalent to climax) that prevailed under grazing by buffalo, elk, antelope, etc. that characterized most of the grassland" (Lewis 1969). Dansereau (1957) considered natural biotic influences, such as the effect of buffalo herds on the North American grassland or of herds of elephants and other herbivores on African savannahs, to be part of succession and important determinants of climax. Larson (1940) concluded the bison and associated large and small herbivores were an indispensable part of the climax on the North American prairies. Sims *et al.* (1978) agreed that the grazed situation was more natural than the ungrazed for ten North American grassland types. Fire as well as grazing is often excluded from exclosures and other isolated areas used as indicators of climax, yet fire is a well-documented part of grassland ecosystems (Anderson 1982).

(2) Contemporary ecologists (Connell and Slatyer 1977, Westoby

1980) are seriously questioning the generality of Clementsian succession. They recognize several pathways of vegetation change, and attach more importance to the mere presence of a species in an early seral stage as an indication of its chance of maintaining a niche in the community as the botanical composition changes through time.

(3) Many factors other than grazing are responsible for changes in botanical composition of rangelands. Weather effects on botanical composition may be much greater than those of grazing, particularly if relative aboveground biomass of constituent species in the plant community is used as an estimate of condition. Biomass responds directly to precipitation amount and distribution and is highly variable from year to year; this is especially obvious if there are annual species in the community (Samuel 1987). On ungrazed shadscale/squirreltail vegetation in southern Idaho, dominance switched from shadscale in 1951 to squirreltail in 1976, with a short-lived eruption of globemallow in the early 1960's, all without the influence of grazing (Sharp and Sanders 1978). Norton (1978) discovered that cover trends in the grazed areas at the Desert Experimental Range in Utah were almost identical to changes in exclosures over a period of 40 years. During that time dominance on shrubby sites switched from shadscale to winterfat, the more palatable of the two species, at all three levels of grazing intensity. Studies such as these raise doubts about the pre-eminent role of grazing in determining species composition on the range.

(4) Changes produced by grazing may be reversed when grazing is excluded on grassland, but reversal is less likely on grass-shrub vegetation in semi-arid rangeland (Herbel 1984, Tueller and Platou 1987). It may be a thousand years before the creosote bush that took over the desert grasslands of West Texas is eventually replaced via a successional pathway, or mesquite is eventually superseded by perennial grasslands in the Southwest, or the juniper stands that have moved into sagebrush-grass ranges are forced to retire in Clementsian style (Herbel 1984). Once the competitive balance has shifted in favor of the woody species, one must wait until the new generation of shrub plants has reached an age of natural senescence before other perennial plants have an opportunity to replace them. During this time the seed reserves of desirable herbaceous perennials may have been depleted or exhausted, while the regenerative potential of the dominant shrub species has remained high. On some Australian rangelands, the original vegetation has been so completely replaced by spinifex that it may never recover (Newman 1974).

Greater exposure of soil surface associated with the loss of perennial grasses may increase erosion and set back the site recovery time even further. Charley and Cowling (1968) concluded overgrazing of saltbush

(*Atriplex*) rangelands in Australia had so altered the nutrient cycle in the soil that it will take years for saltbush to reestablish itself.

(5) Climax vegetation is not always the most desirable for livestock production. Shrub-grass rangelands produce far more forage at low than at high successional stages (Herbel 1984, West 1984). Forage production on severely disturbed mixed-grass prairie in Wyoming, in an early seral stage ten years after disturbance, was about 50% greater than that on near-climax range (M. J. Samuel, unpublished data).

As range managers in other parts of the world have tried to adapt the American concepts of range management to their local conditions they have encountered many of the problems mentioned above. One solution that may be instructive is the range condition and trend model developed by Foran *et al.* (1978) for a grassveld type in South Africa (Fig. 3). In this model, condition declines because of under-utilization as well as over-utilization. Frequency of fire, or incidence and severity of drought, as well as level of grazing pressure, can be expressed on the ordinate. Once the linkage between range condition and ecological climax has been broken, the precise location of the goal line along the successional axis can be adjusted depending on the management objectives; it need not be at ecological climax. This model readily accomodates the presence of introduced species such as crested wheatgrass, because the most favorable species composition for forage production need not be a member of the indigenous climax community.

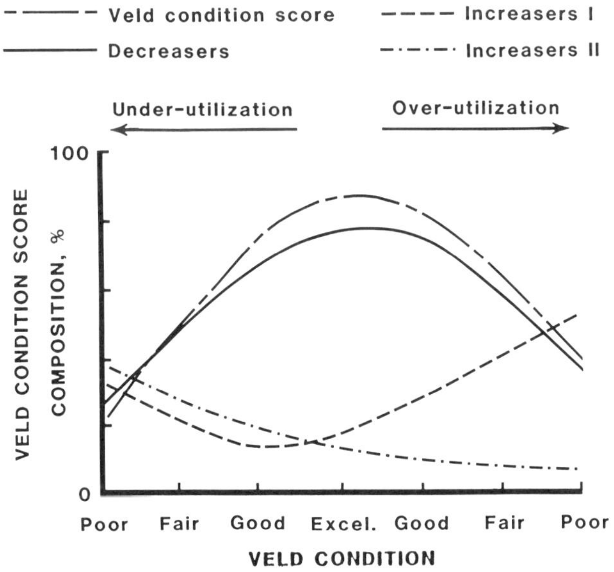

Figure 3. Model of vegetation change in which either under- or over-utilization forces condition class from excellent to poor (Foran *et al.* 1978).

Any useful concept of range condition must consider the desired use of the range, the goals to be achieved, and the criteria for evaluating progress toward those goals (Tueller and Platou 1987). Rather than struggling to attain a poorly-defined and perhaps unattainable "climax" community, one might choose to establish and maintain a stable plant community yielding a desirable product at an economically viable level. Such a community appears to have developed after 80 to 90 years of sheep grazing on Utah mountain ranges (Bowns and Bagley 1986).

20.4.2 Grazing pressure and botanical composition

Branson (1985), after an extensive review of grazing literature, concluded "there was a drastic deterioration of ranges late in the last century and continuing into this century, followed by some impressive improvements. . . . Some of the changes are undesirable and appear to be irreversible . . . and other undesirable changes have unknown degrees of permanency. . . . In the Great Basin, it was concluded pristine conditions cannot be restored, even by the removal of domestic animals, but that a balance between grasses and shrubs can stabilize the plant communities." He continues, "In the case of extreme drought . . . changes in vegetation may be more drastic than those caused by overgrazing. . . . Moist years following 1960 caused remarkable improvements in range vegetation."

Changes in botanical composition of rough fescue grassland over 30 years under no grazing and four levels of continuous grazing (Willms *et al.* 1985) illustrate the difficulties of interpreting the impact of grazing on vegetation (Table 1). Initial and final composition are averages of six years each, to smooth out fluctuations produced by variance in annual precipita-

Table 1. Original botanical composition (cover) of rough fescue grassland in Alberta, and composition after 32 years at four levels of continuous grazing and protection from grazing.

Species or group	1949–1954	1976–1981; AUM/ha 0.0	1.2	1.6	2.4	4.8
	---% of total cover---					
Parry danthonia	45.8	23.4	24.5	32.7	48.0	35.3
Idaho fescue	5.1	5.0	5.2	5.6	12.5	11.9
Rough fescue	13.4	43.8	37.7	20.7	7.9	2.5
Sedges	8.0	2.8	2.7	3.8	6.1	9.7
Other graminoids	10.7	5.1	2.2	4.0	2.1	3.3
Forbs	11.7	11.6	18.5	20.4	18.0	31.8
Shrubs	5.2	8.3	9.2	12.8	5.4	5.5

tion. Rough fescue declined sharply with increasing grazing pressure but actually increased under no grazing or the two lightest grazing pressures. Idaho fescue, sedges, and total forbs increased as rough fescue decreased. But what do these changes mean? Against what standard should final composition be compared? Composition in 1949—1954 is the result of subjecting the original vegetation of the area to moderate grazing by horses or cattle 1884 to 1943, then grazing it only during the dormant season 1944 to 1949. Therefore it may or may not represent "climax". It definitely does not represent stability, at least not when grazing is completely excluded, as indicated by comparison of 1949—1954 composition to that of 1976—1981. But 1976—1981 composition in the exclosures also is suspect as a standard; almost certainly, it has departed significantly from a climax formed under grazing by bison and other large herbivores. Composition after grazing for 32 years at a stocking rate of 1.6 AUM/ha appears to resemble the original composition more closely than does the composition under any other treatment. This might indicate that vegetation composition will stabilize at a stocking rate of about 1.6 AUM/ha. But is this generally true, or only for the weather patterns experienced from 1949 to 1981? Perhaps the most significant point illustrated by this data is that it is very difficult to isolate the causes of changes in botanical composition and more difficult to determine what standard such changes should be measured against.

Olson *et al.* (1985), noting the interaction between precipitation and stocking rate, developed surfaces describing the response of cover of different plant species to current and preceding year's precipitation under four grazing intensities (0.76, 0.92 or 1.24 ha per animal unit month or ungrazed). Data was taken 1932 to 1956 at the same site used by Houston and Woodward (1966). Plant species were divided into types according to their response to precipitation. Cover of Type 1 species increased in a wet year following a dry year or vice versa. Type 2 species increased following any combination of extremely high or extremely low precipitation in two successive years. Cover of Type 1 and 2 species decreased under average or consistent precipitation which favored other species. Type 3 species increased whenever precipitation in the current or preceding year was high and the effect was cumulative. Type 4 species increased as current year's precipitation increased, without regard to preceding year's precipitation. Response surfaces of some species differed with grazing intensity, and moderate grazing appeared to be optimum for several important species, including western wheatgrass, blue grama, buffalograss, and needleleaf sedge.

The effect of stocking rate on cover of key plant species on buffalograss-blue grama range in Kansas was determined by precipitation in the current and preceding years (Launchbaugh 1967). The drought of 1952—1956, with annual precipitation averaging 27% below normal, produced large reductions in cover of buffalograss and blue grama under heavy stocking, small reductions under moderate stocking, and almost none under light stocking (Fig. 4). Cover of buffalograss was about 72% at the beginning of the study; by 1965, after 20 years of grazing, it was 53, 47 and 36% under heavy, moderate and light stocking, respectively. Hyder *et al.* (1975), considering similar range, stated "Repeated heavy grazing failed to produce consistent effects from year to year . . . most of the significant effects . . . were derived from isolated events — interactions between weather and grazing."

In general, botanical composition may deteriorate significantly under "heavy" grazing but remain unchanged or improve under "light" grazing (to simplify discussion, movement away from assumed botanical composition at climax will be referred to as "deterioration" or "decrease in range condition" and movement toward climax as "improvement" or "increase in range condition"). Examples have been reported from blue grama-prairie sandreed-needleandthread range in Colorado (Sims *et al.* 1976); blue grama-buffalograss in Colorado (Klipple and Costello 1960); blue grama-western wheatgrass-needleandthread in Montana (Houston and Woodward 1966) and in North Dakota (Sarvis 1941); blue grama-western wheatgrass in Wyoming (Hart *et al.* 1986); Texas wintergrass-buffalo grass-curly mesquite in Texas (Heitschmidt *et al.* 1985); bluebunch wheatgrass-Sandberg bluegrass in Oregon (Skovlin *et al.* 1976); the puna or high-elevation grasslands of Peru (Wilcox *et al.* 1987); and the flooding pampa of Argentina (Sala *et al.* 1986). Under grazing, the vegetation composition

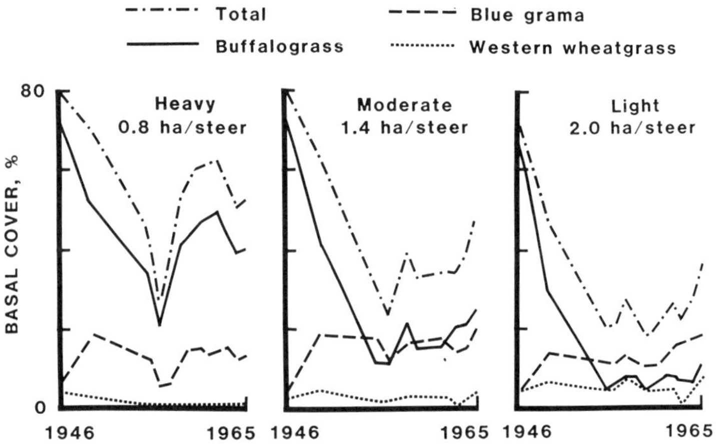

Figure 4. Changes in basal cover of buffalograss-blue grama range in response to stocking rate and yearly variation in precipitation (Launchbaugh 1967).

of riparian communities usually diverges from that at climax (Kauffman and Krueger 1984, Skovlin 1984, Platts and Raleigh 1984), but the number of plant species is often greater in grazed than in ungrazed riparian sites (Kaufmann *et al.* 1983a and b).

20.4.3 Grazing systems and botanical composition

The commercial range livestock industry generally endorses the principle of periodically resting a pasture or range, and the maxim of occasional deferment is even appearing in the grazing guides of Federal land management agencies. Yet research has produced no consensus on the relative merits of different grazing systems or strategies. Reviews of grazing systems research (Driscoll 1967, Hickey 1969, Shiflet and Heady 1971, Herbel 1974, Gammon 1978a) indicated range condition often increased but could be unaffected or decreased by rotation or deferred grazing as compared to season-long or continuous grazing. Such surveys can suggest trends, but unfortunately include good studies and bad and confound the effects of systems on range condition with the effects of range type, stocking rate, general level of management, and precipitation and other weather factors. Hickey (1969) reported that over half of the studies he reviewed (1) ran out of forage on rotation and abandoned it, (2) fed supplemental feed on some treatments, (3) changed class of livestock, stocking rate, or grazing period on some treatments but not on others, or (4) presented conclusions without supporting data.

Errors of interpretation creep in and become lodged in the public consciousness. Ratliff *et al.* (1972) claimed "a 10% increase in (carrying capacity) after one year of rest-rotation grazing", but had no control for comparison and ignored the fact that carrying capacity is not an objective measurement but a management decision. Rogler (1951) said of Sarvis's (1941) study "the rotation pasture grazed at one head to 5 acres (2 ha) . . . did not show the adverse effects evident in the pasture grazed continuously at the same rate." In fact, the rotation pasture was grazed at one head per 7 acres (2.8 ha), and the continuous pasture grazed at that rate was in equivalent condition.

A grazing system might be expected to improve or maintain range condition only if it produces a pattern of grazing resembling that under which the range evolved (Platou and Tueller 1985). High-intensity short-duration grazing resembles the way in which bison and pronghorn antelope grazed the Great Plains grasslands; rest-rotation grazing resembles the way in which mule deer, pronghorn antelope and bighorn sheep grazed the Great Basin shrub-steppe. Rest-rotation has not demonstrated any advantage over continuous grazing on the Great Plains, while high-intensity short-duration grazing may damage bunchgrasses in the Great Basin.

Rotation grazing might be expected to improve vigor and productivity of a desirable plant species if rotation reduces frequency or intensity of defoliation of that species during a critical period in its life cycle. Gammon and Roberts (1978a, b and c) found no significant difference in defoliation frequency and intensity between continuous and 6-paddock rotational grazing. Rogler (1951) reported 59% utilization under 3-paddock rest rotation grazing vs. 70% under continuous grazing, but Fisher and Marion (1951) reported utilization was less under continuous. Engle *et al.* (1983) reported defoliation was heavier under short-duration than under repeated seasonal grazing, and Test (1984) found little difference in frequency and intensity of defoliation of blue grama and western wheatgrass between continuous and 8-paddock short-duration rotation grazing.

Van Poollen and Lacey (1979) noted that rotational grazing achieves rest on part of the system by temporarily overstocking the rest or by reducing stocking rate, and implied that reduced stocking would be more effective in improving range condition. Cook (1966) and Hughes (1979) concluded rest cannot always compensate for the temporary overuse or use during a critical season which is an inescapable part of any rotation system. Hyder and Bement (1977) speculated that deferment was more favorable to range condition than was rotation grazing because deferment did not require as great a degree of overuse during the growing season. Stroud *et al.* (1985) reported that simulated rotational grazing of western wheatgrass was followed by a greater reduction in tiller numbers, aboveground and root biomass, and root carbohydrate reserves than was simulated continuous grazing. Lewis (1969) and Owensby *et al.* (1975) presented evidence that resting is generally unnecessary to maintain plant productivity if proper stocking rate is maintained. West (1964) noted that the greatest benefit to range condition results from maximum dispersal of grazing animals during the critical season, as is achieved in well-distributed continuous or season-long grazing.

Interactions of weather, system and stocking rate make it difficult to interpret the effects of systems. Arnold (1955) and Hyder *et al.* (1975) cautioned that if weather in the last year of a study is especially favorable to one system at the expense of the others, improvement under the favored system will be inflated.

Simulation modeling may provide a tool for understanding the short- and long-term effects of grazing, weather, soil and their complex interactions. Goodall (1979) reviewed the progress of rangeland modeling and concluded that many models adequately simulate effects which become apparent within a single growing season. Simulation of succession has been less successful. The model of Wilkins and Norton (1974) considers the impact of the preceding as well as the current year's grazing on productivity of each species or species group. The US/IBP Desert Biome Programme developed a model which assigns reserve carbon to storage organs of perennial plants, then uses this reserve to initiate growth the following season (Goodall 1979). Grazing modifies the level of reserve carbon of a species in the current year, thus modifying growth rate in the succeeding year. The SPUR model employs a similar approach (Wight 1983) as do several other models reviewed by Wight (1983) and Goodall (1979). Such models provide estimates of changes in annual production of major plant species or species group, but provide no information on changes in frequency, cover or plant demography. Noy-Meir (1979) emphasized the importance of spatial heterogeneity and immigration in re-establishing plant populations after they have been severely reduced by grazing or other stress.

Gammon (1978a) suggests that botanical composition under continuous grazing may or may not differ from that under a specialized grazing system for two apparently contradictory reasons. If the rotation schedule is too rigid, tied to the calendar rather than phenology and forage supply, or if the increased stocking rate recommended for some systems is rigidly maintained regardless of weather, excessive range damage and declining animal weights will result. On the other hand, stocking rate may be too low to produce any effect on vegetation under any system, or different stocking rates on different systems may hopelessly confound system with grazing pressure.

Recent studies confirm that rotation or deferred grazing usually has little effect on botanical composition of rangelands. Howell (1978) reported that use of short-duration grazing systems in Orange Free State, Republic of South Africa "resulted in the return and increase of rare and unknown grasses," but the importance of such species is doubtful. Walker and Scott (1968) reported tall grasses were more dominant under rotational grazing while short grasses were more dominant under continuous grazing in Tanzania, but the differences had no effect on cattle gains under moderate or heavy stocking. Denny *et al.* (1977) and Denny and Steyn (1977) reported no differences in botanical composition between 4-, 12- and 16-paddock rotational systems in Zimbabwe. Clatworthy (1984) reported "rapid rotational grazing . . . failed to produce the marked improvements in the grass cover so often claimed" in six years of grazing in Zimbabwe.

Martin and Cable (1974) found that winter rest produced no higher range condition than did continuous grazing in New Mexico, and summer rest produced lower condition when rest began so late that repeated grazing of new spring growth occurred. No differences were seen in pinyon-juniper vegetation of southern Wyoming after 5 years of season-long, 2-pasture deferred, or 4-pasture rest rotation grazing (Gibbens and Fisser 1975). Laycock and Conrad (1981) reported no differences among summer-long, summer-long in alternate years, and 3-paddock rest-rotation grazing in species composition or cover after sever years of grazing on Utah sagebrush-grass range. Smoliak (1960) saw no improvement in range condition after seven years of 2-pasture deferred-rotation grazing in Alberta.

Reece (1986) noted declines in prairie sandreed after four years of 7-paddock short-duration grazing in Nebraska, and a 41% reduction of carbohydrate reserves in the stem bases of blue grama below levels of plants grown in exclosures. One year of rest after three years of rotation grazing allowed plants to recover 46% of the difference in reserves. Frequency of major plant species differed little between continuous and deferred-rotation grazing on two major range sites in Texas (Heitschmidt *et al.* 1985). Texas wintergrass occurred more frequently under continuous grazing on clay loam sites but less frequently on loamy bottomland; the reverse was true for sideoats grama. Heitschmidt (1986) reported no evidence of changes in species composition under simulated 14- and 42-paddock short-duration rotation systems as compared to moderately-stocked continuous grazing in Texas, even though stocking rate was 16 to 63% higher under rotation. Dahl (1986) concluded "our attempts to evaluate the role of (short-duration grazing in Texas) in . . . rapid range improvement have been disappointing." No improvement in condition was found after four years of 16-paddock short-duration grazing compared to continuous grazing. Hart *et al.* (1986) and Samuel *et al.* (1987) noted no differences in species composition between season-long, rotationally deferred, and short-duration rotation grazing after five years in Wyoming. On grassland in Oregon's Blue Mountains (Skovlin *et al.* 1976), composition changed from year to year but changes were nearly identical from 1954 to 1966 under season-long and deferred rotation grazing. Responses of understory vegetation in ponderosa pine forest were similar.

20.5 Conclusions

The conflicting recommendations from the scientific and political communities concerning grazing management on rangelands may be more a fault of the current paradigms of range science than of grazing as such. Too often livestock grazing is viewed as an unnatural intrusion into the grassland ecosystem, when in fact it may be a continuation of the influences under which the ecosystem developed.

Clark (1956) concluded "(Following) a replacement of wild by domesticated grazing animals, the real change may have been small; where the new grazing pressure clearly was greater than before, or was virtually a new phenomenon, more substantial changes have occurred . . . (On the Great Plains) the weight of opinion is that the bison did not graze as heavily (as cattle and sheep), but it is equally true that pronghorn-grazing was probably much heavier than has been assumed . . . most of the kinds of changes ascribed to cattle- or sheep-grazing pressure must have occurred without it . . . it is possible to see in the depletion of productive forage, rainwash and gullying, the spread of shrubs at the expense of grass, or the increase of rodents or insects, not an example of destructive cultural exploitation of a valuable natural resource, but an expectable phase . . . which might have been observed many times in recent centuries. (In contrast), the original California grassland was 'grazed' only by deer, rabbits and rodents . . . with the invasion of man, his plants and his animals, came a change in the plant cover almost as substantial as that achieved by plowing or cutting and burning of forest."

Too often grazing by livestock is blamed for vegetation changes that are more directly related to weather and climate. Drought periods and times of low forage production are normal features of rangelands that may be associated with successional changes in the vegetation. Grazing during a drought and the immediate post-drought recovery season is a crucial interaction that demands more research in order to understand the processes involved and to plan appropriate management strategies. Drought provides an essential test of a range management plan. If it doesn't work during a drought, it doesn't work.

Plant cover and demography, which fluctuate less in response to short-term stresses, can provide a more reliable guide to vegetation status and successional trends than composition based on production estimates. Range analysis should separate data collection for carrying capacity (principally a function of forage biomass and the desired level of animal performance) from data collection for determining condition and trend (an expression of successional status).

Range condition classification based on ecological climax may be unrelated to suitability and long-term carrying capacity for livestock grazing or most other uses on most public land in the western United States. Range condition is a management concept that cannot be forced into a mold of natural ecological succession, except perhaps in certain situations such as the North American prairies. In order for land management to be justified by ecological arguments it is necessary to define range condition and trend in terms of specific vegetation types; their response to climate, grazing and other driving variables; and management goals, rather than in reference to presumed climax.

Provided grazing management (grazing pressure and animal distribution; grazing systems appear to have minor impact) is designed to conform to the ecological characteristics of the site, is implemented in order to achieve clearly defined objectives, and is operated with the normal flexibility required in a ranch enterprise, there is a sound ecological basis for development and maintenance of productive, stable rangeland ecosystems. Such ecosystems almost certainly will not be at climax; net productivity of climax ecosystems often is very near zero. They are more likely to be what Tueller (1975) calls disclimax; "a relatively long term (decades) steady state exhibiting maintenance of a given species composition and regulated by contemporary environmental factors." Grazing management must be considered as one of the contemporary environmental factors.

20.6 References

Agnew, W., D. W. Uresk, and R. M. Hansen. 1986. Flora and fauna associated with prairie dog colonies and adjacent ungrazed mixed-grass prairie in western South Dakota. J. Range Manage. 39: 135—139.

Albertson, F. W., and J. H. Weaver. 1942. History of the native vegetation of western Kansas during seven years of continuous drought. Ecol. Monogr. 12: 23—51.

Anderson, D. M. 1982. Seasonal grazing of semidesert tobosa rangeland in southern New Mexico. p. 137—142. *In*: D. D. Briske, and M. M. Kothmann (eds.) Proc. Natl. Conf. Grazing Mgt. Tech. Dept. Range Sci., Texas A & M Univ., College Station.

Anderson, J. E., and M. L. Shumar. 1986. Impacts of black-tailed jackrabbits at peak population densities on sagebrush-steppe vegetation. J. Range Manage. 39: 152—156.

Anderson, R. C. 1982. An evolutionary model summarizing the roles of fire, climate, and grazing animals in the origin and maintenance of grasslands: an end paper. p. 297—308. *In*: J. R. Estes, R. J. Tyrl, and J. N. Brunken (eds.) Grasses and grasslands — systematics and ecology. Univ. of Oklahoma Press, Norman.

Archer, S., and J. K. Detling. 1984. The effects of defoliation and competition on regrowth of tillers of two North American mixed-grass prairie graminoids. Oikos 43: 351—357.

Arnold, G. W. 1977. Effects of herbivores on arid and semi-arid rangelands — defoliation and growth of forage plants. p. 57—72. *In*: The impact of herbivores on arid and semi-arid rangelands. Proc. 2nd US/Austr. Rangel. Panel. Austr. Rangeland Soc., Perth.

Arnold, G. W., and J. L. Hill. 1972. Chemical factors affecting selection of food plants by ruminants. p. 71—101. *In*: J. B. Harborne (ed.) Phytochemical Ecology. Academic Press, New York.

Arnold, J. F. 1955. Plant life-form classification and its use in evaluating range conditions and trend. J. Range Manage. 8: 176—181.

Bartolome, J. W. 1984. Impacts of grazing intensity and grazing systems on vegetation composition and production. p. 917—925. *In*: NRC/NAS. Developing strategies for rangeland management. Westview Press, Boulder, CO.

Belsky, A. J. 1986a. Population and community processes in a mosaic grassland in the Serengeti, Tanzania. J. Ecol. 74: 841—856.

Belsky, A. J. 1986b. Revegetation of artificial disturbances in grasslands of the Serengeti National Park, Tanzania. I. Colonization of grazed and ungrazed plots. J. Ecol. 74: 419—437.

Blackburn, W. H. 1984. Impacts of grazing intensity and specialized grazing systems on watershed characteristics and responses. p. 927—983. *In*: NRC/NAS. Developing strategies for rangeland management. Westview Press, Boulder, CO.

Blocker, H. D. 1977. The impact of invertebrates as herbivores on arid and semi-arid rangeland in the United States. p. 357—376. *In*: The impact of herbivores on arid and semi-arid rangelands. Proc. 2nd US/Austr. Rangel. Panel. Austr. Rangeland Soc., Perth.

Bowns, J. E., and C. F. Bagley. 1986. Vegetation responses to long-term sheep grazing on mountain ranges. J. Range Manage. 39: 431—434.

Branson, F. A. 1985. Vegetation changes on western rangelands. Soc. Range Manage., Denver, CO.

Briske, D. D. 1987. Plant response to defoliation: morphological considerations and allocation priorities. Proc. 2nd Intl. Rangel. Cong.

Bryson, R. A. 1974. A perspective on climatic change. Science 184: 753—760.

status resulting from overgrazing and their consequences in plant communities of semi-arid areas. Proc. Aust. Ecol. Soc. 3: 28—28.

Clark, A. H. 1956. The impact of exotic invasion on the remaining New World mid-latitude grasslands. p. 737—762. *In*: W. L. Thomas, Jr., C. O. Sauer, M. Bates, and L. Mumford (eds.) Man's role in changing the face of the earth. Univ. Chicago Press, Chicago.

Clary, W. P., and R. C. Holmgren. 1987. Difficulties in interpretation of long-term vegetation trends in response to livestock grazing. p. 154—161. *In*: F. D. Provenza, J. T. Flinders, and E. D. McArthur (eds.) Proc. Symp. on Plant-Herbivore Interactions. Intermountain Res. Sta., For. Serv., USDA, Ogden, UT.

Clatworthy, J. N. 1984. The Charter Estate grazing trial. Results of the botanical analysis. Zimbabwe Agric. J. 81(2): 49—52.

Clements, F. E. 1928. Plant succession and indicators. H. W. Wilson & Co., New York.

Clements, F. E. 1934. The relict method in dynamic ecology. J. Ecol. 22: 39—68.

Connell, J. H., and R. O. Slatyer. 1977. Mechanisms of success in natural communities and their role in community stability and organization. Amer. Natur. 111: 1119—1144.

Cook, C. W. 1966. Carbohydrate reserves in plants. Utah Agr. Exp. Sta., Utah Resources Ser. 31.

Cook, C. W. 1971. Effects of season and intensity of use on desert vegetation. Utah Agr. Exp. Sta. Bull. 483.

Cowles, H. C. 1899. The ecological relations of the vegetation on the sand dunes of Lake Michigan. Bot. Gaz. 27: 95—117.

Cowling, S. W. 1977. Effects of herbivores on nutrient cycling and distribution in rangeland ecosystems. p. 277—298. *In*: The impact of herbivores on arid and semi-arid rangelands. Proc. 2nd US/Aust. Rangel. Panel. Austr. Rangel. Soc., Perth.

Crider, F. J. 1955. Root growth stoppage resulting from defoliation of grass. USDA Tech. Bull. 1102.

Dahl, B. E. 1986. The West Texas experience in short duration grazing. p. 103—116. *In*: J. A. Tiedeman (ed.) Short duration grazing. Coop. Ext. Serv., Washington State Univ., Pullman, WA.

Dansereau, P. 1957. Biogeography: an ecological perspective. Ronald Press, New York.

Denny, R. P., D. L. Barnes, and T. C. D. Kennan. 1977. Trials of multi-paddock grazing systems on veld. 1. An exploratory trial of systems involving twelve paddocks and one herd. Rhod. J. Agric. Res. 15: 11—23.

Denny, R. P., and J. S. H. Steyn. 1977. Trials of multi-paddock grazing systems on veld. 2. A comparison of a 16-paddocks-to-one-herd system with a four-paddocks-to-one-herd system using breeding cows. Rhod. J. Agric. Res. 15: 119—127.

Deregibus, V. A., R. A. Sanchez, J. J. Casal, and M. J. Trlica. 1985. Tillering responses to enrichment of red light beneath the canopy in a humid natural grassland. J. Appl. Ecol. 22: 199—206.

Detling, J. K., M. I. Dyer, and D. T. Winn. 1979a. Effect of simulated grasshoppers grazing on CO2 exchange rates of western wheatgrass leaves. J. Econ. Entomol. 72: 403—406.

Detling, J. K., M. I. Dyer, and D. T. Winn. 1979b. Net photosynthesis, root respiration, and regrowth of *Bouteloua gracilis* following simulated grazing. Oecologia 41: 127—134.

Donaldson, C. H. 1967. The immediate effects of the 1964/66 drought on the vegetation of specific study areas in the Vryberg District. Proc. Grassl. Soc. S. Africa 2: 137—141.

Dorn, R. D. 1986. The Wyoming landscape, 1805—1878. Mountain West Publ., Cheyenne, WY.

Driscoll, R. S. 1967. Managing public rangelands: effective livestock grazing practices and systems for National Forests and National Grasslands. USDA Forest Serv. Agr. Info. Bull. 315.

Dyksterhuis, E. J. 1949. Condition and management of rangeland based on quantitative ecology. J. Range Manage. 2: 104—115.

Ellison, L. 1960. The influence of grazing on plant succession of rangelands. Bot. Rev. 26: 1—78.

Enevoldsen, M. E., and J. K. Lewis. 1978. Effect of range site and range condition on height and location of the shoot apex in vegetative shoots of western wheatgrass. p. 387—391. *In*: D. N. Hyder (ed.) Proc. 1st Intl. Rangel. Cong. Soc. Range Manage., Denver, CO.

Engle, D. M., J. K. Lewis, and L. S. Bilger. 1983. Tiller defoliation patterns under high performance short duration and repeated seasonal grazing systems. SRM Abstr. 36: 147.

Fisher, C. E., and P. T. Marion. 1951. Continuous and rotation grazing on buffalo and tobosa grassland. J. Range Manage. 4: 48—51.

Foran, B. D., N. M. Tainton, and P. de V. Booysen. 1978. The development of a method for assessing veld condition in three grassland types in Natal. Proc. Grassl. Soc. S. Africa 13: 27—33.

Frith, H. J. 1970. The herbivorous wild animals. p. 74—83. *In*: R. M. Moore (ed.) Australian grasslands. Australian Natl. Univ. Press, Canberra.

Fremont, J. C. 1843. A report on an exploration of the country lying between the Missouri River and the Rocky Mountains, on the line of the Kansas and Great Platte Rivers. Senate Doc. 243, 27th Congress. Washington, DC.

Gammon, D. M. 1978a. A review of experiments comparing systems of grazing management on natural pastures. Proc. Grassl. Soc. S. Africa 13: 75—82.

Gammon, D. M. 1978b. Patterns of defoliation during continuous and rotational grazing of rangeland by cattle. p. 603—605. *In*: D. N. Hyder (ed.) Proc. 1st Intl. Rangel. Cong. Soc. Range Manag., Denver, CO.

Gammon, D. M., and B. R. Roberts. 1978a. Patterns of defoliation during continuous and rotational grazing of the Matopos sandveld of Rhodesia. I. Selectivity of grazing. Rhod. J. Agr. Res. 16: 117—131.

Gammon, D. M., and B. R. Roberts. 1978b. Patterns of defoliation during continuous and rotational grazing of the Matopos sandveld of Rhodesia. 2. Severity of defoliation. Rhod. J. Agr. Res. 16: 133—145.

Gammon, D. M., and B. R. Roberts. 1978c. Patterns of defoliation during continuous and rotational grazing of the Matopos sandveld of Rhodesia. 3. Frequency of defoliation. Rhod. J. Agr. Res. 16: 147—164.

Gamogoun, N. D., R. P. Smith, M. K. Wood, and R. D. Pieper. 1984. Soil, vegetation, and hydrologic responses to grazing management at Fort Stanton, New Mexico. J. Range Manage. 37: 538—541.

Gibbens, R. P., and H. G. Fisser. 1975. Influence of grazing management systems on vegetation in the Red Desert region of Wyoming. Wyoming Agr. Exp. Sta. Sci. Monog. 29.

Gifford, G. F., and R. H. Hawkins. 1978. Hydrologic impact of grazing on infiltration: a critical review. Water Resources Res. 14: 305—313.

Goodall, D. W. 1979. The modeling of arid ecosystem dynamics. p. 385—409. *In*: D. W. Goodall, and R. A. Perry (eds.) Arid-land ecosystems: structure, functioning and management, vol. 2. Cambridge Univ. Press, Cambridge.

Haley, J. E. 1936. Charles Goodnight. Houghton Mifflin Co., New York.

Hart, R. H., and E. F. Balla. 1982. Forage production and removal from western and crested wheatgrasses under grazing. J. Range Manage. 35: 362—366.

Hart, R. H., and G. E. Carlson. 1975. Agricultural implications of climatic change: agronomic implications: forages. p. 4—163 to 4—173. *In*: Impacts of climatic change on the biosphere. CIAP Monogr. 5, Part 2 — Climatic effects. Dept. of Transportation, Washington, DC.

Hart, R. H., G. E. Schuman, M. J. Samuel, and M. A. Smith. 1986. Time-controlled grazing research in Wyoming. p. 73—82. *In*: J. A. Tiedeman (ed.) Short duration grazing. Coop. Ext. Serv., Washington State Univ., Pullman, WA.

Heady, H. F. 1975. Structure and function of climax. p. 73—80. *In*: D. N. Hyder (ed.) Arid shrublands. Proc. 3rd Wkshp. US/Aust. Rangel. Panel. Soc. Range Manage., Denver, CO.

Heitschmidt, R. K. 1986. Short duration grazing at the Texas Experimental Ranch. p. 91—102. *In*: J. A. Tiedeman, (ed.) Short duration grazing. Coop. Ext. Serv., Washington State Univ., Pullman.

Heitschmidt, R. K., S. L. Dowhower, R. A. Gordon and D. L. Price. 1985. Response of vegetation to livestock grazing at the Texas Experimental Ranch. Texas Agr. Exp. Sta. Bull. 1515.

Herbel, C. H. 1974. A review of research related to development of grazing systems on native ranges of the western Unived States. p. 138—149. *In*: K. W. Kreitlow, and R. H. Hart (eds.) Plant morphogenesis as the basis for scientific management of range resources. USDA Misc. Publ. 1271.

Herbel, C. H. 1984. Successional patterns and productivity potentials of the range vegetation in the warm, arid portions of the Southwestern United States. p. 1333—1365. *In*: NRC/NAS. Developing strategies for rangeland management. Westview Press, Boulder, CO.

Hickey, W. C. Jr. 1969. A discussion of grazing management systems and some pertinent literature (Abstracts and Excerpts) 1895—1966. USDA Forest Serv. Rep. (unnumbered).

Hodgkinson, K. C., and H. G. Baas Becking. 1977. Effect of defoliation on root growth of some arid zone perennial plants. Aust. J. Agric. Res. 29: 31—42.

Hormay, A. L. 1970. Principles of rest-rotation grazing and multiple-use land management. USDA Forest Serv. Training Text 4.

Hornaday, W. T. 1889. Extermination of the American bison. Rep. Natl. Museum (1886—1887). US Govt. Printing Off., Washington, DC.

Houston, W. R., and R. R. Woodward. 1966. Effects of stocking rates on range vegetation and beef cattle production in the Northern Great Plains. USDA Tech. Bull. 1357.

Howell, L. N. 1978. Development of multi-camp grazing systems in the southern Orange Free State, Republic of South Africa. J. Range Manage. 31: 459—465.

Hughes, L. E. 1979. Rest-rotation grazing on the Arizona strip: an observation. Rangelands 1: 106—108.

Humphrey, R. R. 1947. Range forage evaluation by the range condition method. J. For. 45: 10—16.

Hutchings, S. S., and G. Stewart. 1953. Increasing forage yields and sheep production on intermountain winter ranges. USDA Circ. 925.

Hyder, D. N. 1972. Defoliation in relation to vegetative growth. p. 304—317. *In*: The biology and utilization of grasses. Academic Press, New York.

Hyder, D. N., and R. E. Bement. 1977. The status of seasons of grazing and rest in grazing management. p. 73—82. *In*: The impact of herbivores on arid and semi-arid rangeland. Proc. 2nd US/Australia Rangel. Panel. Aust. Rangel. Soc., Perth.

Hyder, D. N., R. E. Bement, E. E. Remmenga, and D. F. Hervey. 1975. Ecological responses of native plants and guidelines for management of shortgrass range. US Dept. Agric. Tech. Bull. 1503.

Jameson, D. A. 1963. Responses of individual plants to harvesting. Bot. Rev. 29: 532—594.

Jeffries, R. L. 1984. The phenotype: its development, physiological constraints and environmental signals. p. 347—358. *In*: R. Dirzo, and J. Sarukhan (eds.) Perspectives on plant population ecology. Sinaver Assoc., Inc., Sunderland, MA.

Kauffman, J. B., and W. C. Krueger. 1984. Livestock impacts on riparian ecosystems and streamside management implications . . . a review. J. Range Manage. 37: 430—438.

Kauffman, J. B., W. C. Krueger, and M. Vavra. 1983a. Effects of cattle grazing on riparian plant communities. J. Range Manage. 36: 685—691.

Kauffman, J. B., W. C. Krueger, and M. Vavra. 1983b. Impacts of cattle on streambanks in northeastern Oregon. J. Range Manage. 36: 683—685.

Klemmedsen, J. O. 1977. Physical effects of herbivores on arid and semi-arid rangeland ecosystems. p. 187—210. *In*: The impact of herbivores on arid and semi-arid rangelands. Proc. 2nd US/Australia Rangel. Panel. Aust. Rangel. Soc., Perth.

Klipple, G. E., and D. F. Costello. 1960. Vegetation and cattle responses to different intensities of grazing on shortgrass ranges on the Central Great Plains. USDA Tech. Bull. 1216.

Lacey, J. R., and H. W. van Poollen. 1981. Comparison of herbage production on moderately grazed and ungrazed western ranges. J. Range Manage. 34: 210—212.

Larson, F. 1940. The role of bison in maintaining the short grass plains. Ecol. 21: 113—121.

Launchbaugh, J. L. 1967. Vegetation relationships associated with intensity of summer grazing on a clay upland range site in the Kansas 20- to 24-inch precipitation zone. Kansas Agr. Exp. Sta. Tech. Bull. 154.

Laycock, W. A. 1978. Co-evolution of poisonous plants and large herbivores on rangelands. J. Range Manage. 31: 335—342.

Laycock, W. A., and P. W. Conrad. 1967. Effect of grazing on soil compaction as measured by bulk density on a high elevation cattle range. J. Range Manage. 20: 136—140.

Lee, K. E. 1977. Physical effects of herbivores on arid and semi-arid rangeland ecosystems. p. 173—186. *In*: The impact of herbivores on arid and semi-arid rangelands. Proc. 2nd US/Aust. Rangel. Panel. Aust. Rangel. Soc., Perth.

Le Houerou, H. N., and C. H. Hoste. 1977. Rangeland production and annual rainfall relations in the Mediterranean Basin and in the African Sahelo-Sudanian zone. J. Range Manage. 30: 181—189.

Lewis, J. K. 1969. Range management viewed in the ecosystem framework. p. 97—187. *In*: G. M. Van Dyne (ed.) The ecosystem concept in natural resource management. Academic Press. New York.

Lewis, M. 1814 (re-issued 1961). The Lewis and Clark expedition. J. B. Lippincott Co., Philadelphia.

Lorenz, R. J. 1974. Effects of grazing on microenvironment of United States rangelands. p. 186—196. *In*: K. W. Kreitlow, and R. H. Hart (eds.) Plant morphogenesis as the basis for scientific management of range resources. USDA Misc. Publ. 1271.

Madany, M. H., and N. E. West. 1984. Vegetation of two relict mesas in Zion National Park. J. Range Manage. 37: 456—461.

Marshall, J. K. 1974. Effects of stock on micro-environments in Australian rangelands. p. 167—185. *In*: K. W. Kreitlow, and R. H. Hart (eds.) Plant morphogenesis as the basis for scientific management of range resources. USDA Misc. Publ. 1271.

Martin, S. C., and D. R. Cable. 1974. Managing semidesert grass-shrub ranges: vegetation responses to precipitation, grazing, soil texture, and mesquite control. U.S. Dept. Agric. Forest Serv. Tech. Bull. 1480.

McNaughton, S. J. 1983. Compensatory plant growth as a response to herbivory. Oikos 40: 329—336.

McNaughton, S. J. 1984. Grazing lawns: animal herds, plant form, and coevolution. Amer. Natur. 24: 863—886.

McNaughton, S. J., M. B. Coughenour, and L. L. Wallace. 1982. Interactive processes in grassland ecosystems. p. 167—193. *In*: J. R. Estes, R. J. Tyrl, and J. N. Brunker (eds.) Grasses and grasslands — systematics and ecology. Univ. of Oklahoma Press. Norman, OK.

McNaughton, S. J., and J. L. Tarrants. 1983. Grass leaf silicification: natural selection for an inducible defense against herbivores. Proc. Natl. Acad. Sci. 80: 790—791.

Menke, J. W., and M. J. Trlica. 1981. Carbohydrate reserve, phenology and growth cycles of nine Colorado species. J. Range Manage. 34: 269—277.

Menke, J. W., and M. J. Trlica. 1983. Effects of single and sequential defoliations on the carbohydrate reserves of four range species. J. Range Manage. 36: 70—74.

Morfi, Fray Juan Augustin de. 1777. Viaje de Indios y Diario del Neuvo Mexico. 1935 ed., V. A. Robles ed. Mexico, D. F.

Morris, R. M. 1969. The pattern of grazing in continuously grazed swards. J. Brit. Grassl. Soc. 24: 65—76.

Newman, J. C. 1974. Effects of past grazing in derermining range management principles in Australia. p. 197—206. *In*: K. W. Kreitlow, and R. H. Hart (eds.) Plant morphogenesis as the basis for scientific management of range resources. USDA Misc. Publ. 1271.

Newsome, R. E., and L. K. Corbett. 1977. The effects of native, feral and domestic animals on the productivity of Australian rangelands. p. 331—356. *In*: The impact of herbivores on arid and semi-arid rangelands. Proc. 2nd US/Aust. Rangel. Panel. Aust. Rangel. Soc., Perth.

Norton, B. E. 1978. The impact of sheep grazing on long-term successional trends in salt desert shrub vegetation of southwestern Utah. p. 610—612. *In*: D. N. Hyder (ed.) Proc. 1st Intl. Rangel. Cong. Soc. Range Manage., Denver, CO.

Norwine, J. 1978. Twentieth-century semi-arid climates and climatic fluctuations in Texas and northeastern Mexico. J. Arid Environ. 1: 313—325.

Nowak, R. S., and M. M. Caldwell. 1984. A test of compensatory photosynthesis in the field: implications for herbivory tolerance. Oecologia 61: 311—318.

Noy-Meir, I. 1973. Desert ecosystems: environment and producers. Ann. Rev. Ecol. & Syst. 4: 25—51.

Noy-Meir, I. 1979. Spatial effects in modeling of arid ecosystems. p. 411—432. *In*: D. W.Goodall, and R. A. Perry (eds.) Arid-land ecosystems: structure, functioning and management, vol. 2. Cambridge Univ. Press, Cambridge.

Olson, K. C., R. S. White, and B. W. Sindelar. 1985. Response of vegetation of the Northern Great Plains to precipitation amount and grazing intensity. J. Range Manage. 38: 357—361.

Owensby, C. E., E. F. Smith, and H. Dieta. 1975. Intensive early stocking of bluestem range. Proc. Bluestem Range Clinic, Kansas Agr. Exp. Sta. Rep. Prog. 224: 21—26.

Packard, R. L. 1977. Effects of herbivores on seed usage, dispersal and reproduction with reference to mammals. p. 211—226. *In*: The impact of herbivores on arid and semi-arid rangelands. Proc. 2nd US/Aust. Rangel. Panel. Aust. Rangel. Soc., Perth.

Paige, K. N., and T. G. Whitham. 1987. Overcompensation in response to mammalian herbivory. Amer. Natur. 129: 407—416.

Painter, E. L., and J. K. Detling. 1981. Effects of defoliation on net photosynthesis and regrowth of western wheatgrass. J. Range Manage. 34: 68—71.

Perry, R. A. 1970. Arid shrublands and grasslands. p. 246—259. *In*: Moore, R. M. (ed.) Australian grasslands. Australian Natl. Univ. Press, Canberra.

Pieper, R. D. 1977. Effects of herbivores on nutrient cycling and distribution. p. 249—276. *In*: The impact of herbivores on arid and semi-arid rangelands. Proc. 2nd US/Aust. Rangel. Panel. Aust. Rangel. Soc., Perth.

Pieper, R. D., G. B. Donart, E. E. Parker, and J. D. Wallace. 1978. Livestock and vegetational response to continuous and 4-pasture, 1-herd grazing systems in New Mexico. p. 560—562. *In*: D. N. Hyder (ed.) Proc. 1st Intl. Rangel. Cong. Soc. Range Manage., Denver, CO.

Platou, K. A., and P. T. Tueller. 1985. Evolutionary implications for grazing management systems. Rangel. 7: 57—61.

Platts, W. S., and R. F. Raleigh. 1984. Impacts of grazing on wetlands and riparian habitat. p. 1105—1117. *In*: NRC/NAS. Developing strategies for rangeland management. Westview Press, Boulder, CO.

Poulton, C. E. 1959. Ecology for the land manager. A clarification of ecological principles, concepts, and philosophies. Oregon State Univ. Range Sci. Mimeo.

Pound, R., and F. E. Clements. 1898. The vegetation regions of the prairie province. Bot. Gaz. 25: 381—394.

Pound, R., and F. E. Clements. 1900. The phytogeography of Nebraska. Bot. Seminar Univ. Nebraska, 2nd ed.

Ratliff, R. D., J. N. Reppert, and R. J. McConnen. 1972. Rest-rotation grazing at Harvey Valley . . . range health, cattle gains, costs. USDA Forest Serv. Res. Paper PSW-77.

Rattray, J. W. 1960. The grass cover of Africa. FAO Agric. Studies No. 49. Food & Agric. Org. of United Nations, Rome.

Rauzi, F., and R. L. Lang. 1967. Effect of grazing intensity on vegetation and sheep gains on shortgrass rangeland. Wyoming Agr. Exp. Sta. Sci. Monogr. 4.

Reardon, P. O., and L. B. Merrill. 1978. Response of sideoats grama grown in different soils to addition of thiamine and bovine saliva. p. 396—397. *In*: D. N. Hyder (ed.) Proc. 1st Intl. Rangel. Congr. Soc. Range Manage., Denver, CO.

Reece, P. E. 1986. Short duration grazing research and case studies in Nebraska. p. 39—71. *In*: J. A. Tiedeman (ed.) Short duration grazing. Coop. Ext. Serv., Washington State Univ., Pullman.

Reed, M. J., and R. A. Peterson. 1961. Vegetation, soil, and cattle responses to grazing on Northern Great Plains range. USDA Forest Serv. Tech. Bull. 1252.

Reichardt, P., T. Clausen, and J. Bryant. 1987. Plant secondary metabolites as feeding deterrents to vertebrate herbivores. p. 37—42. *In*: F. D. Provenza, J. T. Flinders, and E. D. McArthur (eds.) Proc. Symp. on Plant-Herbivore Interactions. Intermountain Res. Sta., Forest Serv., USDA, Ogden, UT.

Reichman, O. J., I. Prakash, and V. Roig. 1979. Food selection and consumption. p. 681—716. *In*: D. W. Goodall, and R. A. Perry (eds.) Arid-land ecosystems: structure, functioning and management, vol. 1. Cambridge Univ. Press, Cambridge.

Reitan, C. H., and C. R. Green. 1968. Weather and climate of desert environments. p. 19—92. *In*: W. G. McGinnies, B. J. Goldman, and P. Paylore (eds.) Deserts of the world. Univ. of Arizona Press, Tucson.

Rogler, G. A. 1951. A twenty-five year comparison of continuous and rotation grazing in the northern plains. J. Range Manage. 4: 35—41.

Roseveare, G. M. 1948. The grasslands of Latin America. Imp. Bur. Pastures & Field Crops Bull. 36. William Lewis Printers Ltd., Cardiff.

Ruess, R. W., and S. J. McNaughton. 1984. Urea as a promotive coupler of plant-herbivore interactions. Oecologia 63: 331—337.

Ryle, G. J. A., and C. E. Powell. 1975. Defoliation and regrowth in the Graminaceous plant: the role of current assimilate. Ann. Bot. 39: 297—310.

St. Clair, L. L., B. L. Webb, J. R. Johansen, and G. T. Nebeker. 1984. Cryptogamic soil crusts: enhancement of seedling establishment in disturbed and undisturbed areas. Reclam. Reveg. Res. 3: 129—136.

Sala, O. E., M. Oesterheld, R. J. C. Leon, and A. Soriano. 1986. Grazing effects upon plant community structure in subhumid grasslands of Argentina. Vegetatio 67: 27—32.

Sampson, A. W. 1917. Succession as a factor in range management. J. For. 15: 593—596.

Sampson, A. W. 1951. A twenty-five year comparison of continuous and rotational grazing in North America. J. Range Manage. 4: 19—24.

Samuel, M. J. 1987. Botanical composition changes from six years of low-rate nitrogen fertilization. p. 65. Proc. 2nd Intl. Rangel. Congr.

Samuel, M. J., R. H. Hart, and P. S. Test. 1987. Vegetation changes from five years of grazing systems. SRM Abstr. 40: 188.

Sarvis, J. T. 1941. Grazing investigations on the Northern Great Plains. North Dakota Agr. Exp. Sta. Bull. 308.

Savory, A. 1983. The Savory Grazing Method or Holistic Resource Management. Rangel. 5: 155—159.

Schmutz, E. M. 1978. Let's put range into range management. Rangeman's J. 5: 185—187.

Senft, R. L. 1987. Domestic herbivore foraging tactics and landscape pattern. p. 137—140. *In*: F. D., Provenza, J. T. Flinders, and E. D. McArthur (eds.) Proc. Symp. on Plant-Herbivore Interactions. Intermountain Res. Sta., For. Serv., USDA, Ogden, UT.

Shantz, H. L., and B. L. Turner. 1958. Photographic documentation of vegetational changes in Africa over a third of a century. Univ. Arizona Coll. Agric. Rep. 169.

Sharon, D. 1972. The spottiness of rainfall in a desert area. J. Hydrol. 17: 161—175.

Sharp, L. A., and K. D. Sanders. 1978. Rangeland resources of Idaho: a basis for development and improvement. Univ. of Idaho For. Wildl. & Range Exp. Sta. Contrib. 141.

Shiflet, T. N., and H. F. Heady. 1971. Specialized grazing systems: their place in range management. USDA Soil Cons. Serv. SCS-TP-152.

Sims, P. L., B. E. Dahl, and A. H. Denham. 1976. Vegetation and livestock response at three grazing intensities on sandhill rangeland in eastern Colorado. Colorado State Univ. Exp. Sta. Tech. Bull. 130.

Sims, P. L., and D. D. Dwyer. 1965. Pattern of retrogression of native vegetation in north central Oklahoma. J. Range Manage. 18: 20—25.

Sims, P. L., J. S. Singh, and W. K. Laurenroth. 1978. The structure and function of ten western North American grasslands. I. Abiotic and vegetational characteristics. J. Ecol. 66: 251—285.

Sims, P. L., R. E. Sosebee, and D. M. Engle. 1982. Plant and vegetation responses to grazing management. p. 4—31. *In*: D. D. Briske, and M. M. Kothmann (eds.) Proc. Natl. Conf. Grazing Mgt. Tech. Texas A & M Univ., College Station, TX.

Skiles, J. W. 1984. A review of animal preference. p. 153—211. *In*: NRC/NAS. Developing strategies for rangeland management. Westview Press, Boulder, CO.

Skovlin, J. M. 1984. Impacts of grazing on wetlands and riparian habitat: a review of our knowledge. p. 1001—1103. *In*: NRC/NAS. Developing strategies for rangeland management. Westview Press, Boulder, CO.

Skovlin, J. M., R. W. Harris, G. S. Strickler, and G. A. Garrison. 1976. Effects of cattle grazing methods on ponderosa pine-bunchgrass range in the Pacific Northwest. USDA For. Serv. Tech. Bull. 1531.

Smith, E. L. 1979. A critical evaluation of the range condition concept. p. 266—267. *In*: D. N. Hyder (ed.) Proc. 1st Intl. Rangel. Congr. Soc. Range Management, Denver, CO.

Smith, J. G. 1899. Grazing problems in the Southwest and how to meet them. USDA Div. Agrost. Bull. 16.

Smith, J. S. 1827. Letter to William Clark. Missouri Republican 11 Oct. 1827. St. Louis, MO.

Smoliak, S. 1960. Effects of deferred-rotation and continuous grazing on yearling steer gains and shortgrass prairie vegetation of south-eastern Alberta. J. Range Manage. 13: 239—243.

Smoliak, S. 1986. Influence of climate conditions on production of *Stipa-Bouteloua* prairie over a 50-year period. J. Range Manage. 39: 100—103.

Stout, D. G., and B. Brooke. 1985. Rhizomes and roots below clipped pinegrass tillers have a higher percent carbohydrate when attached to other nonclipped tillers. J. Range Manage. 38: 276—277.

Stroud, D. O., R. H. Hart, M. J. Samuel, and J. D. Rodgers. 1985. Western wheatgrass responses to simulated grazing. J. Range Manage. 38: 103—108.

Svejcar, T., and S. Christiansen. 1987. Grazing effects on water relations of Caucasian bluestem. J. Range Manage. 40: 15—18.

Test, P. S. 1984. Vegetation and livestock response to three grazing systems — continuous, rotationally deferred and short-duration rotation. Ph.D. Diss., Univ. of Wyoming. Diss. Abstr. Intl. 46(2B): 385.

Thurow, T. L., W. H. Blackburn, and C. A. Taylor. 1986. Hydrologic characteristics of vegetation types as affected by livestock grazing systems, Edwards Plateau, Texas. J. Range Manage. 39: 505—509.

Trlica, M. J. 1977. Effects of frequency and intensity of defoliation on primary producers of arid and semi-arid rangelands. p. 27—56. *In*: The impact of herbivores on arid and semi-arid rangeland. Proc. 2nd US/Australia Rangel. Panel. Aust. Rangel. Soc., Perth.

Trlica, M. J., and C. W. Cook. 1971. Defoliation effects on carbohydrate reserves of desert species. J. Range Manage. 24: 418—425.

Trlica, M. J., and J. S. Singh. 1979. Translocation of assimilates and creation, distribution, and utilization of reserves. p. 537—571. *In*: R. A. Perry, and D. W. Goodall (eds.) Arid-land ecosystems: structure, functioning, and management, vol. 1. Cambridge Univ. Press, Cambridge.

Tueller, P. T. 1975. Secondary succession, disclimax, and range condition standards in desert shrub vegetation. p. 56—65. *In*: D. N. Hyder (ed.) Arid shrublands. Proc. 3rd Wkshp. US/Aust. Rangel. Panel. Soc. Range Manage., Denver, CO.

Tueller, P. T., and K. A. Platou. 1987. Range vegetation condition classification problems. Proc. 2nd Intl. Rangel. Cong., p. 540—541.

Vale, T. R. 1975. Presettlement vegetation in the sagebrush-grass area of the intermountain West. J. Range Manage. 28: 32—36.

Van Poolen, H. W., and J. R. Lacey. 1979. Herbage response to grazing systems and stocking intensities. J. Range Manage. 32: 250—253.

Vorster, L. F. 1975. The influence of prolonged seasonal defoliation on veld yields. Proc. Grassl. Soc. S. Africa 10: 119—122.

Wagner, F. H. 1976. Integrating and control mechanisms in arid and semi-arid ecosystems: considerations for impact assessment. *In*: Proc. Symp. Biol. Eval. Environ. Impact. Council on Environ. Quality, Washington, DC.

Wahl, E. W., and T. L. Lawson. 1970. The climate of the midnineteenth-century United States compared to the current normals. Mon. Weather Rev. 98: 259—265.

Walker, B., and G. D. Scott. 1968. Grazing experiments at Ukiriguru, Tanzania. I. Comparisons of rotational and continuous systems on natural pastures of hardpan soils. E. Afr. Agric. For. J. 34: 224—234.

Wallen, C. C., and M. D. Gwynne. 1978. Drought — a challenge to range management. p. 21—31. *In*: D. N. Hyder (ed.) Proc. 1st Intl. Rangel. Cong. Soc. Range Manage., Denver, CO.

Walter, H. 1963. Productivity of vegetation in arid countries, the savannah problem and bush encroachment after overgrazing. Paper RT.9/III/2, Intl. Union Cons. Nature & Natural Resources. Morges, Switzerland.

Weaver, J. E., and F. W. Albertson. 1936. Effects of the great drought on the prairies of Iowa, Nebraska, and Kansas. Ecology 17: 567—639.

Weaver, J. E., and F. W. Albertson. 1944. Nature and degree of recovery of grassland from the great drought of 1933 to 1940. Ecol. Monogr. 14: 393—479.

Weaver, J. E., and W. W. Hanson. 1941. Native midwestern pastures — their origin, composition, and degeneration. Nebraska Conservation Bull. 22.

Webb, W., S. Szarek, W. Laurenroth, R. Kinerson, and M. Smith. 1978. Primary productivity and water use in native forest, grassland, and desert ecosystems. Ecology 59: 1239—1247.

West, N. E. 1984. Successional patterns and productivity potentials of pinyon-juniper ecosystems. p. 1301—1332. *In*: NRC/NAS. Developing strategies for rangeland management. Westview Press, Boulder, CO.

West, O. 1964. Pasture management and improvement in Rhodesia. Rhod. Agric. J. 61: 119—124.

Westoby, M. 1980. Elements of a theory of vegetation dynamics in arid rangelands. Israel J. Bot. 28: 169—194.

White, L. M. 1973. Carbohydrate reserves of grasses: a review. J. Range Manage. 26: 13—18.

Whitman, W. C. 1974. Influence of grazing on the microclimate of mixed grass prairie. p. 207—218. *In*: K. W. Kreitlow, and R. H. Hart (eds.) Plant morphogenesis as the basis for scientific management of range resources. USDA Misc. Publ. 1271.

Whyte, R. O. 1957. The grassland and fodder resources of India. Indian Council of Agric. Res. Sci. Monog. 22. Indian Council of Agric. Res., New Delhi.

Wight, J. R. (ed.) 1983. SPUR — Simulation of Production and Utilization of Rangelands: a rangeland model for management and research. USDA Misc. Publ. 1431.

Wilcox, B. P., F. C. Bryant, and V. B. Fraga. 1987. An evaluation of range condition on one range site in the Andes of Central Peru. J. Range Manage. 40: 41—45.

Wilkin, D. C., and B. E. Norton. 1974. Resource management. US/IBP Desert Biome Res. Memo. RM 74—67. Utah State Univ., Logan.

Williams, O. B. 1977. Reproductive wastage in rangeland plants, with particular reference to the role of mammals. p. 227—248. *In*: The impact of herbivores on arid and semi-arid rangelands. Proc. 2nd US/Aust. Rangel. Panel. Aust. Rangel. Soc., Perth.

Willms, W., A. W. Bailey, and A. McLean. 1980. Effect of burning or clipping *Agropyron spicatum* in the autumn on the spring foraging behavior of mule deer and cattle. J. Appl. Ecol. 17: 69—84.

Willms, W. D., S. Smoliak, and J. F. Dormaar. 1985. Effects of stocking rate on a rough fescue grassland vegetation. J. Range Manage. 38: 220—225.

Wilson, A. D., and G. J. Tupper. 1982. Concepts and factors applicable to the measurement of range condition. J. Range Manage. 35: 684—689.

Wood, M. K., G. B. Donart, and M. Weltz. 1986. Comparative infiltration rates and sediment production on fertilized and grazed blue grama rangeland. J. Range Manage. 39: 371—374.

21. Understanding fire ecology for range management

ARTHUR W. BAILEY

21.1. Abstract

Fire is a natural part of most rangeland ecosystems because of the periodic availability of flammable fine fuels, dry weather and ignition sources. The expanding human population is tending to limit the role of fire in many areas because herbivores have eaten the fuel source or prescribed fires are perceived to be too risky to humans or their animals.

The principle rangeland ecosystems of the world include tundra, desert, temperate arid forest, grassland, sclerophyllic woodland, tropical savanna and various kinds of temporary rangeland. Fire occurs in all of these ecosystems but it's frequency and effect ranges from minor importance in most tundra and desert ecosystems to major ecological importance in humid grasslands and savannas.

Rangeland plants are adapted to normal burning patterns. The high frequency of fire in historical times has required that range vegetation have mechanisms enabling it to survive fire. The plants avoid, escape or endure fire.

Burning has a variety of effects on ecosystems. The fire regime in particular influences the response of plants and animals. Fire is primarily used on rangeland to manipulate plant populations, maintain habitat for animals, improve forage quality, improve forage production, and fireproof vegetation preventing damaging wildfires.

Prescribed burning has a role to play for many rangeland managers. The steps required to conduct any prescribed fire are preburn planning, the burning operation and postburn management.

21.2. Introduction

Rangelands occupy about 40% of the earth's land surface (Branson,

Tueller, P. T. (ed.), Vegetation science applications for rangeland analysis and management.
ISBN 90-6193-195-9.
© 1988, *Kluwer Academic Publishers, Dordrecht, Boston, London.*

Gifford and Owen 1972) and they predominate in the arid to semiarid regions of the world. Soil moisture is often limiting, droughts are frequent and are highly variable in intensity when they strike. Consequently vegetative cover is essential to minimize soil erosion and maximize the infiltration of water (Warren 1987) to maintain productive plant populations. It seems an anomaly that fire, such an historically important environmental factor on many of the world's rangelands, should still be considered basically detrimental by many environmentalists. Where fire has been excluded from some range ecosystems, major vegetative and edaphic changes are taking place (Wright and Bailey 1982). Fire is being prevented on many rangelands because of (1) concerted effort to prevent the spread of wildfire with it's excessive risk to human life, possessions, animals and forage supply, (2) removal of fuel by plowing or cropping, and (3) grazing. The lack of fires in fire dependent grasslands frequently results in encroachment of shrubs and trees and may be followed by degradation of the soil as it changes over many years from a grassland soil to a forest soil.

The promotion and propaganda by forestry interests around the world that advocates the prevention of all fire creates problems for range managers in dealing with the lay public, other scientists and with research funding agencies. There is still considerable misunderstanding about the natural role of fire in rangeland ecosystems compared to it's role in forests. For instance, in tropical rainforests fire rarely occurs and in closed canopy coniferous forests there are many small fires and they burn only small areas whereas a few large fires cause major conflagrations (Martin *et al.* 1976, Hemstrom 1979). Inspite of the recognition of the place of fire in some ecosystems by resource managers and environmentalists alike, the pressure by foresters and others for the removal of fire from most commercial forests has a negative effect on the lay public. This spillover causes the public and the politicians to question the use of prescribed burning on rangelands.

Fire was common on many rangelands because all three criteria required for burning were met; these requirements are an ignition source, a suitable fuel, and suitable weather. Lightning and man were the normal ignition source, rangeland herbage and woody components were the fuels and there was usually no shortage of dry, warm, windy weather.

What happens when fire burns across a range? Fire is basically a rapid oxidation process. Organic materials are very rapidly broken down into carbon dioxide and water releasing energy as heat during the process. The quantity and duration of heat released as a fire passes over an area has a major effect on living organisms, surface soil and the amount of fuel available for another fire. As a flame passes, fine, dry porous fuels are the first to be consumed, followed by some consumption of more dense woody fuels if conditions permit. The depth of the fire front is usually not great in most range fires because the fine, grassy fuels that predominate are readily consumed. Fire temperatures in herbaceous fuels are often in the range of 100 to 400C (Bailey and Anderson 1980) but the duration of high temperature is short since fine fuels are consumed rapidly. When woody fuels burn, temperature ranges are often in the order of 500 to 1000C; the duration of high temperature is longer since denser fuels burn more slowly. The fire leaves a blackened surface behind it of charred, partially burned organic materials and a less favourable landscape for animals that require shelter. Soil surface temperatures usually rise, the rate of evaporation rises, chemical processes near the soil surface that respond to higher temperatures increase (Sharrow and Wright 1977) and rate of soil erosion may or may not rise. What happens next depends on the environmental conditions at the time of the fire. For example, the potential deleterious effects on the environment of a moderate intensity headfire in some rangelands in Africa are much greater if the fire occurred at the beginning of a six month dry season than on damp soil at the beginning of the rainy season. Similarly if a fire of a given intensity occurred on frozen soil in Canada at the beginning of spring, it would likely not be as potentially damaging to vegetation as if it took place six weeks later in the middle of the growing season.

21.3. Historical

Fire was as natural a component of rangelands as the vegetation itself. Much of the natural vegetation evolved under a fire regime and was well adapted to enduring it. Fire was far older than mankind and has been a major agent in the environment since the earth's formative stages (Hall 1984).

The common sources of ignition were lightning and man although other sources such as burning coal banks (Nelson and England 1971) and volcanic activity (Wein 1978) occasionally caused fires. It is not clear from the literature how much of the burning on rangelands was caused by man and how much was caused by lightning. In earliest times lightning was the only major cause of fires on rangeland and it has always been one of the causes of major conflagrations. It was usually not in early man's best interest to be caught in a high intensity fire or to burn under circumstances that would seriously damage productivity of the ecosystems that he depended upon for sustenance. Only recently have anthropologists accepted that hunter-gatherers and nomadic pastoralists were capable of affecting environmental change (Lewis 1982) through the controlled use of fire to better their environment. There are examples from northern Canada (Lewis 1982), southern Africa (Hall 1984), and northern Australia (Lewis 1985) of native peoples burning the vegetation in order to use up fuels thus preventing more serious conflagrations at times of the year and under weather and fuel conditions that would cause degradation to the environment and/or endanger their lives. Biswell *et al.* (1973) has used similar arguments to justify the use of prescribed burning of Ponderosa pine forests in the United States in order to thin the stands and reduce fuel loadings. Reducing fuel loads in these forests also lowers the risk of wildfires, increases the growth rate of surviving trees as well as the productivity and species diversity of understory rangeland vegetation.

Hall (1984) divided the fires used by primitive man into two general types. "Extensive regimes" were used in grassland and heath in order to remove litter and encourage seasonal regeneration while "intensive regimes" were associated with the clearing of savannas and forest for cultivation of crops and livestock grazing. Man has been introducing fire into more open vegetation types (rangelands) for about 150,000 years in Southern Africa and has been working with more "intensive regimes" since about the 3rd century AD. Lewis (1982, 1985) found exceptionally high sophistication and ecological understanding amongst native peoples (hunter-gatherers) in Canada and Australia who were using what Hall (1984) would describe as "extensive fire regimes". These native peoples appreciated the need to keep fire under control and many of them understood the responses of plants and animals to burning at specific seasons, intensities and frequencies. Lewis (1982) quoted a 76 year old Cree indian as follows "Fire had to be controlled. You couldn't just start a fire anywhere, anytime. Fire can do a lot of harm or a lot of good. You have to know how to control it . . . . It has been a long time since my father and my uncles used to burn each spring. But we were told to stop. The Mounties (Royal Canadian Mounted Police) arrested some people . . . . The country has changed from what it used to be — brush and trees where there used to be lots of meadows and not so many animals as before".

During the 19th and 20th century, fire fell into disfavour on rangelands particularly where Europeans had settled and brought along their traditional bias against the use of fire. European man had not used controlled fires in land management for about 1000 years when he went out to discover new lands and colonize them. He found it peculiar that the natives frequently burned these lands. He neither understood why the natives were burning nor did he care to learn why. As settlement took place the first natural resource to be managed was often the forest. Fire was a major competitor with man for the mature trees desired by loggers.

The traditional forestry bias of all fires being bad was reinforced by the homesteader who saw the range fire as a serious danger to his family, livestock and to their forage supply. Some of the earliest legislation in western Canada was related to fires. Railroad caused fires were common and in 1902 about 1750 miles of firebreaks were constructed (Nelson and England 1971) particularly across the unbroken open prairie grassland range (steppe) that extended from Winnipeg to Calgary. In the scientific community of the western world it has only been in the past several decades that the attitude towards fire has begun to change (Wright and Bailey 1982). Currently range scientists and resource managers alike show great interest in learning more about fire. This renewed interest is frequently due to the recognition of the role fire can have in correcting serious brush encroachment problems.

The semidesert grass-shrub vegetation of the southwestern United States and northern Mexico can be used as an example of what has likely happened in many other semiarid rangelands of the world. Originally the occasional fires, when combined with drought, plant competition, rodents and lagomorphs probably played a distinct role in controlling shrubs (Wright and Bailey 1982), thus favoring grasses. In most areas this is no longer possible because there have been widespread changes from semi-desert grassland to shrubland (Branson 1985) caused by the longest history of livestock grazing in North America. The original plant cover of a century ago was mostly grass whereas now it is mostly shrubs (York and Dick-Peddie 1969). The A horizon has been eroded away in many areas preventing the reestablishment of grasses, thus eliminating a source of fuels needed for fires that are able to control the shrubs. Similar changes have occurred in other areas of the world. For instance Booysen and Tainton (1984) indicated that fires had harmful effects on susceptible shrubs in the Karoo biome of Southern Africa, another semidesert grass-shrub vegetation. There are cycles of wet years in the Karoo as elsewhere and grassy fuels would have accumulated under those circumstances. Roberts (1981) indicated that the Karoo ecosystem was generally fragile, particularly susceptible to overgrazing and erosion, and it had the highest rate of soil loss in Southern Africa. He also noted that excessive soil loss following over a century of continuous sheep grazing had often removed the topsoil and left the subsoil exposed reducing the possibility of natural revegetation.

It would appear that the semiarid grassland rangelands have suffered extensively from overgrazing over the past centuries. With the reduced grass fuel loads and reduced grass-shrub competition, fires have become less frequent. The consequence has been an increase in shrubs, reduction or disappearance of perennial grasses, increased erosion and often a permanent change to the ecosystem due to loss of the surface soil.

In many parts of the third world, the human population has expanded so much in this century that all forage is required by livestock and insufficient carryover remains for controlled burning. Overgrazing and the lack of fire has led to marked shifts in species composition, an increase in woody species at the expense of grasses and forbs. There has also been an increase in desertification because of the excessive herbivore populations and periodic cultivation (Rapp *et al.* 1976).

Attitudes today towards fire on rangeland varies greatly from country to country and region to region. Few condone rampant, uncontrolled burning as was commonly practiced during the settlement era. On the other hand, the extremely negative attitudes held by some professionals and the public alike towards any rangeland fire is also unwarranted. The general attitude that has come from promotion campaigns of forestry interests that all trees are "good trees" is as unfounded as the idea that all range fires are "bad fires". In many countries there is now a growing awareness that prescribed burning of rangeland under predetermined conditions to meet specific management objectives has a place in land management.

21.4. Fire terminology

Fire is often considered to either burn or not burn, no consideration is given to the fact that there are "good fires", "bad fires", "harmful fires", "beneficial fires" to name but a few terms that are used by lay people to describe the effects of fire. All fires have both negative and positive attributes because the effect of fire on an ecosystem is complex.

Those who study fire know that not all burns have the same effect. There is a rapidly evolving terminology to describe fire behavior and fire effects. Much of the terminology has developed in forestry and it has been reviewed by Chandler *et al.* (1983). Trollope (1981) has reviewed the literature and proposed appropriate terminology for rangelands. Many of the definitions presented are from Trollope (1981). Fire ecology is the study of the response of the biotic and abiotic components of the ecosystem to the season, frequency, type and intensity of fire. There are three general types of fires. A ground fire burns below the surface usually in deep layers of organic matter; it is usually a smoldering fire (i.e., a peat fire). A surface fire sweeps over the ground consuming litter, dead fuels, living herbs and scorching tree trunks (Daubenmire 1959) but not burning into the crowns of trees. A surface fire may or may not kill tree trunks and shrub stems. A crown fire burns the canopies of trees; it is normally sustained by an accompanying surface fire. A headfire is either a surface fire or a crown fire that is burning with the wind. A backfire is usually a surface fire burning against the wind. Backfires usually burn at a constant rate independent of wind velocity whereas the rate of spread of headfires is very dependent on wind speed. Fire regime refers to the season and frequency of burning and the type and intensity of the fire. Fire severity refers to the effect of the fire on the ecosystem (Albini 1976). Fire intensity is the release of heat energy per unit time per unit length of fire front (Byram 1959). It refers to the rate of heat release during a fire. The preferred term in Canadian forest fire research and management is frontal fire intensity, which distinguishes between line fire intensity and area fire intensity or reaction intensity (Alexander 1982). The area fire intensity is frontal fire intensity (kw/m) divided by active horizontal flame depth (m). Fire behavior refers to the release of heat energy during combustion as described by fire intensity, rate of spread of the flame front, flame characteristics and other related phenomena. Residence time refers to the length of time for a flame front to pass a given point (Alexander 1982). Fuel load is the mass of fuel per unit area available for combustion during a fire (Luke and McArthur 1978). Fire return interval refers to the interval of time between fires at one point or small stand (Romme 1980). The fire return interval is highly area-dependent so the size of the area must be given when calculating this parameter.

21.5. Rangeland ecosystems

There is great diversity in the vegetation of the World's rangelands. Most rangelands fall outside of the categories of closed canopy forest, intensively cropped agricultural land, and urban-industrial complex.

Temperature and moisture are major factors influencing the broad groupings of vegetation. Rodin, Bazilevich and Rozov (1975) distinguished five thermal zones: arctic, boreal, temperate, subtropical and tropical. They also divided these thermal zones into climatic regions on the basis of moisture. The arctic and boreal thermal zones were recognized as being in a humid climatic region but the remaining thermal zones were each divided into humid, semiarid and arid climatic regions. This concept can be applied worldwide recognizing that in each climatic region, edaphic, topographic, fire and faunal factors will have an influence on distribution and productivity of the vegetation. The concept of dividing the thermal zones into three climatic zones based upon moisture is important to fire ecology because in general the humid areas have sufficient fuel to support a fire, semiarid areas occasionally have sufficient fuel for a fire to carry and arid areas rarely have sufficient fuel for a fire to burn; nevertheless some isolated arid areas do accumulate sufficient fuels for fire as a result of favourable soil moisture conditions.

In order to understand the fire ecology of rangelands it is necessary to relate fire regimes to the vegetation of a region. Daubenmire (1978) and Walter (1985) have been consulted in the development of the following descriptions of some world ecosystems.

21.5.1. Tundra

The tundra occurs at high altitudes and high latitudes where the heat supply is too low to permit the tree life form (Daubenmire 1978). It is a vegetation consisting of perennial herbs, shrubs and/or cryptograms. Close to the forest at lower latitudes or lower elevations there are more shrubs and the aboveground biomass ranges from 1000 to 8000 kg/ha because of the greater heat supply. In more extreme environments like the polar deserts the above ground biomass may only be about 60 kg/ha (Wielgolaski 1975). There are usually few days during the average year when the herbaceous and woody vegetation is dry enough to burn in the tundra region. Although there is little opportunity for fire on polar deserts because of a lack of fuel, there is sufficient fuel and fires do occasionally occur in the more dense tundra vegetation of lower latitudes and altitudes. The fires are generally small, usually occur during drought and are often caused by lightning (Wein and Bliss 1973; Wein 1976, Wein and Shilts 1976). Wein and Bliss (1973) found that fire in the *Eriophorum vaginatum* tussock community, which grows on wet peak substrate in the tundra-forest transition, caused the loss of few species but the charred organic surface absorbed more energy and the active layer increased through summer melting of the permafrost. This continued for several years. Annual production had almost recovered after two years and nitrogen content was higher in the burned areas; there was no consistent effect on other common nutrients.

Fire will likely continue to play a minor role on tundra. It is important for nutrient recycling where organic matter accumulates. One of the more common sources of fire in tundra will probably continue to be the summer burns that sweep into tundra from the adjacent boreal or subalpine forests. Since the tundra adjacent to these forests has higher than average fuel mass and is as affected by the occasional prolonged drought as in the forest, then lightning and man caused fires will continue.

21.5.2. Desert

Deserts are areas with low precipitation in relation to available heat so potential evaporation is much higher than annual precipitation. Deserts can be subdivided into semiarid, arid and extremely arid (Walter 1985). The vegetation growing on zonal soils is dominated by shrubs with annual herbs as an understory. The negligible representation of perennial grasses on zonal soils is a key character when differentiating desert from more mesic vegetation (Daubenmire 1978).

Fuel supply is often the key limiting factor to fire on deserts. The extremely arid deserts never have sufficient fuel; the semiarid and occasionally the arid desert areas will have sufficient fuel following seasons of abnormally high precipitation events.

Humphrey (1974) indicated that fire is a rarity in North American deserts despite the high frequency of lightning in some areas. When fire does occur, it removes most of the litter, a key factor in the infiltration of soil moisture and a particularly serious problem on deserts where rainfall events are rare, and are often of high intensity when they do occur.

21.5.3. Temperate arid forests

The pines and oaks usually dominate in the arid forests of North America and Eurasia (Daubenmire 1978, Walter 1985). They generally lie between steppe (or chaparral) and more mesic closed canopied forests. In North America, Ponderosa pine, Pinyon-Juniper and oak forests are prevalent (Daubenmire 1978, Wright and Bailey 1982). The understory is often composed of shrubs and perennial grasses.

Fire is an important natural factor in these forests and the vegetation is well adapted to the low and moderate intensity fires that swept through the forest every few years. Historically, fire was common here because there were sufficient fuel loads during many years, dry weather and dry fuels were common and lightning or native peoples were the ignition source (Arno 1976, Barrett 1980, Lewis 1980). Since the climate is relatively dry, fire, drought and competition often worked together to influence the species composition and structure of these forests. Fire frequency was probably greater in Ponderosa pine forest (about every 5—20 years) than in pinyon-juniper (about every 10—70 years) (Wright and Bailey 1982) because the fuel load is about double under Ponderosa pine. In the oak forests of the western United States and in Mediterranean areas, repeated burning tends to convert these woodlands into a fire disclimax composed of species primarily having the characteristics of chaparral (Daubenmire 1978, Walter 1985). This may be related to a shift in microclimate; the prolonged periods of reduced litter accompanied by reduced moisture infiltration may negatively affect the soil moisture regime and thus favor chaparral species that are better adapted to the drier conditions.

21.5.4. Grassland

21.5.4.1. Upland grassland. The upland grasslands of the world generally occur in semiarid to humid areas. There are many names in the literature referring to the various kinds of grasslands. The term steppe refers to grasslands outside of the tropics. The temperate zone steppes include the grasslands of North America and Eurasia, the pampas of South America, and the tussock grasslands of New Zealand (Daubenmire 1978, Walter 1985). The grasslands of tropical areas are often closely associated with savannas although some have been derived from forests. Grasslands are generally in areas of summer rainfall but there are exceptions such as the sagebrush-grass region of western North America where there is a winter precipitation pattern. A cold winter is common on steppes of the mid-continent. Soils are often high in organic matter at the surface and there is usually a lime layer at depth.

Herbaceous perennial grasses are predominant in pristine steppe vegetation and they may be rhizomatous or tufted. Forbs tend to be minor in the drier areas and more abundant in the more humid areas. Grasslands are generally well adapted to fire with the greatest adaptation being amongst species growing in more humid areas where fire frequency has been the highest. Since there is a range in vegetation types from humid to semiarid within the grassland biome, there is also a range in the fuel load available for burning. Fire frequency cannot be as high in the more arid areas because of insufficient fuel to carry the fire. Consequently some of the species suited to semiarid environments are not as well adapted to fire as those found in more humid grasslands. For example, *Artemisia tridentata*, the dominant shrub of the semiarid sagebrush-grass region, is frequently killed by fire and does not resprout whereas every woody species that grows in the humid *Festuca* grasslands of the Northern Great Plains of Canada has the ability to resprout. In sagebrush-grass fuel loads were low and the fire frequency was probably only once every 25 to 100 years whereas fuel loads were higher in humid grasslands such as the *Festuca* grasslands or the tallgrass prairie of the eastern plains of North America and fire frequency was probably about once every 2 to 5 years (Wright and Bailey 1982). Where the fires have been stopped on humid grasslands, tree and shrub encroachment is often a serious problem (Bailey and Wroe 1974, Bragg and Hulbert 1976, Wright, Bunting and Neuenschwander 1976).

21.5.4.2. Herbaceous wetland. There are wetlands scattered throughout the world that are dominated by herbaceous vegetation. There is considerable variation in species composition depending upon the continent, climate and edaphic conditions. Within a region, wetland vegetation is usually controlled by water depth and duration and by soil and water chemistry rather than by climate (Walker and Coupland 1970). Fire frequency in wetlands is variable from region to region due to variability in climate, fuel loads and burning opportunities.

The cycles of prolonged dry weather that lower water levels exposes greater quantities of dry fuels and may result in unusually severe burns. Prescribed burning can be used in this vegetation to remove moribund, decaying vegetation, alter species composition including retarding tree invasion, and alter physical structure of the standing vegetation thus influencing the habitat and food sources for animals that depend upon these rangelands.

21.5.5. Sclerophyllic woodland

In the Mediterranean winter rain regions of the world where there is a cool moist winter, a dry hot summer and sporadic frost, sclerophyllic woodland predominates. These are mostly shrublands and are called chaparral in North America, garrigue in France, fynbos in South Africa and mallee in Australia. The species diversity is often very great.

These woodlands are dominated by many genera and species of shrubs. Most shrubs are deep-rooted hardwoods, sclerophyllous (thick, tough, hard leaves), often evergreen and their canopies frequently cover the ground completely. There is usually only a sparse understory of annuals or geophytes (Daubenmire 1978).

Litter accumulates in the woodlands from undecayed leaves, dead branches and stems over a period of years resulting in the mature vegetation becoming highly flammable during the dry hot summers. Major intense fires in chaparral usually occur in stands greater than 30 years of age (Wright and Bailey 1982). If there is sufficient dead wood remaining after fire and herbaceous species, mainly forbs, germinate and grow abundantly, then reburns are possible over the next few years. Once the resprouting and seedling chaparral shrubs have recaptivated the site these young vigorous stands are essentially fireproof until sufficient litter has accumulated to again carry a fire.

Most species in these woodlands are well adapted to fire because of the long evolutionary history of burning. Many woody species sprout from adventitious buds with some developing lignotubers at the summit of taproots, other species recolonize by both resprouting and germinating from seed, while still others reproduce only by seed. Herbaceous species of annual grasses and forbs usually predominate for a few years after fire and then the woody plants become dominant shading out the understory species (Daubenmire 1978).

In many parts of the Mediterranean region the original schlerophyllous forest, which included *Quercus ilex*, has been deforested and abused by severe grazing with resultant soil erosion and is now in various stages of degradation (Naveh 1975, Walter 1985). Similarly, repeated burning of arid oak forests in western North America tends to convert this woodland into a fire climax composed of species having the characteristics of chaparral (Daubenmire 1978).

21.5.6. Tropical savanna

This vegetation occupies all of the tropical area in which the natural vegetation on zonal soils includes a layer of perennial grass and a scattering of trees or shrubs growing singly or in groves (Daubenmire 1978). Moisture is usually sufficient to permit trees and shrubs to maintain populations but not enough to exclude the perennial grasses. There is generally a prolonged dry season when the grass layer becomes dry, highly flammable and very susceptible to burning. Fire is an integral part of the savanna environment and both woody and herbaceous species are adapted to this factor. Many trees are able to endure low and even moderate intensity surface fires but high intensity fires and crown fires will topkill most trees. The trees that are topkilled usually sprout from dormant basal buds. The tropical savanna is one of the most frequently burned rangeland regions in the world.

Woody species may start to increase in density when grazing or drought reduces the competitive ability of the grass. As this develops the fire frequency is reduced, and if nothing is done, the trees may shade out the understory, removing the fine fuels and preventing subsequent surface fires. Alternatively grazing and browsing may be so severe as to remove all vegetation which results in intolerable levels of soil being lost and eventually the development of a desert.

21.5.7. Temporary rangeland

There are many kinds of temporary rangelands in the world. Most have been created from forest and are kept as rangeland for the purpose of grazing livestock or providing wildlife with a suitable habitat. There are many fuel types and varying fire management programs to maintain this vegetation in the desired state. Two examples will be presented.

21.5.7.1. Humid grasslands. The humid grasslands around the world are usually of a fire climax nature because there is sufficient annual precipitation to permit tree invasion and there is also enough grass growth each year to enable annual burning under all but the most severe grazing practices. Native peoples and lightning would both have been important sources of ignition. Fires and periodic heavy grazing by herds of wild herbivores had a major influence on the evolution of these grasslands. Now modern man is experiencing problems with tree and shrub encroachment into these grasslands because of changes to the environment that kept them in such a productive state. Consequently, tree and shrub invasion is occurring in the forest-steppe area of Eurasia (Walter 1985), the aspen parkland of Canada (Bailey and Wroe 1974), the tall grass prairie of the central United States (Bragg and Hulbert 1976) and in the highland sourveld in southern Africa when prescribed fire is not regularly imposed.

21.5.7.2. Early successional forest. The closed canopy forests of the world are rarely considered rangeland even though they do fit the definition of rangeland in many cases because they provide the forage and shelter for certain herbivores. These natural grazers or browsers depend upon the forest understory vegetation for their sustenance and cover.

Forest fire or other perturbations such as clearcut logging that removes the tree overstory provides an opportunity for succession of existing and invader plant species to expand rapidly. The luxurious growth of understory herbs and shrubs favors herbivores. Their populations usually multiply quickly in response to this more favorable food supply. Prescribed burning can be used to retard forest succession and is often the major tool at the disposal of the peasant farmers, hunter-gatherers and resource managers who wish to favor the understory species over forest trees and thus perpetuate the food and cover requirements of herbivores. Fire is frequently used in conjunction with grazing and browsing favoring herbivores of interest to man.

There are many tropical areas where the forest vegetation has been destroyed and subsequent seral vegetation is burned at frequent intervals (Daubenmire 1978). A savanna physiognomy develops and is maintained by management. This vegetation is frequently called derived savanna.

21.6. Plant adaptations to fire

Range plants have adapted to burning over time. Fires have occurred repeatedly at various intensities, climatic and edaphic conditions and stages of growth. Therefore species found on today's rangelands have adapted strategies which ensure their survival under most wildfire and prescribe fire conditions. Nevertheless there are only a few plants that cannot be killed by fire at some point in their life cycle. Physiologically active plants are usually more susceptible to fire than are dormant plants. Thus burning during the growing season generally kills or damages more plants than during the dormant season. The most harmful wildfires in a region normally occur during the growing season, killing some living plants. They may remove all fuels leaving the soil exposed to erosion and there will be a prolonged period of recovery for surviving plants. Most prescribed fires are set when the desired rangeland plant species are dormant; some fires are intentionally set when certain desired species are dormant and other undesirable species are physiologically active.

Range plants generally avoid, escape or endure fire. These three general reactions to fire have been described by Frost (1984) as avoidance in space, escape in time and fire tolerance. Plants that avoid fire do so by having buried seed, buried vegetative tissue such as perennating parts, grow where there is a lack of fuel or where weather conditions do not permit burning. Some of the species most adapted to fire in the Mediterranean region are annuals that grow at a time of the year when fires are rare and then disperse seed which is able to bury itself (Naveh 1975). Amongst these annual species, those that grow and set seed most rapidly when the vegetation is still green are most effective at surviving late growing season fires.

Plants that escape fire do so by seasonal dormancy or by being in the seed stage. Most rangeland fires occur during the dry season. This is when most perennial plants are dormant in the tropics. In temperate regions rangeland prescribed fires usually are set soon after snow melt at the beginning of spring or after growing ceases in autumn.

Species that endure fire usually have some mechanism to protect the perennating buds from heat damage. Many conifers and savanna trees have thick bark that protects the bole from low and moderate intensity fires. The branches are often elevated on a tall, fire-resistant stick, the bole, enabling the tree to escape the flames. Nevertheless intense burns such as crown fires will kill the trees. Woody species that are topkilled by fire are often still quite fire tolerant because they have a large, vigorous root system, may or may not have lignotubers, but do have the ability to coppice after the fire, usually from buried dormant buds. For example, in western Canada *Symphoricarpos occidentalis* had five times as many live stems and had completely replaced canopy cover three months after it was destroyed by a spring fire (Anderson and Bailey 1979). The majority of forbs and grasses fall into the fire tolerant category because of the location of perennating buds at or below ground level. Repeated burning usually favors forbs and grasses over shrubs and trees for this reason. Within the fire tolerant category of plant species there is an enormous range of variation in ability to endure season of burning, fire intensity and frequency of burning. Conrad and Poulton (1966) and Wright (1971) found that the more densely tufted perennial grasses were more susceptible to injury from burning or mowing than loosely tufted grasses.

Perennial grasses and forbs that have meristems below ground are usually resistant to fire regardless of whether they are tufted or rhizomatous species. The key characteristic seems to be the location of the meristem. Tufted perennials that have meristems located at ground level are usually resistant to moderate intensity fires, whereas those that have meristems just above the ground surface can only endure low intensity fires. Wet, cold or frozen soils are capable of absorbing great quantities of heat and often tend to protect buried perennating plant parts from the full intensity of a fire.

Raunkiaer (1934) recognized the importance of life forms in plants and classified them according to their ability to survive the unfavourable season. The phanerophytes and chamaephytes have aerial perennating parts (buds) and would be susceptible to fire. The hemicryptophytes have perennating parts in or just below the soil surface. The perennial grasses are hemicryptophytes (Vogl 1974). The cryptophytes have buds below the surface of the soil; many forbs are included in this category (Rowe 1983). The therophytes are annuals and have the characteristics to survive fire similar to the hemicryptophytes if the seeds lie on the surface of the ground or similar to the cryptophytes if the seeds become buried.

Kruger (1977) found that the vegetation of two fynbos (chaparral) stands regenerated vegetatively after a fire in the following manner:

Life form	Species regenerating vegetatively (%)
Phanerophyte	28
Chamaephyte	40
Hemicryptophyte	92
Geophyte	100

Geophytes escaped the fire because of their buried perennating buds and most hemicryptophytes also escaped by having perennating buds protected from heat damage.

The classification scheme of Raunkiaer (1934) has been modified and adjusted by a number of scientists to attempt to specifically design a concept that best fits the response of plants to fire rather than to climate. A relatively recent proposal was developed by Rowe (1983) for application to the boreal forest. The concept has merit for adaptation to rangelands although no attempt has been made to do so in this chapter. He proposed the following modes of persistence to fire:

1. Disseminule based
   1.1. Invaders: highly dispersive, pioneering fugitives with short-lived disseminules.
   1.2. Evaders: species with relatively long-lived propagules that are stored in soils or in canopy.
   1.3. Avoiders: shade-tolerant species that slowly reinvade burned areas, late successional, often with symbiotic requirements.
2. Vegetative based
   2.1. Resisters: shade-intolerant species whose adult stage can survive low-severity fires.
   2.2. Endurers: resprouting species with shallow or deep perennating buds.

All life forms of plants occur on rangelands but not in equal abundance. Some of the least common plant life forms are the phanerophytes and chamaephytes that are described by Rowe (1983) as avoiders. Rangelands that have extensive fire histories have mostly crytophytes, hemicryptophytes and therophytes or according to Rowe (1983), invaders, evaders, resisters and endurers. The type of fire, fire intensity and frequency influence the abundance of the various categories of plants found on rangelands.

21.7. When a rangeland burns

Fire can burn many rangelands at essentially any time of the year. Rangelands are normally found in areas that have extensive dry periods. Dry grass and dead brush are the usual fuels of fire. Thus many rangelands are the perfect setting for wildfire. The circumstances surrounding a wildfire must be differentiated from those of a planned fire, a prescribed burn. The extensive and damaging wildfires that give resource burning a bad name, usually occur when wind speeds are high, rates of spread are rapid, fuel moisture conditions are extremely low and fire intensities are high. The wildfire does not necessarily select the season of the year when plants are quite tolerant of fire. The potential for erosion also varies by season with the least amount of erosion likely to occur at a time of year when there will be a rapid vegetative growth soon after burning.

Fire generally sweeps across rangeland leaving litter consumed or scorched, living plant tissue either dead or alive; those plants able to survive fire usually having a competitive advantage over species that must reestablish from seed. Rangeland fires, with some rather spectacular exceptions such as chaparral, are normally carried along by a continuous mat of fine fuels. The behavior and effectiveness of the fire is closely tied to relative humidity since fine fuels are so responsive to atmospheric change in moisture content. As a fire front passes over a rangeland, temperatures rise and living plant tissue is killed when the temperature rises sufficiently for a long enough period of time. In fine fuels, the duration of high temperature is short and if plants have developed mechanisms to endure high temperature or to escape the heat effects, then the plants will survive.

A reasonable estimate of the temperature lethal to plant tissue is 60C (Hare 1961; Kayll 1966) but this is only an average and the actual value varies widely. The death of plant tissue is an exponential function between temperature and time and it may take from about 2 to 60 minutes to kill plant tissue depending upon moisture content (Wright 1970).

Seeds are usually more tolerant to heat than is living plant tissue (Daubenmire 1968) with those having a lower moisture content being more resistant to heat injury (McKell *et al.* 1962). Thus the more mature the seed is, the more able it is to resist heat injury. Seeds still in the inflorescence are likely to be burned in a fire (Bentley and Fenner 1958) but seeds lying on the ground usually survive (Daubenmire 1968) and those buried in the soil have the greatest chance of survival (Naveh 1975).

Heat is the immediate cause of death to living plant tissue but plants rarely live alone. They live in communities of many individuals and the surviving individuals will respond to an environmental disturbance factor, like fire and secondary succession will begin. Repeated burning will often increase the species diversity. This is in part due to removal of litter allowing lower growing species to establish, and often enabling those species adapted to more xeric conditions to invade or expand (Anderson and Bailey 1980).

The response of vegetation to fire heavily depends upon fire frequency, seasonality and intensity. Whatever the characteristics of the fire regime, the plants will respond to it. Generally herbaceous plants are favoured over woody plants because of the location of the perennating buds at or below ground level. Fire commonly favours forbs over grasses (Daubenmire 1968) although not always and in arid and semiarid rangelands annual grasses are often favoured over perennial grasses (Hopkins *et al.* 1948, West 1965, Naveh 1975).

21.8. Manipulation of rangeland vegetation with fire

There is currently great interest in understanding the fire ecology of rangelands because of the realization that 1. there is always going to be fire on rangelands whether as prescribed fire or as wildfire; 2. fire is a natural environmental phenomenon that normally does not produce serious residual effects; 3. fire alone or fire and grazing together are management tools that have the potential to manipulate rangeland vegetation favoring optimum productivity of the desired resources. Normally the prime product of interest is mass of herbivore whether it be domestic cattle, impala, wapiti, sheep, goat or, other but it may also include production of trophy animals, antler production, water, recreation, wildlife, plant and wood products.

Fire is used on rangelands for the following five basic purposes:

1. Modify species composition to favor preferred species over others through 1.1. brush control, 1.2. grass or forb control, 1.3. seedbed creation favoring seedling establishment, 1.4. modifying species diversity.
2. Maintain or change a physical vegetation structure to provide a desired habitat for animals.
3. Improvement of forage quality through 3.1. removal of litter, 3.2. reduction of structural components, 3.3. increased mineral constituents of forage.
4. Modify forage production through 4.1. increasing production, 4.2. shifting the height of the main canopy to grow within reach of the herbivore, 4.3. increasing forage production by shifting species composition to favor forage species over non-forage species.
5. Fireproof stands of vegetation by using prescribed fire under predetermined conditions to remove excess fuel thus decreasing the risk of wildfire.

21.8.1. Manipulating plants

Burning usually increases species diversity (Anderson and Bailey 1980) which in itself will normally favor an increased diversity of fauna because of the presence of more plant species and a more complex mosaic of habitats.

Fire and herbivores have influenced the species composition of forest-steppe and tropical savanna for centuries. Burning tends to favor herbaceous plants over woody species and is an important natural factor in maintaining a balance between grasslands and forests, herbs and trees. In the forest-steppe areas of Europe (Walter 1985) and North America (Wright and Bailey 1982), lack of burning results in an invasion of forest into grassland. Fire and herbivory together can be used to reverse this trend (Bailey 1986).

In tropical savannas the relationships are more complex. Increasing human and animal populations have placed excessive demands on the vegetation. This has resulted in a lack of fuel in many areas which effectively eliminates the opportunity for these savannas to burn. Tree and shrub encroachment has occurred and now competition for light and moisture favors trees and shrubs over grasses. Soil moisture deficiencies are a normal component of savannas which limit tree density. However, when grass competition is reduced by prolonged overgrazing, the extra soil moisture then made available favors further tree expansion or establishment usually preventing grass reestablishment. Trees in savannas are generally well adapted to fire but high fuel loads at the bases of the trees, coupled with high winds and low relative humidity will cause high fire intensity that can result in considerable topkill. Fire intensity is generally highly variable from place to place within the drier parts of savannas because the grasses are usually tufted, often growing well spaced from one another and larger bare patches are common. Fire is one of several factors that may operate together in a complex manner to influence species composition of a site. Burning, grazing, drought, competition and rodents interact together and can be used to control mesquite (*Prosopis glandulosa* var *glandulosa*) more effectively than fire alone on the semiarid rangelands of the southern United States (Wright and Bailey 1982).

21.8.2. Maintaining habitat for animals

Fire can be used to provide the habitat and ecological diversity required by some animal species (Bigalke and Willan 1984). Brynard (1972) and Vesey-Fitzgerald (1972) also described the role of fire in providing habitat for wildlife in parts of Africa. Mentis and Rowe-Rowe (1979) and Mentis and Bigalke (1979) found that certain francolin species in the Drakensberg Mountains of southern Africa are favored by a patchwork pattern of recently burned and unburned grassland. Gullion (1985, 1986) described the need of wildlife dwelling in the aspen forest ecosystems of North America for a diversity of ages of stands to optimize their habitat. The normal original cause of perturbation of the aspen forest was wildfire but (Gullion 1986) indicated that in 1986 the Minnesota legislature had appropriated US $1 million to simply fell and leave overmature aspen, to stimulate regeneration of future stands for the benefit of wildlife. This gives an indication of how far man is prepared to go to undo problems caused by the lack of natural fire.

Many of the cervids and antelope are edge animals that feed in openings and find shelter in the forest. An optimization of the edge environment provides them with the necessary access to both forage and escape cover. Similarly, wetland dwelling animals are often favored by a diversity of habitats as found in recent and older aged burns. Decadent vegetation that dominates in older marshes is virtually a desert for many animals.

21.8.3. Forage quality

In higher precipitation areas like the humid grasslands, wetlands, and humid savannas the perennial grass species are usually productive and very tall at maturity. As the stems grow each one requires more structural components to support the added height while the proportion of major nutrients like protein and phosphorus declines. At maturity and during the following year the litter from these grasses is of very low forage quality; it is very fibrous, unpalatable and acts as a barrier to grazing of green growth causing selective (patchy) grazing. Prescribed burning is often used at frequent intervals to remove this unpalatable litter in the Tall Grass Prairie of the midwestern U.S. (Kucera and Ehrenreich 1962, Smith and Owensby 1972), on Mesquite-Tobosa grass range in Texas (Heirman and Wright 1973) and in the Highland Sourveld of southern Africa. Burning may be required annually or once ever few years to eliminate the litter and reduce the effects of selective grazing.

Green vegetation that grows after fire is often utilized by herbivores in preference to the green growth on adjacent unburned areas because of a decrease in the proportion of structural components and an increase in mineral constituents (Daubenmire 1968, Willms *et al.* 1981). The woody sprouts that emerge following fire are also usually more preferred than those of unburned areas. There is some indication, however, that secondary compounds may increase in sprouts of some species after fire and result in a reduced preference.

21.8.4. Forage production

The C4 grasses in particular may produce more forage for several years after burning whereas normally any burning of areas dominated by C3 grasses causes a short-term reduction in forage productivity. In the Tall Grass Prairie of Wisconsin (Anderson 1972) and in the tobosa grass (*Hilaria mutica*) dominated areas of Mixed Prairie in Texas (Heirman and Wright 1973) a greater forage production was realized for several years after a spring burn.

In savanna, derived savanna, forest edge and climax forest ecosystems around the world, prescribed burning is used to favor herbaceous species over woody species, to topkill tall shrubs and trees enabling subsequent sprouting of fire adapted species (Booysen and Tainton 1984, Bailey 1986). The woody sprouts are within the reach of herbivory for several years. Although the total browse productivity may not be altered, the quantity available to herbivores is often substantially greater. Often the key to success or failure of the process lies not in the fire regime but in the season, intensity and duration of herbivore use of young woody sprouts. Post burn management is frequently more critical to the maintenance of the desired vegetation than the actual burning process itself.

21.8.5. Fireproofing

Rangelands in the humid to semiarid areas will usually be capable of burning given a sufficient period of years to accumulate fuels. The other criteria are dry weather and an ignition source. Any ignition source has the potential to set off a wildfire. However, a controlled burn can remove some of the fuel. When this happens the fuel is no longer available for another fire. Prescribed fire can be used under less hazardous conditions to reduce the fuel loading limiting the potential deleterious effects of wildfires.

There is a need in many countries to plan more prescribed burns for the purpose of fuel hazard reduction. It is unfortunate that it is usually easier to justify nearly unlimited budgets for fire fighting but give little or no support to the long-term planning and execution of fuel hazard reduction burning.

The long, hot, dry summer in mediterranean climates follows the cool, rainy winter growing season. These climatic areas require that more attention be paid to the potential use of controlled burning for fuel hazard reduction purposes. The grasslands, sclerophyllic woodlands, such as chaparral, and arid forests like ponderosa pine and certain oak forests of North America and Europe all need to be examined from time to time for fuel hazard reduction purposes.

21.8.6. Problems caused by burning

Fire is rarely feasible on many semiarid rangelands because overgrazing and/or climatic constraints restrict the buildup of fuel sufficient for burning to rare cycles of wet years. Where burning is still practiced there is a tendency to leave the soil exposed for long periods increasing the risk of erosion to unacceptable levels. Erosion is often not serious when prescribed burning is conducted early in the growing season in areas that have had recent rains. There is a brief period of exposure before spring growth provides a protective cover against raindrop impact and wind erosion.

Fire consumes herbaceous fuels. The dead herbage (litter) is often of the quality of straw in humid grasslands, savannas and wetlands; the loss of this forage supply is usually considered quite acceptable. In more arid areas of both the southern and northern hemispheres the grasses tend to cure on the stem. They are normally of higher forage quality than in humid areas and can produce liveweight gain of about 0.5 to 1.5 kg/day depending on growth stage and season of the year. The burning of these grasses is usually ill advised unless there are important overriding reasons such as brush control.

Woody species on rangeland are often adapted to fire and will resprout or germinate from seed after burning. Unless followup treatments are applied, the number of stems per unit area increases substantially for a period of years following burning. A single burn often leaves the rangeland with a higher density of woody stems than in unburned areas.

Fire is a powerful force and each prescribed burn carries with it the risk of escape. Spotfires are common when burning certain plant species. Most herbaceous species do not cause firebrands but many woody species do. Flaming or smoldering firebrands rise in the updraft from the fire, are caught and carried by wind currents for some distance before landing. They can ignite dry fuels beyond the firebreak causing wildfires. Dry dung, dry organic soils and porous wood can smolder for hours or days depending upon the circumstances. A knowledge of these risks is essential in successfully conducting a prescribed burn.

21.9. Prescribed burning

Prescribed burning has a place as a management tool on many rangelands but there are limitations. Fire needs to be used to accomplish specific management objectives. It is important to clearly define objectives. The greatest challenge to achieve this is to understand 1. what fire is able to do, 2. how to conduct a burn safely and effectively, 3. how to put out the fire, 4. what management to use after the fire to maintain the vegetation and the plant structure in the desired state. There are a number of steps to be followed in any prescribed burning operation and they include 1. preburn planning, 2. the burning operation, and 3. postburn management (Bailey 1986, 1987).

21.9.1. Preburn planning

21.9.1.1. Objectives. Prescribed burning can be used to realize one or more objectives depending upon management goals. The first step is to establish realistic objectives. The realism is derived from experience gained in using fire on a particular rangeland in cooperation and coordination with the current resource management practices for the region. A cardinal rule in developing objectives is to realize what fire can and cannot be expected to accomplish and then to evaluate if the postburn management is capable of maintaining the vegetation in an acceptable state.

21.9.1.2. Legislation and regulations. There is legislation and regulations in most areas governing the use of fire. Burning is risky particularly for the amateur so regulations are required. Some of the earliest legislation during the colonization era in the Northwest Territories of western Canada in present day Alberta and Saskatchewan dealt with the use of fire by settlers (Nelson and England 1971). Apparently, too many amateurs were using fire without proper safeguards causing undue risk to human life, livestock, property and forage supplies. Yet this was the same region in which the indian had used fire effectively for countless generations prior to the settlement of the region by Europeans.

Regulations governing the use of fire usually apply to a local area. They reflect the local knowledge and burning conditions. It is strongly advised that a beginner use fire conservatively in order to become aquainted with the practical limitations and strengths of the practice. This would include the use of wider than normal firebreaks, accurate spotweather forecasts, and an excessive amount of equipment and personnel on standby in case of an emergency.

21.9.1.3. Fuel loads. Fire will not burn without sufficient dry fuel. Fire spread also requires that there be a uniform distribution of fuel across a landscape; any large gaps without fuel will disrupt the movement of the fire front. A reasonable estimate of a minimal fuel load required for burning grassland is 1000 kg/ha of dry herbaceous fuels that is compacted near the ground as found after spring snowmelt. Taller herbaceous plants that have very little fuel lying on the ground actually burn much like a crown fire. They require about 2000 kg/ha of dry fuel and greater wind speeds for continuous fire coverage. When there are many bare patches, burning will be discontinuous, irregular and a higher proportion of the landscape will be left as unburned islands. Similarly if there is great diversity in topography there will also be a patchwork of unburned areas. The quantity of woody fuels required for burning varies greatly from one vegetation type to another. There are a host of factors responsible for this including proportion of dead to live fuel, vertical and horizontal distribution, wood density and flammability. The amount of dry, dead surface fuel is a key factor and if there is little, then a large quantity of standing dead woody fuel is required to carry a crown fire rather than a surface fire.

The rate of biomass accumulation following a prescribed burn can be forecast and is related to 1. quantity of fuel remaining after fire, 2. annual rate of accumulation of dead biomass, 3. flammability of live biomass. This knowledge can be used to predict fire frequency.

21.9.1.4. Season to burn. The season of burning varies around the world with climate, vegetation and objectives. The tropical savannas, tropical grasslands and areas with a mediterranean climate have about a six month dry season each year when fuels will burn. In contrast, some northern areas have a dry season of only a few days in spring soon after snowmelt or in autumn following frosts. In the far north, these periods may be restricted to the occasional midsummer drought that results in fuels being sufficiently dry to burn in wildfires of boreal forest and adjacent tundra.

Summer burning is usually not recommended because of the risk of very dry soil and fuel conditions, excessive consumption of litter and surface soil organic matter, and possible damage to physiologically active plants. There is also the possibility of prolonged exposure of bare soil to wind and water erosion.

21.9.1.5. Weather requirements. An accurate local spot weather forecast is highly recommended before, during and after a prescribed burn. A set of criteria required to burn the vegetation in the local area needs to be developed and communicated to the forecaster. Only certain wind directions, speeds, relative humidity and temperatures will be acceptable, plan accordingly. Relative humidity, temperature, windspeed all have a reasonably predictable diurnal cycle and vary from area to area depending upon local climate.

A suitable burning prescription must be developed for each vegetation type and it will include acceptable relative humidity, windspeed and air temperature. For example, experience has shown that the following weather conditions are effective in the aspen parkland of central Alberta, Canada where burns are conducted in spring:

Fuel type	Minimum temperature (C)	Range of windspeed (kph)	Maximum relative humidity (%)	No. drying days after snowmelt
grassland	7	3—20	62	1—3
deciduous shrubland	13	3—20	50	3—5
deciduous forest	18	6—20	30	8—10

Recent research in Israel has shown, however, that prescribed burning of annual grassland can be conducted in the dry midsummer period at 65—70% relative humidity and 2—4 kph wind. However, the hot sun (28—34C) must dry the surface of herbaceous fuels for at least an hour before the fire will readily ignite and burn from plant to plant.

21.9.1.6. Burning plan. Arrange with a weather forecaster to provide an accurate spot weather forecast for the area to be burned. Plan to have a forecast for the day before the burn, the day of the burn and the day after the burn, and for as many days thereafter as there are smoldering hot spots. Important information will include 1. stability of the current weather system, 2. cloud cover, 3. maximum temperature, 4. minimum relative humidity, 5. wind direction, 6. windspeed and gusts.

All steps need to be planned in advance. There is not time the day of the prescribed fire to do anything but conduct the burn according to a detailed plan. There will be restrictions on any burning plan. Fuel load must meet preset conditions, burning will be conducted at a specified time of year when plants are in a given physiological state and it will also be conducted within preset weather and fuel moisture conditions. An ignition pattern will be planned for the specific area to be burned. The distribution of ignition and suppression crews will be planned in advance. Emergency suppression tactics will also be planned in advance. Actual small practice fires should be set to provide an inexperienced crew with actual ignition and suppression experience before setting the main prescribed burn.

21.9.1.7. Firebreaks. Every prescribed burn should be conducted within an area surrounded by firebreaks. These firebreaks normally vary in width from a cattle trail as used to burn some grasslands to a 100m wide tilled bare strip as required to burn certain fuel types that are high in volatile oils or other fatty compounds that cast firebrands. The prescription for burning a patchy fuel bed that also has volatile fuels such as sagebrush (*Artemisia* spp.) will require unusually dry and windy conditions. More than 200m wide firebreaks may occasionally be required under these circumstances.

The firebreak can be prepared well ahead of the actual burn and can be planned around certain topographic features and vegetation types.

21.9.1.8. Manpower. Burning is labour intensive and is limited in its potential use in developed countries because of the high cost of manpower. However, well planned prescribed burns can often be done in a matter of hours and be fitted in with other tasks.

Prescribed burning should not be done by a committee. One compassionate dictator, the fire boss, needs to be in absolute charge of the complete operation. On a large burn with an experienced crew, the fire boss may delegate limited authority to an ignition boss, in charge of an ignition crew, and a suppression boss, in charge of the suppression crew which prevents the fire from getting out of control if there is an excursion beyond the firebreak. The potential trouble spots where the fire can get away are usually predictable, are usually related to topography and local wind currents, as well as to concentrations of fuels. The fire boss will be knowledgeable about what fire can do under specific fuel and weather conditions and will be able to direct people accordingly.

21.9.1.9. Equipment. Communication is most important on a burn. All ignition and suppression activities need to be coordinated through the fire boss. Light, highly portable two-way radios that can be strapped to a belt and leave the hands free are useful provided they operate reliably over the distances required.

We use drip torches or propane torches to ignite the fire but others use systems of ignition ranging in sophistication from helicopter mounted equipment, fuel soaked rags, carrying burning grass on a fork, to a match. The ignition device need not necessarily be expensive or sophisticated but it does need to be effective and reliable.

Suppression equipment must be on hand in sufficient quantity and capability to be able to put the fire out under any foreseeable circumstance. It too can vary from wet sacks, flappers and man-carried hand pumps for spraying water to machine mounted sprayers and aerial water bombers, depending upon the circumstances. The equipment needed must suit the circumstances of the burn. One of the major flaws of amateur planners is to not have sufficient water on hand at the site as a backup supply in case of an emergency. Although most fires can be planned and executed to use little water to suppress the burn, sufficient equipment, manpower and water must be readily available at the site of the fire.

21.9.2. The burning operation

The spot weather forecast the day of the burn will assist the fireboss in reaching the decision to proceed. No burning will be done unless the weather is within the preset prescription. Final checks will be made of all personnel and all equipment to ensure that everything is functioning normally. Then the burning plan will be put into effect. One normally starts to burn with a small test ignition in one corner to determine if the fire will carry satisfactorily through the fuels and accomplish the objectives under the specific weather conditions for that day. Once this has been completed a decision is made whether to proceed. If the decision is positive, the ignition crew sets the fire as rapidly as possible. The suppression crew is moved into positions of known risk to prepare for fire excursions. There will be continuous communication between fireboss and crew to coordinate the fire. If problems arise, the fireboss is notified, attempts are made to put out the fire burning beyond the firebreak and the fireboss makes the decision about possible reallocation of suppression forces. Any spotfire that develops outside the firebreak must be put out immediately before it becomes a large fire. If the suppression crew cannot contain a wildfire beyond the firebreak quickly, call the local authorities for reinforcements before a larger fire has time to develop. Time is of the essence in controlling a fire.

Most well planned and well executed prescribed burning operations are conducted effectively and in safety and the public never hears about them. There is however the occasional exception when a fire gets out of control and it tends to mask and negate the many previous successful applications of prescribed burning. Good public relations is a necessity if prescribed burning is to be conducted successfully in populated areas.

Once the burning operation has been completed the crew is not disbanded until there has been a thorough check of the entire burn perimeter. The amount of smoldering material is evaluated for it's potential of causing the fire to escape. This check must be done beyond the firebreak in search of smoldering fuels. Thoroughly soak hot spots beyond or near the firebreak until they are cold.

21.9.3. Management after the burn

The key to success of most prescribed fires on rangeland lies with the effectiveness of postburn management because many plants are in a weakened condition after the fire and are susceptible to further injury by grazing animals. Burning is but a tool in an overall management plan. If no such plan exists then one should not be surprised if the long-term objectives for burning are not realized.

Ash can be used as a seedbed for certain herb and shrub species. The seed should be broadcast as soon as feasible after the fire and certainly before the first rains which will cause a crust to form making seed penetration difficult. Heavy concentrations of herbivores for a brief period may result in better establishment than if no herbivores are allowed to graze the site until much later (FitzGerald and Bailey 1983).

Succulent fresh green forbs, grasses, sedges and woody sprouts that grow after a burn are usually very palatable. Heavy continuous grazing can destroy the regrowth that comes up on recently burned areas. On the other hand, herbivores can be used to successfully manipulate the vegetation to favor the desired species. We have used cattle in a short duration grazing system to control palatable woody species thus favoring the desired herbaceous plants (FitzGerald and Bailey 1984, Bailey 1986) but a different intensity of grazing using the same regime would have permitted the maintenance of a herb-woody species mixture beneficial to wildlife.

Large concentrations of wild or domestic herbivores on small burned patches can do great damage to the vegetation. They can effectively eliminate most palatable species giving the unpalatable vegetation a competitive edge. If there is a possibility of excessive use of postburn vegetation by herbivores, take it into account in preburn planning. For wild herbivores this normally requires that burns be of sufficient size to prevent overuse of succulent forage growing on a new burn. In the case of domestic herbivores, the entire management unit, i.e., the whole field, needs to be burned so that livestock do not have the opportunity to select and overgraze the burn leaving the unburned area underutilized.

21.10. References

Albini, F. A. 1976. Estimating wildfire behaviour and effects. USDA For. Serv. Gen. Tech. Rep. INT-30, Intermt. For. and Range Exp. Stn., Ogden, Utah.

Alexander, M. E. 1982. Calculating and interpreting forest fire intensities. Can. J. Bot. 60(4): 349—357.

Anderson, H. G., and A. W. Bailey. 1980. Effects of annual burning on grassland in the Aspen Parkland of east-central Alberta. Can. J. Bot. 58(8): 985—996.

Anderson, M. L., and A. W. Bailey. 1979. Effect of fire on a *Symphoricarpos occidentalis* shrub community in central Alberta. Can. J. Bot. 57(24): 2819—2823.

Anderson, R. C. 1972. The use of fire as a management tool on the Curtis prairie. Proc. Tall Timbers Fire Ecol. Conf. 12: 23—35.

Arno, S. F. 1976. The historical role of fire on the Bitterroot National Forest. USDA For. Serv. Res. Pap. INT-187. Intermt. For. and Range Exp. Stn., Ogden, Utah.

Bailey, A. W. 1986. Prescribed burning for range and wildlife management. Univ. Alberta, Agriculture-Forestry Bull. 9(3): 10—14.

Bailey, A. W. 1986. Woodland to grassland: fire and grazing versus mechanical clearing in the Canadian aspen parkland. p. 592—593. *In*: P. J. Joss, P. W. Lynch, and O. B. Williams. Rangelands: a resource under siege. 2nd Int. Rangel. Congr. Aust. Acad. Sci. Canberra.

Bailey, A. W. 1987. Planning and conducting a burning operation. p. 646—652. *In*: L. S. Pope. Beef cattle science handbook, Vol. 21, Int. Stockman's School. Spilman Press, Sacramento.

Bailey, A. W., and M. L. Anderson. 1980. Fire temperatures in grass, shrub and aspen forest communities of central Alberta. J. Range Manage. 33(1): 37—40.

Bailey, A. W., and R. A. Wroe. 1974. Aspen invasion in a portion of the Alberta parklands. J. Range Manage. 27(4): 263—266.

Barrett, S. W. 1980. Indian fires in the pre-settlement forests of western Montana. p. 35—41. *In*: M. A. Stokes, and J. H. Dieterich. Proc. of the fire history workshop, Oct. 20—24, 1980, Tucson, Ariz. U.S.D.A. For. Gen. Tech. Rep. RM-81.

Bentley, J. R., and R. L. Fenner. 1958. Soil temperatures during burning related to postfire seedbeds on woodland range. J. For. 56: 737—740.

Bigalke, R. C., and K. Willan. 1984. Effects of fire regime on faunal composition and dynamics. p. 253—271. *In*: P. de V. Booysen, and N. M. Tainton. Ecological effects of fire in South African ecosystems. Springer-Verlag. Berlin.

Biswell, H. H., H. R. Kallander, R. Komarek, R. J. Vogl, and H. Weaver. 1973. Ponderosa fire management. Misc. Pub. No. 2, Tall Timbers Res. Stn., Tallahassee, Fla.

Booyson, P. de V., and N. M. Tainton. 1984. Ecological Effects of Fire in South African Ecosystems. Springer-Verlag, Berlin.

Bragg, T. B., and L. C. Hulbert. 1976. Woody plant invasion of unburned Kansas bluestem prairie. J. Range Manage. 29: 19—24.

Branson, F. A. 1985. Vegetation changes on Western Rangelands. Range Mono. 2, Soc. Range Manage., Denver.

Branson, F. A., G. F. Gifford, and J. R. Owen. 1972. Rangeland hydrology. Range Sci. Ser. 1, Soc. Range Manage., Denver.

Brynard, A. M. 1972. Controlled burning in the Kruger National Park — history and development of a veld burning policy. Proc. Tall Timbers Fire Ecol. Conf. 11: 219—231.

Byram, G. M. 1959. Combustion of forest fuels. *In*: K. P. Davis (ed.) Forest Fire Control and Use. McGraw Hill Book Co., New York.

Chandler, C., P. Cheney, P. Thomas, L. Trabaud, and D. Williams. 1983. Fire in Forestry, Vol. I and II. Wiley, New York.

Conrad, E. C., and C. E. Poulton. 1966. Effect of wildfire on Idaho fescue and bluebunch wheatgrass. J. Range Manage. 19: 138—141.

Daubemire, R. F. 1959. Plants and Environment. Wiley, New York.

Daubemire, R. F. 1968. Ecology of fire in grasslands. Adv. Ecol. Res. 5: 209—266.

Daubemire, R. 1978. Plant Geography. Academic Press, New York.

FitzGerald, R. D., and A. W. Bailey. 1983. Influence of grazing with cattle on establishment of forage in burned aspen brushland. Proc. 14th Int. Grassl. Congr. p. 564—566. Westview Press, Boulder.

FitzGerald, R. D., and A. W. Bailey. 1984. Control of aspen regrowth by grazing with cattle. J. Range Manage. 37: 156—158.

Frost, P. G. H. 1984. The responses and survival of organisms in fire-prone environments. *In*: P. de V. Booyson, and N. M. Tainton. Ecological Effects of Fire in South African Ecosystems. Springer-Verlag, Berlin.

Gullion, G. W. 1985. Aspen management — an opportunity for maximum integration of wood fiber and wildlife benefit. Trans. N. Amer. Wildl. and Nat. Res. Conf. 50: 249—261.

Gullion, G. W. 1986. Northern forest management for wildlife. Forest Industry Lecture Series No. 17, Faculty of Agriculture and Forestry, Univ. Alberta. Edmonton.

Hall, M. 1984. Man's historical and traditional use of fire in southern Africa. *In*: P. de V. Booyson, and N. M. Tainton. Ecological Effects of Fire in South African Ecosystems. Springer-Verlag, Berlin.

Hare, R. C. 1961. Heat effects on living plants. USDA For. Serv. Occas. Pap. S-183. Southern For. Exp. Stn., New Orleans, La.

Heirman, A. L., and H. A. Wright. 1973. Fire in the medium fuels of West Texas. J. Range Manage. 26: 331—335.

Hemstrom, M. A. 1979. A recent disturbance history of forest ecosystems at Mount Rainier National Park. Ph.D. Diss., Oregon State Univ., Corvallis.

Hopkins, H., F. W. Albertson, and A. Riegel. 1948. Some effects of burning upon a prairie in west central Kansas. Kansas Acad. of Sci. Trans. 51: 131—141.

Humphrey, R. R. 1974. Fire in deserts and desert grassland of North America. p. 366—400. *In*: T. T. Kozlowski, and C. F. Ahlgren (ed.) Fire and Ecosystems. Academic Press, New York.

Kayll, A. J. 1966. A technique for studying the fire tolerance of living tree trunks. Can. Dept. For. Pub. No. 1012. Ottawa, Ont.

Kruger, F. J. 1977. Ecology of Cape fynbos in relation to fire. *In*: H. A. Mooney, and C. E. Conrad (ed.) Proc. Sym. Evironmental consequences of fire and fuel management in Mediterranean ecosystems. USDA For. Serv. Gen. Tech. Rep. WO-3, Washington, D.C.

Kucera, C. L., and J. H. Ehrenreich. 1962. Some aspects of annual burning on central Missouri prairie. Ecol. 43: 334—336.

Lewis, H. T. 1980. Hunter-gatherers and problems for fire history. p. 115—119. *In*: M. A. Stokes, and J. H. Dieterich. Proc. of the fire history workshop, Oct. 20—24, 1980, Tucson, Ariz. U.S.D.A. For. Serv. Gen. Tech. Rep. RM-81.

Lewis, H. T. 1982. A time for burning. Occ. Publ. No. 17. Bor. Inst. Nor. Studies., University of Alberta, Edmonton, Alberta, Canada.

Lewis, H. T. 1985. Why indians burned: specific versus general reasons. *In*: J. E. Lotan, B. M. Kilgore, W. C. Fischer, and W. R. Mutch. Proc.: Symposium and Workshop on Wilderness Fire. USDA For. Serv., Intermt. For. and Range Exp. Stn., Ogden, Utah.

Luke, R. H., and A. G. McArthur. 1978. Bushfires in Australia. Aust. Gov. Pub. Serv., Canberra.

Martin, R. E., D. D. Robinson, and W. H. Schaeffer. 1976. Fire in the Pacific Northwest — perspectives and problems. Proc. Tall Timbers Fire Ecol. Conf. 15: 1—23.

McKell, C. M., A. M. Wilson, and B. L. Kay. 1962. Effective burning of rangelands infested with Medusahead. Weeds 10: 125—131.

Mentis, M. T., and R. C. Bigalke. 1979. Some effects of fire on two grassland francolins in the Natal Drakensberg. S. Afr. J. Wildl. Res. 9: 1—8.

Mentis, M. T., and D. T. Rowe-Rowe. 1979. Fire and faunal abundance and diversity in the Natal Drakensberg. Proc. Grassl. Soc. Sth. Afr. 14: 75—77.

Naveh, Z. 1975. The evolutionary significance of fire in the Mediterranean region. Israel J. Bot. 24(1): 51—52.

Nelson, J. G., and R. E. England: 1971. Some comments on the causes and effects of fire in the northern grasslands area of Canada and the nearby United States, CA. 1750—1900. Can. Geogr. 15(4): 295—306.

Rapp, A., H. N. Le Houerou, and B. Lundholm (eds.) 1976. Can desert encroachment be stopped? Ecological Bull. 24, Swedish Nat. Sci. Res. Council, Stockholm.

Raunkier, C. 1934. The Life Forms of Plants and Statistical Plant Geography. Clarendon Press, Oxford.

Roberts, B. R. 1981. Sweet and mixed grassveld. p. 383—391. *In*: N. M. Tainton (ed.) Veld and Pasture Management in South Africa. Shuter and Shooter, Pietermaritzburg.

Rodin, L. E., N. I. Bazilevich, and N. N. Rozov. 1976. Productivity of the world's main ecosystems. p. 13—26. Proc. Symp: Productivity of World Ecosystems. U.S. Nat. Com. for the IBP. Nat. Acad. of Sci., Washington, D.C.

Romme, W. H. 1980. Fire history terminology: report of the Ad Hoc committee. p. 135—137. *In*: Proc. Fire History Workshop, Oct. 20—24, 1980. U.S.D.A. For. Serv. Gen. Tech. Rep. RM-81.

Rowe, J. S. 1983. Concepts of fire effects on plant individuals and species. *In*: R. W. Wein, and D. A. MacLean. The Role of Fire in Northern Circumpolar Ecosystems. Wiley, New York.

Sharrow, S. H., and H. A. Wright. 1977. Effects of fire, ash, and litter on soil nitrate, temperature, moisture and tobosagrass production in the Rolling Plains. J. Range Manage. 30: 266—270.

Smith, E. F., and C. E. Owensby. 1972. Effects of fire on true prairie grasslands. Proc. Tall Timbers Fire Ecol. Conf. 12: 9—22.

Trollope, W. S. W. 1981. Recommended terms, definitions, and units to be used in fire ecology in South Africa. Proc. Grassld. Soc. Sth. Afr. 16: 107—107.

Vesey-Fitzgerald, D. 1972. Fire and animal impact on vegetation in Tanzania National Parks. Proc. Tall Timbers Fire Ecol. Conf. 11: 297—318.

Vogl, R. J. 1974. Effects of fire on grasslands. p. 139—194. *In*: T. T. Kozlowski, and C. E. Ahlgren. Fire and Ecosystems. Academic Press, New York.

Walker, B. H., and R. T. Coupland. 1970. Herbaceous wetland vegetation in the aspen grove and grassland regions of Saskatchewan. Can. J. Botany 48: 1861—1878.

Walter, H. 1985. Vegetation of the Earth, 3rd ed. Springer-Verlag, Berlin.

Warren, S. D. 1987. Soil hydrologic response to intensive rotation grazing: A state of knowledge. p. 488—501. *In*: Y. S. Folk. Int. Conf. on Infiltration Development and Application, University of Hawaii, Honolulu, Hawaii.

Wein, R. W. 1976. Frequency and characteristics of arctic tundra fires. Arctic 29: 213—222.

Wein, R. W. 1978. The role of fire in the degradation of ecosystems. *In*: M. W. Holdgate, and M. J. Woodman. The Breakdown and Restoration of Ecosystems. Phlenum Publishing Corp.

Wein, R. W., and L. C. Bliss. 1973. Changes in arctic *Eriophorum* tussock communities following fire. Ecol. 54: 845—852.

Wein, R. W., and W. W. Shilts. 1976. Tundra fires in the District of Keewatin. Geological Survey of Canada, Paper 76-1A: 511—515.

West, O. 1965. Fire in vegetation and its use in pasture management with special reference to tropical and subtropical Africa. p. 1—53. Mimeogr. Publ. No. 1. Comm. Bur. Pastures and Field Crops, Hurley, Burkshire.

Wielgolaski, F. E. 1975. Productivity of tundra ecosystems. p. 1—12. *In*: Proc. Symp. Productivity of World Ecosystems. U.S. Nat. Com. for the IBP., Nat. Acad. Sci., Washington, D.C.

Wright, H. A. 1971. Why squirreltail is more tolerant to burning than needle-and thread. J. Range Manage. 24: 277—284.

Wright, H. A., and A. W. Bailey. 1982. Fire ecology of the United States and southern Canada. Wiley, New York.

Wright, H. A., S. C. Bunting, and L. F. Neuenschwander. 1976. Effect of fire on honey mesquite. J. Range Manage. 29: 467—471.

York, J. C., and W. A. Dick-Peddie. 1969. Vegetation changes in southern New Mexico during the past hundred years. p. 155—166. *In*: W. G. McGinnies, and B. J. Goldman (ed.) Arid lands in perspective. Univ. Arizona Press, Tucson.

22. Reclamation of drastically disturbed rangelands

EDWARD F. REDENTE AND EDWARD J. DEPUIT

22.1. Abstract

Successional ecology is a discipline of vegetation science integral to the reclamation of rangelands drastically disturbed by mining activity. This chapter is structured around the successional processes first defined by Clements in 1916: migration, ecesis, reaction, competition and stabilization. Initial reclamation success is dependent upon adequate and appropriate supply of propagules. Propagule migration can be artificially enhanced through seeding and planting activities, but the plant community that initially establishes is also dependent upon natural migration along with biotic and abiotic conditions that exist at the site. Following propagule migration, plant establishment must take place before succession can proceed, and this process is influenced through the improvement of edaphic and topographic conditions and by the selection of environmentally adapted species. Once plants become established the reaction process is important, as organisms inhabiting the site modify the abiotic and biotic environment. These modifications of the environment (above- and belowground) are well documented, and may favor the subsequent establishment of later successional species. Belowground changes during the reaction process lead to the immobilization of inorganic compounds, reductions in nutrient availability and increased diversity of the decomposer subsystem. These changes have direct influence on plant community dynamics and structure. Competition can also play an important role in determining structure of primary producers. Competition can be mediated through the modification of resource availability or by direct manipulation of species that are planted or become established on a site. Reductions in plant species diversity have been shown to occur on reclaimed sites with high nutrient or water additions. The final step in the successional process is stabilization. Since successional studies on drastically disturbed rangelands have been conducted over relatively short periods of time, the

Tueller, P. T. (ed.), Vegetation science applications for rangeland analysis and management.
ISBN 90-6193-195-9.
© 1988, *Kluwer Academic Publishers, Dordrecht, Boston, London.*

process of stabilization has not been observed and is little understood. Approaches to accelerate stabilization have been applied in recent years, but the ultimate effect of these treatments is generally unknown.

22.2. Introduction

22.2.1. Mining disturbance

The extraction of fossil fuels and other minerals from rangelands in the United States and throughout the world can affect the rangeland resource both beneficially and adversely. Mining and subsequent reclamation can improve the physical and chemical characteristics of the soil, remove toxic elements that inhibit plant growth, and establish plant communities that provide increased quality and quantity of forage above pre-mining conditions. Mining, however, can adversely affect rangelands if reclamation practices are not properly applied. Soil productivity can be severely reduced if overburden with inimical physical or chemical attributes becomes part of the plant growth medium.

The disturbance of rangelands by mining activity can be termed drastic when mining removes the vegetation and animal components of the ecosystem and disrupts soils to the point of reducing or eliminating the soil flora and fauna. In addition, the physical and chemical properties of the soil may be altered to the point of not being suitable for plant growth. Although the extent of such drastic disturbance on rangelands is and will continue to be limited, mitigation of impacts through reclamation has become an ethical, ecological and legal mandate. Successful reclamation must be soundly based upon vegetation science in concert with a broad array of other physical and biological sciences.

22.2.2. Reclamation process

There are two primary goals associated with the reclamation of mining disturbances on rangelands. The first goal is to stabilize the disturbed site with vegetation cover to provide immediate control of wind and water erosion. A longer-term goal is to establish a diverse, effective and self-sustaining plant community. This second goal is the most challenging to the reclamation specialist, and requires an adequate understanding of the structure and function of both the pre-mining and the disturbed ecosystem. Vegetation science has played an important role in this regard by contributing basic knowledge in plant ecology and plant-soil interrelationships.

Successional ecology is a part of vegetation science that provides an invaluable foundation for building a successful reclamation program. The successional process is the driving force that advances the ecosystem that was artificially established during reclamation toward a self-sustaining and naturally functioning condition. Because of this, knowledge of succession must be incorporated into the reclamation process at every conceivable opportunity.

The authors agree with MacMahon (1983) when he stated "It was Clements (1916 and 1928) who went furthest to propose mechanisms for the successional phenomenon. His argument, in essence, was that a site which underwent development to a *climax formation* involved the *initiation* of plant establishment on that site and a series of subsequent processes, for which he coined specific terms."

Clement's six basic processes are all a function of vegetation except for the first one: (1) nudation, (2) migration, (3) ecesis, (4) competition, (5) reaction, and (6) stabilization. Each of these processes, beginning with migration, form the framework for the rest of this chapter. Each process is described and its importance in the reclamation process is discussed.

22.3. Vegetation science and ecosystem development on disturbed rangelands

22.3.1. Migration process

Propagule supply. The migration process described by Clements (1916) has two components. First, following disturbance or nudation there will be a residual supply of propagules remaining in the soil. Second, there will be a supply of propagules that will migrate to the disturbed site either by chance or highly directed movement (MacMahon 1983).

Invasion by sexually reproducing plant species is dependent upon the dispersal and establishment of their seeds. Adequacy of seed dispersal is dependent upon seed size, weight, shape, and amount produced (Praeger 1923). However, adequate dispersal and the capacity for successful establishment may be negatively correlated. Webb (1966) indicated that while large seeds may be advantageous for establishment, they may confer disadvantages with regard to dispersal mechanisms. Colie (1940) reported that on recently disturbed areas, the new vegetation resulted largely from wind dispersed seeds, with a decreasing gradient from the seed sources. Seed dispersal within the soil profile also may be important and is accomplished through the actions of soil macro-organisms and surface disturbance by animals or physical processes of the soil such as cracking (Young 1988).

In the context of reclamation, both residuals and migrants may be important sources of propagules. Drastically disturbed lands that are topsoiled will have residuals re-introduced to the site during the topsoiling process. Conversely, non-topsoiled areas will be void of residual propagules. Under both scenarios, the process of migration will be in operation and provide a source of propagules from the surrounding plant community. This movement of propagules will be through wind, water, and animal transport.

Reclamation success is dependent upon the migration of propagules from both the short- and long-term perspectives. Artificial seeding of disturbed lands represents, in essence, an effort to replace migration and influence the composition of would be migrant species. Research on disturbed rangelands has often demonstrated more rapid rates of succession on seeded than on non-seeded sites (e.g. Smith and DePuit 1987). The propagule supply introduced during seeding usually influences initial plant community composition to a greater extent than residuals or natural migrants, unless the seeded species fail to establish. A complete or partial seeding failure will result in a greater establishment of early successional species, and the continued presence of these species will be dependent upon abiotic conditions of the site and competition from later successional plants (Redente and Hargis 1985).

Influence of propagule supply. It is difficult to distinguish the influence of propagule supply on succession from that of other successional processes, such as ecesis, reaction and competition. For example, Clements (1916) indicated that it makes no difference whether propagules occupy a site as residuals or migrants, for they will have no role in succession if establishment does not occur. Only limited research has been conducted on reclaimed rangelands to quantify the influence of propagule supply in the successional process.

A study by Doerr *et al.* (1983) examined the influence of propagule supply on early plant community structure. An area was seeded in northwest Colorado using three separate seed mixtures (native, introduced and native + introduced). The ratio among life forms in each mixture was varied to manipulate structure of the established stand. Two life form ratios were used for all seed mixtures. The first (LF-1) had approximately twice the seeding rate of grasses and forbs than did the second (LF-2). Conversely, LF-2 had two to three times the seeding rate of shrubs than did LF-1. The results of Figure 1 represent an average across the three mixtures after the second, third and fourth growing seasons.

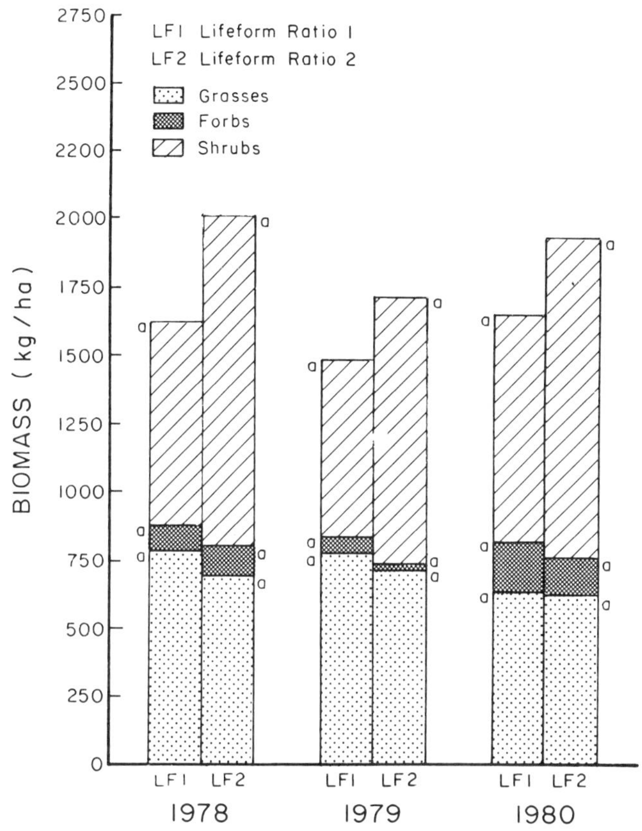

Figure 1. Mean biomass of grasses, forbs, and shrubs for seeding Life Form Ratios 1 and 2 in 1978, 1979 and 1980. Means with same letters within year and life form are not significantly different ($P \leqslant 0.10$). From Doerr *et al.* 1983.

The high grass-forb seeding rate (LF-1) resulted in first-yar (1977) grass densities of 44 plants/m^2 compared to densities of 38 plants/m^2 with the low grass-forb seeding rate (LF-2). Similarly, grass and forb biomass was slightly greater on LF-1 plots than on LF-2 plots in 1978 through 1980 (Figure 1). Conversely, LF-2 produced consistently greater shrub biomass than LF-1. The biomass differences between the two life form ratio treatments, while not statistically significant, suggest that composition changes can occur in the plant community when propagule supply is altered.

This study on propagule supply along with others (Cook *et al.* 1967; DePuit and Coenenberg 1980) indicate that successional plant expression will change with variations of seeding rate. However, these changes will not necessarily be major nor proportional to the change in seeding rate. Consequently, initial vegetation structure is also influenced by other factors in the successional process, as discussed in later portions of this chapter.

The influence of propagule supply on plant community structure was also evaluated by Redente and Hargis (1985). A diverse mixture of 17 grasses, forbs and shrubs were seeded on topsoiled coal mine spoil in northwest Colorado. At the end of the first growing season, non-seeded annual species composed 77 to 85 percent of the plant biomass. The dominant species was blue mustard (*Chorispora tenella*), and its presence caused severe seedling mortality of many seeded species. During the second growing season, dominance of volunteer species shifted to downy brome (*Bromus tectorum*), blue mustard and prickly lettuce (*Lactuca scariola*). Volunteer annual species composed 40—70% of the density and biomass during the second growing season. The trend of decreasing composition of volunteer annual species and increasing composition of seeded perennial species continued until the end of five growing seasons, when volunteer species composition was less than two percent. This study dramatically illustrates the importance of naturally and artificially introduced propagules to plant community dynamics early in succession, and reflects the unpredictability associated with the migration process.

22.3.2. Ecesis process

Once migration has occurred, the ability of a plant to become established is dependent upon its ability to withstand the conditions of the habitat. Establishment (i.e., ecesis) implies that the species completes all of its life processes from germination to reproduction. Werner (1977) attributed unsuccessful establishment to failure at one of the phenological stages: seed germination, juvenile survival, flowering, or seed set. Kerster (1968) reported that colonizing species must have the capability for surviving environmental stresses over several growing seasons in addition to withstanding conditions that inhibit seedling establishment.

In a reseeded plant community, naturally migrating species must be capable of invading the established community and then persisting and reproducing over time (McIntosh 1970). Piemeisel (1951) noted that invading annuals only passively populate openings which have been created by disturbances in perennial communities. Establishment in a seeded community will be highly dependent upon availability of areas with environmental conditions that enable invading species not only to establish, but to compete successfully. Harper *et al.* (1965) referred to these areas as "safe microsites." The availability of safe microsites and timing of migration appear to play significant roles in the ecesis process.

Species differ in their rate of establishment and may be classified as non-aggressive, aggressive or very aggressive. Aggressiveness during establishment is associated with seedling emergence and survival, and with the growth rate of seedlings (Griggs 1956). The existence of microsites also becomes important during ecesis. Harper *et al.* (1965) stressed the importance of microenvironment as it relates to germination and establishment. Evans *et al.* (1969) attributed seedling establishment to availability of microsites which met the specific seed germination and seedling growth requirements of invading species.

Because of limited water and other environmental constraints, the ecesis process comprises a high degree of uncertainty to succession on disturbed arid and semiarid rangelands in the west. Once initial plant establishment has occurred, however, succession often may proceed more rapidly in arid than in more humid environments (MacMahon 1981). Practices to promote plant establishment thus assume critical importance in successional augmentation. Such practices relate both to proper site preparation and to strategies and methods of plant establishment.

Site preparation. Initial modifications of soils and topography can serve at least two broad functions in modifying succession (DePuit 1986). First, the direction of succession may be partially influenced by the types of edaphic (e.g., Schafer 1982, 1984) and topographic (e.g., Wollenhaupt and Richardson 1982) conditions created. Second, plant establishment and soil development may be accelerated through improvements in edaphic and topographic conditions, and increased soil stabilization. The second function is of particular importance to the ecesis process.

Regrading may be accomplished to yield macrotopographies conducive to soil stabilization and water conservation (Verma and Thames 1978). Various manipulations of microtopography have also been found to enhance soil stabilization, conserve limited water and, sometimes, improve soil chemical attributes on disturbed lands. Examples of such practices include contour furrowing, trenching and terracing, pitting, microimprinting, and various primary tillage practices (Verma and Thames 1978; Dollhopf *et al.* 1985; Dixon 1980).

Various types of organic materials may be added to soils for purposes of erosion control, water conservation and enhanced pedogenesis. Organic mulches have been used most extensively for these purposes (Kay 1978), although other organic materials, such as sewage sludge (Sabey *et al.* 1986), may also have value in specific situations. Alternative approaches include use of preparatory or stubble mulch crops (Schuman *et al.* 1980; Day *et al.* 1980), which assume the functions of pioneer species in secondary succession.

Soil physiochemical problems may sometimes be corrected by direct application of the above organic and other types of soil amendments. Calcareous or acid-forming amendments may sometimes treat salt-affected soils (Sandoval and Gould 1978), while various forms of lime may ameliorate acidic soil conditions (Mays and Bengtson 1978). However, the longevity of such treatments has often been questioned. The best approach for problem soils or spoils, if feasible, may be to isolate them beneath a sufficiently thick layer of non-toxic soil material. The proper salvage and reapplication of topsoil comprises one of the most important site preparation practices for successional acceleration, since it often changes the mode of succession from primary to secondary. Considerable progress in topsoiling procedures has been made in recent years (see reviews by Hargis and Redente 1984; Schuman and Power 1981).

Disturbed rangeland soils may have serious deficiencies or imbalances in plant nutrients. In such cases supplemental fertilization may enhance initial plant establishment (Bauer *et al.* 1978; DePuit and Coenenberg 1983). However, the effectiveness of fertilization in promoting establishment may decline with increasing aridity (Aldon 1978), and over-fertilization can occasionally create problems in terms of vegetation composition (DePuit and Coenenberg 1979), soil C:N relationships (Schafer and Nielsen 1978) or microbial recolonization (Klein *et al.* 1984). Therefore, decisions on initial fertilization must be made judiciously.

Lastly, proper site preparation must provide a soil microenvironment conducive to plant establishment under the specific seeding or transplanting methods to be practiced. Richardson (1979) reviewed principles and practices of primary and secondary tillage necessary for seeding operations, and McKell *et al.* (1979) discussed practices appropriate for certain types of transplantation. In brief, transplant and seedbed preparation must be accomplished to reduce compaction and increase aeration, provide for good propagule:soil contact, reduce erosion and conserve water, control competition from indigenous plants, and yield a surface suitable for efficient operation of seeding or transplanting equipment.

The benefits of proper site preparation for plant establishment are well documented in the reclamation literature. A recent study by Smith *et al.* (1985) illustrates the importance of several facets of site preparation. This study evaluated effects of varied rates of N fertilization and soil organic matter (sawmill wood residue) amendment on revegetation of abandoned bentonite mine spoil in Wyoming. Figure 2 presents selected canopy cover data for one of the seed mixtures evaluated during the second growing season. Without wood residue amendment, establishment of seeded perennial grasses was minimal irrespective of fertilization treatment, primarily because of the poor hydraulic characteristics of the clayey spoil. Cover increased progressively and significantly with wood residue amend-

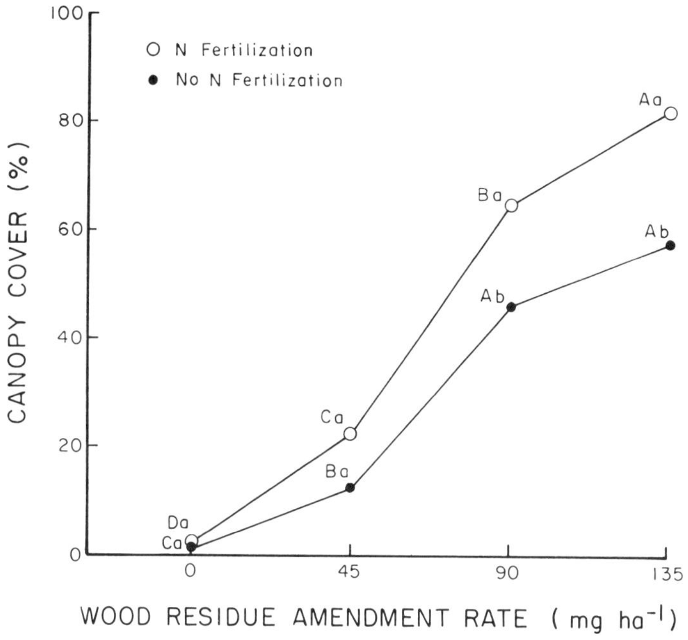

Figure 2. Canopy cover of a mixture of introduced perennial grasses as influenced by varied rates of wood residue amendment, and no N fertilization vs. N fertilization during the second growing season on bentonite mine spoils. Means across residue rates with different upper case letters and between fertilization treatments with different lower case letters are significantly different ($P \leqslant 0.05$). Derived from Smith (1984).

ment, to the intermediate residue rate without N fertilization and to the heaviest residue rate with N fertilization. This response was primarily attributed to improved water infiltration with residue amendment. Grasses were significantly stimulated by N fertilization only at the two heaviest residue rates, presumably because of higher fertilizer-use efficiency with higher water availability. These results dramatically demonstrate the value that proper site preparation can have to plant establishment. Establishment failed without amendment to improve the poor physical characteristics of the bentonite spoil. With residue amendment, revegetation succeeded and was further enhanced by fertilization.

Establishment strategies and methods. Successful establishment depends upon suitable approaches and techniques for introducing plants to disturbed lands after site preparation has been completed. At this point two basic questions must be addressed: (1) what species should be selected for introduction, and (2) what methods should be selected for plant establishment. Several reviews have been published on species selection (Thornburg 1982; Wasser 1982; Brown and Wiesner 1984; Carlson and MacArthur 1985) and seeding/transplanting methods (Cook *et al.* 1974;

Packer and Aldon 1978; McKell *et al.* 1979; U.S. Forest Service 1979; Long *et al.* 1984; DePuit 1986), the contents of which will not be reiterated here. It is important to realize, however, that successful plant introduction depends not only upon adherence to standard agricultural principles, but upon ecological propriety as well.

As reviewed by DePuit (1986), several autecological and synecological attributes of species must be considered in selection of plant materials for revegetation. Species must be adapted for establishment, growth and survival under ambient environmental conditions and the type of management proposed. They must also be capable of rapid establishment under either ambient or temporarily modified environmental conditions. They must be synecologically acceptable in terms of both co-existance with other desirable species and successional niche. Lastly, species should have utilitarian value in terms of ecological function or support of desired land uses.

An implicit, primary goal of species selection is to introduce plants that will accelerate ecesis and, thereby, the overall successional process (Moore *et al.* 1977). Successional studies on disturbed rangelands (Sindelar 1985; and others) have often noted initial dominance of species with the ruderal or competitor primary plant strategies proposed by Grime (1979). Since invasion by ruderals often occurs naturally, revegetation efforts usually strive to introduce and establish highly competitive species, such as perennial grasses that largely bypass the expression of ruderals in succession. Indeed, initial success has usually been highest with the latter species on disturbed lands (Ries and DePuit 1984). However, this success has sometimes been to the detriment of less initially vigorous, slower establishing species (such as stress tolerators) that may be quite important later in succession or in terms of plant community diversity. A number of revegetation strategies may be employed to overcome this problem.

DePuit (1982) suggested that establishment of less initially vigorous species might be enhanced by either excluding or proportionately reducing the seeding rates of highly competitive species in mixtures. While application of these approaches has sometimes been partially successful (e.g. Nilson *et al.* 1985; Kiger *et al.* 1987), the former approach carries the risk of inadequate overall plant establishment, and the latter has not always proven effective (Doerr *et al.* 1983). Multiple seeding dates have sometimes proven an effective strategy for establishment of less vigorous species, whereby such species are seeded first and allowed to become established before interseeding of more competitive species (Coenenberg 1982).

Methods of seeding, transplanting or initial plant culture may be modified to specifically favor difficult-to-establish species and enhance diversity. Research in Montana, for example, has suggested higher initial vegetation diversity with broadcast than with drill seeding, and lower diversity with heavier seeding rates (DePuit *et al.* 1980). Results of a recent Wyoming study (Powell *et al.* 1987) have indicated that the cultural practice of irrigation may enhance establishment of shrubs and perennial forbs under arid conditions. Figure 3 presents selected results of this study indicating that limited first-year irrigation (i.e., 3 months during the spring-early summer) significantly increased established densities of shrubs and forbs over non-irrigation. This occurred despite the concurrent stimulation of dominant perennial grasses under irrigation, suggesting that with proper cultural practices both overall ecesis and diversity may be enhanced. Similar relationships have been demonstrated elsewhere with other types

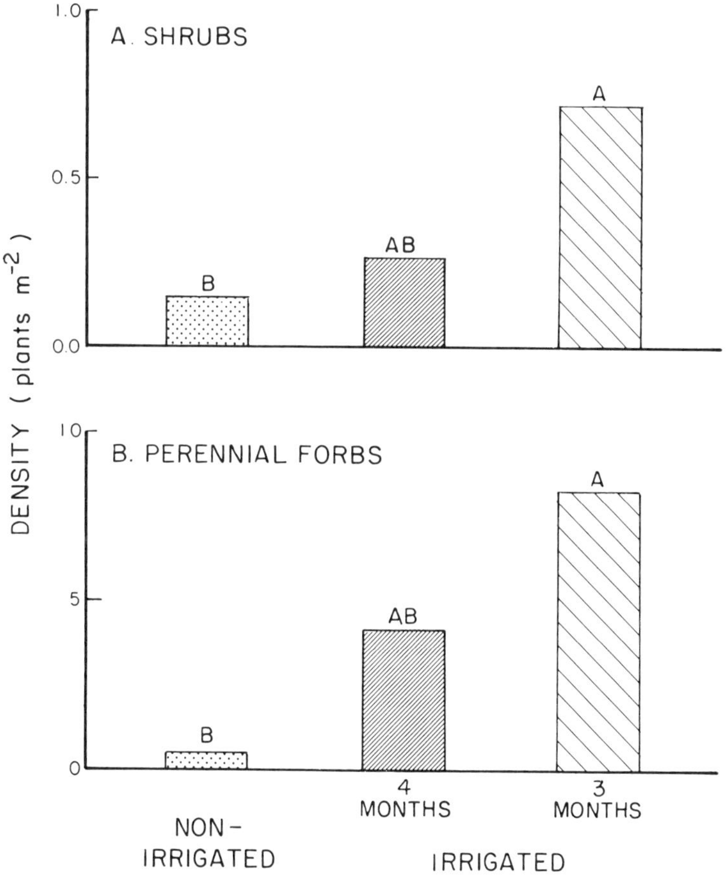

Figure 3. Effects of two durations of first season irrigation on third season density of shrubs and perennial forbs. Means with same letter are not significantly different ($P \leqslant 0.05$).

of species (e.g., Young and Rennick 1982; DePuit *et al.* 1982) or with other types of cultural practices (e.g. fertilization, DePuit and Coenenberg 1979).

Biotic interactions. Another area of importance in the ecesis process relates to plant-microbe interactions. Numerous studies have shown that the rhizosphere microflora population is an active component of the ecosystem that provides substantial amounts of nutrients for primary producers. The rhizosphere microflora may also act as a buffer allowing plants to better cope with environmental stress during and after establishment.

The interdependence of higher plants and mycorrhizal fungi have profound implications for plant species establishment and succession on severely disturbed soils in semi-arid environments (Reeves *et al.* 1979; Miller 1979). The suggestion has been made that the type and rate of succession may depend on the presence of plant-associated mycorrhizal fungi (Reeves *et al.* 1979; Janos 1980; Allen 1984; Allen and Allen 1984; Reeves 1985). The number of mycorrhizal propagules has been shown to be severely reduced with soil disturbance (Moorman and Reeves 1979; Allen and Allen 1980; Doerr *et al.* 1984). If this reduction is so great that the probability for a plant to become mycorrhizal is reduced, then the establishment of obligatory mycorrhizal species will be retarded. Rather, a succession of ruderals and facultatively mycorrhizal species will occur. The facultatively mycorrhizal species may serve as hosts for the propagation of mycorrhizal fungi, thus providing inoculum for later, obligatory mycorrhizal species (Reeves 1985).

Research in secondary succession on sagebrush communities has related the establishment of primary producers to observed patterns of soil biological activity. Mycorrhizal inoculum potential (MIP), which is a measure of the soil's potential to support mycorrhiza formation, and dehydrogenase activity, which is an index of the capacity of the soil microflora to process carbon (Skujins 1978), were positively correlated to dominance by middle and late successional species on natural and artificially reclaimed sites (Redente *et al.* 1984; Biondini *et al.* 1984, 1985).

Reclamation practices can significantly affect the ability of soil to inoculate plants with mycorrhizal fungi. One common practice associated with mining is the storage of topsoil for later use. Topsoil storage may be as short as one year or as long as the life of the mining operation (e.g. 20 years). Reeves *et al.* (1984) have shown significant and predictable reductions in the number of viable vesicular-arbuscular mycorrhizal (VAM) fungi in topsoil stored for just four years when the soil was not planted. When topsoil is planted with species able to host mycorrhizal fungi, the number of viable propagules of VAM fungi is initially increased in the rooting zone of the plants, and then stabilizes. The effective root zone appears to be 90 cm of stored topsoil. Thus, the number of VAM fungi will not increase in deeper portions of the stockpiled topsoil. Data from depths greater than 90 cm of the planted portion of the stockpile indicate a reduction of VAM fungi that parallels that found on the non-planted half of the stockpile.

If reclamation practices include the use of stored topsoil as the principal plant growth medium, upper layers of the planted stockpile should be used as inoculum to insure that a viable population of VAM fungi is present after prolonged storage. This soil can be mixed with lower layers of stored topsoil or spread as a thin veneer to enhance populations of mycorrhizal fungi and establishment of obligatory mycorrhizal species.

22.3.3. Reaction process

One of the driving forces of succession is the influence plants have on their habitat. Plants increase shade, alter vapor pressure deficit within their canopy, add organic matter to the soil, modify fluctuations in soil temperature, and alter soil structure and chemistry through root growth (Barbour *et al.* 1980; MacMahon 1983). The sum of these changes represents the reaction process; in other words, the site itself reacts to the biota and becomes a different habitat (MacMahon 1983). As this process occurs, migrants which at first were unable to establish in the immediate post-nudation environment may eventually encounter an altered environment where ecesis is possible.

In recent years there has been increasing interest shown in changes that take place belowground during the reaction process. As succession proceeds there appears to be an increase in the immobilization of inorganic compounds, a reduction in nutrient availability, and an increase in the diversity and coupling of the decomposer subsystem. These changes are believed to ultimately influence the dynamics of successional plant communities on drastically disturbed rangelands.

Immobilization of inorganic compounds and changes in nutrient availability. The relationships between nutrient cycling and succession were formally linked in a conceptual model by Odum (1969). Odum theorized that changes in both nutrient pools and nutrient cycling take place in the course of succession. Odum's model predicted that with the advance of succession:

(1) Most nutrients become immobilized in organic compounds, and the nutrient supply changes from extrabiotic to intrabiotic,

(2) Nutrient cycling closes and the nutrient exchange rates are reduced,

and

(3) The role of detritus in nutrient regeneration becomes more important.

These predictions are consistent with the behavior of some ecosystems but not with others (MacMahon 1980; Connell and Slatyer 1977). Vitousek and Reiners (1975) showed that increases in nutrient retention by the ecosystem only occur during the period of increased net primary production. During this stage, immobilization dominates over mineralization (Reiners 1981).

Nutrient cycling rates have been shown to be highest in the intermediate stages of succession and decline as the ecosystem reaches maturity (Titlyanova 1982). Changes in plant internal cycling of nutrients during succession have been examined. Reiners (1981) and Vitousek (1982) proposed that an increase in internal recycling of nutrients may be a mechanism by which late successional species compensate for reduced availability of mineralized nutrients in the soil as succession advances.

In summary, it can be stated that as the reaction process of succession occurs there is evidence that nutrients become immobilized in organic compounds and the internal recycling of nutrients becomes more important. These changes provide a different habitat for plant species to survive. Plant species that can establish and persist under these altered conditions of nutrient availability are later successional species that can tolerate lower nutrient availability and possess a greater ability to internally recycle nutrients that are in limited supply.

Importance of the decomposer subsystem. The studies noted above relate secondary succession during the reaction process to changes in plant utilization patterns and nutrient cycling. Although such theoretical concepts of microbial succession and growth strategies have been forwarded, few experiments have directly assessed the role of soil biota in succession (Swift and Heal 1986). Studies suggest that changes in microbial and faunal activity can have an important impact on succession by affecting patterns of nutrient cycling. For example, changes in nitrification potentials, N_2-fixation and soil enzymatic activities during succession have been documented in many ecosystems (Rice 1974, 1984; Robertson and Vitousek 1982; Redente *et al.* 1984; Klein *et al.* 1985; Biondini *et al.* 1985).

Research on secondary succession in sagebrush communities has related the composition of primary producers to observed patterns of soil biological activity. Carbon cycling and the potential to form mycorrhizae were positively correlated with the dominance of middle- and late-successional plant species on reclaimed rangelands (Redente *et al.* 1984;

Biondini *et al.* 1984, 1985). Heil and Dighton (1986) suggested that changes in the decomposer subsystem during the reaction process can determine whether secondary succession repeats previous successional stages. However, no direct evidence has been provided to support this important point.

The role of decomposers in plant community succession, whether it be during the establishment, competition or reaction process, is extremely important. However, many questions still go unanswered, especially as to the importance of soil microflora and fauna in affecting resource availability during plant community development and how microbial activity can be modified by applied reclamation methods.

Klein *et al.* (1984) have studied the influence of reclamation methods such as irrigation and fertilization on microbial activity. Results from studies such as these provide some insight into changes that occur during the reaction process. In their studies, Klein *et al.* (1984) found that irrigation for two growing seasons following seeding continued to show effects on belowground processes five years later. Treatments that were initially irrigated maintained higher levels of zymogenous dehydrogenase, phosphatase activity, N_2-fixation potential, NH_4^+ and NO_2^- oxidation potential, and organic matter compared to non-irrigated treatments five years following irrigation.

The use of N fertilizer on disturbed lands has also shown long-term effects on belowground processes. With added N, increased NH_4^+ and NO_2^- oxidation potentials, were observed and decreases in fungal hyphal lengths were evident five years after fertilization (Klein *et al.* 1984). This suggests that with fertilization the development of the microbial community can be retarded, an observation also made by Fresquez and Lindemann (1982). In the study by Klein *et al.* (1984), the decreases in fungal development in relation to added N could be the result of a series of possible interactions. More N could have influenced plant root exudation patterns or could have been an indirect effect of shifted carbon-nitrogen ratios which influenced the fungal-bacterial balance in decomposition. Increased N availability could also have led to the functioning of a smaller, yet more active, fungal population with a higher turnover rate. Further research, however, is required to better explain these results, which could lead to an improved understanding of reclamation effects on the reaction process and secondary succession in general.

22.3.4. Competition process

Competition, relates to the interaction among plants to obtain needed resources, and may induce changes in plant community structure during and after the ecesis phase of succession (Samuel and DePuit 1987; Harper 1977; Grime 1979). Welden and Slauson (1986) recently made a distinction between the intensity and effects of competition on community structure, and felt the two may not always be correlated. In augmented succession on disturbed lands, we are basically attempting to mediate the effects of competition on community structure and function. This mediation may be achieved by either altering the resources for which plants compete, or by directly manipulating the plants themselves.

Resource availability. Risser (1969) reviewed the manifold resources for which rangeland plants compete. The relationship between the abundance of these resources and the importance of competition in changing community structure has been the subject of considerable debate (Welden and Slauson 1986). While some argue differently (e.g., Wiens 1977), many ecologists (e.g., Huston 1979; Grime 1979; Wilson and Keddy 1986) feel that as resources become more abundant, differences in competitive ability among plants are increasingly manifested and, consequently, more strongly influence community structure. Recent revegetation research supports this hypothesis. Reductions in vegetation diversity have been noted concommitant with improved soil quality, high nutrient abundance or higher water availability (Biondini and Redente 1986; DePuit *et al.* 1982; DePuit and Coenenberg 1979). This relationship may be caused, in part, by increased expression of competitive exclusion under elevated resource abundance. Figure 4, derived from the fertilization study of DePuit and Coenenberg (1979), illustrates this relationship. While total biomass of perennial grasses increased progressively with increasing fertilization, total legume biomass declined. This relationship was primarily caused by plant density responses, which were identical to those of total biomass. Individual plant vigor (i.e., biomass) of perennial grasses was similarly stimulated by all rates of fertilization and legume vigor declined only at the heaviest fertilization rate. Individual legume species responses (DePuit and Coenenberg, 1979) were generally similar to those of total legume biomass and density in Figure 4; however, only a few perennial grass species were consistantly stimulated by increasing fertilization. Because of this, total vegetation diversity declined at the heaviest fertilization rate (Figure 4).

These relationships have important implications. They suggest that competition may play a greater role in mediating community structure in resource-rich than in resource-poor environments. If this is true, efforts to influence competition may be of far greater importance to successional augmentation in productive environments. However, the above results indicate that competitive relationships can be altered by changing resource

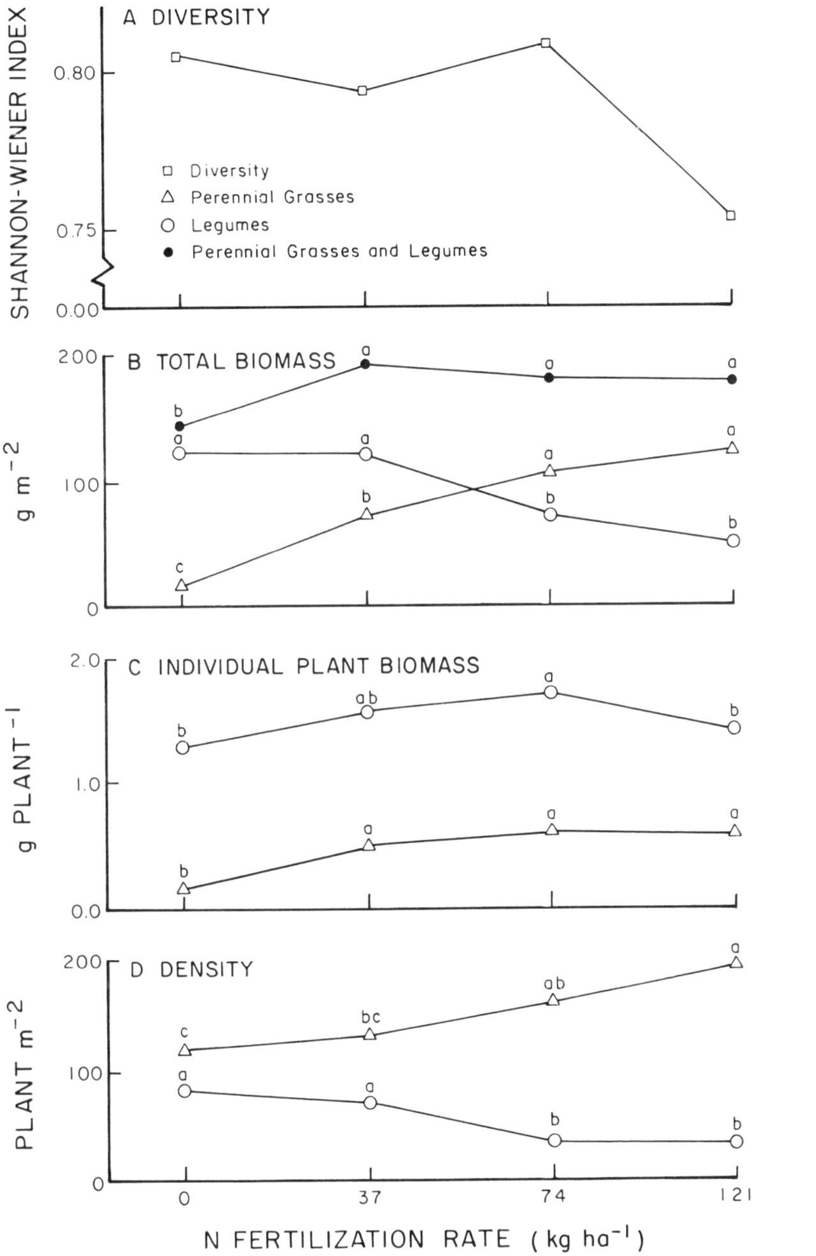

Figure 4. Effects of varied N fertilization rates on diversity, aboveground biomass and density of seeded perennial grasses and legumes. Within plant classes, fertilization rate means with same letter are not significantly different ($P \leqslant 0.05$).

availability even in less productive environments. They also suggest that if vegetation diversity is an important goal, cultural practices to increase resource abundance (e.g., fertilization and irrigation) should be applied only at low to moderate levels to maximize both productivity and diversity. Over-improvement of resource availability may accentuate effects of competition by allowing highly responsive species to express their competitive advantage over less responsive species.

Direct manipulations of plants. Competition and its influence on community structure over time can also be mediated by direct manipulations of plants themselves. As noted previously, differences in competitive ability among species (when known or suspected) should be considered when mixtures of species are composed for introduction to disturbed sites. Examples of apparent competitive exclusion in improperly composed species mixtures are common (e.g., Schuman *et al.* 1982; DePuit *et al.* 1978). Figure 5, derived from data of Samuel and DePuit (1986), illustrates the differences in competitive compatibility that can exist among species. In southeastern Montana, intermediate wheatgrass (*Agropyron intermedium*) is a vigorous and highly effective interspecific competitor, as evidenced by its reduced biomass with intraspecific competition in monoculture and suppression of western (*A. smithii*) and, to a lesser extent, thickspike (*A. dasystachyum*) wheatgrasses in two-species combinations.

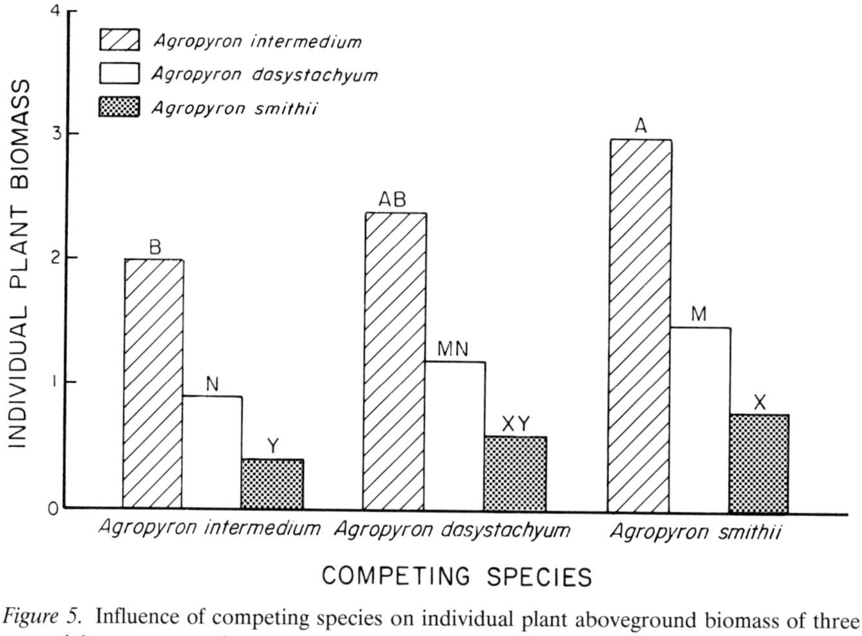

Figure 5. Influence of competing species on individual plant aboveground biomass of three perennial grasses sown in monoculture and in two-species combinations. Within a species, values across competing species treatments with same letter are not significantly different ($P \leqslant 0.05$) Derived from Samuel and DePuit 1986.

Conversely, western wheatgrass was the weakest interspecific competitor among the three species, while thickspike wheatgrass was intermediate in competitive ability. In this case, maximum compatibility would be achieved with thickspike-western and, secondarily, thickspike-intermediate wheatgrass combinations; concurrent seeding of intermediate and western wheatgrasses could result in significantly reduced western wheatgrass vigor due to intermediate wheatgrass competition.

The influence of competition on community structure can be modified by a number of direct manipulations of plants after initial establishment. Plant species often differ dramatically in their response to applied management practices. If the degree and nature (i.e., positive, neutral or negative) of response to a specific management practice varies among species in a community, the practice may alter competitive relationships among species and, therefore, induce changes in structure. For example, varied grazing systems have been suggested as means of altering competitive relationships among species on revegetated mined lands and thereby inducing changes in composition (DePuit and Coenenberg 1978), as have use of selective herbicides, prescribed burning or mowing techniques (Coenenberg 1982). Unfortunately, information on the effectiveness of such treatments is extremely limited on disturbed lands. Introduction of new species to a community through interseeding or interplanting techniques (Brock 1982) comprises another alternative for changing competitive relationships and composition later in succession, and sometimes may be practiced in combination with management practices to increase effectiveness. For example, prior mechanical or herbicidal treatment to reduce competition has sometimes been shown to improve interseeding or interplanting success on rangelands (Vallentine 1980), and this approach may have utility on disturbed lands as well.

22.3.5. Stabilization process

The successional processes described previously tend to drive recovering ecosystems toward a state of dynamic equilibrium (Whitford and Elkins 1986), during which changes in structure and species composition decline and nutrient cycling becomes more closed. While such trends have in fact been observed in successional studies on disturbed rangelands (DePuit and Redente 1987; Smith and DePuit 1987; MacKey and DePuit 1985), results have indicated that full stabilization may not occur over short time frames (i.e., years to decades) under either natural or augmented succession. Ecological stabilization has yet to be observed on drastically disturbed rangelands.

Approaches to accelerate progression toward stabilization involve many of the management practices, described previously, that are applied to influence reaction and competition processes. As reviewed by DePuit and Redente (1987), however, the ultimate ramifications of these techniques on ecosystem stabilization are largely unknown. We are hampered by lack of information on the longer-term effects of practices applied early in the successional sere. Until such information becomes available, the stabilization process on disturbed rangelands must remain little-understood.

22.4. Conclusions

The use of vegetation sciences, specifically plant ecology, physiology and soil-plant relationships, in the reclamation of drastically disturbed rangelands is essential. Without the use of these disciplines, it is unlikely that reclamation would be successful in establishing properly functioning and self-sustaining plant communities within an acceptable time frame. Successional ecology is a field of plant ecology of overriding importance to the reclamation process. The processes of succession determine the rate and nature of initial revegetation, and continue to influence plant community development long after reclamation activities are complete.

Each step of the successional process as described by Clements (1916) can be directly manipulated through reclamation practices. Through these manipulatory activities, succession can be accelerated and the structure of the plant community altered. We presently best-understand, and consequently can most effectively modify, the initial processes of succession, i.e. migration and ecesis. Our understanding of the subsequent processes of competition, reaction and, particularly, stabilization is seriously lacking. As our knowledge of basic successional relationships on disturbed lands increases, our ability to accelerate and direct succession through applied science should improve correspondingly.

22.5. References

Aldon, E. F. 1978. Reclamation of coal-mined land in the Southwest. J. Soil and Water Conserv. 33: 75—78.

Allen, E. B. 1984. VA Mycorrhizae and colonizing annuals: Implications for growth, competition and succession. p. 41—51. *In*: S. E. Williams, and M. F. Allen (eds.) VA Mycorrhizae and reclamation of arid and semi-arid lands, Wyo. Agric. Exp. Sta. Sci. Rept. No. SA1261, Univ. of Wyo., Laramie.

Allen, E. B., and M. F. Allen. 1980. Natural re-establishment of vesicular-arbuscular mycorrhizae following stripmine reclamation in Wyoming. J. App. Ecol. 17: 139—147.

Allen, E. B., and M. F. Allen. 1984. Competition between plants at different successional stages: Mycorrhizae as regulators. Can. J. Bot. 62: 2625—2629.

Barbour, Michael G., J. H. Burk, and W. D. Pitts. 1980. Terrestrial plant ecology. The Benjamin/Cummings Pub. Co. Menlo Park, CA.

Bauer, A., W. A. Berg, and W. L. Gould. 1978. Correction of nutrient deficiencies and toxicities in strip-mined lands in semiarid and arid regions. p. 451—466. *In*: F. W. Schaller, and P. Sutton (eds.) Reclamation of drastically disturbed lands. ASA/CSSA/ SSSA, Madison, WI.

Biondini, M. E., C. D. Bonham, and E. F. Redente. 1984. Relationships between induced successional patterns and soil biological activity of reclaimed areas. Recl. Reveg. Res. 3: 323—342.

Biondini, M. E., C. D. Bonham, and E. F. Redente. 1985. Secondary successional patterns in a sagebrush (*Artemisia tridentata*) community as they relate to soil disturbance and soil biological activity. Vegetatio 60: 25—36.

Biondini, M. E., and E. F. Redente. 1986. Interactive effect of stimulus and stress on plant community diversity in reclaimed lands. Reclam. Reveg. Res. 4: 211—222.

Brock, J. H. 1982. Interseeding and interplanting techniques for southwestern reclamation. p. 192—195. *In*: E. F. Aldon, and W. R. Oaks (eds.) Proceedings: Symposium on Reclamation of Mined Lands in the Southwest. Soil Conserv. Soc. Amer., Albuquerque, NM.

Brown, G. A., and L. E. Wiesner. 1984. Selecting species for revegetation: a guide for disturbed lands in the western coal region. Mt. Agric. Exp. Sta. Special Rep. No. 3, Bozeman, MT.

Carlson, J. R., and E. D. McArthur (ed.) 1985. Symposium: Range plant improvement. p. 107—220. *In*: Proceedings: Selected Papers Presented at the 38th Annual Meeting of the Society for Range Management, Salt Lake City, Utah. Soc. Range Manage., Denver, CO.

Clements, F. E. 1916. Plant succession: An analysis of the development of vegetation. Carnegie Institution Publication 242. Washington, D.C., USA.

Clements, F. E. 192. Plant succession and indicators. Hafner Publishing Company, New York.

Coenenberg, J. C. 1982. Methods for establishment of diverse native plant communities at the Rosebud Mine. p. B6(1)—B6(20). *In*: Proceedings: Symposium on Surface Coal Mining and Reclamation in the Northern Great Plains. Mt. Agric. Exp. Sta. Res. Rep. 194, Bozeman, MT.

Colie, T. S. 1940. Soil changes associated with loblolly pine succession on abandoned agricultural land of the Piedmont Plateau. Duke University School of Forestry Bull. 5: 1—85.

Connell, J. H., and R. O. Slatyer. 1977. Mechanisms of succession in natural communities and their role in community stability and organization. Am. Nat. 111: 1119—1144.

Cook, C. W., R. M. Hyde, and P. L. Sims. 1974. Revegetation guidelines for surface mined areas. Range Sci. Dept., Sci. Ser. No. 16, Col. State Univ., Fort Collins, CO.

Cook, C. W., L. A. Stoddart, and P. L. Sims. 1967. Effects of season, spacing, and intensity of seeding on the development of foothill range grass stands. Utah Agr. Exp. Sta. Bull. 455.

Day, A. D., T. C. Tucker, and J. L. Thames. 1980. Wheat establishment for mulch on coal mine soil in a semiarid environment. Desert Plants 2: 123—126.

DePuit, E. J. 1986. Western revegetation in perspective: past progress, present status and future needs. p. 6—34. *In*: M. A. Schuster, and R. H. Zuck (eds.) Proceedings: High Altitude Revegetation Workshop No. 7. Colorado Water Resources Res. Inst. Inf. Ser. No. 58, Fort Collins, CO.

DePuit, E. J. 1982. Cool-season perenial grass establishment on Northern Great Plains mined lands: status of current technology. p. B1(1)—B1(24). *In*: Preoceedings: Symposium on Surface Coal Mining and Reclamation in the Northern Great Plains, Mt. Agric. Exp. Sta. Res. Rep. 194, Bozeman, MT.

DePuit, E. J., and J. G. Coenenberg. 1979. Responses of revegetated coal strip mine spoils to variable fertilization rates, longevity of fertilization program and season of seeding. Mt. Agric. Exp. Sta. Res. Rep. 150, Bozeman, MT.

DePuit, E. J., and J. G. Coenenberg. 1978. Plant response and forage quality for livestock grazing on coal mine spoils in Montana. p. 892—898. *In*: M. K. Wali (ed.) Ecology and Coal Resource Development. Pergamon Press, NY.

DePuit, E. J., and E. F. Redente (*In Press*). 1987. Ecosystem dynamics on reconstructed semiarid lands. *In*: E. B. Allen (ed.) The reconstruction of disturbed arid ecosystems. Westview Press.

DePuit, E. J., J. G. Coenenberg, and C. L. Skilbred. 1980. Establishment of diverse native plant communities on coal surface mined lands in Montana as influenced by seeding method, mixture and rate. Mt. Agric. Exp. Sta. Res. Rep. 163, Bozeman, MT.

DePuit, E. J., J. G. Coenenberg, and W. H. Willmuth. 1978. Research on revegetation of surface mined lands at Colstrip, Montana: progress report 1975—1977. Mt. Agric. Exp. Sta. Res. Rep. 127, Bozeman, MT.

DePuit, E. J., C. L. Skilbred, and J. G. Coenenberg. 1982. Effects of two years of irrigation on revegetation of coal surface-mined land in southeastern Montana. J. Range Manage. 35: 67—74.

Dixon, R. M. 1980. Conservation tillage for mined land revegetation in southwestern United States. p. 9(1)—9(16). *In*: Proceedings: Workshop on Vegetative Reclamation of Mine Wastes and Tailings in the Southwest, Univ. Arizona, Tucson, AZ.

Doerr, Ted B., E. F. Redente, and F. Brent Reeves. 1984. Effects of soil disturbance on plant succession and levels of mycorrhizal fungi on a sagebrush-grassland community. J. Range Manage. 37: 135—139.

Doerr, T. B., E. F. Redente, and T. E. Sievers. 1983. Effect of cultural practices on seeded plant communities on intensively disturbed soils. J. Range Manage. 36: 423—428.

Dollhopf, D. J., D. W. Hedberg, S. A. Young, J. D. Goering, W. M. Schafer, and C. J. Levine. 1985. Effects of surface manipulation on mined land reclamation. Mt. Agric. Exp. Sta. Special Rep. 18, Bozeman, MT.

Evans, R. A., J. A. Young, and R. E. Eckert, Jr. 1969. Herbaceous weed control and revegetation of semi-arid rangelands in western U.S. Outlook Agric. 6: 60—66.

Farmer, E. E., R. W. Brown, B. Z. Richardson, and P. E. Packer. 1974. Revegetation research at the Decker Coal Mine in Southwestern Montana. U.S.D.A. For. Serv. Res. Paper INT-162, Ogden, UT.

Fresquez, P. R., and W. C. Lindemann. 1982. Soil and rhizosphere microorganisms in amended coal mine spoils. Soil Sci. Soc. Am. J. 46: 751—755.

Griggs, R. F. 1956. Competition and succession on a rocky fell field. Ecology 37: 8—28.

Grime, J. P. 1979. Plant strategies and vegetation processes. John Wiley & Sons, New York, NY.

Hargis, N. E., and E. F. Redente. 1984. Soil handling for surface mine reclamation. J. Soil and Water Conserv. 39: 300—305.

Harper, J. L. 1977. Population Biology of Plants. Academic Press. London.

Harper, J. L., L. T. Williams, and G. R. Sagar. 1965. The behaviorof seeds in soil. I. The heterogenity of soil surfaces and its role in determining the establishment of plants. J. Ecol. 53: 273—286.

Heal, O. W. and J. Dighton. 1986. Nutrient cycling and decomposition in natural terrestrial ecosystems. p. 14—73. *In*: M. J. Mitchell, and J. P. Nakas (eds.) Microflora and faunal interactions in natural and agro-ecosystems. Martin Nijhoff/Dr. W. Junk Publishers, Dordrecht.

Huston, M. 1979. A general hypothesis of species diversity. Amer. Naturalist 113: 81—101.

Janos, D. P. 1980. Mycorrhizae influence tropical succession. Biotropica 12: 1—95.

Kay, B. L. 1978. Mulch and chemical stabilizers for land reclamation in dry regions. p. 467—484. *In*: F. W. Schaller, and P. Sutton (eds.) Reclamation of drastically disturbed lands. ASA/CSSA/SSSA, Madison, WI.

Kerster, H. W. 1968. Population age structure in the prairie forb, *Liatris aspura*. Bioscience 18: 430—432.

Kiger, J. A., W. A. Berg, J. T. Herron, C. M. Phillips, and R. G. Atkinson. 1987. Shrub establishment in the mountain shrub zone. p. L3(1)—L3(6). *In*: Proceedings, Fourth Annual Meeting, Amer. Soc. Surface Mining and Reclam., Billings, MT.

Klein, D. A., W. Metzger, and B. Crews. 1984. Soil microorganisms and plant community development in disturbed ecosystems. p. 37—50. *In*: E. F. Redente, and C. W. Cook (eds.) Ecological studies of natural and established ecosystems on energy related disturbances in Colodado. U.S. DOE Research Report No. DE-AC02-76EVO4018, Range Sci. Dept., Colo. State Univ., Fort Collins, CO.

Klein, D. A., D. L. Sorensen, and E. F. Redente. 1985. Soil enzymes: a predictor of reclamation potential and progress. p. 141—171. *In*: R. L. Tate III, and D. A. Klein (eds.) Soil reclamation processes: Microbiological analyses and applications. Marcel Dekker, New York.

Long, S. G., J. K. Burrell, N. M. Laurenson, and J. M. Nyenhuis. 1984. Manual of revegetation techniques. U.S.D.A. For. Serv., Missoula Equip. Devel. Center 8471—2601, Missoula, MT.

Mackey, C. V., and E. J. DePuit. 1985. Natural revegetation of surface-deposited spent oil shale in Colorado. Recl. Reveg. Res. 4: 1—16.

MacMahon, J. A. 1981. Successional processes: Comparisons among biomes with special reference to probable roles of and influences on animals. p. 277—304. *In*: H. Shugart, D. Botkin, and D. West (eds.) Forest succession: Concept and application. Springer-Verlag, New York, USA.

MacMahon, James A. 1983. Nothing succeeds like succession: Ecology and the human lot. 67th Faculty Honor Lecture, Utah State University Press. Logan, UT.

Mays, D. A., and G. W. Bengtson. 1978. Lime and fertilizer use in land reclamation. p. 307—328. *In*: F. W. Schaller, and P. Sutton (eds.) Reclamation of drastically disturbed lands, ASA/CSSA/SSSA, Madison, WI.

McIntosh, R. P. 1970. Community, competition, and adaptation. Q. Rev. Biol. 45: 259—279.

McKell, C. M., and others. 1979. Selection, Propagation and Field Establishment of Native Plant Species on Disturbed Arid Lands. Ut. Agric. Exp. Sta. Bull. 500, Logan, UT.

Miller, R. M. 1979. Some occurrences of vesicular-arbuscular mycorrhiza in natural and disturbed ecosystems of the Red Desert. Can. J. Bot. 57: 619—623.

Moore, R. T., S. L. Ellis, and D. R. Duba. 1977. Advantages of natural successional processes on western reclaimed lands. p. 274—282. *In*: Proceedings: Fifth Symposium on Surface Mining and Reclamation, Nat. Coal Assoc., Louisville, KY.

Moorman, T.. B., and F. B. Reeves. 1979. The role of endomycorrhizae in revegetation practices in the semi-arid West. II. A bioassay to determine the effect of land disturbance on endomycorrhizal populations. Am. J. Bot. 66: 14—18.

Nilson, D. J., R. L. Williamson, J. C. Thompson, and J. E. Schultz. 1985. Techniques used to establish, maintain and enhance grassland seasonal variety at the Glenharold Mine in North Dakota. p. 144—152. *In*: Proceedings, Second Annual Meeting, Amer. Soc. Surface Mining and Reclam., Denver, CO.

Odum, E. P. 1969. The strategy of ecosystem development. Science 164: 262—270.

Packer, P. E., and E. F. Aldon. 1978. Revegetation techniques for dry regions. p. 425—450. *In*: F. W. Schaller, and P. Sutton (eds.) Reclamation of drastically disturbed lands. ASA/CSSA/SSSA, Madison, WI.

Piemeisel, R. L. 1951. Causes affecting change and rate of change in a vegetation of annuals in Idaho. Ecology 32: 53—72.

Powell, K. B., R. B. Vincent, E. J. DePuit, and F. E. Parady. 1987. Temporary irrigation for revegetation of mined lands under cool desert conditions. p. D4(1)—D4(8). *In*: Proceedings, Fourth Annual Meeting, Amer. Soc. for Surface Mining and Reclamation, Billings, MT.

Praeger, R. 1923. Dispersal and distribution. J. Ecol. 11: 114—123.

Redente, E. F., and N. E. Hargis. 1985. An evaluation of soil thickness and manipulation of soil and spoil for reclaiming mined land in northwest Colorado. Recl. and Reveg. Res. 4: 17—29.

Redente, E. F., J. M. Stark, M. E. Biondini, and T. A. Oliver. 1984. Vegetation structure and succession as they relate to soil disturbance and retorted oil shale. p. 1—35. *In*: E. F. Redente, and C. W. Cook (eds.) Ecological studies of natural and established ecosystems on energy related disturbances in Colorado. U.S. DOE Research Report No. DE-AC02-76EV04018, Range Sci. Dept., Colo. State Univ., Fort Collins, CO.

Reeves, F. B. 1985. Survival of VA mycorrhizal fungi — interactions of secondary succession, mycorrhizal dependency in plants, and resource competition. p. 10—113. *In*: R. Molina (ed.) 6th North American Conference on Mycorrhizae. Forest Res. Lab., Oregon State Univ., Corvallis.

Reeves, F. Brent, Jan Acsai, Saleh Solhaibani, and Andy Park. 1984. Importance of mycorrhizal fungi in revegetating disturbed soils and retorted oil shale. p. 51—59. *In*: E. F. Redente, and C. W. Cook (eds.) Ecological studies of natural and established ecosystems on energy related disturbances in Colorado. U.S. DOE Research Report No. DE-AC02-76EV04018, Range Sci. Dept., Colo. State Univ., Fort Collins, CO.

Reeves, F. B., D. Wagner, T. Moorman, and J. Kiel. 1979. The role of endomycorrhizae in revegetation practices in the semi-arid West. I. A comparison of incidence of mycorrhizae in severely disturbed vs. natural environments. Am. J. Bot. 66: 6—13.

Reiners, W. A. 1981. Nitrogen cycling in relation to ecosystem succession. p. 507—528. *In*: F. E. Clark, and T. Rosswall (eds.) Terrestrial nitrogen cycles. Processes, ecosystem strategies and management impacts. Volume 33, Ecol. Bull., Stockholm.

Rice, E. L. 1974. Inhibition of nitrification by vegetation; increases during succession and pronounced inhibition by climax ecosystems. p. 77—103. *In*: E. L. Rice (ed.) Allelopathy. Academic Press, New York.

Rice, E. L. 1984. Allelopathy. 2nd Edition. Academic Press, New York.

Richardson, B. Z. 1979. Site preparation. p. 27—34. *In*: User guide to vegetation: Mining and reclamation in the West. U.S. For. Serv. Gen. Tech. Rep. INT-64, Ogden, UT.

Ries, R. E., and E. J. DePuit. 1984. Perennial grasses for mined land. J. Soil and Water Conserv. 39: 26—29.

Risser, P. G. 1969. Competitive relationships among herbaceous plants and their influences on ecosystem function in grasslands. p. 153—171. *In*; R. D. Dix, and R. G. Beidleman (eds.) The grassland ecosystem: a preliminary synthesis. Range Sci. Ser. No. 2, Co. State Univ., Fort Collins, CO.

Robertson, G. P., and P. M. Vitousek. 1982. Nitrification potentials in primary and secondary succession. Ecology 62: 376—386.

Sabey, B. E., K. Topper, and G. Voos. 1986. Use of sewage sludge for reclamation of coal mine spoils. p. 158—186. *In*: M. A. Schuster, and R. H. Zuck (eds.) Proceedings: High Altitude Revegetation Workshop No. 7. Colorado Water Resources Res. Inst. Inf. Ser. No. 58, Fort Collins, CO.

Samuel, M. J., and E. J. DePuit (*In Press*). 1987. Competition and plant establishment. *In*: Proceedings, Symposium on Seed and Seedbed Ecology of Rangeland Plants, USDA/ARS, Tucson, AZ.

Sandoval, F. M., and W. Gould. 1978. Improvement of saline and sodium-affected disturbed lands. p. 485—504. *In*: F. W. Schaller, and P. Sutton (eds.) Reclamation of drastically disturbed lands. ASA/CSSA/SSSA, Madison, WI.

Schafer, W. M. 1984. Minesoil restoration and maturity: a guide for managing minesoil development. p. 172—185. *In*: Proceedings: Symposium on Surface Coal Mining and Reclamation in the Great Plains, Mt. State Univ., Bozeman, MT.

Schafer, W. M. 1982. Changes in land capability class resulting from mining in the Northern Great Plains. p. C1(1)-C1(10). *In*: Proceedings: Symposium on Surface Coal Mining and Reclamation in the Northern Great Plains, Mt. Agric. Exp. Sta. Res. Rep. 194, Bozeman, MT.

Schafer, W. M., and G. A. Nielssen. 1978. Soil development and plant succession on 1 to 50 year old strip mine spoils in southeastern Montana. p. 541—549. *In*: M. K. Wali (ed.) Ecology and coal resource development. Pergamon Press, NY.

Schuman, G. E., and J. F. Power. 1981. Topsoil Management on mined lands. J. Soil and Water Conserv. 36: 77—78.

Schuman, G. E., F. Rauzi, and D. T. Booth. 1982. Production and competition of crested wheatgrass-native grass mixtures. Agron. J. 74: 23—26.

Schuman, G. E., E. M. Taylor, F. Rauzi, and G. S., Howard. 1980. Standing stubble vs. crimped straw mulch for establishing grass on mined lands. J. Soil and Water Conserv. 35: 25—27.

Sindelar, B. W. 1985. Vegetation development on surface mined lands in eastern Montana. p. 6/1—30. *In*: Ecological studies of disturbed landscapes, U.S. Dept. Energy, Washington, D.C.

Skujins, J. J. 1978. Dehydrogenase: an indicator of biological activities in arid soils. Bull. Ecol. Res. Comm., Stockholm, 17: 235—241.

Smith, J. A. 1984. Wood residue and fertilizer amendments for reclamation of orphan bentonite mine spoils. Unpub. M.S. Thesis, Dept. Range Manage., Univ. Wyo., Laramie.

Smith, J. A., G. E. Schuman, E. J. DePuit, and T. A. Sedbrook. 1985. Wood residue and fertilizer amendment of bentonite mine spoils: I. Spoil and general vegetation responses. J. Environ. Qual. 14: 575—580.

Smith, P. W., and E. J. DePuit. (*Submitted*). 1987. Plant community development on abandoned petroleum drill sites in northwestern Wyoming. J. Range Manage.

Swift, M. J., and O. W. Heal. 1986. Theoretical considerations of microbial succession and growth strategies: intellectual exercise or practical necessity? p. 115—131. *In*: V. Jensen, A. Kjoller, and L. H. Sorensen (eds.) Microbial communities in soil. Elsevier Applied Science Publishers, London.

Thornburg, A. A. 1982. Plant Materials for Use on Surface Mined Lands in Arid and Semiarid Regions. U.S.D.A., Soil Conserv. Serv., SCS-TP-157, EPA-600/7-79-134, Washington, D.C.

Titylanova, A. A. 1982. Ecosystem succession and biological turnover. Vegetatio 50: 43—51.

U.S. Department of Agriculture. 1979. User Guide to Vegetation. U.S.D.A. For. Serv. Gen. Tech. Rep. INT-64, Ogden, UT.

Vallentine, J. F. 1989. Range development and improvements (2nd Edition). Brigham Young Univ. Press, Provo, UT.

Verma, T. R., and J. L. Thames. 1978. Grading and shaping for erosion control and vegetative establishment in dry regions. p. 399—410. *In*: F. W. Schaller, and P. Sutton (eds.) Reclamation of drastically disturbed lands. ASA/CSSA/SSSA, Madison, WI.

Vitousek, P. 1982. Nutrient cycling and nutrient use efficiency. Am. Nat. 119: 553—572.

Vitousek, P. M., and W. A. Reiners. 1975. Ecosystem succession and nutrient retention: a hypothesis. BioScience 25: 376—381.

Webb, D. A. 1966. Dispersal and establishment: What do we really know? p. 93—102. *In*: J. G. Hanhec (ed.) Reproductive biology and taxonomy of vascular plants. Pergamon Press, NY.

Welden, C. W., and C. W. Slauson. 1986. The intensity of competition versus its importance: an overlooked distinction and some implications. Quart. Rev. Biol. 61: 23—44.

Werner, Patricia A. 1977. Colonization success of a "biennial" plant species: Experimental field studies of species cohabitation and replacement. Ecology 58: 840—849.

Whitford, W. G., and N. Z. Elkins. 1986. The importance of soil ecology and the ecosystem perspective in surface mine reclamation. p. 151—187. *In*: C. C. Reith, and L. D. Potter (eds.) Principles and methods of reclamation science. Univ. New mexico Press, Albuquerque, NM.

Wiens, J. A. 1977. On competition and variable environments. Amer. Sci. 65: 590—597.

Wilson, S. D., and P. A. Keddy. 1986. Species competitive ability and position along a natural stress/disturbance gradient. Ecol. 67: 1236—1242.

Wollenhaupt, N. C., and J. L. Richardson. 1982. The role of topography in revegetation of disturbed lands. p. C2(1)—C2(11). *In*: Proceedings: Symposium on Surface Coal Mining and Reclamation in the Northern Great Plains, Mt. Agric. Exp. Sta. Res. Rep. 194, Bozeman, MT.

Young, James A. 1988. Seedbeds as selective factors in the species composition of rangeland communities. *In*: P. T. Tueller (ed.) Vegetation science applications for rangelands analysis and management. Kluwer Academic Publishers, The Netherlands.

Young, S. A., and R. B. Rennick. 1982. Establishment of Seeded Species on Coal Minesoils Following Temporary Irrigation. Mt. Agric. Exp. Sta. Res. Rep. 200, Bozeman, MT.

23. Rangeland vegetation as wildlife habitat

JOHN G. KIE AND JACK WARD THOMAS

23.1 Abstract

Managing rangelands as wildlife habitat requires knowledge of both plant ecology and wildlife biology. A common understanding of principles fundamental to both disciplines is also necessary. Furthermore, it should be recognized that vegetation, and wildlife habitat values, can change daily, seasonally, annually, and over longer periods of time. These changes need to be considered when formulating management and monitoring plans. The management process itself consists of four distinct steps: defining management goals, measuring habitat conditions and assessing wildlife habitat values, prescribing and carrying out management actions, and monitoring the results. Wildlife-habitat relationships models can be useful in this context.

23.2 Introduction

Rangelands support plant communities which are dominated to varying degrees by grasses, forbs, or shrubs. The predominate worldwide use of rangelands by man is to support domestic livestock. Livestock may also be present on other grazing land, for example in hardwood or coniferous forest communities. Regardless of where grazing by domestic livestock is occurring, those communities also provide habitat for numerous species of wildlife. And in many cases, rangelands are managed primarily for wildlife, particularly ungulate species.

Many techniques are available to land managers wanting to improve habitat for wildlife. If the management goal is to improve foraging habitat for mule deer (*Odocoileus hemionus*), for example, one might consider prescribed burning of decadent, over-mature stands of shrubs to encourage sprouting and germination of new, more palatable seedlings. Mechanical or

Tueller, P. T. (ed.), Vegetation science applications for rangeland analysis and management.
ISBN 90-6193-195-9.
© 1988, *Kluwer Academic Publishers, Dordrecht, Boston, London.*

chemical control of less desirable plant species, and their replacement through planting or seeding of preferred species might be considered. However, constraints imposed by other resource uses such as grazing by domestic livestock, watershed protection, recreation, and aesthetic concerns, and by costs involved frequently limit the practical use of such direct habitat improvement techniques to a relatively small scale. On the other hand, manipulation of vegetation for other major land uses such as livestock grazing, timber harvesting, and silvicultural practices impact much larger areas. As a result, coordination of wildlife habitat management goals with those larger scale, and usually better funded, management activities has a much greater potential for habitat improvement than direct improvement techniques used for wildlife enhancement alone. Similarly, while it is possible to intensively measure and assess wildlife habitat quality at the microplot level, rarely do managers have the personnel and financial resources to do so. It is usually preferable to make full use of existing habitat classification, measurement, and monitoring schemes developed and conducted for range management or forestry purposes in evaluating wildlife habitat.

Traditionally, the primary wildlife concerns of range managers have been directed towards predators of domestic livestock, gallinaceous birds, and wild ungulates. Today, managers of public and private rangelands must increasingly consider not only these species but others as well. For example, management of public rangelands and other grazing lands in the United States is guided by numerous Federal and State laws. Those laws frequently require range managers to account for the impacts of livestock grazing on the entire spectrum of wildlife species that occur within the area of concern.

Because of increased public scrutiny and concern reflected by legislative constraints, land managers are asking more and tougher questions about how range management activitics will affect wildlife. Too often, the wildlife biologist will respond by saying "Don't graze it," or of equally little help, "I don't know, I need more data." Neither answer is of much help to the range manager who cannot delay a decision. In such cases, the usual outcome is to proceed with management activities without the participation of the wildlife biologist. The resultant effect on wildlife is almost always worse than envisioned.

Why are wildlife biologists hesitant to make specific recommendations? There are good reasons. The literature on range-wildlife interactions fills volumes. Yet, much of it presents information about one specific aspect of a problem based on research conducted at one location over a short period of time. However, many wildlife species are cosmopolitan in their distribution, and their response to habitat change resulting from livestock grazing can vary widely from one ecological situation to another. In addition, annual variations in weather can be extreme, and vegetation conditions can change daily, seasonally, annually, and over longer time periods. These may lead to perturbations in plant and animal communities that can be fully understood only after many years of detailed study. Therefore, it is not surprising that much of the existing literature is replete with seeming contradictions. Most reported studies are indeed valid for the locations and years on which they are based, but may simply not be applicable over wider geographic or temporal scales. To develop comprehensive theories of habitat-wildlife interactions for rangelands, and subsequently provide detailed management guidelines applicable to all situations, is obviously beyond the scope of this chapter. Some such guidelines do exist for specific ecosystems, for example those for managing wildlife habitat on rangelands in the Great Basin of southeastern Oregon (Thomas and Maser 1986). Additional guidelines will inevitably be developed for other areas.

What we attempt to do here is: (1) to present some basic principles of rangeland wildlife management, (2) to classify types of changes in rangeland vegetation and their effects on wildlife habitat, and (3) to discuss how rangeland habitats may be assessed and managed for wildlife.

23.3 Principles of rangeland wildlife management

Professionals active in resource management come from varied educational backgrounds such as range management, ecology, geology, wildlife biology, fisheries biology, engineering, animal husbandry, and landscape architecture. In order to work together, they need a common vocabulary and understanding of range management principles, plant and animal ecology, and wildlife management. Discussed by Odum (1963), Leopold (1933), and Hylander (1966), these definitions are mainly from Thomas (1979) and Maser and Thomas (1983).

Each animal species requires for its existence three elements: food, water, and cover. Food and water are necessary to sustain the basic functions of maintenance, growth, and reproduction. Cover is vegetation used by wildlife for protection from predators (hiding or escape cover), or to ameliorate extremes of weather (thermal cover) (Thomas 1979).

Habitat is the place where an animal finds the required arrangement of food, cover, and water to meet its biological needs. The role a particular wildlife species plays in the environment is referred to as its ecological niche.

A plant community type is a unique combination of plants that occurs in particular locations under certain environmental influences. The plant community type reflects the environmental influences of the site, such as soil, temperature, elevation, solar radiation, slope, aspect, and rainfall as they influence vegetation (Daubenmire 1976).

Plant communities are defined and named in terms of dominant overstory and understory species of climax vegetation. Several plant community types may be included. A plant community evolves through a general series of conditions as it progresses from bare ground to a climax stage. This process is called primary succession, and the various stages are known as successional stages. In rangeland communities in which the vegetation is manipulated, singly or in combination, by fire, mechanical means, herbicides, and ungulate grazing, secondary succession occurs. In some cases with repeated, frequent disturbance, this process is so radically modified that a relatively stable state evolves and is maintained by people or their domestic animals. This is called a stage of disclimax or anthrogenic (human generated) subclimax (Odum 1971, Tueller 1973).

Each combination of a plant community and a successional stage or structural condition produces a unique set of habitats. The wildlife supported by these habitats make up the attendant animal community. The animals fill various ecological niches and, in turn, influence the plant community. Individual species may use a particular habitat on a seasonal or yearlong basis. See also the reviews of Meslow and Wight (1975) and Thomas *et al.* (1975).

Some of the ecological terms already discussed have counterparts that are more commonly used in forest and range management. It is important to understand the relationship of these terms to the language of ecology and wildlife biology.

A site is an area considered in terms of its environment, particularly as this determines the type and quality of the vegetation the area can support. Sites are qualitatively classified into site types by their climate, soil, and vegetation. Site type is analogous to habitat type or plant community.

Each site is occupied by one or more stands. A stand is a subset of a plant community that has sufficient uniformity of composition, size, density, age, spatial arrangement, and condition to distinguish it from adjacent stands (see chapter by West in this book). Stands are the common basis on which prescriptions to manipulate vegetation are considered.

The juxtaposition of plant communities, successional stages, or stand conditions within communities produces edge. The area where communities or successional stages overlap or produce a distinct combination of plants or structure is called the ecotone. Edges and ecotones are rich habitat for many species of wildlife because of the attributes of the edge itself plus those of the adjoining communities or successional stages (Leopold 1933).

The influence of this phenomenon on animal populations is called edge effect.

Increasing the amount of edge increases habitat richness, which is an index to the potential number of wildlife species resident within an area. It should be kept in mind, however, that some species of wildlife require large, unbroken blocks of habitat and that such species may be lost from a dramatically altered habitat composed of many small stands and much edge. Interspersion is a measure of the degree to which plant communities or successional stages mix. An increase in interspersion increases the amount of edge. In turn, this may increase diversity or the variety that exists in plant and animal communities (Patton 1975). A rangeland with a high degree of diversity of communities and successional stages provides habitat for a wide variety of wildlife (Odum 1971).

A rangeland ecosystem is a dynamic complex of plant and animal communities that, along with the abiotic environment, comprises one functioning whole. Any change in vegetative structure or composition will favor some wildlife species while adversely affecting others. Such changes can affect the number and type of wildlife species and their use of habitat. The key to using the relationship between wildlife and habitat as a predictive range management tool is the synthesis of existing knowledge into some useable format (see Thomas 1979, Verner and Boss 1980, Maser and Thomas 1983, Hoover and Wills 1984, Brown 1985a, Brown 1985b, and Cooperrider *et al.* 1986 as examples).

23.4 Vegetation change and its effect on wildlife habitat

The concept of plant succession, the replacement of one assemblage of plant species with another and then another through time until some stable climax community is reached, was proposed by Clements (1916). As one group of plant species is replaced by another, the value of the community as habitat to any particular species of wildlife changes. Although succession represents change in plant community structure and value as wildlife habitat, other changes occur in vegetation communities, and succession should be viewed in context with those changes.

Heady (1973) proposed a useful classification of vegetation changes that included daily, seasonal (within-year), annual (between-year), and successional changes, as well as long-term changes brought about by introduction of new species, evolution, climatic and geographical shifts. Although long-term changes mentioned by Heady (1973) have little applicability to the day-to-day problems facing range and wildlife managers, the daily, seasonal, annual, and successional changes do have an impact on the way rangelands should be evaluated, managed, and monitored.

23.4.1 Daily changes in vegetation

The daily changes that occur in plant communities mostly involve physiological responses of individual plants. Although plant structure, and hence value as wildlife cover, rarely changes on a daily basis, plant water and carbohydrate cycles may affect forage values. For example, oryx (no scientific name given in original article) in eastern Africa may exist in xeric environments by feeding on grasses and shrubs only at night when they are highest in moisture content (Taylor 1969). Because daily changes are primarily the result of diel cycles in plant physiology, little opportunity exists for manipulating them to achieve specific wildlife habitat management goals. Daily changes in vegetation may be sufficient to warrant consideration in monitoring programs, however. For instance in the previous example, sampling for forage moisture content should be done at the same time each day.

23.4.2 Seasonal changes in vegetation

Seasonal variation results from changes in plant phenology and is well exemplified by vegetation structure in the California annual grasslands of the United States. Seeds from annual plants germinate in fall following rainfall that exceeds 15 mm (Heady 1977). Plants grow slowly during the cool, wet winter months, and then increase rapidly in size during the warm spring months. By early summer, rainfall has ceased in the Mediterranean climate, and growth of most annual plants has stopped. A few summer-maturing annual plants appear during the dry season such as turkey mullein (*Eremocarpus setigerus*), which produces seeds that are important to mourning doves (*Zenaidura macroura*) and other wildlife species, both seed-eating birds and small mammals. Not only does plant community structure change, so does forage quantity and quality for both domestic and wild ungulates. Forage quality is high but quantity is low from the period of germination through the cool winter months. As plant growth rates increase during spring, quantity increases while quality remains high. As plants mature and cure during the summer, quantity may still be high, but quality declines rapidly (Kie 1988). For example, crude protein in soft chess (*Bromus mollis*), a common grass species, may range from a high of over 22% during early-leaf stage in February to a low of less than 4% after seed cast in September (Gordon and Sampson 1939).

Similar changes can also occur over the course of a single growing season in perennial plant communities. In montane meadows in the Sierra Nevada of California in the United States, corn lily (*Veratrum californicum*) begins growth from an underground root stock in early summer and provides valuable hiding cover for mule deer fawns when they are young and most susceptible to predation from mountain lions (*Felis concolor*), coyotes (*Canis latrans*), and black bears (*Ursus americanus*). As the corn lily plants mature, they lose their value as cover through the natural weathering process even in the absence of grazing by livestock (Loft *et al.* 1987). By the time corn lily loses its value as hiding cover through natural processes, the fawns are old enough and agile enough to escape most predators. Hence, the loss of hiding cover is not as critical as it would have been earlier in the summer. However, excessively heavy livestock grazing accelerates the loss of cover early in the summer, and may result in increased susceptibility of fawns to predation.

Grazing by domestic livestock, and to a lesser extent by wild ungulates, can have profound effects on vegetation structure over a short time period through the removal of biomass and by changing species composition – even within a single growing season. In the annual grasslands of California, cattle remove the taller grasses and encourage the growth of lower-stature forbs such as filaree (*Erodium* spp.) and summer-maturing forbs such as turkey mullein (Duncan 1976, Kie 1988). These and other plant species are valuable as forage for many species of wildlife. In most cases, the loss of herbaceous cover is somewhat offset by the production of preferred forbs as forage if cover is available in nearby shrub-dominated communities.

23.4.3 Annual changes in vegetation

Continuing with the annual grasslands in California as an example of vegetation changes between years, years when fall rains which initiate germination are followed by consistent precipitation, the herbaceous plant community is composed mostly of rapidly growing grasses (Pitt and Heady 1978). However, if rainfall sufficient for seed germination is followed by a period of reduced precipitation, deeper rooted forbs survive better than do grasses and the herbaceous community is dominated by those species (Duncan and Woodmansee 1975).

As with daily changes, annual changes are the primary result of abiotic factors, and there is little opportunity to influence them through management practices. However, the manager needs to recognize the existence of such potential changes in plant community structure and composition between years and adjust plans accordingly. For example, a level of livestock grazing designed to leave adequate hiding cover for mule deer fawns during years of average precipitation may be excessive during years of drought. Close attention to annual differences in weather and resulting plant production, coupled with flexibility in livestock management programs, are necessary to insure the maximum probability that wildlife habitat management goals are met.

23.4.4 Successional changes in vegetation

The concept of ecological succession is well covered in all basic ecology (Odum 1971, Krebs 1972, Ricklefs 1973) and introductory range management textbooks (Stoddart *et al.* 1975). Although the concept of terrestrial plant succession was first proposed by Clements (1916), succession in freshwater fish communities was proposed by Shelford (1911a, 1911b). The process of succession also applies to terrestrial wildlife species groups. As one group of plant species is replaced by another, the value to any particular species of wildlife as foraging habitat changes. Also, as early seral stage plant species, usually low in stature, are replaced by often taller mid-successional and late-successional stage species, cover values change as well. The result of these changes in habitat values is a succession in assemblages of wildlife species (Lyon 1987).

Rangelands exist in any number of successional stages and structural conditions because of the influence of fire, mechanical disturbance (plowing, disking, chaining), herbicide treatments, and grazing by wild and domestic herbivores. Plant communities respond to ungulate grazing in a more or less predictable way depending on the plant species present and type of ungulates (Dyksterhuis 1949, Stoddart *et al.* 1975). Some plant species are dominant in climax communities because they are superior competitors in the absence of disturbance. These species are often the most palatable to livestock and the most susceptible to grazing pressure, and they begin to decline in vigor and abundance with defoliation caused by grazing. These species are referred to as decreasers (Dyksterhuis 1949, Stoddart *et al.* 1975). As the decreasers decline in importance in the community, other plants which are present at climax but which may be somewhat less palatable, become more abundant with relief from competition. These species are called increasers. If grazing intensity is sufficiently heavy and it occurs over a long enough period of time, the increasers also begin to decline in abundance, and new plant species that are well adapted to repeated, heavy defoliation appear in the community. These are referred to as invaders. The relationship between these three types of plant species can be expressed along a retrogressive successional gradient (Fig. 1).

Range condition is traditionally judged by the relative contribution of

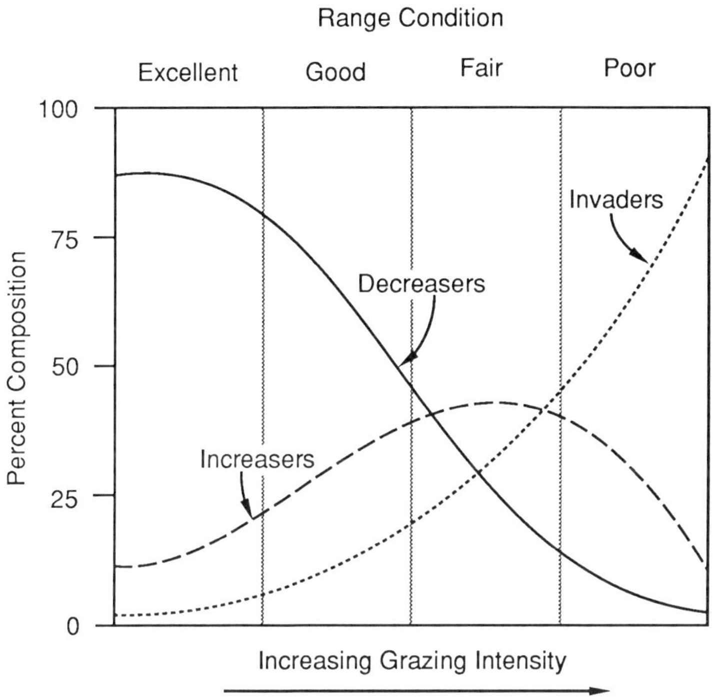

Figure 1. The relationship between decreaser, increaser, and invader plant species, grazing intensity, and traditional definitions of range condition (after Dyksterhuis 1949, Stoddart *et al.* 1975).

these three types of plants to total cover, based on concepts developed by Dyksterhuis (1949). Sites dominated by decreasers are generally classified in excellent condition, while those made up primarily by invaders are judged to be in poor condition (Fig. 1). However, the terms excellent, good, fair, and poor all depend on management objectives (Smith 1978). For example, range dominated by increasers, or even invaders, may be better habitat for mule deer than range dominated by decreasers. The reader should see the last section on managing rangeland vegetation as wildlife habitat.

Because different successional stages differ in condition and value as habitat for any given species of wildlife, the mix of wildlife species depends a great deal on successional stage. Johnston and Odum (1956) expressed the abundance of passerine birds as a function of community type along a progressive successional gradient in the Piedmont region of Georgia in the United States (Fig. 2). Using terminology familiar to the range manager, red-eyed vireos (*Vireo olivaceus*) and wood thrushes (*Hylocichla mustelina*) would be classified as decreasers, pine warblers (*Dendroica*

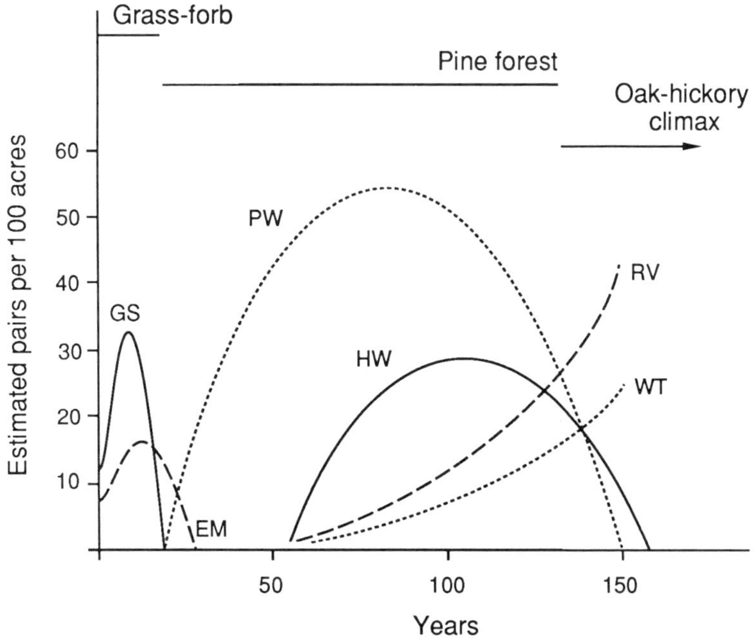

Figure 2. Abundance of selected bird species along a successional gradient in the Piedmont region of Georgia (data from Johnston and Odum 1956: 54). (GS = grasshopper sparrow, EM = eastern meadowlark, PW = pine warbler, HW = hooded warbler, RV = red-eyed vireo, WT = wood thrush).

pinus) and hooded warblers (*Wilsonia citrina*) would be grouped as increasers, and grasshopper sparrows (*Ammodramus savannarum*) and eastern meadowlarks (*Sturnella magna*) would be invading species.

In a similar manner, roe deer (*Capreolus capreolus*) are common in England in open stands of larch with an understory of grasses, sedges, and bracken fern. As the forest becomes more closed with a heather understory, red deer (*Cervus elaphus*) become more numerous and roe deer decline. Finally, in a mature forest with patchy openings, fallow deer (*Cervus dama*) become the dominant deer species (Batcheler 1960).

Hanley and Page (1982) reported on the effects of livestock grazing in the Great Basin in the United States on habitat structure and rodent populations. They found that with livestock grazing, microtine rodents such as montane voles (*Microtus montanus*) and long-tailed voles (*M. longicaudus*) acted as decreaser species. Other species acted as decreasers in some habitats and increasers in others. Least chipmunks (*Eutamias minimus*), Great Basin pocket mice (*Perognathus parvus*), and deermice (*Peromyscus maniculatus*) responded as decreasers in xeric habitats dominated by shadscale (*Atriplex confertifolia*), greasewood (*Sarcobatus vermiculatus*), low sagebrush (*Artemisia arbuscula*), and big sagebrush (*A.

tridentata). In more mesic habitats, those dominated by silver sagebrush (*A. cana*), baltic rush (*Juncus balticus*), aspen (*Populus tremuloides*), and Nevada bluegrass (*Poa nevadensis*), those same rodent species responded as increasers.

All organisms fit into a trophic pyramid either as producers (plants), primary consumers (herbivores), or secondary and higher-order consumers (predators). As plant communities undergo successional changes, resulting changes can occur upwards through the pyramid. Andelt *et al.* (1987) reviewed the effects of successional changes in vegetation on the diets of coyotes (*Canis latrans*) in south Texas in the United States. In the late 1950's, prickly pear cactus (*Opuntia lindheimeri*), a low-successional stage species, was common over much the study area and was a codominant in the prickly pear-shortgrass habitat type. Coyotes ate substantial amounts of prickly pear fruit during the months of July, August, and September. In addition, the prickly pear-shortgrass community provided preferred cover for southern plains wood rats (*Neotoma micropus*) and coyotes preyed extensively on them during winter. As a result of above-average precipitation and reductions in livestock grazing pressure, prickly pear became rare by the early 1970's because it could not compete with other plant species. The former prickly pear-shortgrass community was reclassified along with another habitat type by Drawe *et al.* (1978). As a result of the loss of prickly-pear habitat, wood rats became much less abundant. Both prickly pear fruit and wood rats have been rare occurrences in coyote diets since those changes occurred (Andelt *et al.* 1987).

Vegetational changes resulting in shifts in small mammal populations can, in turn, affect raptorial birds. Winter (1981) found that great gray owls (*Strix nebulosa*) in the central Sierra Nevada of California in the United States preyed on both pocket gophers (*Thomomys* spp.) and microtine rodents (*Microtus* spp.). On a study area grazed by cattle on the Stanislaus National Forest, total biomass of great gray owl diets averaged 85% gophers and 11% microtines. In an ungrazed study area of comparable habitats within Yosemite National Park, diets averaged 54% gophers and 43% microtine rodents. Winter (1981) further suggested that the food value of gophers was lower in terms of gross food yield per unit of effort because they were more difficult to catch. Some studies have found that gophers are more abundant on heavily grazed rangelands (Buechner 1942), although such is not always the case (Turner *et al.* 1973). There is a clear relationship between the removal of herbaceous cover by livestock and the decline of microtine rodents (Hanley and Page 1982).

The numbers and distribution of Swainson's hawks (*Buteo swainsoni*) have decreased significantly throughout a major portion of the species range in the western United States (Sharp 1986). Retrogressive successional changes in plant communities brought about as a result of heavy grazing by livestock have resulted in the reduction of native bunchgrasses (primarily *Agropyron* spp. and *Festuca* spp.) and the increase of sagebrush (*Artemisia* spp.) and rabbitbrush (*Chrysothamnus* spp.) in Butte Valley in northeastern California (Woodbridge, personal communication, Sharp 1986). Few microtine rodents are available as a prey base in sagebrush-rabbitbrush habitats, and Swainson's hawks now spend 95% of their foraging time in mowed alfalfa fields, feeding primarily on Belding's ground squirrels (*Spermophilus beldingi*) (Woodbridge 1986).

23.5 Measurement and analysis of rangeland vegetation as wildlife habitat

Techniques used in the measurement and analysis of rangeland vegetation are well documented in handbooks, textbooks, and other documents. Publications by the U.S. National Research Council of the National Academy of Sciences (NRC 1962) and the USDA Forest Service (USFS 1962, 1970), although somewhat dated, provide much useful information on basic range research methodology that is useful in considering wildlife habitat. More recent treatments include Chambers and Brown (1983) and the comprehensive textbook of Cook and Stubbendieck (1986). Measurement of wildlife habitat variables are covered by several chapters in Schemnitz (1980) and by Hays *et al.* (1981).

We define several terms here. Measurement is the determination or characterization of plant or plant community variables. Those variables might include characteristics of individual plants such as height, weight, crown cover, foliage density, or the chemical makeup of different plant parts. They might include the first-order community variables of average plant height, number of species, frequency, density, cover (traditionally a vertical projection of plant crowns, but might also include a description of the vertical profile of the plants projected horizontally), biomass or standing crop, or annual production. And they might include second and higher-order variables such as species composition, dominance, pattern, diversity, amount of edge, community coefficient, or continuum coefficient.

Analysis is the process of relating physical measurements to the basic habitat needs of food, cover, and water for specific wildlife species. Plant communities composed primarily of seral grasses and forbs provide optimum habitat for some wildlife species, while those consisting of climax vegetation provide optimum habitat for a largely different group of species. A plant community characterized by particular attributes of vertical plant profile and foliage density may provide optimum hiding or escape cover for mule deer but inadequate cover for the larger elk (*Cervus elaphus*) (Thomas *et al.* 1979b, Leckenby 1984). Plants containing a given level of nitrogen may or may not provide adequate levels of digestible protein for a ruminant depending on presence of secondary plant compounds that inhibit digestion (Robbins *et al.* 1975, Mould and Robbins 1982) as well as the specific nutritional requirements of the particular ruminant species, age, and sex involved.

Monitoring is repeated measurement and analysis over time. Monitoring occurs because of the need for information about temporal changes in wildlife habitat values (over daily, seasonal, annual, or longer time periods), or the need to determine and follow wildlife habitat response to management practices.

In planning schemes to measure, analyze, and monitor rangeland vegetation as wildlife habitat, several considerations come into play. Of overriding concern is the need to clearly define one's objectives. A research scientist studying some aspect of wildlife habitat may need very precise and accurate estimates of a variety of variables. Such micromeasurement is invariably time-consuming, labor-intensive, and costly. For example, a study of the effects of livestock grazing on deer hiding-cover values in a riparian zone might involve the collection of information on shrub density, canopy cover, and the density of their foliage, as well as similar information about the herbaceous understory. These measurements would likely be required from a large number of plots to yield the power sufficient for detecting subtle changes in hiding-cover values. In addition, since these variables are likely to change within and between seasons, and from one year to the next, repeated sampling will likely be necessary over long periods of time.

A resource manager, on the other hand, may need far less detailed information on which to base management decisions. A series of photos taken from permanent locations in the riparian zone during the same time of the year over several years may be entirely adequate to determine if any changes in deer hiding-cover values are occurring and what, if anything, needs to be done in modifying livestock grazing practices to insure deer habitat goals are being met. In most cases the needs of resource managers can be addressed through information gained at the macromeasurement level.

Another consideration involves the stratification, mapping, and classification of sites into habitat types. We cannot overstress that it is far preferable to use existing habitat classification systems, maps, and geographic information systems than to develop new ones if at all possible. The development and verification of such information is extremely costly, and seldom would such costs be justifiable. Also, because wildlife habitat quality is so dependent on other land uses such as logging and livestock grazing, developing a new habitat classification system would almost preclude the use of existing resource data sets (Thomas 1979b). For example, detailed plant productivity information may be available from livestock managers based on existing range site classification systems. Developing a new wildlife habitat classification scheme would not only involve great initial cost, but repeated costs to obtain parallel plant production data. Where existing classification systems are either too narrowly or too broadly defined, it may be possible to group or further subdivide habitat types to suit particular purposes (Thomas *et al.* 1979a).

Although the use of existing data collection programs is encouraged, there are wildlife habitat variables that are unique and require separate efforts. In the terminology of the plant ecologist or range manager, cover refers to the percentage of the area occupied by various plant species (Cook and Stubbendieck 1986). It is obtained by projecting the crowns of the individual plants downward onto a horizontal plane. As previously defined here, cover to a wildlife biologist is the vegetation used by wildlife for protection from predators or to ameliorate extremes of weather. To measure cover in this sense, the canopies of individual plants are often projected onto a vertical rather than a horizontal plane (Nudds 1977).

For example, in studying the effects of cattle grazing on the structural profile of mule deer hiding cover, Loft *et al.* (1987) measured hiding cover at eight compass bearings from the center of each of several 5.65 m radius circular plots. A 1-m^2 grid, divided into 0.01-m^2 cells, was observed from 0.75 m above ground at the center of the plot. The percentage of the grid obscured by vegetation and debris was recorded as hiding cover at height strata of 0—0.5 m, 0.5—1.0 m, and 1.0—1.5 m above the ground. Hiding cover was considered most important for fawns at the two lower strata and for adult deer at the two higher strata.

Thomas *et al.* (1976) defined elk hiding cover as vegetation capable of hiding 90% of an elk from human view at 61 m or less. The actual distance was called the sight distance. These and other measures of hiding cover are discussed by Gysel and Lyon (1980), Hays *et al.* (1981), and Nudds (1977).

The concept of habitat edge has been considered by wildlife biologists in two primary ways. The first is as a habitat attribute for individual species such as elk (Leckenby 1984, Reynolds 1966, Harper 1969, Thomas *et al.* 1979c). The second is as a generalized measure of diversity or habitat heterogeneity. This concept and techniques for measurement are described by Patton (1975) and Thomas *et al.* (1979c).

Although the amount of habitat edge is a variable that is infrequently included with other resource data collection systems, it can be readily calculated from existing geographic information systems. Edge can be measured directly from a map, or distribution of the distances from a random sample of points to the nearest habitat edge can be plotted and compared between areas. Wildlife use of or preference for edges can also be estimated. However, Donavin Leckenby of the Oregon Department of Fish and Wildlife (personal communication) has pointed out that a random distribution of points in any habitat mosaic is not uniform, but that most points lie close to habitat edges. Therefore, animal preference for habitat edges can be examined only by looking at the distribution of animal locations in conjunction with the distribution of a random set of points. Edge effect cannot be assumed merely because the animals in question are most frequently encountered in or near edges. Only if the distribution of animal locations is skewed towards zero to a greater extent than the distribution of random points can preference for edges be inferred.

23.6 Managing rangeland vegetation as wildlife habitat

There are four steps to consider in managing rangeland vegetation for wildlife habitat (Fig. 3). The first step is to clearly define one's management goals. Too often, management goals are vaguely worded with no clear objective. An example of a vague management goal is to use prescribed fire to improve deer habitat. More specific objectives include habitat responses (to encourage the germination, growth, and survival of a specific species and numbers of seedlings per unit area), deer use (to increase deer use of burned areas by a specified number of deer days per unit area over a specified time period), or deer population response (to increase the size of the herd by a specified amount).

The second step is to measure current habitat conditions and assess their value to wildlife. Measurement and analysis of habitat conditions require careful attention to sound sampling design and statistical considerations. Designation of desired statistical power should be part of the design of the sampling scheme. The intensity and detail of evaluation is determined by the manager's need for information and the cost of obtaining the data. Each increment in statistical reliability results in increased cost. Such trade-offs should be carefully considered prior to deciding on a sampling scheme.

The third step is to prescribe management actions to achieve the desired goals. Direct habitat manipulation may be necessary in some cases, and might include the use of prescribed fire, mechanical treatments, planting and seeding, and structural developments. Such direct techniques are costly and, as a result, rarely can they be carried out on a scale large

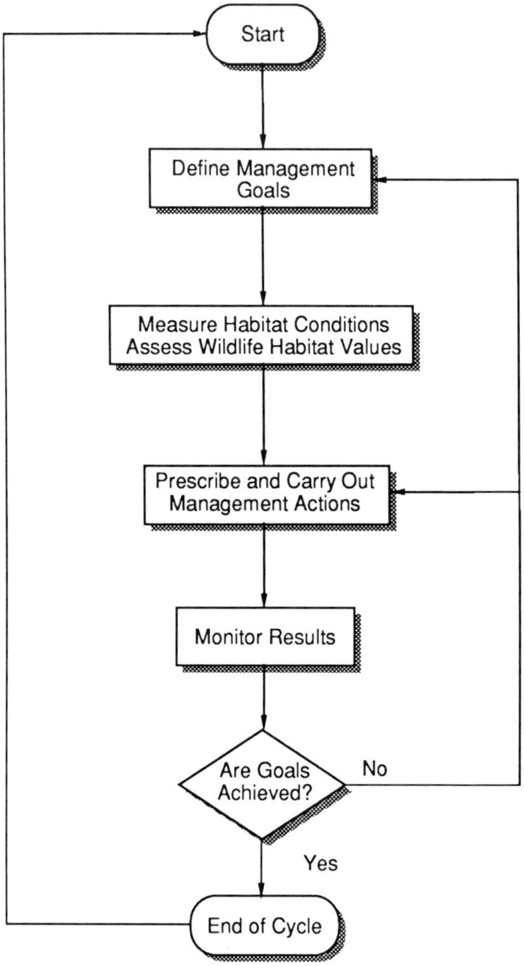

Figure 3. Flowchart for managing rangeland vegetation as wildlife habitat, showing a single management cycle.

enough to fully meet wildlife habitat management objectives. Indirect habitat improvement through coordination with other resource uses such as livestock grazing and timber harvesting provides a much greater opportunity to benefit wildlife habitat over large areas. Conversely, lack of coordination can have equally as extensive adverse effects.

The fourth step, often overlooked or ignored, is to monitor the effects of changes in management and the degree to which wildlife habitat management objectives are being met. Any monitoring program should be directly related to what were defined as the management goals or objectives. If the purpose of a prescribed burn was to encourage the growth of new shrub seedlings, then shrub seedling density and growth rates need to be measured periodically. Better yet, if information on the density of shrubs is available from existing resource inventories, for example those carried out for livestock management purposes, then that information should be used. Such habitat-based measurements are often the easiest to obtain and favor defining habitat management goals in terms of plant community response. More meaningful measures of habitat improvement success would include either increases in deer use of burned habitats or, ultimately, increases in deer population size as a result of habitat improvement activities. Unfortunately, the value of information collected in monitoring is proportional to its cost.

Information-handling systems on which to base resource management decisions have been used in one form or another for years. The four steps outlined above follow in chronological order, with a continuous loop feeding back from step four (monitoring) to step three (management), and in some cases back to step one (definition of new management goals) (Fig. 3). Information systems are designed to facilitate this process.

An example of such a system is the procedure developed by Dyksterhuis (1949). This procedure is used extensively by the United States Soil Conservation Service (SCS) for evaluating range condition (Fig. 1). Range condition is based on how close existing vegetation approximates the potential of an undisturbed, climax community. For purposes of livestock management, this definition has served well in many areas of the United States. However, the SCS system cannot be used on most seeded stands, or other sites containing large numbers of introduced plant species such as the annual grasslands of California (Smith 1978, also see chapter by Smith in this book).

More importantly, as we have seen, different wildlife species differ in their habitat needs. A site rated in poor condition by the SCS system may provide optimum habitat for a wildlife species dependent on seral stage vegetation, while a site rated as excellent may provide no habitat at all for the same species. Descriptions of range condition, then, should be dependent on management goals (Smith 1978).

Similar to principles suggested by Smith (1978), Schlatterer (1986) proposed a system of classifying range condition based on management objectives for soil stability and protection, forage species composition, and specific management goals for other resource values such as big game cover, riparian areas, or fish habitat. Overall range condition would be judged satisfactory only if a site were judged satisfactory under all objectives, or if one or more objectives were not being met, then there was an improving trend in that area. Such a system clearly has benefit over one that defines range condition in a narrow sense.

Finally, the system of classifying wildlife habitats according to potential natural vegetation and seral stage in coniferous forests developed by Thomas (1979) and adopted for other areas (see for example, Verner and Boss 1980), has also been applied to rangeland vegetation in southeastern Oregon (Maser *et al.* 1984, Thomas and Maser 1986). Such models can be invaluable for managing rangelands as wildlife habitat. It should be remembered, however, that predictions of wildlife response to management activities are only as good as the models on which they are based (Lyon 1987).

23.7 References

Andelt, W. F., J. G. Kie, F. F. Knowlton, and K. Cardwell. 1987. Variation in coyote diets associated with season and successional changes in vegetation. J. Wildl. Manage. 51: 273—277.

Batcheler, C. L. 1960. A study of the relations between roe, red, and fallow deer with special reference to Drummond Hill Forest, Scotland. J. Anim. Ecology 29: 375—384.

Brown, E. R. (ed.) 1985a. Management of wildlife and fish habitats in forests of western Oregon and Washington. Part I — Chapter narratives. USDA Forest Service, Pacific Northwest Region, Portland, Oregon.

Brown, E. R. (ed.) 1985b. Management of wildlife and fish habitats in forests of western Oregon and Washington. Part II — Appendices. USDA Forest Service, Pacific Northwest Region, Portland, Oregon.

Buechner, H. K. 1942. Interrelationships between the pocket gopher and land use. J. Mammal. 23: 346—348.

Chambers, J. C., and R. W. Brown. 1983. Methods for vegetation sampling and analysis on revegetated mined lands. General Technical Report INT-151. USDA Forest Service, Ogden, Utah.

Clements, F. E. 1916. Plant succession: analysis of the development of vegetation. Carnegie Inst. Wash. Publ. 242: 1—512.

Cook, C. W., and J. Stubbendieck. 1986. Range research: basic problems and techniques. Soc. Range Manage. Denver, Colorado.

Cooperrider, A. Y., R. J. Boyd, and H. R. Stuart (eds.) 1986. Inventory and monitoring of wildlife habitat. U.S. Dept. of Interior, Bureau of Land Management, Denver. CO.

Daubenmire, R. 1976. The use of vegetation in assessing the productivity of forest lands. Bot. Rev. 42: 115—143.

Drawe, D. L., A. D. Chamrad, and T. W. Box. 1978. Plant communities of the Welder Wildlife Refuge. Rob and Bessie Welder Wildl. Found. Contrib. 5, Ser. B. Sinton, Texas.

Duncan, D. A. 1976. Frequent mowing increases turkey mullein on California foothill rangeland. Calif. Fish and Game 62: 82—84.

Duncan, D. A., and R. G. Woodmansee. 1975. Forecasting forage yield from precipitation in California's annual rangeland. J. Range Manage. 28: 327—329.

Dyksterhuis, E. J. 1949. Condition and management of rangeland based on quantitative ecology. J. Range Manage. 2: 104—115.

Gordon, A., and A. W. Sampson. 1939. Composition of common California foothill plants as a factor in range management. University of California, College of Agriculture Bulletin 627.

Gysel, L. W., and L. J. Lyon. 1980. Habitat analysis and evaluation. p. 305—327. *In*: S. D.

Schemnitz (ed.) Wildlife management techniques manual, 4th ed. The Wildlife Society, Washington, D.C.
Hanley, T. A., and J. L. Page. 1982. Differential effects of livestock use on habitat structure and rodent populations in Great Basin communities. Calif. Fish and Game 68: 160—174.
Harper, J. A. 1969. Relation of elk to reforestation in the Pacific Northwest. p. 67—71. *In*: H. C. Black (ed.) Wildlife and reforestation in the Pacific Northwest. School of Forestry, Oregon State University, Corvallis.
Hays, R. L., C. Summers, and W. Seitz. 1981. Estimating wildlife habitat variables. FWS/OBS-81/47. U.S. Fish and Wildlife Service, Washington, D.C.
Heady, H. F. 1973. Structure and function of climax. p. 73—80. *In*: D. N. Hyder (ed.) Proc. 3rd Workshop U.S./Australia Rangelands Panel. Society for Range Management, Denver, Colorado.
Heady, H. F. 1977. Valley grassland. p. 491—514. *In*: M. G. Barbour, and J. Major (eds.) Terrestrial vegetation of California. Wiley-Interscience Publ. New York.
Hoover, R. L., and D. L. Wills (eds.) 1984. Managing forested lands for wildlife. Colorado Division of Wildlife, in cooperation with USDA Forest Service, Denver, Colorado.
Hylander, C. J. 1966. Wildlife communities from the tundra to the tropics in North America. Houghton Mifflin Co., Boston, Massachusetts.
Johnston, D. W., and E. P. Odum. 1956. Breeding bird populations in relation to plant succession on the Piedmont of Georgia. Ecology 37: 50—62.
Kie, J. G. (*In Press*). 1988. Annual grassland. *In*: K. E. Mayer and W. F. Laudenslayer, Jr. (eds.) A Guide to the Wildlife Habitats of California. California Dept. Forestry, Sacramento, California.
Krebs, C. J. 1972. Ecology — The experimental analysis of distribution and abundance. Harper and Row Publishers, New York.
Leckenby, D. A. 1984. Elk use and availability of cover and forage components in the Blue Mountains, northwest Oregon, 1976—1982. Wildlife Research Report No. 14. Oregon Dept. Fish and Wildlife, Portland, Oregon.
Leckenby, D. A., D. P. Sheehy, C. H. Nellis, R. J. Scherzinger, I. D. Luman, W. Elmore, J. C. Lemos, L. Doughty, and C. E. Trainer. 1982. Wildlife habitats in managed rangelands — The Great Basin of southeastern Oregon. Mule deer. General Technical Report PNW-139. USDA Forest Service, Portland, Oregon.
Leopold, A. 1933. Game management. Charles Scribner Sons, New York.
Loft, E. R., J. W. Menke, J. G. Kie, and R. C. Bertram. 1987. Influence of cattle stocking rate on the structural profile of deer hiding cover. J. Wildl. Manage. 51: 655—664.
Lyon, L. J. 1987. Succession and wildlife management. Paper presented at the Symposium on land classifications based on vegetation: applications for resource management, Univ. of Idaho, Moscow, 17—19 November.
Maser, C., and J. W. Thomas. 1983. Wildlife habitats in managed rangelands — The Great Basin of southeastern Oregon. Introduction. General Technical Report PNW-160. USDA Forest Service, Portland, Oregon.
Maser, C., J. W. Thomas, and R. G. Anderson. 1984. Wildlife habitats in managed rangelands — The Great Basin of southeastern Oregon. The relationship of terrestrial vertebrates to plant communities. General Technical Report PNW-172. USDA Forest Service, Portland, Oregon.
Meslow, E. C., and H. W. Wight. 1975. Avifauna and succession in Douglas-fir forests of the Pacific northwest. p. 266—271. *In*: D. R. Smith (ed.) Proceedings of the symposium on the management of forest and range habitats for nongame birds. General Technical Report WO-1. USDA Forest Service, Washington, D.C.
Mould, E. T., and C. T. Robbins. 1982. Digestive capabilities in elk compared to white-tailed deer. J. Wildl. Manage. 46: 22—29.

NRC. 1962. Range research — basic problems and techniques. National Academy Sciences, National Research Council Publ. 890, Washington, D.C.

Nudds, T. D. 1977. Quantifying the vegetative structure of wildlife cover. Wildl. Soc. Bull. 5: 113—117.

Odum, E. P. 1963. Ecology. Holt, Reinhart, and Winston, Inc. New York.

Odum, E. P. 1971. Fundamentals of ecology, 3rd ed. W. B. Saunders Co., Philadelphia, Pennsylvania.

Patton, D. R. 1975. A diversity index for quantifying habitat "edge". Wildl. Soc. Bull. 3: 171—173.

Pitt, M. D., and H. F. Heady. 1978. Responses of annual vegetation to temperature and rainfall patterns in northern California. Ecology 59: 336—350.

Reynolds, H. G. 1966. Use of openings in spruce-fir forests of Arizona by elk, deer, and cattle. Research Note RM-66. USDA Forest Service, Fort Collins, Colorado.

Ricklefs, R. E. 1973. Ecology. Chiron Press, Newton, Massachusetts.

Robbins, C. T., P. J. Van Soest, W. W. Mautz, and A. N. Moen. 1975. Feed analyses and digestion with reference to white-tailed deer. J. Wildl. Manage. 39: 67—79.

Schemnitz, S. D. (ed.) 1980. Wildlife management techniques manual, 4th ed. The Wildlife Society. Washington, D.C.

Schlatterer, E. R. 1986. Background, present status, and future of the evaluation of soil condition in rangeland monitoring in the Forest Service. p. 41—46. *In*: E. L. Smith *et al.* (Compilers) Use of cover, soil, and weather data in rangeland monitoring — symposium proceedings. 39th Annual Meeting, Society Range Manage. 12 February, Kissimmee, Florida.

Sharp, B. 1986. Management guidelines for the Swainson's hawk, Region 1. U.S. Fish and Wildlife Service, Portland, Oregon.

Shelford, V. E. 1911a. Ecological succession: stream fishes and the method of physiographic analysis. Biol. Bull. 21: 9—34.

Shelford, V. E. 1911b. Ecological succession: pond fishes. Biol. Bull. 21: 127—151.

Smith, E. L. 1978. A critical evaluation of the range condition concept. p. 266—267. *In*: International Rangelands Congress. Society for Range Management, Denver, CO.

Stoddart, L. A., A. D. Smith, and T. W. Box. 1975. Range management, 3rd ed. McGraw-Hill Book Co., New York.

Taylor, C. R. 1969. The eland and the oryx. Scientific American 220: 88—95.

Thomas, J. W. (ed.) 1979. Wildlife habitats in managed forests — The Blue Mountains of Oregon and Washington. USDA Agricultural Handbook 553.

Thomas, J. W., G. L. Crouch, R. S. Bumstead, and L. D. Bryant. 1975. Silvicultural options and habitat values in coniferous forest. p. 272—287. *In*: D. R. Smith (ed.) Proceedings of the symposium on the management of forest and range habitats for nongame birds. General Technical Report WO-1. USDA Forest Service, Washington, D.C.

Thomas, J. W., R. J. Miller, H. Black, J. E. Rodiek, and C. Maser. 1976. Guidelines for maintaining and enhancing wildlife habitat in forest management in the Blue Mountains of Washington and Oregon. Trans. N. Amer. Wildl. Nat. Res. Conf. 41: 452—476.

Thomas, J. W., R. J. Miller, C. Maser, R. G. Anderson, and B. E. Carter. 1979a. Plant communities and successional stages. p. 22—39. *In*: J. W. Thomas (ed.) Wildlife habitats in managed forests — The Blue Mountains of Oregon and Washington. USDA Agricultural Handbook 553.

Thomas, J. W., H. C. Black, Jr., R. J. Scherzinger, and R. J. Pedersen. 1979b. Deer and elk. p. 104—127. *In*: J. W. Thomas (ed.) Wildlife habitats in managed forests — The Blue Mountains of Oregon and Washington. USDA Agricultural Handbook 553.

Thomas, J. W., C. Maser, and J. E. Rodiek. 1979c. Wildlife habitats in managed rangelands — The Great Basin of southeastern Oregon. Edges. General Technical Report PNW-85. USDA Forest Service, Portland, Oregon.

Thomas, J. W., and C. Maser (eds.) 1986. Wildlife habitats in managed rangelands — The Great Basin of southeastern Oregon. U.S. Department of Agriculture, Forest Service, and U.S. Department of Interior, Bureau of Land Management.

Tueller, P. T. 1973. Secondary succession, disclimax, and range condition standards in desert shrub vegetation. p. 57—65. *In*: D. N. Hyder (ed.) Proc. 3rd Workshop U.S./ Australia Rangelands Panel. Society for Range Management, Denver, Colorado.

Turner, G. T., R. M. Hansen, V. H. Reid, H. P. Tietjen, and A. L. Ward. 1973. Pocket gophers and Colorado mountain rangeland. Colorado State Univ. Bull. 554S, Fort Collins, Colorado.

USFS. 1962. Range research methods — a symposium. USDA Forest Service, Misc. Publ. 940, Denver, Colorado.

USFS. 1970. Range and wildlife habitat evaluation — a research symposium. USDA Forest Service, Misc. Publ. 1147, Flagstaff and Tempe, Arizona.

Verner, J., and A. S. Boss. 1980. California wildlife and their habitats: western Sierra Nevada. General Technical Report PSW-37. USDA Forest Service, Berkeley, California.

Winter, J. 1981. Some aspects of the ecology of the great gray owl in the central Sierra Nevada. USDA Forest Service, Stanislaus National Forest, Sonora, California.

Woodbridge, B. 1986. Biology and management of Swainson's hawks in the Butte Valley, California. USDA Forest Service, Klamath National Forest, Goosenest Ranger District, Macdoel, California.

24. Revegetation of arid and semiarid rangelands

BRUCE A. ROUNDY AND C. A. CALL

24.1 Abstract

Lack of range management and subsequent deterioration of many arid and semiarid grazing lands throughout the world have created a desire to revegetate to recover the productive potential of these lands. Since revegetation is such an intensive and expensive practice, it should be justified in terms of management goals and ecological impacts. Revegetation considerations are site-specific and include: determination of site potential, selection of plant materials, manipulation of competing vegetation and seedbed preparation, selection of seeding and planting methods, and subsequent management of revegetated and associated areas.

24.2 Introduction

Arid and semiarid rangelands throughout the world support a diversity of uses, including: grazing by domestic and wild herbivores, water storage and distribution, energy and mineral development, fuelwood gathering, subsistence crop production, and recreation. Increased population pressures, indifference, and mismanagement of the aforementioned activities have resulted in the degradation of these fragile rangelands (Walker 1979). Many of these degraded rangelands have potential for improvement by natural or artificial revegetation, depending upon: the kinds and amounts of vegetation remaining, climatic conditions, the feasibility of using grazing management practices and/or range improvement practices to accelerate successional processes, the expected recovery rate, and the cost of alternative approaches (Box 1984, Vallentine 1980).

Natural revegetation implies improved management, particularly of grazing, to restore vigor and accelerate the spread of remaining desirable plants (Vallentine 1980). Although vegetation response to improved

Tueller, P. T. (ed.), Vegetation science applications for rangeland analysis and management.
ISBN 90-6193-195-9.
© 1988, *Kluwer Academic Publishers, Dordrecht, Boston, London.*

management varies from site to site, a minimum of 15 percent desirable perennial species in the vegetal cover is often used as an index to indicate the potential for successful natural improvement on semiarid ranges in western North America (Garza 1986, Vallentine 1980). Vegetation manipulation (chemical, mechanical, biological and/or prescribed burning) in combination with grazing management has resulted in the improvement of many shrub-infested rangelands with sufficient remnant perennial grasses (Scifres 1986). Destruction or suppression of woody vegetation stimulates herbaceous vegetation succession, the direction of which is shaped by grazing management. Grazing management by itself, however, does not automatically result in a timely return to vegetation similar to the original plant community under arid and semiarid conditions. Exclusion of livestock from depleted sagebrush (*Artemisia*) — grass ranges in the Intermountain Region of the western United States for 13 years (West *et al.* 1984) or 38 years (Holmgren 1976) did not improve species composition or forage production. Even under good management, limited precipitation on more arid salt desert shrub ranges in the same region may only allow seedling recruitment of desirable perennial species once or twice over a 15-year period (Bleak *et al.* 1965). Consequently, manipulation of competing vegetation and artificial revegetation may be the only recourse if rapid range improvement is desired.

Artificial revegetation involves the establishment of adapted species by planting harvested seed or by transplanting seedlings or vegetal segments (Sampson 1952). An existing plant community may be completely removed and replaced by another community through some sort of a type conversion. In many instances, this method of revegetation works against normal successional processes by attempting to hold the plant community in some artificial stage (Box 1984). Under other circumstances with less intensive land preparation (overseeding), the development of desired plant communities depends upon the integration of artificial revegetation (establishment of seeded species) with natural revegetation (recovery of remaining desirable species) (Vallentine 1980).

Vegetation response to improved management has been addressed in other chapters in this volume (see Hart and Norton, and Bailey); therefore, this discussion will focus on artificial revegetation of arid and semiarid rangelands, with emphasis on research efforts in the western United States.

Interest in revegetation developed following the downward trend in range conditions associated with the relatively recent intensive use and settlement of arid and semiarid lands in the western United States. Over a period of just a few years in the latter part of the nineteenth century, ranchers and scientists observed a tremendous decline in the grass component of many of these rangelands (Bentley 1898, Davy 1902, Griffiths

1901, 1902, 1904, Rowley 1985, Young and Sparks 1985) mainly due to free range use and associated overstocking. Plant scientists suggested replanting the range with useful species and pleaded for better grazing management. Over a 3-year period, 1500 seeding trails were conducted by the Division of Agrostology (Stoddart and Smith 1943). These and subsequent trials by the U.S. Forest Service were largely unsuccessful because seed available for planting was mainly from forage plants adapted to humid regions (Stoddart *et al.* 1975). In ensuing years, cultivation and abandonment (Lorenz 1986), fire suppression (Humphrey 1958), establishment of highly adapted exotic weeds (Young *et al.* 1972), continued overstocking of many areas (Cox *et al.* 1982, Young *et al.* 1979), and subtle climatic changes (Hastings and Turner 1965, Neilson 1986) all contributed to changes in vegetation composition and production and soil stability on arid and semiarid rangelands in the western United States. Generally, herbaceous forage species decreased, native woody species and exotic annual herbs increased, and soil erosion potential increased. As concern for the condition of the range increased, extensive revegetation trials were conducted to evaluate the adaptability of many different species for conservation and forage on various range sites.

Evident in this effort was the continuing search for an ideal species that would cure all range deterioration problems. Crested wheatgrass (*Agropyron desertorum, A. cristatum*), an introduction from western Siberia, satisfied many of the criteria expected of an ideal species (Keller 1979). This cool-season bunchgrass had high seed production, viability and germinability, high seedling drought tolerance, and high grazing tolerance compared to most native grasses (Johnson 1986). Once its adaption to the sagebrush zone of the Intermountain West was discovered, and the rangeland drill was developed to seed it, millions of hectares were seeded in what could be referred to as the "Golden Age" of range revegetation (Young and Evans 1986). About the same time, several South African lovegrasses (*Eragrostis* spp.) were successfully seeded in the southwestern United States (Cox and Ruyle 1986). These successes renewed interest in the search for, and the development of, better plant materials and planting methods for forage improvement, wildlife habitat improvement, and soil conservation.

Outside the United States, most rangeland revegetation efforts have emphasized forage improvement for domestic livestock until recent years. The cumulative effects of severe droughts and overgrazing by exotic herbivores (cattle, sheep, rabbits) resulted in the degradation of arid and semiarid rangelands in Australia by the end of the nineteenth century (Noble *et al.* 1984). Revegetation research since that time has investigated the planting of native and introduced species in conjunction with surface modification treatments (furrowing, pitting, water spreading) to improve forage production. In the 1930's and 1940's scientists in Soviet Central Asia failed to establish artificial plant communities (agrophytocenoses) to meet the year-round diet requirements of sheep (Nechaeva 1985). Revegetation efforts were successful in the 1950's after Soviet researchers identified and worked within the ecological constraints of the region. During the 1970's and 1980's, revegetation research in arid and semiarid regions of developing countries has been directed at slowing the multi-faceted desertification process. Researchers have identified the need to establish multipurpose plants to provide fodder for livestock, fuelwood for cooking and heating, and protective cover for soil conservation (Goodin and Northington 1985, Child *et al.* 1984).

Research over the years in the western United States as well as in other countries has resulted in a rather consistent list of considerations and guidelines for the revegetation of arid and semiarid rangelands (Stewart 1950, Plummer *et al.* 1968, Vallentine 1980, Jordan 1981, Nechaeva 1985, Silcock 1986). These considerations — revegetation planning, site selection, plant materials selection, vegetation manipulation and seedbed preparation, planting methods, and post-revegetation management — are discussed in the following sections of this chapter.

24.3 Revegetation planning

Artificial revegetation should not be viewed as a panacea for production or conservation problems, nor should it be regarded as an alternative to good management (Kennedy and Doten 1901, Heady 1975, Vallentine 1980, Noble *et al.* 1984). Revegetation must be considered one of the tools of management, and land managers should be realistic in their expectations and approach to the application of treatments. Realistic goal setting and planning are not possible without reliable estimates of the production potential or ecological potential of the range resource for various uses (Scifres 1986). Estimates of production potential and ecological potential are developed from a comprehensive inventory and evaluation of resource capabilities and limitations. If a revegetation program is adopted before a comprehensive survey and evaluation of vegetation and soil resources or a projection of responses, it may eliminate the consideration of more appropriate alternatives (Scifres *et al.* 1985). Many revegetation practices and plant species are site specific, and should be used on a prescription basis for each range site rather than a general approach for large areas.

Any revegetation treatment or associated range improvement treatment will initiate a dynamic successional response, and the consequence of this response must be understood and directed toward desirable management goals (Young 1983). Ideally, the land manager should itemize and predict outputs available from different successional stages, and be able to move succession forward or backward to achieve the desired combination of goods and services (Box 1984). In many instances, a high successional stage, near climax, is not always the most productive or suitable stage for multiple uses on arid and semiarid rangelands. In the Intermountain Region of the western United States, pinyon-juniper (*Pinus-Juniperus*) ranges near climax do not provide ample perennial grass forage for cattle, and sagebrush — grasslands near climax do not provide effective habitat for mule deer (*Odocoiles hemionus*) winter range when dominated by bluebunch wheatgrass (*Pseudoroegneria spicata*).

Revegetation planning, from the development of realistic and preceptive management goals through site and plant materials selection, vegetation manipulation and seedbed preparation, planting, and post-revegetation management, should be integrated; and the costs and benefits to livestock, wildlife, watersheds, recreation, and/or ecological integrity should be assessed (Herbel 1984). Direct costs (manipulation of undesirable plants, land preparation, seeds and seeding, fencing, etc.) and indirect costs (risk of failure, nonuse during the establishment period, and interest on direct costs and nonuse) should not exceed the value of the land and the potential increase in productivity or stability (Vallentine 1980, Nielsen 1984, Workman 1986). For optimum allocation of limited funds, emphasis should be placed on sites and practices with the greatest potential for increasing net benefits in the shortest amount of time. Under favorable conditions, seeding improved species on a semiarid range site with high potential can increase forage production 100 to 1000 percent within 1 to 3 years (Herbel 1984). However, even if revegetation costs must be recovered over many years, long-term soil conservation benefits may justify the project (Pratt and Gwyne 1977, Box 1984).

24.4 Site selection

Revegetating arid and semiarid rangelands is generally a difficult endeavor because of limiting climatic, soil, and topographic conditions. Unless these conditions and their interactions are understood, it may not be possible to accurately evaluate range sites as to their potential and need for revegetation.

Average annual precipitation values are commonly used to characterize range sites and describe the revegetation potential of a site. Seedings are usually successful in areas receiving 600 to 800 mm of annual precipitation (upper end of the precipitation regime for semiarid areas), and are generally not recommended in areas receiving less than 250 mm of annual precipitation (boundary between arid and semiarid areas) (Heady 1975, Stoddart *et al.* 1975, Walker 1979, Vallentine 1980). However, no definite limits can be set for seedings with respect to amounts of precipitation because the effectiveness of precipitation in plant establishment and maintenance is determined by seasonal distribution and intensity of precipitation and by associated temperatures and wind velocities which regulate evaporative losses (Vallentine 1980, Jordan 1981). The Koppen climatic mapping system, based on amount and seasonal distribution of rainfall and average annual temperatures, has been presented as a model to set lower limits for seeding semiarid rangelands in the western United States (Jordan 1983). These climatic variables are related to soil texture, and germination rates, drought tolerance, and seedling vigor of seeded species.

In South Africa, the lower limits of annual precipitation for veld revegetation are 300 mm in winter rainfall areas; 400 mm and 500 mm, respectively, where annual evaporation is less than and greater than 1,778 mm in all-year rainfall areas; and 500 mm and 600 mm, respectively, where annual evaporation is less than and greater than 2,032 mm in summer rainfall areas (Tainton 1981). Even where annual precipitation is 300 to 400 mm in some regions of Australia, irregularity of seasonal distribution and high evaporation rates limit revegetation potential (Noble *et al.* 1984).

As the lower limit for successful seeding is approached, establishment becomes risky, production declines, and management becomes precarious (Keller 1979). On arid sites, seeded grasses often establish in above average precipitation periods or with the use of special techniques that concentrate and/or conserve soil water, but in many instances, stands are short-lived and low-producing as a result of drought and grazing (Bleak *et al.* 1965, Vallentine 1980). Transplanting adapted, drought-tolerant shrubs may be the only viable revegetation strategy in many instances (Institute for Land Rehabilitation 1979, Van Epps and McKell 1980).

Soil characteristics such as texture, depth, and salinity influence the effectiveness of precipitation on range sites. Loamy soils with high infiltration rates, a high water holding capacitv, and good aeration, should have the greatest potential for plant establishment and production (Cluff *et al.* 1983). Sandy soils have high infiltration rates but a low water holding capacity and readily dry at the surface (Vallentine 1980). Silty soils have a high water holding capacity but low infiltration rates and a tendancy to form restrictive crusts (Eckert *et al.* 1978). Clayey soils also have a high water holding capacity but wet up slowly and dry hard enough to restrict seedling root growth and shoot emergence (Cox 1984). Shallow soils, cobbly soils, and those with restrictive layers, have limited water holding capacity and limited production potential. Saline and sodic soils, common in valley bottoms of arid and semiarid regions, may have low water availability due to the dispersion of soil particles in the surface layer (reducing infiltration) and more negative osmotic potentials in the rooting zone. At high salinity levels, toxic concentrations of ions may restrict germination and growth (Roundy *et al.* 1984). When seeding fine-textured soils or saline and sodic soils, precipitation should be 20 percent higher than levels implied earlier for medium textured soils, in order to compensate for poorer infiltration rates and greater evaporative losses (Jordan 1983).

Topographic features, primarily slope and exposure, create different microclimates, and influence soil development, plant species composition, and the use of land preparation and planting treatments. Shallow soils and soils with high drainage have limited soil moisture on steep slopes, and this is compounded by greater evaporative losses on south and west aspects in the Northern Hemisphere. Level to gently sloping lands have the greatest revegetation potential and lowest erosion hazard. Except for watershed protection and where special treatments (terracing, contour furrowing, hand or aerial seeding, and mulching) can be used effectively, complete seedbed preparation and drill seeding are generally economically infeasible and hazardous on slopes greater than 30 percent (Vallentine 1980, Tainton 1981).

The presence and vigor of certain indicator species growing on a site can be useful in determining revegetation potential, since plants integrate the previously described site factors — precipitation, temperature, soil, and topography (Stoddart *et al.* 1975). For example, in the Intermountain Region of the western United States, big sagebrush (*Artemisia tridentata*) sites with large plants are generally suitable for revegetation, but blackbrush (*Coleogyne ramosissima*) sites at the southern limits of big sagebrush are not suitable (Jordan 1983). Thus, a site may be dominated by several undesirable species, but one or two species may indicate that the site has a deep soil with a favorable moisture regime and a high potential for revegetation. As a rule, sites with high potential are revegetated first, since they offer the greatest chances for plant establishment and production, and the most favorable return on the investment (Heady 1975, Keller 1979, Vallentine 1980).

24.5 Plant materials selection

Plant materials are selected for rangeland revegetation on the basis of their ability to establish and persist under the climatic, edaphic, and management conditions on the site of interest (Thornburg and Fuchs 1978, Low and Berlin 1984). Establishment and persistence characteristics of importance when environmental conditions are limiting include: germinability, seedling vigor, regeneration strategy, and tolerance to drought, temperature extremes, salinity, and herbivory. Management considerations for species selection include productivity, palatability, seasonal nutrient content, potential toxicity in relation to phenology, compatibility with other species, rate of growth, growth habit, and ability to control erosion. The selection of proper plant materials implies an understanding of the autecology of the plants involved, and the ability to modify the environmental conditions of the site to meet the autecological characteristics of the desired plants (Vories and Sims 1977, Noble *et al.* 1984). Unfortunately, a detailed knowledge of the autecology of many range species is lacking (West 1968), and many recommendations are largely based on site-specific trials. Numerous publications are available which give plant recommendations for different climatic, edaphic and management conditions throughout the world (Hafenrichter *et al.* 1968, Plummer *et al.* 1968, Monsen 1975, Institute for Land Rehabilitation 1979, Keller 1979, Vallentine 1980, Jordan 1981, Cox *et al.* 1982, Thornburg 1982, Wasser and Shoemaker 1982, Goodin and Northington 1985, Nechaeva 1985, Barrett-Lennard 1986).

Historically, perennial grasses have been dominantly used in range revegetation programs due to greater seed availability, ease of seeding, and suitability for livestock production and soil conservation. During recent years increased emphasis has been placed on the selection and development of legumes (Barnes *et al.* 1985, Rumbaugh and Townsend 1985), non-leguminous forbs (Shaw and Monsen 1983), and shrubs (Van Epps and McKell 1978, McArthur *et al.* 1985, Goodin and Northington 1985). Mixtures of grasses, forbs, and shrubs are generally preferred to single species plantings because a diverse community increases ecological stability and affords a greater variety of food and shelter for wild and domestic animals (Brown and Hallman 1984). Multiple species seedings may also result in greater plant establishment in a heterogeneous seedbed, and may be more resistant to insects and pathogens than a monoculture (Keller 1979). When water is not limiting, legumes may contribute nitrogen and increase production of grasses in a mixed stand (Bleak 1968). However, differential germinability, palatability, grazing tolerance, competitiveness, and adaptability of mixed species may result in dominance by a single species anyway (Harris and Dobrowolski 1986) or may require careful management to maintain the diversity.

Native species are a logical starting place for the selection of adapted plant materials, but they may not be the best species under certain circumstances (Thornburg and Fuchs 1978, Johnson *et al.* 1981). Disturbance may have altered environmental conditions of the site in such a way that highly-successional native species are unable to establish. Many plants are adapted and highly desirable for forage or conservation but are difficult to establish due to low germination and seedling vigor. For example, Indian ricegrass (*Oryzopsis hymenoides*) and fourwing saltbush (*Atriplex canescens*) are both excellent forage plants adapted to dry sandy areas. However, the seed coat and embryo dormancy of Indian ricegrass (McDonald and Kah 1977) and the highly variable germination requirements of different ecotypes and collections of fourwing saltbush (Young *et al.* 1984) limit the ability to successfully and consistently seed these species. Species that are known to be strongly ecotypic, such as fourwing saltbush, should only be used when local seed or seed for very similar sites to the one being seeded are available. Plant materials from other parts of the world with similar climates should be evaluated in plant selection and improvement programs. Cox *et al.* (1987) have shown that an understanding of the distribution of warm season grasses in relation to climate and soils in their native Africa can be used to select areas of greatest adaptation for these species in the United States, and avoid costly seeding failures.

In selection programs, plant materials (ecotypes or accessions) from a wide range of climatic and edaphic conditions are assembled at a common location and evaluated for several establishment and production characteristics (seedling establishment, forage yield and quality, seed production, pest resistance) (Shiflet and MacLauchlan 1986). Promising accessions are further evaluated under different growing conditions at multiple locations. Superior accessions are selected and released as cultivars assumed to be adapted to a wide range of environmental conditions within a particular region. The success of this selection method is highly dependent upon assembling a diverse collection of germplasm from several sources. If the desired gene combinations are absent initially, then they will be absent when the plant material is released several years later (Hussey 1985).

More effective improvement of range plants requires a basic understanding of previously mentioned adaptive characteristics and the development of procedures to identify and create the proper combinations of genes to facilitate plant establishment and persistence. Although morphological characteristics of seedlings account for differences in drought tolerance and establishment of some grasses (Hyder 1974) the interaction of these characteristics with the environment generally precludes their use as reliable indicators of drought tolerance or ability to establish (Johnson *et al.* 1981). An approach has been developed for the genetic improvement of drought tolerance in cool-season perennial grasses adapted to semiarid rangelands in western North America and Central Asia (Johnson *et al.*

1981). Drought stress is characterized in terms of intensity and duration of water deficit. Since the effect of drought stress is particularly pronounced during germination, emergence, and early seedling growth, superior establishment characteristics such as early root initiation and elongation and seedling recovery after exposure to drought are designated as breeding goals. A diverse collection of germplasm is assembled because improvement in drought tolerance probably involves many genes with small effects. Laboratory and field screening techniques assess plant performance under drought stress and isolate plants that have desired characteristics for the breeding of improved cultivars. Seedling response is not necessarily indicative of mature plant response (Johnson *et al.* 1981), and therefore, drought tolerance must be considered throughout the phenology of the plant (Jordan 1983). Further research on mechanisms of tolerance to drought (Johnson *et al.* 1983) and herbivory (Caldwell 1984), as well as plant competitive ability with respect to acquisition of limited resources (Fowler 1986, Caldwell 1987) will enhance the development of improved range plants.

New developments in breeding technology and biotechnology hold great promise for improving rangeland vegetation. Using interspecific hybridization, genetic resources from widely divergent taxa can be combined into new cultivars, or in some cases, new species (Asay and Knowles 1985). 'Hycrest' crested wheatgrass, an induced tetraploid (*Agropyron cristatum*) × natural tetraploid (*A. desertorum*) hybrid, has been consistently more vigorous and productive than commonly used cultivars on semiarid range sites in the Intermountain region of the western United States. A quackgrass (*Elytrigia repens*) × bluebunch wheatgrass (*Pseudoroegneria spicata*) hybrid has forage quality aspects of the bluebunch wheatgrass parent and persistence, productivity, and salinity tolerance aspects of the quackgrass parent. Biotechnology may benefit range vegetation through genetic improvement of plants by using a combination of plant breeding, tissue culture, and recombinant DNA technology (McKell 1985). However, before any intensive programs can be undertaken with range plants, a clear understanding of the biology, diversity, and genetic relationships of those plants is necessary.

24.6 Vegetation manipulation and seedbed preparation

Vegetation manipulation is usually a prerequisite to revegetation since most degraded rangelands are usually dominated by less desirable species. Numerous studies have described the competitive interactions between undesirable and desirable plants for limited resources such as soil water and nutrients (Robertson and Pearce 1945, Evans 1961, Harris and Wilson 1970, Cable 1977, Ueckert 1979), and the competitive role of established woody and herbaceous species in the failures of many early revegetation attempts in arid and semiarid regions (Stoddart and Smith 1943, Nechaeva 1985, Silcock 1986). The interactive mechanisms of belowground competition for limited resources, however, are not completely understood (Caldwell 1987). Recent research in sagebrush-grassland communities in the Intermountain Region of the western United States has quantified differences in the abilities of crested wheatgrass and bluebunch wheatgrass to compete with big sagebrush for phophorus (Caldwell *et al.* 1985). Future research of this nature will allow us to better understand vegetation dynamics following manipulation treatments, and to better select plant species to structure ecologically stable and desirable artificial plant communities.

Most vegetation manipulation treatments do not remove entire populations of undesirable species, and thus, treated areas are constantly disposed to reinvasion by undesirable plants (Scifres 1986). When only partially controlled, remaining plants respond dynamically by taking advantage of the environmental potential released by the reduction in plant density (McKell and Chilcote 1957, Young and Evans 1974, Roundy *et al.* 1981). Depending upon the species and growing conditions, undesirable plants may rapidly increase their densities by crown or root sprouting or by seedling recruitment from new, large seed crops or persistent seedbanks. Vegetation manipulation treatments create a temporary shift in plant succession, and this shift is given a time duration by revegetation and grazing management programs (Young 1983). Once desirable perennial species gain dominance in a manipulated community, they should supress seedlings of undesirable plants, but they will not eliminate them from the community.

Complete eradication of undesirable species may not be desirable for certain management objectives. Numerous brush species may be of little value for grazing by domestic livestock, but in certain amounts may be useful for recreational services, watershed protection, and/or wildlife habitat (Scifres 1980). Conversions from brush to grassland may benefit stock production but reduce habitat for wildlife species such as the black rhino (*Diceros bicornis*) in East Africa (Pratt and Gwynn 1977). In some cases, replacement of woody species with herbaceous species directly benefits wildlife species by providing forage for them or indirectly benefits them by providing forage for domestic stock away from important wildlife habitats (Urness 1986).

Rangeland vegetation can be manipulated by chemical, mechanical, prescribed burning, or biological methods. Each method has advantages and disadvantages (depending on site conditions, undesirable species present, and the degree of infestation), but the major objective in all cases is to substitute desirable plants for undesirable plants (Herbel 1983). Many of these methods depend on natural revegetation for vegetation improvement, but some are effective seedbed preparation techniques that precede or are performed simultaneously with artificial seeding. More detailed descriptions of these methods and their applications are available elsewhere (Bovey 1977, De Loach 1980, Scifres 1980, Vallentine 1980, McDaniel 1983).

A variety of herbicides and application methods are available for vegetation manipulation and chemical seedbed preparation (Evans *et al.* 1969, Scifres 1977, Bovey *et al.* 1981). Herbicides provide a means of selective plant control when desirable plants are present, leave a protective mulch cover over a firm seedbed for better plant establishment and erosion control, and can be used on areas too rough, steep, or erosive for traditional mechanical treatment (Vallentine 1980). In the western United States, degraded big sagebrush communities that are not invaded by downy brome (*Bromus tectorum*, a highly competitive introduced annual grass) can be sprayed with 2,4-D [(2,4-dichlorophenoxy) acetic acid], and a heavy-duty rangeland drill can be used to seed perennial grasses into the standing dead brush (Evans *et al.* 1979). For degraded big sagebrush communities in which perennial grasses have been replaced by downy brome, it is possible to combine in sequence the application of 2,4-D for big sagebrush control, an atrazine [2-chloro-4-(ethylamino)-6-(isopropylamino)-s-triazine] fallow treatment for downy brome control, and the drill seeding of perennial grasses (Evans and Young 1977). An atrazine fallow is also an excellent land preparation treatment when transplanting seedlings of desirable shrub species on downy brome infested areas (Christensen *et al.* 1974). Spraying short-lived herbicides such as paraquat (1-1-dimethyl-4, 4′-bipyridinium ion) or glyphosate [isopropylamine salt of N-(phosphonomethyl) glycine] in bands and drill seeding down the center of each band has been successfully used to modify species composition in semiarid plant communities (Vallentine 1980). Suppression of resident vegetation reduces competition for limited resources, and remaining vegetation between seeded rows creates a favorable microclimate for establishing seedlings.

Mechanical vegetation manipulation methods vary considerably in their effectiveness to control herbaceous and woody species and prepare a suitable seedbed. Equipment is generally designed either to remove the top growth of undesirable plants (mowing, roller chopping, and rotobeating) or to remove plants by completely uprooting them (grubbing, bulldozing, chaining, cabling, railing, root plowing, disk plowing, raking, and others)

(Larson 1980, Scifres 1980). In general, top removal methods are used to stimulate the release of native forage with minimal soil disturbance and minimal cost, but they may have a relatively short treatment life because of the regrowth potential of many woody species (Scifres *et al.* 1985). Large disk plows cut and underturn herbaceous and small woody vegetation into the soil, and can effectively reduce undesirable plant competition and prepare seedbeds when applied early in the growing season and again during mid-season (Davis 1983). Chaining or cabling twice in opposite directions is useful for releasing desirable undergrowth from mature, even-aged stands of nonsprouting woody species. Seed can also be broadcast from tractors on the first pass and covered with loosened soil on the second pass. Modified chains, with disks or steel bars welded to links, have been developed for greater brush removal and soil scarification (Larson 1980). When used in sequence, complete plant removal methods such as chaining, root plowing, and raking provide highly effective brush control and prepare a clean, relatively firm seedbed for improved forage grasses (Scifres 1986). Type conversions using these costly methods are usually only feasible on semiarid range sites with high production potential.

Controlled burning can extend the treatment lives of previously mentioned chemical and mechanical manipulation techniques and prepare firm seedbeds with an ash mulch when fine fuel and continuity are sufficient (Scifres 1980, Wright and Bailey 1982). Increased soil temperatures, soil water, and nutrient availability after a burn may enhance the establishment of seeded species and increase the production and quality of vegetation that emerges following burning. Drill seeding or some type of mechanical treatment to cover broadcast seed in the loose ash are recommended; and the burn should be reseeded as soon as possible, before resident plants reestablish (Vallentine 1980).

Biological methods of manipulating undesirable vegetation range from the use of natural enemies (insects and pathogens) to selective grazing by large herbivores. Although spectacular results have been attained with some host-specific insects (Huffaker 1951; DeLoach 1978, 1980), their use involves highly complex strategies. Goats have been used extensively in Texas and other areas of the southwestern United States to control low-growing shrubs and regrowth of manipulated woody species (Scifres 1980). However, careful grazing management is essential to prevent goats from damaging desirable grasses and forbs when the availability of preferred woody species becomes limited. Late-fall grazing by sheep can control the increase of big sagebrush on seeded range in the Intermountain Region of the western United States (Frischknecht and Harris 1973). Daily protein supplementation will keep sheep from losing excessive weight and will encourage greater use of sagebrush.

Integrated management systems have been emphasized in recent years to more efficiently control undesirable plants and establish desirable plants by using a combination of previously described methods (Evans 1982, Scifres 1986). Each method has characteristic strengths and limitations depending upon site conditions and management objectives. Combinations of methods in complementary sequences, such as tandem roller chopping, seeding perennial grasses, and prescribed burning in mixed brush communities in South Texas can increase the biological effectiveness and life of the treatments (Thomas 1982, Scifres *et al.* 1985). Tandem roller chopping temporarily suppresses undesirable woody species and prepares a suitable seedbed for plant establishment. Seeded perennial grasses provide forage for livestock and fine fuel to carry a fire. Periodic prescribed burning suppresses the regrowth of woody species and extends the life of the seeding treatment. The herbicide-seeding system for improving degraded big sagebrush communities in the western United States described above is another good example of integrated management.

In addition to removing or suppressing competing vegetation, seedbed preparation practices attempt to ameliorate harsh environmental conditions in the surface soil (rapid drying, unfavorable temperatures, and impenetrable crusts) on arid and semiarid rangelands. The seedbed must be suitable for providing seed coverage, to: increase seed-soil contact and the imbibition of available water (Hadas and Russo 1974 a, b), buffer temperature extremes that occur at the surface (Evans *et al.* 1970), and provide leverage for radicle growth. The effects of seedbed modifications are site-specific and highly dependent on soil texture, structure, and surface morphology. Disturbance of soils with good aggregate stability by plowing, furrowing or other mechanical methods may create microsites or "safe sites" (Harper *et al.* 1965) suitable for seed germination and plant establishment and persistence. However, silty soils with low organic matter content and low aggregate stability can slake and flow after disturbance and wetting, burying seeds too deep for emergence and reforming an impenetrable vesicular crust (Wood *et al.* 1978, 1982).

Problems associated with mechanical seedbed preparation include loose, soft soil which interferes with proper seed placement and has low water holding capacity; and reduced surface litter which can increase erosion hazards (Vallentine 1980). Soil firming with cultipackers, rollers, or roller choppers may be required to hold more moisture near the surface, help control depth of seeding, and provide anchorage for seedling root systems (Stuth and Dahl 1974, Brown and Hallman 1984). Strip cropping and preparatory crop methods have been used with some success on semiarid sites that are susceptible to wind or water erosion (Bement *et al.* 1965, Stroh and Sundberg 1971). For strip cropping, mechanically fallowed strips 10 meters wide are seeded to perennial grasses between untreated strips of similar width. After seeded strips establish, untreated strips are fallowed and seeded in the same manner. The preparatory crop method involves the planting of a residue-producing crop (an adapted annual grain or forage species) the season before seeding desired perennial species and seeding directly into the residue without further seedbed preparation. The application of mulch materials (hay, straw, wood chips, synthetics, etc.) in revegetation programs on arid and semiarid rangelands is not considered practical except for the stabilization of critical areas (Jordan 1981).

As annual and seasonal precipitation decreases, there is a need for seedbed modifications to increase available moisture (Pratt 1963). The main approach for increasing effectiveness of precipitation is that of creating soil depressions to collect water, increase infiltration and decrease evaporation. Methods such as pitting (Jordan 1981, Noble *et al.* 1984), furrowing (Wight *et al.* 1978, Nechaeva 1985) and land imprinting (Dixon 1980) have enhanced establishment, depending on the precipitation during the year of treatment. These methods of water collection to increase establishment may be unnecessary in years of high precipitation and inadequate during drought years, but may make a difference in years of moderate to low precipitation.

Other mechanical surface modification techniques, water spreading (Miller *et al.* 1969, French 1978, Nobel *et al.* 1984) or water harvesting (Matlock and Dutt 1986, Medina and Garza 1987), have been used to establish productive vegetation in dry areas. These systems divert runoff water from stream channels or gullies, following heavy rains or snowmelt, onto adjacent natural or seeded areas where diversion dykes spread water uniformly and control flow velocity for greater infiltration. Data from several systems across the western United States indicate that in areas receiving less than 230 mm of mean annual precipitation the small forage response from water spreading does not justify the installation of a system (Miller *et al.* 1969).

24.7 Seeding and planting methods

Regardless of the type of seeding or planting method, or equipment used, the purpose is the same: to place the seed or plant in the soil at the correct time and place with the necessary ammendments to facilitate establishment and growth. Seeds are either placed in a furrow and covered with soil by drilling (also referred to as direct seeding) or scattered on the soil surface by broadcasting (includes aerial seeding). Although more costly and time consuming than seeding, transplanting containerized or bare root plants or plant propagules (rhizomes, sprouting root segments, bulbs, and corms) can improve the establishment and survival of species with poor germination and seedling development characteristics. More detailed descriptions of seeding and planting methods and associated equipment can be found elsewhere (Abernathy and Herbel 1973, Larson 1980, Vallentine 1980, Malcolm and Allen 1981, Young and McKenzie 1982, Nechaeva 1985).

Time of planting depends on many factors, including climate, site conditions and accessibility, and establishment requirements of selected species. The most desirable time to revegetate rangeland is immediately prior to the season of most reliable rainfall, and when temperatures are favorable for plant establishment (Herbel 1983). Favorable growing conditions in semiarid areas may be limited to 30 days or less (Vallentine 1980). Both precipitation and temperature may be favorable for germination only 3 to 7 days out of the entire year in certain arid regions (Jordan 1983). Transplanting pre-hardened plants provides more flexibility in timing of planting in arid areas; but under exceptionally dry conditions, transplants may require supplemental water for several weeks to ensure survival (Institute for Land Rehabilitation 1979). Time of planting is also important when both warm-and cool-season species are desired on the site. Both seasonal types can be seeded at the same time, but climatic patterns in the seeding year will often favor one type to the detriment of the other (Brown and Hallman 1984). Another method is to seed warm-season species one year and then interseed cool-season species into an established warm-season stand the following year. Competition from resident vegetation is temporarily reduced by scalping out wide, shallow furrows and drilling in the center of the furrows (Larson 1980). Resident vegetation remaining between the furrows creates a favorable microenvironment for seedling establishment.

The proper depth of seed placement for germination and seedling emergence depends on the period of available water in the seedbed, the germination requirements of the seed, the size of the seed and its associated seedling vigor. The period of water availability increases with depth in the seedbed but so does the energy requirement for emergence (Keller 1979). Larger seeds generally are able to emerge from greater depths and are associated with greater seedling vigor than smaller seeds (Kneebone 1972, McKell 1972). General recommendations are to plant very small seeds such as those of the lovegrasses at 6 mm, medium sized seeds like those of crested wheatgrass at 12 mm, and larger seeds such as those of tall wheatgrass (*Agropyron elongatum*) at 25 mm (Vallentine 1980).

Planting depth requirements influence the selection of seeding methods.

Drilling is generally considered superior to broadcasting because it covers the seed to a proper depth, provided the seedbed is fairly firm and uniform in topography. However, broadcasting can be effective when followed by some mechanical treatment (cultipacking, chaining) that provides suitable coverage. Trampling of broadcast seeds by animals can increase seedling establishment (Owens and Brzostowski 1967, Howell 1976) but may be impractical for treating large areas. When broadcast onto loose or rough seedbeds, soil sloughing or soil movement following rainfall events may provide the shallow coverage required by small seeds (Jordan 1981). The problem of soil drying around small seeds near the soil surface can be reduced by planting seeds in deep furrows (McGinnies 1984) or in open, small-diameter holes punched in the soil to a greater depth than for conventional seeding (Hauser 1982).

Rate of seeding will be influenced by the quality of the seed, the condition of the seedbed, the method of seeding, and the potential for reinvasion by undesirable species (Keller 1979). To insure quality of seed, seed lots should be quantified in terms of germinability, viability, and purity. Seeds of many species from wildland communities may have low viability or adaptive dormancy mechanisms that delay germination until environmental conditions are more favorable for seedling establishment. In artificial seedings, however, high germinability is desired for establishment of seeded species during the first favorable season after seeding before undesirable species invade and dominate the site. Thus, increased seeding rates may be required to overcome low viability, and presowing treatments such as scarification, stratification, storage until afterripening requirements are met or leaching of inhibitors may be required to enhance germination. Combining pretreatments and seeding rate recommendations to include seeds with a range of germination requirements may allow more successful seeding in areas with erratic environmental conditions. Seed germination of domestic and wildland species have been studied extensively and some excellent texts, reviews, and manuals are available (Kozlowski 1972; Heydecker 1973; McDonough 1977; Koller and Hadas 1982; Mayer and Polijakoff-Mayber 1982; Bewley and Black 1982, 1983; Young *et al.* 1983; Fenner 1985; Young and Young 1986).

Another seed quality factor of concern is the presence of appendages (awns, hairs, bristles) that can interfere with proper seed metering. These inert materials can be removed by mechanical and aerodynamic conditioning (Dewald *et al.* 1983) and flame processing (Pogue 1983) or covered with coating materials (Oaks 1982) to improve uniformity of seed metering. Variations in the seeding rates of many chaffy-seeded species have been largely overcome by the development of a semi-circular seedbox and auger agitator system that is presently available from several commercial manufacturers (Wiedemann and Cross 1981). Since seed appendages may have important functions in the establishment of some plants, specialized seeding techniques are being developed to plant intact diaspores (Booth 1987).

Generally, seeding rates should be based on pure live seed (PLS), or the portion of the seed lot that is germinable seed of the desired species. Although seeding rates vary with species and specific site conditions, the general recommendation for drilling medium-sized seed on upland sites in the western United States is 215 PLS per square meter (Vallentine 1980). When seeding on poorly prepared seedbeds, or broadcast seeding, the seeding rate may have to be increased 25 to 75 percent on semiarid sites (Plummer *et al.* 1955, Keller 1979). Heavier seeding rates may also be required to suppress seedlings of undesirable species (Plummer *et al.* 1955). Seeding rates lower than the recommended average may be adequate for species which spread rapidly by rhizomes or tiller extensively (Vallentine 1980).

Diverse mixtures of plant species have often been recommended in revegetation programs to support multiple uses and to increase chances of establishment on varied microhabitats of a given area (Monsen 1975, Plummer 1977). Even though diversity of growth form provides for stratification of aboveground and belowground biomass and differential use of resources in natural communities (Harper 1977), it is often difficult to overcome interspecific competition during the initial establishment of mixtures of grasses, forbs, and shrubs on arid and semiarid rangelands. Many forbs and shrubs are slow to establish and may not be able to compete with more aggressive seeded grasses and undesirable plants (Hubbard 1957, Nord *et al.* 1971, Monsen 1975). Forbs and shrubs have been seeded in alternate rows with grasses (Keller 1979, Nechaeva 1985) and have been established within grass stands by interseeding and transplanting (Stevens *et al.* 1981, McKell 1986, Petersen *et al.* 1986). Under harsh environments (Plummer *et al.* 1968) and competitive environments (Petersen *et al.* 1986), transplanting of shrubs was more successful than direct seeding. Seeding or transplanting in straight, evenly-spaced rows may not achieve the desired stand heterogeneity found under natural conditions. Natural communities are differentiated according to microsite conditions and have few distinct boundaries. A more heterogeneous planting may be achieved by seeding and transplanting species at different rates in various areas (patch planting) (Brown and Hallman 1984).

Disturbance or degradation of rangelands can reduce levels of soil microorganisms that form essential symbiotic associations with many species used in subsequent revegetation programs. Successful establishment and production of legumes in range seedings depends on effective nodulation and nitrogen fixation by *Rhizobium* bacteria (Roughley 1985). Similarly, establishment and growth of many non-leguminous woody species (in the genera *Casuarina, Ceanothus, Cercocarpus, Discaria, Elaeagnus, Purshia, Shepherdia*, and others) depends on effective nodulation and nitrogen fixation by actinomycetes (*Frankia*) (Torrey 1978). Most herbaceous and woody plants of arid and semiarid rangelands form mycorrhizae with endo- or ectomycorrhizal fungi that improve host plant nutrient and water relations in limiting environments (Trappe 1981). In order to assure nodulation and nitrogen fixation, legume seed should be inoculated with a specific, effective strain of *Rhizobium*. Actinomycete and mycorrhizal fungi propagules are not readily adapted for seed inoculation at the time of sowing. Transplants, inoculated in the nursery, can be outplanted with mycorrhizal or actinorhizal associations already established.

24.8 Post planting evaluation and management

Plantings should be evaluated for success as early as possible, but not before plants have had sufficient time to establish. In many instances, only tentative information can be obtained the first season, because seedlings or transplants have not been subjected to the stresses of winter (Cook and Stubbendieck 1986). Thus, measurement of stand establishment of perennial plants on arid and semiarid areas should be started near the end of the second growing season. Several estimates of relative success have been proposed for different areas, and usually include one or more of the following observations or measurements: number of plants per unit area, distribution of plants, apparent vigor, and stage of plant development (Vallentine 1980). On seeded foothill range (280 to 330 mm annual precipitation) in the Intermountain Region of the western United States, relative success based on the number of seeded plants per square meter is as follows: excellent stand (8.0 or more), good stand (5.4 to 8.0), fair stand (2.7 to 5.4), and fair stand (less than 2.7) (Cook *et al.* 1967). Unfortunately, most revegetation trials are short-term in nature, paying little attention to long-term plant persistence. Long-term revegetation studies in arid regions have documented that several perennial grasses and shurbs may have good initial establishment on several sites but do not persist past 10 to 20 years (Bleak *et al.* 1965, Judd and Judd 1976, Jordan 1983). Other revegetation studies in more favorable semiarid regions indicate that stands of perennial grasses remained productive up to 35 years (Cook 1966, Hull and Klomp 1966).

Vegetation manipulation and revegetation treatments provide the basis for improving rangeland and increasing output, but subsequent management greatly influences the longevity of the plant community and determines the efficiency of translating the increased output into a merchandisable product (Scifres 1986). Due to the variability in vegetation establishment and production that exists from location to location and during different years at the same location, land managers need to develop flexible management strategies based on the capabilities and limitations of the planted species (Sharp 1983). Management objectives for revegetated arid and semiarid rangelands may include: maintenance of planted species and soil stability, increased livestock production, improved wildlife habitat, increased fuelwood production, aesthetic improvement, or a combination of these.

Where providing forage for livestock is an important objective, it is the manipulation of grazing animals that will be of primary concern in maintaining the forage species. A new seeding should be protected from grazing to allow seedlings to become firmly rooted, to establish significant secondary growth, and to produce a seed reserve (Silcock 1986). The length of protection usually varies from one to two growing seasons, depending upon site conditions, species seeded, competition from undesirable vegetation, and initial stand development (Keller 1979, Vallentine 1980). Subsequent grazing management is based upon an understanding of plant response to the intensity, frequency, and time (season) of defoliation (Caldwell 1984, Kothmann 1984), and the monitoring of these plant responses. Periodic outbreaks of rodents, lagomorphs, insects, and other pests may complicate the management of revegetated areas because these sources of defoliation are less well controlled than those of livestock (Sharp 1983).

Vegetation manipulation treatments, primarily selective herbicides, may be required if undesirable species invade and reduce the productivity of desirable species. Before application of a manipulation treatment, the land manager should consider possible changes in competitive interactions during the establishment period. Depending upon the growth characteristics of the species, many grass seedings may be weedy the first year and become dominated by the seeded species the second year without vegetation manipulation beyond seedbed preparation (Vallentine 1980).

The use of fertilizers to increase seedling establishment and growth on extensive revegetated areas in arid and semiarid regions is a questionable practice (Heady 1975). Variations in climate, topography, soil type, and vegetation make it difficult to generalize about plant responses to added nutrients. With the exception of mesic meadows, irrigated pastures, and areas receiving 380 mm of annual precipitation or more, it is probably uneconomical to fertilize extensive plantings on arid and most semiarid rangelands (Vallentine 1980, Herbel 1983). However, fertilization and other intensive management practices such as irrigation may be required to promote plant establishment and growth on drastically disturbed areas (see Redente and Depuitt, this volume).

24.9 References

Abernathy, G. H., and C. H. Herbel. 1973. A brush eradicating, basin pitting and seeding machine for arid and semiarid rangeland. J. Range Manage. 26: 189—192.

Asay, K. H., and R. P. Knowles. 1985. Current status and future of introduced wheatgrasses and wildryes for rangeland improvement. p. 109—116. *In*: J. R. Carlson, and E. D. McArthur (eds.) Symposium: Range plant improvement. Proceedings selected papers presented at 38th annual meeting, Society for Range Management. Denver, Colorado.

Barnes, R. F., P. R. Ball, R. W. Brougham, G. C. Marten, and D. J. Minson (eds.) 1985. Forage legumes for energy-efficient animal production. USDA Agric. Res. Serv., Washington, D.C.

Barrett-Lennard, E. G., C. V. Malcolm, W. R. Stern, and S. M. Wilkins (eds.) 1986. Forage and fuel production from salt affected wasteland. Reclamation and Revegetation Research. Vol. 5. Elsevier, New York.

Bement, R. E., R. D. Barmington, A. C. Everson, L. O. Hylton, and E. E. Remmenga. 1965. Seeding of abandoned croplands in the Central Great Plains. J. Range Manage. 18: 53—59.

Bentley, H. L. 1898. Cattle ranges of the Southwest. USDA Farmers' Bulletin 72.

Bewley, J. D., and M. Black. 1982. Physiology and biochemistry of seeds in relation to germination. Vol. 2. Springer-Verlag, New York.

Bewley, J. D., and M. Black. 1983. Physiology and biochemistry of seeds in relation to germination. Vol. 1. Springer-Verlag, New York.

Bleak, A. T. 1968. Growth and yield of legumes in mixtures with grasses on a mountain range. J. Range Manage. 21: 259—261.

Bleak, A. T., N. C. Frischknecht, A. P. Plummer, and R. E. Eckert. 1965. Problems in artificial and natural revegetation of the arid shadscale vegetation zone of Utah and Nevada. J. Range Manage. 18: 59—65.

Booth, D. T. 1987. Diaspores of rangeland plants: Ecology and management. p. 202—211. *In*: G. W. Frasier, and R. A. Evans. Proceedings of symposium "Seed and seedbed ecology of rangeland plants." USDA, Agric. Res. Serv., National Technical Information Service, Springfield, Virginia.

Bovey, R. W. 1977. Response of selected woody plants in the U.S. to herbicides. USDA, Agric. Res. Serv., Agric. Handbook No. 493, U.S. Government Printing Office, Washington, D.C.

Bovey, R. W., R. E. Meyer, and J. R. Baur. 1981. Potential herbicides for brush control. J. Range Manage. 34: 144—148.

Box, T. W. 1984. Role of land treatments on public and private lands. p. 1397—1419. *In*: Developing strategies for rangeland management. National Research Council/National Academy of Sciences. Westview Press, Boulder, Colorado.

Brown, D., and R. G. Hallman. 1984. Reclaiming disturbed lands. USDA-Forest Service, Equipment Development Center, Missoula, Montana.

Cable, D. R. 1977. Seasonal use of soil water by mature velvet mesquite. J. Range Manage. 30: 4—11.

Caldwell, M. M. 1984. Plant requirements for prudent grazing. p. 117—152. *In*: Developing strategies for rangeland management. National Research Council/National Academy of Sciences. Westview Press, Boulder, Colorado.

Caldwell, M. M. 1987. Competition between root systems in natural communities. p. 167—185. *In*: P. J. Gregory, J. V. Lake, and D. A. Rose (eds.) Root development and function. Society for Experimental Biology Series 30. Cambridge University Press.

Caldwell, M. M., D. M. Eissenstat, J. H. Richards, and M. F. Allen. 1985. Competition for phosphorus: Differential uptake from dual isotopelabeled soil interspaces between shrub and grass. Science 229: 384—386.

Child, R. D., H. F. Heady, W. C. Hickey, R. A. Peterson, and R. D. Pieper. 1984. Arid and semiarid lands: Sustainable use and management in developing countries. Winrock International, Morrilton, Arkansas.

Christensen, M. D., J. A. Young, and R. A. Evans. 1974. Control of annual grasses and revegetation in ponderosa pine woodlands. J. Range Manage. 27: 143—145.

Cluff, G. J., J. A. Young, and R. A. Evans. 1983. Edaphic factors influencing the control of Wyoming big sagebrush and seedling establishment of crested wheatgrass. J. Range Management. 36: 786—792.

Cook, C. W. 1966. Development and use of foothill ranges in Utah. Utah Agric. Exp. Sta. Bull. 461. Utah State University, Logan.

Cook, C. W., L. A. Stoddart, and P. L. Sims. 1967. Effects of season, spacing, and intensity of seeding on the development of foothill range grass stands. Utah Agric. Exp. Sta. Bull. 467. Utah State University, Logan.

Cook, C. W., and J. Stubbendieck (eds.) 1986. Range research: Basic problems and techniques. Society for Range Management. Denver, Colorado.

Cox, J. R. 1984. Temperature, timing of precipitation and soil texture effects on germination, emergence and seedling survival of South African lovegrasses. J. South African Bot. 50: 159—170.

Cox, J. R., M. H. Martin-R., F. A. Ibarra-F., J. H. Fourie, N. F. G. Rethman, and D. G. Wilcox. 1987. Effects of climate and soils on the distribution of four African lovegrasses. p. 225—241. *In*: G. W. Frasier, and R. A. Evans (eds.) Proceedings of symposium "Seed and seedbed ecology of rangeland plants." USDA, Agric. Res. Serv., National Technical Information Service, Springfield, Virigina.

Cox, J. R., H. L. Morton, T. N. Johnsen, Jr., G. L. Jordan, S. C. Martin and L. C. Fierro. 1982. Vegetation restoration in the Chihuahuan and Sonoran Deserts of North America. ARM-W-28, Agric. Res. Serv., Western Region, Oakland, California.

Cox, J. R., and G. B. Ruyle. 1986. Influence of climatic and edaphic factors on the distribution of *Eragrostis lehmanniana* Nees in Arizona, USA. J. Grassland Soc., South Africa 3: 25—29.

Davis, W. F. 1983. Mechanical control of sagebrush. p. 49—53. *In*: S. B. Monsen, and N. Shaw (compilers) Managing intermountain rangelands: Improvement of range and wildlife habitat. USDA For. Serv. Gen. Tech. Rep. INT-157, Ogden, Utah.

Davy, J. B. 1902. Stock ranges of northwestern California. USDA Bureau of Plant Industry Bulletin 12.

DeLoach, C. J. 1978. Considerations in introducing foreign biotic agents to control native weeds of rangelands. p. 39—50. *In*: T. E. Freeman (ed.) Proceedings 4th International Symp. on Biological Control of Weeds. Institute of Food and Agricultural Sciences, University of Florida, Gainesville.

DeLoach, J. C. 1980. Prognosis for biological control of weeds of southwestern U.S. Rangelands. p. 175—199. *In*: Proceedings 5th International Symp. Biological Control of Weeds, Brisbane, Australia. USDA, Agric. Res. Serv., Temple, Texas.

Dewald, C. L., A. E. Beisel, and S. Cowles. 1983. The Woodward chaffy seed conditioning system. p. 69—80. *In*: H. T. Wiedemann, and J. F. Cadenhead (compilers) Range and pasture seeding in the Southern Great Plains. Texas A & M University Agric. Research and Exten. Center, Vernon.

Dixon, R. M. 1980. Rangeland imprinter. p. 8—10. *In*: 34th annual report, vegetative rehabilitation and equipment workshop. USDA For. Serv. Equipment Development Center, Missoula, Montana.

Eckert, R. E., Jr., M. M. Wood, W. H. Blackburn, F. F. Peterson, J. L. Stephens, and M. S. Meurisse. 1978. Effects of surface-soil morphology on improvement and management of some arid and semiarid rangelands. p. 299—302. Proceedings First International Rangeland Congress.

Evans, R. A. 1961. Effect of different densities of downy brome (*Bromus tectorum*) on growth and survival of crested wheatgrass (*Agropyron desertorum*) in the greenhouse. Weeds 9: 216—223.

Evans, R. A. 1982. Integrated pest management on western rangelands. Council for Agricultural Science and Technology. Report No. 93.

Evans, R. A., H. R. Holbo, R. E. Eckert, Jr., and J. A. Young. 1970. Functional environment of downy brome communities in relation to weed control and revegetation. Weed Sci. 18: 154—162.

Evans, D. D., T. W. Sammis, and D. R. Cable. 1981. Actual evapotranspiration under desert conditions. p. 195—218. *In*: D. D. Evans, and J. L. Thames (eds.) Water in desert ecosystems. Dowden, Hutchinson & Ross, Inc., Stroudsburg, Pennsylvania.

Evans, R. A., and J. A. Young. 1977. Weed control — revegetation system for big sagebrush — downy brome rangelands. J. Range Manage. 30: 331—336.

Evans, R. A., J. A. Young, and R. E. Eckert, Jr. 1969. Herbaceous weed control and revegetation of semiarid rangelands in western United States. Outlook on Agriculture 6: 60—66.

Evans, R. A., J. A. Young, and R. E. Eckert, Jr. 1979. Use of herbicides as a management tool. p. 110—116. *In*: The sagebrush ecosystem: A symposium. Utah State University, Logan.

Fenner, M. 1985. Seed ecology. Chapman and Hall, New York.

Fowler, N. 1986. The role of competition in plant communities in arid and semiarid regions. Ann. Rev. Ecol. Syst. 17: 89—110.

French, N. 1978. Water spreading around the world. p. 314—317. *In*: D. N. Hyder (ed.) Proceedings First International Rangeland Congress. Society for Range Management, Denver, Colorado.

Frischknecht, N. C., and L. E. Harris. 1973. Sheep can control sagebrush on seeded range. Utah Sci. 34: 27—30.

Garza, H. M. 1986. Tecnologia forrajera. *In*: J. G. Medina Torres, and L. N. Beltran (eds.) Metodologia de planeacion integral de los recursos naturales. Universidad Autonoma Agraria Antonio Narro. Saltillo, Mexico.

Goodin, J. R., and D. K. Northington (eds.) 1985. Plant resources of arid and semiarid lands. A global perspective. Academic Press, New York.

Griffiths, D. 1901. Range improvement in Arizona. USDA Bureau.of Plant Industry Bulletin 4.

Griffiths, D. 1902. Forage conditions on the northern border of the Great Basin. USDA Bureau of Plant Industry Bulletin 15.

Griffiths, D. 1904. Range investigations in Arizona. USDA Bureau of Plant Industry Bulletin 67.

Hadas, A., and D. Russo. 1974a. Water uptake by seeds as affected by water stress, capillary conduotivity and seed-soil contact. I. Experimental study. Agron. J. 66: 643—646.

Hadas, A., and D. Russo. 1974b. Water uptake by seeds as affected by water stress, capillary conductivity and seed-soil contact. II. Analysis of experimental data. Agron. J. 66. 647 652.

Hafenrichter, A. L., J. L. Schwendiman, H. L. Harris, R. S. MacLauchlan, and H. W. Miller.

1968. Grasses and legumes for soil conservation in the Pacific Northwest and Great Basin states. USDA Handbook 339. Washington, D.C.
Harper, J. L. 1977. Population biology of plants. Academic Press, New York.
Harper, J. L., J. T. Williams, and G. R. Sagai. 1965. The behavior of seeds in the soil. I. The heterogeneity of soil surfaces and its role in determining the establishment of plants from seed. J. Ecol. 53: 273—386.
Harris, G. A., and J. P. Dobrowolski. Population dynamics of seeded species on northeast Washington semiarid sites, 1948—1983. J. Range Manage. 34: 46—51.
Harris, G. A., and A. M. Wilson. 1970. Competition for moisture among seedlings of annual and perennial grasses as influenced by root elongation at low temperature. Ecol. 51: 530—534.
Hastings, J. R., and R. M. Turner. 1965. The changing mile. University of Arizona Press, Tucson, Arizona.
Hauser, V. L. 1982. Punch planting to establish grass. J. Range Manage. 35: 294—297.
Heady, H. F. 1975. Rangeland Management. McGraw-Hill, New York.
Herbel, C. H. 1983. Principles of intensive range improvements. J. Range Manage. 36: 140—144.
Herbel, C. H. 1984. Manipulative range improvements: Summary and recommendations. p. 1167—1178. *In*: Developing strategies for rangeland management. National Research Council/National Academy of Sciences. Westview Press, Boulder, Colorado.
Heydecker, W. (ed.) 1973. Seed ecology. Pennsylvania University Press, University Park, Pennsylvania.
Holmgren, R. C. 1976. Preliminary summary of vegetation changes on 17 Nevada range sites under conditions of grazing and no grazing by livestock for 38 years. USDA For. Serv. Mimeo Rep., Ogden, Utah.
Howell, D. 1976. Observations on the role of grazing animals in revegetating problem patches of veld. Proc. Grassland Soc. South Africa 11: 59—63.
Hubbard, R. L. 1957. The effects of plant competition on the growth and survival of bitterbrush seedlings. J. Range Manage. 10: 135—137.
Huffaker, C. B. 1951. The return of native perennial bunchgrass following the removal of Klamath weed (*Hyoericum perforatum* Linn.) by imported beetles. Ecol. 32: 443—458.
Hull, A. C., and G. J. Klomp. 1966. Longevity of crested wheatgrass in the sagebrush-grass type in southern Idaho. J. Range Manage. 19: 5—11.
Humphrey, R. R. 1958. The desert grassland. Bot. Rev. 24: 193—252.
Hussey, M. A. 1985. Plant materials: Old and new. p. 186—189. *In*: Proc. of the International Ranchers Roundup. Texas Agric. Exten. Serv. Texas A & M University, College Station.
Institute for Land Rehabilitation. 1979. Selection, propagation and field establishment of native plant species on disturbed arid lands. Utah Agric. Exp. Sta. Bulletin 500. Utah State University, Logan.
Johnson, D. A., R. A. Richard, and N. C. Turner. 1983. Yield, water relations, gas exchange and surface reflectances of near-isogenic wheat lines differing in glaucousness. Crop Sci. 23: 318—325.
Johnson, D. A., M. D. Rumbaugh, and K. H. Asay. 1981. Plant improvement for semiarid rangelands: possibilities for drought resistance and nitrogen fixation. Plant and Soil 58: 279—303.
Johnson, K. L. (ed.) 1986. Crested wheatgrass: its values, problems, and myths; symposium proceedings. Utah State University, Logan.
Jordan, G. L. 1981. Range seeding and brush management on Arizona rangelands. University of Arizona Agric. Exp. Sta. T81121. Tucson, Arizona.
Jordan, G. L. 1983. Planting limitations for arid, semiarid and salt-desert shrublands. p. 11—16. *In*: S. B. Monsen, and N. Shaw (compilers) Managing intermountain rangelands:

Improvement of range and wildlife habitat. USDA For. Serv. Gen. Tech. Rep. INT-157, Ogden, Utah.

Keller, W. 1979. Species and methods for seeding the sagebrush ecosystem. p. 129—163. *In*: The sagebrush ecosystem; a symposium. Utah State University, Logan.

Kennedy, P. B., and S. B. Doten. 1901. A preliminary report on the summer ranges of western Nevada sheep. Nevada State University Agric. Exp. Sta. Bulletin 51. Reno, Nevada.

Judd, B. I., and L. W. Judd. 1976. Plant survival in the arid Southwest 30 years after seeding. J. Range Manage. 29: 248—251.

Kneebone, W. R. 1972. Breeding for seedling vigor. p. 90—100. *In*: U. B. Youngner, and C. M. McKell (eds.) The biology and utilization of grasses. Academic Press, New York.

Koller, D., and A. Hadas. 1982. Water relations in the germination of seed. p. 402—431. *In*: O. L. Lange, P. S. Nobel, C. B. Osmond, and H. Ziegler (eds.) Physiological plant ecology II. Water relations and carbon assimilation. Encyclopedia of plant physiology. New Series, Vol. 12B. Springer-Verlag, New York.

Kothmann, M. M. 1984. Concepts and principles underlying grazing systems: A discussant paper. p. 903—916. *In*: T. T. Kozlowski (ed.) Developing strategies for rangeland management. National Research Council/National Academy of Sciences. Westview Press, Boulder, Colorado. 1972. Seed biology. Academic Press, New York.

Larson, J. E. 1980. Revegetation equipment catalog. USDA For. Serv. Equipment Development Center, Missoula, Montana.

Lorenz, R. J. 1986. Introduction and early use of crested wheatgrass in the Northern Great Plains. p. 9—20. *In*: K. L. Johnson (ed.) Crested wheatgrass: its values, problems and myths; symposium proceedings. Utah State University, Logan.

Low, B. S., and J. A. Berlin. 1984. Natural selection and the management of rangelands. p. 1179—1234. *In*: Developing strategies for rangeland management. National Research Council/National Academy of Sciences. Westview Press, Boulder, Colorado.

Malcolm, C. V., and R. J. Allen. 1981. The Mallen niche seeder for plant establishment on difficult sites. Aust. Rangeland J. 3: 106—109.

Matlock, W. G., and G. R. Dutt. 1986. A primer on water harvesting and runoff farming. Agricultural Engineering Dept., College of Agriculture, University of Arizona, Tucson.

Mayer, A. M., and A. Poljakoff-Mayber. 1982. The germination of seeds. Third edition. Pergamon Press, New York.

McArthur, E. D., B. L. Welch, and D. L. Nelson. 1985. Developing improved cultivars of sagebrushes and other composite shrubs. p. 188—196. *In*: J. R. Carlson, and E. D. McArthur (eds.) Symposium: Range plant improvement. Proceedings selected papers presented at 38th annual meeting, Society for Range Management. Denver, Colorado.

McDaniel, K. C. (ed.) 1983. Proceedings Brush Management Symposium. Society for Range Management. Denver, Colorado.

McDonough, W. T. 1977. Seed physiology. p. 155—184. *In*: R. E. Sosebee (ed.) Rangeland plant physiology. Society for Range Management. Denver, Colorado.

McGinnies, W. J. 1984. Seedbed preparation — The forgotten step in range seeding. p. 23—25. *In*: 38th annual report, vegetative rehabilitation and equipment workshop. USDA For. Serv. Equipment Development Center, Missoula, Montana.

McDonald, M. B., Jr., and A. A. Kah. 1977. Factors determining germination of Indian ricegrass seeds. Agron. J. 69: 558—563.

McKell, C. M. 1972. Seedling vigor and seedling establishment. p. 74—89. *In*: V. B. Youngner, and C. M. McKell. The biology and utilization of grasses. Academic Press, New York.

McKell, C. M. 1985. Biotechnology in range vegetation. p. 116—123. *In*: Proc. of the National Range Conference. USDA, Washington, D.C.

McKell, C. M. 1986. The role of shrubs in diversifying a crested wheatgrass monoculture. p.

109—116. *In*: K. L. Johnson (ed.) Crested wheatgrass: Its values, problems, and myths; symposium proceedings. Utah State University, Logan.

McKell, C. M., and W. W. Chilcote. 1957. Response of rabbitbrush following removal of competing vegetation. J. Range Manage. 10: 228—230.

Medina T., J. G., and H. Garza. 1987. Range seeding research in Northern Mexico. p. 246—259. *In*: G. W. Frasier, and R. A. Evans. Proceedings of symposium "Seed and seedbed ecology of rangeland plants." U.S. Dept. Agric., Agric. Res. Serv., National Technical Information Service, Springfield, Virginia.

Miller, R. F., I. S. McQueen, F. A. Branson, L. M. Shown and W. L. Buller, 1969. An evaluation of range floodwater spreaders. J. Range Manage. 22: 246—257.

Monson, S. B. 1975. Selecting plants to rehabilitate disturbed areas. p. 76—90. *In*: R. S. Campbell, and C. H. Herbel (eds.) Improved range plants. Range Symp. Ser. 1, Soc. Range Manage. Denver, Colorado.

Nechaeva, N. T. (ed.) 1985. Improvement of desert ranges in Soviet Central Asia. Harwood Academic Publishers, New York.

Nielsen, D. B. 1984. Economic factors to be considered in sagebrush/grassland management. p. 1373—1386. *In*: Developing strategies for rangeland management. National Research Council/National Academy of Sciences. Westview Press, Boulder, Colorado.

Noble, J. C., G. M. Cunningham, and W. E. Mulham. 1984. Rehabilitation of degraded land. p. 171—186. *In*: G. N. Harrington, A. D. Wilson, and M. D. Young (eds.) Management of Australia's rangelands. Commonwealth Scientific and Industrial Research Organization, East Melbourne, Australia.

Nord, E. C., P. F. Hartless, and W. D. Nettleton. 1971. Effects of several factors on saltbush establishment in California. J. Range Mange. 24: 216—223.

Oaks, W. R. 1982. Seed coating hard to drill seeds. p. 21—22. *In*: 36th annual report, vegetative rehabilitation and equipment workshop. USDA For. Serv. Equipment Development Center, Missoula, Montana.

Owens, M. A., and H. W. Brzostowski. 1967. Grass establishment under semi-arid conditions in Central Tanganyika. Trop. Agric. Trin. 44: 275—291.

Petersen, J. L., D. N. Ueckert, and R. L. Potter. 1986. Cultural practices for establishing fourwing saltbush within perennial grass stands. J. Range Manage. 39: 460—463.

Plummer, A. P. 1977. Plant and treatment for revegetation of disturbed sites in the Intermountain area. p. 155—173. *In*: R. A. Wright (ed.) the reclamation of disturbed arid lands. University of New Mexico Press, Albuquerque.

Plummer, A. P., D. R. Christensen, and S. B. Monsen. 1968. Restoring big-game range in Utah. Publication No. 68—73. Utah Department Natural Resources, Division of Fish and Game. Salt Lake City, Utah.

Plummer, A. P., A. C. Hull, G. Stewart, and J. H. Robertson. 1955. Seeding rangelands in Utah, Nevada, and southern Idaho, and western Wyoming. USDA Agric. Handbook 71. Washington, D.C.

Pouge, G. 1983. Flamecoat seed treatment process. p. 65—68. *In*: H. T. Wiedemann, and J. F. Cadenhead (compilers) Range and pasture seeding in the Southern Great Plains. Texas A & M University Agric. Research and Exten. Center, Vernon.

Pratt, D. J. 1963. Reseeding denuded land in Baringo District, Kenya. E. African Agric. and For. J. July 1963: 78—91.

Pratt, D. J., and M. D. Gwynne (eds.) 1977. Rangeland management and ecology in East Africa. Robert Krieger Publishing Co., New York.

Robertson, J. H., and C. K. Pearse. 1945. Artificial reseeding and the closed community. Northwest Sci. 14: 58—66.

Roughley, R. J. 1985. Nitrogen fixation by the legume — *Rhizobium* symbiosis: The role of the bacteria. p. 140—145. *In*: R. F. Barnes, P. R. Ball, R. W. Brougham, G. C. Marten, and D. J. Minson (eds.) Forage legumes for energy — efficient animal production. USDA Agric. Res. Serv., Washington, D.C.

Roundy, B. A., R. A. Evans, and J. A. Young. 1984. Surface soil and seedbed ecology in salt-desert plant communities. p. 66—74. *In*: A. R. Tiedemann, E. D. McArthur, H. C. Stutz, R. Stevens, and K. L. Johnson (compilers) Proceedings symposium on the biology of *Atriplex* and related chenopods. Gen. Tech. Rep. INT-172. USDA For. Serv., Intermountain Exp. Sta., Ogden, Utah.

Roundy, B. A., J. A. Young, and R. A. Evans. 1981. Phenology of salt rabbitbrush (*Chrysothamus nauseous* spp. *consimilis*) and greasewood (*Sarcobatus vermiculatus*). Weed Sci. 29: 48—454.

Rowley, W. D. 1985. U.S. Forest Service grazing and rangelands: A history. Texas A & M University Press, College Station.

Rumbaugh, M. D., and C. E. Townsend. 1985. Range legume selection and breeding in North America. p. 137—147. *In*: J. R. Carlson, and E. D. McArthur (eds.) Symposium: Range plant improvement. Proceedings selected papers presented at 38th annual meeting, Society for Range Management. Denver, Colorado.

Sampson, A. W. 1952. Range management principles and practices. John Wiley & Sons, New York.

Scifres, C. J. 1977. Herbicides and the range ecosystem: Residues, research and the role of range men. J. Range Manage. 30: 86—91.

Scifres, C. J. 1980. Brush management principles and practices. Texas A & M University Press, College Station.

Scifres, C. J. 1986. Integrated management systems for improvement of rangeland. p. 227—259. *In*: M. A. Sprague, and G. B. Triplett (eds.) Notillage and surface-tillage agriculture: The tillage revolution. John Wiley & Sons, New York.

Scifres, C. J., W. T. Hamilton, J. R. Conner, J. M. Inglis, G. A. Rasmussen, R. P. Smith, J. W. Stuth, and T. G. Welch. 1985. Integrated brush management systems for south Texas: Development and implementation. B-1493, Texas Agric. Exp. Sta., College Station, Texas.

Sharp, L. 1983. Management of seeded rangeland to maintain forage plants. p. 180—183. *In*: S. B. Monsen, and N. Shaw (compilers) Managing intermountain rangelands: Improvement of range and wildlife habitat. USDA For. Serv. Gen. Tech. Rep. INT-157. Ogden, Utah.

Shaw, N. and S. B. Monsen. 1983. Non-leguminous forbs for rangeland sites. p. 123—131. *In*: S. B. Monsen, and N. Shaw (compilers) Managing intermountain rangelands: Improvement of range and wildlife habitat. USDA For Serv. Gen. Tech. Rep. INT-157. Ogen, Utah.

Shiflet, T. N., and R. S. MacLauchlan. 1986. Selecting plants for revegetation of rangelands in the United States. p. 295—296. *In*: P. J. Joss, P. W. Lynch, and O. B. Williams (eds.) Rangelands: a resource under siege. Proc. Second International Rangeland Congress. Australian Academy of Science, Canberra.

Silcock, R. G. 1986. Strategies for plant introduction and reseeding of semiarid pastoral lands. p. 268—274. *In*: P. J. Joss, P. W. Lynch, and O. B. Williams (ed.) Rangelands: a resource under siege. Proceedings Second International Rangeland Congress. Australian Academy of Science, Canberra.

Stevens, R., W. L. Moden, and D. W. McKenzie. 1981. Interseeding and transplanting shrubs and forbs into grass communities. Rangelands 3: 55—58.

Stewart, G. 1950. Reseeding research in the Intermountain region. J. Range Manage. 3: 52—59.

Stoddart, L. A., and A. D. Smith. 1943. Range Management, First edition. McGraw-Hill, New York.

Stoddart, L. A., A. D. Smith, and T. W. Box. 1975. Range Management, Third Edition. McGraw-Hill, New York.

Stroh, J. R., and V. P. Sundberg. 1971. Emergence of grass seedlings under crop residue culture. J. Range Mange. 21: 226—230.

Stuth, J. W., and B. E. Dahl. 1974. Evaluation of rangeland seedings following mechanical brush control in Texas. J. Range Manage. 27: 146—149.

Tainton, A. M. (ed.) 1981. Veld and pasture management in South Africa. Shuter & Shooter, Pietermaritzburg, South Africa.

Thomas, R. A. 1982. Rollerchopping — A valuable range improvement tool. p. 172—180. *In*: Proceedings of the Intermountain Rancher's Roundup. Texas Agric. Exten. Serv., Texas A & M University, College Station.

Thornburg, A. A. 1982. Plant materials for use on surface-mined lands in arid and semiarid regions. SCS — TP — 157. USDA Soil Cons. Serv., Washington, D.C.

Thornburg, A. A., and S. H. Fuchs. 1978. Plant materials and requirements for growth in dry regions. p. 411—424. *In*: F. W. Schaller., and P. Sutton (eds.) Reclamation of drastically disturbed lands. American Society of Agronomy, Madison, Wisconsin.

Torrey, J. G. 1978. Nitrogen fixation by actinomycete — nodulated angiosperum. BioScience 28: 586—592.

Trappe, J. M. 1981. Mycorrhizae and productivity of arid and semiarid rangelands. p. 581—599. *In*: J. T. Manassah, and E. J. Briskey. Advances in food-producing systems for arid and semiarid lands. Part A. Kuwait Foundation for Advancement of Sciences. Academic Press, New York.

Ueckert, D. N. 1979. Broom snakeweed: effect on shortgrass forage and soil water depletion. J. Range Mange. 32: 216—220.

Urness, P. J. 1986. Value of crested wheatgrass for big game. p. 147—153. *In*: K. L. Johnson (ed.) Crested wheatgrass: its values, problems and myths; symposium proceedings. Utah State University, Logan.

Vallentine, J. F. 1980. Range development and improvements. Second edition. Brigham Young University Press, Provo, Utah.

Van Epps, G. A., and C. M. McKell. 1978. Major criteria and procedures for selecting and establishing range shrubs as rehabilitators of disturbed lands. p. 352—354. *In*: D. N. Hyder (ed.) Proceedings First International Rangeland Congress. Society for Range Management. Denver, Colorado.

Van Epps, G. A., and C. M. McKell. 1980. Revegetation of disturbed sites in the salt desert range of the Intermountain West. Land Rehabilitation Series No. 5, Utah Agric. Exp. Sta., Logan.

Vories, K. C., and P. L. Sims. 1977. The plant information network, Vol. I: A user's guide. FWS/OBS-77/38, Fish and Wildlife Serv., U.S. Dept. Interior, Washington, D.C.

Walker, B. H. 1979. Principles for semiarid ecosystems. p. 379—388. *In*: B. H. Walker (ed.) Management of semiarid ecosystems. Elsevier, Amsterdam.

Wasser, C. H., and J. Shoemaker. 1982. Ecology and culture of selected species useful in revegetating disturbed lands in the West. U.S. Dept. Interior, Fish and Wildlife Service. Government Printing Office, Washington, D.C.

West, N. E. 1968. Outline for autecological studies of range grasses. J. Range Manage. 21: 102—105.

West, N. E., F. D. Provenza, P. S. Johnson, and M. K. Owens. 1984. Vegetation change after 13 years of livestock grazing exclusion on sagebrush semidesert in West Central Utah. J. Range Mange. 37: 262—264.

Wiedemann, H. J., and B. T. Cross. 1981. Rangeland seeder development using semicircular seedbox and auger agitator seed metering concept. J. Range Manage. 34: 340—342.

Wight, J. R., F. H. Siddoway, R. J. Kartchner, and J. L. Bishop. 1978. Contour furrows seeded with grasses and legumes increase production on semiarid range sites. p. 643—644. *In*: D. N. Hyder (ed.) Proceedings First International Rangeland Congress. Soc. Range Manage. Denver, Colorado.

Wood, M. C., W. H. Blackburn, R. E. Eckert, Jr., and F. F. Peterson. 1978. Interrelations of the physical properties of coppice dune and vesicular dune interspace soils with grass seedling emergence. J. Range Manage. 31: 189—192.

Wood, M. K., R. E. Eckert, Jr., W. H. Blackburn, and F. F. Peterson. 1982. Influence of crusting soil surfaces on emergence and establishment of crested wheatgrass, squirreltail, Thurber needlegrass and fourwing saltbush. J. Range Manage. 35: 282—287.

Workman, J. P. 1986. Range economics. MacMillan Publ. Co., New York.

Wright, H., and A. Bailey. 1982. Fire ecology in the United States and Canada. Wiley Interscience, New York.

Young, J. A. 1983. Principles of weed control and plant manipulation. p. 6—10. *In*: S. B. Monsen, and N. Shaw (compilers) Managing intermountain rangelands: Improvement of range and wildlife habitat. USDA For. Serv. Gen. Tech. Rep. INT-157, Ogden, Utah.

Young, J. A., R. E. Eckert, Jr., and R. A. Evans. 1979. Historical perspectives regarding the sagebrush ecosystem. p. 1—13. *In*: The sagebrush ecosystem: a symposium. Utah State University, Logan.

Young, J. A., and R. A. Evans. 1974. Population dynamics of green rabbitbrush in disturbed big sagebrush communities. J. Range Manage. 27: 127—132.

Young, J. A., and R. A. Evans. 1986. History of crested wheatgrass in the Intermountain area. p. 21—25. *In*: K. L. Johnson (ed.) Crested wheatgrass: its values, problems and myths; symposium proceedings. Utah State University, Logan.

Young, J. A., R. A. Evans, and J. J. Major. 1972. Alien plants in the Great Basin. J. Range Manage. 25: 194—201.

Young, J. A., and D. McKenzie. 1982. Rangeland drill. Rangelands 4: 108—113.

Young, J. A., R. A. Evans, B. Roundy, and G. Cluff. 1983. Moisture stress and seed germination. USDA, Agric. Res. Serv. Reviews and Manuals No. 36. Oakland, California.

Young, J. A., R. A. Evans, B. A. Roundy, and G. J. Cluff. 1984. Ecology of seed germination in representative Chenopodiaceae. p. 159—165. *In*: A. R. Tiedemann, E. D. McArthur, H. C. Stutz, R. Stevens, and K. L. Johnson (compilers) Proc.: symposium on the biology of *Atriplex* and related chenopods. Gen. Tech. Rep. INT-172. USDA Forest Service, Intermountain Exp. Sta., Ogden, Utah.

Young, J. A., and B. A. Sparks. 1985. Cattle in the cold desert. Utah State University Press, Logan.

Young, J. A., and C. G. Young. 1986. Collecting, processing and germinating seeds of wildland plants. Timber Press, Portland, Oregon.

Index